Sustainable Development and Biodiversity

Volume 36

Sustainable Development Goals are best achieved by mechanisms such as research, innovation, and knowledge sharing. This book series aims to help researchers by reporting recent progress and providing complete, comprehensive, and broad subject-based reviews about all aspects of sustainable development and ecological biodiversity.

The series explores linkages of biodiversity with delivery of various ecosystem services and it offers a discourse in understanding the biotic and abiotic interactions, ecosystem dynamics, biological invasion, ecological restoration and remediation, diversity of habitats and conservation strategies. It is a broad scoped collection of volumes, addressing relationship between ecosystem processes and biodiversity. It aims to support the global efforts towards achieving sustainable development goals by enriching the scientific literature. The books in the series brings out the latest reading material for botanists, environmentalists, marine biologists, conservationists, policy makers and NGOs working for environment protection.

We welcome volumes on the themes --

Agroecosystems, Agroforestry, Biodiversity, Biodiversity conservation, Conservation of ecosystem, Ecosystem, Endangered species, Forest conservation, Genetic diversity, Global climate change, Hotspots, Impact assessment, Invasive species, Livelihood of people, Plant biotechnology, Plant resource utilization, Sustainability of the environment, Sustainable management of forests, Sustainable use of terrestrial ecosystems and plants, Traditional methods, Urban horticulture.

Hanuman Singh Jatav • Vishnu D. Rajput •
Tatiana Minkina • Eric D. Van Hullebusch •
Asik Dutta
Editors

Agroforestry to Combat Global Challenges

Current Prospects and Future Challenges

Springer

Editors
Hanuman Singh Jatav
Department of Soil Science & Agricultural Chemistry
College of Agriculture, Sri Karan Narendra Agriculture University, Jobner
Fatehpur-Shekhawati, Sikar
Rajasthan, India

Vishnu D. Rajput
Academy of Biology and Biotechnology
Southern Federal University
Rostov-on-Don, Russia

Tatiana Minkina
Academy of Biology and Biotechnology, Department of Soil Science
Southern Federal University
Rostov-on-Don, Russia

Eric D. Van Hullebusch
Institut de physique du globe de Paris
Université Paris Cité
Paris, France

Asik Dutta
ICAR-Indian Institute of Pulses Research
Kanpur, India

ISSN 2352-474X ISSN 2352-4758 (electronic)
Sustainable Development and Biodiversity
ISBN 978-981-99-7281-4 ISBN 978-981-99-7282-1 (eBook)
https://doi.org/10.1007/978-981-99-7282-1

This Springer imprint is published by the registered company Springer Nature Singapore Pte Ltd.
The registered company address is: 152 Beach Road, #21-01/04 Gateway East, Singapore 189721, Singapore

Paper in this product is recyclable

Preface

Globally agriculture is facing an unprecedented task of feeding the ever-increasing population, and the situation is further complicated by climate change, especially in the past decades. Population boom along with faulty agricultural practices is making the task of the farming community more challenging and cumbersome. Land use change is a win-win strategy which not only sustains the system productivity but also helps to sequester atmospheric C combating both hunger and climate change at the same time. Organic agriculture, natural farming, and agro-forestry all are highly eco-friendly measures of recovering the ecosystem but, among all the aforementioned, agro-forestry is the best as it is the amalgamation of all the three components of farms, namely crops, trees, livestock. Internationally, the role of agro-forestry system has been acclaimed by all in Kyoto protocol, and all the aspects are discussed for maximizing the benefits.

Mono-cropping without the use of any organic amendments, excessive tillage without any ground coverage, depleting water table caused large-scale land degradation leaving behind only barren land suitable for nothing. But this barren land provides an excellent site for the regeneration of trees, and within short time the soil can regain its productive ability with a provision for nutritional and economical security for the locals of that region. As per different group of researchers across the globe, agro-forestry system can restore 27 ± 14 tons CO_2 per hectare per year of which 70% to be stored in biomass and rest amount in soil. Studies in North America reported that agro-forestry system can annually sequester 548.4 Tg carbon per enough to offset 34% of US emissions from coal, oil, and gas. As a positive outcome of C-sequestration, the productive capacity and microbial biodiversity of the soil increased with concomitant reduction in erosion, better deposition of organic matter, ground water recharge and less pollution in lieu of the conventional system. So, undoubtedly this integrated approach has a special significance largely in the developing countries in mitigating climate change and securing a nutritionally rich future.

Apart from bringing down the gloomy impact of climate change agro-forestry measures have multiple benefits. Biodiversity conservation, protection of heavy winds, stabilizing the market volatility, and ensuring stable economical return for the growers and better nutrient cycling can be achieved by this system. Research suggests perennial vegetation can trap 80% or more nutrients lost through runoff and harbour array of microbes in its rhizosphere. Steady deposition of mineral nutrients, differential rooting pattern of the component crops, and deposition of leaf litter enrich the soil and improve the productivity. In mono-cropping system, landowners of farms are engaged in seasonal farming activities, whereas agroforestry system provides round the year employment opportunity in diverse works with higher family wage. However, achieving the goal is not an easy walk, and there are many bottlenecks in these paths like initial cost involvement, lack of suitable tree availability with scanty knowledge about the management conditions, logistic management, and marketing. Therefore, an endeavour must be made from all sectors via public-private mode to harness the utmost potential using technological and management advancements. Also, additional information about the ecological production potential of the system with subsidiary factors like site and soil characteristics, climatic conditions, market demand, and in-depth knowledge about the suitability of species in a particular area must be gathered to multiply the effectiveness of the agroforestry system to combat global challenges.

Sikar, Rajasthan, India — Hanuman Singh Jatav
Rostov-on-Don, Russia — Vishnu D. Rajput
Rostov-on-Don, Russia — Tatiana Minkina
Paris, France — Eric D. Van Hullebusch
Kanpur, India — Asik Dutta

Acknowledgement

Every accomplishment task deserved a token of thanks for those who helped directly or indirectly to complete that work. It gives me pleasure to remember and express my feelings to those, who helped me in the form of motivation to compile this outcome. I Dr. Hanuman Singh Jatav as an author of this book *Agroforestry to Combat Global Challenges: Current Prospects and Future Challenges* acknowledge token of thanks to Prof. Balraj Singh, Hon'ble Vice-Chancellor, Sri Karan Narendra Agriculture University, Jobner-Jaipur, for his motivation to compile such work. I am deeply indebted to Dr. Amitava Rakshit, Associate Professor, Banaras Hindu University, Varanasi for his lifelong motivation to compile the work. I also extend my thanks to Prof. K.K. Sharma (HOD-SSAC), and Dr. S.K. Dadhich for their direct and indirect support.

It is very difficult to acknowledge the token of words to Prof. S.R. Dhaka (Dean-COA Fatehpur- Sikar) who not only supported but also guided me to compile this upshot for the students. I would like to thank from the bottom of my heart Dr. Sanjay K. Attar, Prof. Athar Uddin, Dr. Muddser A. Khan, Dr. K.C. Verma, Dr. Kailash Chandra, Dr. Mujahid Khan, Dr. Champa L. Khatik, Dr. Shubita Kumawat, Dr. Ramu Meena, Dr. Ajeet Singh, and Dr. Vishnu D. Rajput for the direct and indirect support to compile this outcome.

At the outset, being a student of the great institution Banaras Hindu University, Varanasi, I am thankful to the Prof. A.K. Ghosh, Prof. Nirmal De, Prof. Surendra Singh (FISSS), Prof. Priyankar Raha, Prof. P.K. Sharma, Dr. Y.V. Singh, and Dr. Ashish Latare for their support and guidance during my studies.

I am grateful to my schoolteacher Shri Surendra Sarthak Ji, Shri Ghanshyam Ji, Shri Harkesh Ji, and Shri Hanuman Ji. I am also delighted to offer thanks to my friends Dr. Sunil Kumar, Dr. Dinesh Jinger, Dr. Sonu Kumar, Dr. Surendra Singh Jatav, Dharamveer Verma, and Rajat.

I don't have words for the encouragement led by beloved Nikki (OTA Sahab) who has always been with me in my ups and downs in compiling this work.

I am also thankful to Prof. Bijay Singh for his regular efforts in compiling and critical evaluation of this book. I am also thankful to Prof. S.K. Singh, and

Dr. Vishnu D. Rajput for their continued checking of the content in making very less scientific or grammar errors. I also extend my sincere gratitude to my family members who always support me in any difficult situation.

The grace of the Lord Hanuman Ji, Goddess Maa Sharde along with my parents Smt. Supedi Devi and Shri Babulal Jatav who always blessed me and gave me power and patience to overcome difficulties that came my way in the accomplishment of this endeavour. I cannot dare to say thanks but only pray to bless me always. VDR and TM acknowlege support by the Ministry of Science and Higher Education of the Russian Federation (no. FENW-2023-0008) and the Strategic Academic Leadership Program of the Southern Federal University ("Priority 2030").

Soil Science and Agril. Chemistry, H. S. Jatav
COA, Fatehpur, Rajasthan, India

SKNAU, Jobner, Rajasthan, India

Contents

Editors and Contributors

Editors

Hanuman Singh Jatav is an Assistant Professor in the Department of Soil Science and Agricultural Chemistry, Sri Karan Narendra Agriculture University, Jobner Jaipur (Rajasthan), India. He has more than 6 years of teaching and 12 years of research experience in the field of Sewage Sludge, Biochar, Micronutrients, Heavy Metals, Soil Fertility Management including M. Sc. (Ag.) and Ph.D. He has completed B.Sc. (Hons.) Agriculture from Maharana Pratap University of Agriculture and Technology, Udaipur-Rajasthan, India. Dr. Jatav has completed his M.Sc. (Ag.) and Ph.D. in Soil Science and Agricultural Chemistry from the Institute of Agricultural Sciences, Banaras Hindu University, Varanasi-U.P. India under guidance of Prof. Satish Kumar Singh [FISSS, FISAC]. At present, Dr. Jatav is an emerging person in the field of research transforming himself to an excellent scientific level by enhancing research capacity by promoting environmental sustainability. He has published more than 120 paper(s) including research papers, review papers, books, book chapters, and technical papers from various national and international repute Journals and Publishers. He is serving as an Academic Editor and member of Subject expertise in many internationally acclaimed journals for the last 7 years and serving as an active member of more than 20 national and international scientific societies. Dr. Jatav has been awarded Certificate of Gratitude as an Ambassador by IUCN: International Union for Conservation of Nature-Gland, Switzerland in the World Conservation Congress during Global Youth Summit. Considering Dr. Jatav has been awarded as Outstanding Assistant Professor of the Sri Karan Narendra Agriculture University, Jobner Jaipur (Rajasthan) India.

Vishnu D. Rajput is working as a Leading Researcher (Associate Professor), Head of "International Lab on Nanobiotechnology & Soil Health" at Southern Federal University, Rostov-on-Don, Russia. His ongoing research is based on toxic effects of bulk- and nano-forms of metals and investigating the bioaccumulation, bio/geo-transformations, uptake, translocation, and toxic effects of bulk- and nano-forms of metals on plant physiology, morphology, anatomy, the ultrastructure of cellular and subcellular organelles, cytomorphometric modifications, and DNA damage. He has also discovered the possible remediation approaches such using biochar/nano-biochar-based sorbents and nanotechnological approaches. With long experience and experimental work, Dr. Rajput comprehensively detailed the state of research in environmental science in regard to "how nanoparticles/heavy metals interact with plants, soil, microbial community, and the larger environment as well as possible remediation technology using nanoparticles/nano-biochar". He has published (total of 351 scientific publications) 204 peer-reviewed full-length articles, 18 books, 70 chapters (Scopus indexed), and 44 conference articles. He is an editorial board member of various high-impacted journals. He is a leader of several grants including Mega-Grant, BRICS, and performer/co-PI of several national and international projects. Dr. Rajput has received "Certificate for Appreciation 2019" 2021, "Certificate of Honor 2020", Diploma Award 2021, Letter of Gratitude 2022, 2023 by Southern Federal University, Russia, for outstanding contribution in academic, creative research, and publication activities. He has also received "Highly Qualified Specialist" status by Russian government and included in the list of top 2% of Russian scientists.

Tatiana Minkina is the Head of Soil science and land evaluation department of Southern federal university. The Head of International Master's Degree Educational Program "Management and estimation of land resources" (2015–2022 years, accreditation by ACQUIN). Her area of scientific interest is Soil Science, Biogeochemistry of Trace Elements, Environmental Soil Chemistry, Soil monitoring, assessment, modelling, and remediation using physicochemical treatment methods. She was awarded with Diploma of the Ministry of Education and Science of the Russian Federation for many years of long-term work for development and improvement of the educational process, significant contribution to the training of highly qualified specialists. Currently, she is handling projects funded by Russian Scientific foundation, Ministry of Education and Science of the Russian Federation, Russian foundation of Basic Researches. Member of Expert Group of Russian Academy of Science; the International Committee on Contamination Land; Eurasian Soil Science Societies; the International Committee on Protection of the Environment; the International Scientific Committee of the International Conferences on Biogeochemistry of Trace Elements. She has over 750 total scientific publications.

Eric D. van Hullebusch received his Ph.D. (Aquatic Chemistry and Microbiology) from Université de Limoges (France) in 2002. From November 2002 until October 2004 he was a Marie Curie Postdoctoral fellow at Wageningen University and Research (the Netherlands) where his research focused on the optimization of anaerobic granular sludge reactors by studying the speciation, bioavailability, and dosing strategies of trace metals. In 2005, he was appointed as associate professor in biogeochemistry of engineered ecosystems at Université Paris-Est (France). In 2012, he obtained his Habilitation qualification in Environmental Sciences from Université Paris-Est (France). The title of his Habilitation thesis is "Biofilms in the environment: from anaerobic wastewater treatment to material bioweathering". From September 2016 until August 2018, he worked at IHE Delft as chair professor in

Environmental Science and Technology and head of the Pollution Prevention and Resource Recovery chair group. In September 2018, he joined Institut de Physique du Globe de Paris (France) as full professor in Biogeochemistry of engineered ecosystems. Prof. Eric D. van Hullebusch has published more than 280 peer-reviewed papers in the field of biogeochemistry of metals and metalloids in engineered ecosystems and soil remediation. His research is currently focusing on the implementation of biohydrometallurgical approaches for the recovery of technology critical elements as well as polluted soils remediation.

Asik Dutta is from West Bengal, India, and he completed his graduation from Bidhan Chandra Krishi Viswavidyalaya and joined ICAR-Indian Agricultural research Institute, New Delhi for M.Sc. (Soil Science and Agricultural Chemistry). After Completion of post-graduation, he joined the same institute for Ph.D. He was selected in Agricultural Research Services (Govt. of India) in 2018. Currently, he is posted at ICAR-Indian Institute of Pulses Research, Kanpur (UP). His area of work is phosphorus chemistry and nutrient cycling in pulse-based cropping system under climate change scenario. Along with this, he is also working on micronutrient bio-fortification in lentil and urad bean. He has published his research in journals of national and international repute.

Contributors

Ghazanfer Abbas Department of Silviculture and NRM, Forest College and Research Institute, Mettupalayam, India

Avantika Adhruj Indian Institute of Forest Management, Bhopal, India

Maryam Ali Sustainable Development Study Center, Government College University, Lahore, Pakistan

J. Aliyu Department of Soil Science, Ahmadu Bello University, Kaduna, Nigeria

K. S. Anjali Department of Silviculture and NRM, Forest College and Research Institute, Mettupalayam, India

M. N. Ashwath Department of Forest Biology and Tree Improvement, Kerala Agricultural University, Thrissur, Kerala, India

Kodchire Avinash ICFRE-Institute of Forest Biodiversity, Hyderabad, Telangana, India

Ayush Bahuguna Institute of Agricultural Sciences, Banaras Hindu University, Varanasi, India

A. Balasubramanian Department of Silviculture and NRM, Forest College and Research Institute, Mettupalayam, India

Th. Bhaigyabati Department of Botany, Pravabati College, Mayang Imphal, Manipur, India

Ankita Bhardwaj Department of Silviculture and Agroforestry, Kerala Agricultural University, Thrissur, Kerala, India

Sumanta Bhattacharya MAKAUT, Howrah, West Bengal, India

Urjashi Bhattacharya School of Agriculture and Allied Sciences, The Neotia University, Jhinga, Sarisha, West Bengal, India

Tanmaya Kumar Bhoi Forest Protection Division, ICFRE-Arid Forestry Research Institute (ICFRE-AFRI), Jodhpur, Rajasthan, India

Rohit Bishist Department of Silviculture and Forestry, College of Forestry, Dr YS Parmar University of Horticulture and Forestry Nauni, Solan, HP, India

Quan Cao College of Animal Science and Veterinary Medicine, Qinghai University, Xining, China

S. Chakravarty MS Swaminathan School of Agriculture, Centurion University of Technology and Management, Paralakhemundi, Odisha, India

Saroj Chauhan Department of Agricultural Entomology, Institute of Agriculture and Natural Sciences, Deen Dayal Upadhyaya Gorakhpur University, Gorakhpur, India

Deepali Chittora Microbial Research Laboratory, Department of Botany, University College of Science, Mohanlal Sukhadia University, Udaipur, India

A. Dabral Department of Forestry, North Eastern Hill University, Tura, Meghalaya, India

Bishnuprasad Dash Department of Soil Science, M.S. Swaminathan School of Agriculture, Centurian University of Technology and Management, Paralakhemundi & Jatni, Odisha, India

Umakanta Dash ICAR-Indian Agricultural Research Institute, Dirpai, Assam, India

Akoijam Ranjita Devi SRM College of Agricultural Sciences, SRM Institute of Science and Technology, Chengalpattu, Tamil Nadu, India

Yendrembam K. Devi Department of Entomology, Lovely Professional University, Phagwara, Punjab, India

Vinod Kumar Dubey Department of Entomology, Dr. Rajendra Prasad Central Agricultural University, Pusa, Samastipur, Bihar, India

Asik Dutta ICAR-Indian Institute of Pulses Research, Kanpur, India

Dinkar Jagannath Gaikwad M. S. Swaminathan School of Agriculture, Centurion University of Technology and Management, Paralakhemundi, India

Krishan Lal Gautam Department of Silviculture and Forestry, College of Forestry, Dr YS Parmar University of Horticulture and Forestry Nauni, Solan, HP, India

Pushpa Gehlot Microbial Research Laboratory, Department of Botany, University College of Science, Mohanlal Sukhadia University, Udaipur, India

Mainak Ghosh Department of Agronomy, Bihar Agricultural University, Bhagalpur, Bihar, India

Varsha Giri Silviculture and Forest Management Division, ICFRE-Arid Forestry Research Institute (ICFRE-AFRI), Jodhpur, Rajasthan, India

Harun I. Gitari School of Agriculture and Enterprise Development, Kenyatta University, Nairobi, Kenya

Sarbasree Goswami Institute of Agricultural Sciences, Banaras Hindu University, Varanasi, India

Suraj Goswami Indian Institute of Forest Management, Bhopal, India

Hitesh Gupta College of Forestry, Ranichauri, Uttarakhand, India
V.C.S.G. Uttarakhand University of Horticulture and Forestry, Bharsar, Uttarakhand, India

Dimjaneng Haokip College of Forestry and College of Agricultural Sciences, FEEDS Group of Institution, Hengbung, Kangpokpi, Manipur, India

Akbar Hossain Division of Soil Science, Bangladesh Wheat and Maize Research Institute, Dinajpur, Bangladesh

Tripta Jain Microbial Research Laboratory, Department of Botany, University College of Science, Mohanlal Sukhadia University, Udaipur, India

M. Jaison Department of Agricultural Chemistry and Soil Science, Bidhan Chandra Krishi Viswavidyalaya, Mohanpur, Nadia, West Bengal, India

Hanuman Singh Jatav Department of Soil Science and Agricultural Chemistry, S. K.N. Agriculture University, Jobaner, Jaipur, Rajasthan, India

Department of Soil science and Agricultural Chemistry, Sri Karan Narendra Agriculture University, Jaipur, Rajasthan, India

Surendra Singh Jatav Institute of Agricultural Sciences, Banaras Hindu University, Varanasi, India

I. A. Jimoh Department of Geography, Gombe State University, Gombe, Nigeria

S. Senjam Jinus School of Horticulture, and School of Agricultural Sciences, Pandit Deen Dayal Upadhyay Institute of Agricultural Sciences, Utlou, Bishnupur and Kameng Khunjao Campus, Lamlai, Manipur, India

Athokpam Kalpana School of Horticulture, and School of Agricultural Sciences, Pandit Deen Dayal Upadhyay Institute of Agricultural Sciences, Utlou, Bishnupur and Kameng Khunjao Campus, Lamlai, Manipur, India

Abha Manohar Kandileri M. S. Swaminathan School of Agriculture, Centurion University of Technology and Management, Paralakhemundi, India

Sahely Kanthal School of Agriculture, Swami Vivekananda University, Kolkata, India

J. Komal Basic Seed Multiplication and Training Centre, Central Silk Board, Kharsawan, Jharkhand, India

S. Navaneetha Krishnan Department of Silviculture and NRM, Forest College and Research Institute, Mettupalayam, India

Tulika Kumari Dr. Rajendra Prasad Central Agricultural University, Pusa, Samastipur, Bihar, India

Sanjay Kumar Department of Agronomy, Bihar Agricultural University, Bhagalpur, Bihar, India

Vijay Kumar Department of Silviculture and Agroforestry, Kerala Agricultural University, Thrissur, Kerala, India

Suprabuddha Kundu School of Agriculture, Swami Vivekananda University, Kolkata, India

Kanhaiya Lal Department of Agronomy, Dr. Rajendra Prasad Central Agricultural University, Pusa, Samastipur, Bihar, India

B. Lalramhlimi Department of Vegetable Science, Bidhan Chandra Krishi Viswavidyalaya, Nadia, West Bengal, India

Ningthaipuilu Longmei College of Forestry and College of Agricultural Sciences, FEEDS Group of Institution, Hengbung, Kangpokpi, Manipur, India

Deepak Kumar Mahanta Forest Entomology Discipline, Forest Protection Division, Indian Council of Forestry Research and Education (ICFRE)-Forest Research Institute (ICFRE-FRI), Dehradun, Uttarakhand, India

Sagar Maitra M. S. Swaminathan School of Agriculture, Centurion University of Technology and Management, Paralakhemundi, India

Prasanta Kumar Majhi Regional Research and Technology Transfer Station (RRTTS), Odisha University of Agriculture and Technology (OUAT), Keonjhar, Odisha, India

V. Manimaran J K K Munarajah College of Agricultural Sciences, Erode, India

K. A. Manohar MS Swaminathan School of Agriculture, Centurion University of Technology and Management, Paralakhemundi, Odisha, India

Chowlani Manpoong Department of Forestry, Wildlife and Environmental Sciences, Guru Ghasidas Vishwavidyalaya, Bilaspur, Chhattisgarh, India

Bhanu Raj Meena Ramjas College, University of Delhi, Delhi, India

Rajesh Kumar Meena Dr. Rajendra Prasad Central Agricultural University, Pusa, Samastipur, Bihar, India

Sayon Mukherjee Institute of Agricultural Sciences, Banaras Hindu University, Varanasi, India

Siddhartha Mukherjee Faculty Centre of Agriculture, Rural and Tribal Development (ARTD), Ramakrishna Mission Vivekananda Educational and Research Institute, Ranchi, Jharkhand, India

Vandana Mutum School of Horticulture, and School of Agricultural Sciences, Pandit Deen Dayal Upadhyay Institute of Agricultural Sciences, Utlou, Bishnupur and Kameng Khunjao Campus, Lamlai, Manipur, India

Ankit Pandey Department of Forestry, Wildlife and Environmental Sciences, Guru Ghasidas Vishwavidyalaya, Bilaspur, Chhattisgarh, India

Astha Pandey Institute of Agricultural Sciences, Banaras Hindu University, Varanasi, India

K. S. Pant Department of Silviculture and Forestry, College of Forestry, Dr YS Parmar University of Horticulture and Forestry Nauni, Solan, HP, India

Vinita Partel Jawaharlal Nehru Krishi Vishwa Vidyalaya, Jabalpur, MP, India

Ruby Patel ICFRE-Institute of Forest Biodiversity, Hyderabad, Telangana, India

Abhik Patra Institute of Agricultural Sciences, Banaras Hindu University, Varanasi, India

Sarthak Pattanayak Krishi Vigyan Kendra, Odisha University of Agriculture and Technology, Puri, Odisha, India

Prem Prakash Department of Silviculture and Forestry, College of Forestry, Dr. YS Parmar University of Horticulture and Forestry Nauni, Solan, HP, India

Biswajit Pramanick Department of Agronomy, Dr. Rajendra Prasad Central Agricultural University, Samastipur, Bihar, India

C. N. Hari Prasath Department of Silviculture and NRM, Forest College and Research Institute, Mettupalayam, India

Adyasha Priyadarshini Institute of Agricultural Sciences, Banaras Hindu University, Varanasi, India

PriyaKumari Department of Agronomy, Bihar Agricultural University, Bhagalpur, Bihar, India

Om Prakash Raigar Department of Plant Breeding and Genetics, Punjab Agricultural University, Ludhiana, India

Manish Raj Department of Agronomy, Bihar Agricultural University, Bhagalpur, Bihar, India

Roma Rani Ramjas College, University of Delhi, Delhi, India

Arvind Rathore Indian Institute of Forest Management, Bhopal, India

B. Venkateshwar Reddy Forest College and Research Institute, Hyderabad, India

Gorantla Prathap Reddy Institute of Agricultural Sciences, Banaras Hindu University, Varanasi, India

Esmaeil Rezaei-Chiyaneh Department of Plant Production and Genetics, Faculty of Agriculture, Urmia University, Urmia, Iran

Vinay Rojaria Department of Genetics and Plant Breeding, BRD PG College, Deoria, India

Lalichetti Sagar M. S. Swaminathan School of Agriculture, Centurion University of Technology and Management, Paralakhemundi, India

Upasana Sahoo M. S. Swaminathan School of Agriculture, Centurion University of Technology and Management, Paralakhemundi, India

Biswabara Sahu Department of Soil Science and Agricultural Chemistry, Faculty of Agriculture, Siksha O Anusandhan Deemed to be University, Bhubaneswar, Orissa, India

Shubhashree Sahu College of Forestry, Ranichauri, Uttarakhand, India
V.C.S.G. Uttarakhand University of Horticulture and Forestry, Bharsar, Uttarakhand, India

Masina Sairam M. S. Swaminathan School of Agriculture, Centurion University of Technology and Management, Paralakhemundi, India

Ipsita Samal ICAR-National Research Centre on Litchi, Mushahari, Ramna, Muzaffarpur, Bihar, India

B. C. Sarkar Department of Forestry, North Eastern Hill University, Tura, Meghalaya, India

Sudip Sarkar ICAR-Research Complex for Eastern Region, Patna, India

Tanmoy Sarkar School of Agriculture, Swami Vivekananda University, Kolkata, India

Satdev Department of Soil Science and Agricultural Chemistry, Bihar Agricultural University, Bhagalpur, Bihar, India

Sudip Sengupta School of Agriculture, Swami Vivekananda University, Kolkata, India

Laila Shahzad Sustainable Development Study Center, Government College University, Lahore, Pakistan

ShailjaKumari Department of Soil Science and Agricultural Chemistry, Bihar Agricultural University, Bhagalpur, Bihar, India

Tanmoy Shankar M. S. Swaminathan School of Agriculture, Centurion University of Technology and Management, Paralakhemundi, India

Faiza Sharif Sustainable Development Study Center, Government College University, Lahore, Pakistan

Harish Sharma Department of Silviculture and Forestry, College of Forestry, Dr YS Parmar University of Horticulture and Forestry Nauni, Solan, HP, India

Kanika Sharma Microbial Research Laboratory, Department of Botany, University College of Science, Mohanlal Sukhadia University, Udaipur, India

Sanjukta Samurailatpam Sharma Institute of Bioresources and Sustainable Development, Regional Centre, Gangtok, Sikkim, India

Yuying Shen College of Pastoral Agriculture Science and Technology, Lanzhou University, Lanzhou, China

Susmita Shil Department of Silviculture and Agroforestry, Kerala Agricultural University, Thrissur, Kerala, India

G. Shukla Department of Forestry, Uttar Banga Krishi Viswavidyalaya, Pundibari, Cooch Behar, West Bengal, India

Hausila Prasad Singh Department of Genetics and Plant Breeding, BRD PG College, Deoria, India

Noopur Singh Department of Genetics and Plant Breeding, Institute of Agriculture and Natural Sciences, Deen Dayal Upadhyaya Gorakhpur University, Gorakhpur, India

Puja Singh Department of Soil Science and Agricultural Chemistry, Horticulture College, Khuntpani, Chaibasa, Birsa Agriculture University, Ranchi, Jharkhand, India

S. L. Singh Dr. Rajendra Prasad Central Agricultural University, Samstipur, Bihar, India

Sangeeta Singh Forest Protection Division, ICFRE-Arid Forestry Research Institute (ICFRE-AFRI), Jodhpur, Rajasthan, India

Satish Kumar Singh Institute of Agricultural Sciences, Banaras Hindu University, Varanasi, India

Rashmi Sinha Dr. Rajendra Prasad Central Agricultural University, Samstipur, Bihar, India

Trisha Sinha Department of Plant Physiology, Gopal Narayan Singh University, Sasaram, Bihar, India

Vishnu K. Solanki Jawaharlal Nehru Krishi Vishwa Vidyalaya, Jabalpur, MP, India

Sushant Department of Agronomy, Bihar Agricultural University, Bhagalpur, Bihar, India

Ningombam Swapana Department of Chemistry, Manipur Technical University, Takyelpat, Imphal, Manipur, India

G. Swathiga Department of Silviculture and NRM, Forest College and Research Institute, Mettupalayam, India

Prabhat Tiwari Department of Silviculture and Agroforestry, Rani Lakshmi Bai Central Agriculture University, Jhansi, Uttar Pradesh, India

J. Vanitha SRM College of Agricultural Sciences, SRM Institute of Science and Technology, Chengalpattu, Tamil Nadu, India

Preeti Vats Department of Forest Product and Wood Utilization, Kerala Agricultural University, Thrissur, Kerala, India

Vipula Vyas Forest Protection Division, ICFRE-Arid Forestry Research Institute (ICFRE-AFRI), Jodhpur, Rajasthan, India

Anam Waheed Sustainable Development Study Center, Government College University, Lahore, Pakistan

Zikui Wang College of Pastoral Agriculture Science and Technology, Lanzhou University, Lanzhou, China

Jyoti Yadav Microbial Research Laboratory, Department of Botany, University College of Science, Mohanlal Sukhadia University, Udaipur, India

Simran Yadav Indian Institute of Forest Management, Bhopal, India

Part I
Agro-Forestry-Inception for Restoring Soil Health (Physical, Chemical and Biological Properties)

Chapter 1
Soil Fertility and Soil Biodiversity Health Under Different Agroforestry Systems

Laila Shahzad, Anam Waheed, Faiza Sharif, and Maryam Ali

Abstract The rising climate crisis and the upsurge of Anthropocene badly impact the agriculture in the entire globe, thus impacting crop production and food security. Poor agriculture practices expose soil to the factors causing soil damages that led to the infertility of soil. Agroforestry is a sustainable option in all such scenarios not only to acquire more crop outputs but also to raise trees, livestock, and crop on the same piece of land and get maximum benefits. The current chapter covers how different agroforestry systems play a decisive role in climate change mitigation. As trees deposit carbon stock to their woody biomass and also fix it to the soil, the soil biota take advantage from it. It is a fact that agroforestry not only reduces greenhouse gas emissions but also can be adopted as a best climate adaptation practice to make agriculture risk resilient. The agroforestry systems need no or less chemical fertilizer and other expensive inputs, also offering more productivity supporting soil biota, carbon sequestration, wind shielding, etc. Therefore, farmer and landowner need to be aware about the agroforestry system profits and adopt agroforestry in agricultural practices to become ecologically and economically sustainable.

Keywords Agroforestry · Fertility · Biodiversity · Carbon sequestration · Soil · Sustainability · Climate change

1.1 Introduction

Pedosphere is a vital component of ecosystem, like oceans, contributing as the biggest carbon sink. The climate-agriculture relationship is directly correlated; however the sudden changes in climate will adversely impact agro-productivity and agro-biodiversity (Arora 2019). Apart from climate, agroecosystems are at the verge of breakdown because of anthropogenic activities. The change in climate is behind the boom of causative process that gave rise to Anthropocene (Nair et al.

L. Shahzad (✉) · A. Waheed · F. Sharif · M. Ali
Sustainable Development Study Center, Government College University, Lahore, Pakistan
e-mail: lailashahzad@gcu.edu.pk

H. S. Jatav et al. (eds.), *Agroforestry to Combat Global Challenges*, Sustainable Development and Biodiversity 36, https://doi.org/10.1007/978-981-99-7282-1_1

2021). In agriculture, land use management and practices are unsustainable, the reason of degradation of the agroecosystems and exposed system to climate change. The poor agricultural practices effects the soil fertility though world's majority of population relay on improved systems for their better livilihood. By crops, the nutrients are continuously removed from soil and cause deficiency of organic matter. The deficiency of important nutrients like NPK, zinc, etc. enhances infertility in soil, leading to low crop yield. However, the organic matter balance may be regained by some means of applications, in which fertilizers are common. But the burning of harvested crop residue ultimately again lessens the soil nutrients. Trees are a vital component to reduce vulnerability of agricultural systems and increase resilience against climate change. Planting trees and promoting forestry were found to be the most helpful and cheap sources to mitigate carbon dioxide and changing climate. With the series of benefits, trees not only increase the nutrient in soil but also gave sustainability and referred as an "evergreen agricultural" component by the International Council for Research in Agroforestry (ICRAF) (Mbow et al. 2014). As a potential, trees provide a constructive impact on soil overall properties and improve crop quality. In such prospective, the tree-soil relationship was found to be the best application to make agro-systems less vulnerable and more resilient against climate change (Dollinger and Jose 2018).

1.2 Agroforestry and Climate Change Mitigation

Apart from many traditional agricultural practices, agroforestry was still devised to be the most beneficial application to maintain soil fertility and provisioning soil biota in agroecosystems. Agroforestry works with the aim to optimize the interaction between the biological systems and physical component of environment to achieve productivity from land (Reppin et al. 2020). A well-managed agroforestry system is the best practice to mitigate climate change and make agro-systems more diversified and sustainable.

Like forest trees, the trees planted in agroforestry are the best sequesters for carbon and depositors for greenhouse gas emissions. Agroforestry is the integration of shrubs or trees to the crop land and animal-farm systems to achieve economic, social, and environmental benefits. Agroforestry is multifunctional; it not only sequesters the carbon emissions but also aims to benefit the farmer in many ways (Tschora and Cherubini 2020). The benefits may include risk reduction, higher yields, more pollinators, and capacity development for adaptation against climate change, with farmers getting more outcomes (Amare et al. 2019). The in-cooperation of agriculture with trees could be better understand by the tree cover around/in the agricultural land cause the wind breakdown and crop cover, which not only guard the top soil from erosion but also functioned as the cover crop in harsh climatic conditions. By agroforestry it is not necessary to convert the whole land into forest. The practice is done at the edge of the crop field in the form of riparian buffers or in the form of alley cropping by planting trees in rows between the fields (De Stefano

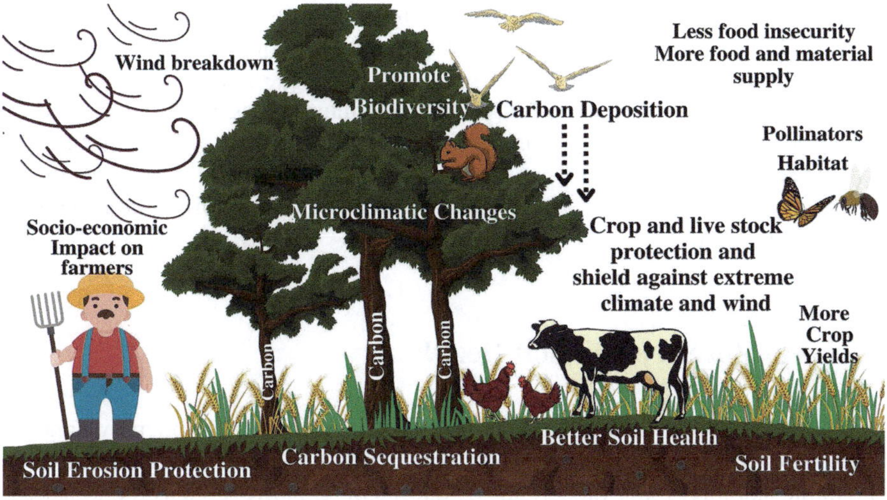

Fig. 1.1 Agroforestry and its foremost contributions

and Jacobson 2018). Even small number of farmers if forested their fields it will potentially significant to sequester carbon. A small piece of land converted to forested areas eventually play a significant role to sequester carbon.

Like riparian buffer and alley cropping, silvopasture system is also a significant mitigatory agroforestry practice against climate change. In silvopasture systems, trees are added to pasture land and sequester carbon from pasture and soils, also helping to reduce methane emission. This system also provides shade-induced area which gave microclimate changes to the place (Mayerfeld et al. 2016). All the agroforestry practices are significant in mitigating climate-induced risks and provide production and financial benefits to the farmers and increase livelihood of small-holder farmers. The trees are vital component for agriculture with huge economic impacts to the individual farmer and landowners as well. This tree crop integration system works differently in different regions of the world and plays an important role to abate negative climate impacts and diversify crops and carbon storage compared to system without trees.

Figure 1.1 is designed to explain the major benefits of integrating forest with agriculture. The tree covers before in and around the field protect the land with the damages caused by wind, provide home for soil biota, and prevent soil from erosion. The tree cover absorbs carbon emissions from air and deposits it into its woody biomass which can be locked up for a long period of time. The nutrients are balanced by the trees, and soil gets healthier and fertility increased. The trees also provide shelter and shade to the livestock and protect them from extreme weather conditions. The crop yields are improved and greener promoting pollinators to the place and more biodiversity to the fields. All of these gave advantage to the farmer or landowner and help in their socio-economic prosperity.

The changing climate is not a minor term; it's a major devastating change which needs comprehensive approaches to tackle it. Agroforestry is an integrated approach applied to reduce climate-related threats, improve landscape resilience, increase biological movements, provide favorable settings, and sequester carbon responsible for healthy soil (Toppo and Raj 2018). This technique helps land managers to achieve productivity and profitability, with less intensive agricultural practices, at the same time. Different agroforestry approaches apply in different settings to achieve social, environmental, and economic remunerations and to make landscape sustainable and healthier as well (Trozzo et al. 2014). These different settings are discussed in Table 1.1.

1.3 Agroforestry and Soil Health

In general trees are beneficial, both in rural and urban lands. They are providing best environmental benefits and ecosystem services to the surroundings by providing economic stability, food, timber, shade, and income. However, specific to agriculture, soil health, and fertility, trees provide positive impact on soil chemical and physical properties by optimizing nutrient cycling (Udawatta et al. 2017). A good soil is having good retention and infiltration properties. The more biological activities occur in the soil, the healthier the soil structure and fertility increases (Ollinaho and Kröger 2021). The good soil health achieved by soil protection occurs by working of different agroforestry systems together, providing benefits like crop covering, low tilling, crop rotation, and management of nutrients in the soil (Weerasekara et al. 2016).

Integrating trees with field soil, change soil properties in such a way that under trees the soil pH changes, cation exchange capacity enhances, and nutrient supply increases (Pinho et al. 2012). Soil under tree cover has higher microbial biomass and mineralizable potassium, nitrogen, phosphorus, and calcium, compared to field without tree crown (Geris et al. 2015). But if trees got fires, the changes will be relapse. Another advantage of tree-field integration is that it attracts the biota to the area, and the bird's dropping and cow dung improve the soil nutrient (Udawatta et al. 2021).

Another reason for adding up nutrient to soil by trees is the leaf litter. Although it sounds minor, it impacted a lot to enrich soil with nutrients. The dead organic matter of trees like falling leaf, fruits, flowers, branches, etc. enrich the top soil layer with nutrients expose to crop growth. The agroforestry systems are designed as continuous cycle so that they can be effectively used repeatedly by resource sharing between tree and soil compartments. The tree crown also provides shade helping to maintain moisture in the soil by lessening evapo-transpiration from the soil (Tsufac et al. 2019).

Like tree shoot, the tree roots are also step forward in providing benefits to agroforestry. The roots provide carbon enrichment to the soil by improving soil structure, by root turnover and discontinuing the nutrient leaching (Shi et al. 2018). Roots are also popular for their nutrient pumping in tree-based cropping systems. The tree roots extend deep down the soil, where the roots of crop cannot reach to take

Table 1.1 Description of different agroforestry types with climate change risk factors and adaptation services provided by different agroforestry techniques

Agroforestry practice type	Description	Benefits	Climate change risks	Adaptation services	References
Wind break (shelter belts)	The shrubs or trees are planted in the form of rows to protect crop from winds and sometimes livestock and people also (Fig. 1.3 A)	Shelter wind-sensitive crops Enhance C-storage Control snow dispersal Resist wind erosion Improved habitat for insects and wildlife Barrier to odor, dust, and chemicals Decrease animal stress Increase crop outputs	Frequent droughts More intensified storms Alteration in growing seasons Temperature extremes Winter storms Harvest failure due to other risks	Windbreaks reduce evapotranspiration Protection from wind damage Reduce climate stresses Crop diversification	Smith et al. (2021) Thevs et al. (2022) Buvaneswaran et al. (2018) Akinwalere and Okunlola (2019)
Alley cropping (intercropping)	The shrub or trees are planted in the form of alleys or rows between which crops are being produced (Fig. 1.3 B)	More C-storage in woody biomass Long-term crops Reduce erosion Minimize surface water runoff Enhanced microclimate Improved nutrient cycle Reduce offsite movement of nutrients Improve crop quantity and quality	Frequent rainfalls Intense precipitations Change cropping season due to temperature and precipitation Increase pest attack and diseases Crop failure by more risks	Slow water runoff Reduce flooding and soil erosion Crop protection by creating microclimate Pest control Home for beneficial insects Increase crop diversity	Wolz et al. (2018) Cary and Frey (2020) Dupraz et al. (2018) Ashraf et al. (2019) Sharma et al. (2020)
Silvopasture	Integration of livestock pasture with tree plantation (Fig. 1.3 C)	Escalate C-storage Minimize fuel usage Decrease nutrient loss Support livestock health Enhance livestock productivity Produce long-term and high-value products Produce livestock	Temperature increase Abrupt temperature extremes Crop failure	Minimize heat stress on livestock Provide shade and shelter Decrease crop loss by crop diversification	Castillo et al. (2020) Tourangeau and Sherren (2021)

(continued)

Table 1.1 (continued)

Agroforestry practice type	Description	Benefits	Climate change risks	Adaptation services	References
		diversification and more plant yields in less space and time			
Multistory cropping (forest farming)	Integration of understory and overstory shrubs or trees, grown together on the same landscape (Fig. 1.3 D)	Intensify crop diversification by allowing mixed and multiple heighted compatible crops at same place Upsurge C-storage Increase soil quality Upgrade nutrient cycling	Growing seasons have changed due to unexpected temperature extremes and unpredictable precipitations	Support cropping and provide microclimate to the crops	Riyadh et al. (2021) Negash et al. (2013)
Riparian forest buffer	Group of shrubs, trees, or vegetation adjacent to the stream, lake, and wetland (Fig. 1.3 E)	Manage stream flows Minimize NPS pollution Support terrestrial and aquatic habitat life Increase C-storage	Pest attack Insects and associated diseases	Reduce surface runoff Avoid flooding Prevent water pollution	Ellis (2020) Yonce et al. (2021)

Above-discussed different agroforestry techniques are used to maintain soil health and soil fertility, improve water and air quality, increase livestock and crop production, and provide home for livestock and wildlife and more revenue. The table also provides the view to understand different climate-induced risks and associated adaptation services provided by different agroforestry practices up nutrients from deeper soil, dumping them in the crop surface layer. This nutrient pumping depends on the tree type chosen for agroforestry, the topography of the land, and the moisture content of the soil (Isaac and Borden 2019).

For soil fertility, among all the agroforestry practices, silvopasture practice is the most effective one for farmers. This practice gave more diversity to the system via livestock, crops, and trees. This diversity associates the three components to benefit each other. For example, trees served as food for livestock and also provide shade; on the other hand, livestock fertilize trees, and likewise, crops serve as food for livestock and nutrients for trees (Mayerfeld et al. 2016). This interdependent link between trees, crops, and livestock makes silvopasture more diverse and a more soil-friendly mechanism for sustainable agriculture (Tsufac et al. 2019).

Briefly, agroforestry causes improvement in soil fertility by organic matter, biological N-fixation, minimized nutrient loss, discontinuing soil erosion, improved water holding capacity, nutrient recycling, and enhanced soil physical and chemical properties. Therefore, the agroforestry systems are supposed to be the best sustainable land use practice for the agriculture. The system enhances sustainability by biodiversity conservation, improved water and air quality, healthier soil, and climate mitigation by carbon sequestration. Although agroforestry is not the only way to have soil fertility, it is a significant component to support and maintain good soil health and structure (Elagib and Al-Saidi 2020).

All is not an ideal practice; the soil fertility may also decline, although using best agroforestry systems. Using tree-based cropping but facing reduced soil fertility is credited to poor agricultural practices. Many farmers do trees with their crops, but some physical, chemical, and geographic restrictions failed the agroforestry practice. Farmers claim slashing and burning of land and contour ploughing as physical; using of pesticides, insecticide, etc. as chemical, and the hilly topographic feature of land as geographical constraints in agroforestry systems (Baig et al. 2021).

1.4 Agroforestry and Soil Biodiversity

The tree-based cropping systems have a very high potential of diversifying soil organisms. The agroforestry systems provide habitat to the soil biota. The community residing in soil is a different world. There are a vast variety of microbial communities doing their vital roles in different nutrient cycles and supply, mineralization, chemical degradation, healthy soil components, and soil fertility. Not only the microfauna but also the macrofauna in the soil are also contributing their best to support healthy soil with agroforestry systems (Isaac and Borden 2019). These tree soil biota work together to improve soil quality and control soil from pest, diseases, and contamination as well.

Cropping without trees does not provide the nutrient enrichment and life support system to soil biota. Although using external efforts to the soil like fertilizers may enhance the functions for time, it did not provide long-lasting and sustainable aid (Biasi et al. 2017). The tree-soil association is very compromising and promising to

Fig. 1.2 Different agroforestry types under climatic stress

achieve sustainable outputs. Trees maintain micro-systems in soils for soil biodiversity. Tree roots produce carbon sugars utilized by microbes as energy; the tree litter also contributes to add on organic matter to the soil contributing in biota life support (Korboulewsky et al. 2016). Not only this, trees attract different birds and animals; the more the birds, the more will be their droppings and more contribution in the soil organic matter. Trees can also maintain the moisture content in soil, supporting microbial communities at apex. Meanwhile, the soil biological components, which are involved in decomposition, ecosystem engineering, nutrient transformation, and bio-controlling, supply nutrients to trees and crops, providing multiple benefits in return (Tedersoo et al. 2016). The carbon from litter and other means are transformed into soil organic matter by decomposer via series of enzymatic activities and mediate the nutrient cycle in the soil. The microbes with roots form biological aggregates which provide home for other soil biota. The soil creatures also defend by competition, parasitism, and predation against pests and diseases in the soil (Cherubin et al. 2019). All these factors promote soil biota to create strong communities in the soil, and by tree-soil interaction, they stand against pressure causing decline in their population. Use of agroforestry to enhance resilience is not new but a unique option for agricultural landscape management. In the context of biological diversity and life support, agroforestry works fabulously. For instance, habitat fragmentation in an area obstructs the species to move, relocate, or adapt against climate risks. In that certain scenario, agroforestry worked as a stepping stone or corridors in landscape and provided connectivity and enabled species to travel and breed (Mantyka-pringle et al. 2012). Figure 1.2 demonstrates the overall attributes of integrating agriculture

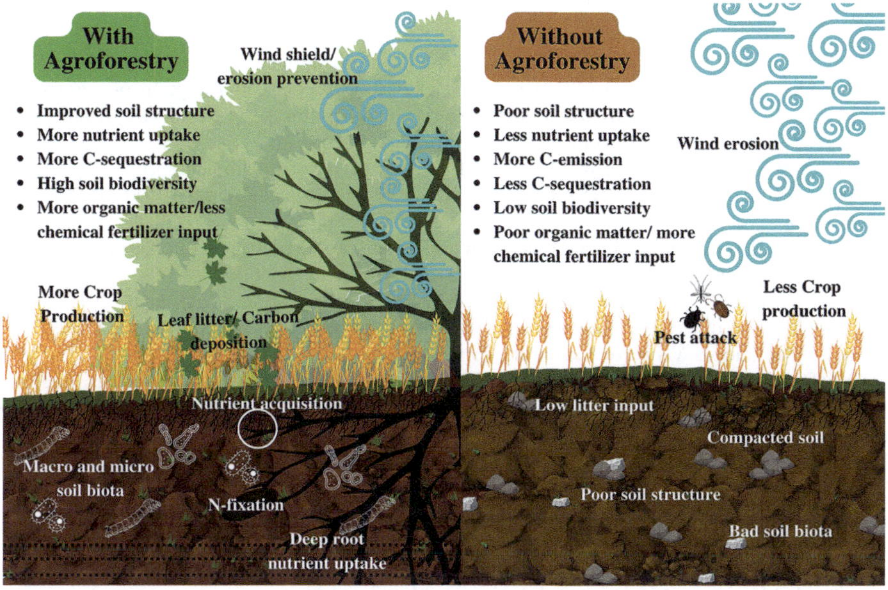

Fig. 1.3 Perspective on agroforestry encompassing land use changes

with trees to get more outputs with fewer inputs. Perspective on agroforestry encompasses land use changes which has been depicted in Fig. 1.3.

1.5 Conclusion

Farmers' livelihood is threatened by changing climate; the agroforestry systems have a real potential to mitigate climate change impacts and to resist associated risks. The trees have real power to preserve and strengthen the agricultural resources leading to food security and sustainability. It is a viable process that could be used anywhere, but the state of the art is to have knowledge about tree species, where and which type, tree-crop synergistic relationship, specie behavior against climate change, and farmer willingness, before practicing agroforestry. The knowledge gaps needed to be overcome on this scenario, especially on important cropping systems so that the agroforestry practices strengthen under scientific knowledge without the power of government or influential actors. Although the benefits and socioeconomic profits from trees are witnessed by farmers themselves, farmers still feel hesitant to invest on trees. So, by knowledge, awareness, and financial aid, farmers could achieve their best by agroforestry practices and combat on-field climate change impacts. The chapter was designed to address the tree-based farming systems and its ability to mitigate climate and insight of agroforestry science on which agroforestry future research can be built.

References

Akinwalere BO, Okunlola JO (2019) Roles of extension Services in Adoption of agroforestry practices among farmers in South West Nigeria. Int J Sustain Agric Res 6(1):1–7. https://doi.org/10.18488/journal.70.2019.61.1.7

Amare D, Wondie M, Mekuria W, Darr D (2019) Agroforestry of smallholder farmers in Ethiopia: practices and benefits. Small-Scale For 18(1):39–56. https://doi.org/10.1007/s11842-018-9405-6

Arora NK (2019) Impact of climate change on agriculture production and its sustainable solutions. Environ Sustain 2(2):95–96. https://doi.org/10.1007/s42398-019-00078-w

Ashraf M, Sanusi R, Zulkifli R, Tohiran KA, Moslim R, Ashton-Butt A, Azhar B (2019) Alley-cropping system increases vegetation heterogeneity and moderates extreme microclimates in oil palm plantations. Agric For Meteorol 276:107632

Baig MB, Burgess PJ, Fike JH (2021) Agroforestry for healthy ecosystems: constraints, improvement strategies and extension in Pakistan. Agrofor Syst 95(5):995–1013. https://doi.org/10.1007/s10457-019-00467-4

Biasi R, Brunori E, Ferrara C, Salvati L (2017) Towards sustainable rural landscapes? A multivariate analysis of the structure of traditional tree cropping systems along a human pressure gradient in a Mediterranean region. Agrofor Syst 91(6):1199–1217. https://doi.org/10.1007/s10457-016-0006-0

Buvaneswaran C, Vinoth KK, Velumani R, Senthilkumar S (2018) Potential of agroforestry to provide wood resources to Central Asia. J Tree Sci 37(1):11–18

Cary MA, Frey GE (2020) Alley cropping as an alternative under changing climate and risk scenarios: a Monte-Carlo simulation approach. Agric Syst 185:102938. https://doi.org/10.1016/j.agsy.2020.102938

Castillo MS, Tiezzi F, Franzluebbers AJ (2020) Tree species effects on understory forage productivity and microclimate in a silvopasture of the southeastern USA. Agric Ecosyst Environ 295:106917. https://doi.org/10.1016/j.agee.2020.106917

Cherubin MR, Chavarro-Bermeo JP, Silva-Olaya AM (2019) Agroforestry systems improve soil physical quality in northwestern Colombian Amazon. Agrofor Syst 93(5):1741–1753. https://doi.org/10.1007/s10457-018-0282-y

De Stefano A, Jacobson MG (2018) Soil carbon sequestration in agroforestry systems: a meta-analysis. Agrofor Syst 92(2):285–299. https://doi.org/10.1007/s10457-017-0147-

Dollinger J, Jose S (2018) Agroforestry for soil health. Agrofor Syst 92(2):213–219. https://doi.org/10.1007/s10457-018-0223-9

Dupraz C, Blitz-Frayret C, Lecomte I, Molto Q, Reyes F, Gosme M (2018) Influence of latitude on the light availability for intercrops in an agroforestry alley-cropping system. Agrofor Syst 92(4):1019–1033. https://doi.org/10.1007/s10457-018-0214-x

Elagib NA, Al-Saidi M (2020) Balancing the benefits from the water–energy–land–food nexus through agroforestry in the Sahel. Sci Total Environ 742:140509. https://doi.org/10.1016/j.scitotenv.2020.140509

Ellis CJ (2020) Microclimatic refugia in riparian woodland: a climate change adaptation strategy. For Ecol Manag 462:118006

Geris J, Tetzlaff D, McDonnell J, Soulsby C (2015) The relative role of soil type and tree cover on water storage and transmission in northern headwater catchments. Hydrol Process 29(7):1844–1860

Isaac ME, Borden KA (2019) Nutrient acquisition strategies in agroforestry systems. Plant Soil 444(1):1–19. https://doi.org/10.1007/s11104-019-04232-5

Korboulewsky N, Perez G, Chauvat M (2016) How tree diversity affects soil fauna diversity: a review. Soil Biol Biochem 94:94–106. https://doi.org/10.1016/j.soilbio.2015.11.024

Mantyka-pringle CS, Martin TG, Rhodes JR (2012) Interactions between climate and habitat loss effects on biodiversity: a systematic review and meta-analysis. Glob Chang Biol 18(4):1239–1252. https://doi.org/10.1111/j.1365-2486.2011.02593.x

Mayerfeld D, Rickenbach M, Rissman A (2016) Overcoming history: attitudes of resource professionals and farmers toward silvopasture in Southwest Wisconsin. Agrofor Syst 90(5): 723–736. https://doi.org/10.1007/s10457-016-9954-7

Mbow C, Smith P, Skole D, Duguma L, Bustamante M (2014) Achieving mitigation and adaptation to climate change through sustainable agroforestry practices in Africa. Curr Opin Environ Sustain 6:8–14. https://doi.org/10.1016/j.cosust.2013.09.002

Nair PK, Kumar BM, Nair VD (2021) Agroforestry and land Management in the Future. In: An introduction to agroforestry. Springer, Cham, pp 609–614

Negash M, Starr M, Kanninen M, Berhe L (2013) Allometric equations for estimating aboveground biomass of Coffea arabica L. grown in the Rift Valley escarpment of Ethiopia. Agrofor Syst 87(4):953–966. https://doi.org/10.1007/s10457-013-9611-3

Ollinaho OI, Kröger M (2021) Agroforestry transitions: the good, the bad and the ugly. J Rural Stud 82:210–221. https://doi.org/10.1016/j.jrurstud.2021.01.016

Pinho RC, Miller RP, Alfaia SS (2012) Agroforestry and the improvement of soil fertility: a view from Amazonia. Appl Environ Soil Sci 2012:616383. https://doi.org/10.1155/2012/616383

Reppin S, Kuyah S, de Neergaard A, Oelofse M, Rosenstock TS (2020) Contribution of agroforestry to climate change mitigation and livelihoods in Western Kenya. Agrofor Syst 94(1): 203–220. https://doi.org/10.1007/s10457-019-00383-7

Riyadh ZA, Rahman MA, Saha SR, Ahamed T, Current D (2021) Adaptation of agroforestry as a climate smart agriculture technology in Bangladesh. Int J Agric Res Innov Technol 11(1): 49–59. https://doi.org/10.3329/ijarit.v11i1.54466

Sharma H, Kumar D, Pant KS (2020) Agroforestry-a sustainable tool for climate change adaptation and mitigation. Agric Rev 41(4):364–371. https://arccjournals.com/journal/agricultural-reviews/R-1924

Shi L, Feng W, Xu J, Kuzyakov Y (2018) Agroforestry systems: meta-analysis of soil carbon stocks, sequestration processes, and future potentials. Land Degrad Dev 29(11):3886–3897. https://doi.org/10.1002/ldr.3136

Smith MM, Bentrup G, Kellerman T, MacFarland K, Straight R, Ameyaw L (2021) Windbreaks in the United States: a systematic review of producer-reported benefits, challenges, management activities and drivers of adoption. Agric Syst 187:103032. https://doi.org/10.1016/j.agsy.2020.103032

Tedersoo L, Bahram M, Cajthaml T, Põlme S, Hiiesalu I, Anslan S, Harend H, Buegger F, Pritsch K, Koricheva J, Abarenkov K (2016) Tree diversity and species identity effects on soil fungi, protists and animals are context dependent. ISME J 10(2):346–362. https://doi.org/10.1038/ismej.2015.116

Thevs N, Aliev K, Emileva B, Yuldasheva D, Eshchanova G, Welp M (2022) Potential of agroforestry to provide wood resources to Central Asia. Forests 13(8):1193. https://doi.org/10.3390/f13081193

Toppo P, Raj A (2018) Role of agroforestry in climate change mitigation. J Pharmacogn Phytochem 7(2):241–243

Tourangeau W, Sherren K (2021) Why is grazing management being overlooked in climate adaptation policy? Agroecol Sustain Food Syst 45(6):843–867. https://doi.org/10.1080/21683565.2020.1870645

Trozzo KE, Munsell JF, Chamberlain JL, Aust WM (2014) Potential adoption of agroforestry riparian buffers based on landowner and streamside characteristics. J Soil Water Conserv 69(2): 140–150. https://doi.org/10.2489/jswc.69.2.140

Tschora H, Cherubini F (2020) Co-benefits and trade-offs of agroforestry for climate change mitigation and other sustainability goals in West Africa. Glob Ecol Conserv 22:e00919. https://doi.org/10.1016/j.gecco.2020.e00919

Tsufac AR, Yerima BPK, Awazi NP (2019) Assessing the role of agroforestry in soil fertility improvement in Mbelenka-Lebialem, Southwest Cameroon. Int J Glob Sustain 3(1):115–135. https://doi.org/10.9734/bpi/ciees/v7/11772D

Udawatta RP, Gantzer CJ, Jose S (2017) Agroforestry practices and soil ecosystem services. In: Soil health and intensification of agroecosystems. Academic Press, Cambridge, pp 305–333

Udawatta RP, Rankoth LM, Jose S (2021) Agroforestry for biodiversity conservation. In: Agroforestry and ecosystem services. Springer, Cham, pp 245–274

Weerasekara C, Udawatta RP, Jose S, Kremer RJ, Weerasekara C (2016) Soil quality differences in a row-crop watershed with agroforestry and grass buffers. Agrofor Syst 90(5):829–838. https://link.springer.com/article/10.1007/s10457-016-9903-5

Wolz KJ, Lovell ST, Branham BE, Eddy WC, Keeley K, Revord RS, Wander MM, Yang WH, DeLucia EH (2018) Frontiers in alley cropping: transformative solutions for temperate agriculture. Glob Chang Biol 24(3):883–894. https://doi.org/10.1111/gcb.13986

Yonce HN, Sarkar S, Butcher JB, Johnson TE, Julius SH, LeDuc SD (2021) Forest riparian buffers reduce timber harvesting effects on stream temperature, but additional climate adaptation strategies are likely needed under future conditions. J Water Climate Change 12(5): 1404–1419. https://doi.org/10.2166/wcc.2020.031

Chapter 2
Agroforestry: A Resource Conserving Technology for Efficient Utilization of Agricultural Inputs, Leads to Food and Environmental Security

Sagar Maitra, Urjashi Bhattacharya, Biswajit Pramanick, Lalichetti Sagar, Dinkar Jagannath Gaikwad, Sarthak Pattanayak, Masina Sairam, Upasana Sahoo, Hanuman Singh Jatav, Harun I. Gitari, Tanmoy Shankar, Abha Manohar Kandileri, Esmaeil Rezaei-Chiyaneh, and Akbar Hossain

Abstract The agroforestry system (AFS) is established for on-farm and off-farm revenues both from tree and crop production by utilizing natural resources. In AFS, trees plantation on farms affords various livelihood welfares and environmental sustainability. Earlier and contemporary evidence undoubtedly highlights that AFS is a viable land-use option for alleviating poverty and compromises several

---

S. Maitra (✉) · L. Sagar · D. J. Gaikwad · M. Sairam · U. Sahoo · T. Shankar · A. M. Kandileri
M. S. Swaminathan School of Agriculture, Centurion University of Technology and Management, Paralakhemundi, India
e-mail: sagar.maitra@cutm.ac.in

U. Bhattacharya
School of Agriculture and Allied Sciences, The Neotia University, Jhinga, Sarisha, West Bengal, India

B. Pramanick
Department of Agronomy, Dr. Rajendra Prasad Central Agricultural University, Samastipur, Bihar, India

S. Pattanayak
Krishi Vigyan Kendra, Odisha University of Agriculture and Technology, Puri, Odisha, India

H. S. Jatav
Department of Soil Science and Agricultural Chemistry, S.K.N. Agriculture University, Jobaner, Jaipur, Rajasthan, India

H. I. Gitari
School of Agriculture and Enterprise Development, Kenyatta University, Nairobi, Kenya

E. Rezaei-Chiyaneh
Department of Plant Production and Genetics, Faculty of Agriculture, Urmia University, Urmia, Iran

A. Hossain
Division of Soil Science, Bangladesh Wheat and Maize Research Institute, Dinajpur, Bangladesh

H. S. Jatav et al. (eds.), *Agroforestry to Combat Global Challenges*, Sustainable Development and Biodiversity 36, https://doi.org/10.1007/978-981-99-7282-1_2

environmental benefits including carbon sequestration, conservation of biodiversity, improves soil health and enhances the quality of water and air. Considering the food security of the increasing population and the future projected climate change, AFS may be useful as a cost-effective resource to boost-up the sustainability of food production, and also mitigate the adverse of climate change. For developing countries both in the tropical and subtropical regions, AFS may be considered one of the suitable options because of its multifaceted benefits focusing on agricultural sustainability. The present review deals with the principles, practices, and advantages of AFS system for food and environmental security and agricultural sustainability.

Keywords Agroforestry · Food security · Environment · Security · C-sequestration · Advantages

2.1 Introduction

The agroforestry system (AFS) is an ancient practice and probably it is as old as agriculture itself. Many centuries witnessed agroforestry in different countries. The home-gardening, a form of agroforestry, is an old practice connected with the farmers of the humid tropics since 10,000 BC (Nair and Kumar 2006). Similarly, the introduction of domestic animals into the forest was common to the people of Europe during 4000 BC (Nair 2014). But during the second half of the twentieth century, the agroforestry system lost its importance due to policy matters adopted by different countries. In the developing world, different countries focused on self-sufficiency in food production and invited industrialized agriculture with new crop ideotypes/cultivars, chemical fertilizers and pesticides, creation of irrigation facilities and arrangement of farm mechanization. These practices were recognized as the Green Revolution (GR) and technologies were known as Green Revolution Technologies (GRTs). Thus, monoculture started to dominate replacing ancient practices. The new technologies adopted also resulted in quick enhancement of crop productivity and the developing countries witnessed a boost in food-grain production at that time it was believed that modern technology-based agriculture was the suitable option for alleviation of hunger. Simultaneously, some poor farmers in tropical and sub-tropical regions continued their old practices of crop cultivation under mixed stand due to different reasons, especially non-capability to invest much favouring high-energy modern agriculture.

On other words it may be mentioned that poor farmers did not adopt modern agriculture due to their economic inefficiency and the GRTs did not address the issues of the economically weaker section of the farmers; therefore, they preferred to sustain with mixed cultivation. But after a few decades, it was further noticed that the supply-driven and high-input-based agriculture failed to assure agricultural sustainability and created degradation of the agro-ecosystem (Lichtfouse et al. 2009; Maitra et al. 2019). Moreover, random deforestation in the tropics caused due to the expansion of farmland for food-grain production ultimately created enormous social and environmental problems. Considering the above limitations of GRTs, researchers started to search for suitable options for farming which will be sustainable, socially acceptable and economically viable. For the developing countries of the tropical and subtropical regions, AFS may be considered as one of the suitable options because of its multifaceted benefits focusing on agricultural sustainability. In support of AFS, the World Agroforestry Centre was established in 1977 in Nairobi, Kenya (which was earlier known as the International Council for Research in Agroforestry) and provided efforts to institutionalize agroforestry as a multidisciplinary science (Nair and Garrity 2012). Gradually, agroforestry gained the attention of researchers and policymakers of different countries. Even in the industrialized and developed world inclusive of Europe, Australia and New Zealand attention was given to AFS (Polglase et al. 2008; Rigueiro-Rodriguez et al. 2008; Mosquera et al. 2012). The present review article focused on the principles, practices and advantages of the AFS system for food and environmental security as well as agricultural sustainability.

2.2 Agroforestry as Low Input Agriculture

During the present time agriculture is facing a tremendous challenge in developing countries with an ever-increasing target for production with gradual progression for the fulfilment of present and future needs with gradually declining resources. Degradation and shrinkage of natural resources are major concerns to achieving the present target as well as ensuring production sustainability in the future. The second half of the previous century witnessed a quantum jump in food production through the adoption of GRTs. But there is no doubt that input-driven GRTs caused a lot of harm within a few decades, triggering the degradation of natural resources with ecological unbalance (Maitra et al. 2018). In this regard, low input technologies are regaining importance because these have less carbon footprint and thus lead to a greener earth. AFS is an age-old system with a promise of safer ecology and several other benefits in terms of conservation of natural resources, (especially soil) maintenance of proper nutrient cycle and C sequestration, enhancement of soil fertility, better management of marginal lands, improvement of biodiversity and higher ecological services (Fahad et al. 2022). Further, AFS is regarded as low-input agriculture because unlike industrialized agriculture in AFS, demand is less for high-energy chemical inputs, assured supply of irrigation and mechanized operations with the combustion of fossil fuel. Under arid and semi-arid conditions in tropical countries, high energy inputs may not yield satisfactorily; whereas, low input AFS can exhibit multifaceted benefits as mentioned confirming its journey towards a safer planet (Nöldeke et al. 2021). Now, in the present context of global warming and climate change (Bhadra et al. 2021), it is the appropriate time to relook into the hidden treasures of ancient practices where forests can be nurtured in parity of the landscape along with the production of foods and economic crops under AFS aiming fulfilment of some of the SDGs and livelihood security.

2.2.1 Concept and Principles of Agroforestry

In general, AFS is the purposeful growing of crops and trees with an interaction among themselves targeting a wide range of outputs (in terms of food grains, commodities, or products) along with superior environmental impacts and improved ecological services (Nair 2014). AFS is a combination of trees and/or shrubs and crops or animals accommodated spatially or temporally. Such arrangements result in noteworthy interactions between the woody and the non-woody constituents producing more than one output for economic gain and creating complex ecological combinations (Luedeling et al. 2016; Muschler 2015). Nair (2014) suggested four 'I's to describe the spirit of AFS and these are intentional, intensive, interactive and integrated. He further described that the word 'intentional' is indicative of intentional design and management of the system and 'intensive' denotes intensive management to achieve more benefits. These two components are 'interactive' and 'integrated' to harness the goal of the system. Sometimes AFS systems are categorized by three attributes, namely, productivity, sustainability and adaptability. Considering all the above concepts and ideas, International Council for Research in Agroforestry mentioned that AFS is 'a dynamic, ecologically sound system of natural resource management; by integrating trees on farms and in the agricultural landscape, it helps diversify and sustain production for enhanced economic, environmental and social benefits' (Nair 2014).

2.3 Ecological Base of Agroforestry

Agroforestry is a concept which is grounded on the evidence that better utilization of resources than monoculture and in this direction structural and functional complexities are created in land-use system and more output along with ecosystem services and nutrient cycling (Fahad et al. 2022). Undoubtedly, above- and below-ground diversity enables a relationship with forests and landscapes and thus ensures a multi-functional land-use system. The multi-species homegardens and silvopastoral systems are common examples where ecological sustainability is prominently visible. The complexity may result in beneficial impacts like increased productivity of diversified crops, nutrients cycling and C sequestration, enhancement of soil fertility and congenial micro-climate for different species (Nöldeke et al. 2021). However, the negative influences may also be observed in terms of allelopathy, competition among species and incidence of more pests and diseases due to micro-climate and the presence of alternate hosts in AFS (Nair et al. 2008). Further, Nair et al. (2008) suggested four ecological properties to understand the system design and management of AFS and these are: spatial and temporal heterogeneity, disturbance, perennials, and structural and functional complexity (Fig. 2.1). These ecological properties together create a more complex energy flow within AFS.

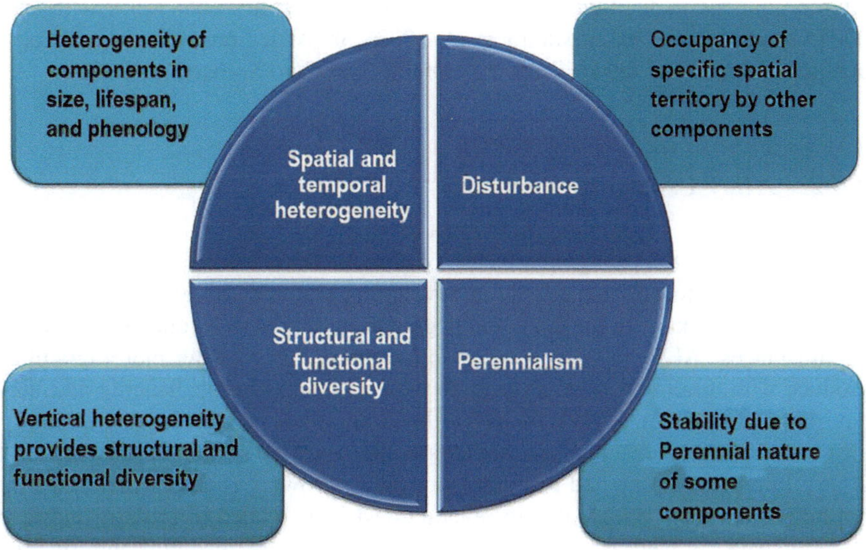

Fig. 2.1 Ecological properties of agroforestry to understand system design and management

2.4 Practices and Systems

There are numerous examples of AFS systems existing in the tropical and temperate regions which can be categorized into the following groups. Different characteristics of the various AFS systems are also mentioned.

2.4.1 Improved Fallows

The word 'fallow' means a temporal sequence of crop or tree species on a given piece of land (Muschler 2015); however, the improved fallows are the thoughtful planting of legumes targeting the improvement of soil fertility (Buresh and Cooper 1999; Jena et al. 2022). Green manuring is practised in suitable crop rotation in improved fallows with different economic crops. But the choice of woody species in improved fallows is a new practice in AFS systems (Kang 1993). The practice is also known as 'managed fallow' to differentiate the system from natural fallows. The annual crop species chosen for improved fallows are from the genera of *Cajanus*, *Crotalaria*, *Mimosa*, *Sesbania* and *Tephrosia* and tree species considered are *Acioabarteri*, *Anthonotha macrophylla*, *Alchornea cordifolia*, *Casuarina species*, *Gliricidiasepium*, and *Leucaena leucocephala* (Muschler 2015). The woody plants are allowed to grow during short fallow periods (less than three years) and these improve soil along with some economic products (Nair et al. 2010). Improved management practices may also be included like inoculation of N-fixer microorganisms to facilitate crop growth and soil fertility enhancement (Stamets 2005; Maitra et al. 2022; Maitra et al. 2023). In Africa and Latin America, the practice of improved fallow is more common (Kass and Somarriba 1999).

2.4.2 Alley Cropping

At present, under the changing scenario of climatic aberrations, it is very difficult to attain food and nutritional security in developing countries because of the growing population, shrinkage in the land, and degradation of natural resources. Under these circumstances, AFS may give enough opportunity for growing more agricultural produce with minimum or no damage to the ecosystem, even enhancing a healthier environment (Xu et al. 2019). In AFS, alley cropping is an excellent opportunity to grow crops and to ensure some benefits of the forestry system (Solaimalai et al. 2005). In erosion-prone and resource-poor areas, growing food or other crops may not be economically viable due to several constraints related to ecology and facilities, but AFS, as well as alley cropping, can fulfil short-, medium- and long-term requirements.

Alley cropping, also known as hedgerow intercropping, is the growing of food, forage or specialty crops in the space available between tree rows and it is a type of intercropping (Maitra et al. 2021). In alley cropping, annual crops, perennial shrubs, and tree crops are included which give yield output periodically. Generally, in alley cropping, annual food crops are cultivated in alleys comprised of hedgerows of shrubs or trees (Wolz and DeLucia 2018). Mainly in Indian conditions, alley cropping is practised targeting grain and forage needs. In the resource-poor marginal lands, investment in agriculture is less and farmers do not use costly inputs. Under this situation, alley cropping may be considered a suitable option for efficient land use and sustainable agricultural production.

Alley cropping creates a balanced diversification of farms with scope for short- and long-term cash flow from marginal lands. Control of soil erosion and fertility enhancement are additional benefits with desirable yield output from crops. The greater resource utilization (land, soil moisture and sunlight) is one of the important benefits of alley cropping. In the semi-arid tropics, alley cropping created competition among the species cultivated for soil moisture which resulted in reduced yields (Singh et al. 1989). Alley cropping with perennial pigeon pea + groundnut recorded 64.9% more pigeon pea equivalent yield than when pigeon pea was cultivated with sorghum alley cropping system (Solaimalai et al. 2005). The yield performance of cowpea in alleys of *Gliricidiasepium* and *Leucaena leucocephala* was better than pure cropping of cowpea in Guyana and alley cropping was recommended to enhance annual crop yield with reduced chemical N fertilizers (Chesney et al. 2010). Annual and specialty crops provide frequent cash flow whereas medium and long-duration income comes from trees as farm-grown fuel-wood and timber (Nair 2014). In alley cropping, the soil surface is covered by annual crops and trees make a network of roots, and thus soil erosion is reduced.

Planting of tree crops *Gliricidiamaculata* and *Leucaena leucocephalain* parallel alleys on 8–10% sloppy land 5–10 m wide is suggested in the eastern region of India to check soil erosion. In low rainfall areas trench planting and in high rainfall areas ridge planting is recommended for hedgerows (Madhu et al. 2019). The area between two alleys can be used for the growing of annual crops. *Gliricidia* alley cropping system can reduce runoff by 27.9–28.2% and soil loss by 49.3–51.1% over no alley cropping. Similarly, *Leucaena* alley with a miniature trench can reduce runoff by 18.3–18.7% and soil loss by 37.2–43.0%. Further, alley cropping can check the nutrient loss from the top soil. *Gliricidia* alley can conserve organic C, N, P and K by 63.4, 5.0, 0.3 and 2.4 kg ha^{-1}, respectively. However, *Leucaena* alley can conserve organic C, N, P and K by 57.7, 4.6, 0.3 and 2.2 kg ha^{-1}, respectively, over no alley cropping (Madhu et al. 2019). In Kenya, AFS enhanced carbon storage in the dry region (Reppin et al. 2019). Alley cropping with *Gliricidiamaculataand* *Leucaena leucocephalain* the eastern region of India can store 4.26 and 4.52% more soil organic carbon, respectively, than no alley cropping system in sloppy areas. Moreover, AFS as well as alley cropping supports habitat for different birds and bats and thus further enriches biodiversity (Harvey and Villalobos 2007). AFS, as well as alley cropping (a type of AFS), is considered one of the suitable options to contribute to climate change mitigation and adaptation advantages (Zomer et al. 2016). Reppin et al. (2019) reported that in Kenya, the adoption of AFS increased carbon storage in soil and upgraded the livelihoods of smallholder farmers. Presently it is considered one of the important adaptation strategies to minimize the adverse effects of climate change (Kristjanson et al. 2012).

2.4.3 Homegarden, Shaded Perennials and Multi-Strata Systems

Homegarden is broadly used to mean various practices adopted in tropical and sub-tropical regions ranging from raising vegetables in the backyard system to multi-strata systems (Nair 2014). Homegarden is adopted by small and marginal farmers for subsistence based on the ecological and socio-economic perspectives, preference of the grower and dietary habit, and demand in the local market. A combination of fruit plants, food crops and vegetables is common in homegardens. In other words, it may be stated that homegardens provide food and nutritional security to smallholders in the tropics and subtropics. The inclusion of legume species further improves soil quality (Blaser et al. 2013).

Shaded-perennial system of AFS is to some extent like a homegarden, and in the tropics perennials chosen for the purpose are areca nut, oil palm, rubber, coconut, cacao, cashew, tea black pepper and so on (Nair 2014). However, the focus is given more to perennials ensuring efficient nutrient cycling and storage with reduced erosion and leaching.

2.4.4 Silvopastoral Systems

The phrase 'silvopastoral system' is indicative of a production system comprised of forest or tree or shrub and animal rearing. In both temperate and tropical countries, the silvopastoral system is the major form of AFS system. Among the tree-types fodder and timber-producing plant species are chosen in the system. In temperate regions of Australia and New Zealand, the USA and southern Europe, commercial trees are grown along with the provision of pasture and animals (Mackay-Smith et al. 2021). Mostly, trees do not produce any fodder and during the lean period of tree plantations, pasture and animals are included for augmentation of income. But in the tropics, unlike the temperate regions tree species are considered and fodder-producing perennials or woodlots are added with animals (Soni et al. 2016). Sometimes fodder perennials are grown as live fences and pruned or cut periodically to meet the requirement. In developing countries, the silvopastoral system is characterized by extensive grazing by a herd of animals as it is common in sub-Saharan Africa. Further, the silvopastoral system has enough potential for mitigation of climate change (addition of manures and grazing) and carbon sequestration (Eichhorn et al. 2006; Porqueddu and Franca 2013; Seddaiu et al. 2013). Other advantages are soil conservation, providing shade to cattle and water quality improvement.

2.4.5 Protective Systems

Protective AFS systems include trees and shrubs by planting windbreaks, riparian buffers, soil conservation hedges, and similar plantations to obtain ecosystem-protection benefits. Windbreaks are inclusive of hedgerows, timber belts and shelterbelts, and these are planted to provide protection to crops from winds and to manage livestock operations (Nair 2014). Moreover, windbreaks facilitate pollination as pollinators and other wildlife shelter on these (Ouin and Burel 2002; Brandle et al. 2009; Le et al. 2008). Livestock also gets the benefit of windbreak as these provide feed and reduce animal stress. However, timber breaks are created to obtain forestry products and these are well managed. Shelterbelts are a type of windbreak generally planted along the coasts to restrict encroachment by the sea. Further, shelterbelts protect the crops from damage caused by saline water and salt stress (Zhang et al. 2016). The strips of permanent vegetation are known as riparian and upland buffers comprised of grasses, shrubs and trees planted and managed simultaneously and riparian buffers are created in between natural waterbodies and cropland for reduction of run-off of water and nutrients, erosion, non-point source of pollution and improvement of wild habitat (Hartel et al. 2014). The shrubs and trees of riparian and upland buffers are pruned frequently and used for incorporation in the crop field and as animal feed (Papanastasis et al. 2008).

2.4.6 *Tree Woodlots and Specialty Crops*

Tree woodlots are the tree crops cultivated at the boundary or degraded lands for fodder, fire-woods, poles and small timbers as well as bioenergy production. Tree woodlots are planted for recovery of saline, alkaline and acidic lands, and mined lands. Fallow regeneration is also another programme adopted in Sub-Saharan Africa. Specialty crops are ornamentals, other high-value crops and honey, etc., which are very common in Canada (Nair et al. 2010). Bee farming is also adopted in Cuba. In AFS system bee farming is chosen because of the abundance of floral diversity and less use of pesticides (Muschler 2015). The specialty crops are selected based on local demand and significance. Aquaculture and bee farming are common features of AFS in Costa Rica and Cuba (Muschler 2015). The large tree woodlots contribute to the mitigation of climate change and C sequestration and specialty crops provide regular income and cashflow.

2.5 Animals in Agroforestry

Livestock is responsible for the contribution of greenhouse gas emissions (GHGs) to the atmosphere. Food and Agriculture Organisation (FAO) estimated that livestock directly or indirectly produces 18% of anthropogenic GHGs (which is equivalent to 7100 T_g CO_2) (Nair et al. 2010). Livestock constitutes an essential component of different AFS systems. In northern America, the silvopasture system is a common practice (Garrett 2009), however, in tropical countries also tree fodder is considered to feed the animals (Nair 2014). The traditional European AFS systems comprised livestock as the main component (Kadirvel et al. 2003; Mosquera et al. 2012; Plieninger et al. 2015; Moreno et al. 2018). In east Africa, livestock-based AFS system in the in-focus area of the integrated farming system where the leguminous fodder tree *Calliandra calothyrsus* is included to obtain green forage in addition to crop residues, Napier and natural grasses (Dawson et al. 2014) and 500 Calliandra trees can provide enough year-round feed to a dairy animal. Increased production from dairy animals and small ruminants was also noted due to the livestock-based AFS system adopted by smallholders in Africa (Franzel et al. 2014). The impact of livestock and AFS system on global warming has not been studied, but it can be mentioned that the inclusion of animals in AFS can reduce the carbon footprint in livestock farming practised in developing countries (Nair et al. 2010).

2.6 Geographical Distribution of Agroforestry Systems

There are limitations in the availability of areas under AFS systems adopted worldwide. Montagnini and Nair (2004) mentioned that 'with no reliable estimates on the extent of area and the gross variability expected in terms of tree species and soil attributes, it is an "almost insurmountable" task to estimate C stocks in AFS'. Based on information available, Nair et al. (2019) stated that there were 1023 million ha of land under AFS system. But the World Agroforestry Centre and collaborators mentioned that about 1.5 billion farmers in the world adopted AFS on around 1 billion ha (Zomer et al. 2009). The area with potential to adopt AFS like the degraded forest is not included in the estimate. In this regard, it may be mentioned that IPCC (2000) stated that about 630 m ha of land worldwide which was unproductive could be brought under AFS (Nair et al. 2010). Further, Nair et al. (2008) claimed correctly that farming and forestry are in close association in many places and interwoven to serve common targets and ecological benefits. Presently, in the context of climate change and food and nutritional security, AFS established its superiority over monoculture targeting sustainable agriculture and fulfilment of sustainable development goals (SDGs).

2.7 Sustainable Agroforestry System Design

Unprecedented challenges associated with food generation throughout the world are more aggravated by climate change, rising population, overexploitation of natural resources and loss of biodiversity. So, in the foreground of getting sustained yield to feed the projected 9.8 billion human population by 2050 (UN 2015) and to get multiple outputs along with environmental conservation and stability from limited sources of land, the age-old concept of AFS where trees, crops and livestock are grown together in a single unit of land became an imperative sustainable concept to adapt.

Agroforestry acts as a compact system of highly interrelated and interactive biological, chemical and physical processes. More than 46% of all farmland is surrounded by more than 10% tree cover (Zomer et al. 2016). In developing countries, the smallholder farmers are now taking up AFS practices more as a nonmainstream approach to cultivation with a target of achieving food security with maximum co-benefits. United Nations set up 17 SDGs in 2015, as a resolution called 'The 2030 Agenda' to reduce hunger and poverty to achieve a sustainable future (Fig. 2.2). SDGs are more pertinent for small-scale farmers who comprise of 90% of the developing world (Arze del Granado 2012).

The AFS has the potential to increase yield by twice that of the original yield if the system is properly managed especially when location specificity is taken into consideration (Waldron et al. 2012). The reasons for the increase in yield are attributed to various beneficial effects of trees like better infiltration rate, enhancing

Fig. 2.2 Sustainable Development Goals by UN (2015)

soil nutrients, preventing soil erosion, etc. It provides resilience to climatic extremes such as drought, flood, sudden high temperature and sudden low temperature, through better canopy cover, decrease in evaporation rate, increased infiltration of water in the soil, regulating temperature in its microclimate as well as soil, which coincides with the SDGs on climate action (World Bank 2015). It has nearly 200 million tonnes of carbon added to agricultural lands annually through AFS (Zomer et al. 2016). Trees pose as a buffer for food, fibre, fuel, fodder, and income when the crop fails. Small-scale farmers are always negatively affected by poor supply chain linkages and market fluctuations. Food sovereignty in local areas, through AFS, maintains the equity and dignity of the farmers which are often threatened in general, hence, meeting the motto of some of the SDGs like decent work and economic growth, reducing inequality and partnerships for the goals (Chappell et al. 2013). The need for deforestation is exponentially reduced since the trees from agroforests provide sufficient, cheap and easily available fuel-wood (affordable and clean energy), for which previously the women used to walk long distances for collection (Sharma et al. 2016). Now, the women can concentrate on education and taking care of their children or can focus on other sources of income targeting SDGs like good health and well-being, quality education and gender equality (Sharma et al. 2016). Further, it creates a huge ecological niche of habitats on-farm, harbouring rich biodiversity with interconnected complex biological cycles (life on land as stated in SDG 15).

According to the triple-bottom-line approach of any sustainable agricultural system, designs should satisfy the following objectives:

- Food and fibre demand
- Efficient use of on-farm resources with a view to conserving the environment and its natural cycles

Fig. 2.3 The triple bottom-line approach to sustainability

- Securing the profit of the agricultural system
- Improving the quality of life of people

A properly designed AFS will always fit into the triple-bottom-line approach of sustainable agriculture (Pereira and Martin 2021) as shown in Fig. 2.3.

2.7.1 Planet

Tending to the environment is the prime concern nowadays of any food production system provided the current climate change scenario. The advantages of practising AFSs which bring resilience to climatic variations are as follows.

2.7.1.1 Soil Conservation and Soil Fertility Improvement

Trees have the immense capacity of mining nutrients from subsoil which otherwise are most of the time unavailable to crops (since their roots extend to a lesser depth than trees), and they deposit the nutrients on the topsoil in the form of leaf litters or other organic biomass (Thorup-Kristensen et al. 2020). In the process, the organic nitrogen and phosphorus present within the biomass is mineralized in the form of plant-available nitrates and orthophosphates within the soil solution, improving the nutrient supply to both the crops and the trees. Agroforests act as a natural sink of atmospheric carbon dioxide and hence trap carbon within the terrestrial ecosystem.

Tropical rainforests can sequester 160 t of carbon ha^{-1} in aboveground biomass (Houghton et al. 1997).

Trees absorb methane emitted by paddy fields when grown together (Singh et al. 2012). The biomass, various root exudates, leaf litters, and metabolites released in huge amounts from trees act as a cementing material for soil aggregates making it more stable. Stable aggregates give rise to better soil structures together both the canopy cover and the organic residues check the temperature, from excessively rising during summer or reducing during winter. Leguminous trees can fix approximately 25–250 kg N ha^{-1} year^{-1} Giller and Wilson (1991). In an AFS, nutrient cycling and soil protection take place in a spontaneous and self-sustained manner.

2.7.1.2 Biodiversity Conservation

Growing different plant species of multi-strata on a single unit of the land itself increases the biodiversity of fauna and flora both above and below the soil surface (Nair 2013). AFS is a highly recommended alternative for the popular slash and burn cultivation, practised in tropical areas. Complex agroforests act as habitats for various birds, animals which might act as predators of other harmful insects and animals feeding on crops. The presence of trees alters and modifies the microclimate of crops and helps in curbing weeds, and harmful insects' pathogens by releasing various allelochemicals into the environment. The soil is the abode of various macro-fauna (ex-earthworms), micro-fauna (like nematodes, protozoa), microflora (bacteria, fungi, actinomycetes) aiding in nutrient cycling and bio-geo-chemical cycles (Nair et al. 2010). Many landraces, indigenous species and ethno-cultivars are conserved for a particular region. It forms a highly interactive matrix of different species which prevents habitats from loss and degradation basically mimicking the natural forest ecosystem. In rubber agroforests of Sumatra, plant species diversity is said to be 300 ha^{-1} which in a sole rubber plantation is 5 plant species ha^{-1} (Sanchez et al. 1997). The species diversity in the agroforest is found to be close to the species diversity of adjacent forests having approximately 420 plant species ha^{-1}. Bird species found in mature *damur* (*Shoreajavanica*) agroforests is 50% of rainforests (Thiollay et al. 1995).

2.7.2 *Profit*

Local site-specific AFSs, where farmers can grow indigenous high-value trees along with basic crops and the in situ procurement of inputs from within the system, will give them complete authority on farm inputs as well as production. The synergistic benefits of growing high-value crops along with low-value crops are risk management factors during the stress period. Better soil fertility adds to natural resources investment capital improving the productivity of the entire system. Growing of multipurpose trees like *Gliricidiasepium*, *Leucaena leucocephala*, *Casuarina equisetifolia*, *Acacia nilotica*, *Azadirachta indica*, etc., providing timber, firewood, fuel-wood, fruits, essential oils and other high-value products fetches additional income for the farmers apart from cash inflows from the basic crops. Trees reduce the dependence of the small-scale farmers on basic crops, in the years of crop failure.

The multiple arrays of products obtained from farm diversification intensify the small holder's income. Especially for smallholder farmers, it is suggested to grow high-value plants (multipurpose trees) on small scale and for large holdings, low-value crops like millets, maize, etc., depending on location specificity. In Africa, high-value crops are grown along with low-value crops. In 100 m^2 area, the farmers are growing French beans under contract farming which has an assured export market in Europe (Sanchez et al. 1997). It has been estimated that in Niger, Africa, a single tree can generate $1.40 per year by its various benefits of restored soil fertility, firewood, timber, high-value fruits, fodder, essential oils and various other products (Larwanou and Adam 2008) which would fetch an additional income of $ 56 per year and total income of $ 280 million.

2.7.3 People

A family forms the basic structure of a society, so to alleviate families from the current hunger and poverty crisis a sustainable low-cost, environmentally friendly agricultural system is of foremost importance (FAO 2012). When most of the farmers throughout the world are small-scale farmers, improving their income and livelihood by AFS is the grass-root level approach to eliminating poverty and hunger, and ensuring food and nutritional security. Trees generally require less labour for its maintenance which reduces labour costs as well as gives the farmers ample opportunity to work elsewhere for generating off-farm income. This is a nature-based system, where the use of external inputs like fertilizers and pesticides is minimum, leading to crop quality improvement. The perennial nature of the trees has the production cycle and provides income for the farm family for the long run. The stability of farm income and employment generated from AFS improves the standard of living (Fig. 2.4).

2.7.4 Designing Agroforestry Systems for Sustainability

The nature and arrangement of components of an AFS determine its sustainability for a particular area. More the complexity and species diversity of an AFS system, the better it upholds the productivity (food fibre, timber and other products) and protection service functions (prevention of soil erosion, reclamation of land, soil water conservation). AFS through an artificial ecosystem but the temporal and spatial arrangement of different trees with crops and livestock and the synergistic linkages of ecological pathways force it to behave like a natural ecosystem,

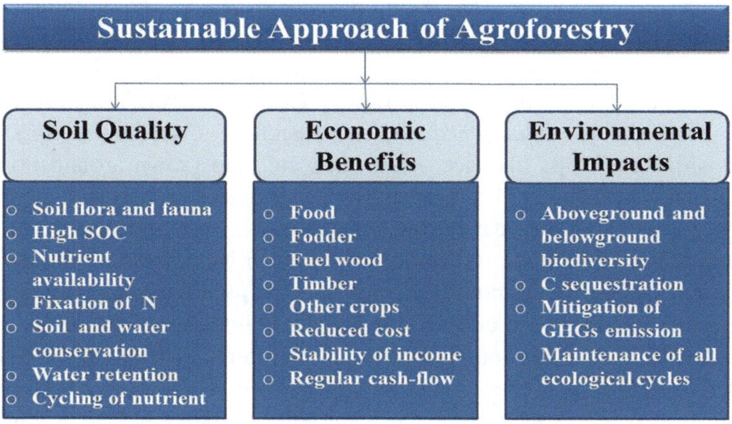

Fig. 2.4 A sustainable approach to agroforestry

transforming it into a win-win situation. Examples of certain practices exclusively designed to improve the sustainability of AFS systems throughout the world are as follows:

- The inclusion of trees in pasture lands of Florida checked the phosphorus loss from course soil to nearby water bodies, making the silvopastoral system sustainable by retaining nutrients within the soil (Jose and Dollinger 2019).
- In the temperate zones, trees are grown in single or double rows in wide spaces along with crops on the same land. Though the trees take up 10% of land area its multifaceted services and products surpass the loss and make the system ecologically and economically sustainable (Swieter et al. 2022).
- Growing of "fertilizer trees" like *Gliricidia, Sesbania*, *Tephrosia* in high plant densities in fallow lands of dry areas is a common practice. The organic biomass added to the soil increases soil fertility and nutrient use efficiency of the land without any addition of inorganic fertilizers within 2–3 years as compared to natural fallows which takes a time span of 15–20 years (Buresh and Cooper 1999).
- Planting *Faiderbia albida* in degraded dry soils of Africa is giving promising results in soil fertility restoration, improving the scope of the land for growing basic crops (Nair and Garitty 2012).
- Incorporating nitrogen fixing trees and shrubs, like *Gliricidiasepium* in 4–8 m apart rows in maize fields, is practised by a large number of small-scale farmers in Africa and other humid tropical areas (Liyanage et al. 1994; Pandey and Rai 2007). In India, incorporating perennial trees like *Grevellia robusta*, *Dalbergia latifolia*, etc. (exclusive to the Western Ghats region) in coffee plantations (shade tolerant crop) is widely practised in the multi-strata system. The dominance of perennial trees over annual crops results in a higher accumulation of nutrients within trees which in turn reduces the chances of soil erosion and leaching of

nutrients (Mohan Kumar 2007; Koutouleas et al. 2022). The growing of indigenous trees leads to the conservation of native landraces.

- A homestead garden having a combination of trees and crops with multiple strata is very common in humid regions of India. Coconut trees (*Cocos nucifera*), fruit trees like mango *(Mangifera indica*), jackfruit (*Artocarpus* sp.), banana (*Musa*sp); food crops like rice, sugarcane, etc., are grown around homesteads (Jamnadass et al. 2013). This kind of practice provides cash as well as nutritional security throughout the year to small-scale farmers around India.
- In high wind-prone areas, windbreaks and shelter breaks around the agricultural lands and pasture land should be grown. It helps in reducing wind speed by 20 times its height on the leeward side and 5–10 times its height on the windward side and checks soil erosion. The trees grown around the field act as a source of fodder, fruits, timber, fuel-wood, etc. (Brandle et al. 2004; Van Thuyet et al. 2014). It also serves as a home for local wildlife, birds and other animals and saves the agricultural land from snow and hailstorms.
- Multiple rows of trees should be planted around agricultural lands in coastal areas to protect crops from land inundation and salt water damage. This will eventually improve the sustainability of the agro-ecosystem which would otherwise be fragile due to sea encroachment. The trees act as guards against further soil degradation in coastal areas (Maity et al. 2018; Susware et al. 2021). In areas of heavy metal toxicity, land can be reclaimed by well-designed AFS systems. Augmenting trees such as *Populus* sp., *Salix* sp., *Sebertia acuminata*, etc., with crops, namely, sunflower, hemp, sugarcane, etc., depending on the climate situations of the particular place, can transform a land unsuitable for production into a sustainable fertile land giving guaranteed production from crops and other multiple benefits from trees (Sade 2020).
- In waterlogged areas where cultivation of normal crops is unsuitable can be reclaimed by growing a huge number of bio-drainage trees like *Eucalyptus* sp. for initial years, then using the land for further cultivation of food crops in between the trees in later years, when the water tables underground have receded (Maitra and Zaman 2017).
- *Prosopis cineraria* in pasture land is said to improve the organic carbon, available nitrogen, phosphorus, and potassium of the soil at the same time securing huge fodder, timber, fuelwood and food for the farmers (Yadav et al. 2011). *Acacia nilotica* based silvipastoral system especially growing conservative, less competitive grasses like *Eulaliopsis binata* is commonly practised in Shivalik foothills for rehabilitating degraded soils (Samra and Singh et al. 2000). Multipurpose trees like *Azadirachta indica*, *Eucalyptus* sp., etc., help in ameliorating sodic soils having high pH and electrical conductivity by their extensive root system, which breaks the soil compaction, improves infiltration of rainwater, eventually leaching the salts and bringing the pH to the desired level so that crops can be grown. Bamboo (*Dendrocalamus hamiltonii*) should be introduced in different AFS systems since it has a remarkable ability to improve potassium levels in the soil.

2.8 Role of Agroforestry

2.8.1 Food and Nutritional Security

An increase in demand for food, due to a rapid increase in population pressure consequently, raised concerns over food and nutritional security in the world (Arora 2019). This was aggravated by urbanization, resulting in over-pressurizing arable land through unprecedented use of agrochemicals and exploitation of scarce natural resources under changing climate (Godfray and Garnett 2014). Moreover, the lack of access to an adequate amount of high-quality food may lead to nutritional insecurity and malnutrition (Adesogan et al. 2020). Under these circumstances, AFS plays a very crucial role in doubling food production over the coming years which is attributable to its multifunctional approach, making it ecologically sustainable (Shukal et al. 2018).

AFS is a multifunctional landscape approach that supports diverse food systems, wherein woody perennials are grown in association with crops, pastures, livestock, etc. (Saikia et al. 2017; Mathuia et al. 2016). AFS involves the integration of more than two components, thus playing a complementary role to overall yield enhancement globally (Shin et al. 2020); and this was attributed due to the pumping of nutrients and water by deep-rooted trees from deeper layers, contributing to effective recycling of nutrients in the soil and improving soil fertility whose depletion was otherwise one of the major causes for food insecurity (Pierret et al. 2016). However, the role of forests in augmenting organic matter reduces evaporation, improves infiltration and reduces soil erosion, marked its scope towards amelioration of degraded lands (Sarkar et al. 2019; Lal 2015), which is attributed to enhanced biological nitrogen fixation, efficient nutrient cycling and deep capture of nutrients (Kumar 2016). Although resource conservation is not the primary goal of the AFS system due to its rich biodiversity, it plays an important role in resource conservation by improving soil water, soil organic matter (SOM) and nitrogen availability thus gaining sustainable productivity enhancement in associated crops (Fahad et al. 2022).

Several AFS components also have insect-pest repellent properties which in turn significantly minimized the damage due to pests and diseases creating conditions favourable for plant growth (Pumariino et al. 2015). However, healthy soil certainly enriches the nutrient content in the food thus minimizing malnutrition and building up nutritional security among the population (Berkhout et al. 2019; Bilali et al. 2018). In light of these facts, it is evident that the role of AFS is not restricted to the enhancement of yield alone and hence could be a befitting approach in achieving food and nutritional security.

2.8.2 Soil Conservation and Reclamation

Soil degradation is a serious threat to food and environmental security (Trivino et al. 2016; Bindraban et al. 2012). This was aggravated by field crop's ability to hold soil intact, resulting in accelerated soil and nutrient erosion (Jnr 2014; Trnka et al. 2016). In this context, the presence of trees as a component in AFS was perceived as a silver lining to address the issues of soil conservation effectively (Nainwal et al. 2016; Bashir et al. 2018). Owing to the deep root system, trees along with deep-rooted annuals explore a greater soil volume, thus enhancing the stability of soil aggregates to resist erosion (Ola et al. 2015). Besides, trees act as a barrier, slowing down runoff water velocities by resisting the movement of water and preventing the loss of suspended sediments and this character was enhanced due to the involvement of annual crops and animals (Piyuh et al. 2018; Udawatta et al. 2010; Jianbo et al. 2018). Tress as a component of AFS also bestowed the soil with a developed organic layer that tweaks the stability of soil aggregates ensuing in increased infiltration and reduced runoff (Vermang et al. 2015; Hosseini et al. 2016). Further, this was supported by increased soil organic carbon (SOC) under tree-based horticultural cropping system over crop and uncultivated land, respectively (Mandal et al. 2018), and similarly, a significant increment was observed in organic matter content and electrical conductivity without affecting the soil pH under eucalyptus based agri-Silvi-horticultural system (Johar et al. 2017, b). However, considering the role of the canopy cover of trees along with thick organic cover on the soil, turn down the impact of a raindrop on soil, thus eliciting the potential of AFS in nutrient and soil conservation (Zhu et al. 2020; Mahmud and Islam 2017). This is further supported by improved soil infiltration under single trees over open areas or trees with a termite mound in association in semiarid regions of Burkina Faso (Tobella et al. 2014).

Soil salinization is one of the major threats in the world. This was mainly aggravated due to an increase in groundwater level with reduced transpiration which was ascribed to the substitution of forests with annual crops (Banyal et al. 2017). Besides, the contributions of seawater and arid environments towards the development of soil salinity were noteworthy (Szymkiewicz et al. 2018). The aftermath of salinity is directly linked with significant yield reduction (Morniruzzaman and Shamim 2015; Chauhan et al. 2016); due to minimized water and nutrient intake by plants owing to increased osmotic pressure at higher pH (Munns et al. 2019). Moreover, the domination of sodium salts results in the deterioration of the physical properties of soil (Choudhary and Kharche 2018). The role of trees in minimizing the deposition of salt on the soil surface layer is influenced by the type of litter facilitating improved water infiltration for successful leaching of surface salts into the soil profile (Behera et al. 2015; Gelaye et al. 2019). Similarly, in a study conducted to assess chemical properties under different land-use systems, AFS-based systems (eucalyptus + turmeric) was found more efficient in influencing the soil chemical properties (Pandey et al. 2019). Further, this was supported by a study conducted in Punjab wherein lower pH values were recorded under the paddy-wheat cropping system over poplar-based AFS treatments, comparatively (Sharma et al. 2015). Most of the popular tree species in an agroforest were accomplished with higher salt tolerance like *Prosopis*, *Casuarina*, *Salix*, etc.. Moreover, when these tree species were added by adopting salt-tolerant annuals could show extended tolerance towards soil salinity (Sera 2017; Yaish and Kumar 2015). Concomitantly, before decomposition tree and crop residues on the surface serve as a mulch minimizing the accumulation of salts and promoting better surface characteristics (Aragues et al. 2014; Memon et al. 2017). Every so often in various types of research, certain tree species were found to exude acidic components, thereby significantly taking part in reclaiming saline soils (Behera et al. 2015). Therefore, the cumulative role of all components made the role of AFS inevitable in conserving soil and its reclamation ensuring sustainability (Alao and Shuaibu 2013).

2.8.3 Climate Change Adaptation and Mitigation

Globally, rapid climate change in the past few decades has a serious impact on agriculture questioning food security (Yadav et al. 2018). Consequently, its adaptation in addition to ideal mitigation strategies has become imperative for successful crop production (Wahid et al. 2015; Onu and Ikehi 2016). In this context, the role of AFS in adaptation and mitigation of climate change has been widely acknowledged (Zoysa and Inoue 2014). Its role in climate change adaptation and mitigation is attributed mainly due to the inclusion of trees in crop and pasture lands (Ogle et al. 2019; Bhatt 2018). Several studies showed that annual crops were more sensitive to extreme weather events than perennials (Johansson et al. 2013). Indeed, the integration of perennials with annuals were also found to be beneficial in coping with extreme weather conditions (Scaven and Rafferty 2013). The rich diversity in an AFS system largely involves trees with a mechanism to adapt extreme weather conditions at an expense of its carbon gain and growth (Nair et al. 2019; Kaul et al. 2011). However, improved organic matter accumulation through the continuous addition of litter and shade due to extensive tree canopy minimized evapotranspiration and thereby enhanced the ability of annuals in adapting to drought and heat stress (Wang et al. 2016). Subsequently, well-developed tree crown and deep root system add up its ability to counteract runoff and at the same time play an important role in modifying soil temperature (Zhang et al. 2019). Tree height and canopy growth further determine its capacity to resist fast-moving winds through shelterbelts and windbreaks consequently minimizing impact of direct sunlight and evaporation playing a significant role in moisture conservation (Dafa and Nawal 2016; Alemu 2016). As AFS is an integration of many diversified components, it thus creates many production opportunities and improves the risk-bearing ability under fluctuating climate (Prasad et al. 2016; Brown et al. 2018).

Accelerated increase in GHGs in the atmosphere cannot be combated efficiently for a long time merely through adaptation alone (Kweku et al. 2017). Further, it significantly contributed to global warming, a major threat to the ecosystem (Latake and Pawar 2015). To withstand this, there is a need for an approach that ameliorates

GHG from the atmosphere (Singh 2013; Saklani and Khurana 2019). In this context, AFS gained immense popularity because of its built-in mechanism to sequester carbon dioxide (Nair and Nair 2014). However, trees in this system sequester atmospheric carbon into its woody biomass both above and below the ground, substantially higher than that of herbaceous vegetation (Mayrinck et al. 2019). Above ground, sequestration was attributed due to increasing above-ground biomass production due to excessive branching and leaf-bearing ability, while prolific root growth exploring larger volumes of soil and litter-fall contributed to below ground sequestration in tree-based AFS systems (Chauhan et al. 2011; Meena et al. 2020). Besides, this system also contributes to the increase in the availability of fuel-wood serving primarily helping in the conservation of natural forests (Gronau et al. 2018) and minimizing burning of fossil fuel (Adhikari and Ozarska 2018).

2.8.4 Agroforestry and Ecosystem Services

The AFS system involves the integration of trees with annual crops and livestock ought to have potential to ecosystem services, namely, increases soil nutrient inputs, improves the sink capacity for enhanced carbon sequestration, improves water quality, enhances conservation of biodiversity, spikes soil productivity, protects and reduces nutrient losses. Ecosystem services are identified as recognized benefits that were believed to be achieved by the adoption of different AFS practices. Here, we have classified these services into five different headings, namely, increased soil nutrient inputs; soil productivity, protection and reduced nutrient loss; carbon sequestration; water quality enhancement and environment amelioration and biodiversity conservation.

2.8.4.1 Increased Soil Nutrient Inputs

Biological Nitrogen Fixation

Nitrogen is one of the most vital plant nutrients that plays a key role in plant growth and development (Leghari et al. 2016). Globally, rise in population concomitantly raises the demand for higher food production which could possibly be achieved through the attainment of the potential yield of nutrient responsive genotypes by supplying nitrogen countering its demand (Imran et al. 2015). Conventionally, this demand was met largely by mineral fertilizers and to some extent by biological nitrogen fixation (BNF) (Gendy et al. 2013). However, mineral fertilizers assure the timely supply of nitrogen at adequate amounts but concomitantly triggers environmental pollution (Ahman and Zhang 2018). Besides, indiscriminate use of fertilizers also contributes to eutrophication, species extinction, climate change, ozone depletion, etc. (Skiba and Rees 2014). Henceforth, the importance of BNF can be emphasized to achieve environmental security and improve soil health (Weisany et al. 2013).

Biological nitrogen fixation is an important biochemical process involving conversion of atmospheric nitrogen to ammonia by certain bacteria, actinomycetes, blue-green algae, fungi and yeast (Bhat et al. 2015; Chakdar et al. 2012). The atmosphere constitutes more than 78% of nitrogen by volume, thus integration of nitrogen-fixing trees/crops with crops acts as an alternate strategy to increase nitrogen availability without the addition of synthetic fertilizers (Islam et al. 2013). Subsequently, provide nitrogen as required to the non-nitrogen fixing crops within the AFS system (Verma et al. 2018). Further, the ability of nitrogen-fixing trees in fixing atmospheric nitrogen was usually governed by several factors, namely, effectiveness of microbial strains, soil available phosphorus, soil nitrogen content, soil moisture and soil pH (Hamza et al. 2017). Effectiveness of microbial strains significantly influences the plant nodulation and the amount of nitrogen fixed which was observed by Ndoye et al. (2015) in a glasshouse study resulting in a higher number of nodules in plants inoculated with rhizobial strains (ORS 3474, ORS 3593) in *Acacia senegal.* A field study conducted at Eastern Cameroon by 15 N dilution method (Sarr et al. 2016) estimated the increase in nitrogen significantly by 83% in *Pueraria phaseoloides* inoculated with *Bradyrhizobium strain* S3-4. Henceforth, this role enhanced complementarity between the components of AFS for nitrogen and helped the non-legume components to thrive well in nitrogen-deficient soils with extended scope in the reclamation of degraded soil (Stamford et al. 2013).

Deep Nitrate Capture

Nitrate form of nitrogen is the major source of nitrogen that determines the crop yields (Liu et al. 2014). It is susceptible to leaching, resulting in turmoil of nitrogen availability in the crop root zone, collaterally increasing the use of N-based fertilizers (Syahrawati et al. 2018). Nitrate leaching is an important component of N cycle in agriculture, with excess irrigation; nitrate leaching does occur which is compensated by using chemical fertilizers affecting farmer and environment simultaneously (Abrol et al. 2012; Nganchamung et al. 2017). Capturing the nitrate from the subsoil and pumping it back into the root zone offers a wide range of benefits to the ecosystem (Quemada et al. 2013). The role of AFS in deep nitrate capture was mainly attributed due to the deep roots of trees and to its SOM contribution (Zhu et al. 2020). Deep nitrate capture is one of the processes of nitrogen input besides biological nitrogen fixation (Querne et al. 2017). Since annual crops within the system have shallow root systems and cannot reach deep into the soil to utilize the nitrate lost through leaching, this is sufficed by the extended role potential of deep-rooted trees in nitrate recovery (Lan et al. 2012; Shah et al. 2017). Kevin et al. (2018) in an extensive research from 2013 to 2016 estimated the reduction in nitrogen losses after converting from row crop agriculture including maize-soybean rotation to alley cropping with mixed fruit and nut trees at a pomology research farm, and University of Illinois resulted in a significant reduction in nitrate leaching by 82–91% under alley cropping.

Further, an experiment was conducted for 6 long years in Brazil to estimate the nitrogen, calcium and potassium potential uptake by *Eucalyptus grandis* by injecting NO_3, 15N, Rb^+ and Sr^+ tracers through plastic tubes at various depths which revealed that the relative uptake potential of nitrate is more from deep soils. The nitrate that was captured from the deeper soil layers was utilized by the annual crops in the system after being added to the soil as leaf litter, root, leaf and branch pruning (Kaushal et al. 2012). According to Wang et al. (2011), the reduction in total subsurface nitrate flow estimated to ranging between 45 to 64 kg ha^{-1} in the mono-cropping system and 16 to 48 kg ha^{-1} in AFS, resulting in a long time for the citrus tree to reclaim the leached nitrogen. This highlights the role of adopting a land-use system that involves the integration of shallow-rooted crops with deep-rooted trees to enhance water and nutrient pumping thereby promoting the restoration of nitrate loss beyond the crop root zone (Ropokis et al. 2019).

Biomass Transfer

Biomass transfer is the process of the addition of plant biomass into the crop field to improve the soil organic matter content (Daniel et al. 2013). The woody perennials accumulate biomass for years mainly through enhanced carbon sequestration which was nutritiously enriched by BNF or deep nutrient capture, respectively (Moussa et al. 2015). This nutrient-rich biomass from multipurpose trees was added to crops under the AFS aim to supply essential nutrients for crop production upon decomposition (Hasan et al. 2019). The addition of leaves and twigs were the key sources of biomass transfer that occurs frequently in an AFS. However, the results of Oyebamiji et al. (2016) were quite appealing to this context since the investigation revealed that the maize sown in plots incorporated with *Albizia lebbeck* performed well both in terms of growth and yield attributes due to increased mineral nitrogen in soil resulting from the decomposition of added litter material. Moreover, the addition of litter contributes to build-up soil organic matter content in the soil surface which thereby serves as a reserve for consistent nutrient supply in the soil (Novara et al. 2015). The rate of decomposition and mineralization of the added litter further influences the utilization of biomass transferred by the crops growing under tree alleys (Maitra 2023). Research conducted under different silvopastoral systems found that organic matter within 0–2 cm was significantly improved under the semi dehiscent trees and thereby releases nutrients upon decomposition of fallen tissues (Lana et al. 2018). Similarly, the litter quality and individual nutrient concentrations influence the quantity of nutrient added into the soil such that the nitrogen and phosphorus uptake was found to be best under *Dalbergia sissoo* while potassium uptake was best under *Azadirachta indica* (Hossain et al. 2011). Hence, organic residue contribution from integrated trees within the AFS system act as a source of low-cost inputs in readily available form attributing for better plant growth.

Enhanced Nutrient Cycling

The role of AFS in nutrient cycling is more significant than other land use systems due to higher rates of turnover and lower rates of losses from the system (Seneviratne et al. 2015). This enhancement of nutrient cycling was attributed to this system due to enhanced interception of nutrients by deep-rooted trees and subsequent addition of the captured nutrients on the surface by added litter (Egnell et al. 2015). Further, due to the integration of trees, the loss of nutrients by harvest and dependence on external inputs were minimized (Pinho et al. 2012). In a view to restocking soil fertility, Yamada et al. (2016) recommended the integration of fast-growing trees. In research conducted using the 15N dilution technique, Augustin et al. (2017) reported that the age of the tree will not affect N-fixation instead it will be stored in standing litter, leaves, stem, branches, etc., sufficiently contributing to nutrient recycling in Acacia magnesium in the Philippines. Singh (2009) found a total addition of 176 kg/ ha N, 21.7 kg/ha P, 133 kg/ha K, 368 kg/ha Ca and 55.4 kg/ha S with an addition of 20.1 Mg/ha of dry litter from poplar plantations and subsequently with the increase in age of the nutrient concentration decreases. Similarly, Aguiar et al. (2010) reported significant addition of calcium (436 ± 18 kg ha^{-1}) and phosphorus (56 ± 3 kg ha^{-1}) with the surface incorporation of Acacia branches into the soil. Innangi (2017) revealed that the new shed litter of alder contains 30.3 mg/g of N on a dry weight basis. This process of capture and cycling nutrients is collectively termed nutrient pumping (Moore et al. 2013).

Soil Organic Nitrogen

Nitrogen is the most important nutrient required by the crop and being a constituent of biologically important molecules, namely, proteins, amino acids, porphyrins, nucleic acids, etc., it is required by the crop in relatively higher concentrations (Leghari et al. 2016). Besides, weathering cannot contribute to the addition of nitrogen to the soil and thereby SOM turned out to be the potential source of nitrogen in the soil (Schmidt et al. 2011). The inherent quality of materials constituted for the formation of SOM plays a major role in the mineralization and availability of nitrogen to the soil. This was in confirmation with the results of an experiment at Makoka Research Station, at Southern Malawi which reported inclusion of *Glyceridia* in the cropping system increased the POM-N by 86% (Beedy et al. 2010). Plant litter usually consists of water-soluble compounds, glycoproteins, polymer carbohydrates, lignin, lipids, waxes, etc. (Gispert et al. 2017). However, a high concentration of lignin in SOM reduced the rate of decomposition affecting the N mineralization. On the evaluation of lignin and cellulose dynamics in *Cistus incanus* L., *Myrtus communis* L. and *Quercus ilex* L. during litter decomposition found that an increase in lignin and cellulose content within the biomass raised the nitrogen immobilization (Fioretto et al. 2005).

Soil Organic Phosphorus

Phosphorus availability is the most essential and largely influences the productivity of an ecosystem (Liu et al. 2021). Due to rising concerns about inorganic phosphorus application on soil and water quality, it was imperative to integrate organic phosphorus with inorganic sources to increase its availability (Darch et al. 2014). AFS's role in the addition of organic phosphorus was highly attributed by heavy litter addition subsequently transferring the nutrient-rich biomass (Hossain et al. 2011). This added biomass acts as a reserve for organic phosphorus, namely, phosphates, monoesters and diesters which upon mineralization adds up to the inorganic phosphorus content (Vincent et al. 2010). In an investigation, it was estimated that continuous addition of litter for a span of three years increased the organic phosphorus by 16% in the surface soil (0–2 cm) (Vincent et al. 2010).

2.8.4.2 Soil Productivity, Protection and Reduced Nutrient Loss

The decrease in nutrient loss from the soil was highly accounted for the role of the AFS system in minimizing soil leaching, erosion and runoff (Zhu et al. 2020). Soil erosion involves the removal of topsoil due to wind or water by the influence of natural processes and human activities (Balasubramanian 2017). Primarily, the reduction in erosion was attributed to the ability of tree roots in exploring deeper soil layers and holding the fragile soil tight which helps to impart enhanced soil compaction within the system (Vannoppen et al. 2017). Above-ground functions of trees, namely, rainfall interception by tree canopies minimized the impact of a raindrop on soil also attribute to erosion control (Zhujun et al. 2015). Besides, frequent litter addition leads to the building up of SOM, thus enhancing the ability of soil to resist erosion by improving stability within the soil aggregates (Zhang et al. 2018). On steep slopes too planting of trees on terraces act as a biological barrier to soil erosion and helps in cutting off the velocity of runoff (Jia et al. 2020). Similarly, in an investigation conducted for 8 years at e Yangtze River basin, China, found a remarkable reduction of runoff (63 to 70.8%) and sediment (231.2–242.8 t/ha) under hedgerows of both false indigo and vetiver, respectively (Lin et al. 2009). Further, this extensive influence of AFS systems on both physical and chemical properties of soil is attributed to improving the innate ability of the soil to protect nutrients against loss (Chandra et al. 2016), hence highlighting its role in addressing one of the major concerns most efficiently which is otherwise exaggerated by intensive agricultural practice (Ncube 2020). This agreed with the results of Tully et al. (2012) and according to him, there was a linear decline in N losses from the surface due to an increase in above-ground biomass with the integration of shade trees with coffee plantations. Moreover, AFS involves more diverse cropping systems and acts as supplemental habitat for natural forests protecting both plant and animal biodiversity through reduced clearance of natural forests (Udawatta et al. 2015). This was further regulated by soil micro and macro-fauna affecting both nutrient release and carbon conservation from the accumulated biomass protecting the productive capacity of the underneath soils (Naresh et al. 2016). Collectively, all these ecosystem services provided by a well-managed AFS ultimately aim to improve soil fertility, thus complementing the yield of the annual crops cultivated under this system (Prasad et al. 2016). Although poor light interception through tree canopies limits the photosynthetic accumulation of the intercrop, this limitation was complemented by improved soil fertility (Qiao et al. 2019).

2.8.4.3 Carbon Sequestration

Agroforestry is considered one of the most plausible ways to sequester more carbon in both aboveground and underground conditions (Pramanick et al. 2021). The agroforestry area is much higher in developing countries (about 1 billion ha) compared to the industry-intensive developed country. Agroforestry systems (AFSs) are known to sequester atmospheric carbon in an efficient way since their apparent ability to sequester more carbon (C) from the atmosphere during their growth and development. The estimates of C stored in AFSs range from 0.29 to 15.21 Mg ha^{-1} $year^{-1}$ aboveground, and 30 to 300 Mg C ha^{-1} up to 1 m depth in the soil (Nair et al. 2010). Many recent studies suggested that tree-based AFSs can store more C in the deeper layers of the soil as compared to treeless AFSs. More surprisingly, it was also revealed that the C_3 plants could contribute to more C-build-up in soils as compared to C_4 plants in the deeper layers (Nair et al. 2010). However, the magnitude of C-sequestration under AFSs massively depends on the ecology as well as the management of the systems. The AFSs practitioners in the world are now gaining more and more importance as far as C-trading is concerned (Pramanick et al. 2023). Many previous studies reported on the C-sequestration potential of varied AFSs, namely, silvi-pasture, agri-silviculture, agri-horticulture and agri-horti-silviculture. It was found from various studies that the AFS can increase ~25% of C-sequestration than non-AFS (Kiran Kumara et al. 2023). The best-suited AFS is agro-horticulture which can contribute to ~38 Mg C ha^{-1} accounting for ~30% more C-stock comparing conventional systems. C-sequestration can be increased to the tune of about 37% when grassland is converted to AFS. In contrast, generally, ~25% decline in the C-sequestration is observed when a forest gets converted to agroforestry (Kiran Kumara et al. 2023).

2.9 Future Perspective of Agroforestry

Generally, agriculture and forestry are considered as the two different domains, but many times these are interrelated on landscape and stake common targets and ecological fundamentals. The multitude of AFS is based on ecological values and ecosystem services with a focus on contribution to the accomplishment of many regional developmental goals. Worldwide, AFS is considered an underexploited system. Further, enough potential is there for the flourish of AFS in tropical as well as in the temperate zone. In tropical regions, it can ensure agricultural sustainability with the fulfilment of major SDGs. Past researches carried out during the last four decades created an enormous impression, particularly in the context of food and nutritional security in the arena of combatting ill effects of climate change. As the AFSs are having strong and scientific bases, there are enough opportunities for the flourishing of the systems. The technologies and research results evolved in a direction, which need dissemination in a broader perspective. The adoption of AFS systems should be intensified considering their suitability in different landscapes for agricultural sustainability, and in this direction there is an urgent need for policy interventions for lowering carbon footprint in fragile ecological conditions and other suitable areas.

2.10 Conclusions

The agroforestry system (AFS) gained attention in the 1980s when the adverse impacts of GRTs were realized worldwide and alternative thoughts of sustainable agriculture were seeded. There is no doubt that the research works carried out globally on AFS influenced a lot to set a future direction to attain agricultural sustainability as a suitable option wherever applicable. Recent researches clearly emphasized the multifaceted roles of AFS like food and nutritional security, beneficial environmental impacts and ecosystem services along with greater ecosystem services. Under the present context of climate change impacts and carbon footprint in agriculture, the focus should be given on eco-friendly farming practices that can lead to evergreen agriculture and AFS has enough potential to ensure agricultural sustainability. There is an urgent need for enhancement of agricultural production in the populous countries of tropics and AFS can play a magnificent role as it has a wide range of promising approaches like soil fertility improvement, proper nutrient cycling, C sequestration, checking erosion of soil and nutrient loss, better management of marginal lands and greater ecosystem services.

References

Abrol YP, Pandey R, Raghuram N, Ahmad A (2012) Nitrogen cycle sustainability and sustainable technologies for nitrogen fertilizer and energy management. J Ind Inst Sci 92:17–36

Adesogan A, Havelaar AH, Mckune SL, Eilitta M, Dahl GE (2020) Animal source foods: sustainability problem or malnutrition and sustainability solution? Perspective matters. Global Food Sec 25:100225. https://doi.org/10.1016/j.gfs.2019.100325

Adhikari S, Ozarska B (2018) Minimizing environmental impacts of timber products through the production process "from sawmill to final products". Environ Syst Res 7:6

Aguiar AF, Bicudo SJ, Sobrinho JR, Martins RL, Coelho KP, Moura EG (2010) Nutrient recycling and physical indicators of an alley cropping system in a sandy loam soil in the pre-Amazon region of Brazil. Nut Cyc Agroecosyst 86:189–198. https://doi.org/10.1007/s10705-009-9283-6

Ahman AMK, Zhang D (2018) Effects of fertilizer broadcasting on the excessive use of inorganic fertilizers and environmental sustainability. Sustainability 10:1–15. https://doi.org/10.3390/su10030759

Alao JS, Shuaibu BR (2013) Agroforestry practices and concepts in sustainable land-use systems in Nigeria. J Hort Forest 5:156–159

Alemu MM (2016) Ecological benefits of trees as windbreaks and shelterbelts. Int J Ecosystem 6: 10–13. https://doi.org/10.5923/j.ije.20160601.02

Aragues R, Medina ET, Claveria I (2014) Effectiveness of inorganic and organic mulching for soil salinity and sodicity control in a grapevine orchard drip-irrigated with moderately saline waters. Span J Agric Res 12:501–508. https://doi.org/10.5424/sjar/2014122-5466

Arora KN (2019) Impact of climate change on agriculture production and its sustainable solutions. Environ Sustain 2:95–96

Arze del Granado F, Coady D, Gillingham R (2012) The unequal benefits of fuel subsidies: a review of evidence from developing countries. World Dev 40:2234–2248

Augustin R, Mercado J, Noordwijk VM, Cadisch G (2017) Positive nitrogen balance of *Acacia mangium* woodlots as fallows in The Philippines based on 15N natural abundance data of N2 fixation. Agroforest Syst 81:221–233. https://doi.org/10.1007/s10457-010-9309-8

Balasubramanian A (2017) Soil erosion-causes and effects. University of Mysore, pp 1–7. https://doi.org/10.13140/rg.2.2.26247.39841

Banyal R, Kumar R, Kumar M, Yadav KR, Dagar CJ (2017) Agroforestry for rehabilitation and sustenance of saline ecologies. Nature Singapore, Springer, pp 413–454. https://doi.org/10.1007/978-981-10-7650-3_16

Bashir S, Javed A, Bibi I, Ahmad N (2018) In: Sabir M, Akhtar J, Hakeem KR (eds) Soil and water conservation. In soil science concepts and applications. University of Agriculture, Faisalabad, Pakistan, pp 263–284

Beedy TL, Snapp SS, Akinnifesi FK, Sileshi GW (2010) Impact of *Gliricidiasepium* intercropping on soil organic matter fractions in a maize-based cropping system. Agric Ecosys Environ 138: 139–146

Behera L, Nayak RM, Patel D, Mehta A, Sinha KS, Gunaga R (2015) Agroforestry practices for physiological amelioration of salt affected soils. J Plant Stress Physiol 1:12–17. https://doi.org/10.5455/jpsp.2015-06-06

Berkhout ED, Malan M, Kram T (2019) Better soils for healthier lives. An econometric assessment of the link between soil nutrients and malnutrition in sub-Saharan Africa. PLoS One 14:1–26

Bhadra P, Maitra S, Shankar T, Hossain A, Praharaj S, Aftab T (2021) Climate change impact on plants: Plant responses and adaptations. In: Aftab T, Roychoudhury A (eds) Plant perspectives to global climate changes. Elsevier Inc., Academic Press, pp 1–24. https://doi.org/10.1016/B978-0-323-85665-2.00004-2

Bhat AT, Ahmad L, Ganai AM, Haq US, Khan AO (2015) Nitrogen fixing biofertilizers; mechanism and growth promotion: a review. J Pure Appl Microbiol 9:1675–1690

Bhatt PR (2018) Climate change assessment, impacts of global warming, projections and mitigation of ghg emissions endorsing green energy. Int Edu Sci Res J 4:33–48

Bilali EH, Callenius C, Strassner C, Probst L (2018) Food and nutrition security and sustainability transitions in food systems. Food Energy Sec: 8(2): 1–20. https://doi.org/10.1002/fes3.154

Bindraban PS, Velde M, Ye L, DVM B, Materechera S, Kiba DI, Tamene L, Ragnarsdottir VK, Jongschaap R, Hoogmoed M, Hoogmoed W, Beek VC, Lynden VJ (2012) Assessing the impact of soil degradation on food production. Curr Opin Environ Sustain 4:478–488

Blaser WJ, Sitters J, Hart SP, Edwards PJ, Olde VH (2013) Facilitative or competitive effects of woody plants on understorey vegetation depend on N-fixation, canopy shape and rainfall. J Ecol 101:1598–1603

Brandle J, Hodges L, Zhou X (2004) Windbreaks in north American agricultural systems. Agrofor Syst 61:65–78. https://doi.org/10.1023/B:AGFO.0000028990.31801.62

Brandle JR, Hodges L, Tyndall J, Sudmeyer RA (2009) Windbreak practices. In: Garrett HE (ed) North American agroforestry: an integrated science and practice, 2nd edn. Am Soc Agron, Madison, WI, pp 75–104

Brown ES, Miller CD, Ordone JP, Baylis K (2018) Evidence for the impacts of agroforestry on agricultural productivity, ecosystem services, and human Well-being in high-income countries: a systematic map protocol. Environ Evidence 7:4. https://doi.org/10.1186/s13750-018-0136-0

Buresh RJ, Cooper PJM (1999) The science and practice of improved fallows. Agroforest Syst 47: 13–58

Chakdar H, Jadhav DS, Dhar WD, Pabbi S (2012) Potential applications of blue green algae. J Sci Industrial Res 71:13–20

Chandra LR, Gupta S, Pande V, Singh N (2016) Impact of forest vegetation on soil characteristics: a correlation between soil biological and physico-chemical properties. Biotechnology 6:188. https://doi.org/10.1007/s13205-016-0510-y

Chappell MJ, Wittman H, Bacon CM, Ferguson BG, Barrios LG, Barrios RG, Perfecto I (2013) Food sovereignty: an alternative paradigm for poverty reduction and biodiversity conservation in Latin America. F1000 Res 61:1–17

Chauhan JS, Prasad SR, Pal S, Choudhury PR, Bhaskar UK (2016) Seed production of field crops in India: quality assurance, status, impact and way forward. Ind J Agric Sci 86:563–579

Chauhan KS, Gupta N, Walia R, Yadav S, Chauhan R, Mangat SP (2011) Biomass and carbon sequestration potential of poplar-wheat intercropping system in irrigated agro-ecosystem in India. J Agric Sci Technol:575–586

Chesney PE, Simpson LA, Cumberbatch RN, Homenauth O, Benjamin F (2010) Cowpea yield performance in an alley cropping practice on an acid infertile soil at Ebini, Guyana. Open Agric J 4:80–84

Choudhary OP, Kharche VK (2018) Soil salinity and sodicity. In: Soil Science, Indian Society of Soil Science, New Delhi, India, pp 353–384

Dafa SMA, Nawal ANK (2016) The influence of optical porosity of tree windbreaks on windward wind speed, erosive force and sand deposition. J Forest Environ Sci 32:212–218. https://doi.org/10.7747/JFES.2016.32.2.212

Daniel MD, Tiemann LK, Grandy AS (2013) Does agricultural crop diversity enhance soil microbial biomass and organic matter dynamics-a meta-analysis. Ecol Appl 24:560–570. https://doi.org/10.1890/13-0616.1

Darch T, Blackwell MSA, Hawkins JMB, Haygarth PM, Chadwick D (2014) A meta-analysis of organic and inorganic phosphorus in organic fertilizers, soils and water; implications for water quality. Critical Rev Environ Sci Tech 44:1–57

Dawson IK, Carsan S, Franzel S, Kindt R, Breugel P, Graudal L, Lilleso JPB, Orwa C, Jamnadass R (2014) Agroforestry, livestock, fodder production and climate change adaptation and mitigation in East Africa: issues and options. World Agroforestry Centre, Nairobi, p 178. https://doi.org/10.5716/WP14050.PDF

Egnell G, Jurevics A, Peichl M (2015) Negative effects of stem and stump harvest and deep soil cultivation on the soil carbon and nitrogen pools are mitigated by enhanced tree growth. Forest Ecol Manag 338:57–67

Eichhorn MP, Paris P, Herzog F, Incoll LD, Liagre F, Mantzanas K, Mayus M, Moreno G, Papanastasis VP, Pilbeam DJ, Pisanelli A, Dupraz C (2006) Silvoarable systems in Europe-past, present and future prospects. Agroforest Syst 67:29–50

Fahad S, Chavan SB, Chichaghare AR, Uthappa AR, Kumar M, Kakade V, Pradhan A, Jinger D, Rawale G, Yadav DK, Kumar V, Farooq TH, Ali B, Sawant AV, Shah Saud S, Chen S, Poczai P (2022) Agroforestry systems for soil health improvement and maintenance. Sustainability 14: 14877. https://doi.org/10.3390/su142214877

FAO (2012) Towards the future we want: End hunger and make the transition to sustainable agricultural and food systems, available online: https://www.fao.org/3/an894e/an894e00.pdf, accessed 29 August 2021, p 28

Fioretto A, Nardo CD, Papa S, Fuggi A (2005) Lignin and cellulose degradation and nitrogen dynamics during decomposition of three leaf litter species in a Mediterranean ecosystem. Soil Biol Biochem 37:1083–1091

Franzel S, Carsan S, Lukuyu B, Sinja J, Wambugu C (2014) Fodder trees for improving livestock productivity and smallholder livelihoods in Africa. Curr Opinion Environ Sustain 6:98–103

Garrett HE (2009) North American agroforestry: an integrated science and practice, 2nd edn. American Society of Agronomy, Madison, WI

Gelaye KK, Zehetner F, Loiskandl W, Klik A (2019) Effects of soil texture and groundwater level on leaching of salt from saline fields in kesem irrigation scheme, Ethiopia. Soil Water Res 14: 221–228

Gendy HSA, Ahl S, Mahmoud AA, Mohamed YH (2013) Effect of nitrogen sources, bio-fertilizers and their interaction on the growth, seed yield and chemical composition of guar plants. Life Sci J 10:389–402

Giller KE, Wilson KJ (1991) Nitrogen fixation in tropical cropping systems. CAB Int

Gispert M, Pardini G, Emran M, Doni S, Masciandaro G (2017) Seasonal evolution of soil organic matter, glomalin and enzymes and potential for C storage after land abandonment and renaturalization processes in soils of NE Spain. Catena 162:402–413

Godfray JCH, Garnett T (2014) Food security and sustainable intensification. Phil Trans R Soc A 369:2012–0273

Gronau S, Winter E, Grote U (2018) Papyrus, forest resources and rural livelihoods: a village computable general equilibrium analysis from northern Zambia. Nat Resour 9:268–296

Hamza TA, Hussein Z, Mitku R, Ayalew P, Belayneh T (2017) Isolation and characterization of nitrogen fixing bacteria from rhizosphere soil collected from shell mele agricultural center, Southern Ethiopia. J Agric Sci Food Technol 3:117–124

Hartel T, Hanspach J, Abson D, Mathe O, Moga C, Fischer J (2014) Bird communities in traditional wood-pastures with changing management in Eastern Europe. Basic App Ecol 15:385–395

Harvey CA, Villalobos JAG (2007) Agroforestry systems conserve species-rich but modified assemblages of tropical birds and bats. Biodivers Conserv 16:2257–2292. https://doi.org/10.1007/s10531-007-9194-2

Hasan MK, Islam MT, Roshni NA, Akter R, Hossain MB (2019) Effect of agroforest tree leaf biomass on yield and yield contributing characters of wheat *cv. Akbar*. Prog Agric 30:32–38. https://doi.org/10.3329/pa.v30i1.42206

Hossain MD, Siddique MR, Rahman MD, Hossain MD, Hasan MD (2011) Nutrient dynamics associated with leaf litter decomposition of three agroforestry tree species (*Azadirachta indica*, *Dalbergia sissoo*, and *Melia azedarach*) of Bangladesh. J Forest Res 22:577–582

Hosseini M, Keize JJ, Pelayo GO, Prats AS, Ritsema C, Geissen V (2016) Effect of fire frequency on runoff, soil erosion, and loss of organic matter at the micro-plot scale in north-Central Portugal. Geoderma 269:126–137. https://doi.org/10.1016/j.geoderma.2016.02.004

Houghton JT, Gylvan MFL, Griggs DJ, Maskell K (1997) Stabilization, of atmospheric greenhouse gases: physical, biological and socioeconomic implications. In: IPCC Technical Paper III. IPCC Secretariat, Geneva, p 48

Imran S, Arif M, Khan A, Khan AM, Shah W, Latif A (2015) Effect of nitrogen levels and plant population on yield and yield components of maize. Adv Crop Sci Technol 3:1–7. https://doi.org/10.4172/2329-8863.1000170

Innangi M (2017) Dynamics of organic matter in leaf litter and topsoil within an Italian Alder (*Alnus cordata* (Loisel.) Desf.). Ecosys Forest 8:240

IPCC (2000) The IPCC assessment reports. Available from: www.ipcc.ch/. Accessed on 25 December 2022

Islam MD, Sultana T, Joe MM, Yim W, Cho CJ, Sa A (2013) Nitrogen-fixing bacteria with multiple plant growth-promoting activities enhance growth of tomato and red pepper. J Basic Microbiol 53:1004–1015. https://doi.org/10.1002/jobm.201200141

Jamnadass R, Place F, Torquebiau E, Malézieux E, Iiyama M, Sileshi GW, Kehlenbeck K, Masters E, McMullin S, Weber JC, Dawson IK (2013) Agroforestry, food and nutritional security. ICRAF working paper no. 170. World Agroforestry Centre, Nairobi. https://doi.org/10.5716/WP13054.PDF

Jena J, Maitra S, Hossain A, Pramanick B, Gitari HI, Praharaj S, Shankar T, Palai JB, Rathore A, Mandal TK, Jatav HS (2022) Role of legumes in cropping system for soil ecosystem improvement. Ecosystem services: types, management and benefits. Nova Science Publishers, Inc, New York, pp 1–22

Jia C, Sun B, Xinxiao Y, Yang X (2020) Analysis of runoff and sediment losses from a sloped roadbed under variable rainfall intensities and vegetation conditions. Sustainability 12:2077. https://doi.org/10.3390/su12052077

Jianbo L, Guangyao G, Shuai W, Lei J, Xing W, Bojie F (2018) The effects of vegetation on runoff and soil loss: multidimensional structure analysis and scale characteristics. J Geograph Sci 28: 59–78. https://doi.org/10.1007/s11442-018-1459-z

Jnr DS (2014) Land degradation and agriculture in the Sahel of Africa: causes, impacts and recommendations. J Agric Sci App 3:67–73

Johansson J, Bolmgren K, Jonzen N (2013) Climate change and the optimal flowering time of annual plants in seasonal environments. Glob Chang Biol 19:197–207. https://doi.org/10.1111/gcb.12006

Johar V, Dhillon RS, Bangarwa KS, Bharadwaj KK (2017) Effect of eucalyptus based Agri-silvi-horticultural system on growth and yield of wheat in North-Western region of Haryana, India. Int J Curr Microbiol Appl Sci 6:2702–2711

Jose S, Dollinger J (2019) Silvopasture: a sustainable livestock production system. Agroforest Syst 93:1–9. https://doi.org/10.1007/s10457-019-00366-8

Kadirvel G, Bhatt BP, Das A (2003) Agroforestry: a promising farming Systenl for livestock production in NEH region. In: Bhatt BP, Bujarbaruah KM, Sharma YP, Patiram (eds) Proc. approaches for increasing agricultural productivity in hill and Moulltain 'Ecosystem. ICAR Research Complex for NEH Region, Umiam, Meghalaya, India, pp 139–147

Kang BT (1993) Potential for sustainable agroforestry and alley farming in tropical Africa. In: soil tillage in Africa, needs and challenges, FAO soil bulletin, p 69, Available online: http://www.fao.org/3/t1696e/t1696e03.htm#TopOfPage, Accessed on 25 December 2022

Kass DCL, Somarriba E (1999) Traditional fallows in Latin America. Agrofor Syst 147:13–36

Kaul M, Mohren JMG, Dadhwal VK (2011) Phytomass carbon pool of trees and forests in India. Clim Chang 2011(108):243–259. https://doi.org/10.1007/s10584-010-9986-3

Kaushal R, Verma SK, Chaturvedi PO, Alam MN (2012) Leaf litter decomposition and nutrient dynamics in four multipurpose tree species. Range Manag Agroforest 33:20–27

Kevin JW, Bruce EB, Evan HD (2018) Reduced nitrogen losses after conversion of row crop agriculture to alley cropping with mixed fruit and nut trees. Agric Ecosys Environ 258:172–181

Kiran Kumara TM, Suresh P, Prem C, Ankita K (2023) Carbon sequestration potential of agroforestry systems in Indian agricultural landscape: a meta-analysis. Ecosyst Serv 62:101537. https://doi.org/10.1016/j.ecoser.2023.101537

Koutouleas A, Sarzynski T, Bordeaux M, Bosselmann AS, Campa C, Etienne H, Turreira-García N, Rigal C, Vaast P, Ramalho JC, Marraccini P, Ræbild A (2022) Shaded-coffee: a nature-based strategy for coffee production under climate change? A Review. Front Sustain Food Syst 6: 877476. https://doi.org/10.3389/fsufs.2022.877476

Kristjanson P, Neufeldt H, Gassner A, Mango J, Kyazze FB, Desta S, Sayula G, Thiede B, Foch W, Thornton PK, Coe R (2012) Are food insecure smallholder households making changes in their farming practices? Evidence form East Africa. Food Sec 4:381–397. https://doi.org/10.1007/s12571-012-0194-z

Kumar V (2016) Multifunctional agroforestry systems in tropics region. Nat Environ Poll Technol 15:365–376

Kweku WD, Bismark O, Maxwell A, Desmond AD, Danso BK, Mensah OAE, Quachie TA, Adormaa BB (2017) Greenhouse effect: greenhouse gases and their impact on global warming. J Sci Res Rep 17:1–9. https://doi.org/10.9734/JSRR/2017/39630

Lal R (2015) Restoring soil quality to mitigate soil degradation. Sustainability 7:5875–5895. https://doi.org/10.3390/su7055875

Lan ZM, Lin XJ, Wang F, Zhang H, Chen CR (2012) Phosphorus availability and rice grain yield in a paddy soil in response to long-term fertilization. Biol Fertil Soils 48:579–588. https://doi.org/10.1007/s00374-011-0650-5

Lana MQ, Lana RMQ, Lemes EM, Reis GL, Moreira GHFA (2018) Influence of native or exotic trees on soil fertility in decades of silvopastoral system at the Brazilian Savannah biome. Agroforest Sys 92:415–424. https://doi.org/10.1007/s10457-016-9998-8

Larwanou M, Adam T (2008) Impacts de la regeneration naturelle assistee au Niger: Etude de quelquescas dans les Regions de Maradi et Zinder. In: Synthese de 11 memoires d'étudiants de 3eme cycle. de l'Universite Abdou Moumouni de Niamey, Niger. Photocopy, p 49

Latake TP, Pawar P (2015) The greenhouse effect and its impacts on the environment. Int J Inn Res Creative Technol 1:333–337

Le D, Le L, Coeur D, Thenail C, Burel F, Baudry J (2008) New hedgerows in replanting programmes: assessment of their ecological quality and their maintenance on farms. In: Berlan-Darque M, Terrasson D, Luginbiuhl Y (eds) Landscape: from knowledge to action. Versailles, Editions Quae, pp 177–191

Leghari JS, Wahocho AN, Laghari MG, Laghari AH, Bhabhan MG, Talpur HK, Bhutto AT, Wahocho AS, Lashari AA (2016) Role of nitrogen for plant growth and development: a review. Adv Envir Biol 10:209–218

Lichtfouse E, Navarrete M, Debaeke P, Souchere V, Alberola C, Menassieu J (2009) Agronomy for sustainable agriculture. A review. Agron Sustain Dev 29:1–6

Lin C, Tu S, Huang J, Chen Y (2009) The effect of plant hedgerows on the spatial distribution of soil erosion and soil fertility on sloping farmland in the purple-soil area of China. Soil Till Res 105:307–312

Liu WC, Sung Y, Chen CB, Lai YH (2014) Effects of nitrogen fertilizers on the growth and nitrate content of lettuce (*Lactuca sativa* L.). Int J Environ Res Pub Health 11:4427–4440. https://doi.org/10.3390/ijerph110404427

Liu Y, Bing H, Wu Y, Zhu H, Tian X, Wang Z, Chang R (2021) Nitrogen addition promotes soil phosphorus availability in the subalpine forest of the eastern Tibetan plateau. J Soils Sediments 22(1):1–11

M de Liyanage S, SKA D, HPS J (1994) Biological nitrogen fixation in four Gliricidiasepium genotypes. Plant Soil 161:267–274

Luedeling E, Smethurst PJ, Baudron F, Bayala J, Huth NI, Noordwijk M (2016) Field-scale modeling of tree–crop interactions: challenges and development needs. Agric Syst 142:51–69. https://doi.org/10.1016/j.agsy.2015.11.005

Mackay-Smith TH, Burkitt L, Reid J, López IF, Phillips C (2021) A framework for reviewing Silvopastoralism: a New Zealand Hill country case study. Land 10:1386. https://doi.org/10.3390/land10121386

Madhu M, Hombegowda HC, Beer K, Adhikary PP, Jakhar P, Sahoo DC, Dash CJ, Kumar G, Naik GB (2019) Status of natural resources and resource conservation technologies in eastern region of India. In: Resource conservation in eastern region of India: Lead papers of FFCSWR, 2019. Indian Association of Soil and Water Conservationists, Dehradun, India, pp 60–82

Mahmud K, Isalam A (2017) Reclamation of degraded land through agroforestry introduction in Bangladesh. J Nat Sci Res 5:56–62

Maitra S, Brestic M, Bhadra P, Shankar T, Praharaj S, Palai JB, Shah MMR, Barek V, Ondrisik P, Skalický M et al (2022) Bioinoculants—natural biological resources for sustainable plant production. Microorganisms 10:51. https://doi.org/10.3390/microorganisms10010051

Maitra S, Hossain A, Brestic M, Skalicky M, Ondrisik P, Gitari H, Brahmachari K, Shankar T, Bhadra P, Palai JB, Jana J, Bhattacharya U, Duvvada SK, Lalichetti S, Sairam M (2021) Intercropping–a low input agricultural strategy for food and environmental security. Agron 11:343. https://doi.org/10.3390/agronomy11020343

Maitra S, Palai JB, Manasa P, Prasanna KD (2019) Potential of intercropping system in sustaining crop productivity. Int J Agric Environ Biotechnol 12:39–45

Maitra S, Praharaj S, Brestic M, Sahoo RK, Sagar L, Shankar T, Palai JB, Sahoo U, Sairam M, Pramanick B, Nath S, Venugopalan VK, Skalický M, Hossain A (2023) *Rhizobium* as biotechnological tools for green solutions: an environment-friendly approach for sustainable crop production in the modern era of climate change. Curr Microbiol 80:219. https://doi.org/10.1007/s00284-023-03317-w

Maitra S, Zaman A (2017) Bio-drainage in waterlogged areas for agricultural sustainability. Int J Biores Sci 4(2):79–84

Maitra S, Zaman A, Mandal TK, Palai JB (2018) Green manures in agriculture: a review. J Pharmacog Phytochem 7(5):1319–1327

Maitra S (2023) Alley cropping. In: Intercropping systems. NIPA Genx Electronic Resources and Solutions P. Ltd., India, pp 91–102

Maity PK, Das S, Das R (2018) Remedial measures for saline water ingression in coastal aquifers of South West Bengal in India. MOJ Ecol Environ Sci 3(1):00061. https://doi.org/10.15406/mojes.2018.03.00061

Mandal A, Toor AS, Dhaliwal SS (2018) Effect of land-uses on physico-chemical properties and nutrient status of surface (0-15 cm) and sub-surface (15-30 cm) layers in soils of South-Western Punjab, India. Int J Curr Microbiol Appl Sci 7:2659–2671

Mathuia RK, Sagarka BK, Panara DM (2016) Fodder production through agroforestry: a boon for profitable dairy farming. Inn J Agric Sci 4:13–19

Mayrinck CR, Laroque PC, Amichev YB, Rees VK (2019) Above- and below-ground carbon sequestration in shelterbelt trees in Canada: a review. Forests 10:922. https://doi.org/10.3390/f10100922

Meena SR, Kumar S, Yadav SG (2020) Soil carbon sequestration in crop production. Nature Singapore, Springer, pp 1–39. https://doi.org/10.1007/978-981-13-8660-2_1

Memon SM, Jun Z, Jun G, Ullah F, Hassan M, Ara S, Changying J (2017) Comprehensive review for the effects of ridge furrow plastic mulching on crop yield and water use efficiency under different crops. Int Agric Eng J 26:58–66

Mohan Kumar B (2007) Agroforestry systems and practices of Kerala. In: Puri S, Panwar P (eds) Agroforestry systems and practices of India. New India Publishing Agency, New Delhi, pp 459–483

Montagnini F, Nair PKR (2004) Carbon sequestration: an underexploited environmental benefit of agroforestry systems. Agroforest Sys 61:281–295

Moore CM, Mills MM, Arrigo KR, Rank BI, Boyd PW, Galbraith ED, Geide RJ, Guieu C, Jaccar SL, Jickell TD, Roch J, Lento TM, Mahowald NM, Maranon E, Marinov I, Moore JK, Nakatsuka T, Oschlies A, Saito MA, Thingstad TF, Tsuda A, Ulloa O (2013) Processes and patterns of oceanic nutrient limitation. Nat Geosci 6:701–710. https://doi.org/10.1038/NGEO1765

Moreno G, Aviron S, Berg S, Crous DJ, Franca A, Garcia DS, Hartel T, Mirck J, Pantera A, Palma JHN, Paulo JA, Re GA, Sanna F, Thenail C, Varga A, Viaud V, Burgess PJ (2018) Agroforestry systems of high nature and cultural value in Europe: provision of commercial goods and other ecosystem services. Agroforest Syst 92:877–891. https://doi.org/10.1007/s10457-017-0126-1

Morniruzzaman M, Shamim MHA (2015) Effect of mulching on soil salinity and yield of sweet gourd. Int J Sustain Crop Prod 10:16–20

Mosquera L, Moreno G, Pardini A, McAdam JH, Papanastasis V, Burgess PJ, Lamersdorf M, Castro FL, Rigueiro RA (2012) Past, present and future of agroforestry systems in Europe. In: Agroforestry the future of global land use. Springer Netherlands, pp 285–312

Moussa M, Mahamane L, Saadou M (2015) Allometric equations for biomass estimation of woody species and organic soil carbon stocks of agroforestry systems in West African Africa: state of current knowledge. Int J Res Agric Forest 2:17–33

Munns R, Passioura BJ, Colmer DT, Byrt SC (2019) Osmotic adjustment and energy limitations to plant growth in saline soil. New Phytol 225:1–6. https://doi.org/10.1111/nph.15862

Muschler RG (2015) Agroforestry: essential for sustainable and climate-smart land use? In: Pancel L, Köhl M (eds) Tropical forestry handbook. Springer, Berlin, Heidelberg. https://doi.org/10.1007/978-3-642-41554-8_300-1

Nainwal CR, Katiyar SR, Singh D, Sharma SK, Singh S, Tewari SK (2016) Agroforestry as a reservoir for biodiversity conservation, soil fertility and mitigation of climatic hazards. In: Holistic Development of Agroforestry. Jaya Publishing House, Delhi, pp 203–211

Nair PKR (2013) An introduction to agroforestry. Springer Science & Business Media

Nair PKR (2014) Agroforestry: practices and systems. In: Vanalfen NK (ed) Encyclopedia of agriculture and food systems. Elsevier Inc, p 1. https://doi.org/10.1016/B978-0-444-52512-3.00021-8

Nair PKR, Garrity DP (2012) Agroforestry research and development-the way forward. In: Nair PKR, Garrity DP (eds) Agroforestry: the future of global land use. Springer, Dordrecht, The Netherlands, pp 515–531

Nair PKR, Gordon AM, Mosquera-Losada MR (2008) Agroforestry. In: Jorgensen SE, Faith BD (eds) Encyclopedia of ecology. Elsevier, Oxford, UK, pp 101–110

Nair PKR, Kumar BM (2006) Introduction tropical Homegardens: a time-tested example of sustainable agroforestry. Springer, Dordrecht, The Netherlands, pp 1–10

Nair PKR, Kumar MB, Nair DV (2019) Agroforestry as a strategy for carbon sequestration. J Plant Nut Soil Sci 172:10–23. https://doi.org/10.1002/jpln.200800030

Nair PKR, Nair DV (2014) 'Solid–fluid–gas': the state of knowledge on carbon-sequestration potential of agroforestry. Curr Opin Environ Sustain 6:22–27

Nair PKR, Nair VD, Mohan KB, Julia M, Showalter JM (2010) Carbon sequestration in agroforestry systems. Adv Agron 108:237–307. https://doi.org/10.1016/S0065-2113(10)08005-3

Naresh RK, Dwivedi A, Gupta RK, Rathore RS, Dhaliwal SS, Singh SP, Kumar P, Kumar R, Singh V, Singh V, Singh O (2016) Influence of conservation agriculture practices on physical, chemical and biological properties of soil and soil organic carbon dynamics in the subtropical climatic conditions: a review. J Pure Appl Microbiol 10:1061–1080

Ncube D (2020) The importance of contract farming to small-scale farmers in Africa and the implications for policy: a review scenario. Open Agric J 14:59–86. https://doi.org/10.2174/1874331502014010059

Ndoye F, Kane A, Diedhiou AG, Bakhoum N, Fall D, Sadio O, Oureye M, Noba K, Diouf D (2015) Effects of dual inoculation with arbuscular mycorrhizal fungi and rhizobia on *Acacia Senegal* (L.) Wild. Seedling growth and soil enzyme activities in Senegal. Int J Biosci 6:36–48

Nganchamung T, Robson GM, Siriwong W (2017) Chemical fertilizer use and acute health effects among chili farmers in ubonratchathani province, Thailand. J Health Res 31:427–435. https://doi.org/10.14456/jhr.2017.53

Nöldeke B, Winter E, Laumonier Y, Simamora T (2021) Simulating agroforestry adoption in rural Indonesia: the potential of trees on farms for livelihoods and environment. Land 10:385. https://doi.org/10.3390/land10040385

Novara A, Ruhl J, Mantia LT, Gristina L, Bella SL, Tuttolomondo T (2015) Litter contribution to soil organic carbon in the processes of agriculture abandon. Solid Earth 6:425–432. https://doi.org/10.5194/se-6-425-2015

Ogle MS, Alsaker C, Baldock J, Bernoux M, Breidt JF, Conkey MB, Regina K, Amabile VGG (2019) Climate and soil characteristics determine where no-till management can store carbon in soils and mitigate greenhouse gas emissions. Sci Res 9:11665. https://doi.org/10.1038/s41598-019-47861-7

Ola A, Dodd CI, Quinton NJ (2015) Can we manipulate root system architecture to control soil erosion, vol 1. J Europ Geosci Union, pp 603–612

Onu MF, Ikehi EM (2016) Mitigation and adaptation strategies to the effects of climate change on the environment and agriculture in Nigeria. J Agric Veterinary Sci 9:26–29. https://doi.org/10.9790/2380-0904022629

Ouin A, Burel F (2002) Influence of herbaceous elements on butterfly diversity in hedgerow agricultural landscapes. Agric Ecosyst Environ 93:45–53

Oyebamiji NA, Adesoji AG, Aduradola AM (2016) Influence of leguminous trees biomass and nitrogen fertilizer on growth and yield of maize. J Agric Environ 12:119–127

Pandey CB, Rai RB (2007) Nitrogen cycling in Gliricidia*(Gliricidiasepium)* alley cropping in humid tropics. Trop Ecol 48(1):87–97

Pandey V, Gautam P, Singh AP (2019) Assessment of chemical properties of soil under different land use systems in a mollisol. Int J Curr Microbiol App Sci 8:1165–1175

Papanastasis VP, Yiakoulaki MD, Decandia M, Dini PO (2008) Integrating woody species into livestock feeding in the Mediterranean areas of Europe. Animal Feed Sci Technol 140:1–17

Pereira THM, Martin HC (2021) People, planet, and profit: a bibliometric analysis of triple bottom line theory. J Manage Sustain 11:64–73

Pierret A, Maeght JL, Clement C, Montoroi JP, Hartmann C, Gonkhamdee S (2016) Understanding deep roots and their functions in ecosystems: an advocacy for more unconventional research. Ann Bot 118:621–635. https://doi.org/10.1093/aob/mcw130

Pinho CR, Miller PR, Alfaia SS (2012) Agroforestry and the improvement of soil fertility: a view from Amazonia. App Environ Soil Sci **2012**:616383. https://doi.org/10.1155/2012/616383

Piyuh M, Arun N, Rashmi YJ, Pankaj T (2018) A study of managing crop damage by wild animals in Himachal Pradesh. Int J Agric Sci 10:6438–6442

Plieninger T, Hartel T, Martin LB, Beaufoy G, Bergmeier E, Kirby K, Montero MJ, Moreno G, Oteros RE, Van UJ (2015) Wood-pastures of Europe: geographic coverage, social—ecological values, conservation management, and policy implications. Biol Cons 190:70–79

Polglase PJ, Paul KI, Hawkins C, Siggins A, Turner J, Booth T, Crawford D, Jovanovic T, Hobbs T, Opie K, Almeida A, Carter J (2008) Regional opportunities for agroforestry systems in Australia. RIRDC Pub 8:98

Porqueddu C, Franca A (2013) Sardinian agro-silvo-pastoral systems: management and constraints. In: Proceedings of MONTADOS and DEHESAS as high nature value farming systems: implications for classification and policy support. ICAAM int conference 6-8 February, p 77

Pramanick B, Kumar M, Singh SK, Sapna K, Maitra S (2021) Soil-centric approaches towards climate resilient agriculture. In: Rakshit A, Singh S, Abhilash P, Biswas A (eds) Soil science: fundamentals to recent advances. Springer, Singapore. https://doi.org/10.1007/978-981-16-0917-6_17

Pramanick B, Mahapatra BS, Datta D, Dey P, Singh SP, Kumar A, Paramanik B, Awasthi N (2023) An innovative approach to improve oil production and quality of mustard (*Brassica juncea* L.) with multi-nutrient-rich polyhalite. Heliyon 9(3):e13997. https://doi.org/10.1016/j.heliyon.2023.e13997

Prasad R, Dhyani SK, Newaj R, Kumar S, Tripath VD (2016) Contribution of advanced agroforestry research in sustaining soil quality for increased food production and food security. J Soil Water Conserv 15:31–39

Pumariino L, Sileshi WG, Gripenberg S, Kaartinen R, Barrios E, Muchane NM, Midega C, Jonsson M (2015) Effects of agroforestry on pest, disease and weed control: a meta-analysis. Basic Appl Ecol 16:1–10. https://doi.org/10.1016/j.baae.2015.08.006

Qiao X, Sai L, Chen X, Xue L, Lei J (2019) Impact of fruit-tree shade intensity on the growth, yield, and quality of intercropped wheat. PLoS One 14:e0203238

Quemada M, Baranski M, Lange MNJ, Vallejo A, Cooper MJ (2013) Meta-analysis of strategies to control nitrate leaching in irrigated agricultural systems and their effects on crop yield. Agric Ecosystem Environ 174:1–10. https://doi.org/10.1016/j.agee.2013.04.018

Querne A, Laclau BP, Dufour L, Wery J, Dupraz C (2017) Effects of walnut trees on biological nitrogen fixation and yield of intercropped alfalfa in a Mediterranean agroforestry system. Europ J Agron 84:35–46. https://doi.org/10.1016/j.eja.2016.12.001

Reppin S, Kuyah S, de Neergaard A, Oelofse M, Rosenstock TS (2019) Contribution of agroforestry to climate change mitigation and livelihoods in Western Kenya. Agrofor Syst 94:203–220. https://doi.org/10.1007/s10457-019-00383-7

Rigueiro-Rodriguez A, McAdam JH, Mosquera-Losada MR (2008) Agroforestry in Europe. Springer, Dordrecht, Netherlands
Ropokis A, Ntatsi G, Kittas C, Katsoulas N, Savvas D (2019) Effects of temperature and grafting on yield, nutrient uptake, and water use efficiency of a hydroponic sweet pepper crop. Agron 9:110. https://doi.org/10.3390/agronomy9020110
Sade AG (2020) Role of agroforestry in phytoremediation of contaminated soils: review. Int J Curr Res Aca Rev 8(9):29–37
Saikia P, Kumar A, Khan ML (2017) Agroforestry: a sustainable land-use system for livelihood security and climate change mitigation. In: Climate Change and Agroforestry. New India Publishing Agency, New Delhi, India, pp 61–70
Saklani N, Khurana A (2019) Global warming: effect on living organisms, causes and its solutions. Int J Eng Manage Res 9:24–26. https://doi.org/10.31033/ijemr.9.5.4
Samra JS, Singh SC (2000) Silvipasture systems for soil, water and nutrient conservation on degraded lands of Shivalik foothills (subtropical northern India). Ind J Soil Con 28:35–42
Sanchez PA, Buresh RJ, Leakey RRB (1997) Trees, soils and food security. Philos. Trans. R. Soc., B 352:949–961
Sarkar AK, Kundu DK, Ghosh GK (2019) Sustainable farmer-friendly technologies for management of natural resources in eastern India. Ind J Fertil 15:1406–1428
Sarr SP, Okon WJ, Begoude BAD, Araki S, Ambang Z, Shibata M, Funakawa S (2016) Symbiotic N2-fixation estimated by the 15N tracer technique and growth of *Pueraria phaseoloides* (Roxb.) Benth. In: Inoculated with Bradyrhizobium strain in field conditions. Hindawi Publishing Corporation Scientifica, pp 1–10
Scaven LV, Rafferty EN (2013) Physiological effects of climate warming on flowering plants and insect pollinators and potential consequences for their interactions. Curr Zool 59:418–426. https://doi.org/10.1093/czoolo/59.3.418
Schmidt MW, Torn MS, Abiven S, Dittmar T, Guggenberger G, Janssens IA, Kleber M, Kögel-Knabner I, Lehmann J, Manning DA, Nannipieri P (2011) Persistence of soil organic matter as an ecosystem property. Nature 478:49–56
Seddaiu G, Porcu G, Ledda L, Roggero PP, Agnelli A, Corti G (2013) Soil organic matter content and composition as influenced by soil management in a semi-arid Mediterranean Agrosilvo-pastoral system. Agric Ecosys Environ 167:1–11
Seneviratne HTJ, Sumanasekar VDW, Dissanayake DMMR (2015) Role of agroforestry in achieving food and nutritional security, climate change mitigation and environmental resilience: a review. In: NBRO symposium. 2015. Innovations for resilient environmental. Accessed on 17 October, 2020, https://www.nbro.gov.lk/images/content_image/pdf/symposia/33.pdf
Sera B (2017) Salt-tolerant trees usable for central European cities-a review. Hort Sci 44:43–48. https://doi.org/10.17221/201/2015-HORTSCI
Shah S, Hookway S, Pullen H, Clarke T, Wilkinson S, Reeve V, Fletcher MJ (2017) The role of cover crops in reducing nitrate leaching and increasing soil organic matter. Aspects Appl Biol 134:243–252
Sharma N, Bohra B, Pragya N, Ciannella R, Dobie P, Lehmann S (2016) Bioenergy from agroforestry can lead to improved food security, climate change, soil quality and rural development. Food Energy 5:165–183
Sharma S, Singh B, Sikka R (2015) Changes in some physico-chemical characteristics of soils under poplar-based agroforestry. Agric Res J52:19–22. https://doi.org/10.5958/2395-146X.2015.00030.7
Shin S, Soe KT, Lee H, Kim TH, Lee S, Park MS (2020) A systematic map of agroforestry research focusing on ecosystem services in the Asia-Pacific region. Forests Open Access J 368:1–23. https://doi.org/10.3390/f11040368
Shukal S, Pandey VV, Kumar V (2018) Agroforestry systems stool sustainable rural development, food scarcity and income generation. Ind Forester 44:435–441
Singh AK, Ngachan SV, Munda GC, Mohapatra KP, Choudhury BU, Das A, Srinivasa Rao C, Patel DP, Rajkhowa DJ, Ramkrushna GI, Panwar AS (2012) Carbon Management in Agriculture for

mitigating greenhouse effect. ICAR Research Complex for NEH Region, Umiam-793 103, Meghalaya, India, p 377
Singh B (2009) Return and release of nutrients from poplar litterfall in an agroforestry system under subtropical condition. J Ind Soc Soil Sci 57:214–218
Singh RP, Van den Beldt RJ, Hocking D, Korwar GR (1989) Alley farming in the semi-arid regions of India. In: Kang BT, Reynolds L (eds)., March 10-14, Nigeria Proc. int. workshop on alley farming in the humid and sub-humid tropics, pp 108–122
Singh U (2013) Carbon capture and storage: an effective way to mitigate global warming. Curr Sci 105:914–922
Skiba MU, Rees MR (2014) Nitrous oxide, climate change and agriculture. CAB Rev 9(10):1–7. https://doi.org/10.1079/PAVSNNR20149010
Solaimalai A, Muralidaran C, Subburamu K (2005) Alley cropping in rainfed agro-ecosystem – a review. Agric Rev 26:157–172
Soni ML, Subbulakshmi V, Yadava ND, Tewari JC, Dagar JC (2016) Silvopastoral agroforestry systems: lifeline for dry regions. In: Dagar JC, Tewari JC (eds) Agroforestry research developments. Nova Science Publishers Inc., New York, pp 245–305
Stamets P (2005) Mycelium running. In: How mushrooms can help save the world. Ten Speed Press, Berkeley, p 344
Stamford PN, Junior SDS, Santos SREC, Freitas SDA, Junior LAM, Barros CFM (2013) Cowpea nodulation, biomass yield and nutrient uptake, as affected by biofertilizers and rhizobia, in a sodic soil amended with Acidithiobacillus. Acta Sci Agron 35:453–459. https://doi.org/10.4025/actasciagron.v35i4.16994
Susware NK, Sapkale JB, Rathod BL, Susware VN (2021) Impact of saltwater intrusion on the agricultural land of coastal tract: an investigation along Malvan coast, Maharashtra. J Rs Dev 11(11):17–20
Swieter A, Langhof M, Lamerre J (2022) Competition, stress and benefits: trees and crops in the transition zone of a temperate short rotation alley cropping agroforestry system. J Agron Crop Sci 208(2):209–224
Syahrawati M, Martono E, Putra SN, Purwanto HB (2018) Effects of fertilizer, irrigation level and spider presence on the abundance of herbivore and carnivore in rice cultivation in Yogyakarta. J Appl Sci 6:385–395
Szymkiewicz A, Kawecka GA, Simunek J, Leterme B, Beegum S, Szulc JB, Caceres PM, Langner GW, Jaramillo AR, Jacques D (2018) Simulations of freshwater lens recharge and salt/freshwater interfaces using the HYDRUS and SWI2 packages for MODFLOW. J Hydrol Hydromech 66:246–256. https://doi.org/10.2478/johh-2018-0005
Thiollay JM (1995) The role of traditional agroforests in the conservation of rainforest bird diversity in Sumatra. Conser Biol 9:335
Thorup-Kristensen K, Halberg N, Nicolaisen M, Olesen JE, Crews TE, Hinsinger P, Kirkegaard J, Pierret A, Dresbøll DB (2020) Digging deeper for agricultural resources, the value of deep rooting. Trends Plant Sci 25(4):406–417. https://doi.org/10.1016/j.tplants.2019.12.007
Tobella AB, Reese H, Almaw A, Bayala J, Malmer A, Laudon H, llstedt U (2014) The effect of trees on preferential flow and soil infiltrability in an agroforestry parkland in semiarid Burkina Faso. Water Res 50:3342–3354. https://doi.org/10.1002/2013WR015197
Trivino RR, Cano JLP, Jamieson CB (2016) An innovative agroforestry system for food and fuel production. J Environ Sci Manage 1:74–80
Trnka M, Semeradova D, Novotny I, Dumbrovsky M, Drbal K, Pavlik F, Vopravil J, Stepankova P, Vizina A, Balek J, Hlavinka P, Bartosova L, Zalud Z (2016) Assessing the combined hazards of drought, soil erosion and local flooding on agricultural land: a Czech case study. Clim Res 70: 231–249. https://doi.org/10.3354/cr01421
Tully KL, Lawrence D, Scanlon TM (2012) More trees less loss: nitrogen leaching losses decrease with increasing biomass in coffee agroforests. Agric Ecosystem Environ 161:137–144
Udawatta PR, Garrett EH, Kallenbach LR (2010) Agroforestry and grass buffer effects on water quality in grazed pastures. Agrofor Syst 79:81–87. https://doi.org/10.1007/s10457-010-9288-9

Udawatta RP, Rankothand LM, Jose S (2015) Agrofor Biodiver Sustain 11:2879. https://doi.org/10.3390/su11102879
UN (United Nations) (2015) World Population Prospects: the 2015 Revision. Available online at: https://esa.un.org/unpd/wpp. Accessed 16 May 2021
Van Thuyet D, Van Do T, Sato T, Hung TT (2014) Effects of species and shelterbelt structure on wind speed reduction in shelter. Agrofor Syst 88:237–244. https://doi.org/10.1007/s10457-013-9671-4
Vannoppen W, Baets DS, Keeble J, Dong Y, Poesen J (2017) How do root and soil characteristics affect the erosion-reducing potential of plant species. Ecol Eng 109:186–195. https://doi.org/10.1016/j.ecoleng.2017.08.001
Verma S, Khare N, Daniel S (2018) Effect of nitrogen and Sulphur levels on growth and yield of maize (*Zea mays* L.) under poplar (*Populus deltoides*) based agroforestry system. J Pharm Phytochem 7:2235–2238
Vermang J, Norton DL, Huang C, Cornelis MW, Silva MA, Gabriels D (2015) Characterization of soil surface roughness effects on runoff and soil erosion rates under simulated rainfall. Soil Sci Soc Amer J 79:903–916. https://doi.org/10.2136/sssaj2014.08.0329
Vincent AG, Turner BL, Tanner E (2010) Soil organic phosphorus dynamics following perturbation of litter cycling in a tropical moist forest. Europ J Soil Sci 61:48–57
Wahid F, Sharif M, Khan KM, Ali A, Khattak MA, Saljoqi RA (2015) Addition of rock phosphate to different organic fertilizers influences phosphorus uptake and wheat yield. Printed Portugal 30:90–100
Waldron A, Justicia R, Smith LE, Sanchez M (2012) Conservation through chocolate: a win-win for biodiversity and farmers in Ecuador's lowland tropics. Con Lett 5:213–221
Wang D, Heckathorn AS, Mainali K, Tripathee R (2016) Timing effects of heat-stress on plant ecophysiological characteristics and growth. Front Plant Sci 7:1–11. https://doi.org/10.3389/fpls.2016.01629
Wang Y, Zhang B, Lin L, Zepp H (2011) Agroforestry system reduces subsurface lateral flow and nitrate loss in Jiangxi Province, China. Agric Ecosystem Environ 140:441–453
Weisany W, Raei Y, Allahverdipoor HK (2013) Role of some of mineral nutrients in biological nitrogen fixation. Bulletin Environ Pharm Life Sci 2:77–84
Wolz KJ, DeLucia EH (2018) Alley cropping: global patterns of species composition and function. Agric Ecosyst Environ 252:61–68
World Bank (2015) Ending poverty and hunger by 2030. An agenda for the global food system. World Bank, Washington, DC, USA
Xu H, Bi H, Gao H, Yun L (2019) Alley cropping increases land use efficiency and economic profitability across the combination cultivation period. Agron 9:34. https://doi.org/10.3390/agronomy9010034
Yadav RS, Yadav BL, Chhipa BR, Dhyani SK, Ram M (2011) Soil biological properties under different tree based traditional agroforestry systems in a semi-arid region of Rajasthan, India. Agrofor Syst 81:195–202
Yadav SS, Hegde SV, Habibi BA, Dia M, Verma S (2018) Climate change, agriculture and food security. Food Sec Climate Change:1–24. https://doi.org/10.1002/9781119180661.ch1
Yaish MW, Kumar PP (2015) Salt tolerance research in date palm tree (*Phoenix dactylifera* L.), past, present, and future perspectives. Front. Plant Sci 6:348. https://doi.org/10.3389/fpls.2015.00348
Yamada T, Watanabe K, Okuda T, Sugimoto T, Azlin NY (2016) Growth and survival of trees planted in an oil palm plantation: implications for restoration of biodiversity. J Trop Forest Sci 28:97–105
Zhang C, Stratopoulos FML, Pretzsch H, Rotzer T (2019) How do tilia cordata greenspire trees cope with drought stress regarding their biomass allocation and ecosystem services. Forest 10: 676. https://doi.org/10.3390/f10080676
Zhang H, Goll SD, Manzoni S, Ciais P, Guenet B, Huang Y (2018) Modeling the effects of litter stoichiometry and soil mineral N availability on soil organic matter formation using

CENTURY-CUE (v1.0). Geosci Model Dev 11:4779–4796. https://doi.org/10.5194/gmd-11-4779-2018

Zhang JG, Lei JQ, Wang YD, Zhao Y, Xu XW (2016) Survival and growth of three afforestation species under high saline drip irrigation in the Taklimakan Desert. China Ecosphere 7(5): e01285. https://doi.org/10.1002/ecs2.1285

Zhu X, Liu W, Chen J, Bruijnzeel AL, Mao Z, Yang X, Cardinael R, Meng RF, Sidle CR, Seitz S, Nair DV, Nanko K, Zou X, Chen C, Jiang JX (2020) Reductions in water, soil and nutrient losses and pesticide pollution in agroforestry practices: a review of evidence and processes. Plant Soil 453:45–86. https://doi.org/10.1007/s11104-019-04377-3

Zhujun G, Xiaoxia W, Zhou F, Azofeifa SAG, Feng J, Nianwei Q (2015) Analysis of time scale influences on water and soil conservation effects for trees on experimental plots using vegetation fractional coverage. For Sci 61:67–75. https://doi.org/10.5849/forsci.13-028

Zomer RJ, Neufeldt H, Xu J, Ahrends A, Bossio D, Trabucco A, Van NM, Wang M (2016) Global tree cover and biomass carbon on agricultural land: the contribution of agroforestry to global and national carbon budgets. Sci Rep 6:29987. https://doi.org/10.1038/srep29987

Zomer RJ, Trabucco A, Coe R, Place F (2009) Trees on Farm: An Analysis of Global Extent and Geographical Patterns of Agroforestry, ICRAF Working Paper no. 89. World Agroforestry Centre, Nairobi, Kenya, pp 88–89

Zoysa DM, Inoue M (2014) Climate change impacts, agroforestry adaptation and policy environment in Sri Lanka. Open J Forest 4:439–456. https://doi.org/10.4236/ojf.2014.45049

Chapter 3
The Tree-Crop Interface: Soil Moisture Relations

Zikui Wang, Quan Cao, and Yuying Shen

Abstract Understanding the soil moisture dynamics and evapotranspiration (ET) patterns in the tree-crop interface is crucial for effective design and management of agroforestry systems, particularly in water-limited environments. In this chapter, we conducted a comprehensive analysis of soil moisture conditions, ET, and their key influencing factors in agroforestry. Subsequently, we developed an integrated ET and soil water balance model to unravel the underlying mechanisms governing soil moisture dynamics. Furthermore, we investigated the soil moisture conditions and ET relationships in an apple tree-cocksfoot agroforestry system established in the Loess Plateau of China. Our findings indicate that the ET and soil water balance coupling model successfully captured the intricate water cycling processes within agroforestry systems. This model holds promise for evaluating water utilization efficiency and informing the design and management of agroforestry systems across diverse environmental contexts. We emphasize the significance of appropriate field management practices, such as maintaining low coverage of understorey crop species, to mitigate the potential negative impacts of interspecific competition on tree performance in agroforestry under water-limited environments. We encourage the adoption of agroforestry systems incorporating well-managed understorey crops to enhance water productivity while providing valuable ecological services.

Keywords Agroforestry · Soil moisture · Tree-crop coupling system · ET partitioning · Water sustainability

Z. Wang (✉) · Y. Shen
College of Pastoral Agriculture Science and Technology, Lanzhou University, Lanzhou, China
e-mail: wzk@lzu.edu.cn; yy.shen@lzu.edu.cn

Q. Cao
College of Animal Science and Veterinary Medicine, Qinghai University, Xining, China

H. S. Jatav et al. (eds.), *Agroforestry to Combat Global Challenges*, Sustainable Development and Biodiversity 36, https://doi.org/10.1007/978-981-99-7282-1_3

3.1 Introduction

Diversified agricultural system has long been recognized as a sustainable agricultural approach. However, global agricultural modernization has led to the gradual simplification of agricultural systems, resulting in increased dependence on external inputs such as irrigation, fertilization, and energy, and contributed to a decline in resource use efficiencies (Basso et al. 2021). Furthermore, simplified farming systems are more susceptible to challenges such as pest and weed invasion, heat and drought stress, and greenhouse gas emissions, raising concerns about their sustainability in the face of climate change (zhang et al. 2018; Zimmerer and Vanek 2016). In contrast, a mixed agricultural system, such as intercropping and agroforestry, mimics the function of natural ecosystems and offers potential solutions for designing agroecosystems that rely more on biological processes, while harnessing land productivity and ecosystem services (Frasier et al. 2017).

Agroforestry practices have gained global recognition due to their potential for enhancing production and ecological service of agricultural systems (Smith et al. 2012; Abbas et al. 2017; Garcia et al. 2018). In Europe, agroforestry systems are recognized as a promising approach to integrate traditional agricultural and forestry systems effectively (Kay et al. 2019). Consequently, the "AGFORWARD" project has been implemented to assist smallholders in achieving economic gains while safeguarding the ecological environment. Similarly, in the United States, the integration of agroforestry systems within broader agricultural strategies has played a significant role in enhancing agricultural productivity and ensuring food security (Schoeneberger et al. 2017). Compared to single planting methods of economic forests or crops, agroforestry systems have been shown to effectively enhance the regional ecological environment and achieve simultaneous improvements in economic and ecological benefits (George et al. 2013; Zhu et al. 2018). In China, various types of agroforestry of different combinations of crop and tree species have been implemented. The understorey crops could be grain crops (e.g., cotton and wheat) as studied by Liu et al. (2012), forage crops (e.g., fodder canola and soybean) as investigated by Ling et al. (2017), and cover crops (e.g., cocksfoot) as examined by Wang et al. (2019). Grain and forage crops contribute to enhanced forest productivity, although they require substantial water resources, making them more suitable for arid regions under irrigation or sub-humid regions, as highlighted by Zhang et al. (2014). Conversely, cover crops are predominantly employed in some semi-arid hilly areas to prevent soil erosion, as emphasized by Gao et al. (2014). The selection of appropriate agroforestry systems depends on resource availability and the dynamics of competition or complementarity in resource utilization among different species (Grass et al. 2020; Wang et al. 2021).

Water is one of the main limiting factors for forest and agricultural production. Numerous studies have demonstrated that incorporating economic crops or herbs between trees and rows can enhance rainfall infiltration rates, decrease soil evaporation, and enhance soil water storage (Atucha et al. 2013; Chen et al. 2019). Additionally, the presence of crop canopies between rows can reduce raindrop kinetic energy, mitigating the direct impact of rainfall on the ground and subsequently reducing soil erosion (Chen et al. 2017; Ali et al. 2021). Wallace (1991) highlighted that in arid and semi-arid regions, annual soil evaporation can account for 30–60% of the annual soil rainfall. By incorporating interline canopy shading and reducing wind speed, agroforestry systems significantly mitigate soil water evaporation, thereby facilitating the optimal utilization of water, soil, and other resources within the composite system. A comprehensive understanding of water utilization and soil moisture dynamics in agroforestry is essential for optimizing planting configurations and maximizing both agricultural productivity and ecological benefits.

3.2 Soil Moisture Relations in Tree-Crop Coupling System

3.2.1 Soil Moisture Relations in Agroforestry

Compared to single planting methods, the establishment of a well-designed agroforestry system yields positive effects on the improvement of the soil water environment. The first reason is that incorporating crops into forests can increase infiltration and reduce soil evaporation. In sole planting systems, a significant amount of soil water is lost through non-productive water consumption, such as runoff or inefficient soil evaporation (Celette et al. 2010; Cao et al. 2021). The introduction of biological cover on previously bare surfaces reduces runoff and soil erosion during rainfall events, while the growth of underground roots enhances soil infiltration by creating larger soil pores. This promotes rainfall infiltration into deeper soil layers, increasing water storage capacity (Huang et al. 2014). Consequently, more available soil water becomes accessible to trees and crops during seasonal or prolonged drought periods. For instance, Hernández et al. (2005) found cover crop of clover effectively conserved available soil water without significant competition for water resources with the olive tree. Additionally, organic mulching with cover crop can enhance soil microbial activity and improve the function of soil aggregates, resulting in favourable soil structure and moisture conditions (Siczek and Lipiec 2011). Another reason is that cover crops typically have shallower root systems compared to fruit trees, resulting in distinct vertical divergence in their respective water uptake regions. Hydraulic lifting is a mechanism observed in agroforestry systems with deep-rooted trees and shrubs. It facilitates the efficient use of soil water resources by crops for their growth. This process involves the transfer of water from deep soil layers to the surface through roots. By strategically selecting tree and crop (or grass) layouts, trees can absorb water from deeper soil layers, while crops (or grasses) utilize water from surface soil layers. This strategy minimizes water competition between trees and crops.

However, underwater short age conditions, the effects of agroforestry planting on soil moisture can vary. Soil water competition has been a prominent research focus within agroforestry systems, particularly in semi-arid and arid regions (where seasonal drought or dry years occur), as well as in rain-fed agriculture (Leroy et al. 2009; Zhao et al. 2014). Tree and crop roots are the primary organs responsible for water absorption from the soil. When the root systems of two crops overlap, competition for soil water and other resources becomes unavoidable. In a semi-arid rain-fed region, Du et al. (2015) reported that the cover crop of milk vetch under apricot tree resulted in low soil water as well as nutrient levels compared to mono-cultured orchard, and Fang et al. (2016) validated that crop covering decreased soil water content and the production in an apple orchard.

It should be noted that competitive utilization of soil water between tree and crop species is influenced by many factors such as planting arrangement, tree and crop height difference, and field management measures on water, nutrients and soil tillage. The effective design of the planting arrangement and management of the agroforestry can alleviate interspecific water competition. For instance, Zhang et al. (2014) found that reducing the cotton plant density could alleviate its competition with jujube tree in jujube-cotton agroforestry, and Wang et al. (2017) found that increasing the distance of cotton row to tree row could also alleviate the competition. Similarly, in another study on vineyard agroforestry, Centinari et al. (2013) found that increasing the cut frequency of understorey cover crop could reduce the understorey water consumption. However, the mechanisms contributing to the interaction between evapotranspiration (ET) water transportation and soil moisture dynamics in agroforestry systems remain unclear and warrant further investigation.

3.2.2 ET Partitioning Between Tree and Crop Species

The assessment of water use interactions in tree-crop systems, whether competitive or complementary, can be facilitated by quantifying and partitioning of ET (Padovan et al. 2018). ET refers to the process by which water transitions from a liquid to a gaseous state. Soil evaporation, on the other hand, is a significant hydrological process in which water dissipates from the soil to the atmosphere (Ritchie 1972). However, in terms of agricultural production, soil evaporation represents non-productive water consumption since it cannot be directly utilized by plants (Li et al. 2022). Plant transpiration, on the other hand, is a vital physiological process closely associated with plant growth. Plant roots directly extract water from the soil, facilitating plant growth and development, and ultimately releasing water into the atmosphere through leaf stomata. Achieving full cover is often challenging in young and newly planted forests, leading to substantial water loss through soil evaporation (Wang et al. 2015; Wang et al. 2019). Introducing crop species between tree rows is potential of reducing soil evaporation, however meanwhile, it might contribute to the increase of water use through crop transpiration beneath the tree canopy (Ling et al. 2017).

A study carried out in Kenya demonstrated as high as 35% reduction in soil evaporation in agroforestry systems compared to sole planting, highlighted the role of crop shade in evaporation reduction in forest (Wallace et al. 1999). Huang et al.

(2014) also reported that clover and cocksfoot planted under jujube tree decreased water consumption through soil evaporation and increased rainfall infiltration. However, Zhao et al. (2015) observed that the introduction of agroforestry systems led to alterations in the soil water balance, resulting in a 5–12% increase in system ET. It is found that plant transpiration tends to intensify with higher solar radiation levels, agroforestry primarily modifies the amount of solar radiation intercepted by crops and subsequently affects their evapotranspiration within the forest environment. Meng et al. (2012) investigated crop transpiration in the presence of forests in the North China Plain, and noted a reduction of approximately 12% in crop transpiration under forest conditions. Similar to soil water conditions, proper management of cover crop could also have the potential of balancing water competition and water conservation in agroforestry systems.

3.2.3 Quantification of Water Relations in Tree-Crop Interface

Estimating seasonal or periodic ET in agroforestry systems is commonly achieved by using the soil water balance equation, after measuring the soil moisture in the root layer. However, it is not possible to separate evaporation and transpiration using this method alone. Direct measurement of one of these components is necessary. Micro-lysimeters, a widely adopted technique, can be used to measure the evaporation from soil (Jackson and Wallace 1999). On the other hand, measuring plant transpiration, especially for crop species, presents challenges due to limitations in available instrumentation. As a result, accurately partitioning of ET among tree plants, crop plants, and soil in agroforestry remains a challenging task in field operations. To overcome these challenges, the utilization of theoretical modelling provides an efficient approach for quantifying ET under various canopy and weather conditions, facilitating improvements in planting systems.

The simulation of ET plays a crucial role in facilitating precise water management in agricultural fields. The partitioning of water resources among various plants and soil compartments within a complex planting system is influenced by the interactions between the above-ground canopy and the subsurface root system. Above-ground, the water requirements of plants in such complex populations are determined by meteorological conditions and the spatial distribution of leaves (Ozier-Lafontaine et al., 1998). Below-ground, it primarily depends on the soil water availability and the distribution extent of plant roots (McIntyre et al. 1996). The Penman-Monteith (PM) model was derived by extending the original Penman equation, which utilizes an analogy to electrical resistance to estimate water vapour fluxes from canopies and soil surfaces. According to this analogy, the latent heat of water fluxes is impeded by multiple resistances that characterize the transport processes involved (Monteith 1965). The (PM) model integrates both soil and vegetation components when estimating evapotranspiration, but it overlooks the differences in water vapor transport between these two entities. Building upon the PM model, Shuttleworth and Wallace (1985) recognized that soil and vegetation act as distinct sources of water vapor. Consequently, they developed a dual-source evapotranspiration model that accounts for the evaporation from both soil and transpiration from leaf surfaces of different vegetation types.

Recognizing the presence of multiple plant species within complex populations such as intercropping, Wallace (1997) expanded the dual-source evapotranspiration model. To capture the radiation interception and evapotranspiration patterns in complex populations, a multiple source ET model (or ERIN model) was subsequently presented. The ERIN model has been successfully applied to calculate ET partitioning in several intercrop canopies (Gao et al. 2013; Teh et al. 2001). However, its properness for agroforestry systems might be limited. The main reason is that the multi-source ERIN assumes a mean canopy flow (MCF) dominated by the taller species. This assumption can result in an underestimation of water transporting resistance of understorey crops, and thereby overestimating the crop transpiration. Additionally, in cases where tree coverage is low, the model may also underestimate the boundary layer resistance for soil evaporation.

In order to enhance the simulation of water use processes in agroforestry systems, Wang et al. (2021) recently made improvements to the multi-source ERIN model. The improved model accounted for the impact of significant height differences among species within agroforestry systems on resistance network. Subsequently, the improved multi-source model was combined with a soil water balance model to simulate water cycling in agroforestry. This enhanced model was rigorously validated and demonstrated high accuracy in simulating plant transpiration, soil evaporation, and soil moisture conditions in the tree-crop coupling systems. The model will be presented in the forthcoming chapter.

3.3 Soil Moisture Relations Modelling in Agroforestry

3.3.1 Radiation Partitioning

Radiation drives water evaporating processes, which needs to be firstly figured out. The incident net radiation (R_n, MJ m^{-2} day^{-1}) on the top of tree canopy was determined with the procedures proposed by Allen et al. (1998). Incident radiation transmits through tree canopy, understory crop canopy, and then reaches soil surface. The interception of radiation by each component could be determined by the agroforestry radiation transmission model (Wang et al. 2019). This radiation model incorporates a geometrical approach to simulate radiation transmission within the tree crown, accounting for the spatial heterogeneity of the tree canopy. Additionally, it specifically applies a strip-path radiation transmission model for understory crop strips, guaranteeing the accurate partitioning of radiation between crop and soil. Finally, the fraction of net radiation captured by each component was calculated, and the available radiation for each plant species and soil was calculated as the product of total radiation and the fraction of radiation capture.

3.3.2 ET Partitioning

The multi-source ERIN model was enhanced by Wang et al. (2021) to simulate water processes within agroforestry canopies. Figure 3.1 illustrates the water transportation and the resistances on the way in intercrop canopy, as presented by the ERIN model (Wallace 1997). Species 1 represents the dominant species, while species 2 is the subdominant species. The latent heat for species 1 (λE_1), species 2 (λE_2), and soil (λE_s) are firstly transporting from the crop or soil surface to the level of MCF, and then to the reference level. On their way to the reference layer, the latent heat fluxes encounter various resistances that need to be accounted for. These include the crop and soil surface resistances of r_{s}^{c1}, r_{s}^{c2}, and r_{s}^{s}. Additionally, there are bulk boundary layer resistances for both species r_{a}^{c1} and r_{a}^{c2}, an aerodynamic resistance to the level of MCF r_{a}^{s}, and another aerodynamic resistance from mean canopy to the level of reference height r_{a}^{a}. The total latent heat flux in the canopy is calculated as follows:

$$\lambda \mathrm{ET} = C_{c1}\mathrm{PM}_{c1} + C_{c2}\mathrm{PM}_{c2} + C_{\mathrm{s}}\mathrm{PM}_{\mathrm{s}} \tag{3.1}$$

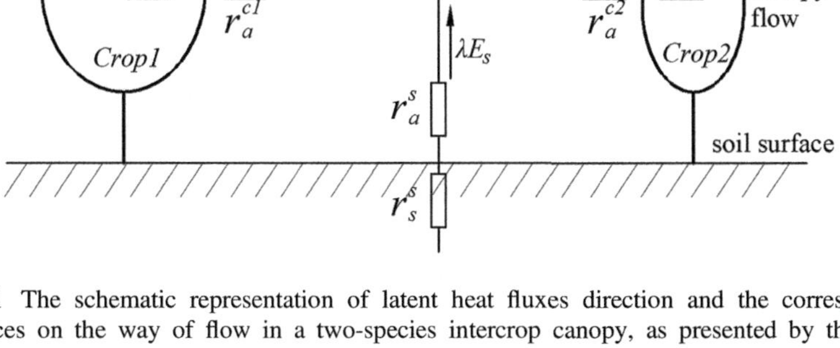

Fig. 3.1 The schematic representation of latent heat fluxes direction and the corresponding resistances on the way of flow in a two-species intercrop canopy, as presented by the ERIN model (Wallace 1997)

$$\mathrm{PM}_i = \frac{\Delta A + \left[\rho c_{\mathrm{p}} D - \Delta r_{\mathrm{a}}^i (R_n - R_{ni})\right] / \left(r_{\mathrm{a}}^a + r_{\mathrm{a}}^i\right)}{\Delta + \gamma\left[1 + r_{\mathrm{s}}^i / \left(r_{\mathrm{a}}^a + r_{\mathrm{a}}^i\right)\right]} \tag{3.2}$$

where, PM_{c1}, PM_{c2}, and PM_{s} are terms similar to the PM model. Definitions for the symbols in eq. (3.2) are referred to Allen et al. (1998). The coefficients C_{c1}, C_{c2}, and C_{s} and intermediate variables R_a and R_i are expressed as follows:

$$C_i = \left(1 + \frac{\sum 1/R_j}{1/R_i + 1/R_a}\right)^{-1} \quad j = [c1, c2, s] \text{ exclude } i \tag{3.3}$$

$$R_a = (\Delta + \gamma) r_{\mathrm{a}}^a \tag{3.4}$$

$$R_i = (\Delta + \gamma) r_{\mathrm{a}}^i + \gamma r_{\mathrm{s}}^i \tag{3.5}$$

Upon obtaining the total ET of the system, the deficit of vapour pressure at the mean canopy level (D_0) can be calculated from measurable vapour pressure deficit at the reference height (D), using eq. (3.6). Subsequently, the latent heat flux for each agroforestry component can be calculated using eq. (3.7):

$$D_0 = D + [\Delta A - (\Delta + \gamma)\lambda E] r_{\mathrm{a}}^a / \rho c_p \tag{3.6}$$

$$\lambda E_i = \frac{\Delta f_i R_n + \rho c_{\mathrm{p}} D_0 / r_{\mathrm{a}}^i}{\Delta + \gamma\left[1 + r_{\mathrm{s}}^i / r_{\mathrm{a}}^i\right]} \tag{3.7}$$

When directly applying the original ERIN model to agroforestry systems, discrepancies were observed between the simulated soil moisture content and ET and the observed data. Wang et al. (2021) identified that the primary cause of these discrepancies should be the inaccurate calculation on aerodynamic resistances of water transportation from the understorey crop canopy and soil surface. For intercropping, the ERIN model estimated the height of MCF based only on the top species 1. Therefore, it is assumed that water vapor from the subdominant species 2 directly flows to the mean canopy level after overcoming its surface resistance ${r_{\mathrm{s}}}^{c2}$ and boundary layer resistance ${r_{\mathrm{a}}}^{c2}$ (Fig. 3.1). However, reality in the agroforestry canopy is that the height of understorey crop is considerably lower than the trees under most circumstances, resulting in a significant vertical distance between the MCF levels of the crop and tree canopies (Fig. 3.2).

In the context of agroforestry canopies, the water evaporated from the understorey crop canopy is firstly impeded by the stomatal resistance (${r_{\mathrm{s}}}^{c}$) and boundary layer resistance (${r_{\mathrm{a}}}^{c}$), and then by an aerodynamic resistance (${r_{\mathrm{a}}}^{a,c}$) from the MCF of crop to the MCF of the tree. Additionally, for water evaporated from soil under the crop strip, the aerodynamic resistance should be the sum of ${r_{\mathrm{a}}}^{s,c}$ (aerodynamic resistance from the soil surface to the MCF of crop) and ${r_{\mathrm{a}}}^{a,c}$. The improved aerodynamic resistances for understorey crop (${r_{\mathrm{a}}}^{c\prime}$) and soil (${r_{\mathrm{a}}}^{s\prime}$) are expressed as:

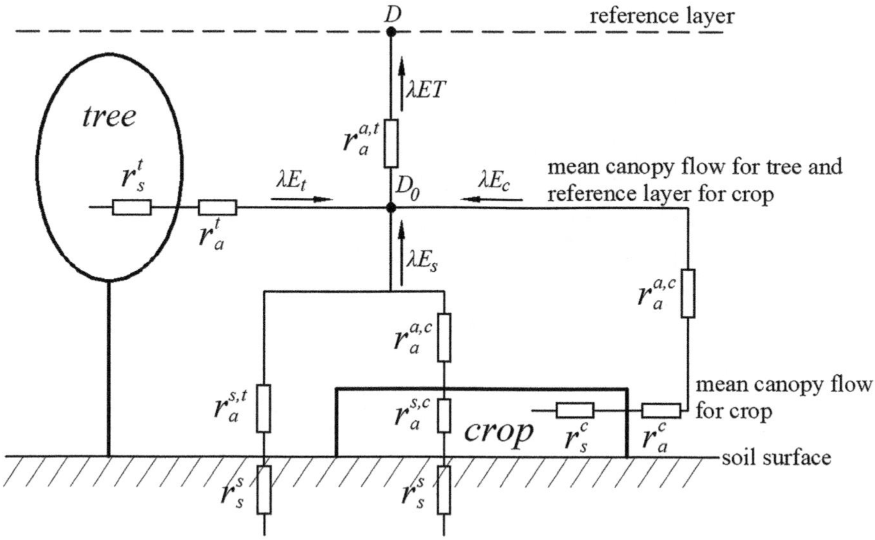

Fig. 3.2 The schematic representation of latent heat fluxes and the network of resistances in an agroforestry system as described by Wang et al. (2021)

$$r_{\mathrm{a}}{}^{c'} = r_{\mathrm{a}}{}^{c} + r_{\mathrm{a}}{}^{a,c} \tag{3.8}$$

$$r_{\mathrm{a}}{}^{s'} = (1-p)r_{\mathrm{a}}{}^{s,t} + p(r_{\mathrm{a}}{}^{s,c} + r_{\mathrm{a}}{}^{a,c}) \tag{3.9}$$

where p is the proportion of understorey land used for planting crop. In the enhanced multi-source model, the calculation procedures of the original ERIN model are strictly followed, except for the calculation of improved aerodynamic resistances. The calculation procedures for different resistances are referred to the studies of Shuttleworth and Wallace (1985), Wallace (1997), and Wang et al. (2021).

3.3.3 Soil Water Balance

Figure 3.3 presents a schematic representation illustrating the partitioning of radiation and the cycling of water in agroforestry. R_n is partitioned into R_{nt} (net radiation absorbed by trees), R_{nc} (net radiation absorbed by crops), and R_{ns} (net radiation absorbed by the soil), which are utilized to drive water transport processes. Precipitation is intercepted by the canopies of both trees and crops, and it can either be stored within the canopy or infiltrate into the soil. Based on field investigations, it has been observed that the roots of crops predominantly occupy the 0–100 cm soil layer. Therefore, we assume that all transpiration of the cocksfoot crop occurs within the 0–100 cm layer beneath the crop strip. The rooting depth of trees, on the other hand, can extend as deep as 300 cm or beyond, and this should be verified through field

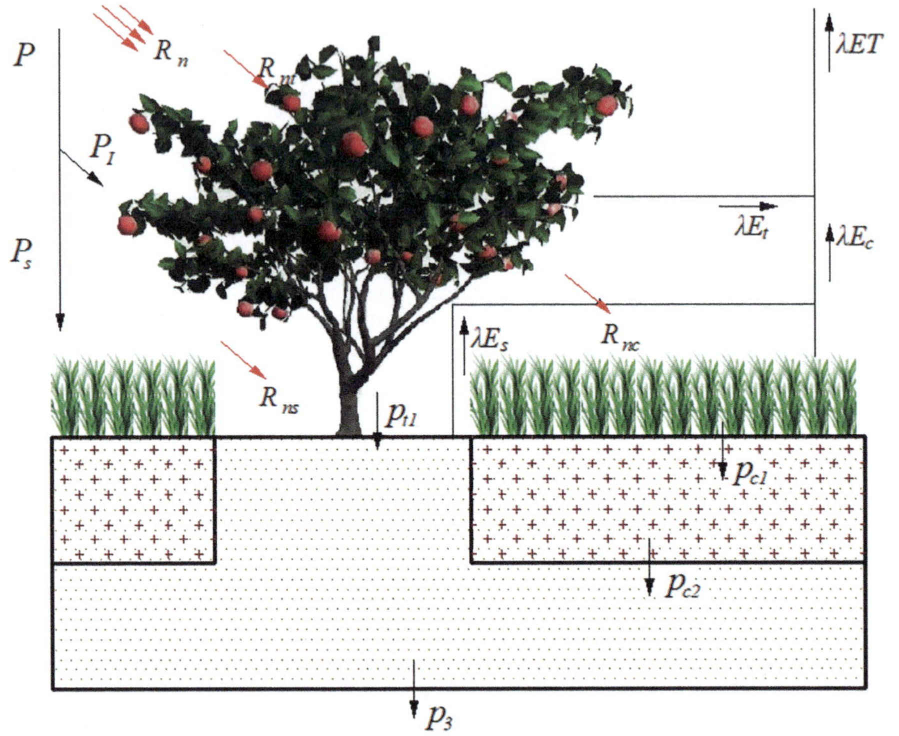

Fig. 3.3 The schematic representation illustrates the partitioning of radiation and the water cycling in an agroforestry system. R_n represents net radiation, P represents precipitation, λE denotes latent heat of water transport, and p represents water percolation into below soil layers. The subscripts t, c, and s indicate tree, crop, and soil, respectively. Area of dots means root area of apple tree while the areas with a cross represent the root zone of understorey crop

investigations for different tree species. Trees utilize soil water from the entire root layer including the area occupied by crop roots, in comparison, crop only depletes soil water from its own root zone. Water in the root layer after rainfall firstly drainages down into the below layers as excessive soil water or in the process of soil water redistribution, and then the stored soil water is consumed through processes such as soil evaporation, crop transpiration, and tree transpiration.

Soil water balance in each layer was expressed following the procedure presented by Teh (2006). Soil water content in layer i at day t,$\Theta_{i,t}$, is calculated as:

$$\Theta_{i,t} = \Theta_{i,t-1} + P_{i-1,t} - P_{i,t} - E_{a(i,t)} - T_{a(i,t)} \tag{3.10}$$

in which, $\Theta_{i,t-1}$ represents the water content in soil layer i at day $t-1$ (mm). $E_{a(i,t)}$ means the daily evaporation from layer i while $T_{a(i,t)}$ denotes the daily transpiration from layer i (mm). $P_{i-1,t}$ represent the percolation of soil water from soil layer $i-1$ into layer i, and $P_{i,t}$ is the outflow of water from layer i $toi+1$ (mm). The soil water deep percolation can be further divided into two parts: $p_{e(i,t)}$ accounts for the loss of excess water, while $p_{d(i,t)}$ corresponds to the soil water redistribution (mm). They can be mathematically expressed as follows:

$$p_{e(i,t)} = \begin{cases} 0 & \text{if } \Theta_{i,t-1} + P_{i-1,t} \leq \Theta_{sat} \\ \Theta_{i,t-1} + P_{i-1,t} - \Theta_{sat} & \text{if } \Theta_{i,t-1} + P_{i-1,t} > \Theta_{sat} \end{cases} \tag{3.11}$$

$$p_{d(i,t)} = \Theta'_{i,t} - \Theta''_{i,t} \tag{3.12}$$

in which, Θ_{sat} is sutured water content; $\Theta'_{i,t}$ and $\Theta''_{i,t}$ represent the water content before and after redistribution, respectively; they are calculated as:

$$\Theta'_{i,t} = \Theta_{i,t-1} + P_{i-1,t} - p_{e(i,t)} \tag{3.13}$$

$$\Theta''_{i,t} = \Theta_{sat} - \frac{\Theta_{sat}}{\delta} \ln\left\{ \frac{\delta K_{sat} \Delta t}{L_i \Theta_{sat}} + \exp\left[\frac{\Theta_{sat}}{\delta} \left(\Theta_{sat} - \Theta'_{i,t} \right) \right] \right\} \tag{3.14}$$

in which, δ represents a coefficient with a value of 13. K_{sat} represents the soil saturated hydraulic conductivity (m day^{-1}). L denotes the calculated depth of the soil layer (m). Δt represents the time step in days. Finally, the potential plant transpiration computed by the improved ERIN model needs to be adjusted based on the real-time soil moisture condition (Doorenbos and Kassam 1979). The reduction factor of transpiration is defined as:

$$R_D = \left(\theta_v - \theta_{v,wp} \right) / \left(\theta_{v,cr} - \theta_{v,wp} \right) \tag{3.15}$$

in which $\theta_{v,wp}$ is the soil wilting point; and $\theta_{v,cr}$ is the critical soil water content which is derived as:

$$\theta_{v,cr} = \theta_{v,wp} + 0.5 \left(\theta_{v,sat} - \theta_{v,wp} \right) \tag{3.16}$$

3.4 Case Study

3.4.1 Field Experiment

A field study was conducted on an apple tree-cocksfoot agroforestry system at the Qingyang Research Station in Gansu, located in northwest China (latitude 35°40'N, longitude 107°51′E), during the growing seasons of 2016–2018. The research site is situated in the south of the Chinse Loess Plateau, has a long-term average air temperature of 9.2 °C, and an annual precipitation of 527.6 mm. The soil at the site is classified as a silty loam.

Fig. 3.4 The apple tree-cocksfoot agroforestry at Qingyang Research Station, which is located on the Chinese Loess Plateau

The apple trees were planted in a north-south orientation, with uniform within-row spacing and inter-row spacing of 4.0 m. Three experimental plots were established to investigate different orchard management treatments: (1) Monoculture of apple orchard which was frequently tilled to keep clean (CT), (2) agroforestry system with cocksfoot strips of 2.4 meter width between tree rows, where cocksfoot vegetation was frequently harvested to make crop had a low coverage (LC, Fig. 3.4), and (3) agroforestry with cocksfoot crop that was less harvested and had a greater cover (GC).

The leaf area index (LAI) of apple trees was determined using an LAI-2000 plant canopy analyser. The LAI of cocksfoot was measured by employing a leaf area meter. Water content in the 0–300 cm soil layer was measured using the oven-drying method. Throughout the growing season, soil moisture was measured continuously using a Diviner 2000 soil water measuring device, covering a depth of 200 cm. Soil evaporation was measured with a micro-lysimeter with a length of 15 cm and internal diameter of 11 cm. Cocksfoot ET was measured with a larger micro-lysimeter with a length of 30 cm and an internal diameter of 25 cm.

3.4.2 Soil Water Dynamics and ET in Apple Tree-Cocksfoot System

In 2016, two distinct drying periods were observed. Soil water content under the cocksfoot cover treatments exhibited more significant depletion compared to the clean tillage (CT) treatment. For example, in the growing season of 2016, soil water content in the 0–80 cm soil layer under the greater coverage (GC) and low coverage (LC) agroforestry treatments decreased by 22.0% and 18.3%, respectively, while the decrease was only 8.0% under CT. Similar trends were observed in soil water content in 2018. In comparison, soil water storage was recharged in 2017, the recharge was also the highest under GC. This indicated that agroforestry increased the soil water depletion and recharge compared to apple tree monoculture, a mechanism contributed a high rainfall capture and use efficiency.

The agroforestry ET and soil water balance model proposed by Wang et al. (2021) was validated using measured soil water content (Fig. 3.5) and ET values (data not shown). The comparison between the measured and simulated values demonstrated a high level of agreement in most cases, indicating a good performance of the model. The simulated dynamics of soil water content closely matched the observed changes in soil water conditions. Notably, the model exhibited better performance in 2018 compared to the preceding two years. The model accurately captured the rising and declining trends in soil water content, demonstrating its ability to simulate the temporal variations in soil water dynamics.

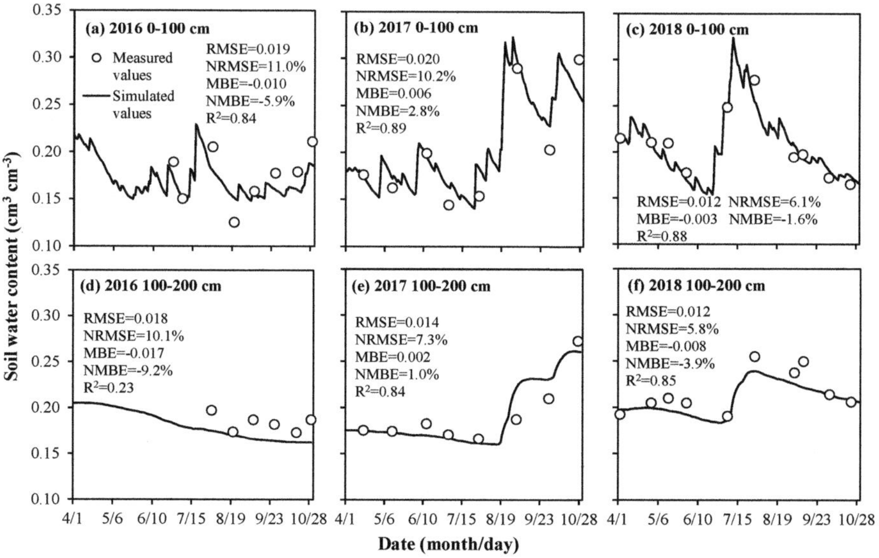

Fig. 3.5 Comparison of simulated soil water dynamics in 0–100 and 100–200 cm soil layers versus the measured ones in apple tree-cocksfoot agroforestry on the Loess Plateau

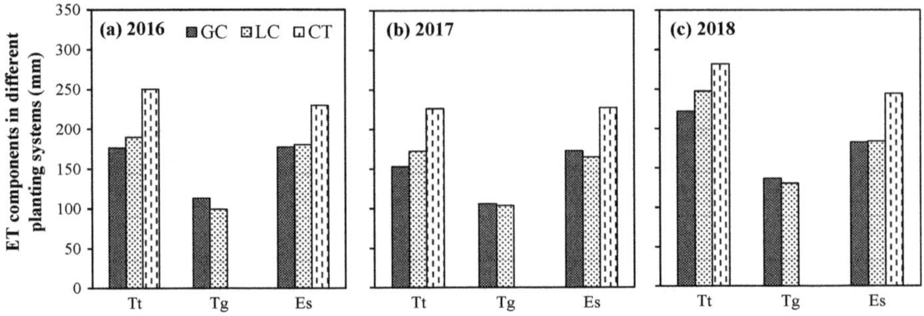

Fig. 3.6 Evapotranspiration partitioning among tree (Tt), crop (Tg), and soil(Es) in an apple tree and cocksfoot agroforestry in the growing seasons of 2016, 2017, and 2018 on the Loess Plateau

Using the validated model, we assessed ET and ET partitioning in the agroforestry. The results revealed a substantial reduction in tree transpiration following the intercropping of cocksfoot, particularly in the dry seasons of 2017. In comparison to the clean tillage (CT) treatment, the apple tree transpiration under the agroforestry with greater cocksfoot coverage (GC) was significantly lower. Furthermore, it was observed that increasing the cutting frequency is efficient in reducing the negative effects of cover crop on apple tree transpiration. This indicates that low coverage of understorey crop helped alleviate the negative impact on tree water use.

In addition to tree transpiration, soil evaporation also exhibited significant reductions compared to the monoculture orchard (Fig. 3.6). The reduction in evaporation under the cocksfoot covered treatments ranged from 21.5 to 22.7% in 2016, 24.1 to 27.5% in 2017, and 24.8 to 25.4% in 2018 when compared to the clean tilled (CT) orchard. Moreover, the total transpiration, which includes both apple tree and cocksfoot transpiration, showed substantial increases in GC and LC. The total system evapotranspiration (ET) in the agroforestry system demonstrated a reduction ranging from 2.2% to 4.9% compared to CT in 2016 and 2017. However, in 2018, the total system ET increased by 2.7–6.7% when compared to CT.

3.4.3 Scenario Analysis

Using the agroforestry ET and soil water model, we conducted a long-term simulation to assess the impacts of apple tree-cocksfoot agroforestry on water use and soil moisture dynamics (Fig. 3.7). Apple tree canopy information was collected by Wang et al. (2010). Our findings indicate that in the early stages (1–8 years of simulation), apple trees in the clean tilled (CT) treatment exhibited higher transpiration rates compared to those in the greater coverage (GC) and lower coverage (LC) treatments. However, as the apple tree LAI increased over time, the presence of cocksfoot did not significantly restrict tree transpiration. Notably, after 2008, tree transpiration in LC surpassed that in CT, while transpiration in GC remained lower than both LC and

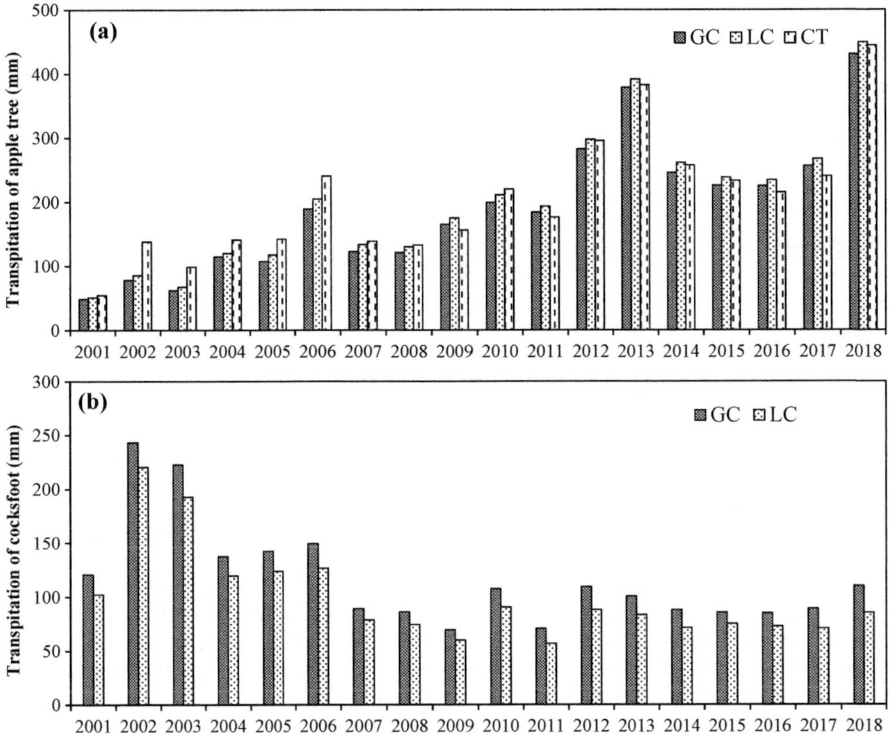

Fig. 3.7 The simulated plant transpiration in an apple tree and cocksfoot agroforestry over an 18-year period on the Loess Plateau of China

CT throughout the entire period. The 18-year averaged apple tree transpiration represented a decrease of 7.3% and 2.1% in GC and LC, respectively, compared to CT. In contrast, the cocksfoot transpiration decreased significantly from 2002 to 2007 and kept a low value thereafter. Furthermore, the total transpiration in agroforestry was largely improved, which in GC was 17.5% higher than that in LC. Importantly, our 18-year simulation demonstrated that although agroforestry increased total water consumption through plant transpiration, it reduced soil evaporation at the same time and did not increase deep soil water depletion, suggesting soil moisture is rechargeable and sustainable.

3.5 Conclusion

The appropriateness of agroforestry is strongly influenced by water availability in arid and semi-arid regions. In this chapter we figured out the soil moisture relations and ET patterns in the tree-crop interface of agroforestry system. We concluded that firstly, the ET and soil water balance coupling model could capture the water cycling processes within agroforestry, which was capable of guiding the design and management of agroforestry systems. Secondly, proper field management practices, such as maintaining low coverage of understorey crop species, should be applied to mitigate the potential negative impacts of crop on tree performance in water-limited environments.

References

Abbas F, Hammad HM, Fahad S, Cerda A, Rizwan M, Farhad W, Ehsan S, Bakhat HF (2017) Agroforestry: a sustainable environmental practice for carbon sequestration under the climate change scenarios-a review. Environ Sci Pollut Res 24:11177–11191

Ali G, Wang Z, Li X, Jin N, Chu H, Jing L (2021) Deep soil water deficit and recovery in alfalfa fields of the loess plateau of China. Field Crop Res 260:107990

Allen RG, Pereira LS, Raes D, Smith M (1998) Crop evapotranspiration—guidelines for computing crop water requirements. In: FAO irrigation and drainage paper no. 56. FAO, Rome

Atucha A, MerwinI A, Brown MG, Gardiazabal F, Mena F, Adriazola C, Lehmann J (2013) Soil erosion, runoff and nutrient losses in an avocado (Perseaamericana mill) hillside orchard under different groundcover management systems. Plant Soil 368(1–2):393–406

Basso B, Jones JW, Antle J, Martinez-Feria RA, Verma B (2021) Enabling circularity in grain production systems with novel technologies and policy. Agric Syst 193:103244

Cao Q, Wang Z, Yang X, Shen Y (2021) The effects of cocksfoot cover crop on soil water balance, evapotranspiration partitioning, and system production in an apple orchard on the loess plateau of China. Soil Tillage Res 205:104788

Celette F, Ripoche A, Gary C (2010) WaLIS—A Simple model to simulate water partitioning in a crop association: the example of an intercropped vineyard. Agric Water Manag 97(11): 1749–1759

Centinari M, Filippetti I, Bauerle T, Allegro G, Valentini G, Poni S (2013) Cover crop water use in relation to vineyard floor management practices. Am J Enol Vitic 64:522–526

Chen C, Liu W, Jiang X, Wu J (2017) Effects of rubber-based agroforestry systems on soil aggregation and associated soil organic carbon: implications for land use. Geoderma 299:13–24

Chen, C., Liu, W., Wu, J., Jiang, X., Zhu, X., 2019. Can intercropping with the cash crop help improve the soil physico-chemical properties of rubber plantations? GEODERMA 335, 149-160.

Doorenbos, J., Kassam A., H., 1979. Yield response to water. FAO Irrigation and Drainage Paper No. 33, FAO, Rome.

Du S, Bai G, Yu J (2015) Soil properties and apricot growth under intercropping and mulching with erect milk vetch in the loess hilly-gully region. Plant Soil 390(1–2):431–442

Fang K, Li H, Wang Z, Du Y, Wang J (2016) Comparative analysis on spatial variability of soil moisture under different land use types in orchard. Sci Horticult 207:65–72

Frasier I, Noellemeyer E, Amiotti N, Quiroga A (2017) Vetch-rye biculture is a sustainable alternative for enhanced nitrogen availability and low leaching losses in a no-till cover crop system. Field Crop Res 214:104–112

Gao X, Wu P, Zhao X, Wang J, Shi Y (2014) Effects of land use on soil moisture variations in a semi-arid catchment: implications for land and agricultural water management. Land Degrad Dev 25(2):163–172

Gao Y, Duan A, Qiu X, Li X, Pauline U, Sun J, Wang H (2013) Modeling evapotranspiration in maize/soybean strip intercropping system with the evaporation and radiation interception by neighboring species model. Agric Water Manag 128:110–119

Garcia L, Celette F, Gary C, Ripoche A, Valdés-Gómez H, Metay A (2018) Management of service crops for the provision of ecosystem services in vineyards: a review. Agric Ecosyst Environ 251:158–170
George SJ, Harper RJ, Hobbs RJ, Tibbett M (2013) A sustainable agricultural landscape for Australia: a review of interlacing carbon sequestration, biodiversity and salinity management in agroforestry systems. Agric Ecosyst Environ 163:28–36
Grass I, Kubitza C, Krishna V V, Corre M D, Mußhoff O, Pütz P, Drescher J, Rembold K, Ariyanti E S, Barnes A D. (2020) Trade-offs between multifunctionality and profit in tropical smallholder landscapes. Nat Commun,11(1):1–13
Hernández AJ, Lacasta C, Pastor J (2005) Effects of different management practices on soil conservation and soil water in a rainfed olive orchard. Agric Water Manag 77(1–3):232–248
Huang J, Wang J, Zhao X, Wu P, Qi Z, Li H (2014) Effects of permanent ground cover on soil moisture in jujube orchards under sloping ground: a simulation study. Agric Water Manag 138: 68–77
Jackson NA, Wallace JS (1999) Soil evaporation measurements in an agroforestry system in Kenya. Agric For Meteorol 94:203–215
Kay S, Graves A, Palma JH, Moreno G, Roces-Díaz JV, Aviron S, Chouvardas D, Crous-Duran J, Ferreiro- Domínguez N, de Jalón SG (2019) Agroforestry is paying off–economic evaluation of ecosystem services in European landscapes with and without agroforestry systems. Ecosyst Serv 36:100896
Leroy C, Sabatier S, Wahyuni NS, Barczi JF (2009) Virtual trees and light capture: a method for optimizing agroforestry stand design. Agrofor Syst 77:37–47
Li Z, Cui S, Zhang Q, Xu G, Feng Q, Chen C, Li Y (2022) Optimizing wheat yield, water, and nitrogen use efficiency with water and nitrogen inputs in China: asynthesis and life cycle assessment. Front Plant Sci 13:930484
Ling Q, Gao X, Zhao X, Huang J, Li H, Li L, Sun W, Wu P (2017) Soil water effects of agroforestry in rainfed jujube (*Ziziphus jujube* Mill.) orchards on loess hillslopes in Northwest China. Agric Ecosyst Environ 247:343–351
Liu JQ, Halik W, Wang GS, Kasim Y (2012) Study on temporal and spatial distribution rules of characteristic fruit industry resources in Xingjiang. Agric Res Arid Areas 30:230–236
McIntyre BD, Riha SJ, Ong CK (1996) Light interception and evapotranspiration in hedgerow agroforestry systems. Agric For Meteorol 81:31–40
Meng P, Zhang JS, Song ZM, Lu MM, Ma XL (2012) Study on evapotranspiration consumption of agroforestry model. For Sci Res 9(3):221–226. (in Chinese with English abstract)
Monteith JL (1965) Evaporation and environment. In: Fogg GE (ed) Symposia of the Society for Experimental Biology: the state and movement of water in living organisms, vol 19. Academic Press, New York, pp 205–234
Ozier-Lafontaine H, Lafolie F, Bruckler L, Tournebize R, Mollier A (1998) Modeling competition for water in intercrops: theory and comparison with field experiments. Plant Soil 204:183–201
Padovan MP, Brook RM, Barrios M, Cruz-Castillo JB, Vilchez-Mendoza SJ, Costa AN, Rapidel B (2018) Water loss by transpiration and soil evaporation in coffee shaded by Tabebuiarosea Bertol and Simaroubaglauca dc. compared to unshaded coffee in sub-optimal environmental conditions. Agric For Meteorol 248:1–14
Ritchie JT (1972) Model for predicting evaporation from a row crop with incomplete cover. Water Resour Res 8:1204–1213
Schoeneberger MM, Bentrup G, Patel-Weynand T (2017) Agroforestry: enhancing resiliency in US agricultural landscapes under changing conditions. In: Gen. Tech. Report WO-96, vol 96. US Department of Agriculture, Forest Service, Washington, DC
Shuttleworth WJ, Wallace JS (1985) Evaporation from sparse crops-an energy combination theory. Q J R Meteorol Soc 111:839–855
Siczek A, Lipiec J (2011) Soybean nodulation and nitrogen fixation in response to soil compaction and surface straw mulching. Soil Tillage Res 114(1):50–56

Smith J, Pearce BD, WoGCe MS (2012) Reconciling productivity with protection of the environment: is temperate agroforestry the answer? Renew Agric Food Syst 28:80–92
Teh CBS (2006) Introduction to mathematical Modeling of crop growth. Brown Walker Press, Boca Raton, Florida, USA
Teh CBS, Simmonds LP, Wheeler TR (2001) Modelling the partitioning of evapotranspiration in a maize–sunflower intercrop. In: The 2nd International Conference on Tropical Climatology, Meteorology and Hydrology. TCMH-2001, Brussels, Belgium
Wallace JS (1991) The measurement and modelling of evaporation from semiarid land. IAHS-AISH 199:131–148
Wallace JS, Jackson NA, Ong CK (1999) Modelling soil evaporation in an agroforestry system in Kenya. Agric For Meteorol 94:189–202
Wallace JS (1997) Evaporation and radiation interception by neighbouring plants. Q J R Meteorol Soc 123:1885–1905
Wang L, Guo X, Shi X (2010) Study on the canopy characteristics of different ages Fujiapple trees in Jinxi loess plateau area. *journal of northwest A&F university*. Nat Sci Ed 28(11):115–120. (in Chinese with English abstract)
Wang Q, Zhang D, Zhang L, Han S, van der Werf W, Evers JB, Su Z, Anten NPR (2017) Spatial configuration drives complementary capture of light of the understory cotton in young jujube plantations. Field Crop Res 213:21–28
Wang Z, Cao Q, Shen Y (2019) Modeling light availability for crop strips planted within apple orchard. Agric Syst 170:28–38
Wang Z, Wu P, Zhao X, Gao Y, Chen X (2015) Water use and crop coefficient of the wheat–maize strip intercropping system for an arid region in northwestern China. Agric Water Manag 161: 77–85
Wang Z, Wu Y, Cao Q, Shen Y, Zhang B (2021) Modeling the coupling processes of evapotranspiration and soil water balance in agroforestry systems. Agric Water Manag 250:106839
Zhang D, Zhang L, Liu J, Han S, Wang Q, Evers J, Liu J, van der Werf W, Li L (2014) Plant density affects light interception and yield in cotton grown as companion crop in young jujube plantations. Field Crop Res 169:132–139
Zhang Q, Bell LW, Shen Y, Whish JPM (2018) Indices of forage nutritional yield and water use efficiency amongst spring-sown annual forage crops in north-West China. Eur J Agron 93:1–10
Zhao X, Huang J, Gao X, Wu P, Wang J (2014) Runoff features of pasture and crop slopes at different rainfall intensities, antecedent moisture contents and gradients on the Chinese loess plateau: a solution of rainfall simulation experiments. Catena 119:90–96
Zhao, P., Li, S., Li, F., Du, T., Tong, L., Kang, S. (2015) Comparison of dual crop coefficient method and Shuttleworth–Wallace model in evapotranspiration partitioning in a vineyard of northwest China. Agric Water Manag 160: 41–56
Zhu X, Liu W, Jiang XJ, Wang P, Li W (2018) Effects of land-use changes on runoff and sediment yield: implications for soil conservation and forest management in Xishuangbanna, Southwest China. Land Degrad Dev 29(9):2962–2974
Zimmerer KS, Vanek SJ (2016) Toward the integrated framework analysis of linkages among agrobiodiversity, livelihood diversification, ecological systems, and sustainability amid global change. Land 5(2):10

Chapter 4
Potential Nutrient Cycling and Management in Agroforestry

Manish Raj, Kanhaiya Lal, Satdev, Priya Kumari, Shailja Kumari, Vinod Kumar Dubey, Sushant, Mainak Ghosh, and Sanjay Kumar

Abstract The nutrient availability in soil system governs an important role in all living species on the earth through food cycle. Inorganic available nutrients influence the growth and development as well as influence the nutrient state in plant acquisition. Agroforestry system (AFS) as an alternative practice of agriculture involves the incorporation of agricultural crop or animal use with trees on arable land use system for sustaining the soil health and productivity. Soil nutrient dynamic is defined as the way of taking up nutrient, retention, transformation (quantity and forms) and cycle over the distance and time due to continuous different bio-chemical processes in agroecosystem. The amount of nutrient released during the decomposition of litter or residue material of agroforestry subsequently satisfies the nutrient need of the plant, while further away the soil releases nutrient losses through various kinds of processes, i.e. erosion, heavy runoff, leaching and gaseous losses, removal by crop, as well as nutrient fixation, etc. The agroforestry system influenced different nutrient cycles of our ecosystem. Agroforestry system relative importance for storage of nutrient as well as increased the SOM (sustain of active SOM) levels in the soil. The agroforestry system may decrease the losses of N through leaching and increase the C immobilization, enhance the cation exchange capacity (CEC) and drastically improve the level of nitrogen, phosphorus and potash. This chapter examines the in situ nutrient dynamics of carbon and essential primary nutrients with its cycle in the agroforestry system (AFS).

M. Raj (✉) · P. Kumari · Sushant · M. Ghosh · S. Kumar
Department of Agronomy, Bihar Agricultural University, Bhagalpur, Bihar, India

K. Lal
Department of Agronomy, Dr. Rajendra Prasad Central Agricultural University, Pusa, Samastipur, Bihar, India

Satdev · S. Kumari
Department of Soil Science and Agricultural Chemistry, Bihar Agricultural University, Bhagalpur, Bihar, India

V. K. Dubey
Department of Entomology, Dr. Rajendra Prasad Central Agricultural University, Pusa, Samastipur, Bihar, India

H. S. Jatav et al. (eds.), *Agroforestry to Combat Global Challenges*, Sustainable Development and Biodiversity 36, https://doi.org/10.1007/978-981-99-7282-1_4

Keywords Carbon and nitrogen dynamic · AFS · SOM · Nutrient

4.1 Introduction

The agroforestry system has been practiced from ancient time where the agricultural crops are diversifying with forest tree or animals on a certain unit of land. This system is one of the sustainable approaches to upsurge the overall production and socio-economic status of the people under a particular edaphic and climatic condition. In Southeast Asia, agroforestry is known by different names like *taungya*, *tumpang sari* and *nônglâmkếthợp* in Burmese, Indonesia and Vietnam, respectively. Worldwide, agroforestry systems are practiced in approximately 1023 Mha of land which is 7.9% of the global land areas (13.0 billion ha). It has been reported that South America has the largest areas of agroforestry, followed by sub-Saharan Africa and then Southeast Asia (Zomer et al. 2009). In India, as per Dhyani and Handa (2013), only 8.2% of India's total land area is covered by agroforestry (25.32 Mha). Among this approximately 78.98% agroforestry system are practiced on cultivated land (20.0 Mha), of which 7.0 Mha are irrigated and 13.0 Mha are rainfed, while the remaining 5.32 Mha comprises shifting cultivation (2.28 Mha) and home gardens and problem soil rehabilitation (2.93 Mha).

Soil nutrient dynamic is the study of variations in nutrient status (quantity and forms) in a soil ecosystem under a certain set of environmental conditions, in both temporal and spatial manners, as a result of ongoing biogeochemical processes at work in the system. Thus it is one of the vital processes in agroforestry to understanding the dynamics of soil nutrient content and forms as well as also affects biomass output, soil nutrient levels and the sustainable management of in situ nutrient cycling.

Nutrient cycling simply states the continuous transfer of nutrients like N, P, K, C, Fe, Zn, Mn, etc. from the soil to plant acquisition and returning back to the soil through various means of activities of soil microflora (bacteria, fungi, algae, etc.) and macrofauna (ants, earthworm, snail, springtails, rats, spider, beetle, etc.), weathering of minerals and other transformation occurring within different components of the ecosystem. The agroforestry systems consist of different types of perennial trees like *Gliricidia*, *Leucaena leucocephala*, *Prosopis juliflora*, *Azadirachta indica*, *Acacia nilotica*, *Casuarina equisetifolia*, *Acacia mearnsii*, etc. and have tap root system that utilized nutrients and water from the deeper layer of the soil more efficiently than herbaceous crops. The herbaceous crops have adventitious root system and utilized water and nutrients only from the shallow to medium depth of the soil.

4.2 Why Study Soil Nutrient Dynamics in AFS

- Estimation of nutrient turnover in agroforestry system
- For evaluating the activity of soil biota population, i.e. microflora (bacteria, fungi, algae, etc.) and fauna (earthworm, springtails, ants, etc.)
- For estimation rate of litter fall and its associated nutrient dynamics
- For assessing the crop residue and litter decomposition rate and also observe its nutrient release pattern
- Effect of root dynamics (fine and coarse) on nutrient supplement

4.3 Key Process Responsible for Soil Nutrient Dynamics in Agroforestry System

4.3.1 Litter Addition

The litter consists of dead leaves, root, twigs, needles, flower, fruit and amorphous material of tree species that contain a considerable amount of different essential nutrients like N, P, K, C, S, Fe, Zn, Mg, etc. necessary for plant growth and development. Apart from this litter production in agroforestry that helps to accumulate the high amount of organic matter on the soil, which helps to compound the soil fertility status by releasing vital nutrients C, N, P, K, Mn and Zn to the soil as well as play a significant role in soil nutrient cycling process (Fig. 4.1). Rawat et al. (2009) reported that the rate of litter production mainly depends on the types of various tree species present in agroforestry and other physical factors like topographic, edaphic, climatic and biotic. Annual crop species or cultivars generally contribute less quantity of litter production as compared to perennial deep-rooted trees in agroforestry systems. Sarvade et al. (2014) noticed that the deciduous species added more in litter fall due to shading of leaves occurring round the year as parallel to the evergreen species. They also mentioned that 25% litter fall is recorded in each season, i.e. spring and summer; however least litter fall is observed in autumn season. Munishamappa et al. (2012), Singh et al. (2011) and Kumar et al. (2011) reported that in summer and spring season wind velocity is more, helping to increase the abscission concentration in leaves.

4.3.2 Decomposition of Litter and Nutrient Release

Litter decomposition is the critical stage that occurred on dead leaves, roots, twigs, branches, reproductive and amorphous material of tree species, barks, etc. and break down into smaller trashes with the passage of time through physical, chemical and biological process by plummeting litter to water, CO_2 and mineral nutrients

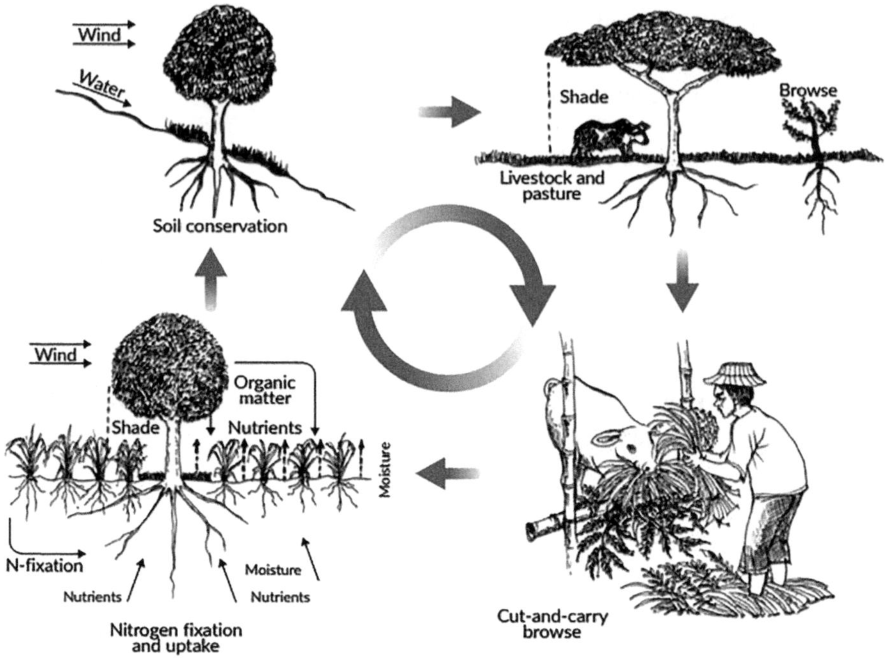

Fig. 4.1 Nutrient cycle in agroforestry (source modified from Xu et al. 2013)

(Lambers et al. 1998). The rate of decomposition is governed by various factors: the type of tree species (litter quality), physical factor (temperature, relative humidity, rainfall, aeration, etc.), biological factors (living entity, i.e. microflora and fauna) and soil nutrient availability (preferably C:N ratio) that confined areas of agroforestry (Zheng 2006). In 2011, Kumar et al. found that tree-based cropping systems provide the favourable environment for soil biota (decomposer) that regulates the litter decomposition process and also govern the nutrient release. Sarvade et al. (2014) observed that the chemical characteristics of the litter, including phosphorus content, C:P ratio, nitrogen content, C:N ratio, phenolics, phenolics to N or P ratio, lignin content and lignin to N ratio. Litter composition, soil type, microbial populations and soil characteristics, all affect how much nutrients are released from the litter. High nitrogen and phosphorus concentrations speed up decomposition, but high C:N and C:P ratios, phenolics, phenolics to N or P ratios, lignin content and lignin to N ratios slow it down (Rawat et al. 2009). Litter from N_2 fixing tree species decays more rapidly than litter from non-N_2 fixing tree and crop species. The rate of disintegration and nutrient discharge from litter is slowed down by a slight drop in the N concentration of the litter, an increase in the lignin content and simultaneously a rise in ambient CO_2 (Kumar 2011).

4.3.3 *Biological Nitrogen Fixation*

In 1901, a Dutch microbiologist Martinus Beijerinck discovered biological nitrogen fixation (BNF) which is governed by a specialized group of prokaryotes. These organisms convert the atmospheric nitrogen to ammonia by using the enzyme nitrogenase (NH_3). Nair et al. (1999) reported that approximately 650 woody trees are identified that belong to 9 different families that are capable of fixing atmospheric N_2 and capturing it into the soil. Among this a large number of trees are confined in a Leguminosae family. In 2012, Nygren et al. observed that globally 246 kg N ha^{-1} $year^{-1}$ are fixed annually in tree-based systems; however the maximum N_2 fixation, i.e. 300–650 kg N ha^{-1} $year^{-1}$, are noticed in improved fallow. N fixation values are varied from plant to plant which are mentioned in Table 4.1.

4.4 Nitrogen (N)

Nitrogen is a crucial element for sustainable systems, which is involved in photosynthesis for sugar formation. It enhances the system productivity by carbon sequestration in the soil and plant system (Lawson et al. 2020; Lu et al. 2021). It is a structural component of amino acid, nucleic acids, chlorophyll and energy compound (ATP). Although nitrogen is about 78.5% of the total atmosphere, it is not available to all living system. Nitrogen is a critical element in agroforestry for sustainable farm management. A self-reliant agroforestry system should be developed. Nitrogen availability in the atmosphere is abundant, but there is limited and deficient nitrogen in the farming system. In the agroecosystem, the supply of nitrogen is available through organic and inorganic nutrient application or by nitrogen fixing plants from atmospheric nitrogen.

4.5 Nitrogen Cycling Pathways

4.5.1 *N Mineralization*

In the agroforestry, component crop and trees continuously adding organic matter (litter fall, roots and straw) are added in the soil and decomposed by soil microbes. Leguminous tree organic matters are N-rich which are converted into ammonical N-form and later to nitrate (Isaac and Borden 2019; Muchane et al. 2020). These inorganic compounds are utilized by the plant (nitrate/ammonia) or lost to below root zone (NO_3^-), surface runoff or atmosphere (N_2, NH_3 and N_2O). Animal waste also incorporated into the soil results in enhancing the soil N. Tree pruning also augments a significant amount of organic C and N in the soil (Young 1989). In Himalayan Bhabar belt, the agri-silviculture system recorded an ammonification rate

Table 4.1 Nitrogen fixed by agroforestry components

S. no.	Plant	Family	Scientific name	N_2 fixed (kg ha^{-1} $year^{-1}$)
Higher tree plant				
1.	Subabul, ipil-ipil	Mimosoideae	*Leucaena leucocephala* (Lam.) de Wit.	100–500
2.	Inga	Do	*Inga jinicuil*	35–40
3.	Apple-ring acacia, ana tree	Do	*Faidherbia (Acacia) albida* (Delile) A. Chev.	20
4.	Black wattle	Do	*Acacia mearnsii* De Wild.	200
5.	Beef wood, Saru	Casuarinaceae	*Casuarina equisetifolia* L.	60–110
6.	Gliricidia	Fabaceae	*Gliricidia sepium* (Jacq.) Kunth ex Walp.	13
Pulse crop				
7.	Cowpea	Fabaceae	*Vigna sinensis* L. Walp.	73–354
8.	Pigeon pea	Do	*Cajanus cajan* L. Millsp.	68–88
9.	Black gram, mung bean	Do	*Vigna mungo* L. Hepper	63–342
10.	Pea	Do	*Pisum sativum* L.	55–77
11.	Soybean	Do	*Glycine max* L. Merr.	60–168
12.	Horse bean	Do	*Vicia faba* L.	45–552
13.	Chickpea	Do	*Cicer arietinum* L.	103
14.	Lentil	Do	*Lens esculenta* Moench	88–114
Fodder crops				
15.	Alfalfa, lucerne	Fabaceae	*Medicago sativa* L.	229–290
16.	Clover	Do	*Trifolium* spp.	128–207
17.	Wild tantan	Do	*Desmanthus virgatus* (L.) Willd.	196–226
Green manure crops				
18.	Sunn hemp	Fabaceae	*Crotalaria juncea* L.	199–223
19.	New dhaincha	Do	*Sesbania rostrata*	70–458
20.	Dhaincha	Do	*Sesbania sesban* (Jacq.) W. Wight	7–18

Source: Nair (1993); Silva and Uchida (2000)

of 6.47 ± 1.47 mg kg^{-1} $month^{-1}$ as compared to agri-horti-silvicultural system (5.67 ± 1.68 mg kg^{-1} $month^{-1}$) (Karki et al. 2021). Soil depth negatively correlated with NH_4–N, NO_3–N and inorganic N activity (Karki et al. 2021; Chen et al. 2005). Various researches reported that ammonification dominates over nitrification in the tree-based system in sub-tropical humid forest region and northeast India (Das et al. 1997; Tanjang et al. 2009; Karki et al. 2021). N mineralization was higher in rainy season than summer and winter, mainly due to the microbial favourable condition (Garkoti et al. 2003; Karki et al. 2021).

4.5.2 Atmospheric N Fixation

Legume family has the capability to fix the atmospheric nitrogen into the soil, resulting to an increased N status in the soil. Trees have different abilities to fix nitrogen, depending upon species, age of trees, climatic condition and symbiotic association. Species like *A. mangium*, *C. equisetifolia* and *L. leucocephala* can fix nitrogen from 100 to 300 kg N ha^{-1} $year^{-1}$, whereas species like *Acacia albida*, *A. raddiana* and *A. senegal* have the ability to fix less nitrogen, i.e. up to 20 kg N ha^{-1} $year^{-1}$ (Steppler and Nair 1987). Generally, agroforestry trees are sown depending upon the need, i.e. aesthetic value or economical purpose. Fixed nitrogen can either be utilized by other plants or help in decomposition and mineralization of the organic matter. Subsequent crop uptakes the fixed nitrogen through mycorrhizal hyphae networks (Jalonen et al. 2009) or N-enriched root exudates (Fustec et al. 2010) released by N_2 fixer.

4.5.3 N Miner from the Subsoil Layer: Tree Roots

The roots of trees grow deep and excavate the nutrients from the subsoil layer or leached nutrients, which help to extensively increase the nutrient cycling below ground area. However, annual crops have shallow roots, which enable them to extract nutrients from the subsoil. Deep-rooted trees should be promoted in the dry region which reduces the competition from crops and shrubs (Pierret et al. 2016; Bordron et al. 2019). Roots can uptake leached N–NO_3^- and reduce up to 30 to 227 kg N ha^{-1} $year^{-1}$ from the subsoil (Bergeron et al. 2011).

4.5.4 Ammonia Mitigation Through the Tree Canopy

Tree can capture ammonia from the atmosphere released by a nearby source in canopy, which reduces the N emission (Mishra et al. 2021). Pasture land, urinated by grazing animals, releases ammonia, which can also be reduced by planting trees. Combined modelling results showed that emissions of NH_3 can be captured by trees for up to 20% of total emissions (Bealeyet al. 2015). Agroforestry is very effective in the sensitive ecosystem zone for capturing pollutants like ammonia (Lawson et al. 2020). In a silvipastral system, trees can capture up to 60% of the livestock emissions (Bealey et al. 2014). Deposited ammonia in the canopy contributes to the soil nitrogen after rain.

4.5.5 N Sequestration in Tree Biomass

In the long term, trees accumulate nitrogen in the biomass, and the capability of different agroforestry systems depends on the type of plant species, soil properties, topography and climatic conditions (Dold et al. 2019; Thomas et al. 2020). Leguminous trees can increase CO_2 sequestration through N forcing into the soil, resulting in high biomass formation. In the Zambia region, Gliricidia-maize cropping system gave stable yield and comparable yield with respect to fertilized plot (Sileshi et al. 2012). Moreover, improved fallow sorghum with Gliricidia gave 55% higher yield as compared to traditional fallow sorghum (Hall et al. 2006). C:N ratio differs drastically in the agroforestry trees. N sequestration by leaf and wood trees ranges from 1.28 to 3.22% and 0.82 to 2.7% of dry mass, respectively (Northup et al. 2005). In *Mangifera indica*-based agroforestry systems, root biomass was significantly higher in agri-horticultural system than home garden (Karki et al. 2021). The addition of nitrogen can sequestrate 13 kg C kg^{-1} N in aboveground woody biomass in temperate regions (Schulte-Uebbing and de Vries 2018).

4.5.6 Atmospheric Losses

The biological process mainly involved in the nitrogen losses by the aerobics class of bacteria reduces the oxides of nitrogen into N_2 and N_2O under anaerobic conditions from the soil to the atmosphere, called denitrification (Firestone 1982). Due to the occurrence of anaerobic conditions mainly by rainfall, this caused a gradual increase in the losses followed by a decline in the emission of N gases in the atmosphere. This happens because of the decline of nitrogen oxides in the vicinity of the soil microbes. Global estimated average N_2O emission was 4.7 ± 1.6 N kg $year^{-1}$ ha^{-1} (Kim and Isaac 2022). Under agroforestry systems, i.e. improved fallow, intercropping and shaded perennial crop system reported that higher N_2O emission is mainly due to nitrogen fixer trees (Kou-Giesbrecht and Menge 2019; Akinnifesi et al. 2010) as well as residual incorporation (Baggs et al. 2006) over conventionally managed fields. Moreover, N-fixing trees have a self-regulating feedback mechanism in which excess N in the soil can reduce the fixation capacity and vice versa (Menge et al. 2015). However, this regulating mechanism can vary from species to species and reduce fixation from 30 to 0 kg N ha^{-1} $year^{-1}$ (Batterman et al. 2013). Another process is called nitrification in which ammonia is oxidized to NO_3^- later on converted by nitrifying bacteria. Nitrous oxide (NO) emission occurs by the process of microbial nitrification and denitrification and chemo-denitrification (Pilegaard 2013). Agroforestry systems emit more NO (6–4 N kg ha^{-1} $year^{-1}$) than forests (2.74 N kg ha^{-1} $year^{-1}$) and upland cropping systems (0.1–0.56 kg N ha^{-1} $year^{-1}$) (Lin et al. 2016). Cocoa agroforestry in Indonesia emitted NO@11–13 mg N m^{-2} h^{-1} (Veldkamp et al. 2008). But Rosenstock et al. (2014) noticed there is no threat of an increase in N_2O emission due to legume-based agroforestry.

4.5.7 *Soil Erosion, Runoff and Leaching Losses*

There is a huge loss of soil that occurs throughout the year but varies in different land management systems. However, the average global soil loss was estimated to be 10.2 Mg ha^{-1} $year^{-1}$ (Yang et al. 2003), with 12.38 to 17.12 kg ha^{-1} of nitrogen lost in different cropping systems, whereas bare land experienced losses of 33.88 N kg ha^{-1} (Zhu et al. 2020). Often, agroforestry systems cause less runoff, overland flow and leaching losses as compared to crop-based cropping systems and fallow land, which are higher than that in forest land (Zhu et al. 2020; Muchane et al. 2020). Meta-analysis data showed that silvoarable agroforestry could reduce runoff and soil and nutrient losses up to 100%, but nitrogen loss was reduced by 45–88% (Zhu et al. 2020). Meanwhile, another scientist reported that there was less chance to leach down N (up to 98%) in the groundwater in an agroforestry system (Pavlidis and Tsihrintzis 2018). Agroforestry system build-up of soil organic matter litter addition which slowly on mineralization release of N, P, and other nutrients (Nair 1987; Palm 1995; Lee and Jose 2003). Moreover, pecan-cotton mixed alley agroforestry system in the USA has less leaching of N and P than treeless pastural systems (Nair and Graetzm 2004; Nair et al. 2007).

4.6 Carbon (C)

Increased atmospheric CO_2 raises global concerns. Atmospheric CO_2 concentrations can be reduced by reducing emissions or by absorbing CO_2. It removes CO_2 from the atmosphere and stores it in terrestrial, marine or aquatic ecosystems. Agroforestry defined as a sustainable land management system use (Bene et al. 1977) has been criticized for some time for using land for production. In 1995, Matta and Jordan highlighted the underground competition that occurred between forest tree and crops for nutrients, water, etc. Perennials stay in place for many years, flushing their waste to the surface of the soil. Most of the carbon enters the ecosystem through leaf photosynthesis. More than half of the assimilated carbon is finally transported via the underground route root growth, root exudate (organic matter) and decomposition of litter. Therefore, the soil contains the largest C stock in the ecosystem. Organic carbon in soil agricultural field levels is depleted. Therefore, agroforestry practice will not only improve the quality of the soil but also increase the amount of carbon sequestration. Jha and Gupta (2002) reported that it is about 1.5 times more organic carbon under poplar and wheat agroforestry than wheat cultivation alone. The agroforestry system also affects the global carbon cycle. They often have higher area equivalent ratios (Graves et al. 2007), reducing the need for further agricultural land expansion and the associated C loss due to changes in land use. In addition, these systems sequester carbon at a higher rate than when trees and crops are grown separately and store carbon in stagnant biomass, or through, for example, defoliation, root turnover, crop residues, etc. introduces carbon into the soil, thereby reducing carbon. Schroeder (1994), Montagnini and Nair (2004) and Upson (2014) reported that atmospheric carbon is one of the essential components to mitigate the effects of global warming. Tropical forests are considered to be one of the most important terrestrial carbon reservoirs (Pan et al. 2011). These forests are particularly threatened by shifting cultivation (Kotto-Same et al. 1997) and soil loss associated with vegetation destruction, leaching and precipitation (Don et al. 2011; Hosonuma et al. 2012).

4.7 Carbon Sequestration in Agroforestry

Tree-based cropping system is a significant leader in terrestrial C sequestration, accounting for around 12% of the total terrestrial C in the world (Dixon 1995). The deeper subterranean horizons are penetrated by the roots of forest trees and perennial crops, which places soil organic carbon (SOC) at these depths and distant from the reach of tillage tools (Lorenz and Lal 2005). The mulch encase the surface of farmed fields as they decay over time and become a component of the SOC pool. AFS not only helps to conserve soil nutrients but also improves soil fertility and carbon sequestration (Montagnini and Nair 2004). It has been recognized as of particular importance carbon sequestration strategy due to its applicability to agricultural land as well as in reforestation programmes (Cairns and Meganck 1994).

4.8 Carbon Cycle

Carbon (C) is the building block of all life on earth. When plants convert CO_2 from the atmosphere into reducing sugars via photosynthesis, carbon cycling in forest environments begins. About half of the total photosynthate that plants normally produce during autotrophic respiration (Ra) is consumed in the synthesis and maintenance of living cells, liberating CO_2 into the atmosphere. The remaining C products are used for net primary production (NPP) such as leaves, branches, stems, roots and plant reproductive organs. Detritus is generated when plants die or lose their leaves, roots or both. Detritus is a substrate that supports microbes and animals that release CO_2 into the atmosphere through heterotrophic (Rh) metabolism. Undisturbed forest ecosystems often exhibit a little net increase in the amount of carbon they exchange with the atmosphere each year. This corresponds to net ecosystem production (NEP). Ecosystems can lose carbon when photosynthesis suddenly declines or when organic matter is removed as a result of disturbance. Soil humus, which has not been oxidized for centuries, represents the largest carbon stock in most ecosystems. It constitutes the most significant long-term C stock in the ecosystem.

4.9 Different Types of Organic Carbon (OC)

4.9.1 Soil Organic Carbon (SOC)

Organic carbon (OC) in soil is a quantifiable component of soil organic matter. Organic matter accounts for just 2–10% of the bulk of most soils but has a major role in the physical, chemical and biological functions of agricultural soils (Stevenson 1986). In a nutshell, SOC affects a variety of agricultural factors, including quality, fertility and productivity. Haynes (2005) noticed that SOC influences the availability of nutrients through mineralization. SOC also affects physical characteristics like aggregate stability and water retention. Therefore, it is crucial to keep soil carbon reserves high (Lal 2014). Bhattacharyya et al. (2011) and Singh et al. (2011) mentioned that specific SOC fractions play a vital role in maintaining soil quality and serve as markers for various management approaches. The SOC dynamic is shown in Fig. 4.2.

4.9.2 Labile and Recalcitrant Soil Organic Carbon

TOC is made up of different fractions of soil organic carbon pools (SOC-pools), including labile and recalcitrant pools. The atmospheric CO_2 content can be considerably impacted by even a slight shift in the SOC-pools. The relative percentage of these pools is a reflection of the soil ecosystem, which includes aerable and non-aerable soils that can have a direct influence on the microbial activity and carbon dynamics in soil. As a comparatively tiny portion of TOC with a very short half-life in soils and a high sensitivity to management difficulties, the labile carbon fractions (LP-C) in soil play a significant role in determining soil quality.

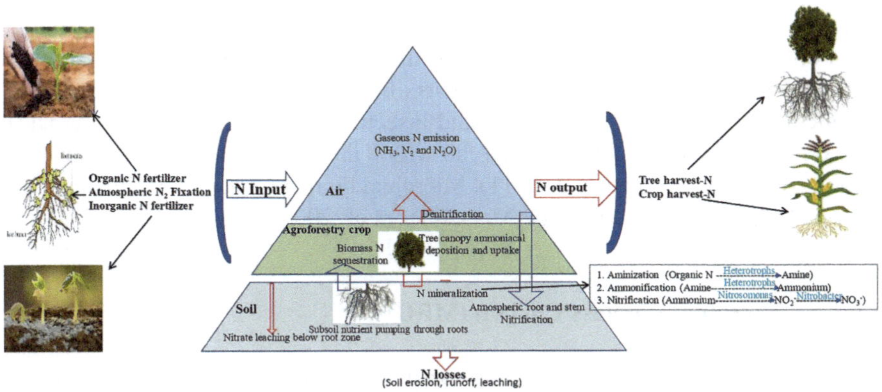

Fig. 4.2 Schematic representation of N dynamics in agroforestry system

In the soil system, there is more recalcitrant carbon (RC), which also has a slower turnover rate. The proportion of long-lived RC is frequently used to estimate long-term C storage. Depending on their state of decomposition and their function in the health and functioning of the soil, these C pools have different chemical compositions. Both the equilibrium of these fractions and the resistance of soil C to microbial attack are influenced by the textural characteristics of the soil.

4.9.3 *Soil Organic Matter*

Soil organic matter (SOM) is predominantly constituted of C, O and H along with trace quantities of Ca, P, N, S, Mg and K prevalent in organic wastes. It is divided into 'living' and 'dead' components, and the former can include comparatively recent inputs like stubble, while the latter might include entirely decomposed parts that are hundreds of years old. About 10% of below-ground SOM is 'living', such as roots, animals and microbes (Fig. 4.3).

SOM occurs in the soil as four separate fractions that vary greatly in size, turnover time and composition.

1. Dispersed organic matter
2. Humus
3. Resistant organic matter
4. Particulate organic matter

4.9.3.1 Total Organic Carbon (TOC)

The term 'TOC' refers to the quantity of carbon in soil that is tied up in organic molecules. These organic components may come from both endogenous and external sources. Examples of organic materials derived from endogenous processes include decomposing organic materials (such as cellulose, hemicellulose, glucose, citric acid, amino, fulvic, humic acid and humin) as well as waste products from the metabolic processes of microorganisms or living things (such as suberans, murein, chitin and glomalin). Manures, composts, biosolids, fertilizers (such as urea), organic dyes (such as X-3B red dye) and insecticides or pesticides (such as DDT) are examples of soil amendments that fall within the category of exogenous organic carbon compounds (Bolan et al. 2011). Microbial biomass is said to make up around 2% of TOC, according to reports (Marschner et al. 2008).

4.9.3.2 Microbial Biomass Carbon (MBC)

The living SOC component is made up mostly of microbial biomass carbon (MBC), which has been extensively investigated. MBC is a significant measurable carbon

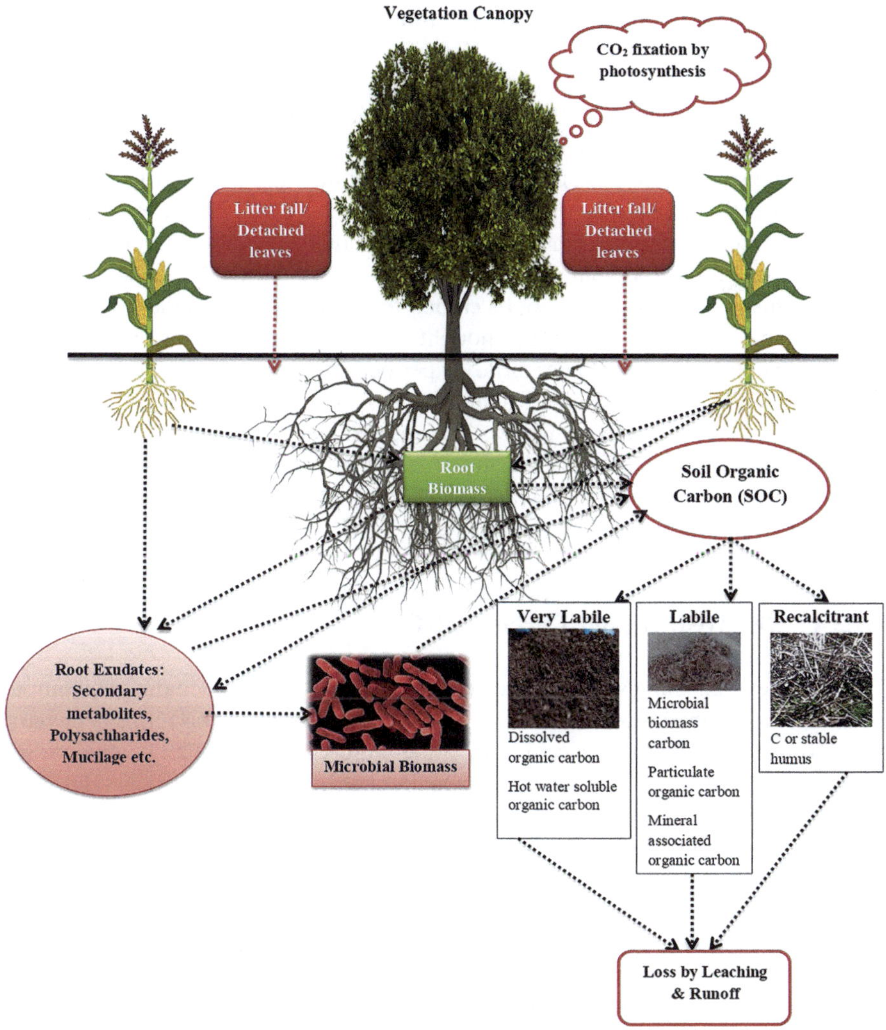

Fig. 4.3 Diagrammatical representation of soil organic carbon dynamics in agroforestry system

fraction and a marker of biological activity in soil in various multi-pool systems of SOC factors (Hanson et al. 2000). According to Franzluebbers et al. in 1999, the overall chloroform fumigation obtaining method, which Jenkinson and Powlson (1976) described, is an effective general definition of MBC, assuming that the carbon extraction via regulated soils is not subtracted from treated soils because doing so might muddy the determination of the difference.

4.9.3.3 Particulate and Mineral-Associated Organic Carbon

The type of carbon present in particulate organic matter (POM) is known as particulate organic carbon or POC. The top soil has a higher concentration of POC than the deeper layers, which are constantly altered by management techniques (Cambardella and Elliot 1992). Microorganisms in the POC prefer different sources, which indirectly alter the community structure (Fierer et al. 2007). By altering the metabolic absorption of substrates, labile POM included in carbon inputs also has an impact on microbial populations (Wang et al. 2014). Mineral-associated organic carbon is made up of the SOM pool's carbon fractions that have undergone physical and chemical stabilization and are thought to be representative of passive carbon pools with comparatively slower turnover periods (Marschner et al. 2008). By using (i) the natural morphological recalcitrance, (ii) associations with the surfaces of minerals or ionized metals as well as (iii) mechanical blockage inside aggregates, SOC elements can be stabilized in soils. Due to their exceedingly poor bioavailability for microbial decomposition and lengthy turnover times, these fractions are highly unstable (Benbi et al. 2014).

4.9.3.4 Dissolved Organic Carbon (DOC)

There are several sources of dissolved organic carbon (DOC), including soil humus, plant litter, microbial biomass and root exudates (Bolan et al. 2004). It is believed to have originated from either recent litter or the more stable SOM that is often present in the lowest portions of organic layers (Qualls et al. 1991). Tipping (1998) proposed the idea that there is a 'potential DOC fraction' in the SOM-pool, which, although not in the soil solution, may be thought of as a portion of the soil solids that might move into the solution given ideal soil circumstances. This potential pool of DOC is controlled and constantly refilled by newly supplied organic wastes, root exudates and microbial biomass. This pool of DOC in the soil is the result of a variety of processes, including physical separation, chemical change, leaching and the creation of soluble humic compounds from additional organic wastes. The actual DOC content may be affected by abiotic variables such as desorption and dissolution from potential DOC pools, but these activities are mostly regulated by soil bacteria (Guggenberger et al. 1998). In order to create and maintain DOC (both prospective and real) in the soil, both biotic and abiotic processes may be at play. The DOC pool can be divided further into a portion that is readily movable and an immobile fraction according on the size of the soil pore (Tipping 1998). The DOC fractions in macro- and meso-sized holes are more susceptible to convective movement than the DOC fractions in micropores, which are typically thought of as static (seepage).

4.9.3.5 Extractable Organic Carbon

It is made up of both biodegradable and non-biodegradable components. The biodegradable DOM has also been separated into labile, semi-labile and non-labile components based on breakdown rates as determined by the quantities of DOC mineralized after a particular amount of time (Marschner and Kalbitz 2003) (Fig. 4.4). Labile carbon, also known as extractable organic carbon, is referred to as a major energy source that can be easily broken down or eaten swiftly (between hours to weeks) by soil microbes. It is also recognized as a short-lived carbon pool. For example, labile carbon molecules include simple sugars like glucose and fructose as well as the by-products of protein breakdown such as amino acids (Gillis and Price 2011; Marschner et al. 2008). These substances are often scarce in most soils because of their quick microbial degradation, although organic additions can be employed to raise the soil's labile carbon component (Gillis and Price 2011).

4.10 Phosphorus

The availability of phosphorus (P) commonly restricts agricultural production in many soils (Hedley et al. 1995). Soil may restrain a large amount of P, but the key criteria for fertility are the soil's capacity to supply adequate solution P for plant growth. However, P availability is reduced due to insufficient crop and soil management. Losses of P occur during cropping in crop extraction and soil erosion, and

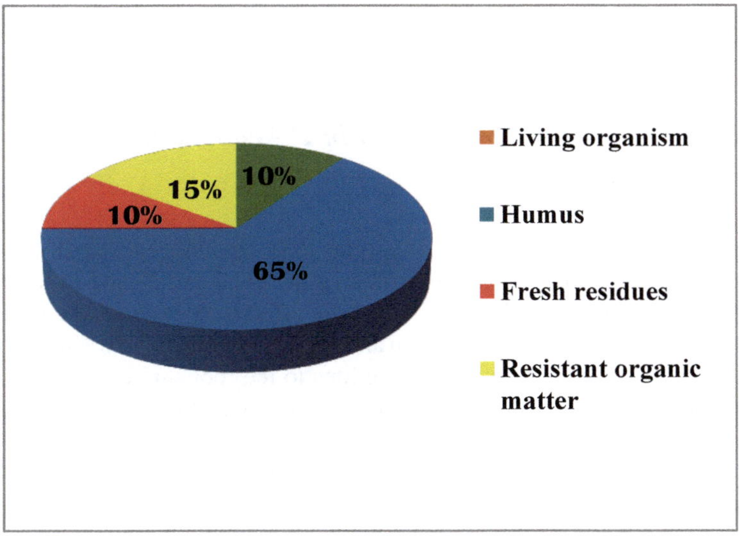

Fig. 4.4 Living organisms make up roughly 10% of the soil organic matter pool, with the majority of it being dead or decomposing (www.agric.wa.gov.au)

these losses can exceed the quantities added through various inputs and by mineral weathering which ultimately results in net reduction of P content. Therefore, this net decrease can be overcome by adding organic and inorganic inputs, and by returning crop residues, but this practice seems to be uneconomical. It is also observed that much of the P applied in the form of inorganic fertilizer gets firmly absorbed or precipitated, and it may not be instantly available to crops.

Agroforestry system in this regard has been reported to be more sustainable for farms where nutrients are in limited amount as forest trees can apprehend more nutrients and use nutrients more efficiently than annual cropping systems (Grierson et al. 2004). In agroforestry, root degradation and litter fall are the key processes that ensure the transfer of P from vegetation to the soil. It was observed that roots under agroforestry systems might report 80% of net primary production and it may contribute up to 1.5–2.0 kg P ha^{-1} $year^{-1}$ through root turnover (Manlay et al. 2002). As a more cost-effective way to increase the availability of P to crops, an agroforestry system successfully focuses on enhancing the use efficiency of soil P. The more substantial roots of trees and shrubs result in enhanced P uptake in comparison with annual crops due to increased exploration of a large soil volume. Compared to annual crops, tree and shrub species produce a greater biomass of leaves and roots, which increases the quantity of P that is recycled back into the soil. One of the many techniques employed in agroforestry to recycle phosphorus from plants to soil is the absorption of plant biomass, either locally or from elsewhere through biomass transfer, which enhances phosphorus availability and, eventually, crop productivity (Niang et al. 1996; Jama et al. 1997; Nziguheba et al. 2000). Several physicochemical processes, as well as the quality of the litter, which is often determined by the C:N:P or C:N ratio, govern the release of P from degrading plant materials. The modulation of these reactions by inclusion of plant residues can promote or hinder P availability (Palm and Rowland 1997).

4.10.1 Role of Root Morphology in P Acquisition in Agroforestry

Hairy roots possessed by many plant species have the ability to rapidly establish a large root system, which helps in maximizing the uptake of P (and water) from the soil. Mycorrhiza in association with root hairs effectively expands the root system, increases the volume of soil exploration and boosts the contact between root and soil, whereas the diffusion path for phosphate ions to the root surface gets shortened in case of closer and more widespread exploration of the soil system. The formation of cluster roots or proteoid by infection of mycorrhizal fungi appears to be an alternative strategy for the enhanced uptake of water and nutrients (Lamont 1986). Some of the important agroforestry plants including *Grevillea robusta* and *Macadamia* spp. contain cluster roots similar to those of the leguminous crop, white lupin. The non-mycorrhizal cluster roots, however, are often highly hairy and have a considerable surface area. They are known to produce huge amounts of phosphatase enzymes and organic acids, which may increase the amount of labile P pools in the soil solution (Grierson 1992; Dinkelaker et al. 1995).

4.10.2 Role of Organic Anions in P Acquisition and Mobilization Under Agroforestry System

Organic anions, such as organic acids, amino acids and phenolics, generated by plant roots can facilitate the release of P from the soil complex. Malate, citrate and oxalate are examples of some low molecular weight organic acids that are known to promote P release by ligand exchange (Gerke 1992). Certain tree species such as *Melaleuca cajuputi* (Watanabe et al. 1998) and *Banksia integrifolia* (Grierson 1992) release large amounts of citrate into their rhizosphere, particularly under condition of low-P or high-Al soils. Through a variety of methods, such as chelation, solubilization and complexation, the ectomycorrhizas of trees have also been identified to create significant amounts of organic acids that aid in raising the quantity of labile P in the soil solution (Malajczuk and Cromack 1982). As observed, it is possible for P to be mobilized by organic anions in the rhizosphere through a pH shift; desorption of P; chelation of metal ions, especially Al and Fe (III); or the development of metal-chelate complexes (Gardner et al. 1982).

References

Akinnifesi FK, Ajayi OC, Sileshi G, Chirwa PW, Chianu J (2010) Fertiliser trees for sustainable food security in the maize-based production systems of east and southern Africa. A review. Agron Sustain Dev 30:615–629

Baggs E, Chebii J, Ndufa J (2006) A short-term investigation of trace gas emissions following tillage and no-tillage of agroforestry residues in western Kenya. Soil Till Res 90:69–76

Batterman SA, Hedin LO, Van Breugel M, Ransijn J, Craven DJ, Hall JS (2013) Key role of symbiotic dinitrogen fixation in tropical forest secondary succession. Nature 502(7470): 224–227

Bealey WJ, Loubet B, Braban CF, Famulari D, Theobald MR, Reis S, Reay DS, Sutton MA (2014) Modelling agro-forestry scenarios for ammonia abatement in the landscape. Environ Res Lett 9(12):125001

Benbi DK, Boparai AK, Brar K (2014) Decomposition of particulate organic matter is more sensitive to temperature than the mineral associated organic matter. Soil Biol Biochem 70: 183–192

Bene JC, Beall HW, Cote A (1977) Tree, food and people: land management in the tropics. IDRC, Ottawa

Bergeron P, Réale D, Humphries M, M.andGarant, D. (2011) Anticipation and tracking of pulsed resources drive population dynamics in eastern chipmunks. Ecology 92(11):2027–2034

Bhattacharyya R, Kundu S, Srivastva AK, Gupta HS, Prakash V, Bhatt JC (2011) Long term fertilization effects on soil organic carbon pools in a sandy loam soil of the Indian sub-Himalayas. Plant Soil 341(1–2):109–124

Bolan NS, Adriano DC, Kunhikrishnan A, James T, McDowell R, Senesi N (2011) Dissolved organic matter: biogeochemistry, dynamics, and environmental significance in soils. Adv Agron 110:1

Bolan NS, Adriano DC, De la Luz M (2004) Dynamics and environmental significance of dissolved organic matter in soil. Doctoral dissertation. Regional Institute Limited

Bordron B, Robin A, Oliveira IR, Guillemot J, Laclau JP, Jourdan C, Nouvellon Y, Abreu-Junior CH, Trivelin PCO, Goncalves JLM, Plassard C, andBouillet, J. P. (2019) Fertilization increases the functional specialization of fine roots in deep soil layers for young Eucalyptus grandis trees. For Ecol Manag 431:6–16

Cairns MA, Meganck RA (1994) Carbon sequestration, biological diversity and sustainable development: integrated forest management. Environ Manag 18(1):13–22

Cambardella CA, Elliot ET (1992) Particulate organic matter changes across a grassland cultivation chronosequence. Soil Sci Soc Am J 56:777–783

Chen E, Zeng D, Singh AN, Chen G (2005) Effects of soil moisture and soil depth on nitrogen mineralization process under Mongolian pine plantations in Zhanggutai sandy land P. R. China. J for Res 16(2):101–104

Das AK, Boral L, Tripathi RS, Pandey HN (1997) Nitrogen mineralization and microbial biomass-N in a subtropical humid forest of Meghalaya, India. Soil Biol Biochem 29(9–10):1609–1612

Dhyani SK, Handa AK (2013) Area under agroforestry in India: an assessment for present status and future perspective. Indian J Agrofor 15(1):1–11

Dinkelaker B, Hengeler C, Marschner H (1995) Distribution and function of proteoid roots and other root clusters. Acta Botanica 108:169–276

Dixon RK (1995) Agroforestry systems: sources or sinks of greenhouse gases ? Agro for Syst 31: 99–116

Dold C, Thomas AL, Ashworth AJ, Philipp D, Brauer DK, Sauer TJ (2019) Carbon sequestration and nitrogen uptake in a temperate silvopasture system. Nutr Cycl Agroecosyst 114(1):85–98

Don A, Schumacher J, Freibauer A (2011) Impact of tropical land-use change on soil organic carbon stocks—a meta analysis. Glob Chang Biol 17:1658–1670

Fierer N, Bradford MA, Jackson RB (2007) Toward an ecological classification of soil bacteria. Ecology 88(6):1354–1364

Firestone MK (1982) Biological Denitrification. In: Stevenson FJ (ed) Nitrogen agricultural soils. Agron. Monogr. 22. *American society agronomy*. Crop Science Society of America and Soil Science Society of American, Madison, pp 289–326

Franzluebbers AJ, Haney RL, Hons FM (1999) Relationships of chloroform fumigation–incubation to soil organic matter pools. Soil Biol Biochem 31(3):395–405

Fustec, J., Lesuffleur, F., Mahieu, S. andCliquet, J. B. (2010).N Rhizodeposition of legumes. A review. Agron Sustain Dev, 30: 57–66

Gardner WK, Barber DA, Parbery DG (1982) The influence of microorganisms on the formation and function of proteoid roots in *L. Albus* L. Aust J Bot 30:303–309

Garkoti SC, Zobel DB, Singh SP (2003) Variation in drought response of Sal (Shorea robusta) seedlings. Tree Physiol 23(15):1021–1030

Gerke J (1992) Phosphate, aluminum and iron in the soil solution of three different soils in relation to varying concentrations of citric acid. Zeitschrift Pflanzenernahrung Bodenkunde 155(4): 339–343

Gillis, J. D. and Price, G. W. (2011). Comparison of a novel model to three conventional models describing carbon mineralization from soil amended with organic residues. Geoderma, 160(3–4): 304–310

Graves AR, Burgess PJ, Palma JHN, Herzog F, Moreno M, Dupraz C, Liagre F, Keesman K, Van der Werf W, Koeffeman de Nooy A, Van den Briel JP (2007) Development and application of bio-economic modelling to compare silvoarable, arable, and forestry systems in three European countries. Ecol Eng 29:434–449

Grierson PF, Smithson P, Nziguheba G, Radersma S, Comerford NB (2004) Phosphorus dynamics and mobilization by plants. In: Below-ground interactions in tropical agroecosystems: concepts

and models with multiple plant components. CABI Books. CABI, pp 127–142. https://doi.org/10.1079/9780851996738.0127
Grierson P (1992) Organic acids in the rhizosphere of *Banksia integrifolia* L. f. Plant Soil 144:259–265
Guggenberger G, Kaiser K, nadZech, W. (1998) Mobilization and immobilization of dissolved organic matter in forest soils. J Plant Nutr Soil Sci 161:401–408
Hall NM, Kaya B, Dick J, Skiba U, Niang A, Tabo R (2006) Effect of improved fallow on crop productivity, soil fertility and climate-forcing gas emissions in semi-arid conditions. Biol Fertil Soils 42(3):224–230
Hanson PJ, Edwards NT, Garten CT, Andrews JA (2000) Separating root and soil microbial contributions to soil respiration: a review of methods and observations. Biogeochemistry 48(1):115–146
Haynes RJ (2005) Labile organic matter fractions as central components of the quality of agricultural soils: an overview. Adv Agron 85:221–268
Hedley MJ, Mortvedt JJ, Bolan NS, Syers JK (1995) Phosphorus fertility management in agroecosystems. In: Tiessen H (ed) Phosphorus in the global environment. Wiley and Sons, Chichester, pp 59–92
Hosonuma N, Herold M, De Sy V, De Fries RS, Brockhaus M, Verchot L, Angelsen A, Romijn E (2012) An assessment of deforestation and forest degradation drivers in developing countries. Environ Res Lett 7(4):044009
Isaac ME, Borden KA (2019) Nutrient acquisition strategies in agroforestry systems. Plant Soil 444(1):1–19
Jalonen R, Nygren P, Sierra J (2009) Transfer of nitrogen from a tropical legume tree to an associated fodder grass via root exudation and common mycelial networks. Plant Cell Environ 32(10):1366–1376
Jama B, Swinkels RA, Buresh RJ (1997) Agronomic and economic evaluation of organic and inorganic sources of phosphorus in western Kenya. Agron J 89:597–604
Jenkinson DS, Powlson DS (1976) The effects of biocidal treatments on metabolism in soil—V: A method for measuring soil biomass. Soil Biol Biochem 8(3):209–213
Jha MN, Gupta MK (2002) Soil nutrient status and economic return under farming and agroforestry land uses. In: National Workshop on Agroforestry: Prospects and Challenges, Coimbatore, 22 November 2002. Proceedings. Coimbatore, IFGTB
Karki H, Bargali K, andBargali, S. S. (2021) Spatial and temporal trends in soil N-mineralization rates under the agroforestry systems in Bhabhar belt of Kumaun Himalaya, India. Agrofor Syst 95(8):1603–1617
Kim DG, Isaac ME (2022) Nitrogen dynamics in agroforestry systems. A review. Agron Sustain Dev 42(4):1–18
Kotto-Same J, Woomer PL, Appolinaire M, Louis Z (1997) Carbon dynamics in slash-and-burn agriculture and land use alternatives of the humid forest zone in Cameroon. Agric Ecosyst Environ 65:245–256
Kou-Giesbrecht S, Menge D (2019) Nitrogen-fixing trees could exacerbate climate change under elevated nitrogen deposition. Nat Commun 10(1):1–8
Kumar BM (2011) Quarter century of agroforestry research in Kerala: an overview. J Trop Agric 49(1–2):1–18
Lal R (2014) Soil carbon management and climate change. In: Hartemink AE, McSweeney K (eds) Soil Carbon. Switzerland, Springer, Cham, pp 339–361
Lambers H, Chapin FS, Pons T (1998) Decomposition. In: Plant physiological ecology. Springer, New York, pp 495–502
Lamont BB (1986) The significance of proteoid roots in proteas. Acta Hortic 185:163–170
Lawson G, Bealey WJ, Dupraz C, andSkiba, U. M. (2020) Agroforestry and opportunities for improved nitrogen management. In just enough nitrogen. Springer, Cham, pp 393–417
Lee KH, Jose S (2003) Soil respiration and microbial biomass in a pecan–cotton alley cropping system in southern USA. Agrofor Syst 58:45–54

Lin HC, Huber JA, Gerl G, Hülsbergen KJ (2016) Nitrogen balances and nitrogen-use efficiency of different organic and conventional farming systems. Nutr Cycling Agroecosyst 105:1–23

Lorenz K, Lal R (2005) The depth distribution of soil organic carbon in relation to land use and management and the potential of carbon sequestration in subsoil horizons. Adv Agron 88:35–66

Lu X, Vitousek PM, Mao Q, Gilliam FS, Luo Y, Turner BL, Mo J (2021) Nitrogen deposition accelerates soil carbon sequestration in tropical forests. Proc Natl Acad Sci 118(16): e2020790118

Malajczuk N, Cromack K (1982) Accumulation of calcium oxalate in the mantle of ectomycorrhizal roots of pinus radiata and eucalyptus marginata. New Phytol 92(4):527–531

Manlay RJ, Kairé M, Masse D, Chotte J-L, Ciornei G, Floret C (2002) Carbon, nitrogen and phosphorus allocation in agro-ecosystems of a west African savanna. Agric Ecosyst Environ 88(3):215–232

Marschner B, Kalbitz K (2003) Controls of bioavailability and biodegradability of dissolved organic matter in soils. Geoderma 113(3–4):211–235

Marschner B, Brodowski S, Dreves A, Gleixner G, Gude A, Grootes PM, Hamer U, Heim A, Jandl G, Ji R, Kaiser K, Kalbitz K, Kramer C, Leinweber P, Rethemeyer J, Schaffer A, Schmidt MWI, Schwark L, Wiesenberg GL (2008) How relevant is recalcitrance for the stabilization of organic matter in soils? J Plant Nutr Soil Sci 171(1):91–110

Matta RP, Jordan CF (1995) Nutrient dynamics during the first three years of an alley cropping agroecosystem in southern USA. Agrofor Syst 30(3):361–362

Menge DNL, Wolf AA, Funk JL (2015) Diversity of nitrogen fixation strategies in Mediterranean legumes. Nat Plants 1:15064

Mishra AK, Sinha B, Kumar R, Barth M, Hakkim H, Kumar V, Datta S, Guenther A, Sinha V (2021) Cropland trees need to be included for accurate model simulations of land-atmosphere heat fluxes, temperature, boundary layer height, and ozone. Sci Total Environ 751:141728

Montagnini F, Nair PKR (2004) Carbon sequestration: an underexploited environmental benefit of agroforestry systems. Agrofor Ecosyst 61:281–295

Muchane MN, Sileshi GW, Gripenberg S, Jonsson M, Pumarino L, Barrios E (2020) Agroforestry boosts soil health in the humid and sub-humid tropics: A meta-analysis. Agric Ecosyst Environ 295:106899

Munishamappa M, Austin D, Muddumadappa N (2012) Carbon sequestration in litter and soil of coffee based agroforestry systems Central Western Ghats Kodagu district of Karnataka. Environ Ecol 30(3B):985–987

Nair P (1993) An introduction to agroforestry. Kluwer Academic Publishers in cooperation with International Centre for Research in Agroforestry

Nair PR, Buresh RJ, Mugendi DN, Latt CR (1999) Nutrient cycling in tropical agroforestry systems: myths and science. In: Agro Forestry in Sustainable Agricultural Systems. CRC Press, Boca Raton, FL, USA

Nair PKR (1987) Soil productivity under agroforestry. In: Gholz HL (ed) Agroforestry: realities, possibilities and potentials. Nijhoff, Dordrecht, pp 21–30

Nair VD, Graetzm DA (2004) Agroforestry as an approach to minimizing nutrient loss from heavily fertilized soils: the Florida experience. Agrofor Syst 61:269–279

Nair VD, Nair PKR, Kalmbacher RS, Ezenwa IV (2007) Reducing nutrient loss from farms through silvopastoral practices in coarse-textured soils of Florida, USA. Ecol Eng 29:192–199

Niang A, Amadalo B, Gathumbi S, Obonyo CO (1996) Maize yield response to green manure application from selected shrubs and tree species in western Kenya: a preliminary assessment. In: Mugah JO (ed) Proceedings of the first Kenya agroforestry conference on people and institutional participation in agroforestry for sustainable development. Kenya Forestry Research Institute (KEFRI), Muguga, Kenya, pp 350–358

Northup BK, Zitzer SF, Archer S, McMurtry CR, Boutton TW (2005) Above-ground biomass and carbon and nitrogen content of woody species in a subtropical thornscrub parkland. J Arid Environ 62(1):23–43

Nygren P, Fernández MP, Harmand JM, Leblanc HA (2012) Symbiotic dinitrogen fixation by trees: an underestimated resource in agroforestry systems? Nutr Cycl Agroecosyst 94:123–160
Nziguheba G, Merckx R, Palm C, Rao MR (2000) Organic residues affect phosphorus availability and maize yields in a nitisol of western Kenya. Biol Fertil Soils 32:328–339
Palm CA, Rowland AP (1997) Chemical characterization of plant quality for decomposition. In: Cadisch G, Giller KE (eds) Driven by nature: plant litter quality and decomposition. CAB International, Wallingford, UK, pp 379–392
Palm CA (1995) Contribution of agroforestry trees to nutrient requirements of intercropped plants. Agrofor Syst 30:105–124
Pan Y, Birdsey RA, Fang J, Houghton R, Kauppi PE, Kurz WA, Phillips OL, Shvidenko A, Lewis SL, Canadell JG, Ciais P, Jackson RB, Pacala SW, Mcguire AD, Piao S, Rautiainen A, Stich S, Hayes D (2011) A large and persistent carbon sink in the world's forests. Science 333(6045): 988–993
Pavlidis G, Tsihrintzis VA (2018) Environmental benefits and control of pollution to surface water and groundwater by agroforestry systems: a review. Water Resour Manag 32:1–29
Pierret A, Maeght JL, Clément C, Montoroi JP, Hartmann C, Gonkhamdee S (2016) Understanding deep roots and their functions in ecosystems: an advocacy for more unconventional research. Ann Bot 118(4):621–635
Pilegaard K (2013) Processes regulating nitric oxide emissions from soils. Phil Trans R Soc B 368: 20130126
Qualls RG, Haines BL, Swank WT (1991) Fluxes of dissolved organic nutrients and humic substances in a deciduous forest. Ecology 72:254–266
Rawat N, Nautiyal BP, Nautiyal MC (2009) Litter production pattern and nutrients discharge from decomposing litter in Himalayan alpine ecosystem. New York Sci J 2(6):54–67
Rosenstock T, Tully K, Arias-Navarro C, Neufeldt H, Butterbach-Bahl K, Verchot L (2014) Agroforestry with N_2-fixing trees: sustainable development's friend or foe? Curr Opin Environ Sustain 6:15–21
Sarvade S, Singh R, Prasad H, Prasad D (2014) Agroforestry practices for improving soil nutrient status. Popular Kheti 2(1):60–64
Schroeder P (1994) Carbon storage benefits of agroforestry system. Agrofor Syst 27(1):89–97
Schulte-Uebbing L, De Vries W (2018) Global-scale impacts of nitrogen deposition on tree carbon sequestration in tropical, temperate, and boreal forests: A meta-analysis. Glob Chang Biol 24(2): e416–e431
Sileshi GW, Debusho LK, Akinnifesi FK (2012) Can integration of legume trees increase yield stability in rainfed maize cropping systems in southern Africa? Agron J 104(5):1392–1398
Silva JA, Uchida RS (2000) Plant nutrient management in Hawaii's soils: approaches for tropical and subtropical agriculture. University of Hawaii
Singh SK, Pandey CB, Sidhu GS, Sarkar D, Sagar R (2011) Concentration and stock of carbon in the soils affected by land uses and climates in the western Himalaya, India. Catena 87(1):78–89
Steppler HA, Nair PKR (1987) Agroforestry: a decade of development. International Council for research in agroforesty
Stevenson FJ (1986) Cycles of soil: carbon, nitrogen, phosphorus, 779 Sulfur, micronutrients. John Wiley, New York, p 380
Tanjang S, Shukla AK, Arunachalam K, Arunachalam A (2009) Mineralization dynamics of nitrogen and phosphorus in areca catechu l.–Based traditional agroforestry system. Commun Soil Sci Plant Anal 40(21–22):3225–3237
Tipping E (1998) Modelling the properties and behavior of dissolved organic matter in soils. Mitt Dtsch Bodenkdl Ges 87:237–252
Thomas AL, Kallenbach R, Sauer TJ, Brauer DK, Burner DM, Coggeshall MV, Dold C, Rogers W, Bardhan S, Jose S (2020) Carbon and nitrogen accumulation within four black walnut alley cropping sites across Missouri and Arkansas, USA. Agrofor Syst 94(5):1625–1638
Upson MA (2014) The carbon storage benefits of agroforestry and farm woodlands. Unpublished PhD thesis. Cranfield University, Cranfield, 319 pp

Veldkamp E, Purbopuspito J, Corre MD, Brumme R, Murdiyarso D (2008) Land use change effects on trace gas fluxes in the forest margins of Central Sulawesi, Indonesia. J Geophys Res: Biogeosci 113(G2):G02003

Wang L, Chen Z, Shang H, Wang J, Zhang PY (2014) Impact of simulated acid rain on soil microbial community function in Masson pine seedlings. Electron J Biotechnol 17:199–203

Watanabe T, Osaki M, Yoshihara T, Tadano T (1998) Distribution and chemical speciation of aluminum in the Al accumulator plant, *Melastoma malabathricum* L. Plant Soil 201:165–173

Xu J, Mercado A, He J, Dawson J (2013) An agroforestry guide for field practitioners. World agroforestry Centre (ICRAF), Kunming, China

Yang D, Kanae S, Oki T, Koike T, Musiake K (2003) Global potential soil erosion with reference to land use and climate changes. Hydrol Process 17:2913–2928

Young A (1989) Agroforestry for soil conservation. CAB International, Wallingford. ISBN 0 85198 648 X

Zheng Z (2006) Litter decomposition and nutrient release in a tropical seasonal rain forest of Xishuangbanna, Southwest China. Biotropica 38(3):342–347

Zhu X, Liu W, Chen J, Bruijnzeel LA, Mao Z, Yang X, Cardinael R, Meng FR, Sidle RC, Seitz S, Nair VD, Nanko K, Zou X, Chen C, Jiang XJ (2020) Reductions in water, soil and nutrient losses and pesticide pollution in agroforestry practices: a review of evidence and processes. Plant Soil 453(1):45–86

Zomer RJ, Trabucco A, Coe R, Place F (2009) Trees on farm: analysis of global extent and geographical patterns of agroforestry, ICRAF Working Paper No. 89. (World Agroforestry Centre (ICRAF)

Chapter 5
Agroforestry-Based Consequences Improve the Soil Health

Jyoti Yadav, Roma Rani, Bhanu Raj Meena, Deepali Chittora, Pushpa Gehlot, Tripta Jain, and Kanika Sharma

Abstract Soil is the upper surface of land which covers the earth in the form of thin layer and, with the optimum quantity of air and water, provides the life and base to all plants; with the extensive availability of the upper surface of soil, once, it takes a longer time for its renovation and to be fertile. Agroforestry is one of the best practices for enhancing soil organic carbon and sustainable ecosystem. Agroforestry has been long in practices/used for improving soil health by combining the trees and shrubs with agricultural crops or livestock. These agricultural practices provide the multiple benefits which have been revealed from the researchers that (1) they provide the vital home of wildlife or in other words the number of wildlife increased; (2) they enhance the farm and land productivity by improving the soil health and animal welfare; (3) they also provide help in mitigation of one of the main global problems, i.e., climate change, (4) managing soil surface quality and water flow, and (5) balancing the nutrient cycling and also in the prevention of soil erosion. Due to the much extensive use of agroforestry in improving soil, improving soil fertility, and environment management, it is needed to focus on agroforestry practices. This chapter focused on the relationship of soil health and agroforestry.

Keywords Soil health · Sustainable ecosystem · Agroforestry · Soil organic carbon

5.1 Introduction

Soil is the essential need for life to be sustained as it is the nourishment provider for a range of organisms, either microorganisms or giant organisms. It is the basis of agriculture and forestry (Stirling et al. 2016). Soil holds nutrients, minerals, and

J. Yadav · D. Chittora · P. Gehlot · T. Jain · K. Sharma
Microbial Research Laboratory, Department of Botany, University College of Science, Mohanlal Sukhadia University, Udaipur, India

R. Rani (✉) · B. R. Meena
Ramjas College, University of Delhi, Delhi, India
e-mail: romarani@ramjas.du.ac.in

H. S. Jatav et al. (eds.), *Agroforestry to Combat Global Challenges*, Sustainable Development and Biodiversity 36, https://doi.org/10.1007/978-981-99-7282-1_5

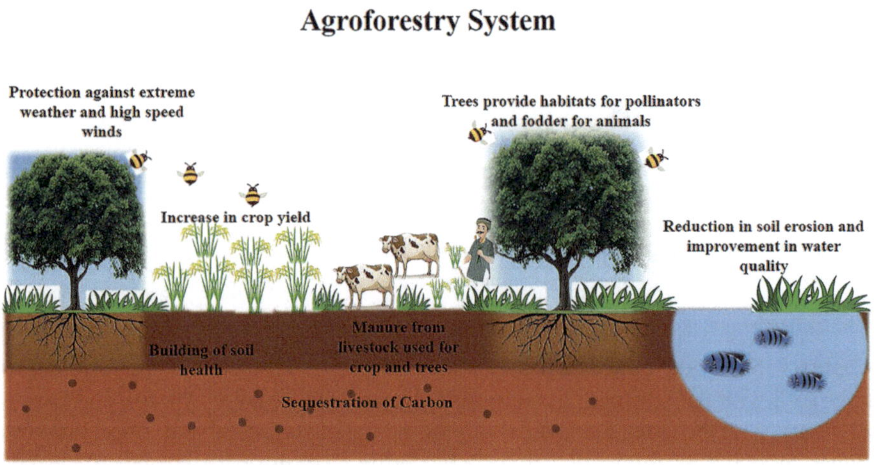

Fig. 5.1 The setup and benefits of agroforestry system

water that are the important factors responsible for agricultural productivity. The content of soil decides its quality that should be capable to sustain the productivity of flora, environmental, and faunal health which is commonly referred as soil health (Yadav et al. 2008). For sustainable agricultural production, soil management is a much needed step. Continuous exploitation of soil without considering or managing its capacity makes its quality or health to deteriorate (Rosa and Sobral 2008). The rapidly increasing demand for food for the growing population of humans and livestock has put immense pressure on natural resources that leads to non-regulated exploitation of soil (Hoppe and Sanders 2014). There are various factors like deforestation, over-grazing, constructions, and unscientific farming practices that contribute direct or indirect pressure on soil health. Other factors that influence the soil health indirectly include climate change and greenhouse effect that alter the temperature and rainfall patterns which might result in heavy rainfall or no rainfall at all. Excessive rainfall could swipe off the nutrient-rich organic matter containing the uppermost layer of soil that is the most fertile part, and it takes years to be that much fertile; this loss of soil quality is called soil erosion (Baig et al. 2013). Soil health plays a major role in agricultural productivity; continuous cultivation makes it necessary to add the lacking constituent again to make it fertile like once it was (Tittonell and Giller 2013). To scientifically manage the soil quality or health, there is a need of a method that is sustainable and ecofriendly, providing long-term multiple benefits and maintaining the soil health. One of such kind of practice is agroforestry. Agroforestry is a sustainable and integrated practice in which trees and shrubs are cultivated with crops and livestock cattle which optimizes the output one can get separately from individual practices (Dollinger and Jose 2018). The setup and benefits of agroforestry system are shown in Fig. 5.1.

5.2 Soil

Soil is the uppermost layer of earth that holds biological diversity of billions of species. It is the reservoir for water and minerals like phosphorous, provides a supportive space for breakdown of organic wastes, and participates in the nutrient cycle (Timmis and Ramos 2021). The process of soil formation takes thousands of years through weathering of large rocks with the help of some abiotic factors like wind, water, and gravity. Soil thickness can vary from few centimeters to many meters depending on weathering process and other factors like deposition, erosion, and patterns of landscape development (Bormann and Likens 2012). Due to combined effect of percolated water and living organism, vertical layers of soil are formed that are called horizons. These horizons combine known as soil profile (McKenzie et al. 2004).

Soil plays various ecological and non-ecological functions. Some of the ecological functions include biomass production, as gene reservoir, and providing protection to humans and environment (Blum 2005). Soil produces renewable energy sources, food, and raw materials that hold various applications. It is a habitat to diverse types of organisms that constitute more variety and number of species than above soil surface put together (Thomas and Kevan 1993). Hence soil is the basis of diverse biodiversity that is also a source of genetic diversity for biotechnological and bioengineering processes (Rastegari et al. 2020). Soil plays an important role in the carbon cycle as it transforms organic carbon into carbon dioxide and other constituents by decomposition with the help of numerous microorganisms present in it and also is a habitat for plants that again capture this carbon dioxide and transform it to organic carbon (Horwath 2015). Soil plays a major role in filtering, buffering, and transforming capacities between atmosphere, groundwater, and plants. There are non-ecological functions that soil plays including providing a physical basis for human activities, forming a source of raw materials, and forming an essential part of landscape in which human civilization lived. Soil preserves paleontological and archeological history of human civilizations that once was there (Pandey 2021).

5.3 Soil Health and Management

Soil health has now been given attention as it is being understood that it not only provides us food and fibers but also maintains the environmental quality and balance. Much of agricultural land has been degraded because of soil erosion, pollution, excessive land cultivation, overgrazing, and desertification induced due to human activities (Jie et al. 2002). Degradation and loss of agricultural land are a big concern among the other human-induced issues like climate change, global warming, and biodiversity losses. Soil health is defined as the capacity of soil to function within natural environmental conditions to sustain plant and animal productivity, maintain air and water quality, as well as support human health (Doran and

Zeiss 2000). Soil quality and biodiversity are influenced by human-caused disturbances. Majorly the soil quality is determined by its inherent physical and chemical properties that are specified by the climate and ecosystem. Other than the natural factors, the quality of soil is also influenced by the kind of management and the decisions about land being used (Morgado et al. 2018). In the past era, soil has been mismanaged to get excess production beyond the capacity of the farmland. The quality of agricultural land and the balance of other ecosystems have been severely affected. Crop cultivation with the use of heavy machineries and continuous same crop cultivation has resulted in decrease of soil quality and even soil loss by erosion (Thapa and Weber 1991). Organic content has been lost in the form of carbon dioxide into the atmosphere. The accelerated growth of human population has increasingly putting pressure on natural resources and soil which results in unmanageable exploitation and quality reduction of resources (Amthor 1995). Thus, we should develop such agricultural land use management practices like agroforestry that restore the lost content of soil to make it healthy and also remain healthy for future generations to come.

5.4 Agroforestry

Agroforestry is a term used for an agricultural land management system in which agricultural field crops and livestock are integrated with trees and shrubs for long-term sustainable farming and improvement of soil health (Sharma et al. 2016). Agroforestry leads to an ecological diverse and productive land with lesser human interference, low cost, and long-term environmental benefits. The concept of agroforestry came into light in the early twentieth century, but written evidences of practice of woody perennials in agriculture are found in Roman times (Rosa-Schleich et al. 2019). Agroforestry is employable in different kinds of ecological units and lifestyles. Proper adaptation of agroforestry is useful to improve the living standard of people by enhancing health and nutrition and accelerating economic growth and environmental sustainability (Montagnini and Metzel 2017). It holds a great promise of sustainable production of fruit crops, medicinal plants, dairy and beef cattle, and organic biomass to regenerate nutrients and minerals for decomposition (Kesavan and Swaminathan 2008). Land use management systems with agroforestry are proven to be efficient in carbon sequestration, enrichment of soil in nutrients and minerals, maintenance of biodiversity, and providing long-term benefit to the environment and the landowner (Udawatta et al. 2021). The basis of the benefits provided by agroforestry is the interaction between the components of the system, i.e., trees, shrubs, crop, and cattle (Mosquera-Losada et al. 2009). Agroforestry encourages the positive interactions like mutualism and commensalism and discourages the negative interactions like competition and amensalism between species (Campbell 2012).

In agroforestry the components are interlinked structurally and functionally such that they can be managed to get the best outcome from the system, for example, trees

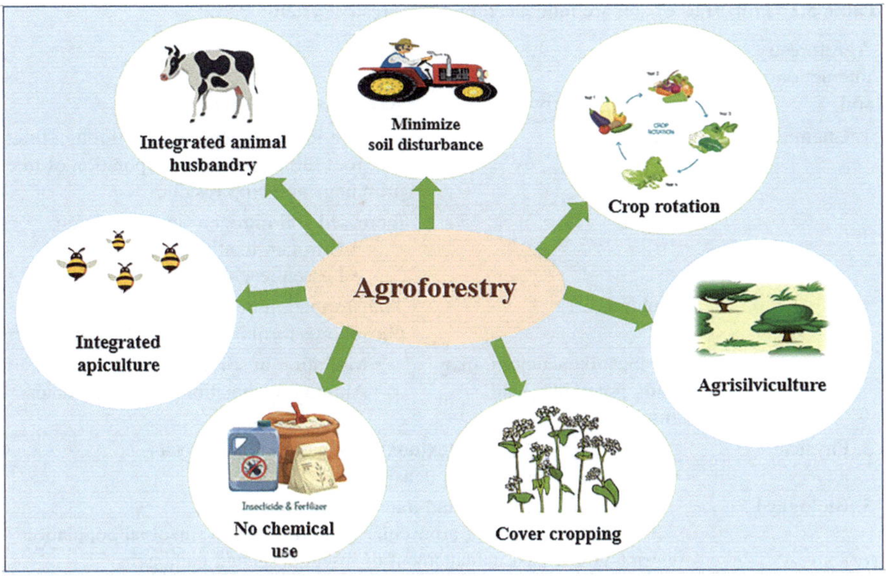

Fig. 5.2 The elements of agroforestry

are regularly cut, and the cuttings are used as mulch to the soil. This encourages the growth of new trees and also prevents the trees to be dark shadowed on undergrown crops and maintains the soil moisture (Nair et al. 2010). The elements of agroforestry are shown in Fig. 5.2.

There are three main types of agroforestry systems (Zerbe 2022):

- Agrisilvicultural system: Integration of crops and trees, such as alley cropping or home gardening
- Silvopastoral system: Integration of forestry and grazing of household cattle on farmlands
- Agrosylvopastoral system: Integration of crop, trees, and cattle as scattered trees on farmland and animal grazing after crop harvest.

5.5 Indicative Factors of Soil Health Influenced by Agroforestry

There are various constituents present in the soil that are important to be considered as indicator of the soil health. The physical, chemical, and biological properties of soil are influenced by agroforestry system given in Table 5.1 (Cardoso et al. 2013). All the properties influence one another as physical properties improve the chemical and biological properties of soil that increase the overall growth of plants (Marinari et al. 2000). Leaves of trees associated with the agricultural land fall off on the soil

Table 5.1 Properties of soil are influenced by agroforestry system

Agroforestry interaction with soil	Nature of property	Agroforestry process
1. Chemical	Carbon	Increase in soil organic matter through litter fall, root turnover, and incorporation of tree prunnings and crop residues
	Nitrogen	Increased soil nitrogen supply through • Nitrogen fixation • Lessen leaching
	Phosphorus	Transformation of less available inorganic phosphorus form into readily absorbable form
	Other ions like calcium, magnesium, potassium, and aluminum	• Migration in soil horizons • Aluminum chelation by organic acids
2. Physical	• Improvement in soil texture, aggregation, and porosity • Reduction in soil bulk density	
3. Biological	• Diversify soil fauna and microbial population • Increase in vascular arbuscular mycorrhizal and rhizobial population • Integrated pest management of insect pests and pathogens	
4. Competitions	Sharing of growth resources between trees and crop	
5. Conservation	Reduction in soil erosion and leaching	

surface that form a protective layer. This layer of litter adds organic matter to the soil after decomposition by microorganisms, reduces loss of moisture content, and also plays an important role in the prevention of soil erosion (Sayer 2006). Some of the physical properties that are indicative of soil health and influenced by agroforestry are soil binding stability, bulk density, water holding capacity, infiltration rate, depth of soil, and structure, texture, and porosity of soil (Bot and Benites 2005). Soil binding or aggregation stability is the binding of primary soil particles around the organic matter or other surrounding particles. It is a good indicator of soil quality (Wei et al. 2006). The structural stability of soil depends on soil aggregation stability; if the binding ability of soil decreases, this can be a strong indication of land degradation. In agroforestry system of land management, trees play an important role in improving soil aggregation stability by adding organic matter to the soil. Several reports have proven that lands with trees have larger soil aggregate when compared to bare farmlands (Somasundaram et al. 2012). One of the other physical properties of soil is the water holding capacity; it is the quantity of water soil can retain that is of crop use. Land systems with trees are proven to be having more water holding capacity and reduced water loss due to evaporation (Waddington et al. 2015). Bulk density of soil is an indicator of soil compaction. It is defined as the dry weight of soil divided by its volume of soil particles and pores between soil particles. The more the value of bulk density, the lesser the soil porosity (Cresswell and Hamilton 2002). Decreased soil porosity leads to poor flow of air and water in the soil and also restricts the plant root growth, hence reducing plant growth. Poor bulk density of soil is the result of practices like use of heavy machinery for field leveling, repetition of crops with more or less equal root structure and depth, not plowing to different depths, application of heavy equipment on wet soil, and adding piles of crop remaining on the field (Kubik 2005). High bulk density and compaction of soil decrease the productivity of soil; hence to reduce this, one must practice reduction in number visits in the field especially when it is wet, plantation of crops with different root lengths, and increasing the organic matter in the soil aggregates (Arvidsson 1999).

Other than the physical properties, there various chemical properties of soil that play an indicator in soil health and also influenced by agroforestry. Soil organic matter and soil organic carbon are reported to be increased in soil due to the addition of litter fall and dead root remains when crop is planted under old plantations (Singh and Sharma 2007). Soil pH is an important factor which influences the surrounding soil microbiota and the growth of plants. Soil pH is a major factor to control the presence of many nutrients and also decides their ionic form which further influences their uptake and growth of plants. The pH range of soil with good amount of nutrients in it is 6 to 7 (Alam et al. 1999). Trees maintain pH by addition of organic matter to the soil surface and releasing organic acid during the process of decomposition of the litter. The acidic pH plays an important role in the process of mineralization and, thus, ensures the availability of nitrogen in the soil (Clarholm and Skyllberg 2013). Salts present in the soil are important in the regulation of soil salinity or electrical conductivity of soil. It is the ability to conduct electrical current; electrical conductivity is used as a measure of soil salinity (Rhoades et al. 1999). The amount of nutrients present in soil very much depends on the tree species in the agroforestry system because the amount of litter fall is different in different tree species with seasonal variations (Murovhi et al. 2012).

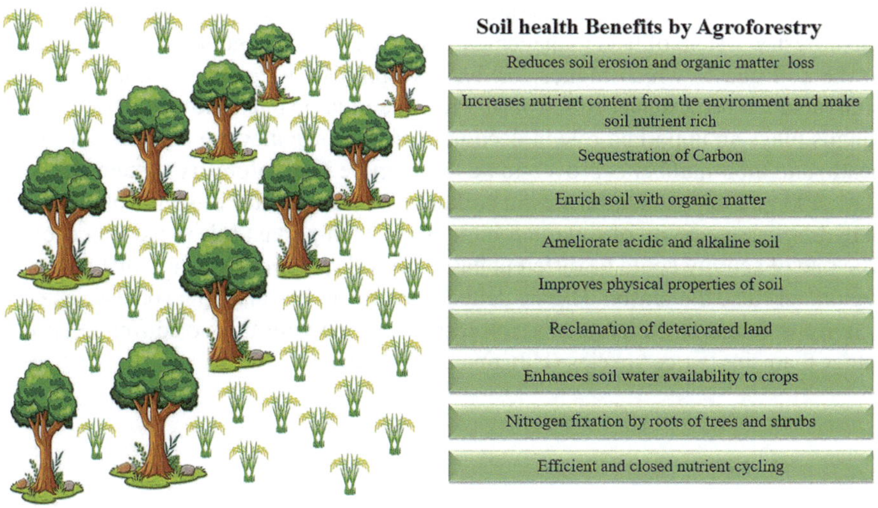

Biological properties of soil include the number of microbes, enzymatic activity in the soil, carbon and nitrogen microbial biomass, soil basal respiration, and microbiota (Salazar et al. 2011). These properties are influenced by organic matter present in the soil and soil organic carbon added by litter fall, decomposition of plant remains, and variety of litter depending on the different types of tree species under the agroforestry system (Gupta et al. 2009). Higher microbial count is reported in cultural agroforestry system, and it has been reported that higher bacterial count is found in the soil containing higher clay and water content, whereas higher fungal count is found in the soil with higher nitrogen content (Tangjang and Arunachalam 2009). The soil pH and organic carbon content are the deciding factors for microbial population in the soil. Higher organic matter and neutral pH in agroforestry favor high number of microbes in the soil (Ramesh et al. 2019). Studies have suggested that soil associated with agroforestry system have more number of total bacteria, anaerobic bacteria, gram-negative bacteria, and mycorrhizal fungi when compared to regular field soil (Unger et al. 2013). More root colonization is found by arbuscular mycorrhiza for *Croton macrostachyus* (45%) tree than *Albizia gummifera* (41%), and significantly higher root colonization of maize crops is grown under the canopy of *Albizia gummifera* and *Croton macrostachyus* trees by AM than outside the canopy (Hailemariam et al. 2013). Various studies have revealed that agrosilviculture-based agroforestry with different tree species showed considerable difference in microbial biomass of carbon, and also variation in microbial biomass of carbon and nitrogen was also varied with the density of the same tree species; the more the density of tree, the more the microbial biomass (Panwar et al. 2017; Rodrigues et al. 2015). The trees that produced high litter fall and root exudates are reported to harbor more microbe population and microbial biomass (Van Der Heijden et al. 2008). Another biological property is soil basal respiration; it is defined as the constant respiration rate originating due to carbon dioxide evolution or oxygen released during the decomposition or mineralization of organic matter. In agroforestry system of land management, much of carbon containing matter is added to the soil due to which higher soil basal respiration is found in these tree-based cultivated land (Cardinael et al. 2020). Presence of large microbiota like bacteria and fungus plays an important role in decomposition; hence they are an indicator of soil heath, and their number is proven to be found more in agroforestry systems (Jeffries et al. 2003). Enzymatic activities in soil are also an indicator of soil health because dehydrogenase activity of soil microbiota plays an important role in nutrient cycling. The rate of activity could be different based on soil type, vegetation of that particular region, and also the season. Higher dehydrogenase and alkaline phosphatase enzyme activity is found to be more in agroforestry systems with trees of different species rather than single tree species (Uthappa et al. 2015).

5.5.1 Agroforestry for Soil Health

5.5.2 Carbon Transformation

Increasing global atmospheric carbon dioxide is a big problem considering global warming as its one of the consequences. Agriculture and forestry can be employed to curb certain levels of carbon dioxide concentrations by incorporating carbon from atmosphere to vegetation biomass (Luckow et al. 2010). The role of trees and forest in carbon cycling is well known; forest is a large sink of carbon. To increase the carbon storage capacity, there is an increasing interest to incorporate agricultural land with trees and reforestation (Murthy et al. 2013). Ever since the industrialization began, atmospheric carbon has increased from 280 mg/L in 1750 to about 392 mg/L in 2012, and at the current rate, it is expected to surpass 400 mg/L by 2015 (Jose and Bardhan 2012). Agroforestry land use system has the greatest potential for carbon sequestration due to large area available to employ agroforestry as a practice (Ramachandran Nair et al. 2009). In comparison to mono-crop field, the area incorporated with trees and shrubs is proven to be more efficient in carbon sequestration both on the aboveground and belowground. Due to much modernization in agriculture, soil carbon pool has been decreased due to deforestation, soil erosion, overexploitation of soil, and unsustainable agricultural practices. The incorporated carbon into vegetation biomass constitutes the soil organic carbon that adds with the litter fall. Studies have found that land cultivation system shifting from without trees to agroforestry systems has improved the levels of soil organic carbons (Lorenz and Lal 2014).

5.5.3 Symbiotic Interaction

The two important mutualistic interactions in agroforestry-based system are nitrogen-fixing microbes and mycorrhizas. Both the relations are based on resource transfer trading. In the case of root nodules, plant trades carbon in return of fixed nitrogen, whereas phosphate, water, and other ions are traded from mycorrhiza. Plant and symbiotically associated microorganism fixing nitrogen biologically are presented in Table 5.2. These interactions contribute to the overall fertilization of the soil.

5.5.4 Control of Soil Erosion

Soil erosion is the main cause of fertile soil loss which affects physical, chemical, and biological properties. There are various control measures to manage erosion, but these methods are not very cost-effective and also are not easy to maintain.

Table 5.2 Plant and symbiotically associated microorganism fixing nitrogen biologically

Plant name	Family	Microorganism associated	References
Silver wattle (*Acacia dealbata*)	Mimosoideae	Rhizobia	Niu (2013)
Black wattle (*Acacia mearnsii*)	Mimosoideae	Bradyrhizobium	Boudiaf et al. (2014)
Gum Arabic tree (*Acacia nilotica*)	Mimosoideae	Rhizobium	Choudhary et al. (2020)
Indian alder (*Alnus nepalensis*)	Betulaceae	*Frankia*	Varghese et al. (2003)
European alder (*Alnus glutinosa*)	Betulaceae	Frankia	Varghese et al. (2003)
Beef wood (*Casuarina equisetifolia* L.)	Casuarinaceae	Frankia	Mink et al. (2016)
Erythrina (*Erythrina poeppigiana*)	Papilionoideae	Bradyrhizobium	Leblanc et al. (2007)
Ana tree (*Faidherbia albida*)	Mimosoideae	Bradyrhizobium	Mokgolodi et al. (2011)
Gliricidia (*Gliricidia sepium*)	Fabaceae	Rhizobium	Samarakoon and Rajapakse (2020)
Inga(*Inga jinicuil*)	Mimosoideae	Rhizobium	Grossman et al. (2006)
New dhaincha (*Sesbania rostrata*)	Fabaceae	Sinorhizobium	Maheshwari et al. (2013)
Indigo (*Indigofera tinctoria* L.)	Fabaceae	Rhizobium	Sarvade et al. (2014)
Horse bean (*Vicia faba* L.)	Fabaceae	Rhizobium	Richardson et al. (1975)
Pigeon pea (*Cajanus cajan* L. Mill sp.)	Fabaceae	Rhizobium	Bopape et al. (2022)
Mung bean (*Vigna mungo* L. Hepper)	Fabaceae	Rhizobium	Veer et al. (2021)
Chickpea (*Cicer arietinum* L.)	Fabaceae	Rhizobium	Singh and Singh (2018)
Peanut, groundnut (*Arachis hypogaea* L.)	Fabaceae	Bradyrhizobium	Zhang et al. (2022)
Pea (*Pisum sativum* L.)	Fabaceae	Rhizobium	Amsalu et al. (2012)
Alfalfa, Lucerne (*Medicago sativa* L.)	Fabaceae	Sinorhizobium	Panahpour (2009)
Clover (*Trifolium* spp.)	Fabaceae	Rhizobium	Rodríguez-Navarro et al. (2022)
Sun hemp (*Crotalaria juncea* L.)	Fabaceae	Methylobacterium	Sarvade et al. (2019)
Wild tantan (*Desmanthus virgatus* (L.) Willd.)	Fabaceae	Rhizobium	Sarvade et al. (2019)

Agroforestry-based cropland ecosystem is predominantly found effective for soil conservation and soil erosion. The forest vegetation along with hedge plants produces a cover effect which decreases soil erosion by creating a barrier.

5.6 Discussion

The environment and soil are dealing with so much pressure to sustain the large population of living organisms; this pressure is somewhere responsible for blind exploitation of natural resources without considering the degrading quality or health of these resources like soil. Good soil health is the important factor to sustain the growing population consistently in long term. It is also the ethical duty of the present population to pass the natural resources to the upcoming generations in good condition. The integration agricultural crops and livestock with trees and shrubs have proven to be more productive than a mere cropland. This system of land cultivation maintains the balance of the ecosystem and enriches and conserves the natural resources like soil and water, making it sustainable in the long term.

5.7 Conclusion and Future Prospective

The agroforestry-based agriculture is efficient in improving soil quality, and its large biodiversity is helpful in the achievement of sustenance in agriculture and environment. This integrated system in which litter is added to the soil by the perennial trees, decomposition is performed by microorganisms, and the nutrients and minerals are recycled again to the soil makes a cycle of agroforestry system. Additionally biological nitrogen fixation and carbon transformation inside the soil also improve the status of the absorbable form of these nutrients and overall quality of soil. Large root system and balanced moisture due to trees make the soil less exposed to soil loss and erosion. Agroforestry system is a sustainable way of agriculture as well as an adaptive method of soil and water conservation with high productivity of soil. Hence, agroforestry system of land use should be adapted widely to achieve sustainable development goals.

References

Alam SM, Naqvi SSM, Ansari R (1999) Impact of soil pH on nutrient uptake by crop plants. In: Handbook of plant and crop stress, vol 2. Routledge, London, pp 51–60

Amthor JS (1995) Terrestrial higher-plant response to increasing atmospheric [CO_2] in relation to the global carbon cycle. Glob Chang Biol 1(4):243–274

Amsalu A, Assefa F, Hailemariam A (2012) Symbiotic and phenotypic characterization of rhizobium isolates of field pea (Pisum sativum L.) Fabaceae, from central and Southern Ethiopia. Ethiop J Biol Sci 11(2):163–179

Arvidsson J (1999) Nutrient uptake and growth of barley as affected by soil compaction. Plant Soil 208(1):9–19

Baig MB, Shahid SA, Straquadine GS (2013) Making rainfed agriculture sustainable through environmental friendly technologies in Pakistan: a review. Int Soil Water Conser Res 1(2): 36–52

Blum WE (2005) Functions of soil for society and the environment. Rev Environ Sci Biotechnol 4(3):75–79

Bopape FL, Beukes CW, Katlego K, Hassen AI, Steenkamp ET, Gwata ET (2022) Symbiotic performance and characterization of pigeon pea (*Cajanus cajan* L. Mill sp.) rhizobia occurring in South African soils. Agriculture 13(1):30

Bormann FH, Likens GE (2012) Pattern and process in a forested ecosystem: disturbance, development and the steady state based on the Hubbard Brook ecosystem study. Springer Science & Business Media, Berlin

Bot A, Benites J (2005) The importance of soil organic matter: key to drought-resistant soil and sustained food production (No. 80). Food & Agriculture Org.

Boudiaf I, Le Roux C, Baudoin E, Galiana A, Beddiar A, Prin Y, Duponnois R (2014) Soil Bradyrhizobium population response to invasion of a natural Quercus suber forest by the introduced nitrogen-fixing tree Acacia mearnsii in El Kala National Park, Algeria. Soil Biol Biochem 70:162–165

Campbell L (2012) Biophysical drivers of tree crop performance in shade agroforestry systems: the case of coffee in Costa Rica. University of Toronto (Canada)

Cardinael R, Chevallier T, Guenet B, Girardin C, Cozzi T, Pouteau V, Chenu C (2020) Organic carbon decomposition rates with depth and contribution of inorganic carbon to CO_2 emissions under a Mediterranean agroforestry system. Eur J Soil Sci 71(5):909–923

Cardoso EJBN, Vasconcellos RLF, Bini D, Miyauchi MYH, Santos CAD, Alves PRL, Paula AMD, Nakatani AS, Pereira JDM, Nogueira MA (2013) Soil health: looking for suitable indicators. What should be considered to assess the effects of use and management on soil health? Sci Agric 70:274–289

Choudhary S, Tak N, Bissa G, Chouhan B, Choudhary P, Sprent JI, James EK, Gehlot HS (2020) The widely distributed legume tree Vachellia (acacia) nilotica subsp. indica is nodulated by genetically diverse Ensifer strains in India. Symbiosis 80(1):15–31

Clarholm M, Skyllberg U (2013) Translocation of metals by trees and fungi regulates pH, soil organic matter turnover and nitrogen availability in acidic forest soils. Soil Biol Biochem 63: 142–153

Cresswell H, Hamilton GJ (2002) Bulk density and pore space relations. In: Soil physical measurement and interpretation for land evaluation, CSIRO Publishing, pp 35–58

Dollinger J, Jose S (2018) Agroforestry for soil health. Agrofor Syst 92(2):213–219

Doran JW, Zeiss MR (2000) Soil health and sustainability: managing the biotic component of soil quality. Appl Soil Ecol 15(1):3–11

Grossman JM, Sheaffer C, Wyse D, Bucciarelli B, Vance C, Graham PH (2006) An assessment of nodulation and nitrogen fixation in inoculated Inga oerstediana, a nitrogen-fixing tree shading organically grown coffee in Chiapas, Mexico. Soil Biol Biochem 38(4):769–784

Gupta N, Kukal SS, Bawa SS, Dhaliwal GS (2009) Soil organic carbon and aggregation under poplar based agroforestry system in relation to tree age and soil type. Agrofor Syst 76(1):27–35

Hailemariam M, Birhane E, Asfaw Z, Zewdie S (2013) Arbuscular mycorrhizal association of indigenous agroforestry tree species and their infective potential with maize in the rift valley, Ethiopia. Agrofor Syst 87(6):1261–1272

Hoppe T, Sanders MPT (2014) Agricultural green gas demonstration projects in The Netherlands. A stakeholder analysis. Environ Eng Manag J 13(12):3083–3096

Horwath W (2015) Carbon cycling: the dynamics and formation of organic matter. In: Soil microbiology, ecology and biochemistry, vol 4, pp 339–382

Jeffries P, Gianinazzi S, Perotto S, Turnau K, Barea JM (2003) The contribution of arbuscular mycorrhizal fungi in sustainable maintenance of plant health and soil fertility. Biol Fertil Soils 37(1):1–16

Jie C, Jing-Zhang C, Man-Zhi T, Zi-tong G (2002) Soil degradation: a global problem endangering sustainable development. J Geogr Sci 12(2):243–252

Jose S, Bardhan S (2012) Agroforestry for biomass production and carbon sequestration: an overview. Agrofor Syst 86(2):105–111

Kesavan PC, Swaminathan M (2008) Strategies and models for agricultural sustainability in developing Asian countries. Philos Trans R Soc Lond B Biol Sci 363(1492):877–891
Kubik R (2005) How to use implements on your small-scale farm. Voyageur Press, Beverly
Leblanc HA, McGraw RL, Nygren P (2007) Dinitrogen-fixation by three neotropical agroforestry tree species under semi-controlled field conditions. Plant Soil 291(1):199–209
Lorenz K, Lal R (2014) Soil organic carbon sequestration in agroforestry systems. A review. Agron Sustain Dev 34(2):443–454
Luckow P, Wise MA, Dooley JJ, Kim SH (2010) Large-scale utilization of biomass energy and carbon dioxide capture and storage in the transport and electricity sectors under stringent CO_2 concentration limit scenarios. Int J Greenhouse Gas Control 4(5):865–877
Maheshwari DK, Agarwal M, Dheeman S, Saraf M (2013) Potential of rhizobia in productivity enhancement of Macrotyloma uniflorum L. and Phaseolus vulgaris L. cultivated in the Western Himalaya. In: Bacteria in agrobiology: crop productivity. Springer, Berlin, Heidelberg, pp 127–165
Marinari S, Masciandaro G, Ceccanti B, Grego S (2000) Influence of organic and mineral fertilisers on soil biological and physical properties. Bioresour Technol 72(1):9–17
McKenzie N, Jacquier D, Isbell R, Brown K (2004) Australian soils and landscapes: an illustrated compendium. CSIRO publishing, Clayton
Mink JN, Holmes WC, Singhurst JR (2016) Casuarina equisetifolia (Casuarinaceae) naturalized in Texas and comments on ecological implications for the Texas coast. Phyton 55:1–8
Mokgolodi NC, Setshogo MP, Shi LL, Liu YJ, Ma C (2011) Achieving food and nutritional security through agroforestry: a case of Faidherbia albida in sub-Saharan Africa. Forestry Studies China 13(2):123–131
Montagnini F, Metzel R (2017) The contribution of agroforestry to sustainable development goal 2: end hunger, achieve food security and improved nutrition, and promote sustainable agriculture. In: Integrating landscapes: agroforestry for biodiversity conservation and food sovereignty. Springer, Cham, pp 11–45
Morgado RG, Loureiro S, González-Alcaraz MN (2018) Changes in soil ecosystem structure and functions due to soil contamination. In: Soil pollution. Academic Press, Cambridge, pp 59–87
Mosquera-Losada MR, McAdam JH, Romero-Franco R, Santiago-Freijanes JJ, Rigueiro-Rodróguez A (2009) Definitions and components of agroforestry practices in Europe. In: Agroforestry in Europe. Springer, Dordrecht, pp 3–19
Murovhi NR, Materechera SA, Mulugeta SD (2012) Seasonal changes in litter fall and its quality from three sub-tropical fruit tree species at Nelspruit, South Africa. Agrofor Syst 86(1):61–71
Murthy IK, Gupta M, Tomar S, Munsi M, Tiwari R, Hegde GT, Ravindranath NH (2013) Carbon sequestration potential of agroforestry systems in India. J Earth Sci Climate Change 4(1):1–7
Nair PR, Nair VD, Kumar BM, Showalter JM (2010) Carbon sequestration in agroforestry systems. Adv Agron 108:237–307
Niu H (2013) Photosynthetic response of nodulated seedlings of three leguminosae species to drought stress. J West China For Sci 42(2):82–87
Panahpour H (2009) Effects of Sinorhizobium meliloti on quality traits of alfalfa varieties (Medicago sativa L.). Iran J Rangel For Plant Breeding Genet Res 16(2):207–217
Pandey D (2021) Impact of climate change on soil functionality. In: Climate change and the microbiome. Springer, Cham, pp 597–604
Panwar P, Chauhan S, Kaushal R, Das DK, ARORA G, Chaturvedi OP, Jain AK, Chaturvedi S, Tewari S (2017) Carbon sequestration potential of poplar-based agroforestry using the CO_2 FIX model in the Indo-Gangetic region of India. Trop Ecol 58(2):1–9
Ramachandran Nair PK, Mohan Kumar B, Nair VD (2009) Agroforestry as a strategy for carbon sequestration. J Plant Nutr Soil Sci 172(1):10–23
Ramesh T, Bolan NS, Kirkham MB, Wijesekara H, Kanchikerimath M, Rao CS, Sandeep S, Rinklebe J, Ok YS, Choudhury BU, Wang H (2019) Soil organic carbon dynamics: impact of land use changes and management practices: a review. Adv Agron 156:1–107

Rastegari AA, Yadav AN, Yadav N (2020) New and future developments in microbial biotechnology and bioengineering: trends of microbial biotechnology for sustainable agriculture and biomedicine systems: diversity and functional perspectives. Elsevier, Amsterdam

Richardson M, Dilworth MJ, Scawen MD (1975) The amino acid sequence of leghaemoglobin I from root nodules of broad bean (Vicia faba L.). FEBS Lett 51(1-2):33–37

Rhoades JD, Chanduvi F, Lesch SM (1999) Soil salinity assessment: methods and interpretation of electrical conductivity measurements (no. 57). Food & Agriculture Org.

Rodríguez-Navarro DN, Lorite MJ, Vera FJT, Camacho M (2022) Selection and characterization of Spanish trifolium-nodulating rhizobia for pasture inoculation. Syst Appl Microbiol 45(2): 126290

Rodrigues RC, Araújo RA, Costa CS, Lima AJ, Oliveira ME, Cutrim JA Jr, Santos FN, Araújo JS, Santos VM, Araújo AS (2015) Soil microbial biomass in an agroforestry system of Northeast Brazil. Trop Grassl Forrajes Trop 3(1):41–48

Rosa DDL, Sobral R (2008) Soil quality and methods for its assessment. In: Land use and soil resources. Springer, Dordrecht, pp 167–200

Rosa-Schleich J, Loos J, Mußhoff O, Tscharntke T (2019) Ecological-economic trade-offs of diversified farming systems–a review. Ecol Econ 160:251–263

Salazar S, Sánchez LE, Alvarez J, Valverde A, Galindo P, Igual JM, Peix A, Santa-Regina I (2011) Correlation among soil enzyme activities under different forest system management practices. Ecol Eng 37(8):1123–1131

Samarakoon SMNS, Rajapakse S (2020) Identification of stress tolerant rhizobial strains inhabiting Gliricidia sepium in Polonnaruwa district, Sri Lanka. Ceylon J Sci 49(1):37

Sarvade S, Gautam DS, Upadhyay VB, Sahu RK, Shrivastava AK, Kaushal R, Singh R, Yewale AG (2019) Agroforestry and soil health: an overview. Scientific Publishers, Jodhpur, pp 276–297

Sarvade S, Singh R, Prasad H, Prasad D (2014) Agroforestry practices for improving soil nutrient status. Popular Kheti 2(1):60–64

Sayer EJ (2006) Using experimental manipulation to assess the roles of leaf litter in the functioning of forest ecosystems. Biol Rev 81(1):1–31

Sharma N, Bohra B, Pragya N, Ciannella R, Dobie P, Lehmann S (2016) Bioenergy from agroforestry can lead to improved food security, climate change, soil quality, and rural development. Food Energy Secur 5(3):165–183

Singh B, Sharma KN (2007) Tree growth and nutrient status of soil in a poplar (Populus deltoides Bartr.)-based agroforestry system in Punjab, India. Agrofor Syst 70(2):125–134

Singh Z, Singh G (2018) Role of rhizobium in chickpea (Cicer arietinum) production-a review. Agric Rev 39(1):31–39

Somasundaram, J., Singh, R. K., Ali, S., Sethy, B. K., Singh, D., Lakaria, B. L., ... & Sinha, N. K. (2012). Soil aggregates and other properties as influenced by different long term land uses under table landscape topography of Chambal region, Rajasthan, India. Indian Journal of Soil Conservation, 40(3), 212–217

Stirling G, Hayden H, Pattison T, Stirling M (2016) Soil health, soil biology, soilborne diseases and sustainable agriculture: a guide. Csiro Publishing

Tangjang S, Arunachalam A (2009) Role of traditional home garden systems in Northeast India. Indian J Tradit Knowl 8:47–50

Thapa GB, Weber KE (1991) Soil erosion in developing countries: a politicoeconomic explanation. Environ Manag 15(4):461–473

Thomas VG, Kevan PG (1993) Basic principles of agroecology and sustainable agriculture. J Agric Environ Ethics 6(1):1–19

Timmis K, Ramos JL (2021) The soil crisis: the need to treat as a global health problem and the pivotal role of microbes in prophylaxis and therapy. Microb Biotechnol 14(3):769–797

Tittonell P, Giller KE (2013) When yield gaps are poverty traps: the paradigm of ecological intensification in African smallholder agriculture. Field Crop Res 143:76–90

Udawatta RP, Anderson SH, Kremer RJ, Garrett HEG (2021) Agroforestry for soil health. In: North American Agroforestry. Wiley, Hoboken, pp 355–386
Unger IM, Goyne KW, Kremer RJ, Kennedy AC (2013) Microbial community diversity in agroforestry and grass vegetative filter strips. Agrofor Syst 87(2):395–402
Uthappa AR, Bana OPS, Kumar M, Kanwal M (2015) Soil physico-bio-chemical properties as influenced by varying tree densities in poplar (Populus deltoides Bartr. Ex marsh.) based agroforestry system. Indian J Agrofor 17(1):81–90
Van Der Heijden MG, Bardgett RD, Van Straalen NM (2008) The unseen majority: soil microbes as drivers of plant diversity and productivity in terrestrial ecosystems. Ecol Lett 11(3):296–310
Varghese R, Chauhan VS, Misra AK (2003) Evolutionary implications of nucleotide sequence relatedness between Alnus nepalensis and Alnus glutinosa and also between corresponding Frankia microsymbionts. In: Frankia Symbiosis. Springer, Dordrecht, pp 219–227
Veer D, Diwakar SK, Tomar SP, Mohasin M, Singh RP (2021) Effect of rhizobium, phosphorus and nitrogen on growth, yield and yield attributes of summer black gram. Pharma Innov J 11(2): 1192–1195
Waddington JM, Morris PJ, Kettridge N, Granath G, Thompson DK, Moore PA (2015) Hydrological feedbacks in northern peatlands. Ecohydrology 8(1):113–127
Wei C, Gao M, Shao J, Xie D, Pan G (2006) Soil aggregate and its response to land management practices. China Particuol 4(05):211–219
Yadav RL, Yadav DV, Duttamajumder SK (2008) Rhizospheric environment and crop productivity: a review. Indian J Agron 53(1):1
Zerbe S (2022) Types of traditional cultural landscapes throughout the world. In: Restoration of multifunctional cultural landscapes. Springer, Cham, pp 19–76
Zhang J, Peng S, Li S, Song J, Brunel B, Wang E, James EK, Chen W, Andrews M (2022) Arachis hypogaea L. from acid soils of Nanyang (China) is frequently associated with Bradyrhizobium guangdongense and occasionally with Bradyrhizobium ottawaense or three Bradyrhizobium genospecies. Microb Ecol 84(2):556–564

Chapter 6
Soil Nutrient Dynamics and Cycling Under Agroforestry

A. Balasubramanian, K. S. Anjali, G. Swathiga, Ghazanfer Abbas, and S. Navaneetha Krishnan

Abstract Agroforestry is the sustainable integration of trees with other agricultural components for multiple benefits. Water, light, and nutrients are apprehended as the basic components of plant growth. Sustainable management of these resources will ensure maximum nutrient and water use efficiency, hence resulting in maximizing the output from the system. Agroforestry systems can promote efficient nutrient cycling than any other agricultural systems by regulating the uptake, recycling, and synchronization. The tree components in an agroforestry are the most integral component of nutrient balance owing to its deep roots of holding nutrients, nutrient addition by litter, and fine root decomposition and symbiotic nitrogen fixation by nitrogen-fixing trees. Agroforestry hence can modify the nutrient dynamics of the system. Hence integrating the right components in the system can reduce the cost of application of external inputs such as fertilizers and results in a system that is self-sustained and productive.

Keywords Agroforestry · Litter · Fine roots · Nutrient balance · Nutrient use efficiency

6.1 Introduction

Agroforestry encompasses a range of land use systems and techniques that involve intentionally cultivating woody perennials such as trees, shrubs, palms, and bamboos alongside agricultural crops and/or animals. These different components are strategically grown together on the same land units, either in specific spatial arrangements or temporal sequences (ICRAF 1996). In many tropical nations, output and food security are seriously threatened by the current agricultural situation's loss of nutrients and organic matter in the soil. Agroforestry encourages a water and nutrient cycle that is more effective than the typical agriculture system. The conjugation of

A. Balasubramanian (✉) · K. S. Anjali · G. Swathiga · G. Abbas · S. N. Krishnan
Department of Silviculture and NRM, Forest College and Research Institute, Mettupalayam, India
e-mail: balasubramanian.a@tnau.ac.in

H. S. Jatav et al. (eds.), *Agroforestry to Combat Global Challenges*, Sustainable Development and Biodiversity 36, https://doi.org/10.1007/978-981-99-7282-1_6

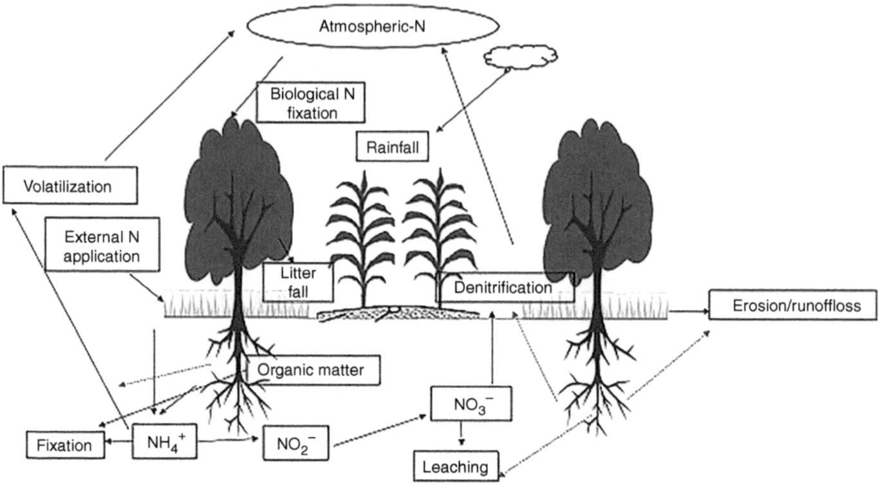

Fig. 6.1 Nutrient dynamics under an agroforestry system

different components in agroforestry will result in a complex system modifying the soil and nutrient dynamics pertinent to the system. The nutrients in the soil are responsible for the overall growth and development of the crop by influencing the productivity and species composition. Hence the shift from optimum physical, chemical, and biological constants of the soil will affect the productivity of the soil, thus influencing the soil and plant nutrient balance.

Soil nutrient dynamics encompass the changes in nutrient status in the soil ecosystem under a given set of environments in temporal and spatial manner due to continuous biogeochemical processes operating within the system. Agroforestry system can alter the soil properties substantially in both surface and sub-surface layers. The intercropping pattern of the trees, management techniques, spatial planning, the quantity and quality of litter, and the rate of decay and decomposition all affect the soil's characteristics in an agroforestry system. The most prominent species in the community, which is frequently the tree component, is mostly responsible for the ecological processes like decomposition, which have overall beneficial impacts from the pumping of nutrients throughout the system. By managing the microclimate of the site and changing the physical composition, infiltration ability, moisture regime, and other physio-chemical aspects of soils, trees as a component of an agroforestry system tend to influence the ecosystem. The tree components in the agroforestry owing to its deep roots are more capable of nutrient and water pumping from the deep layers that are usually not utilized by the herbaceous crops. Trees can also minimize nutrient losses by reducing soil erosion and trapping nutrients that drain out of the crop root zone with the help of their deep and broad root systems. The nutrient dynamics under an agroforestry system has been presented in Fig. 6.1.

Litter fall, root decay, internal recycling of nutrients through crop residue and manure formed within the system, and the recycling of nutrients by trees from deep soil horizons to the soil surface are all examples of nutrient transfers that take place inside the system. The majority of the nutrients that plants absorb return to the soil in the form of decomposed litter, and the patterns of nutrient release are influenced by the climatic conditions and the quality of the litter. This process accounts for around 50% of the nutrients that plants take up. The input of nutrients into the system comes from rainfall, fertilizers, and biological nitrogen fixation by trees, crops, and other organic matter from outside the system, and the output comes from removal of crop, tree, and livestock products, soil erosion, leakage, evaporation, and other processes.

Apart from litter and fine root decomposition, the residues of tending operations such as pruning and pollarding also contribute to nutrient addition. Fine lateral roots are a dominating axis of nutrient acquisition techniques across and between species because they exhibit a wide range of nutrient acquisitive qualities and conservative traits within the zone of tree and crop root interactions. In agroforestry system such as alley cropping, the tree crop interactions in an agroforestry system can be enhanced by planting leguminous trees with crops with increased nitrogen fixation and enhanced productivity.

6.2 Nutrient Recycling Under an Agroforestry System

The nutrient cycling is defined as the cycling of nutrients in the physical environment into the living organisms and recycling back to the physical environment. The nutrient cycling is a vital function of the ecology of a region, and under an environment, the cycle must be stable and balanced. An instability in nutrient balance results in the nutrient loss from the system affecting the sustainability of the system to support life. Nutrient cycling under an agroforestry system can be redefined as the cycling of nutrients within the physical environment and between the components of the system. Due to the continuous accumulation of litter and activities involving decomposition, the biological processes in agroforestry systems increase the system's efficiency in terms of carbon stocking and cycling of nutrients. The movement and interchange of organic and inorganic substances back into the formation of living matter are known as nutrient recycling. Food web mechanisms that break down organic matter into mineral nutrients control the recycling of nutrients.

Nutrient balance (Fig. 6.2) is the difference between nutrient inflow and outflow. The stability of a system is contributed by the state of a positive nutrient balance. For any system where the inflow and outflow are equal, the state of balance is referred to as steady state. On comparing with the tree ecosystems (Fig. 6.3), viz., forest ecosystem, agroforestry, and agricultural ecosystem, regarding the flow of input and output, agroforestry has a predominantly positive effect on the process followed by forest ecosystem and agricultural ecosystem. The roots will hold the soil by nutrient saving, and the litter will contribute to nutrient addition maintaining a

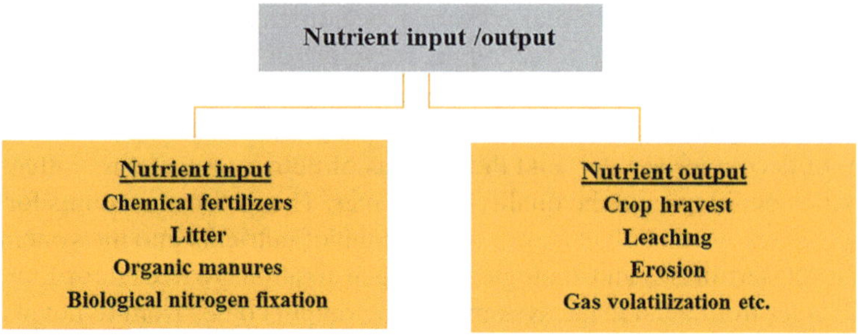

Fig. 6.2 Nutrient balance input and output

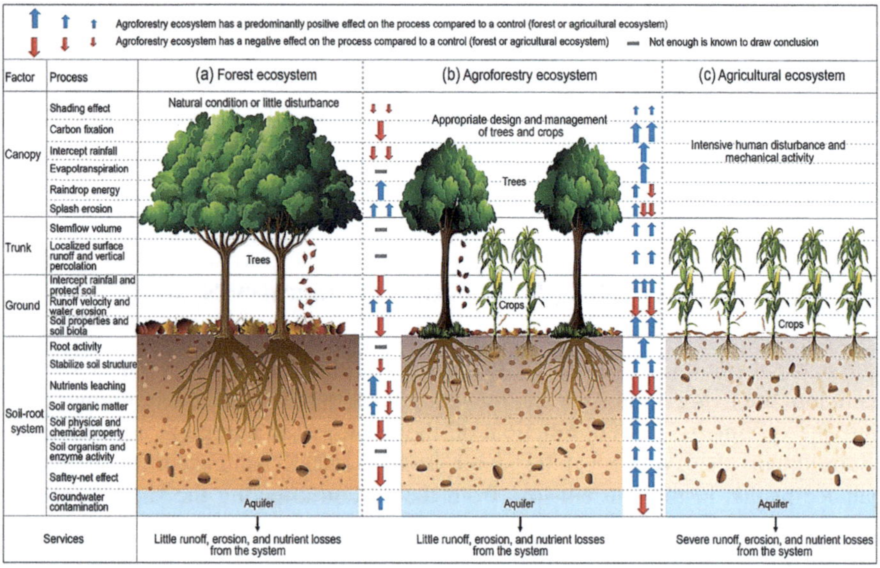

Fig. 6.3 Agroforestry possesses characteristics that can potentially mitigate water, soil, and nutrient losses, in contrast to both natural forests and conventional agricultural systems (source, Zhu et al. 2019)

balance of nutrients in an agroforestry system (Hossain et al. 2011). Trees through their canopy will make a balance of shade and water interception through stem flow and through flow will provide the system the needs for sustainable agricultural practices.

6.3 Major Processes Responsible for Soil Nutrient Dynamics in Agroforestry Systems

6.3.1 Litter Fall

The dead leaves, twig, bark, needles, etc. form the composition of total litter production in the system. Litter production in agroforestry system helps in maintaining or improving the soil fertility through release of critical macronutrients like nitrogen, phosphorus, and potassium to the soil and plays a significant role in soil nutrient recycling process. The addition of organic matter through litter promotes the water holding capacity, water penetration rate, biodiversity, soil microbial activity, and nutrient concentration of the soil, all of which contribute to its better quality (Hossain et al. 2011). In terms of the overall amount of litter generated in the system, leaf litter makes up the majority, and in comparison, it has the greatest nutritional concentrations. It also returns the most nutrients to the soil, making it the main source of soil nutrients (Hossain et al. 2011). Litter fall in soil depends on the nature of plant species, climate, land use, decomposer population, etc. (Tapia-Coral et al. 2005). Major processes responsible for soil nutrient dynamics in agroforestry systems have been presented in Fig. 6.4.

6.3.2 Decomposition

Decomposition, which is a sequence of physical and chemical processes that reduce leaf litter and foliage to their basic chemical components, is the main mechanism in nutrient recycling within ecosystems (Aerts 1997). The decomposable matter on the agroforestry soil floor breaks down into smaller fragments through chemical and physical processes by the activity of microfauna present in the soil (Lorena et al. 2005). This is one of the most significant ecological occurrences because it contributes nutrients to the soil (Dommergues 1987). The availability of nutrients in soil is significantly impacted by the decomposition and mineralization processes carried out by soil microbes, and this is made possible by soil fauna such as non-symbiotic nitrogen-fixing bacteria, phosphate, solubilizing bacteria, and sulfate-oxidizing bacteria. The rate of litter decomposition is influenced by several factors such as quality and quantity of plant component; climatic factors such as temperature, moisture, and aeration; and soil nutrient availability factors including carbon-nitrogen (C : N) ratio, microbial community dynamics, etc.

Fig. 6.4 Major processes responsible for soil nutrient dynamics in agroforestry systems

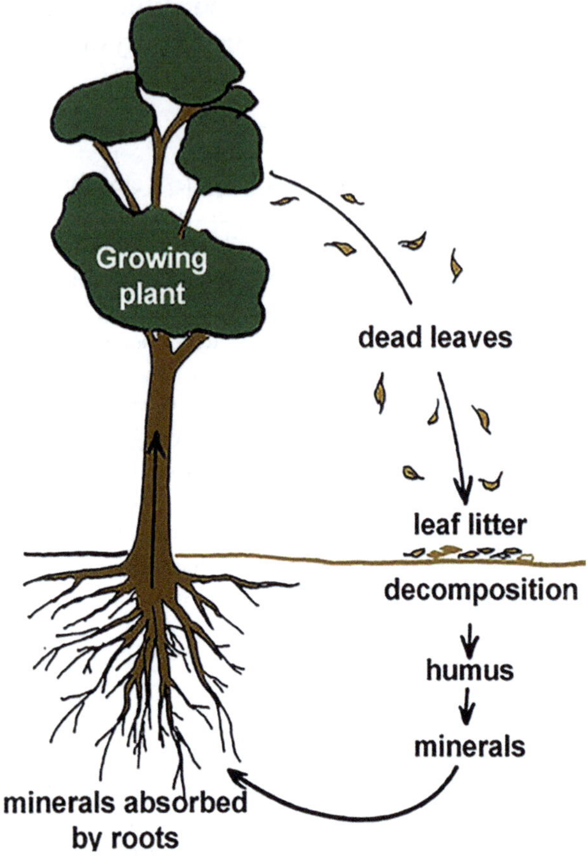

6.4 Biological Nitrogen Fixation Under an Agroforestry System

The process by which inorganic nitrogen is changed into a useful organic form with the support of living things is known as the biological nitrogen fixation mechanism. This may be accomplished in collaboration with both free-living and symbiotic bacteria, as well as with other symbiotic microbes. Nitrogen fixation occurs when atmospheric molecular nitrogen is converted to ammonia. With the help of a group of bacteria called rhizobia, legumes are able to fix atmospheric nitrogen and accumulate it biologically. The bacteria, which are normally free-living in the soil in the native range of a particular legume, inoculate into the tree root hairs and are embodied in small root structures called nodules. The mutualistic association is accomplished by providing energy to the bacteria by the plant to fuel the nitrogen fixation process by which the plant receives nitrogen for growth. The influences of soil biota on soil processes in tree-based ecosystems are given below.

6.4.1 *Microflora*

These groups of soil biota include fungi, bacteria, and actinomycetes. These actively participate in nutrient cycling by catabolizing organic matter and mineralizing and immobilizing soil nutrients. By producing organic substances that bind aggregates and hyphae that entangle particles into aggregates, they improve the structure of the soil.

6.4.2 *Microfauna*

These groups of soil biota include protozoa, nematodes, etc. These actively participate in nutrient cycling by regulating bacterial and fungal population, hence altering nutrient turnover. Through interactions with microflora, they could have an impact on aggregate structure.

6.4.3 *Mesofauna*

Mesofauna includes Acarina, Collembola, enchytraeids, etc. This group regulates fungal and microfaunal populations, alters nutrient turnover, and fragments plant residues. They alter the soil structure by creation of biopores and promoting humification.

6.4.4 *Macrofauna*

Macrofauna includes isopods, centipedes, millipedes, earthworms, etc. These groups fragment plant residues and stimulate microbial activity. These groups mix organic and mineral particles, redistribute organic matter and microorganisms, create biopores, and promote humification.

6.5 Soil Fauna: Small Burrowing Animals

This group enriches soil by breaking down of organic matter and mixing of soil through profile, enhancing decomposition. The interconnecting tunnels created by their activities improve aeration and infiltration. These enhance soil structure by influencing microrelief and soil properties.

6.6 What Are Nitrogen Fixing Trees?

Trees may provide nitrogen to agroforestry systems in two ways: biological nitrogen fixation in the case of nitrogen-fixing tree species (NFTs) and soil deep nutrient collection (Rosenstock et al. 2014). Trees and nutrient cycling under an agroforestry system have been presented in Fig. 6.5. Trees which have the capacity of converting atmospheric nitrogen into plant usable compounds, such as ammonia, are called as nitrogen-fixing trees (NFTs). The role of NFTs in agroforestry includes improvement of soil fertility by nitrogen fixation, and they are incorporated as fodder plants, windbreaks, and plywood and pulpwood trees (Brewbaker 1987; Das and Chaturvedi 2005). The interpolating of N-fixing trees in rows with low crop plants, often known as alley cropping or hedgerow intercropping, can help increase soil fertility. Trees planted as hedge will improve the soil fertility by means of nitrogen fixation and enrichment of fallow land, and the litter provides mulch for crop fertility improvement.

The amount of biological nitrogen fixation by trees, both leguminous and non-leguminous, ranges from 20 to 300 kg N ha^{-1} $year^{-1}$. Nitrogen-fixing trees can supply nitrogen either through symbiotic nitrogen fixation or through litter fall. Some non-NFTs, e.g., cherry, mandarin, etc., also accumulate more nitrogen as NFTs, owing to their greater root volume and ability to capture nutrients from soil surface, where they perform the role of cycling the nutrients apart from fixing them to the system. Apart from leguminous nitrogen fixation, NFTs can fix nitrogen by the process of deep nutrient capture; trees can uptake nutrients at depths by means of deep roots where the crop roots can't reach. In agroforestry systems, these nutrients are leached and lost for crop use, but they are transported to the soil through the decomposition of tree litter, where they constitute an additional nutrient input

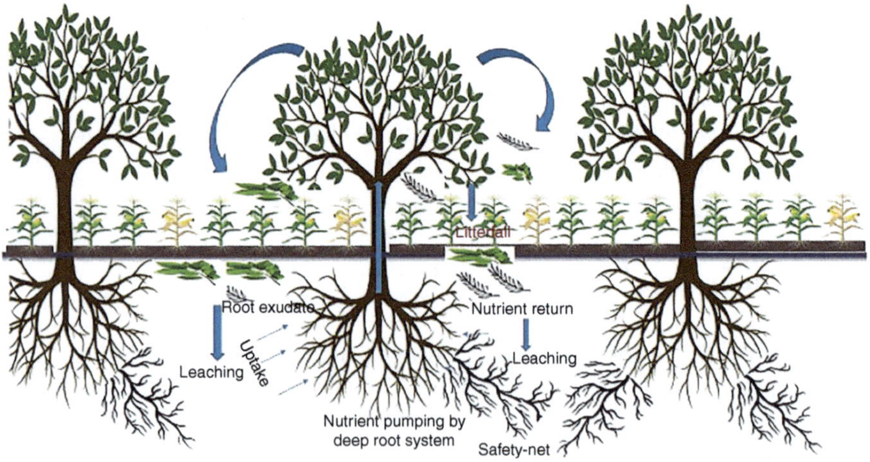

Fig. 6.5 Trees and nutrient cycling under an agroforestry system (Fahad et al. 2022)

Table 6.1 Major nitrogen-fixing trees and its agroforestry utility

Species	Agroforestry utility		
Acacia aneura	Trees and shrubs in pastures		
Acacia farnesiana			Fodder banks
Albizia lebbeck	Trees and shrubs in pastures		
Albizia saman	Trees and shrubs in pastures		
Alnus acuminate	Trees and shrubs in pastures		
Calliandra arborea	Trees and shrubs in pastures		
Calliandra calothyrsus	Trees and shrubs in pastures		Fodder banks
Casuarina cunninghamiana		Live fence posts	
Desmodium velutinum	Trees and shrubs in pastures		Fodder banks
Erythrina variegata	Trees and shrubs in pastures	Live fence posts	Fodder banks
Gliricidia sepium	Trees and shrubs in pastures	Live fence posts	Fodder banks
Pterocarpus hayesii	Trees and shrubs in pastures		

(Galiana et al. 2004). Major nitrogen-fixing trees and its agroforestry utility have been presented in Table 6.1.

6.7 Phosphorus Dynamics Under Agroforestry

Unlike nitrogen, phosphorus is a highly immobile nutrient and is mostly deficiently available to plants. The agroforestry system can also not supply the whole phosphorus requirement for crops. Phosphorus cannot be fixed like nitrogen and is not available to get captured from deep due to the non-availability of phosphorus in subsoil. The phosphorus accumulated in tree biomass will be transferred to the soil by means of litter decomposition, but it is merely nutrient cycling and not an input. During phosphorus cycling non-available form of phosphorus is converted to available form for plants. The phosphorus cycle also affects the biological activity of the system since photosynthesis and microbial activity in decomposing litter require sufficient amounts of phosphorus in certain biochemical forms. Contrary to nitrogen, the phosphorus cycle is closed, indicating that over the course of a few years, neither significant gains nor losses arise from the system. Trees can also solubilize phosphorus by symbiotic association with phosphorus-solubilizing fungi termed mycorrhizae where trees will host the fungi and fungi, in turn, solubilizes phosphorus. Mycorrhizal association between fungi and roots of higher plants is presented in Fig. 6.6. In agroforestry systems with little input, phosphorus is frequently the essential nutrient; as a result, inorganic phosphorus in the form of fertilizers needs to be added to soils that are low in P.

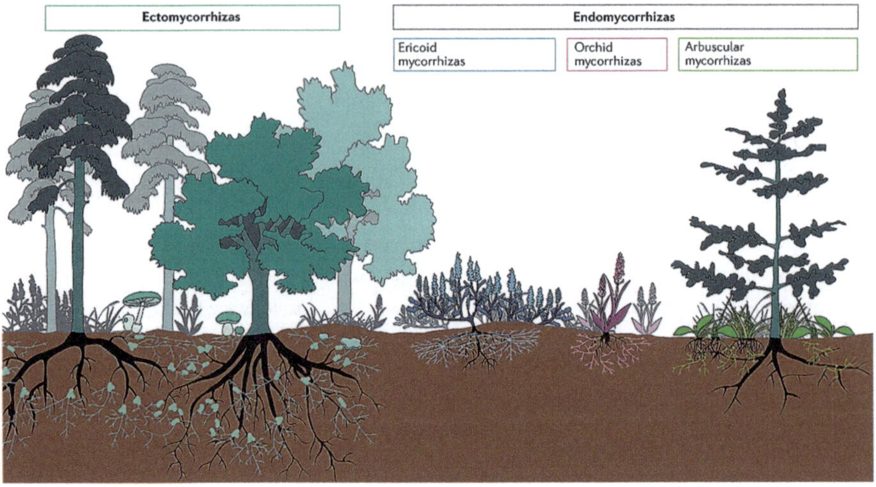

Fig. 6.6 Mycorrhizal association between fungi and roots of higher plants

6.8 Factors Affecting Biomass Decomposition

The factors affecting biomass decomposition include the following.

6.8.1 Substrate Quality

6.8.1.1 Carbon and Nitrogen Content

The rate of decomposition is enhanced by high initial nitrogen content and low carbon:nitrogen (C:N) ratio. High C:N ratio does not provide sufficient nitrogen for the metabolism of decomposing crop.

6.8.1.2 Lignin and Cellulose Content

High lignin and cellulose content in the substrate will increase the stability and reduce/resist the rate of decomposition.

6.8.1.3 Polyphenol

Increased polyphenol content will reduce decomposition as polyphenol forms complexes with protein making inaccessible to microorganisms.

6.8.2 Soil Micro- and Macrofaunal Activity

The microorganisms in the soil play a major role in the decomposition of dead organic matter, nitrogen fixation, improvement in soil aeration, and mixing of soil.

6.8.3 Environmental Parameters

The environmental parameters including moisture and temperature play a major role in influencing biomass decomposition. Prominently, the increased moisture content will increase the decomposition, while the optimum temperature and moisture content affects the populations of the active microbes.

6.9 Root Dynamics Under Agroforestry System

Roots serve as the primary interface between plants and soil for the uptake of water and nutrients. Roots contribute to nutrient dynamics, by returning nutrients to the soil by death, decay, exudation, and leaching. The root biomass is not static (Peter and Lehmann 2000). The roots with diameter less than 5 mm are fine roots distributed in the upper soil layers with 90 percent in the top 30 cm. Fine roots contribute high organic matter with 2–5 times higher than that of the aboveground parts. Eight to 67% of net primary production is made up of fine root production. The detritus input to the soil from the fine roots may be greater than that from the aboveground compartments (Finér and Laine 1998).

The factors affecting fine root decomposition include the following:

(a) **Elevated CO_2**: The fine root production and turnover increase in elevated CO_2.
(b) **Climate change**: When soil moisture and nutrient availability are sufficient, both root growth and death increase as temperature rises (Pregitzer 2002).
(c) **Nutrient availability**: Fine root biomass normally declines with increased N availability, but turnover rises (Nadelhoffer 2000).
(d) **Drought**: In the organic layer of the analyzed spruce stand, soil dryness and frost enhanced fine root mortality by 61% and 29%, respectively (Gaul et al. 2008).
(e) **Changes in water supply**: Rapid rise in root numbers in *Tectona grandis* (Singh and Srivastava 1985) and *Eucalyptus globulus* (Kätterer et al. 1995) following the extinction of drought periods.

6.10 Nutrient Use Efficiency (NUE)

Nutrient use efficiency (NUE) is the proportion of the overall rate of biomass output to the total rate of nutrient intake (Hirose 1975). For many decades, nutrient use efficiency (NUE) has drawn the attention of forest researchers. It was asserted that the more nutrients that were consumed, the less was their NUE, i.e., biomass generated per unit of resource supply declined with an increase in supply. This was based on litter fall as a substitute for growth and net primary productivity (NPP) and litter N content as a replacement for N supply (Vitousek 1982). Nutrient use efficiency of the trees can be maintained by both fertilizer management and irrigation management. By using proper fertilizer scheduling techniques, which provide nutrients at the proper rate, time, and location with proper silviculture techniques, the NUE of the trees may be increased (Panhwar et al. 2019).

The NUE can be quantified based on the potential photosynthetic nutrient use efficiency of leaves (Hiremath 2000), ratio of growth to nutrient uptake (Manschadi et al. 2014), ratio of net primary productivity to soil nutrient supplied (Hirose 2011), litter fall nutrient use efficiency (Vitousek 1982), nitrogen and phosphorus growth efficiency (Berendse and Aerts 1987), product of plant level nutrient efficiency to the uptake efficiency (Xu et al. 2012), and efficiency of recovery (Shaver and Melillo 1984).

The NUE of the trees depends on the factors such as site soil nutrient status and litter dynamics, nutrient loss from the system, nutrient partitioning, soil fertility status, forms of nutrients applied, age of the trees species, and composition of the species.

Another way of improving NUE in trees is applications of biofertilizers for sustainable enhancement of productivity improvement, for instance, improved NUE was obtained in *Pinus wallichiana* on treating with ectomycorrhizal fungus followed by Azotobacter sp., Azospirillum sp., *Pseudomonas fluorescens*, and *Bacillus subtilis* (Asif et al. 2014). Three tropical trees that grow quickly have their potential and cumulative photosynthetic nutrient use efficiency assessed: *Cedrela odorata*, *Cordia alliodora*, and *Hyeronima alchorneoides* where *Cedrela odorata* with shortest-lived leaves was observed with highest potential nitrogen use efficiency, beneficial under high-nutrient environments, whereas *Hyeronima alchorneoides* with longest-lived leaves had the highest cumulative nitrogen and phosphorus use efficiencies which are extremely beneficial under low-nutrient environments (Hiremath 2000).

On comparing deciduous trees, evergreen trees with greater leaf longevity, high cumulative carbon gain over leaf lifetimes (DeLucia and Schlesinger 1995), longer nutrient retention, and lower rates of nutrient losses (Aerts 1995), accompanied by longer nutrient retention and lower rates of nutrient losses while quantifying NUE with respect to potential photosynthetic nutrient use efficiency of leaves (Field and Mooney 1986). On examination, the foliar nutrient dynamics of four deciduous species, *Quercus prinus*, *Quercus alba*, *Acer rubrum*, and *Fagus grandifolia* under different soil nutrient availability, Boerner (1984) observed high nutrient efficiency and improved growth rate under poor site conditions, while Miller et al. (1976) observed a decrease in nutrient use efficiency but an increase in growth rate for conifer stands under high nitrogen availability. In monoculture plantation of Himalayan alder (Sharma 1993) and in the mixed plantation of *Alnus* + cardamom (Sharma et al. 2002), nitrogen as well as phosphorus use efficiency decreased with age; this is in contrary to Jha (2014) in *Tectona grandis*, where NUE increases with age up to 18 years and then reduced up to 30 years.

6.11 Fertilizer Requirements in Trees

The practice of applying nutrients at the proper rate, timing, and location based on the species can be termed fertilizer scheduling (IPNI, 2014). Fertilizer requirements are affected by the type of soil, previous cropping, expected duration of the growth season, stand age, yield, and nutrient demand of the tree. Fertilizer requirement in trees can be monitored by appearance of nutrient deficiency symptoms, soil analysis, and foliar analysis. The same fertilizer requirement can be monitored through mathematical models based on nutrient accumulation and released by fertilized trees with the help of intelligent sensor systems (Lakhiar et al. 2018).

Nitrogen fertilization in boreal conifer forest at a recommended dose of 150 kg ha^{-1} infers high timber yields and cash flows in spruce-dominated stands growing on medium sites than pine stands (Pukkala 2015); this is in contrary to Saarsalmi and Mälkönen (2001) where fertilizing pine trees were more profitable than spruce. Jordan et al. (2003) monitored N fertilizer uptake in *Quercus rubra* and *Quercus coccinea* seedlings by ^{15}N labelled fertilizer and infer a reduced fertilizer uptake and growth owing to soil compaction, while in *Quercus ilex* under favorable nursery conditions, increasing rate of fertilizer 200 mg N $plant^{-1}$ increased nutrient uptake and biomass of the seedlings (Oliet et al. 2009). In *Eucalyptus* plantations, trees respond positively to the increasing rate of fertilizers (fourfold dose of 40 kg ha^{-1} N; 16 kg ha^{-1} P; 53 kg ha^{-1} K) at the initial stages; however the effect goes on decreasing in subsequent years as the tree response to fertilizer doses generally diminish with tree age (Santana et al. 2008), while a split application of fertilizer at rates of 125 and 250 g (16N–7P–7K) per seedling increased growth and biomass accumulation in hybrid eucalyptus (Zeng et al. 2013). Fertilizer dose of 50 N:25 P_2O_5:50 K_2O kg ha^{-1} $year^{-1}$ shows an increment in mean height and diameter, while dose of 100 N:50 P_2O_5:50 K_2O kg ha^{-1} showed an increase in crown width in *Ailanthus triphysa* saplings under an agroforestry system, but the observations were not significant (Kumar et al. 2001). In *Dalbergia sissoo* dual inoculation of biofertilizers, rhizobium, and AM fungi results in an increasing growth and biomass under normal soil, while in alkaline soil blending of micronutrients with dual dose of biofertilizers inferred a favorable outcome (Revathi et al. 2013). The application of 163 kg ha^{-1} urea, 375 kg ha^{-1}mussoorie rock phosphate, 145 kg ha^{-1} muriate of potash, 105 kg ha^{-1} quick lime, and 373 kg ha^{-1} Mg sulfate from two split applications in the first year and four split applications during the second and third years in young plantations of *Tectona grandis* in Kerala is practiced, cited in Kumar (2011), while in Costa Rica application of N-P-K at the rate of 10-30-10 or 12-24-12 at the beginning of the rainy season and an extra N dose during the maximum rainy days up to 3 to 4 years is practiced (Alvarado 2012).

6.12 Fertigation on Tree Growth and Biomass Allocation

Fertigation is referred to a fertilizer application method that uses a drip irrigation system to dissolve fertilizer in irrigation water. Fertigation intensifies the efficiency of fertilizer and water application with the advantage to modify the doses and frequency of water and fertilizer administrations based on plant/tree requirements, influenced by tree age, growth cycle, and weather conditions in which the tree crop is growing. Fertigation has significant impact on tree development and biomass production by synchronizing the application of water and nutrients (Li and Liu 2010) and hence results in a much higher yield and more efficient use of water and nutrients, allowing for the sustainable use of both water and nutrients (Shirazi et al. 2014). Extensive studies on fertigation in forest trees are limited, and some of the research evidences related to tree fertigation are discussed. The N fertigation in *Populus tomentosa* influenced basal area and the aboveground diameter by 38% (N_{115}), 30% (N_{230}), and 32% (N_{345}) (Wang et al. 2015). The estimation of above- and belowground biomass in sweet gum (*Liquidambar styraciflua* L.) and loblolly pine (*Pinus taeda* L.) plantation which is under proper irrigation and fertilization by Coyle et al. 2008 showed increase in mean annual aboveground biomass, for sweet gum ranging from 2.4 to 5 Mg ha^{-1} $year^{-1}$ and for loblolly pine ranging from 5 to 6.9 Mg ha^{-1} $year^{-1}$. Teak (*Tectona grandis* L.) under fertigation on medium black soil, with three fertilizer levels and five fertigation frequencies under drip irrigated system, registered an increase in height, diameter at breast height, basal area, and volume, for fertigation application in six splits a year, and there was an increase in the volume for the fertilizer doses at 200, 43, and 166 kg and 300, 64.5, and 249 kg N/ha, P/ha, and K/ha.

6.13 General Methods of Estimation of Nutrient Dynamics

The nutrient dynamics can be evaluated in terms of litter fall rate, leaf and litter decomposition rate, soil microbial population, soil nutrient status, organic matter content, and aboveground and belowground biomass. The general methods of estimation can be by destructive sampling of trees; soil N, P, K, and organic carbon content estimation; estimation of nutrient content of litter; N, P, and K uptake of trees and crops; etc.; apart from this advanced application, stable isotopes can be also used such as ^{15}N isotope for nutrient uptake dynamics, ^{32}P isotope for root competition, etc. The fine root dynamics or the belowground contribution to nutrient dynamics can be estimated without destructive sampling techniques such as X-ray computed tomography, 3D laser scanning, ground penetrating radar, etc. Various models such as WaNuLCAS, SCUAF, AME, AMAPmod, etc. were also developed in order to justify the tree-soil-crop interactions.

6.14 Conclusion

The system became stable and sustainable if there is a balance in the input with the output, where there are occurrence of nutrient reduction and land quality reduction due to continuous utilization of land for agriculture. Therefore, the need of sustainable agriculture practices is of potent importance for maintaining the soil quality and land integrity for the future. Agroforestry is one such approach with sustainable balance of nutrient input vs output and is the future of green revolution. Thus, integrating with trees, the agricultural system can make use of these benefits from trees and can reduce the intake of inputs such as chemical fertilizers and maximize the output with minimum input cost.

References

Aerts R (1995) The advantages of being evergreen. Trends Ecol Evol 10:402–407

Aerts R (1997) Nitrogen partitioning between resorption and decomposition pathways: a trade-off between nitrogen use efficiency and litter decomposibility? Oikos 3:603–606

Alvarado A (2012) Nutrición y fertilización de *Tectona grandis* Linn. f. In: Alvarado A, Raigosa J (eds) Nutrición y fertilización forestal en regiones tropicales. AsociaciónCostarricense de las Ciencias del Suelo, San José, Costa Rica, pp 313–340

Asif M, Lone S, Baba ZA, Mir NA, Bhat H (2014) Impact of bio-fertilizers on nutrient uptake of blue pine (*Pinus wallichiana*) seedlings under nursery conditions. Ecol Environ Conserv 20 (3):1081–1088

Berendse F, Aerts R (1987) Nitrogen-use-efficiency: a biologically meaningful definition. Funct Ecol 1:293–296

Boerner REJ (1984) Foliar nutrient dynamics and nutrient use efficiency of four deciduous tree species in relation to site fertility. J Appl Ecol 21:1029–1040

Brewbaker JL (1987) Significant nitrogen fixing trees in agroforestry systems. In: Agroforestry: realities, possibilities and potentials. Martinus Nijhoff Publishers, pp 31–45

Coyle DR, Hart ER, McMillin JD, Rule LC, Hall RB (2008) Effects of repeated cottonwood leaf beetle defoliation on Populus growth and economic value over an 8-year harvest rotation. For Ecol Manag 255(8–9):3365–3373

Das DK, Chaturvedi OP (2005) Structure and function of Populus deltoides agroforestry systems in eastern India: 2. Nutrient dynamics. Agrofor Syst 65(3):223–230

DeLucia EH, Schlesinger WH (1995) Photosynthetic rates and nutrient-use efficiency among evergreen and deciduous shrubs in Okefenokee swamp. Int J Plant Sci 156:19–28

Dommergues YR (1987) The role of biological nitrogen fixation in agroforestry. Agroforestry, 245p

Fahad S, Chavan SB, Chichaghare AR, Uthappa AR, Kumar M, Kakade V, Pradhan A, Jinger D, Rawale G, Yadav DK, Kumar V, Farooq TH, Ali B, Sawant AV, Saud S, Chen S, Poczai P (2022) Agroforestry systems for soil health improvement and maintenance. Sustainability 14(22):14877. https://doi.org/10.3390/su142214877

Field CB, Mooney HA (1986) The photosynthesis–nitrogen relationship in wild plants. In: Givnish TJ (ed) On the economy of plant form and function. Cambridge University Press, New York, pp 25–55

Finér L, Laine J (1998) Root dynamics at drained peatland sites of different fertility in southern Finland. Plant Soil 201(1):27–36

Galiana A, Bouillet JP, Ganry F (2004) The importance of biological nitrogen fixation by trees in agroforestry. Oxford & IBH Publishing Co. Pvt. Lrd., pp 185–199

Gaul D, Hertel D, Borken W, Matzner E, Leuschner C (2008) Effects of experimental drought on the fine root system of mature Norway spruce. For Ecol Manage 256(5):1151–1159

Hiremath AJ (2000) Photosynthetic nutrient-use efficiency in three fast-growing tropical trees with differing leaf longevities. Tree Physiol 20(14):937–944

Hirose T (1975) Relations between turnover rate, resource utility, and structure of some plant populations: a study in the matter budgets. J Facul Sci, Univer Tokyo, Sect III: Bot 11:355–407

Hirose T (2011) Nitrogen use efficiency revisited. Oecologia 166(4):863–867

Hossain M, Siddique MRH, Rahman M, Hossain M, Hasan M (2011) Nutrient dynamics associated with leaf litter decomposition of three agroforestry tree species (*Azadirachta indica*, *Dalbergia sissoo*, and *Melia azedarach*) of Bangladesh. J For Res 22(4):577–582

ICRAF (1996) Definition of agroforestry revisited. Agrofor Today 8:5–5

Jha KK (2014) Temporal patterns of storage and flux of N and P in young teak plantations of tropical moist deciduous forest. India J For Res 25(1):75–86

Jordan CF, Montagnini F, Gassman P (2003) Organic farming and agroforestry: Alleycropping for mulch production for organic farms of southeastern United States. Agrofor Syst 61:79–90

Kätterer T, Fabião A, Madeira M, Ribeiro C, Steen E (1995) Fine-root dynamics, soil moisture and soil carbon content in a *Eucalyptus globulus* plantation under different irrigation and fertilisation regimes. For Ecol Manage 74(1–3):1–12

Kumar BM, Thomas J, Fisher RF (2001) *Ailanthus triphysa* at different density and fertiliser levels in Kerala, India: tree growth, light transmittance and understorey ginger yield. Agrofor syst 52(2):133–144

Kumar BM (2011) Quarter century of agroforestry research in Kerala: an overview. J Tropl Agric 49:1–18

Lakhiar IA, Jianmin G, Syed TN, Chandio FA, Buttar NA, Qureshi WA (2018) Monitoring and control systems in agriculture using intelligent sensor techniques: a review of the aeroponic system. J Sensors 2018(1):18

Li J, Liu Y (2010) Water and nitrate distributions as affected by layered textural soil and buried dripline depth under subsurface drip fertigation. Irrig Sci 29:469–478

Lorena CA, Noé VD, Victoria CM, Beatriz BM, Leticia SC, Julia MM (2005) Soil nitrogen in relation to quality and decomposability of plant litter in the Patagonian Monte, Argentina. Plant Ecol 181(1):139–151

Manschadi AM, Kaul HP, Vollmann J, Eitzinger J, Wenzel W (2014) Developing phosphorus-efficient crop varieties—an interdisciplinary research framework. Field Crops Res 162:87–98

Miller HG, Cooper JM, Miller JD (1976) Effect of nitrogen supply on nutrients in litterfall and crown leaching in a stand of Corsican pine. J Appl Ecol 13:233–248

Nadelhoffer KJ (2000) The potential effects of nitrogen deposition on fine-root production in forest ecosystems. New Phytol 147(1):131–139

Oliet JA, Tejada M, Salifu KF, Collazos A, Jacobs DF (2009) Performance and nutrient dynamics of holm oak (*Quercus ilex* L.) seedlings in relation to nursery nutrient loading and post-transplant fertility. Eur J For Res 128(3):253–263

Panhwar QA, Ali A, Naher UA, Memon MY (2019) Fertilizer management strategies for enhancing nutrient use efficiency and sustainable wheat production. In: Organic farming. Woodhead Publishing, pp 17–39
Peter I, Lehmann J (2000) Pruning effects on root distribution and nutrient dynamics in an acacia hedgerow planting in northern Kenya. Agrofor Syst 50(1):59–75
Pregitzer KS (2002) Fine roots of trees: a new perspective. New Phytol 154(2):267–270
Pukkala T (2015) Optimizing continuous cover management of boreal forest when timber prices and tree growth are stochastic. Forest Ecosyst 2(6):1–13
Revathi R, Mohan V, Jha MN (2013) Integrated nutrient management on the growth enhancement of *Dalbergia sissoo* Roxb. seedlings. J Acad Ind Res 1(9):550–557
Rosenstock TS, Tully KL, Arias-Navarro C, Neufeldt H, Butterbach-Bahl K, Verchot LV (2014) Agroforestry with N2-fixing trees: sustainable development's friend or foe? Curr Opin Environ Sustain 6:15–21
Saarsalmi A, Mälkönen E (2001) Forest fertilization research in Finland: a literature review. Scand J Forest Res 16:514–535
Santana RC, Barros NF, Novais RF, Leite HG, Comerford NB (2008) Alocação de nutrientesemplantios de eucalipto no Brasil. R Bras Ci Solo 32:1369–1377
Sharma E (1993) Nutrient dynamics in Himalayan alder plantations. Ann Bot 72(4):329–336
Sharma G, Sharma R, Sharma E, Singh KK (2002) Performance of an age series of Alnus–cardamom plantations in the Sikkim Himalaya: nutrient dynamics. Ann Bot 89(3):273–282
Shaver GR, Melillo JM (1984) Nutrient budgets of marsh plants: efficiency concepts and relation to availability. Ecology 65:1491–1510
Shirazi SM, Yusop Z, Zardari NH, Ismail Z (2014) Effect of irrigation regimes and nitrogen levels on the growth and yield of wheat. Adv Agr 2014:250874. https://doi.org/10.1155/2014/250874
Singh KP, Srivastava SK (1985) Seasonal variations in the spatial distribution of root tips in teak (*Tectona grandis* Linn. F.) plantations in the Varanasi Forest division, India. Plant Soil 84(1): 93–104
Tapia-Coral SC, Luizão FJ, Wandelli E, Fernandes EC (2005) Carbon and nutrient stocks in the litter layer of agroforestry systems in Central Amazonia, Brazil. Agrofor Syst 65:33–42
Vitousek P (1982) Nutrient cycling and nutrient use efficiency. Am Nat 119(4):553–572
Wang Y, Xi BY, Bloomberg M, Moltchanova E, Li GD, Jia LM (2015) Response of diameter growth, biomass allocation and N uptake to N fertigation in a triploid *Populus tomentosa* plantation in the North China plain: ontogenetic shift does not exclude plasticity. Eur J Forest Res 134:889–898. https://doi.org/10.1007/s10342-015-0897-8
Xu G, Fan X, Miller AJ (2012) Plant nitrogen assimilation and use efficiency. Ann Rev Plant Biol 63:153–182
Zeng S, Jacobs DF, Sloan JL, Xue L, Li Y, Chu S (2013) Split fertilizer application affects growth, biomass allocation, and fertilizer uptake efficiency of hybrid eucalyptus. New For 44(5): 703–718
Zhu X, Chen C, Wu J, Yang J, Zhang W, Zou X, Jiang X (2019) Can intercrops improve soil water infiltrability and preferential flow in rubber-based agroforestry system? Soil Tillage Res 191: 327–339

Part II
Addressing Climate Change and Eco-System Services Through Agro-Forestry System

Chapter 7
Benefactions of Agroforestry to Ecosystem Services

B. C. Sarkar, K. A. Manohar, G. Shukla, S. Maitra, A. Dabral, M. Sairam, and S. Chakravarty

Abstract The human population is benefited from Agroforestry systems (AFs) and practices since man settled down and learned the cultivation of crops. However, the recognition for ecosystem services provided by AFs is perceived very recently only. Research evidence envisages as a land-use system agroforestry supports ecosystem services like carbon sequestration, biodiversity, food security, microclimate modification, ground water recharge, etc. Agroforestry (AF) is perceived as a sustainable system than monoculture systems mainly duc to the multiple ecosystem services it provides. The ecosystem services of these land uses are affected by complex tree-crop-animal-environment interactions both above ground and below ground. Deforestation and deterioration of traditional land use systems has caused reduction in area of agroforestry systems and practices. Quantification of ecosystem services provided by these agroforests will help the policy makers to conserve and widen the practices of agroforestry. Adaptation of agroforestry practices has the potential to mitigate the ill effects of climate change as well as it is a way forward to sustainable livelihoods.

Keywords Multi land use system · Biodiversity conservation · Sustainable development goals · Provisional services · Livelihood

B. C. Sarkar
Department of Forestry, North Eastern Hill University, Tura campus, Tura, Meghalaya, India

Faculty of Agricultural Sciences, GLA University, Mathura, 17 KM stone, NH-2 Mathura Delhi Road, Uttar Pradesh, India

K. A. Manohar (✉) · S. Maitra · M. Sairam · S. Chakravarty
MS Swaminathan School of Agriculture, Centurion University of Technology and Management, Paralakhemundi, Odisha, India
e-mail: abha.manohar@cutm.ac.in

G. Shukla · A. Dabral
Department of Forestry, North Eastern Hill University, Tura campus, Tura, Meghalaya, India

H. S. Jatav et al. (eds.), *Agroforestry to Combat Global Challenges*, Sustainable Development and Biodiversity 36, https://doi.org/10.1007/978-981-99-7282-1_7

7.1 Introduction

The importance of ecosystem services in preserving both human well-being and the health of natural ecosystems has been recognized since the publication of the Millennium Ecosystem Assessment (MEA) report (MEA 2005; Kamiyama et al. 2016; Rana et al. 2018). Recent studies have also demonstrated the significance of agroforestry systems (AFs) in delivering ecosystem services, such as reducing the impact of climatic aberrations and improving the soil quality (Jose 2009; Tscharntke et al. 2011; Rathore et al. 2022). Agroforestry (AF) is the land-use systems and methods in which perennial woody plants are taken into the consideration as crops on a piece of land other annual crops and animals are chosen spatially and/or temporally. The ecological and commercial interactions exist among several components considered in AFs (Lundgren and Raintree 1982; Murthy et al. 2016). The multifunctional approach in agroforestry, the deliberate combination of trees and shrubs with crops or livestock, provides ecosystem services. The Ecosystem Service is broadly categorized into four services, that is, provisioning, regulating, culturing, and supporting or protecting services (Fig. 7.1). The provision of tangible and intangible returns to human society such as the livelihood improvement, materials production for an improved livelihood, health, as well as social relationships (Jose 2009; Akinnifesi et al. 2010; Garrity et al. 2010; Mbow et al. 2014; Kamiyama et al. 2016; Moreno et al. 2018; Sida et al. 2018; Thierfelder et al. 2018; Haile et al. 2019;

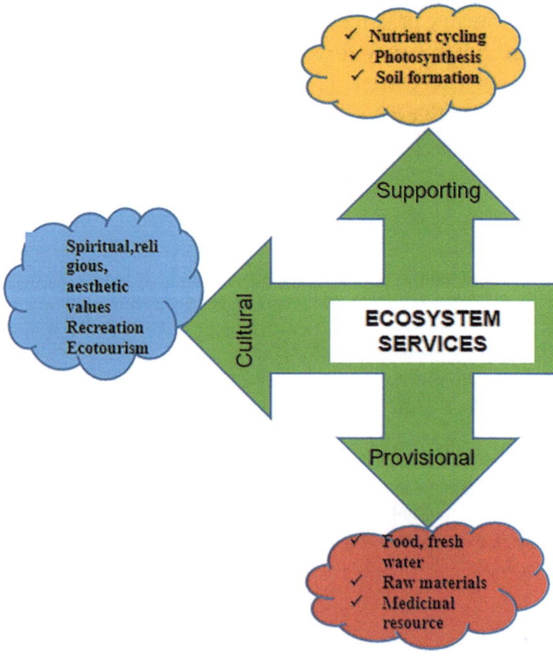

Fig. 7.1 General illustration of the types of ecosystem services (MEA 2005)

Kay et al. 2019; Pradhan et al. 2020). It has been widely noted that agroforestry provides ecosystem benefits to agriculture and related systems as well as at the micro-level of a household. For cultivars that might be susceptible to biotic and abiotic stress in fields, these are protections and hospitable environments; agro-biodiversity conservation; a habitat for small and large animals and a means of promoting plant population gene transfer both inside and outside the garden; provisional, regulating, or supporting services in the form of food, clothing, shelter, materials, and well-being; pollination, soil amelioration, soil conservation, practical nutrient cycling, soil fertility improvement, soil carbon build-up, water retention, air purification, reducing CO_2 emissions, microclimate modification, temperature regulation, providing livelihood opportunities as well as social-cultural preservation, aesthetic or recreational needs, and empowerment or the social position of women (Junqueira et al. 2016; Chakravarty et al. 2017; Vibhuti et al. 2018; George and Christopher 2020; Avilez-López et al. 2020).

7.2 Ecosystem Services in Agroforestry

7.2.1 Provisioning of Agroforestry

The provisioning services include the things that serve humans directly or immediately, such as food, fire, wood, and natural medications, etc. (Table 7.1). The AFs provide tangible benefits to the farmer (land manager), frequently by balancing the production of marketable items with the household's subsistence requirements, mitigating the effects of climate change, and safeguarding soil and water resources (van Noordwijk 2021). Thus, productivity, sustainability, and adaptability are one of the important attributes of the agroforestry system (Nair and Sreedharan 1986). Agriculture is the most prevalent anthropogenic land use, occupying around 38% of the planet's land surface (Foley et al. 2011). However, a key difficulty the globe faces is ensuring that the millions of family's living in poverty have access to enough food and nutrition (Adekunle 2013). Therefore, AFs offer a significant amount of opportunities to discuss the present problems with food, nutrition, energy, employment, and environmental security. Home garden systems ensure that small pieces of land close to the house are used to meet daily demands for staple foods, fibre, fodder, medicine, timber, and fuel (Fig. 7.2). According to numerous studies, an agroforestry system is essential for maintaining sustainability and providing ecosystem services (Jose 2009; Montagnini et al. 2004; Vaast and Somarriba 2014).

The oldest land use practice is home gardening, followed by shifting farming, which has been practiced by many human groups around the world (Kumar and Nair 2004; Galhena et al. 2013; Cerda et al. 2022). It developed over many generations as agriculture gradually became more intensive in return under rising human needs resulting in decline in fertile areas (Nair 2001; Kumar and Nair 2004). Due of the environmental, social, and economic advantages that agroecosystems and traditional agricultural systems, including home gardens, bring, there is growing interest in the

Table 7.1 Provisioning services on home gardens from different regions

Country (region)	Number of Homegardens (HGs) surveyed	Total number of species	Predominant plant species category	Source
Austria (Osttirol)	196	94	Ornamental, spices, fruits	Vogl-Lukasser and Vogl (2004)
Bangladesh (Northern part)	80	62	Edible, medicine, fuel and timber	Roy et al. (2013)
China (Beijing municipality)	104	278	Edible, ornamental, medicine	Clarke et al. (2014)
Ecuador (Amazonian)	138	484	Edible, medicine, ornamental	Caballero-Serrano et al. 2016
Ethiopia (Janithenan District)	48	69	Edible, medicine, ornamental	Mekonnenet al. (2014)
Iran (Bash district)	192	97	Edible, ornamental, medicine	Schadegan et al. (2013)
India (North Bengal)	100	142	Edible, fuel and timber, ornamental	Subba et al. (2015)
Sri Lanka (Western parts)	106	289	Ornamental, food, medicine	Kumari et al. (2009)
Mexico (Tehuacan-Cuicatlan Valley)	30	281	Ornamental, edible, shade, medicinal	Blanckaert et al. (2004)
India (North-Eastern part)	50	122	Edible, medicine, timber, ornamental	Das and Das (2005)

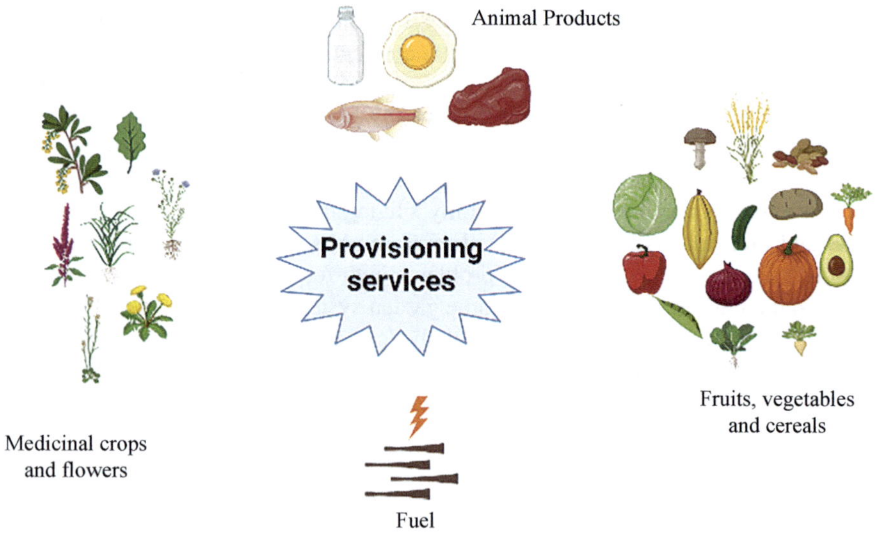

Fig. 7.2 Provisioning services from agroforestry

study of ecosystem services (Calvet-Mir et al. 2012; Caballero-Serrano et al. 2016). Studies in many locations' home gardens have noted remarkable species diversity and their benefits. Caballero-Serrano et al. (2016) reported a total of 142 species as food, fuel 7 species, and fibre 6 species, produced from the home garden of 11 localities of the Sangayparish of the Amazon region of Ecuador. Kala (2010) reported 62% were edible species among a total of 47 species in home gardens of tribal communities of Pachmarhi Biosphere reserve, Madhya Pradesh, India. Growing multiple species simultaneously in home gardens addresses more than only providing resources for food and mechanisms; it also addresses related resilience tactics by minimizing risk and increasing toughness, as is typically seen in single crop farming (Buchmann 2009). In another study, out of 281 species recorded in AFs within 12 use groups 115 were ornamental plants, 92 edible, and 50 medicinal plants in Tehuacan valley, Mexico (Larios et al. 2013). Due to the owner's perception that the provisions (food, medicines) provided by home gardens are most important for his or her well-being and livelihood, even though it provides many services other than provisional services, the invisible nature of provisioning services with home gardens may undermine its regulatory services (Blanckaert et al. 2004; Caballero-Serrano et al. 2016).

7.2.2 *Regulatory Services*

The advantages derived from the regulation of ecological processes are a consequence of regulating ecosystem services (Table 7.2). Because of the several functions that trees serve, agroforestry treatments are the finest management strategies for delivering a range of regulating ecosystem services. There are seven ecosystem services that need to be regulated: C sequestration, improving soil fertility, check in soil erosion, control of flood, wind, pests, and supporting the pollination, etc., as mentioned by MEA 2005 (Fig. 7.3).

7.2.2.1 Carbon Sequestration

According to current estimates, human activities are to blame for the annual emission of 7.9 billion tons of carbon into atmosphere (IPCC 2001). Climate change brought on by increasing CO_2 levels in the atmosphere is anticipated to have a severe impact on fresh water availability, food production, distribution, and the seasonal transmission of infectious vector-borne diseases (UNEP 2002). One strategy to slow down climate change is to reduce anthropogenic carbon emissions; another is to preserve or improve ecosystems' ability to absorb carbon. By repairing degraded landscapes and soils through better management or changing the land's use, it is possible to create an effective terrestrial ecosystem for sequestering carbon dioxide, the main greenhouse gas causing global warming. Switching from conventionally managed pasture lands to improved pasture lands or from degraded crop or

Table 7.2 Role of agroforestry in regulating ecosystems processes (Leeuw et al. 2014)

Regulating services	Livelihood assets		
	Human	Natural	Financial
Microclimate	Improving productivity of drylands, implication on health and nutrition	Providing shade, reducing wind, rain velocity, and momentum, reducing body energy loss from livestock	Impact on income from other services
Air quality	Benefiting health by reducing dust	Lower dust movement and offer soil cover	Indirect impact on income from other services
Macroclimate	Improve productivity of drylands hence implication on health and nutrition	Carbon sequestration	Indirect impact on income from other services
Flood + groundwater Control	Improve productivity of drylands hence implication on health and nutrition	Reduction in runoff amount and speed, improved soil moisture and ground water recharge	Indirect impact on income from other services
Pest and disease control	Health of human and livestock	–	Indirect impact on income from other services

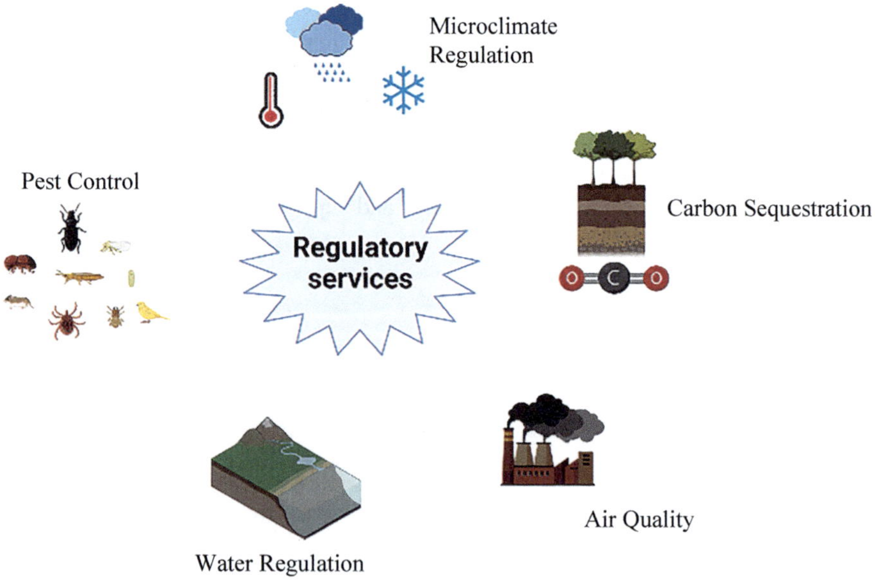

Fig. 7.3 Regulatory services from Agroforestry (Created with BioRender.com)

grasslands to tree plantations and agroforestry systems are particularly best at capturing carbon while still allowing products to be harvested that will benefit the land's users (van Noordwijk et al. 2014).

Carbon is removed from the atmosphere by physical or biological processes and stored in carbon sinks (such as oceans, vegetation, or soils) (Jose 2009). Agroforestry not only provide provision services, but they also supply, safeguard, and enhance conditions for plant kinds that may be vulnerable to biotic and abiotic stress in fields. Due to the potential for C storage by plants and soil, and its applicability on farm lands, and reforestation or afforestation, AFs are significant as a strategy for sequestering carbon. Therefore, complex or traditional agroforestry systems, that is, home gardens, were reported more efficient in carbon sequestration than simpler systems or sole stand. In addition, agroforestry systems can aid carbon sequestration of natural forests by offsetting potential deforestation, for example, five to twenty hectares of deforestation can be offset by one hectare of sustainable agroforestry (Dixon 1995; Nair and Nair 2003; Kumar 2006a, 2006b) and also increase the soil carbon storage (Kumar and Nair 2004; Palm et al. 2004; Nair et al. 2010; Zomer et al. 2016).

The home gardens may be particular in all mechanisms registering their efficiencies, that is, C sequestration in soil and biomass, reduction of fossil fuel consumption by encouraging the output of wood fuel, aid in the storage of carbon stocks in existing forests by reducing extra burden on natural forest, and safeguard better synergy with the Convention on Biological Diversity (CBD). The efficiency of home gardens as carbon sinks depends on factors such as land area, natural site quality, species selection, system management, and the owner's preferences for his garden. Alternatively, a home garden's ability to sequester carbon is a function of the garden's structural design and arrangement as modified by environmental, social, and economic factors (Haile et al. 2008; Newaj and Dhyani 2008; Henry et al. 2009; Saha et al. 2009). Home gardens in Kerala, India, and other nations have been claimed to be stable from an ecological and socioeconomic standpoint and to store more carbon than natural forests (Peyre et al. 2006; Saha et al. 2011; Kassa et al. 2022). Above ground carbon stock in home garden of Kerala estimated up to 16–36 Mg ha^{-1} with smaller gardens reportedly storing more than the larger ones (Kumar and Takeuchi 2009; Saha et al. 2009, 2010). The traditional agroforestry systems around Srinagar, Garhwal Himalayas, were dominated by tree species *Grewia oppositifolia* followed by *Toona ciliata* and carbon stock estimated in these was 19.85–57.45 Mg ha^{-1} with an average of 32.56 Mg ha^{-1} (Kumar et al. 2012). The soil organic carbon estimated in these traditional agroforestry systems of Srinagar was 12.78–146.88 Mg ha^{-1} with an average of 56.74 Mg ha^{-1}. The amount of carbon sequestered by Kashmir Himalayan home gardens with various species composition (salix, poplar, beans, kale, and apple) varied greatly (104.86 Mg ha^{-1}) in comparison to home gardens with poplar, kale, beans, and apple (44.53 Mg ha^{-1}) (Dar et al. 2019). From South-eastern Ethiopia coffee agroforestry, home garden and crop field of Dallo Mena districts reported 426.54, 266.61, 185.26, and 97.56 Mg ha^{-1}, respectively (Mengistu and Asfaw 2019). Estimated carbon storage of 145 Mg ha^{-1} in Panamanian traditional agroforestry systems was lesser than managed forest (335 Mg ha^{-1}) but higher than pastures (46 Mg ha^{-1}) (Kirby and Potvin 2007). In Ethiopia, home gardens were estimated with a higher amount of ecosystem carbon stock (100.4 Mg ha^{-1} against 72.90 Mg C ha^{-1}) estimated in woodlot agroforestry systems (Semere 2019).

In addition to vegetation, the presence of tree species, litter volume, and quality as well as age tree species influenced the soil organic carbon status in AFs; however, the SOC varies with location, geographic position, land use and management systems. Moreover, the studies are meagre to evaluate the benefits of AFs spatially and temporally. Many large cardamom-based traditional AFs in Sikkim Himalayas were reported rich in SOC, however, it remained lesser than natural forest cardamom agroforestry (Sharma et al. 2007, 2008, 2016). Total tree biomass of studied 45 home gardens in Cooch Behar district of West Bengal, India, quantified was 7482.67 Mg with potential to offset 507.94 Mg CO_2. Moreover, these North Bengal home gardens of West Bengal produced 110.86 Mg fuel wood 39.15 Mg fodder per year avoiding deforestation with net gain of 247.06 CO_{2e} annually while the monetary value of carbon offset was US$ 1270 with average carbon credit per household of US$ 28.22 annually (Pala et al. 2019).

7.2.2.2 Water Regulation

Groundwater resources are crucial for supplying over half of the world's drinking water, making them essential for human use (Smith et al. 2017). Water scarcity is a common issue in the world and more than 40% of the people across the world experienced various forms of water shortage (Nordbotten and Celia 2006). The availability of high-quality groundwater is a major concern for the estimated 2.5 billion people who rely mostly on groundwater to meet their daily water needs (WWAP 2015). Urbanization, unsustainable farming methods, and climate change are all projected to have a negative influence on groundwater recharge and water quality, increasing the vulnerability of entire water supply systems (Jiménez Cisneros et al. 2014). Of these, nearly 1.2 billion people experienced physical water scarcity.

Forests have a significant impact on atmospheric moisture fluxes and rainfall patterns over land. Water vapour is released to the atmosphere by the land and ocean surfaces of Earth. Evapotranspiration (ET), which is the process of water evaporating from soil and plant surfaces and being transpired by plants, is facilitated by forests and other vegetation on continental surfaces. Winds move the resulting atmospheric moisture over the planet's continents and oceans.

AFs with the mixed stands of tree species and pasture or agricultural crops were thus added to the list of climate smart technologies. The annual plants in agroforestry systems tend to have shallow roots, but perennial plants, such as trees, have deep root systems that reach deep into the soil, supporting the safety-net concept (van Noordwijk and van de Geijn 1996; van Noordwijk et al. 2015; Bayala and Prieto 2020). Therefore, by adding tree species through supportive root distributions, water sharing mechanisms like hydraulic redistribution and bio irrigation, and intercropping, limiting inputs for crop productivity, namely, soil moisture and essential plant nutrient, may be used more effectively (Maitra et al. 2021). Additionally, widespread mycorrhizal networks and/or decreased runoff enhanced water penetration into the soil (Mao et al. 2012; Prieto et al. 2012; Bayala and Wallace 2015; Brooker et al. 2015).

7.2.2.3 Pest and Pollution Regulation

An essential ecosystem service that contributes to agricultural yield, stabilized production, and the preservation of wild plant populations is insect pollination (Varah et al. 2020). For most flowering plants to reproduce successfully, pollination by animals is necessary. About 22,0000 out of 24,0000 eastimated plant species require animals such as bees, beetles, butterflies, beneficial insects, hummingbirds to complete this crucial duty, have been recorded. This covers over 70% of the crops and their wild relatives used to feed the world's population. These natural pollination services ensure higher productivity of crops on cropping lands, home gardens, rangelands, and forests (Buchmann and Nabhan 1996). The pollination service has obtained specific attention as it ensures more than one-third of global crop output for cross-pollinated crops as a resultant of insect pollination to a certain extent (Klein et al. 2007).

Traditional home gardens are home to many insects, which are crucial for pollination, which helps fruits ripen. In the context, various researchers have claimed that high species variety and close plant connection also lower the danger of pest and diseases in home garden in contrast to monocropping. However, the pest incident has not been recorded or researched in home garden. Regarding the pest control services offered by home gardens, there is no scientific data or research. The variety of natural pollinators for wild plants and domesticated crops is dwindling and over 60 genera of pollinating species are currently categorized under the list of threatened, endangered, or extinct species (Buchmann and Nabhan 1996; Hossain et al. 2021).

7.3 Cultural Services

The Millennium Environment Assessment (2005) narrated cultural ecosystem services vaguely as "non-material benefits people obtain from ecosystems through spiritual enrichment, cognitive development, reflection, recreation, and aesthetic experiences". However, this is controversial or being criticized because it is very difficult to distinguish terms, services, advantages, and values (Milcu et al. 2013). Cultural ecosystem services (CES) have a special role mainly due to their intangibility, emotional and mental benefits, and partly non-consumptive character (Milcu et al. 2013). CES must be measured with non-monetary methods in order to fix their poor quantification and integrate them into Ecosystem services (ES) frameworks (Martín-López et al. 2012; Milcu et al. 2013; Szücs et al. 2015). Cultural values of a

Table 7.3 Prominent home garden cultural services of North Bengal, West Bengal, India, based on field observation (SERB project, Department of Forestry, Uttar Banga Krishi Viswavidyalaya, West Bengal, India)

Cultural value	Home garden of North Bengal
Religious and spiritual significance	Temple, religious plants, animals, etc.
Aesthetic value	Flower garden, pond, greenery, etc.
Health value	Medicinal plants, home grown food, daily activities on garden, etc.

landscape or an ecosystem are important and worth protecting because they are unique, irreplaceable, and have an increasing importance in economic societies (Szücs et al. 2015). Cultural ecosystem services play a special role in their connection and contribution to human well-being. They are on the one hand generally less directly linked to human well-being than provisioning or regulating services. On the other hand, their potential for replacement is low (Plieninger et al. 2013). While degraded provision or regulating services can be replaced by other means, for example, bottled water as a replacement for a contaminated spring, cultural ecosystem services have less possibilities for a substitution (Bieling et al. 2014; Plieninger et al. 2013). Agrodiversity includes cultural memory and many modes of information transmission, such as oral histories, rituals, exchanging first-hand accounts, the arts, and so forth; these are just as crucial as cataloguing species and assembling genetic collections. Additionally, it is well known that placing a focus on replacing local technology and expertise rather than enhancing it, such as in regard to intensifying agricultural production, frequently results in various types of land degradation and a reduction in food security (MEA 2005). Home gardens contribute significantly to social life in neighbourhoods. Every house garden has a level area that is shaded by big trees where kids may play and seniors can socialize when they are free. The gardens are a significant symbol of social status as well (Ahmad et al. 1980). Typically, a live fence or enclosure of shrubs or small trees surrounds traditional home gardens, allowing for simple access to bring water, gather medicinal herbs, and pass through. As far as we know, numerous animal and plant species in India can be found in home gardens and have cultural significance. In diverse AFs, as well as in personal gardens, the species choice, plant geometry and stand, and the level of maintenance differ significantly based on the agroclimatic and soil conditions, market potential of the chosen species, and cultural background of the owner (Saikia and Khan 2009). Home gardens of Darjeeling Himalaya, West Bengal, India, have been presented in Table 7.3 and depicted in Fig. 7.4.

Fig. 7.4 Home gardens of Darjeeling Himalaya, West Bengal, India at different altitudes

7.4 Supporting Services

Agroforestry has the potential to improve livelihood as it provides farmers with a variety of options and opportunities to increase farm output and incomes as well as provides ecosystems with productive and protective forest function (biodiversity, a healthy ecosystem, protection of soil and water resources, terrestrial carbon storage, etc.) while preserving the environment (Sharma et al. 2007). The addition of trees to agricultural settings has the potential to benefit the soil in several ways, both for crop growth and as a habitat for soil creatures. Trees have a range of effects on the soil environment: their leaves deflect rain, release water that their roots have collected from the ground, and shade the understory and soil; their dead or pruned branches cover the soil and nourish it with nutrients. These processes change the temperature, moisture content, erosion susceptibility, nutrient content, and soil biota of the soil (Barrious et al. 2012). Litter is a key element of nutrient cycling in agroforestry systems because it includes a significant quantity of the nutrients required for plant growth (Zheng 2006). Agroforestry systems' floral diversity fosters favourable circumstances for soil microorganisms, which are crucial to the decomposition of litter and the release of nutrients (Kumar 2011). The nitrogen fixing trees most widely used in AFs include *Acacia* spp., *Albizia* spp., *Calliandra calothyrsus*, *Faidherbia albida*, *Flemingia* spp., *Erythrina* spp., *Inga* spp., *Leucaeana* spp., *Gliricidia* spp., and *Sesbania* spp. (Table 7.4), which develop symbiotic associations with N_2-fixing bacteria (Gold 2020; Sileshi et al. 2020). These are specified by their wide adaptation qualities under severe environmental stresses, namely, drought, salinity, erosion, low fertility, and other hostile situations (Ribeiro-Barros et al. 2018). Therefore, the adoption of N-fixing trees to endorse restoration of vegetation

Table 7.4 Important N_2 fixing species (Nair 1993; Silva and Uchida 2000)

Common name	Scientific name	Family	Nitrogen fixed (kg N ha^{-1} $year^{-1}$)
Black wattle	*Acacia mearnsii*	Mimosoideae	200
Beed wood	*Casuarinae quisetifolia*	Casuarinaceae	60–110
Erythrina	*Erythrina poeppigiana*	Palilionoideae	60
Subabool	*Leucaena leucocephala*	Mimosoideae	100–500
Stylo	*Stylosanthes* spp.	Fabaceae	20–263
Horse bean	*Viciafaba*	Fabaceae	45–552
Pigeon pea	*Cajanuscajan*	Fabaceae	68–88
Cowpea	*Vigna sinensis*	Fabaceae	73–354
Black gram	*Vigna mungo*	Fabaceae	63–342
Soybean	*Glycine max*	Fabaceae	60–168
Chickpea	*Cicer arietinum*	Fabaceae	103
Lentil	*Lens esculenta*	Fabaceae	88–114
Groundnut	*Archis hypogaea*	Fabaceae	72–124
Pea	*Pisum sativum*	Fabaceae	55–77
Bean	*Phaseolus vulgaris*	Fabaceae	40–70
Alfalfa	*Medicago sativa*	Fabaceae	228–290
Clover	*Trifolium* spp.	Fabaceae	128–207
Sunn hemp	*Crotalaria juncea*	Fabaceae	199–223
Wild tantan	*Desmanthus virgatus*	Fabaceae	196–226
True indigo	*Indigofera tinctoria*	Fabaceae	79
New dhaincha	*Sesbania rostrata*	Fabaceae	70–458
Dhaincha	*Sesbania sesban*	Fabaceae	7–18
Gliricidia	*Gliricidia sepium*	Mimosoideae	13

cover is a sensible approach for ameliorating soil stabilization and fertility. Agroforestry practices involving tree legumes represents a sustainable cultivation practice for farming communities (Jena et al. 2022).

7.5 Conclusion

Agroforestry methods have been scientifically developed in the current situation to increase the quantity and quality of the same plot of land, yield more, and improve sustainability. The agroforestry systems can help billions of people throughout the world to fulfil their demands, making agroforestry systems the second-largest carbon sink behind forests. In order to maintain ecosystem resilience and integrity while meeting production demands, agroforestry is a viable alternative. The foundation of a sustainable land use that is both socioeconomically and environmentally safe is the spatial and temporal combination of agroforestry components (trees, crops, and animals). The use of agroforestry techniques is a successful way for ecological restoration and climate-smart agriculture. A comparable approach would likewise guarantee the well-being of society, the economy, and the environment.

Acknowledgements Authors acknowledge the Science Engineering Research Board, Government of India, for financial assistance provided through the project entitled Assessment of Ecosystem Services in Home garden Agroforestry systems in Sikkim and Sub-Himalayan region of West Bengal (EMR/2016/ 001369).

References

Adekunle OO (2013) The role of home gardens in household food security in eastern cape: a case study of three villages in Nkonkobe municipality. J Agric Sci 5(10):67–76

Ahmad H, Martadihardja, Suharto (1980) Social and cultural aspects of homegarden. In: Furtado JI (ed) Tropical Ecology and Development. The International Society of Tropical Ecology, Kuala Lumpur

Akinnifesi FK, Ajayi OC, Sileshi G, Chirwa PW, Chianu J (2010) Fertiliser tree for sustainable food security in the maize- based production systems of East and Sothern Africa. A review. Agron Sustain Dev 30:615–629

Avilez-López T, van der Wal H, Aldasoro-Maya E, Rodríguez-Robles U (2020) Homegardens' agrobiodiversity and owner's knowledge of their ecological, economic and socio-cultural multifunctionality: a case study in the lowlands of Tabasco, México. J Ethnobiol Ethnomed 16:42. https://doi.org/10.1186/s13002-020-00392-2

Barrious E, Sileshi GW, Shephard KD, Sinclair F (2012) Agroforestry and soil health: linking trees, soil biota, and ecosystem services. In: Soil ecology and ecosystem services, 1st edn. Chapter: 5.2. Oxford University Press

Bayala J, Prieto I (2020) Water acquisition, sharing and redistribution by roots: applications to agroforestry systems. Plant Soil 453:17–28. https://doi.org/10.1007/s11104-019-04173-z

Bayala J, Wallace JW (2015) The water balance of mixed tree-crop systems. In: Black C, Wilson J, Ong CK (eds) Tree-crop interaction: agroforestry in a changing climate. CABI, London, pp 146–190

Bieling C, Plieninger T, Pirker H, Vogl CR (2014) Linkages between landscapes and human Well-being. An empirical exploration with short interviews. Ecol Econ 105:19–30. https://doi.org/10.1016/j.ecolecon.2014.05.013

Blanckaert I, Swennen RL, Paredes-Flores M, Lopez RR, Saade RL (2004) Floristic composition, plant uses and management practices in homegardens of San Rafael Coxcatlán, Valley of Tehuacán-Cuicatlán, Mexico. J Arid Environ 57:39–62

Brooker RW, Bennett AE, Cong W-F, Daniell TJ, George TS, Hallet PD, Hawes C, Lannetta PPM, Jones HG, Karley AJ, McKenzie BM, Pakeman RJ, Paterson E, Schob C, Squire G, Watson CA, Zhang C, Zhang F (2015) Improving intercropping: a synthesis of research in agronomy, plant physiology and ecology. New Phytol 206:107–117

Buchmann C (2009) Cuban Homegardens and their role in social-ecological resilience. J Hum Ecol 37:705–721

Buchmann SL, Nabhan GP (1996) The pollination crisis: the plight of the 334 honey bee and the decline of other pollinators imperils future harvests. Science 36:22

Caballero-Serrano V, Onaindia M, Alday JG, Caballero D, Carrasco JC, McLaren B (2016) Plant diversity and ecosystem services in Amazonian homegardens of Ecuador. Agric Ecosyst Environ 225:116–125

Calvet-Mir L, Gómez-Baggethun E and Reyes-García V (2012) Beyond food production: ecosystem services provided by Homegardens. A case study in VallFosca, Catalan Pyrenees, Northeastern Spain. Ecol Econ 74: 153–160

Cerda C, Guenat S, Egerer M, Fischer LK (2022) Home food gardening: benefits and barriers during the COVID-19 pandemic in Santiago, Chile. Front Sustain Food Syst 2:6. https://doi.org/10.3389/fsufs.2022.841386

Chakravarty S, Puri A, Subba M, Pala NA, Shukla G (2017) Homegardens: drops to sustainability. In: Dagar JC, Tewari VP (eds) Agroforestry: anecdotal to modern science. Springer Nature, Singapore, pp 517–528

Clarke LW, Li L, Jenereeta GD, Yu Z (2014) Drivers of plant biodiversity and ecosystem service production in home gardens across the Beijing municipality of China. Urban Ecosyst 17:741–760

Dar MUD, Qauser KN, Massoodi TH, Mughal AH, Khan PA (2019) Biomass production and carbon stock potential under home gardens of Kashmir Himalaya. J Plant Dev Sci 11:256–271

Das T, Das AK (2005) Inventorying plant biodiversity in homegarden: a case study in Barak valley, Assam, North East India. Curr Sci 98:155–163

Dixon RK (1995) Agroforestry systems: sources or sinks of greenhouse gases? Agrofor Syst 31:99–116

Foley J, Ramankutty N, Brauman KA, Cassidy ES, Gerber JS, Johnston M, Meiller ND, Connell CO, Ray DK, West PC, Balzer C, Bennett EM, Caroenter SR, Hill J, Monfreda C, Polasky S, Rockström J, Sheehan J, Siebert S, Tilman D, Zaks DPM (2011) Solutions for a cultivated planet. Nature 478:337–342. https://doi.org/10.1038/nature10452

Galhena DH, Freed R, Maredia KM (2013) Home gardens: a promising approach to enhance household food security and wellbeing. Agric Food Secur 2:8. https://doi.org/10.1186/2048-7010-2-8

Garrity DP, Akinnifesi FK, Ajayi OC, Weldesemayat SG, Mowo JG, Kalinganire A, Larwanou M, Bayala J (2010) Evergreen agriculture: a robust approach to sustainable food security in Africa. Food Secur 2:197–214. https://doi.org/10.1007/s12571-010-0070-7

George MV, Christopher G (2020) Structure, diversity and utilization of plant species in tribal homegardens of Kerala, India. Agrofor Syst 94:297–307

Gold MA (2020) Anthromes—temperate and tropical agroforestry. In: Michael I, Goldstein, Dominick A. DellaSala (eds) Encyclopedia of the World's Biomes, pp 107–116. https://doi.org/10.1016/B978-0-12-409548-9.12492-3

Haile KK, Tirivayi N, Tesfaye W (2019) Farmers' willingness to accept payments for ecosystem services on agriculture land: the case of climate-smart agroforestry in Ethiopia. Ecosyst Serv 39: 100964

Haile SG, Nair PKR, Nair VD (2008) Carbon storage of different soil-size fractions in Florida silvopastoral systems. J Environ Qual 37:1789–1797

Henry M, Tittonell P, Manlay RJ, Bernoux M, Albrecht A, Vanlauwe B (2009) Biodiversity, carbon stocks and sequestration potential in above ground biomass in smallholder farming systems of western Kenya. Agric Ecosyst Environ 129:238–252. https://doi.org/10.3390/land10080770

Hossain A, Maitra S, Pramanick B, Bhutia KL, Ahmad Z, Moulik D, Abu Syed M, Shankar T, Adeel M, Hassan MM, Aftab T (2021) Wild relatives of plants as sources for the development of abiotic stress tolerance in plants. In: Aftab T, Roychoudhury A (eds) Plant perspectives to global climate changes. Elsevier Inc., Academic Press, pp 471–518. https://doi.org/10.1016/B978-0-323-85665-2.00011-X

IPCC (2001) Third Assessment Report- Climate Change 2001. In: The scientific basis: summary for policymakers and technical summary. Intergovernmental Panel on Climate Change, Geneva, p 98

Jena J, Maitra S, Hossain A, Pramanick B, Gitari HI, Praharaj S, Shankar T, Palai JB, Rathore A, Mandal TK, Jatav HS (2022) Role of legumes in cropping Systems for Soil Ecosystem Improvement. In: Jatav HS, Rajput VD (eds) Ecosystem services types, Management and Benefits. Nova Science Publishers Inc, New York, USA, pp 1–22. https://doi.org/10.52305/PFZA6988

Jiménez Cisneros BE, Oki T, Arnell NW, Benito W, Cogley JG, Doll P, Jiang T, Mwakalila SS (2014) Freshwater resources. In: Field CB, Barros VR, Dokken DJ, Mach KJ, Mastrandrea MD,

Bilir TE, Chatterjee M, Ebi KL, Estrada YO, Genova RC, Girma B, Kissel ES, Levy AN, MacCracken S, Mastrandrea PR, White LL (eds) Change (2014) impacts, adaptation, and vulnerability, part a: global and sectoral aspects. Contribution of working group ii to the fifth assessment report of the intergovernmental panel on climate change. Cambridge University Press, Cambridge, New York, pp 229–269

Jose S (2009) Agroforestry for ecosystem services and environment benefits: and overview. Agrofor Syst 76:1–10

Junqueira A, Souza N, Stomph T, Almekinders C, Clement C, Struik P (2016) Soil fertility gradients shape the agrobiodiversity of Amazonian homegardens. Agric Ecosyst Environ 221: 270–281

Kala CP (2010) Homegardens and Management of key Species in the Pachmarhi biosphere Reserve of India. J Biodivers 1:111–117

Kamiyama C, Hashimoto S, Kohsaka R, Saito O (2016) Noo-market food provisioning service via homegarden and communal sharing in satoyama socio-ecological production landscapes on Japan's Noto peninsula. Ecosyst Serv 17:185–196

Kassa G, Bekele T, Demissew S, Abebe T (2022) Above- and belowground biomass and biomass carbon stocks in homegarden agroforestry systems of different age groups at three sites of southern and southwestern Ethiopia. Carbon Manage 13(1):531–549. https://doi.org/10.1080/17583004.2022.2133743

Kay S, Graves A, Palma JHN, Moreno G, Roces-Dĩaz JV, Aviron S, Chouvardas D, Crous-Duran J, Ferreiro-Dominguez N, de Jalon SG, Măcicăşan V, Mosquera-Losada MR, Pantera A, Santiago-Freijanes JJ, Szerencsits E, Torralba M, Burgees PJ, Herzog F (2019) Agroforestry is paying off- economic evaluation of ecosystem services in European landscapes with and without agroforestry systems. Ecosyst Serv 36:100896

Kirby KR, Potvin C (2007) Variation in carbon storage among tree species: implications for the management of a small-scale carbon sink project. For Ecol Manag 246:208–221

Klein AM, Vaissiere BE, Cane JH, Steffan-Dewenter I, Cunningham SA, Kremen C, Tscharntke T (2007) Importance of pollinators in changing landscapes for world crops. Proc R Soc B 274: 303–313

Kumar BM (2006a) Agroforestry: the new old paradigm for Asian food security. J Trop Agric 44:1–14

Kumar BM (2006b) Carbon sequestration potential of tropical homegardens. In: Nair PKR, Kumar BM (eds) Tropical homegardens: a time-tested example of sustainable agroforestry. Springer, Netherlands, pp 185–204

Kumar BM (2011) Quarter century of agroforestry research in Kerala: an overview. J Trop Agric 49(1–2):1–18

Kumar M, Anemsh K, Sheikh MA, Raj AJ (2012) Structure and carbon stock potential in traditional agro forestry system of Garhwal Himalaya. J Agric Tech 8(7):2187–2200

Kumar BM, Nair PKR (2004) The enigma of tropical homegardens. Agrofor Syst 61:135–152

Kumar BM, Takeuchi K (2009) Agroforestry in the Western Ghats of peninsular India and the Satoyama landscapes of Japan: a comparison of two sustainable land use systems. Sustain Sci 4: 215–232

Kumari MAS, Kansuntisukmongkol K, Brockelman WY (2009) Plant diversity in Homegardens and its contribution to household economy in suburban areas in Sri Lanka. Environ Nat Resour J 7:12–30

Larios C, Casas A, Vallejo M, Moreno-Calles AI, Blancas J (2013) Plant management and biodiversity conservation in Náhuatl homegardens of the Tehuacán Valley, Mexico. J Ethnobiol Ethnomed 9:74–89

Leeuw JD, Njenga M, Wagner B, Iiyama M (2014) Treesilience: an assessment of the resilience provided by trees in the drylands of eastern Africa. World Agroforestry Centre, Nairobi, Kenya

Lundgren BO, Raintree JB (1982) Sustained agroforestry. In: Nestel B (ed) Agricultural Research for Development : potentials and challenges in Asia. ISNAR, The Hague, pp 37–49

Maitra S, Hossain A, Brestic M, Skalicky M, Ondrisik P, Gitari H, Brahmachari K, Shankar T, Bhadra P, Palai JB, Jana J, Bhattacharya U, Duvvada SK, Lalichetti S, Sairam M (2021) Intercropping–a low input agricultural strategy for food and environmental security. Agron 11:343. https://doi.org/10.3390/agronomy11020343

Mao LL, Zhang LZ, Li WQ, van der Werf W, Sun JH, Spiertz H, Li L (2012) Yield advantage and water saving in maize/pea intercrop. Field Crops Res 138:11–20

Martín-López B, Iniesta-Arandia I, García-Llorente M, Palomo I, Casado-Arzuaga I, Del Amo DG, Gómez-Baggethun E, Oteros-Rozas E, Palacios-Agundez I, Willaarts B, González JA, Santos-Martín F, Onaindia M, López-Santiago C, Montes C (2012) Uncovering ecosystem service bundles through social preferences. PLoS One 7(6):1–11. https://doi.org/10.1371/journal.pone.0038970

Mbow C, Smith P, Skole D, Duguma L, Bustamante M (2014) Achieving mitigation and adaptation to climate change through sustainable agroforestry practices in Africa. Curr Opin Environ Sustain 6:8–14. https://doi.org/10.1016/j.cosust.2013.09.002

MEA, Millennium Ecosystem Assessment (2005) Ecosystems and Human Well-Being: Our Human Planet: Summary for Decision Makers. The Millennium Ecosystem Assessment Series, vol 5. Island Press, Washington DC

Mekonnen EL, Asfaw Z, Zewudie S (2014) Plant species diversity of homegarden agroforestry in Jabithenan district, North-Wertern, Ethiopia. Int J Biodivers Conserv 6(4):301–307

Mengistu B, Asfaw Z (2019) Carbon sequestration in agroforestry practices with relation to other land uses around Dallo Mena districts of bale zone, southeastern Ethiopia. Acad Res J Agric Sci Res 7:218–226

Milcu AI, Hanspach J, Abson D, Fischer J (2013) Cultural ecosystem Services. A literature review and prospects for future research. Ecol Soc 18(3):44. https://doi.org/10.5751/ES-05790-180344

Montagnini F, Cusack D, Petit B, Kanninen M (2004) Environmental services of native tree plantations and agroforestry systems in Central America. J Sustain For 21:51–67

Moreno G, Aviron S, Berg S, Crous-Duran J, Franca A, García de Jalón S, Hartel T, Mirck J, Pantera A, Palma JHN, Paulo JA, Sanna F, Thenial C, Varga A, Viaud V, Burgess PJ (2018) Agroforestry of high nature and culture value in Europe: provision of commercial goods and other ecosystem services. Agrofor Syst 92:877–891

Murthy IK, Dutta S, Varghese V, Joshi PP, Kumar P (2016) Impact of agroforestry systems on ecological and socio-economic systems: a review. Glob J Sci Front Res: H Environ Earth Sci 16(5):15–27

Nair MA, Sreedharan C (1986) Agroforestry farming systems in the homesteads of Kerala, southern India. Agrofor Syst 4:339–363

Nair PKR (1993) An introduction to agroforestry. Kluwer Academic Publishers, Netherlands, p 499p

Nair PKR (2001) Do tropical homegardens elude science, or it is the other way around? Agrofor Syst 53:239–245

Nair PKR, Nair VD (2003) Carbon storage in north American agroforestry systems. In: Kimble J, Heath LS, Birdsey RA, Lal R (eds) The potential of U.S. Forest soils to sequester carbon and mitigate the greenhouse effect. CRC Press, Boca Raton, FL, USA, pp 333–346

Nair PKR, Nair VD, Kumar BM, Showalter JM (2010) Carbon sequestration in agroforestry systems. Adv Agron 108:237–307

Newaj R, Dhyani SK (2008) Agroforestry for carbon sequestration: scope and present status. Indian J Agrofor 10:1–9

Nordbotten JM, Celia MA (2006) An improved analytical solution for interface upconing around a well. Water Resour Res 42. https://doi.org/10.1029/2005WR004738

Pala NA, Sarkar BC, Gokhale Y, Manohar K, Shukla G, Chakravarty S (2019) Potential of homestead gardens to gain from global carbon markets to mitigate climate change. Indian For 146(2):159–164

Palm C, Tomich T, Van Noordwijk M, Vosti S, Alegre J, Gockowski J, Verchot L (2004) Mitigating GHG emissions in the humid tropics: case studies from the alternatives to slash-and-burn program (ASB). Environ Dev Sustain 6:145–162
Peyre A, Guidal A, Wiersum KF, Bongers F (2006) Dynamics of homegarden structure and function in Kerala, India. Agrofor Syst 66:101–115
Plieninger T, Dijks S, Oteros-Rozas E, Bieling C (2013) Assessing, mapping, and quantifying cultural ecosystem services at community level. Land Use Policy 33:118–129. https://doi.org/10.1016/j.landusepol.2012.12.013
Pradhan R, Manohar KA, SarkarBC BJA, Shukla G, Chakravarty S (2020) Ecosystem services of urban green sites-a case study from eastern Himalayan foothills. Trees Forests People 2:100029
Prieto I, Armas C, Pugnaire FI (2012) Water release through plant roots: new insights into its consequences at the plant and ecosystem level. New Phytol 193:830–841
Rana K, Goyal N, Sharma GP (2018) Staging stewards of agro-ecosystems in the ecosystem services framework. Ecosyst Serv 33:89–101
Rathore SS, Babu S, El-Sappah AH, Shekhawat K, Singh VK, Singh RK, Upadhyay PK, Singh R (2022) Integrated agroforestry systems improve soil carbon storage, water productivity, and economic returns in the marginal land of the semi-arid region. Saudi J Biol Sci 29(10):103427. https://doi.org/10.1016/j.sjbs.2022.103427
Ribeiro-Barros A, Silva M, Moura I, Ramalho J, Máguas-Hanson C, Ribeiro N (2018) The potential of tree and shrub legumes in agroforestry systems. In: Amanullah K, Fahad S (eds) Nitrogen in agriculture-update. InTech, Rijeka, Croatia, pp 223–239
Roy B, Md. Rahman H, Fardusi MJ (2013) Status, diversity, and traditional uses of homestead gardens in northern Bangladesh: a means of sustainable biodiversity conservation. SRN Biodivers, pp 1–11
Saha SK, Nair PKR, Nair VD, Kumar BM (2009) Soil carbon stock in relation to plant diversity of homegardens in Kerala, India. Agrofor Syst 76:53–65
Saha SK, Nair PKR, Nair VD, Kumar BM (2010) Carbon storage in relation to soil size-fractions under some tropical tree-based land-use systems. Plant Soil 328:433–446
Saha SK, Stein TV, Nair PKR, Andreu MG (2011) The socioeconomic context of carbon sequestration in agroforestry: a case study from homegardens of Kerala, India. In: Kumar B, Nair P (eds) Carbon sequestration potential of agroforestry systems. Advances in agroforestry, vol 8. Springer, Dordrecht. https://doi.org/10.1007/978-94-007-1630-8_16
Saikia P, Khan ML (2009) Conservation and density of medicinal plants in homegardens of Teok area of Jorhat district Assam. In: Proceeding of the National seminar on agroforestry for socioeconomic development of North Eastern region, 11th and 12th December 2009. NERIST, Nirjuli, Arunachal Pradesh
Schadegan FH, Khoshbakht K, Damghani AM, Veisi H, Liaghat H, Kambouzia J (2013)A Multidisciplinary study of rural Homegardens in Basht, Southwestern Iran. Middle-East J Sci Res 13: 1431–1437. DOI: https://doi.org/10.5829/idosi.mejsr.2013.13.10.1328
Semere M (2019) Biomass and soil carbon stocks Assessment of agroforestry systems and adjacent cultivated land, in ChehaWereda, Gurage zone, Ethiopia. Int J Environ Sci Nat Res 20(4): 556043. https://doi.org/10.19080/IJESNR.2019.20.556043
Sharma E, Sharma R, Sharma G, Rai SC, Sharma P, Chettri N (2008) Values and services of nitrogen-fixing alder based cardamom agroforestry systems in the eastern Himalayas. In: Snelder DJ, Lasco RD (eds) Advances in agroforestry, small tree growing for rural development and environmental services, vol 5. Springer, pp 393–408
Sharma G, Hunsdorfer B, Singh KK (2016) Comparative analysis on the socio-ecological and economic potentials of traditional agroforestry systems in the Sikkim Himalaya. Trop Ecol 57: 751–764
Sharma R, Jianzhu XU, Sharma G (2007) Traditional agroforestry in the eastern Himalayas region; land management system supporting ecosystem. Trop Ecol 48:1–12

Sida TS, Baudron F, Kim H, Giller KE (2018) Climate-smart agroforestry: *Faidherbia albida*trees buffer wheat against climate extremes in the central Rift Valley of Ethiopia. Agric For Meteorol 248:339–347

Sileshi GW, Mafongoya PL, Nath AJ (2020) Agroforestry Systems for improving nutrient recycling and soil fertility on Degraded Lands. In: Agroforestry for degraded landscapes. Springer, Singapore, pp 225–253. https://doi.org/10.1007/978-981-15-4136-0_8

Silva JA, Uchida R (2000) Biological nitrogen fixation: nature's partnership for sustainable agricultural production. In: Plant nutrient management in hawaii's soils, approaches for tropical and subtropical agriculture. University of Hawaii at Manoa, pp 121–126

Smith RG, Knight R, Chen J, Reeves JA, Zebker HA, Farr T, Liu Z (2017) Estimating the permanent loss of groundwater storage in the southern San Joaquin Valley, California. Water Resour Res 53:2133–2148. https://doi.org/10.1002/2016WR019861

Subba M, Pala NA, Shukla G, Chakravarty S (2015) Provisioning type ecosystem services from homegardens in terai and hilly region of West Bengal, India. J Agric Sci Technol 2:32–37

Szücs L, Anders U, Bürger-Arndt R (2015) Assessment and illustration of cultural ecosystem services at the local scale. A retrospective trend analysis. Ecol Indic 50:120–134. https://doi.org/10.1016/j.ecolind.2014.09.015

Thierfelder C, Baudron F, Setimela P, Nyagumbo I, Mapangwa W, Mhlanga B, Lee N, Gérard B (2018) Complementary practice supporting conservation agriculture in southern Africa. A review. Agron Sustain Dev 38:16. https://doi.org/10.1007/s13593-018-0492-8

Tscharntke T, Clough Y, Bhagwat S, Buchori D, Faust H, Hertel D, Hölscher D, Juhrbandt J, Kessler M, Perfecto I, Scherber C, Schroth G, Veldkamp E, Wanger TC (2011) Multifunctional shade-tree management in tropical agroforestry landscapes – a review. J Appl Ecol 48:619–629

UNEP (2002) Environmental threats to children. In: Children in the new millennium: environmental impact on health. United Nations environmental programme, Children's fund and World Health Organization

Vaast P, Somarriba E (2014) Trade-offs between crop intensification and ecosystem services: the role of agroforestry in cocoa cultivation. Agrofor Syst 88:947–956

van Noordwijk M (2021) Agroforestry-based ecosystem services. Land 10:770. https://doi.org/10.3390/land10080770

van Noordwijk M, Bizard V, Wangpakapattanawong P, Tata HL, Villamor GB, Leimona B (2014) Tree cover transitions and food security in Southeast Asia. Glob Food Sec 3:200–208

van Noordwijk M, Lawson G, Hairiah K, Wilson JR (2015) Root distribution of trees and crops: competition and/or complementarity tree–crop interactions. In: Agroforestry in a changing climate CABI, Wallingford, pp 221–257.

vanNoordwijk M, van de Geijn SC (1996) Root, shoot and soil parameters required for process-oriented models of crop growth limited by water or nutrients. Plant Soil 183(1):1–25

Varah A, Jones H, Smith J, Potts SG (2020) Temperate agroforestry systems provide greater pollination service than monoculture. Agric Ecosyst Environ 301:107031. https://doi.org/10.1016/j.agee.2020.107031

Vibhuti, Bargali K, Bargali SS (2018) Effects of homegarden size on floristic composition and diversity along an altitudinal gradient in central Himalaya. India Cur Sci 114(12):2494–2503

Vogl-Lukasser B, Vogl C (2004) Ethnobotanical research in home gardens of small farmers in the alpine region of Osttirol (Austria): an example for bridges built and building bridges. Ethnobot Res Appl 2:111–137

WWAP (2015) Water for a sustainable world. The United Nations World Water Assessment Programme, UNESCO, Paris

Zheng Z (2006) Litter decomposition and nutrient release in a tropical seasonal rain forest of Xishuangbanna. Southwest China Biotropica 38(3):342–347

Zomer RJ, Neufeldt H, Xu J, Ahrends A, Bassio D, Trabucco A, van Noordwijk M, Wang M (2016) Global tree cover and biomass carbon on agricultural land: the contribution of agroforestry to global and national carbon budgets. Sci Rep 6:29987. https://doi.org/10.1038/srep29987

Chapter 8
Agroforestry for Restoring and Improving Soil Health

Ankit Pandey, Prabhat Tiwari, Chowlani Manpoong, and Hanuman Singh Jatav

Abstract Globally, sustainable agricultural systems must be better understood as evidenced by concerns about food security, environmental degradation, and climate change. The demand for food and goods is increasing with rise in population while the productivity of land is declining all over the world. It is predicted that 25% of the world's lands are either highly deteriorated or prone to rapid degradation. Approximately, 12 million hectares of land are degraded every year worldwide due to land degradation. The functioning of soil ecosystems depends mainly on soil biodiversity and soil organic matter content. Inappropriate land use practices, such as deforestation, crop residue clearance, overgrazing, extensive mechanical tillage, and irrigation, are the main factors that contribute to soil nutrient losses and land degradation. Lack of organic matter reduces soil fertility, which ultimately results in reduced agricultural production. There are 175 million acres of degraded land in India. The world's population is expected to reach 9 billion people by 2050, which will necessitate a 60% increase in food production. Many conventional methods have been recommended for preserving soil fertility among which agroforestry is a potential system with multiple benefits. The woody perennial in agroforestry can supply nutritional inputs to crops through biological nitrogen fixation, deep capture, and storage of nutrients in their biomass. Tree roots take up different macro and micronutrients from deeper soil strata, which are then released into the top most layer of soil during the decomposition of roots and litter and have a potential utilization in providing nutrients to agricultural crops. As a land management strategy, agroforestry can simultaneously support household income, food security, soil biodiversity

A. Pandey (✉) · C. Manpoong
Department of Forestry, Wildlife and Environmental Sciences, Guru Ghasidas Vishwavidyalaya, Bilaspur, Chhattisgarh, India

P. Tiwari
Department of Silviculture and Agroforestry, Rani Lakshmi Bai Central Agriculture University, Jhansi, Uttar Pradesh, India

H. S. Jatav
Department of Soil science and Agricultural Chemistry, Sri Karan Narendra Agriculture University, Jaipur, Rajasthan, India

H. S. Jatav et al. (eds.), *Agroforestry to Combat Global Challenges*, Sustainable Development and Biodiversity 36, https://doi.org/10.1007/978-981-99-7282-1_8

preservation, gender equality, and ecosystem services as well as address global issues such as climate change, global warming, and fulfilling the obligations of international agreements such as sustainable development goals (SDGs).

Keywords Agroforestry · Biodiversity · Land degradation · Soil fertility · Soil organic carbon · Global warming · Sustainable development goals

8.1 Introduction

Soil is an important habitat for most of the living organisms including microorganism which regulates the nutrient for enhancing soil fertility and productivity (Laban et al. 2018). Since the last few decades, land managers have realized the value of the soil health and quality for long-term human existence and future sustainability (Anderson and Udawatta 2019; Dollinger and Jose 2018). Soil fertility is the key factor that affects the crop productivity and soil biodiversity. Soil quality is "the capacity of soil to function within ecosystem boundaries to sustain biological productivity, maintain environmental quality and promote plant and animal health" (Soil Science Society of America 2008). The fertility of soil is depleting continuously due to various factors such as erosion, over utilization of land, wrong agricultural practices, and other anthropogenic activities. Intensive agriculture practices such as monocropping and shifting cultivations have immensely contributed to soil degradation. These practices not only damage the soil and its environment but also reduce biodiversity and decline the quantity of nutrients, resulting to low soil productivity (Tsufac et al. 2021; Gupta 2020). Shifting cultivation is a dominant and traditional land use practice in North India, which has led to soil degradation due to loss in soil microbial diversity, reduced soil fertility, enzyme activity, and heavy soil erosion, etc. (Saha et al. 2010). In India, 175 m ha land is subjected to various types of degradation out of which 9.38 m ha is affected by sodicity and alkalinity (Gupta 2020; Uthappa et al. 2015). The land-use systems have a significant impact on the availability and cycling of nutrients and may also have an impact on secondary succession and biomass output (Panwar et al. 2011). Many conventional methods have been recommended to improve the health and quality of the soil and the ecosystem services provided by the soil, thereby increasing the viability of agricultural systems. Agroforestry systems have been widely adopted by the farmers and stakeholders in order to improve the soil quality and increase the farm outputs as well as soil diversity (Akter et al. 2022). These systems include the variety of interactions between trees and crops (Muchane et al. 2020). The integration of trees into croplands is the most crucial step in reducing the risk of degradation. These multi-strata farming agroforestry techniques can improve the less productive areas and provide security against the risks of crop failure in adverse conditions (Sahoo et al. 2020). In a plant and soil system, plant nutrients move continuously and dynamically. Plants absorb nutrients from the soil, use them for metabolic activities, and then return those nutrients to the soil naturally as litter falls as pruning in agroforestry systems, or through root senescence in both managed and unmanaged systems. These plant residues are broken down by microorganisms, which release nutrients into the soil that are once again available for plant uptake (Mao et al. 2011; Panwar et al. 2011; Sharma 2011). The land use systems that include trees, crops, and pastures are crucial for enhancing the quality and fertility of the soil in a number of ways. In contrast to pure agricultural systems, trees in agroforestry systems provide more closed nutrient cycling (Misra 2011; Nainwal et al. 2016). The significant advantage of tree-mediated benefits includes the increased inputs of organic matter, increased biological nitrogen fixation, efficient use of nutrient from deeper soil layer, and enhanced nutrient cycling.

8.2 Agroforestry and Global Scenario

From ancient times, trees were the integral parts on farmland to support agriculture (Pinho et al. 2012; Gupta et al. 2020). Agroforestry is a technique for managing land that can simultaneously raise household income, increase food security, preserve biodiversity, and enhance ecosystem services (Muchane et al. 2020). In addition, agroforestry act is a prominent tool to combat climate change (Muchane et al. 2020). Historically, agroforestry is one of the oldest known farming systems (Nainwal et al. 2016). Human revolution has been initiated from the forest when the human community learnt the art of cultivating crops/plants and domesticated animals for their use. There are countless instances of traditional land-use techniques that involve co-cultivating agricultural plants and trees on the same unit of land throughout the world (Gupta et al. 2020; Garrity et al. 2010). Although agroforestry methods have been used in various forms around the world for centuries, a systematic research on agroforestry started only during the 1970s–1980s in different countries (Marques et al. 2022) with the aim to formulate alternatives to increase agricultural production, restore the degraded lands, and enhance the well-being of small landholders (Gupta et al. 2020; Udawatta et al. 2017; Nainwal et al. 2016; Pinho et al. 2012). Agriculture land having more than 10 percent tree cover have been categorized as agroforestry system by Gupta et al. (2020) and Zomer et al. (2014). Agroforestry provides an alternative and sustainable approach towards the addition of organic matter to the soil by utilizing fast growing and nitrogen fixing plants species (Tsufac et al. 2021). The systematic research on agroforestry were popularized after the establishment of International Council for Research in Agroforestry (ICRAF) which was later renamed as World Agroforestry Centre (WAC) recently (Gupta et al. 2020). In India, for scientific and dedicated research in the field of agroforestry, the National Research Centre for Agroforestry (NRCAF) was established in Jhansi, Uttar Pradesh, in 1998. It was later renamed as Central Agroforestry Research Institute (CAFRI) in 2014. India is the first country in the world with policy on research in agroforestry which is known as "National Agroforestry Policy 2014". Several other projects and schemes were launched for development in the agroforestry sector (Verma et al. 2017).

8.2.1 Characteristics of Good Trees in an Agroforestry System for Soil Improvement

In the agroforestry system, selecting a suitable tree is one of the most important factors for the farmers. The selected trees play a key role in determining the production and providing protection to the farm. It also helps in the management and improvement of soil health (Saha et al. 2010). The following characteristics are given priority while selecting the trees for agroforestry system:

- The tree should be fast growing with deep root system.
- The tree should have high rate of nitrogen fixation capacity.
- Higher production of leafy biomass (avoid thorny species in farmland).
- It should have dense fine root network as well as abundant mycorrhizal association capacity.
- The tree should not have any toxic substance or allelopathic effect in its litter or root exudate.
- It should have the capacity to grow in poor soil, water scarcity, as well as drought condition and should have resistance against a wide range of pest and diseases.
- Low invasiveness and capacity for soil reclamation.
- Trees should have both productive as well as protective functions.

8.3 Agroforestry for Restoration of Degraded Soil

Land degradation is the temporary or long-term reduction in the land's ability in production and providing ecosystem functions and services (Masebo and Menamo 2016; Mahmud and Islam 2017). Negative human activities including overgrazing, over-cultivation, ineffective irrigation systems, deforestation, industrial pollution, and population growth are only a few examples of the natural causes that contribute towards land degradation (Lorenz et al. 2019; Mahmud and Islam 2017; Sharma 2011). Food and agricultural organization (FAO) has recognized 13 different forms of soil degradation that degrades land in which one of the most ubiquitous aspects of worldwide land degradation is soil erosion (Muchane et al. 2020; Gupta et al. 2020; Udawatta et al. 2017: Korneeva 2021). As a result of land abandonment brought on by soil erosion, more grasslands and woodlands were cleared for farming (Lorenz et al. 2019; Udawatta et al. 2017). There are numerous effects of soil erosion both on- and off-site (Sileshi et al. 2020). When compared to conservation agriculture, erosion caused by traditional farming methods causes three times as much land degradation and nearly 75 times as much destruction of natural flora (Udawatta et al. 2017). Muchane et al. (2020) conducted research and concluded that soil erosion rate was significantly lower in agroforestry system than monoculture. Compared to monocultures, agroforestry practices considerably minimize erosion rates (50%), increase infiltration rates (75%), and reduce runoff (57%). Kinama et al. (2007)

concluded that hedgerow cropping of *Senna seima* based system significantly reduces the soil erosion from 100 mg/ha to 2 mg/has as compared to sole cropping system. In general, all agroforestry systems contain at least two components, one of which must be a tree or shrub component. The presence of trees has a lot of benefits, and there are several agroforestry techniques for restoring and boosting land productivity besides fulfilling the demands of farmers (Gupta et al. 2020). There are several agroforestry systems such as agro-silvopasture, silvopasture, and agro-silvopasture (Tsufac et al. 2021) systems that are appropriate for rehabilitation or restoration of degraded lands due to soil erosion, salinization, waterlogging, less soil organic matter, and mining and improving the agricultural crop's resistance to harmful effects of climate variability as well as pests, diseases, and weeds (Nainwal et al. 2016; Anderson and Udwatta 2019; Parewa et al. 2018). Singh et al. (2010) conducted research on soil fertility under various farming systems and concluded that different soil parameters such as organic carbon and available nitrogen, phosphorous, and potassium in soil was enhanced in a poplar-based agroforestry system as compared to a monoculture system. The available nutrient was higher in Subabul, Siris, and Shisham plantation. Thus, tree component in agroforestry systems helps in sustaining the soil health by improving soil physico-chemical and biological properties. The agroforestry system may also reduce runoff by blocking the flow channel, which reduces the amount of sediment that it may carry. Researchers perceived that agroforestry has the potential to restore the degraded soil and regulate ecosystem services. Sileshi et al. (2011) reported a higher water use efficiency in *Leucaena* and maize-based agroforestry system as compared to sole maize cropping system. The most adopted agroforestry systems in tropical and subtropical countries are alley cropping (hedgerow intercropping), home gardens, improved fallow, multipurpose trees on farms and rangelands, shaded perennial crop systems, shelterbelts and windbreaks, silvopasture, and Taungya cultivation (Dollinger and Jose 2018). Uthappa et al. (2015) carried out research in various districts of Bundelkhand region of Madhya Pradesh and Uttar Pradesh and revealed that Aonla based agroforestry farming was profitable in degraded land. Rodriguez et al. (2021) revealed that the agroforestry system helps in reducing land degradation whereas the silvopasture system significantly improves soil conditions as compared to traditional agricultural practices. Thus, the agroforestry system can not only revive the soil ecosystem services but also play a major role in improving soil fertility.

8.4 Soil Fertility Improvement

It is challenging to generalize the impacts of an agroforestry system on soil fertility. Crops continuously remove nutrients from the top layer of soil, leading to deficiencies in macro nutrients such as N, P, K, sulphur, zinc, and micro nutrients such as boron. The lack of organic matter reduces soil fertility, which eventually leads to lower agricultural productivity (Misra 2011; Mahmud and Islam 2017; Gupta 2020). In agroforestry systems, tree and crop roots help in restoring the fertility of the soil in addition to organic matter, nitrogen fixation, soil physical conditions, and nutrient cycling (Sharma 2011). The soil organic carbon stock is essential for the fertility, productivity, quality, and health of the soil. It also provides soil-based ecosystem services like nutrient cycling, provisioning services of food, fresh water, timber, fibre, and regulation of flood control as well as water quantity attenuation (Lorenz et al. 2019). Young (1989) defined soil fertility as the capacity of the soil to support plant growth on a sustainable basis under given condition of the climate and other relevant properties of the land. Agroforestry systems have reported to have comparatively better overall production than systems that only grow crops (Verma et al. 2017). Agroforestry systems, including agri-horticulture, agri-pastoral, agri-silviculture, agri-silvo-pasture, and others, are the most fertile systems to restore soil that has been damaged. The residues of trees, crops, and animals of different agroforestry systems are decomposed with the aid of microorganisms to produce humus, which helps in improving the soil fertility by preventing nutrient loss (Tsufac et al. 2021; Parewa et al. 2018).

8.4.1 Soil Organic Carbon Through Agroforestry

Soil is considered as the third greatest carbon sink and plays a crucial role in carbon sequestration (Handa et al. 2020; Dollinger and Jose 2018) with 1.5–3 times more carbon than vegetation (De Stefano and Jacobson 2018). The soil organic carbon (SOC) content acts as a universal indicator of soil fertility (Sileshi et al. 2020; Anderson and Udwatta 2019) and for monitoring of soil degradation (Lal et al. 2015). Traditional farming methods sometime encourage SOC loss. Tillage operations promote the oxidation of soil organic matter, streamline microbial communities, and quicken erosion, all of which result in lowering SOC pool, worsening soil fertility, and increasing land degradation (Mosier et al. 2021). Trees contribute in addition of organic matter to the soil in several ways, such as through their root decomposition, litter fall, and the incorporation of root exudates into the rhizosphere (Pinho et al. 2012). Saputra et al. (2020) reported that the soil organic carbon in cocoa-based agroforestry system was higher than monoculture system. The agroforestry system has proven beneficial effect on soil organic carbon stocks at varying soil depths (Ramachandran Nair et al. 2009; Handa et al. 2020). Ramachandran Nair et al. (2009) reported a higher soil carbon stock in agroforestry system as compared to sole tree plantations as well as uncultivated/other land uses. This is probably due to the fact that the tree component has longer growing seasons and minimum biomass loss during harvest. The leaf litter and debris of woody perennials that is returned to the soil often increases the soil organic carbon as compared to seasonal crops. Pardon et al. (2017) found maximum SOC concentration in the agroforestry system with an average increase in soil 5300 kg ha^{-1} within the field zone. The tree characteristics (such as species, canopy, density, age, and associated intercrops), system characteristics (such as structure, function, and stability), and system management (such as tillage operation, fertilizer application, harvesting regime, pruning, and holding size) together play a significant role in enhancing the soil carbon sequestration in agroforestry system (Handa et al. 2020). Several studies has revealed a significant increase in soil organic carbon stock when agricultural lands are converted to agri-silviculture and silvo-pasture, pasture/grassland to agri-silvo-pastoral systems, forest to silvo-pasture, forest plantation to silvo-pasture, and uncultivated/other land uses to agri-silviculture (Udwatta and Jose 2012: De Stefano and Jacobson 2018: Gupta et al. 2020). Ramos et al. (2018) reported higher soil carbon stock in cocoa and oil palm-based agroforestry systems in Brazil. Similar findings were reported by Rahman et al. (2022) in home garden-based agroforestry system where most of the species enhanced the SOC status and soil organic matter of the stated home garden as compared to cropland and orchard.

8.4.2 Nitrogen Availability Through Agroforestry

One of the major obstacles to sustainable crop production and plant growth in any region is the lack of primary nutrients, particularly nitrogen (N) (Munroe and Isaac 2014). This lack of primary nutrients and the excessive use of fertilizers also has an adverse effect on the ecosystem (Parewa et al. 2018). In comparison to monoculture, agroforestry greatly enhances the N levels in the soil (Muchane et al. 2020). Agroforestry system perennial crops (trees), whether they are harvested, grazed, or planted in conservation areas, have a great potential for preserving N. In addition, they also have the unique nitrogen absorption mechanisms that reduce the requirement for supplemental nitrogen (Mosier et al. 2021). Incorporation of nitrogen fixing tree species in agroforestry systems enhances the nutrient availability in soil (Sharma 2011; Parewa et al. 2018). Through above-ground inputs like litter fall and pruning as well as below-ground inputs like roots, nitrogen-rich tree biomass adds N to the nitrogen pool. After the processes of decomposition and mineralization, the soil's inorganic nitrogen is then added by these processes (Sarvade et al. 2014; Muchane et al. 2020). In agroforestry, legumes are particularly commonly grown to enhance the production of fodder, and simultaneously they can improve soil fertility, increase soil nitrogen status, enhance below-ground productivity, and consequently increase below-ground C inputs (De Stefano and Jacobson 2018; Solanki et al. 2020). Munroe and Isaac (2014) reported that there are three major ways of interaction between nitrogen fixing trees and non-nitrogen fixing trees: (1) Mineralization and decomposition of organic compound such as litter, pruning, root, and nodule, (2) root to root direct transfer of nitrogen through exudation, and (3) common mycorrhizal networks. Yengwe et al. (2018) estimated the nutritive potential of *F. albida* and maize-based intercropping system and concluded that the addition of litter by *F. albida* could supply more than 18 kg N ha^{-1} $year^{-1}$ and improve the nitrogen availability for maize crop. Augustine et al. (2007) reported higher soil organic matter and N content in the agroforestry as compared to non-agroforestry system.

8.4.3 Phosphorus Availability Through Agroforestry

Phosphorus (P) is the second most important nutrient after nitrogen for plant growth, and is a crucial component of carbon metabolism, energy conservation, and photosynthetic control in plants. It has a key role in the production of ATP, enzymes, DNA, and RNA. Both organic and inorganic forms of P are present in soil, but only inorganic P is available for plant uptake (Gaxiola et al. 2011). Perennial crops enhance soil organic matter, which accelerates P cycling by supplying a P source through decomposition and dissolution of adsorbed inorganic P (Mosier et al. 2021). Woody perennials enhance the P availability and recycling by reducing losses and increasing the presence of plant-available P (Augustine et al. 2007). Many tree species with deep roots system and efficient mycorrhizal assocoation are able to harvest phosphorus and transform it into a form that is useful for plants available form of P (Crews and Brookes 2014). Microbes are essential for converting phosphorus (P) into forms that plants can utilize by releasing metabolites and organic acids that liberate adsorbed, inaccessible phosphorus from minerals and organic materials (Ingle and Padole 2017). Mycorrhizal association found in the roots of higher plants are important to facilitate this process to make phosphorus more easily available to plants and also increase the nutrient uptake of the plant. The processes by which these organisms change the solubility of organic compounds to their soluble form includes several factors such as the production of acids and H_2S under aerobic and anaerobic conditions, respectively, mineralizing organic compounds with the release of inorganic phosphate, and converting insoluble phosphorus to an available form (Sujata and Nibha 2011). Prakash et al. (2018) evaluated phosphorus availability and speciation between different land use systems in India and reported that the poplar-based agroforestry system showed higher levels of organic phosphorus (27%) and soil organic carbon as compared to other land use systems. Chowdhury et al. (2022) reported higher availability of phosphorus in agroforestry system than other land use system. Correa et al. (2015) argued that in eucalyptus-based agroforestry system, P supply may be increased due to the decomposition of litter facilitated by P solubilizing microorganism.

8.5 Nutrient Recovery of Soil Through Agroforestry

Plants absorb nutrients and store them as biomass in various plant parts. Crop harvest is often accompanied by loss of nutrients from the system. Without any external nutrient inputs in the form of inorganic fertilizers, productivity is expected to decline because recycling nutrients through agricultural leftovers cannot make up for these nutrients losses. In an agricultural system, harvesting removes much of the crop biomass (Misra 2011). Plant nutrients are in a stage of continuous and dynamic transfer between soil and plants, where plants absorb nutrients from soil for metabolic activities which is returned to the soil through the addition of organic matter.

Table 8.1 A comparative analysis of major nutrient input by different tree species

Tree	Available N (kg/ha)	Available P (kg/ha)	Available K (kg/ha)	OC (%)
Leucaena leucocephala	150–200	16–18	150–170	1–1.32
Dalbergia sissoo	200–210	14–16	250–280	0.46
Poplar spp.	170–190	15–18	150–160	1.22
Melia azedarach	180–185	14.5–16.14	150–165	0.98–1
Eucalyptus camaldulensis	150–170	10–12	150–158.	0.97
Albizia procera	180–197	14–16.52	150–165	0.54–0.62
Acacia nilotica	200–216	14–15.5	100–150	0.71–0.80

(Source: Sarvade et al. 2019; Uthappa et al. 2015; Sarvade et al. 2014, Misra 2011)

Trees play a significant part in the nutrient cycling by taking back and pumping back the leached nutrients from the soil through deep root system, which serves as a safety net against nutrient losses (Fahad et al. 2022). In agroforestry systems, nutrient are returned to the soil through litter fall, or pruning, and other organic left overs. Trees have the ability to translocate nutrients from deeper soil horizons to the soil surface facilitating plant uptake as indicated in Table 8.1 (Nair et al. 1999: Misra 2011). Hoosbeek et al. (2018) carried out research and revealed that the nutrient content in surface and sub-surface soil under the tree canopy was higher in C and N as compared to the pasture land due to the presence of trees, which ultimately enhanced the availability of nutrient through nutrient cycling. The soil organic matter in the soil surface is decomposed by microorganism (Fungi, bacteria) present in the soil and release the chemically bound immobile nutrient into the soil, thus making the nutrient available to plants.

8.5.1 Leaf Litter Addition

The dead leaves, twigs, bark, needles, etc., form the overall litter in the system, with leaves as the major component. In agroforestry systems, the plant litter containing most of the nutrients is needed for plant growth and is the key component of nutrient cycling. Through the addition of the important nutrients (such as nitrogen, phosphorus, and potassium) to the soil, the agroforestry systems contribute towards the maintenance and improvement of soil fertility and of the nutrient recycling process (Singh 2020). The nature of the tree-crop species and conditions such as climatic, edaphic, topographic, and biotic factors affecting the plant development and phenology are crucial factors for litter production in an agroforestry system (Rawat et al. 2009). Schroth and Krauss (2006) reported that the addition of litter via tree component of agroforestry is essential for preserving soil moisture and maintaining optimum bulk density, which results in an increased nutrient mobilization compared to bare soil. In agroforestry systems, the agricultural crops make up the smallest portion of the overall litter generation as compared to perennial woody tree species (Sarvade et al. 2014). Generally, deciduous plants produce more leaf litter as compared to evergreen plants throughout year. In summer and spring season the plant's contribution towards litter is more as compared to autumn season (Singh et al. 2011; Pragasan and Parthasarathy 2005).

8.5.2 Leaf Litter Decomposition

After the litter formation, decomposition of litter and its rate of decomposition play a significant role in the improvement of soil fertility in terms of nutrient cycling and soil organic matter formation (Bhattarai and Bhatta 2020). Soil fauna such as non-symbiotic nitrogen fixing bacteria, phosphate solubilizing bacteria, thio-sulphate oxidizing bacteria, etc., mediate litter decomposition and its mineralization which in turn enriches nutrients into the soil for the plants (Singh 2020). There are several factors such as litter quality and its quantity, climatic factor (temperature, moisture, etc.), C:N ratio, and microbial community structure which determine the rate of litter decomposition (Sarvade et al. 2014; Dechaine et al. 2005). The term "litter quality" refers to the intrinsic chemical characteristics of the litter, including its nitrogen content (C:N ratio), phosphorus content (C:P ratio), phenolics (C: N or P ratio), lignin content (C:N ratio), and phenolics to P or N ratio. High levels of C:N ratio, C:P ratio, phenolics, phenolics to N or P ratio, lignin content, and lignin to N ratio reduces the rate of decomposition, but high nitrogen and phosphorus content increases the rate of litter decomposition (Nair et al. 1999). Fungi are regarded as the "primary players" in the breakdown of leaf litter due to their capacity to create a variety of extracellular enzymes. Enzymes such as peptidases, urease, and phosphatase, etc., play crucial role for microbial uptake of nitrogen and phosphorus, while others such as phenol oxidase, peroxides, and laccase aid in the catalysis of lignin degradation of leaf litter (Tennakoon et al. 2021). The decomposition rate of litter found in temperate region is slower than the litter found in tropical region (Bhattarai and Bhatta 2020). The rhizosphere of tree species, which is generally rich in microorganism, plays a significant role in nutrient transformation. This is probably due to a increased root exudates status in agroforestry system as compared to seasonal crops. The stages of decomposition as shown in Fig. 8.1 comprises two main simultaneous process: (A) the mineralization and humification of lignin, cellulose, and other substances by microorganisms; (B) the slow mineralization of carbon, nitrogen, and other elements in the soil caused by the leaching of soluble chemicals into the soil (Tennakoon et al. 2021).

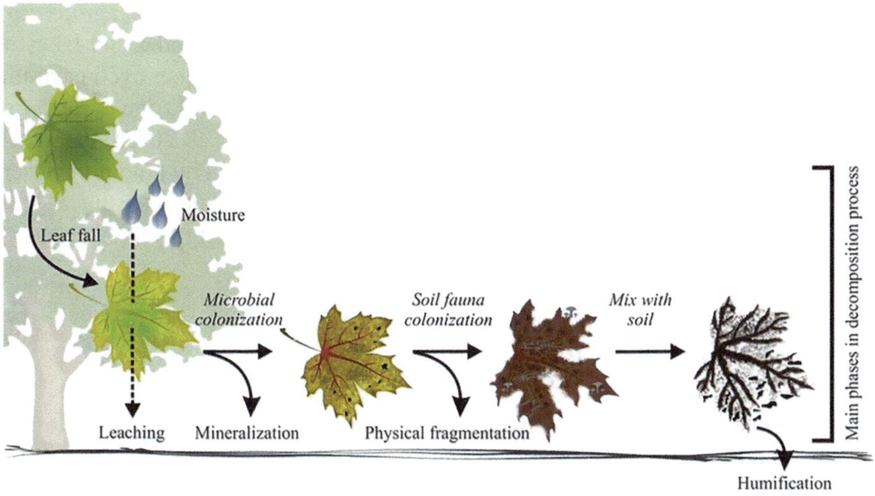

Fig. 8.1 Phases of leaf litter decomposition. (Source: Tennakoon et al. 2021)

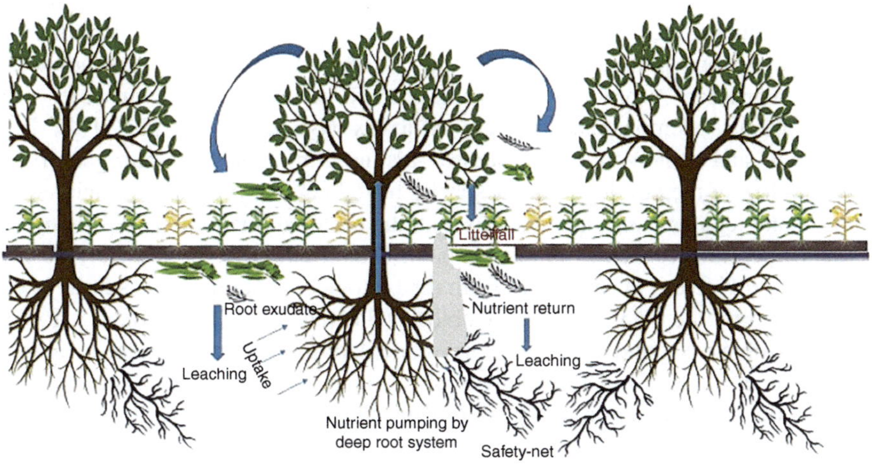

Fig. 8.2 Nutrient pumping through deep root system (Fahad et al. 2022)

8.5.3 Nutrient Pumping

It is a well-recognized fact that because trees are having a deep and spreading root system, they may uptake nutrients and water from deeper soil horizons where herbaceous crop roots often cannot penetrate (Nair et al. 1999). Figure 8.2 depicts the process of absorption of nutrients from a deeper soil profile and its eventual deposition on the surface and subsurface layers through litterfall and other processes by the tree. This process is termed as "nutrient pumping" (Misra 2011). Because of the plastic response of trees to competition with an annual crop, trees in agroforestry systems are likely to have a deeper root system than monocropping and takes up nutrient from deeper soil strata. Closed and more efficient nitrogen cycling results from the nutrient pumping of nitrogen from deep soils to the root zone of deep tree roots leads in closed and more effective N cycling (Fahad et al. 2022). Augustine et al. (2007) reported that *G. sepium* enhanced N content in soil by not only nutrient cycling from decomposing litter and by fixing atmospheric nitrogen but also by taking up nutrient from deeper soil layer than the normal rooting zone of maize.

8.5.4 Biological Nitrogen Fixation

The leguminous trees in agroforestry, enhance the soil fertility by biological nitrogen fixation, addition of organic matter, and recycling of nutrients (Misra 2011: Solanki et al. 2020). In the case of legumes, the mimosoideae and caesalpinioideae subfamilies contain the majority of the nitrogen-fixing tree species, whereas the papilionoideae subfamilies contain the least number of species. In these subfamilies, 98, 60, and 30% of the tested mimosoids, papilionoids, and caesalpinoids, respectively, demonstrated the ability to fix atmospheric nitrogen. Families of non-leguminous plants, such as the ulmaceae, rosaceae, casuarinaceae, chrysobalanaceae, coriariaceae, eleagnaceae, myricaceae, rhamnaceae, and zmiaceae, also demonstrated the ability to fix N_2 (Sarvade et al. 2014). Both symbiotic and non-symbiotic processes are used in biological nitrogen fixing. While a few of the non-leguminous species have Frankia symbionts as their symbionts, many legumes form associations with the bacterial rhizobium. Free-living soil organisms influence non-symbiotic fixation (Misra 2011). Approximately 60% of nitrogen provided to plants comes from the biological fixation whereas 50% of that is provided by the plant bacterial symbiosis (Barea et al. 2005). In India's tropical soils, rhizobia are abundant and capable of nodulating a variety of legume symbioses (Nambiar et al. 1988). The examples of nitrogen fixing bacteria associated with legume trees used in agroforestry system are shown in Table 8.2.

8.6 Agroforestry for Soil Biota

The majority of soil biogeochemical processes depend upon soil microbial communities, which are found in the soil and rhizosphere. The integration of tree in agroforestry not only enhances the physical and chemical properties of soil but also plays a significant role in the enhancement of soil microorganism. The "demand side" for soil resources is represented by root systems, while the "supply side" is influenced by soil microbes and fauna, which play a key role in the degradation and stabilization of soil organic matter as well as the breakdown of litter and other organic components in the soil (Schroth and Krauss 2006). They are crucial for

Table 8.2 Nitrogen fixing bacteria associated with legume trees used in agroforestry system

Bacterial symbionts	Host genus	Source
Azorhizobium caulinodans	*Sesbania* spp.	Dreyfus et al. (1988)
Azorhizobium johannense		Moreira et al. (2002)
Bradyrhizobium sp.	*Acacia* spp.	Dupuy et al. (1994)
B. Japonicam	*Inga* spp.	Leblanc et al. (2005)
Ensifer mexicanus	*Acacia* spp.	Lloret et al. (2007)
Mesorhizobium albiziae	*Albizia* spp.	Wang et al. (2007)
Rhizobium tropici and *Rhizobiumetli*	*Gliricidia* spp.	Hernandez-Lucas et al. (1995)
R. gallicum	*Leucaenaa* spp.	Hernandez-Lucas et al. (1995)
Sinorhizobium kostiense and *Sinorhizobium arbosis*	*Prosopis* spp.	Sprent (2009)

Table 8.3 Comparison of mean densities of different soil biota in soil under agroforestry system and monoculture with calculated response ratio

	Agroforestry	Monocrop	RR
Soil macro fauna			
Earthworm	54.4	17.6	3.1
Beetles	20.9	9.6	2.2
Centipedes	2.7	0.5	5.6
Millipedes	8.1	1.3	6.1
Termites	90.7	81.0	1.1
Ants	23.2	8.6	2.7
Soil mesofauna			
Collembola	3890.1	2000.7	1.9
Mites	5100.7	1860.1	2.7
Soil microfauna			
Non-parasitic nematodes	2922	1288	2.3
Parasitic nematodes	203.7		1

(Source: Barrios et al. 2012)

biogeochemical cycles, nitrogen cycling, mineralization, and nutrient delivery, chemical degradation, maintenance of aboveground biodiversity, soil formation, and maintenance soil health (Udawatta et al. 2019; Jose 2012). When agroforestry system was compared with grazing and row crop management, the soil enzyme activity, microbial diversity, and macro fauna activity were found to be higher under agroforestry system (Udawatta et al. 2017). A major contribution of agroforestry trees significantly contributes to soil ecosystem services as a result of above- and below-ground organic inputs that supply food and nutrition required for soil organisms for nutrient cycling and maintenance of soil fertility (Uthappa et al. 2015).Trees alter the soil environment in a variety of ways, such as by providing shade for the understory and soil, transpiring water taken up by roots from the soil, and covering the soil with dead or pruned leaves/branches to add nutrients (Barrios et al. 2012). A comparative analysis of agroforestry system over sole cropping system (Table 8.3) reveals that the agroforestry system has more macro fauna (termites, earthworm, ants, etc.), miso fauna (Mites, collembola), as well as micro fauna such as nematodes (Nygren et al. 2012). Maurya et al. (2012) reported that the population of *Azospirillum* and *Azotobactor* was higher in agroforestry system whereas lower in rice-wheat and grassland-based cropping system in Uttar Pradesh. Tondoh et al. (2015) revealed that the abundance of earthworms as well as species richness in cocoa-based farming system increased with increasing the plantation age due to adaptation of fauna to degraded land. Marsden et al. (2020) reviewed and found positive effect on abundant fauna in agroforestry as compared to cropland. Bainard et al. (2011) studied and concluded that tree-based intercropping has the highest and diverse Arbuscular mycorrhizal fungal community as compared to traditional cropping system.

8.7 Conclusion

Agroforestry techniques offer the ability to restore ecosystem services, enhance food security, restore livelihoods on degraded land, and decrease pressure on forests. Agroforestry can help to diversify and intensify the existing farming system through integration of trees suitable for local conditions with the agricultural crop as well as animals in the same unit of land. In comparison to agricultural monocultures, agroforestry practices significantly lower the rate of soil erosion, enhance SOC and N status, increase inorganic N availability, and slightly increase inorganic P and pH levels in the soil. The trees in agroforestry help in the reduction of nutrient loss from soil by limiting soil erosion and leaching. Trees also have the capacity to uptake nutrient from a deeper layer of soil, regulate the soil temperature as well as microorganism in soil-based agro-ecosystem. It is assumed that agroforestry systems have a higher capacity to store carbon than pastures or field crops. The study suggests that agroforestry systems can not only ensure sustenance and improvement of physicochemical and biological properties of soil but may also help in the restoration of degraded land and have a huge potential in sustainable production for the farm output, which ultimately ensures a raise in farmer's income too.

References

Akter R, Hasan MK, Kabir KH, Darr D, Roshni NA (2022) Agroforestry systems and their impact on livelihood improvement of tribal farmers in a tropical moist deciduous forest in Bangladesh. Trees For People 9:1–9. https://doi.org/10.1016/j.tfp.2022.100315

Anderson SH, Udawatta RP (2019) Agroforestry: A system for improving soil health. In: Agroforestry for sustainable agriculture. Burleigh Dodds Science Publishing, pp 317–334

Augustine CMJ, Vogt KA, Harrison RB, Hunsaker HM (2007) Nitrogen-fixing trees in small-scale agriculture of mountainous Southeast Guatemala: effects on soil quality and erosion control. J Sustain For 23(4):61–80

Bainard LD, Klironomos JN, Gordon AM (2011) Arbuscular mycorrhizal fungi in tree-based intercropping systems: A review of their abundance and diversity. Pedobiologia 54(2):57–61

Barrios E, Sileshi GW, Shepherd K, Sinclair F (2012) Agroforestry and soil health: linking trees, soil biota and ecosystem services. Soil Ecol Ecosyst Serv 14:315–330

Barea JM, Werner D, Azcón-Guilar C, Azcón R (2005) Interactions of arbuscular mycorrhiza and nitrogen-fixing symbiosis in sustainable agriculture. In: Nitrogen fixation in agriculture, forestry, ecology, and the environment. Springer, Dordrecht, pp 199–222

Bhattarai S, Bhatta B (2020) Leaf-litter decomposition and nutrient dynamics of five selected tropical tree species. Banko Janakari 30(1):32–38

Chowdhury FI, Barua I, Chowdhury AI, Resco de Dios V, Alam MS (2022) Agroforestry shows higher potential than reforestation for soil restoration after slash-and-burn: a case study from Bangladesh. Geol Ecol Landsc 6(1):48–54

Crews TE, Brookes PC (2014) Changes in soil phosphorus forms through time in perennial versus annual agroecosystems. Agric Ecosyst Environ 184:168–181

Correa E, Carvalhais L, Utida M, Oliveira CA, Scotti MR (2015) Effect of plant species on P cycle-related microorganisms associated with litter decomposition and P soil availability: implications for agroforestry management. iForest-Biogeosci For 9(2):294–302. https://doi.org/10.3832/ifor1459-008

De Stefano A, Jacobson MG (2018) Soil carbon sequestration in agroforestry systems: a meta-analysis. Agrofor Syst 92(2):285–299

Dechaine J, Ruan H, Sanchez-de Leon Y, Zou X (2005) Correlation between earthworms and plant litter decomposition in a tropical wet forest of Puerto Rico. Pedobiologia 49(6):601–607

Dreyfus B, Garcia JL, Gillis M (1988) Characterization of *Azorhizobium caulinodans* gen. Nov., sp. nov., a stem-nodulating nitrogen-fixing bacterium isolated from Sesbania rostrata. Int J Syst Evol Microbiol 38(1):89–98

Dollinger J, Jose S (2018) Agroforestry for soil health. Agrofor Syst 92(2):213–219

Dupuy N, Willems A, Pot B, Dewettinck D, Vandenbruaene I, Maestrojuan G, Gillis M (1994) Phenotypic and genotypic characterization of *bradyrhizobiam* nodulating the leguminous tree *Acacia albida*. Int J Syst Evol Microbiol 44(3):461–473

Fahad S, Chavan SB, Chichaghare AR, Uthappa AR, Kumar M, Kakade V, Poczai P (2022) Agroforestry systems for soil health improvement and maintenance. Sustainability 14(22): 14877. https://doi.org/10.3390/su142214877

Garrity DP, Akinnifesi FK, Ajayi OC, Weldesemayat SG, Mowo JG, Kalinganire A, Bayala J (2010) Evergreen agriculture: a robust approach to sustainable food security in Africa. Food Secur 2(3):197–214

Gaxiola RA, Edwards M, Elser JJ (2011) A transgenic approach to enhance phosphorus use efficiency in crops as part of a comprehensive strategy for sustainable agriculture. Chemosphere 84(6):840–845

Gupta VP (2020) Role of agroforestry in soil conservation and soil health management: A review. J Pharmacogn Phytochem 9(4S):555–558

Gupta SR, Dagar JC, Teketay D (2020) Agroforestry for rehabilitation of degraded landscapes: achieving livelihood and environmental security. In: Agroforestry for degraded landscapes: recent advances and emerging challenges, vol 1. Springer Nature Pte Ltd, Singapore, pp 23–68. https://doi.org/10.1007/978-981-15-4136-0

Handa AK, Sirohi C, Arunachalam A, Chavan SB (2020) Agroforestry interventions for carbon sequestration and improving degraded lands. Climate Change Environ Sust 8(1):3–12

Hernandez-Lucas I, Segovia L, Martinez-Romero E, Pueppke SG (1995) Phylogenetic relationships and host range of rhizobium spp. that nodulate *Phaseolus vulgaris* L. Appl Environ Microbiol 61(7):2775–2779

Hoosbeek MR, Remme RP, Rusch GM (2018) Trees enhance soil carbon sequestration and nutrient cycling in a silvopastoral system in South-Western Nicaragukima. Agrofor Syst 92(2):263–273

Ingle KP, Padole DA (2017) Phosphate solubilizing microbes: an overview. Int J Curr Microbiol Appl Sci 6(1):844–852

Jose S (2012) Agroforestry for conserving and enhancing biodiversity. Agrofor Syst 85(1):1–8
Korneeva EA (2021) Economic evaluation of ecological restoration of degraded lands through protective afforestation in the south of the Russian plain. Forests 12(10):1317
Kinama JM, Stigter CJ, Ong CK, Ng'ang'a JK, Gichuki FN (2007) Contour hedgerows and grass strips in erosion and runoff control on sloping land in semi-arid Kenya. Arid Land Res Manag 21(1):1–19
Lal R (2015) Restoring soil quality to mitigate soil degradation. Sustainability 7(5):5875–5895. https://doi.org/10.3390/su7055875
Lorenz K, Lal R, Ehlers K (2019) Soil organic carbon stock as an indicator for monitoring land and soil degradation in relation to United Nations sustainable development goals. Land Degrad Dev 30(7):824–838
Leblanc HA, McGraw RL, Nygren P, Roux CL (2005) Neotropical legume tree Inga edulis forms N2-fixing symbiosis with fast-growing *Bradyrhizobium* strains. Plant Soil 275(1):123–133
Laban P, Metternicht G, Davies J (2018) Soil biodiversity and soil organic carbon: keeping drylands alive. IUCN, Gland, Switzerland, p 10
Lloret L, Ormeno-Orrillo E, Rincon R, Martínez-Romero J, Rogel-Hernández MA, Martínez-Romero E (2007) *Ensifer mexicanus* sp. nov. a new species nodulating *Acacia angustissima* (Mill.) Kuntze in Mexico. Syst Appl Microbiol 30(4):280–290
Maurya BR, Kumar A, Raghuwanshi R, Singh V (2012) Diversity of azotobacter and Azospirillum in rhizosphere of different crop rotations in eastern Uttar Pradesh of India. Res J Microbiol 7(2): 123–130. https://scialert.net/abstract/?doi=jm.2012.123.130
Marques MA, Anjos LHCD, Sanchez Delgado AR (2022) Land recovery and soil management with agroforestry systems. Spanish. J Soil Sci 12(1–9):10457. https://doi.org/10.3389/sjss.2022
Mahmud K, Islam M (2017) Reclamation of degraded land through Agro-forestry introduction in Bangladesh. J Nat Sci Res 7(5):56–62
Mao R, Zeng DH, Li LJ (2011) Fresh root decomposition pattern of two contrasting tree species from temperate agroforestry systems: effects of root diameter and nitrogen enrichment of soil. Plant Soil 347(1):115–123
Marsden C, Martin-Chave A, Cortet J, Hedde M, Capowiez Y (2020) How agroforestry systems influence soil fauna and their functions-a review. Plant Soil 453:29–44
Masebo N, Menamo M (2016) A review paper on the role of agroforestry for rehabilitation of degraded soil. J Biol Agric Healthcare 6(5):128–136
Misra PK (2011) Soil fertility Management in Agro-forestry system. Int J Biotechnol Biochem 7(5): 637–644
Mosier S, Córdova SC, Robertson GP (2021) Restoring soil fertility on degraded lands to meet food, fuel, and climate security needs via perennialization. Front Sust Food Syst 5(1–18): 706142. https://doi.org/10.3389/fsufs.2021.706142
Moreira FMS, Carvalho Y, Gonçalves M, Haukka K, Young PJW, Faria SM, Pedrosa FO (2002) *Azorhizobiumjohannense* sp. nov. and *Sesbania virgata* (Caz.) Pers.: a highly specific symbiosis. In: Nitrogen fixation: from molecules to crop productivity. Springer, Dordrecht, pp 197–197
Muchane MN, Sileshi GW, Gripenberg S, Jonsson M, Pumarino L, Barrios E (2020) Agroforestry boosts soil health in the humid and sub-humid tropics: A meta-analysis. Agric Ecosyst Environ 295:106899. https://doi.org/10.1016/j.agee.2020.106899
Munroe JW, Isaac ME (2014) N2-fixing trees and the transfer of fixed-N for sustainable agroforestry: a review. Agron Sustain Dev 34(2):417–427
Nainwal RC, Katiyar RS, Singh D, Sharma SK, Singh S, Tewari SK (2016) Agroforestry as a reservoir for biodiversity conservation, soil fertility and mitigation of climatic hazard. In: Holistic Development of Agroforestry. Jaya Publishing House, Delhi, India. pp 203–211
Nambiar PTC, Rupela OP, Kumar Rao JVDK (1988) Nodulation and nitrogen fixation in groundnut (*Arachis hypogaea* L), chickpea (*Cicer arietinum* L) and pigeonpea (*Cajanus cajan* L millsp). In: Biological nitrogen fixation: recent developments. Gordon and Breach Science Publishers, Montruex, pp 21–52

Nygren P, Fernandez MP, Harmand JM, Leblanc HA (2012) Symbiotic dinitrogen fixation by trees: an underestimated resource in agroforestry systems. Nutr Cycl Agroecosyst 94(2):123–160
Ramachandran Nair PK, Mohan Kumar B, Nair VD (2009) Agroforestry as a strategy for carbon sequestration. J Plant Nutr Soil Sci 172(1):10–23
Nair PR, Buresh RJ, Mugendi DN, Latt CR (1999) Nutrient cycling in tropical agroforestry systems: myths and science. In: Agro forestry in sustainable agricultural systems. CRC Press, Boca Raton, FL, USA
Panwar P, Pal S, Reza SK, Sharma B (2011) Soil fertility index, soil evaluation factor, and microbial indices under different land uses in acidic soil of humid subtropical India. Commun Soil Sci Plant Anal 42(22):2724–2737
Pardon P, Reubens B, Reheul D, Mertens J, De Frenne P, Coussement T, Verheyen K (2017) Trees increase soil organic carbon and nutrient availability in temperate agroforestry systems. Agric Ecosyst Environ 247:98–111
Parewa HP, Saresh N, Yadav RP, Meena VS, Meena SK, Choudhary A (2018) Revisiting the agroforestry ecosystem for soil sustainability: lessons from the past and objectives for the future. Trop Ecol 59(4):565–573
Pinho RC, Miller RP, Alfaia SS (2012) Agroforestry and the improvement of soil fertility: a view from Amazonia. Appl Environ Soil Sci 2012:1–11
Pragasan LA, Parthasarathy N (2005) Litter production in tropical dry evergreen forests of South India in relation to season, plant life-forms and physiognomic groups. Curr Sci 88:1255–1263
Prakash D, Benbi DK, andSaroa, G. S. (2018) Land-use effects on phosphorus fractions in Indo-Gangetic alluvial soils. Agrofor Syst 92(2):437–448
Rahman S, Ali A, Raihan A (2022) Soil carbon sequestration in agroforestry systems as a mitigation strategy of climate change: A case study from Dinajpur, Bangladesh. Adv Environ Eng Res 3(4):1–13
Rawat N, Nautiyal BP, Nautiyal MC (2009) Litter production pattern and nutrients discharge from decomposing litter in a Himalayan alpine ecosystem. New York Sci J 2(6):54–67
Ramos HMN, Vasconcelos SS, Kato OR, Castellani DC (2018) Above-and belowground carbon stocks of two organic, agroforestry-based oil palm production systems in eastern Amazonia. Agrofor Syst 92(2):221–237
Rodriguez L, Suárez JC, Pulleman M, Guaca L, Rico A, Romero M, Lavelle P (2021) Agroforestry systems in the Colombian Amazon improve the provision of soil ecosystem services. Appl Soil Ecol 164:103933. https://doi.org/10.1016/j.apsoil.2021.103933
Saha R, Ghosh PK, Mishra VK, Majumdar B, Tomar JMS (2010) Can agroforestry be a resource conservation tool to maintain soil health in the fragile ecosystem of north-East India. Outlook Agric 39(3):191–196
Sahoo GR, Wani AM, Satpathy B (2020) Greening wastelands for environmental security through agroforestry. Int J Adv Res Sci Technol 7:2581–9429
Saputra DD, Sari RR, Hairiah K, Roshetko JM, Suprayogo D, van Noordwijk M (2020) Can cocoa agroforestry restore degraded soil structure following conversion from forest to agricultural use. Agrofor Syst 94(6):2261–2276
Sarvade S, Singh R, Prasad H, Prasad D (2014) Agroforestry practices for improving soil nutrient status. Popular Kheti 2(1):60–64
Sarvade S, Gautam DS, Upadhyay VB, Sahu RK, Shrivastava AK, Kaushal R, Yewale AG (2019) Agroforestry and soil health: an overview. Scientific Publishers, Jodhpur, pp 276–297
Schroth G, Krauss U (2006) Biological soil fertility management for tree-crop agroforestry. In: Biological approaches to sustainable soil systems. CRC Press, Boca Raton, pp 291–303
Sileshi GW, Akinnifesi FK, Ajayi OC, Muys B (2011) Integration of legume trees in maize-based cropping systems improves rain use efficiency and yield stability under rain-fed agriculture. Agric Water Manag 98(9):1364–1372
Sileshi GW, Akinnifesi FK, Mafongoya PL, Kuntashula E, Ajayi OC (2020) Potential of *Gliricidia*-based agroforestry Systems for Resource-Limited Agroecosystems. In: Agroforestry for degraded landscapes. Springer, Singapore, pp 255–282

Singh NR (2020) Soil nutrient dynamics in agroforestry systems. Food Sci Rep 1:23–24
Singh B, Gill RIS, Gill PS (2010) Soil fertility under various tree species and poplar-based agroforestry system. J Res, Punjab Agric Univ 47(3–4):160–164
Singh B, Tripathi KP, Singh K (2011) Community structure, diversity, biomass and net production in a rehabilitated subtropical forest in North India. Open J For 1(02):11
Soil Science Society of America (2008) Glossary of Soil Science Terms. Soil Science Society of America, Madison, WI. ISBN 978-0-89118-851-3
Solanki VK, Parte V, Ranawat JS, Bajpai R (2020) Improved nutrient cycling and soil productivity through agroforestry. Int J Curr Microbiol Appl Sci 11:1724–1729
Sprent JI (2009) Legume nodulation: a global perspective. Wiley-Blackwell, Oxford
Sharma KL (2011) Effect of agroforestry systems on soil quality–monitoring and assessment. Central Research Institute for Dryland Agriculture, Santoshnagar, Hyderabad, India
Sujata D, Nibha G (2011) Microbial bioinoculants and their role in plant growth and development. Int J Biotechnol Mol Biol Res 2(13):232–251
Tondoh JE, Kouamé FNG, Guéi AM, Sey B, Koné AW, Gnessougou N (2015) Ecological changes induced by full-sun cocoa farming in Côte d'Ivoire. Global Ecol Conser 3:575–595
Tennakoon DS, Gentekaki E, Jeewon R, Kuo CH, Promputtha I, Hyde KD (2021) Life in leaf litter: fungal community succession during decomposition. Mycosphere 12(1):406–429. https://doi.org/10.5943/mycosphere/12/1/5
Tsufac AR, Awazi NP, Yerima BPK (2021) Characterization of agroforestry systems and their effectiveness in soil fertility enhancement in the south-west region of Cameroon. Curr Res Environ Sustain 3(1–7):100024. https://doi.org/10.1016/j.crsust.2020.100024
Udwatta RP, Jose S (2012) Agroforestry strategies to sequester carbon in temperate North America. Agrofor Syst 86(2):225–242
Udawatta RP, Gantzer CJ, Jose S (2017) Agroforestry practices and soil ecosystem services. In: Soil health and intensification of agroecosystems. Academic Press, pp 305–333
Udawatta RP, Rankoth LM, Jose S (2019) Agroforestry and biodiversity. Sustainability 11(10): 2879. https://doi.org/10.3390/su11102879
Uthappa AR, Chavan SB, Dhyani SK, Handa AK, Newaj R (2015) Trees for soil health and. Indian Farming 65(3):02–05
Verma P, Bijalwan A, Dobriyal MJ, Swamy SL, Thakur TK (2017) A paradigm shifts in agroforestry practices in Uttar Pradesh. Curr Sci 112:509–516
Wang FQ, Wang ET, Liu J, Chen Q, Sui XH, Chen WF, Chen WX (2007) *Mesorhizobiumalbiziae* sp. nov., a novel bacterium that nodulates*Albizia kalkora* in a subtropical region of China. Int J Syst Evol Microbiol 57(6):1192–1199
Yengwe J, Gebremikael MT, Buchan D, Lungu O, De Neve S (2018) Effects of *Faidherbia albida* canopy and leaf litter on soil microbial communities and nitrogen mineralization in selected Zambian soils. Agrofor Syst 92(2):349–363
Young A (1989) Agroforestry for soil management. CAB international with cooperation ICRAF Nairobi, Kenya
Zomer RJ, Trabucco A, Coe R, Place F, van Noordwijk M, Xu JC (2014) Trees on farms: an update and reanalysis of agroforestry's global extent and socio-ecological characteristics, p 33). Working paper 179. (ed WACISARPD WP14064. PDF). Bogor, Indonesia

Chapter 9
Nutrient Acquisition in Agroforestry Ecosystem Services and Soil Health

Vishnu K. Solanki, R. K. Meena, Vinita Parte, and Tulika Kumari

Abstract Agroforestry helps sustain soil nutrients and recycles waste material in a variety of ways to promote the plant growth. The soil nutrients are highly dynamic in nature and can change into several forms that impact the availability to the plants. By controlling recycling and absorption, agroforestry encourages more efficient nutrient cycling than any other system. Agroforestry is more capable of taking up the water from the deep layers and nutrient pumping than any other crops. In any agroforestry systems, litter accumulation and their decomposition processes play a major role for soil fertility improvement. Numerous variables, including land use pattern, soil microbial population, climate, species character, management-related activities, etc., affect the decomposition rate of litter and the rate of nutrient release into the soil. In agroforestry, soil nutrient dynamics influence biomass output, soil nutrient availability, and total nutrient cycling control.

Keywords Agroforestry · Soil nutrients · Dynamics · Recycling · Soil fertility · Decomposition · Accumulation

9.1 Introduction

Agroforestry is a land use system where trees, shrubs, pastures, agricultural crops, and livestock are cultivated on a single land unit. It also involves the interaction between the system's crops and trees. Nair (1984) states that "Agroforestry is a land use that involves deliberate retention, introduction, or mixture of trees or other woody perennials in crop and animal production fields to benefit from the resultant ecological and economic interactions." Agroforestry encompasses not only the combination of crops, trees, and bushes but also livestock, horticultural crops, etc.

V. K. Solanki · V. Parte
Jawaharlal Nehru Krishi Vishwa Vidyalaya, Jabalpur, Madhya Pradesh, India

R. K. Meena (✉) · T. Kumari
Dr. Rajendra Prasad Central Agricultural University, Pusa, Samastipur, Bihar, India
e-mail: rkmeena@rpcau.ac.in

H. S. Jatav et al. (eds.), *Agroforestry to Combat Global Challenges*, Sustainable Development and Biodiversity 36, https://doi.org/10.1007/978-981-99-7282-1_9

Agroforestry systems are more environmentally friendly and able to use various nutrients, moisture, and sunlight, which are primarily found in India. It is typically not feasible to include trees in monocropping systems in agricultural settings (Nelliat et al. 1974). The fundamental goal of any agroforestry system is to maximize positive interactions between different biological and physical environmental components such as crops, trees, shrubs, animals, and land resources, which helps to create a more diverse, productive, and sustainable system. The main aim of any agroforestry system is to improve soil fertilization. Agroforestry not only enhances the biological but also the physical and chemical properties of the soil. Soil conservation is one of the key services where trees in agroforestry systems provide (Misra 2011). Agroforestry has a lot of potential due to its many advantages and broad application in the areas of nutrient cycling, soil conservation, improved soil fertility, productivity enhancement, bio-fuel, bio-energy, bio-drainage, carbon sequestration, microclimate amelioration, etc. Agroforestry provides a great way to combine soil and water conservation (Dhyani et al. 2015). Trees increase the amount of nutrients inside the root zone of crops by pumping up and capturing nutrients from below the root zone, reducing nutrient losses from erosion and leaching, and providing nitrogen through biological N_2 fixation. Trees can increase the availability of nutrients through increased release of nutrients from soil organic matter and through recycling organic materials. Plants used in agroforestry have the ability to absorb both mobile and available nutrients, such as phosphate. Nitrate can accumulate below the rooting depth of annual crops when sub soils have anion exchange sites to absorb the nitrate and when pests and nutrients other than nitrogen restrict the crops' ability to produce. Built-up subsurface nitrate can be effectively removed with *Sesbania sesban*. Sesbania (Daincha), when grown in place of natural grass fallows, is more efficient at retaining subsurface water and reducing nutrient losses. Mineralization of soil organic matter is a good source of nitrogen and phosphorus that are available to plants. The application of organic materials can help recycle phosphorus, which is necessary to meet crops' phosphorus requirements. Agroforestry with continuous crop production on phosphorus-deficient soils typically needs phosphorus inputs, such as fertilizer (Roland et al. 1996).

9.2 Dynamics of Soil Nutrient

Practices in agroforestry and the type of soil can also affect the quantity and quality of litter, as well as its rate of decomposition, intercropping pattern, orientation, and management techniques. In addition to providing food, fuel, wood, building materials, and raw materials for cottage industries and small-scale forest-based businesses, these trees also enrich the soil with vital nutrients (Ghosh et al. 2011). It might be feasible to increase the soil's carbon content by planting trees and crops which is also beneficial in resolving numerous global problems, such as carbon sink, global warming, and climate change (Kumar et al. 2006). Reclamation of salt-affected soils by using agroforestry tree species (Fig. 9.1 shows the sources of

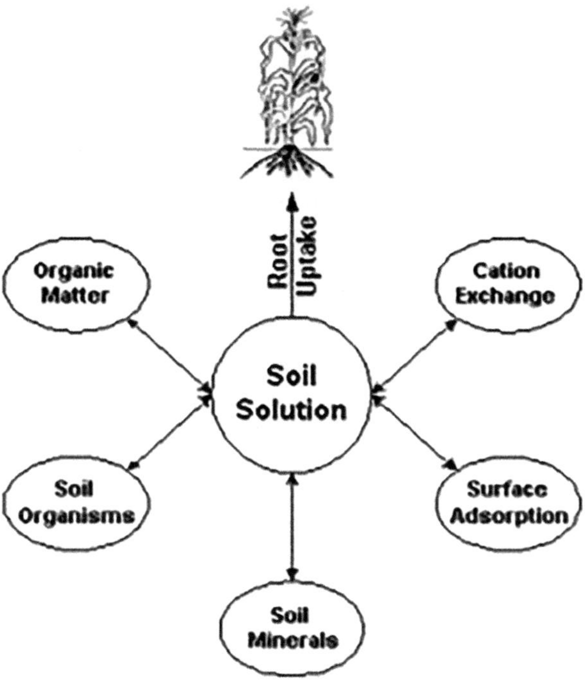

Fig. 9.1 Main sources of plant nutrients that exist in soils (Bierman and Rosen 2019)

plant nutrients in soils). Using leaf litter in agroforestry practices increases the amount of organic matter in the soil. It enhances biological nitrogen fixation in soil and preserves the population dynamics of advantageous microorganisms. Microbiological activity, nutrient cycling, and all soil activities support ecosystem services and functions (Jhariya et al. 2013).

In agroforestry systems, reasons for soil nutrient dynamics are as follows:

- Turnover estimation of nutrients.
- On nutrient supplement impact of fine and coarse roots.
- Rate of litterfall and its associated nutrient dynamics.
- Estimation of leaf litter, litter decomposition, rate of decomposition, and its pattern of nutrient release.
- For assessing the soil microbial population activity.

9.3 Nutrient Cycling of Soil

Through the process of nutrient cycling, nutrients are moved from one component of the plant–soil–animal environment system to another. In agroforestry, rainfall, organic residues, and fertilizers from outside the system, crops and trees, and biological nitrogen fixation all contribute to the addition of nutrients. The harvesting of trees, crops, livestock products, volatilization, leaching, soil erosion, and other processes are the causes of the nutrient losses. The system's produced manure, litter fall, and tree pruning are the sources of the nutrient transfers. Transcend nutrient cycling by improving soil biological and physical processes. Nutrient cycling will help increase soil fertility, which is beneficial for trees and crops, and increase the availability of water and nutrients. Through increased nutrient availability, increased nutrient supply in the crop root zone, and decreased nutrient loss through the control of soil erosion and leaching, trees in agroforestry facilitate the cycling of nutrients to crops (Nair 1979). Through biological nitrogen fixation and nutrient recycling, trees increase the amount of nutrients available to the soil surface. By increasing biological activity, enhancing soil structure, adding leaf litter to the system, and allowing biomass to decompose, it is possible to increase the availability of nutrients for crops. Trees use their deep roots to absorb nutrients that seep out of the crop root zone, thereby reducing nutrient losses and utilizing their extensive root systems to stop soil erosion.

9.4 In Agroforestry Major Processes for Soil Nutrient Dynamics

Tree litter fall: In agroforestry, the majority of the litter produced is made up of leaves, but there are also dead bark, branches, twigs, and other organic materials. Agroforestry systems contribute significantly to soil nutrient recycling and help to maintain and improve soil fertility through the addition of essential elements like potassium, nitrogen, and phosphorus to the soil. Many factors, including the population of decomposers, land use, climate, plant species, etc., also affect soil litter fall and nutrient release from tree and crop residues (Fernandes et al. 1997).

Litter fall decomposition—over time, the accumulation of litter on the agroforestry soil floor began to fragment into smaller pieces, converting the litter to mineral nutrients, water, and carbon dioxide through physical and chemical processes (Lambers et al. 1998). The process of mineralization and decomposition in soil is greatly influenced by soil microbes, and soil fauna including phosphate, non-symbiotic N_2 fixing bacteria, thiosulphate oxidizing bacteria, and solubilizing bacteria help to make soil more nutrient-available (Fig. 9.2 depicts the soil nutrient cycling in the agroforestry system through litter fall and decomposition). The supply of plant nutrients depends on various variables, including the microbial community, the availability of nutrients in the soil, climatic conditions like temperature, moisture, and aeration, the quantity and quality of plant species, etc.

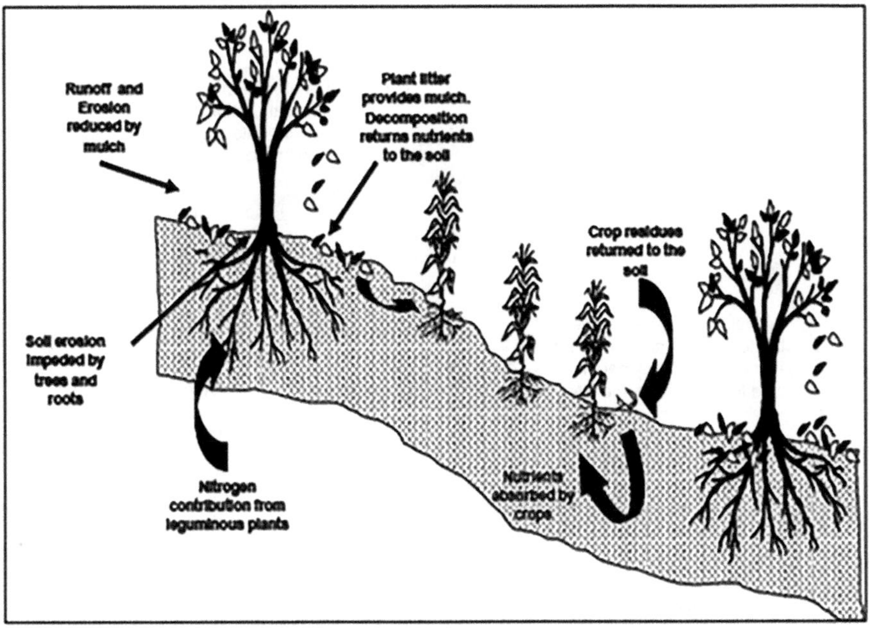

Fig. 9.2 Soil nutrient cycling through litterfall and decomposition in an agroforestry system (Kang et al. 1986)

9.5 Improvement of Soil in Agroforestry

Through leaf litter, woody compound pruning, and atmospheric fixation, agroforestry contributes nutrients. Certain nutrients that are generally believed to be inaccessible to crops may be introduced into the system by tree roots from deeper soil layers that are beneath the area where the roots of the annual crop grow. The topsoil layer and decomposing organic matter, such as fruits, flowers, branches, leaves, etc., aid in enrichment. The two biggest positive effects on the tree's soil may be increased soil structure and nutrient availability (Sanches et al. 1983; Nair 1989). Increasing inputs such as organic matter, atmospheric fixation, and biological nitrogen fixation are signs of improved soil under agroforestry systems. The physical properties of the soil and its ability to hold water for a long time are increased by the addition of organic matter. By modifying nutrient cycling through agroforestry, it is possible to make even more efficient use of nutrients, whether they come from fertilizer or the natural removal process. Compared to pure agriculture systems with agroforestry systems, agroforestry systems support a more closed nutrient cycle. Figure 9.3 depicts the decomposition process in terrestrial ecosystems. The process of soil improvement under agroforestry systems is acknowledged through:

- Increasing inputs such as atmospheric fixation, biological nitrogen fixation, and organic matter.

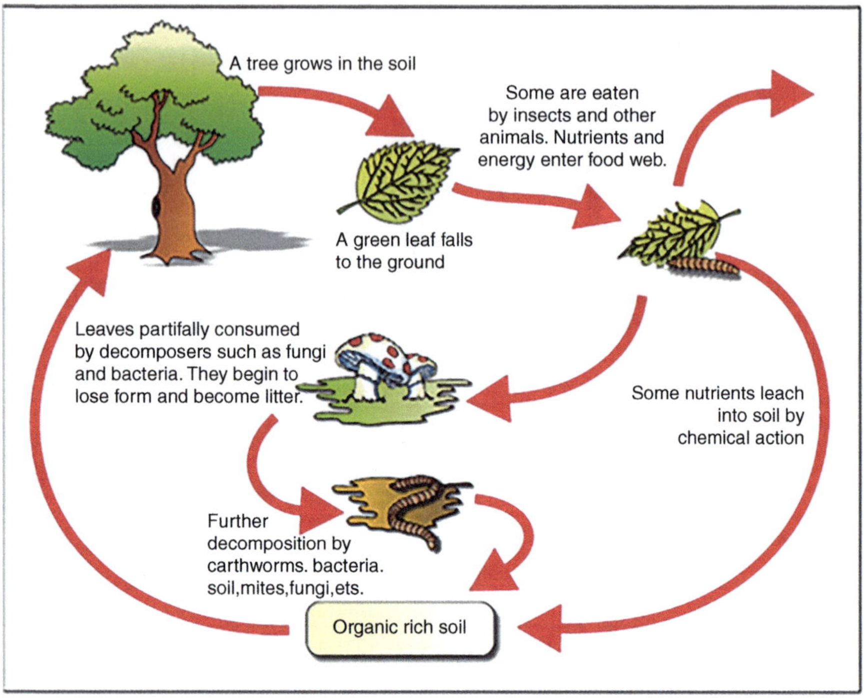

Fig. 9.3 Decomposition process in terrestrial ecosystem (NCERT 2006)

- Reducing losses of organic matter and nutrient.
- Additions of organic water holding capacity are improving and soil physical properties.
- Nutrients are recycled.

9.6 Nutrient Loss Under Agroforestry

Nutrients are uptaken by plants and stored as biomass in the plants and plant parts, resulting in a net loss of nutrients from the system as a result of crop harvesting. It would be expected that inorganic fertilizers, in the absence of any external nutrient inputs, would lower system productivity by recycling nutrients through crop residue. Leaching and erosion, in addition to the removal of biomass, are crucial components of the system in the depletion of nutrients. In agricultural systems, biomass is removed from crops during harvesting (Sanches et al. 1983). In a similar vein, a comprehensive strategy for utilizing trees was discovered from the tree-based system to remove a sustainable quantity of nutrients. Lower levels of organic carbon have generally been found in cultivated soils compared to forest soils (Misra 2011).

9.7 Nutrient Recovery Under Agroforestry

More nutrients are added to the soil through litter fall, which occurs when trees translocate nutrients from deeper soil. Organic residues and leaf shedding also contribute to the deposition of nutrients on the soil surface. Nutrients are released into the soil as a result of the mineralization and breakdown of organic matter. Agroforestry systems release substantially fewer nutrients into the atmosphere than monoculture tree plantations (Young 1987). According to the agroforestry system, soil organic matter can only be preserved by adding biomass from whole trees to the soil (Table 9.1 shows the variation in nutrients in open field and agroforestry systems). To demonstrate that nutrients from tree pruning reach the crop, alley cropping and the biomass transfer system are employed (Szott et al. 1981).

9.8 Soil Enrichment Through Agroforestry

Higher nutrient status and crop fields of soil where trees were previously grown and near trees maintain soil fertility, which is a major trend in agroforestry. Due to their substantial contributions, many tree species are prized in traditional agroforestry. Phosphorus and nitrogen levels in organic matter are higher beneath trees than they are at bare sites (Shankermarayan 1984). Under an agroforestry system with eucalyptus hybrid and *Populus deltoids* canopies, the increase in soil nutrients was 33–83% organic carbon, 38–69% available nitrogen, 3–33% available phosphorus, and 8–24% available potassium (Anonymous 1987). More nutrients are present in the soil under the Prosopis-based agroforestry system than in open fields (Aggarwal 1980). In comparison to open fields, the silvipastoral system has higher levels of organic carbon, nitrogen, and phosphorus (Hazara 1990). A significant rise in nutrient storage in the trees results in improved resource sharing, nutrient cycling, and a tree-based crop system with topsoil compartments (Fig. 9.4). Improved moisture status is attained under trees by lowering evapotranspiration, and more soil organic matter increases the soil's ability to retain moisture, which is known to support improved soil structure. The nutrient pool is enriched by the addition of trees and organic matter, which also checks soil erosion and decreases losses. Sustainable land resource use and tree-based cropping systems can be beneficial.

Table 9.1 Variation of nutrients in agroforestry and open field systems (Misra 2011)

S. no.	Tree species	Organic matter (%)	Nitrogen (%)	Phosphorus (%)
1	*Prosopis cineraria*	250	22.9	633
2	*Prosopis juliflora*	250	10.3	409
3	Open field	203	7.7	370

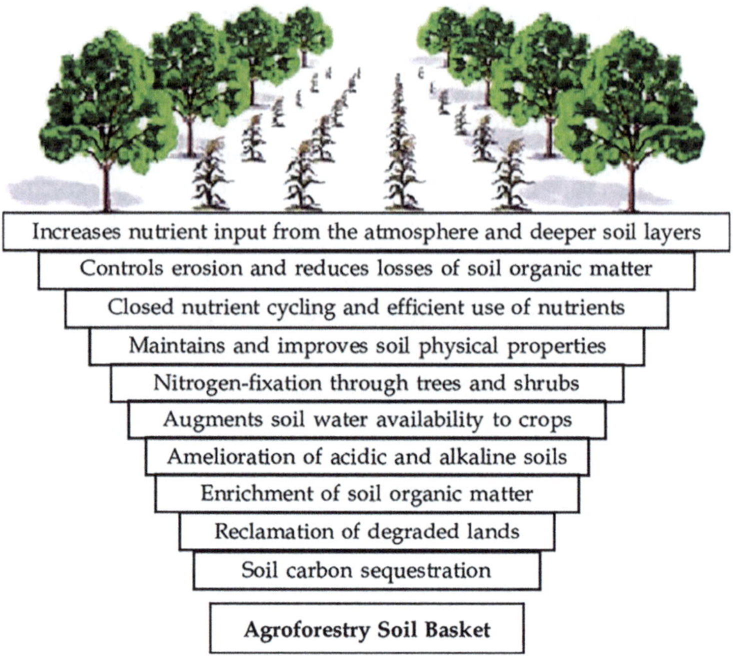

Fig. 9.4 Maintain soil health with the help of agroforestry (Uthappa et al. 2017)

9.9 Nutrient Cycling Under Agroforestry

The nutrient cycle in the system is made up of inputs into gains, outputs from losses, and internal turnover or transfer. Low rates of output and high rates of turnover characterize closed, efficient nutrient cycling systems in forest ecosystems (Fig. 9.5). Normal agricultural systems have low system turnover and relatively high input costs and losses. In agroforestry systems, the cycling of nutrients oscillates between two extremes: more nutrients are recycled by plants before they are eliminated from the system. The transfer of nutrients from one system component to another without compromising the overall productivity of the system's components is the main way that agroforestry differs from land use systems. This allows for higher rates of turnover and the potential for system management. The tree's deep roots allow it to absorb nutrients from soil depths that crop roots are unable to (Raj et al. 2017).

9.10 Nutrient Cycling Through Trees and Agroforestry

- Use of nitrogen-fixing tree species to increase the gains from synthetic fixation.
- Recycling as litter by uptake tree root systems of associated mycorrhiza.

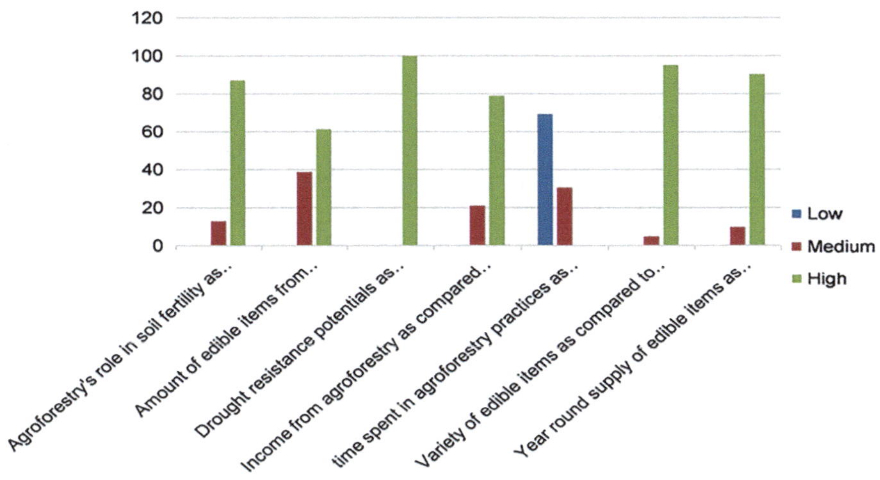

Fig. 9.5 Agroforestry, nutrient cycling, and soil productivity (Solanki et al. 2020)

- As organic residues nutrient deficiencies provide a balanced nutrient supply.
- Nitrogen accumulates in tree fodder.

9.11 Nitrogen-Fixing Trees

- Increase nitrogen inputs to agroforestry systems by using nitrogen-fixing trees that contain a majority of legumes and a minority of non-legumes in quantities comparable to those from herbaceous legumes.
- Substantially improve rates of nitrogen fixation by selection of genotypes of plant and rhizobium inoculation in some cases.
- Both above- and below-ground nitrogen are transferred from nitrogen-fixing through root residues within one season, pruning, litter decomposition, and a longer period via soil organic matter.
- Increasing nitrogen inputs to plant–soil systems does not improve or worsen the performance of nitrogen-fixing trees compared to herbaceous legumes. Non-nitrogen-fixing tree mulches can also be supplies of nitrogen (Raj et al. 2017).

9.12 Some Nitrogen-Fixing Tree Species

Acacia albida, Acacia auriculiformis, Acacia catechu, Acacia aneura, Acacia dealbata, Acacia decurrens, Acacia farnesiana, Acacia implexa, Acacia leucophloea, Bauhinia variegate, Buteamonosperma, Cassia fistula, Cassia siamea, Casuarinaequisetifolia, Dalbergialatifolia, Dalbergiasissoo, Delonixregia,

Gliricidiasepium, Acacia mearnsii, Hardwickia binate, Acacia melanoxylon, Leucaenaleucocephala, Acacia mollissima, Moringaoleifera, Acacia nilotica, Acacia planifrons, Parkinsonia aculeate, Acacia Senegal, Albizia chinensis, Pithecellobium dulce, Albizia lebbeck, Prosopis alba, Albizia procera, Prosopis chilensis, Alnus nepalensis, Prosopis cineraria, Alnus nitida, Samanea saman, Sesbania bispinosa, Saraca indica, Sesbania grandiflora, Tamarindus indica, etc.

9.13 Choose Nitrogen-Fixing Trees

- To fix huge quantities of nitrogen by the potential of nitrogen-fixing tree species.
- Eco-friendly and sustainable viable alternative.
- Restocking of nutrient pumping and soil fertility.

9.14 Help with Nitrogen-Fixing Trees in Agroforestry

Many leguminaceae trees like *Sesbania* spp., *Gliricidia* spp., *Dalbergia sissoo, Acacia nilotica, Leucaena leucocephala,* etc. and several non-legumes trees like *Alnus* spp., *Casurina equisetifolia,* etc. are important to fix about 50 to 500 kg nitrogen per ha (Raj et al. 2017).

9.15 Biological Nitrogen-Fixing Agroforestry

Leguminous trees, which make up the majority of agroforestry trees, recycle nutrients, add organic matter, and improve soil quality by fixing nitrogen biologically. Many tree species like Alnus, Acacia, and Leucaena have annual nitrogen fixations of up to 400–500 kg, 270 kg, and 100–300 kg, respectively, per hectare. The fixed nitrogen contributes to increased soil fertility and may have symbiotic benefits for crop growth. Nitrogen is added when certain tree species are pruned, or a substantial amount of soil organic matter-legumes-is left over after pruning and is not fully absorbed by the first crop, suggesting a longer term nitrogen advantage than an immediate one. The release of nutrients and tree components, such as wood, fruit, twigs, leaves, etc., is distributed over time by the varying rates of decomposition of the various tree components. Biological nitrogen fixation mainly occurs through symbiotic and non-symbiotic relations between plant roots and microorganisms. Certain non-leguminous species have symbiotic relationships with actinomycetes, which is a genus of Frankia bacteria, while a number of legumes have relationships with the bacteria rhizobium. Non-symbiotic fixation can play a major role in natural ecosystems and is influenced by free-living soil organisms from external systems with relatively low nitrogen requirements (Nair 1993a, b).

9.16 Process of Nutrient Pumping

Certain beneficial effects on soils, such as the physical stature enhancement of compact soil layers, the capture of leached nutrients, and the enrichment of soil with carbon through root turnover, are attributed to tree root systems. Trees can absorb water and nutrients from deeper soil layers where the roots of herbaceous crops cannot typically grow because of their spreading and deep roots. Nutrients are taken up by the deeper soil profile and subsequently deposited on the surface layers by tree litter fall and a variety of tree mechanisms collectively referred to as nutrient pumping (Fig. 9.6).

In general, topographic, climatic, soil, and tree species characteristics influence this process because their deep root systems aid in the pumping of nutrients and water. Trees with low-moisture content soils outperform those with high-moisture content soils in this regard. (Makumba et al. 2009; Schroth and Sinclair 2003; Schroth 1999).

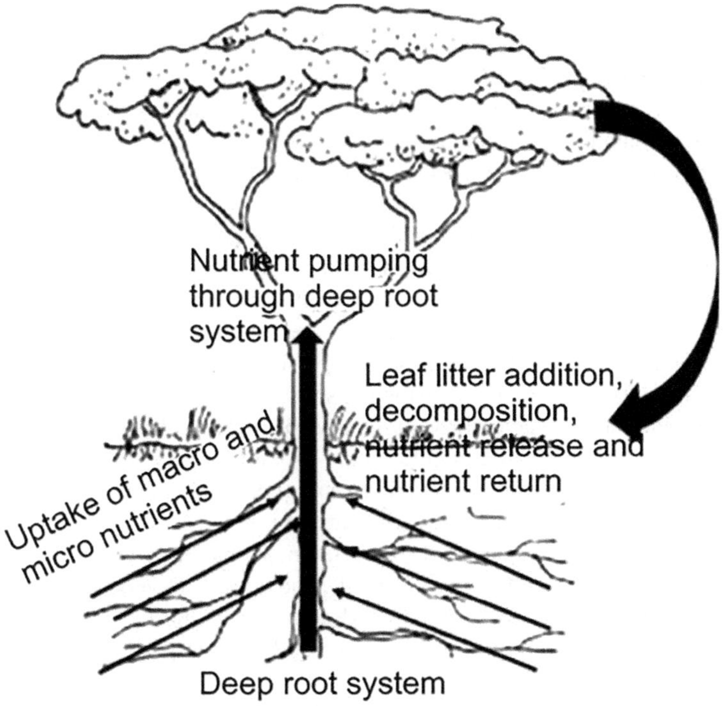

Fig. 9.6 Nutrient pumping through multipurpose trees (MPTs) species' deep roots in agroforestry systems(Sarvade 2019)

9.17 For Agroforestry Desirable Characters in Tree Species

In agroforestry systems, choosing a tree species should take into account a number of desirable traits. Any tree species can be put to multiple uses, but no single species possesses all desirable traits.

- Selecting the tree species should not interfere with soil moisture.
- Selecting the tree species should have very little water requirement for agroforestry.
- For water, trees should not compete with crops.
- Tree species can draw water from deep strata of the soil so tree roots should be deep tap.
- For plant nutrients, tree species should not compete.
- Tree should utilize less plant nutrients.
- In building soil fertility, trees should help.
- Fix atmospheric nitrogen in their roots should be preferred by leguminous tree species.
- The root growth characteristics and root system should explore soil layers.
- For sunlight tree species should not compete.
- On the crops, tree species should not interrupt sunlight falling.
- Tree habit should be light branching.
- Trees promote better crop, pasture growth, and yield and permit the penetration of light into the ground trees.
- If tree species possess dense canopy, they can withstand pruning operation.
- Wider adaptability in selecting tree species.
- For agroforestry, a tree species selected combination must have a wider adaptability.
- Tree species should have soil stabilization attributes and shelter-conferring.
- Some tree species are especially helpful in providing protection for livestock, crops, and soils because of their adaptability and inherent growth habits. Many tree species have been extensively used in soil erosion control, e.g., *Casurinaequisetifolia,* Willows (*Salix* spp.), Poplars (*Populus* spp.), etc. because of their ability to grow in water-logged soils and extensive root system.
- Tree species should have nitrogen fixation attributes and nutrient cycling.
- Trees can play an important role within an agroforestry system leached down through the soil profile, recycling nutrients, and minerals released from weathering parent material like sediments and rocks.
- These nutrients are used in the development and growth of the tree many returning to the topsoil in the form of dead seeds, flowers, twigs, and leaves which are eaten by animals or slowly decompose on the surface.
- In maintaining the nutrient status of the soil by role-play of all trees through recycling.
- A thick mat of leaves on the ground deciduous trees drop most of their leaves in autumn, whereas throughout the year, most evergreen species maintain some level of litterfall.

- Many tree species can convert atmospheric nitrogen into organic nitrogen through complex symbiotic relationships between their fine roots and Rhizobium bacteria for their use.
- On the roots, the bacteria form nodules that can convert nitrogen gas, usable nitrogen for the plant as it is in the atmosphere.
- Some non-leguminous ones and leguminous trees like Casuarina spp. as well as Prosopis, Leucaena, and Acacia fix the atmospheric nitrogen.
- Generally high in nitrogen the litter of these nitrogen-fixing trees, thus increases the nitrogen status of the soil.
- Easily decomposable leaves have the character of selected tree species.
- For agroforestry, the suitable tree species with a fast rate will be one in which fallen leaves decompose.
- The leaves of most of the legume tree species easily, decompose quickly, and small in size, and add a large quantity of nutrients and organic matter to the soil.
- For the agroforestry system, some tree species having broad leaves like banyan, mango, and trek should not be preferred.
- They require a longer time for decomposition and also contain more fiber matter. Broad leaves block their photosynthetic activities when fall on the tender crop plants.

9.18 Plant Nutrient Sources and Losses

Through root uptake, plants receive mineral nutrients from the soil solution. In soil, these soluble nutrient sources are as follows:

- Decomposition of soil microorganisms, animal remains, and plant residues.
- Weathering of soil minerals.
- Application of fertilizers.
- By legumes nitrogen fixation.
- Atmospheric depositions like sulfur and nitrogen from N-fixation or acid rain by lightning discharges.
- From flooding and erosion nutrient-rich sediment deposition.

Mineral nutrients from the soil system may be lost and plant uptake become unavailable. When nutrients leak into groundwater, rivers, and lakes, they cause environmental damage. They are also costly and wasteful. Nutrient losses happen by.

9.18.1 Runoff

Runoff is the term for the loss of dissolved nutrients from the water as it moves through the soil surface.

9.18.2 Erosion

Erosion is the loss of nutrients in or linked to soil particles that are removed from fields due to water and wind movement.

9.18.3 Leaching

Leaching loss of dissolved nutrients out of the field through drain lines or the soil moves down to groundwater.

9.18.4 Atmosphere Gaseous Losses

Mostly losses of various nitrogen types by denitrification and volatilization.

9.18.5 Removal of Trees

Removal of nutrients and plant uptake in harvested goods from the field.

9.19 Increase Nutrient Uptake by Agroforestry

Nutrients recovered from deteriorating rock and lower soil layers, trees can increase the amount of nutrients added to agroforestry systems. Most of the trees species have deep root system, which is often cited as a benefit for agroforestry systems. Trees are able to absorb nutrients from soil depths that agricultural roots are unable to the top 20 cm of the soil are the majority of the feeder roots of many common trees. Although it is difficult that the minerals in weathering rock cannot be identified, uptake from deep soil horizons and weathering rock is quite likely. The fact that some trees can thrive in areas with virtually no soil and that their roots can pierce weathered rock serve as indirect evidence. A major factor is the greater nutrient cycling caused by atmospheric deposition in humid regions compared to dry ones. It includes nutrients that are dissolved in rain and those that are carried in dust (Raj et al. 2017).

Generally speaking, agroforestry techniques improve soil's biological nitrogen fixation, which raises soil organic matter, multiplies the populations of helpful microorganisms, and adds leaf litter (Fig. 9.7). In addition to improving soil nutrient cycling, the additional organic matter serves as a source of energy. Additionally

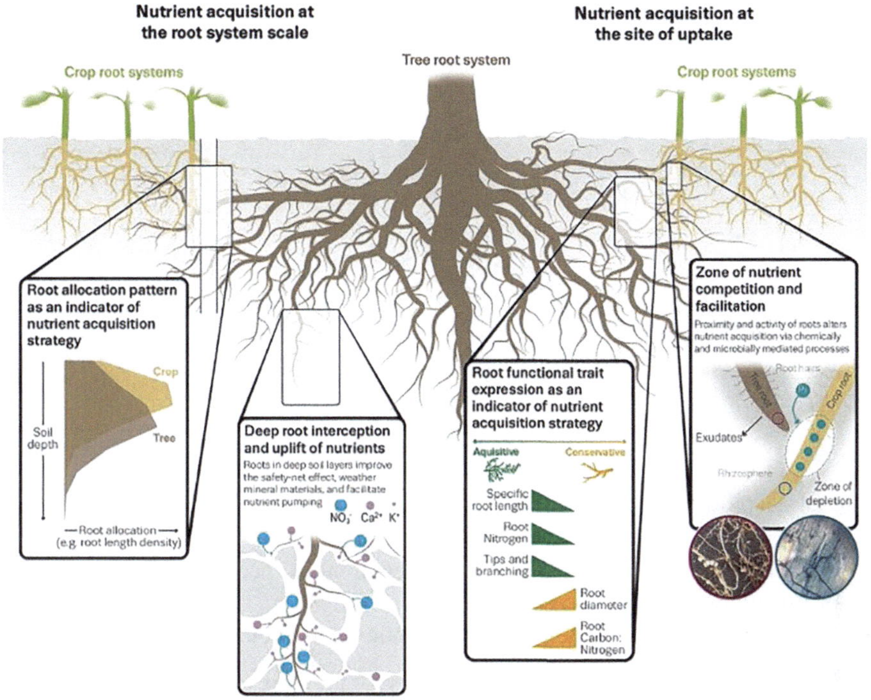

Fig. 9.7 Tree & crop root distribution, function, and processes for nutrients in agroforestry (Marney and Kira 2019)

enhances the system of soil aggregates and modifies the soil microclimate (Raj et al. 2017).

9.20 Through Nitrogen Fixation Agroforestry Systems Increase Inputs to the Plant–Soil System

In agroforestry systems, nitrogen inputs can be significantly increased by adding nitrogen-fixing trees. The biological nitrogen-fixing process involves both symbiotic and non-symbiotic processes. Fixation is carried out independently of plants by non-symbiotic soil organisms (Fig. 9.8). The amounts of fixed nitrogen are negligible compared to the greater needs of agroecosystems, despite being significant in natural ecosystems. The amount of organic matter in the soil and the level of microbiological activity affect its rate (Raj et al. 2017). The process of symbiotic fixation involves the bacterial attachment of nitrogen to plant roots. The mutualisms are as follows:

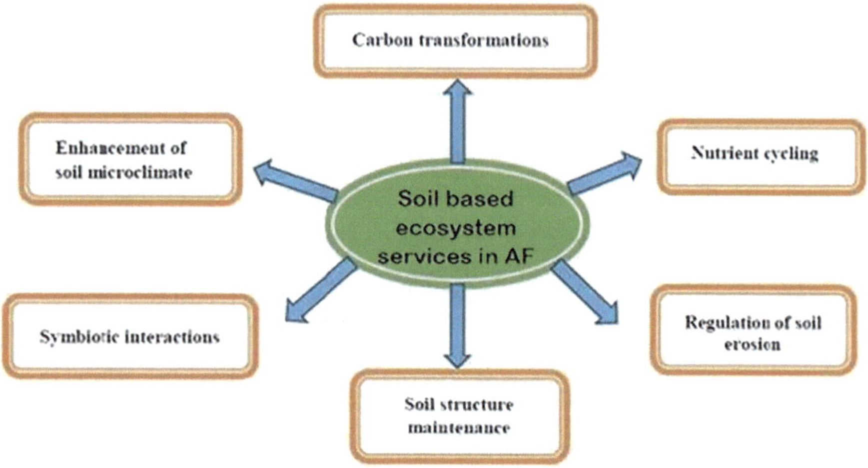

Fig. 9.8 Soil-based ecosystem services in agroforestry (Solanki et al. 2020)

- Forming nodules on root between many leguminous species and bradyrhizobium or rhizobium.
- Between a small number of Frankia and non-leguminous genera.

Many of the chosen shrub and tree species for agroforestry are legumes found in nature, such as Sesbania, Gliricidia, Erythrina, Calliandra, Leucaena, etc., and are fast-growing nitrogen-fixing trees. Additional legumes that fix nitrogen are Prosopis, Inga, Albizia, and several Acacia species. Casuarinaceae is the most commonly used non-leguminous nitrogen fix in the tropics, while Alnus is used in the temperate zone (Raj et al. 2017).

9.21 Erosion Control by Agroforestry

- On erosion factors effect of agroforestry.
- Erosivity of rainfall.
- Erosivity of soil.
- Runoff reduction.
- Cover the ground surface.

9.22 Agroforestry Can Recycle Nutrients

In agroforestry, where trees were previously grown, trees preserve the nutrient status of the soil or its fertility. Due to their substantial contributions, many tree species are prized in traditional agroforestry. A significant increase in the nutrients stored in the trees in the topsoil compartments of tree-based crop systems results in increased efficiency in resource sharing and nutrient cycling. Moisture status was enhanced by lowering evapotranspiration beneath trees by using canopy shade. Better soil structure is being promoted by increased soil organic matter, which enhances the soil's ability to retain moisture. The addition of organic matter in soil from tree species helps reduce soil erosion and similarly improve the nutrient pool. Tree-based cropping systems have the potential to contribute to the sustainable use of land resources. Trees that translocate nutrients from deeper soil through litter fall, leaf shedding, and organic residues contribute significantly to the amount of nutrients added to the soil and on the soil's surface. Agroforestry systems release significantly less nitrogen emissions than single-tree crop plantations. In agroforestry, trees in particular are leguminous, and all trees improve soil via nutrient recycling, the addition of organic matter, and biological nitrogen fixation (Fig. 9.9). Improved soil fertility may be aided by fixed nitrogen, which also has a symbiotic benefit on crop growth. Legumes absorb a significant amount of nitrogen from the first crop or

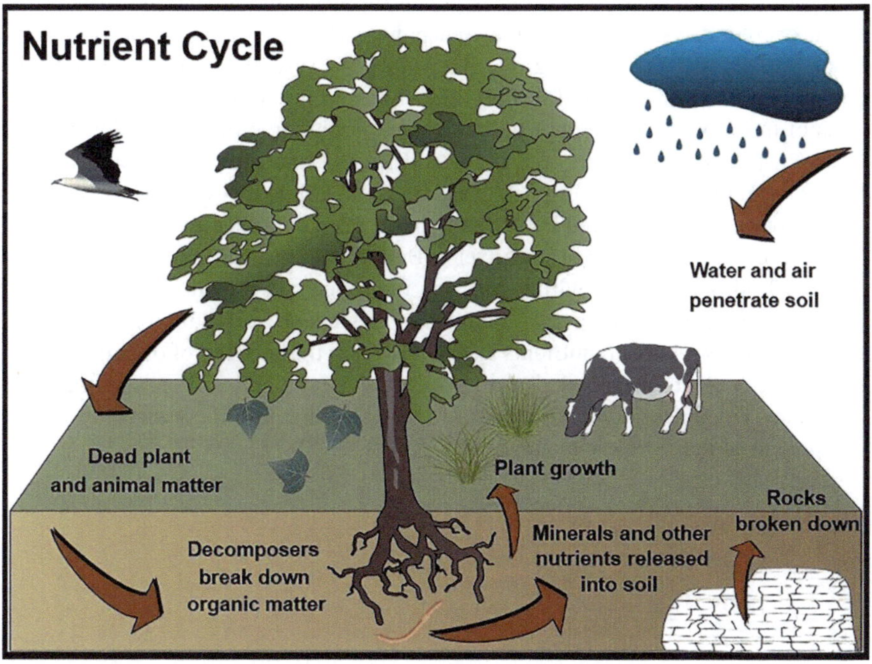

Fig. 9.9 The basic nutrient cycle

from the pruning of certain tree species, but this nitrogen is absorbed over time rather than immediately through the soil's organic matter. Different tree components, like wood, fruit, twigs, leaves, etc., decompose at different rates, which help to distribute the release of nutrients over time.

By increasing the amount of carbon in the soil through root turnover, tree root systems can physically enhance compact soil layers or intercept nutrients that have leached. Because of their deep, spreading roots, trees can absorb water and nutrients from deeper soil layers that herbaceous crop roots are unable to. Nutrient pumping describes the process of removing nutrients from a deeper soil profile and subsequently depositing them on the surface by layers, trees, and other mechanisms. In this process, topographic, climatic, soil, and tree characteristics are all important. Trees aid in the pumping of water and nutrients in low-moisture soils, while deep root systems are characteristic of high-moisture soils (Makumba et al. 2009; Schroth and Sinclair 2003; Schroth 1999).

9.23 How to Improve Nutrient Cycling Efficiency from Management?

Compared to agricultural systems, agroforestry practices result in a more efficient use of soil nutrients through natural processes, whether they are added externally or made available. Increased horizon depth in the soil can enhance nutrient uptake. While deep root systems of trees may reach these sites, the shallow roots of common crops prevent them from doing so. In agroforestry systems, nutrient pumping is a crucial part of increasing soil fertility. By choosing the right species and combinations of trees, one can increase the benefits of symbiotic nitrogen fixation by trees. Nitrogen fixation inputs into the plant–soil system and nitrogen addition through litter or pruning may cause an internal transfer within the system. A significant portion of the nitrogen found in the litter comes from the soil, either from reserves in the soil or from additional fertilizers. Better soil productivity and eventually improved nutrient cycling will result from improved soil organic matter status as a result of management practices. The primary advantage of tree biomass in agroforestry systems is related to nutrients and comes from the addition of organic matter to the soil. It is common knowledge that soil productivity and soil organic matter are important. The potential for reducing nitrogen loss through soil conservation is a key component of agroforestry management (Raj et al. 2017).

9.24 Conclusion

In agroforestry, improving the soil's nutrient status, improving yield productivity, and maintaining the ecosystem through microbial population dynamics are all made possible by carefully pairing multipurpose trees with field crops that have nitrogen-fixing trees. Improvements in the soil beneath agroforestry and trees are closely associated with increases in organic matter, whether in the form of soil carbon or surface litter. There is a lot of potential in the soil for agroforestry systems to raise carbon stocks. Agroforestry is crucial for preserving soil fertility and controlling nutrient cycling in agroecosystems. The decomposition process, together with the timing and quantity of nutrients released to meet the needs of the components, determines the amount of nutrients released from components and the effectiveness of agroforestry. When it comes to nutrient cycling, agroforestry systems encourage more closed-through synchronization, recycling, and uptake than any other type of agricultural system. Through rhizospheric microbial activity, litter fall, tree root regeneration, disintegration, and decomposition contribute continuously to the nutrient supply in the agroforestry system. The type of species, the prevailing climate, and the governing elements of another system all influence this process. Nutrient cycling takes place to varied degrees in different land use systems. Leguminaceae trees are the best component and field crop because of their low output and high nutrient inputs, which boost the efficiency of nutrient cycling. However, this depends on the type of field crops, trees, land types, and combinations of trees and crops. In order to improve soil productivity and nutrient cycling, agroforestry can better utilize nutrients, increase nutrient uptake, and improve the soil's organic matter status.

References

Aggarwal RK (1980) Physiological status of soil under khejri (Prosopis cineraria). In: Mann, Saxena (eds) Khejri (Prosopiscineria) in the Indian desert, pp 32–37

Anonymous (1987) Annual report on AICRAP. G B Pant University of Agriculture and Technology, Pantnagar, Uttar Pradesh

Bierman PM, Rosen CJ (2019) Nutrient cycling and maintaining soil fertility in fruit and vegetable crop systems. By Climate-Woodlands

Dhyani SK, Handa AK, Uma. (2015) Agroforestry research in India: present status and future perspective. In: Dhyani SK, Newaj R, Alam B, Dev I (eds) Organized Agroforestry Research in India: Present Status and Way Forward. Boitech Book Publisher, Delhi

Fernandes ECM, Biot Y, Castilla C, Canto A, Matos JCS (1997) The impact of selective logging and forest conversion for subsistence agriculture and pastures on terrestrial nutrient dynamics in the Amazon. Ciecia Cult 49:34–47

Ghosh SR, Wadud MA, Mondol MA, Rahman GMM (2011) Optimization of plant density of Akashmoni (Acacia auriculiformis) for production of fuel wood in the bunds of cropland. J Agrofor Environ 5(2):1–6

Hazara R (1990) Forage and soil productivity for Subabul agroforestry systems. Proceedings of the International Symposium On National Resource Management sustainable agriculture. Indian Society of Agroforestry, 6–10, Feb, New Delhi
Jhariya MK, Raj A, Sahu KP, Paikra PR (2013) Neem: a tree for solving global problems. Indian J Appl Res 3(10):66–68
Kang BT, van der Kruijs ACBM, Couper DC (1986) Alley cropping for food crop production in humid and subhumid tropics. Paper presented at the Alley Farming Workshop, March 1986, Ibadan, Nigeria
Kumar R, Pandey S, Pandey A (2006) Plant roots and carbon sequestration. Curr Sci 91:885–890
Lambers H, Chapin FS, Pons T (1998) Decomposition. In: Plant physiological ecology. Springer, New York, pp 495–502
Makumba W, Akinnifesi FK, Janssen BH (2009) Spatial rooting patterns of gliricidia, pigeon pea, and maize intercrops and effect on profile soil N and P distribution in southern Malawi. Afr J Agric Res 4(4):278–288
Marney E. Isaac & Kira A. Borden (2019). Nutrient acquisition strategies in agroforestry systems. Plant and soil, 444: 1–19
Misra PK (2011) Soil fertility management in agroforestry system. Int J Biotechnol Biochem 7(5): 637–644
Nair PKR (1979) Agroforestry Research: a retrospective and prospective appraisal. Proceedings of the International Conference on International Cooperation in Agroforestry. ICRAF Nairobi, pp. 275-296
Nair PKR (1984) Soil productivity aspects of agroforestry. ICRAF, Nairobi, p 85
Nair PKR (1989) Agroforestry systems in the tropics. Kluwer, Norwell, MA
Nair PKR (1993a) An introduction to agroforestry. Kluwer Academic Publishers, Dordrecht, p 499
Nair PKR (1993b) An introduction to agroforestry. Kluwer, Dordrecht, p 489
National Council of Educational Research and Training (NCERT) (2006) Ecosystem. In: NCERT class 12 biology, New Delhi, p 244
Nelliat EV, Bavappa KV, Nair PKR (1974) Multistoried cropping. World Crops 26:6
Raj A, Jhariya MK, Toppo P, Oraon PR (2017) Role of agroforestry in nutrient cycling, Kishan World. J Agric Rural Dev 44(7):38–40
Roland J. Buresh, Jama B. and NdufaJ K (1996) Agroforestry trees for nutrient cycling and sustainable management. East Afr Agric For J 62(1&2):115–127
Sanches PA, Villachica JHV, Bandy DE (1983) Soil fertility dynamics after decaying on tropical rain forest in Peru. Soil Sci Soc Am J 47:171–178
Sarvade S (2019) Agroforestry & soil health: an overview. In Book: Agroforestry for climate resilience and rural livelihood
Schroth G (1999) A review of belowground interactions in agroforestry, focusing on mechanisms and management options. Agrofor Syst 43:5–34
Schroth G, Sinclair FL (2003) Trees, crops, and soil fertility concepts and research methods. CABI Publishing, UK/USA, p 423
Shankermarayan KA (1984) Silvipastoral system: a programmatic approach to efficient integrated land management in agroforestry. In: Shankermarayan KA (ed) Arid and semi-arid zones. CAZARI, Jodhpur, pp 137–142
Solanki VK, Parte V, Ranawat JS, Bajpai R (2020) Improved nutrient cycling and soil productivity through agroforestry. Int J Curr Microbiol App Sci Special Issue-11:1724–1729
Szott LT, Fernandes BCM, AndSanches PA (1981) Soil plant interactions in agroforestry systems. For Ecol Manag 45:127–152
Uthappa AR, Chavan S, Handa AK, Chaturvedi OP (2017) Agroforestry- A sustainable solution to address climate change challenges. In Book: Agroforestry for increased production and livelihood security
Young A (1987 The potential of agroforestry for soil conservation

Chapter 10
Carbon Sequestration in Agroforestry: Enhancement of Both Soil Organic and Inorganic Carbon

K. S. Anjali, A. Balasubramanian, Ghazanfer Abbas, C. N. Hari Prasath, S. Navaneetha Krishnan, G. Swathiga, and V. Manimaran

Abstract In India owing to its gradation in climatic conditions from temperate to humid tropics, the agroforestry systems (AFS) and practices are highly diverse within the country. The agroforestry practices followed in India range from intensified simple monoculture systems of planting, such as block and boundary plantations, to more specific, diversified, and complex systems, such as home gardens. In the era of changing climate, the role of trees, and other vegetation, its abatement is of paramount significance. Agroforestry as a sustainable land management system has a major role in carbon conservation and sequestration. Agroforestry practices sequester carbon both above ground as well as belowground. The above ground carbon sequestration by vegetation which sequester atmospheric carbon undergoing various physiological process and conserve it as biomass. The sole terrestrial pool where carbon (C) may be intentionally increased by agroforestry practices is the soil organic carbon (SOC) pool, which has been able to store some carbon for millennia. Agroforestry systems sequester about 2233 g carbon both above and below ground during the period of 50 years, but estimates of the amount of land they occupy globally are highly uncertain.

Keywords Agroforestry · Carbon sequestration · Climate change · Soil organic carbon

---

K. S. Anjali · A. Balasubramanian (✉) · G. Abbas · C. N. Hari Prasath · S. N. Krishnan · G. Swathiga
Department of Silviculture and NRM, Forest College and Research Institute, Mettupalayam, India
e-mail: balasubramanian.a@tnau.ac.in

V. Manimaran
J K K Munarajah College of Agricultural Sciences, Erode, India

H. S. Jatav et al. (eds.), *Agroforestry to Combat Global Challenges*, Sustainable Development and Biodiversity 36, https://doi.org/10.1007/978-981-99-7282-1_10

10.1 Introduction

Agroforestry can be defined as a sustainable land management system in which trees are deliberately integrated with agriculture crops, fodder crops, pastures, and poultry in some form of special arrangement or temporal sequence. Climate change is considered as a potent environmental concern in the twenty-first century. Land Use, Land Use Change, and Forestry (LULUCF) under Kyoto Protocol recommended afforestation and reforestation as the potent Green House Gas (GHG) offset activity to mitigate climate change. But as per the Paris climate agreement, every developed and developing nation shows reduction in carbon emission in order to reduce the global warming of the atmosphere by about 2 °C, whereas the global warming is directly related with carbon emission. Therefore, studies on the carbon sequestration potential of trees are gaining momentum for reducing the global temperature. According to LULUCF, agroforestry became recognized as a carbon sequestration activity under the afforestation and reforestation programmes, and agroforestry systems attracted attention as a carbon sequestration strategy from both industrialized and developing countries. Under the Paris agreement for climate change, the Nationally Determined Contribution (NDC) for India is to sequester 2.5–3.0-billion-ton carbon dioxide equivalent by 2030 (Chavan et al. 2022). A study of the change in forest carbon stock between the years 2019 and 2021 showed an increase of 79.4 million tonnes of carbon (FSI 2021). According to India's Restoration Opportunities Atlas, 87 million hectares (25% of total land) have the potential for carbon reduction through agroforestry (Singh 2021). According to the IPCC's Fifth Assessment Report, agroforestry will have a significant potential to sequester carbon in developing nations by the year 2040 (Nair et al. 2009a, b, c, d; 2010).

The variation in land use sectors have varying carbon emissions and sequestration. Agriculture alone is assumed to be responsible for 10–12% of all global anthropogenic GHG emissions, with non-CO_2 GHG emission estimates of 5120–-6116 Mt CO_2 eq/year in 2005. Agricultural lands frequently undergo extensive management, which presents numerous chances to enhance agronomic techniques, fertilizer and water management, and land use techniques to meet the goals of carbon sequestration. The total carbon sequestration capacity of agricultural lands worldwide is 0.75 to 1 Pg/year, or roughly 50% of the 1.6 to 1.8 Pg/year lost to deforestation and other agricultural activities. Thus, switching from lower biomass land uses like crop fallows, grasslands, etc., to tree-based systems like agroforestry, forests, and plantation forests can result in significant increases in carbon storage. IPCC (2007) defined carbon sequestration as the process of uptake of carbon containing substances, predominantly carbon dioxide, into a secondary reservoir with a long residence time. In agroforestry, carbon accumulation has been estimated to range from 0.29 to 15.2 Mg C ha1 year1 above ground and from 30 to 300 Mg C ha1 year1 for soils that are at least 1 m deep (Nair et al. 2009a, b, c, d).

Above ground carbon sequestration in agroforestry (Fig. 10.1) is by trees and other vegetation (above ground biomass) and below ground carbon sequestration is

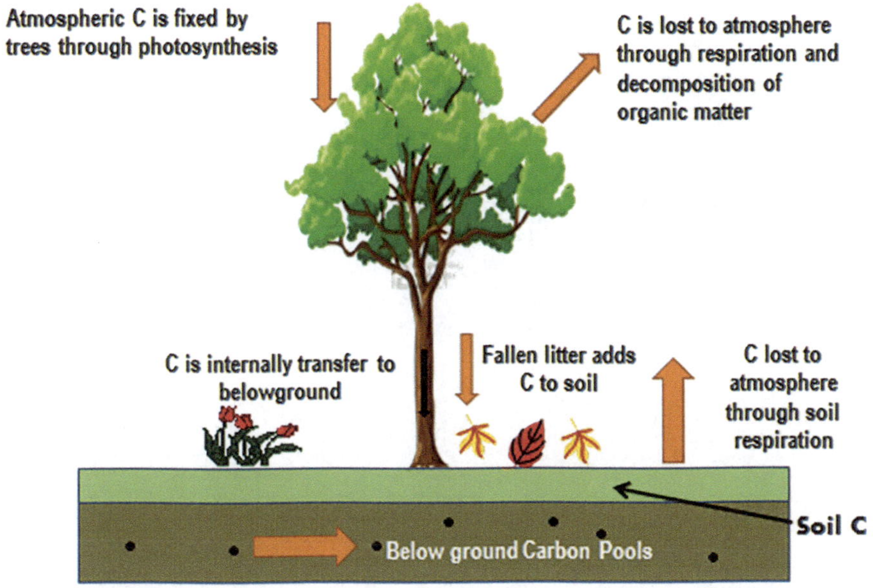

Fig. 10.1 Carbon sequestration in agroforestry

by roots biomass and soil pool. According to the Soil Science Society of America (SSSA), carbon sequestered in the soil is in two ways, direct and indirect (SSSA 2001). Chemical processes that transform carbon dioxide into inorganic soil molecules like calcium and magnesium carbonates directly sequester carbon in the soil. Plants photosynthesize atmospheric carbon dioxide into plant biomass, resulting in indirect carbon sequestration.

10.2 Mechanisms of Carbon Sequestration in Agroforestry Systems

The mechanism of carbon sequestration is through biologically mediated uptake and conversion of atmospheric carbon dioxide into inert, long lived carbon containing substances and hence it is also called as bio sequestration (U.S. DOE 2008). Carbon sequestration in an agroforestry system can be above ground carbon sequestration and below ground carbon sequestration (Fig. 10.2). Over geological time scales of more than 100,000 years, the atmospheric carbon dioxide concentration is regulated by the long-term global carbon cycle, which explains the biogeochemical cycling of atmospheric carbon among surface systems including oceans, the atmosphere, biosphere, and soil.

The carbon dioxide in the atmosphere is fixed in plants through the physiological process of photosynthesis and a small amount of carbon dioxide is released back

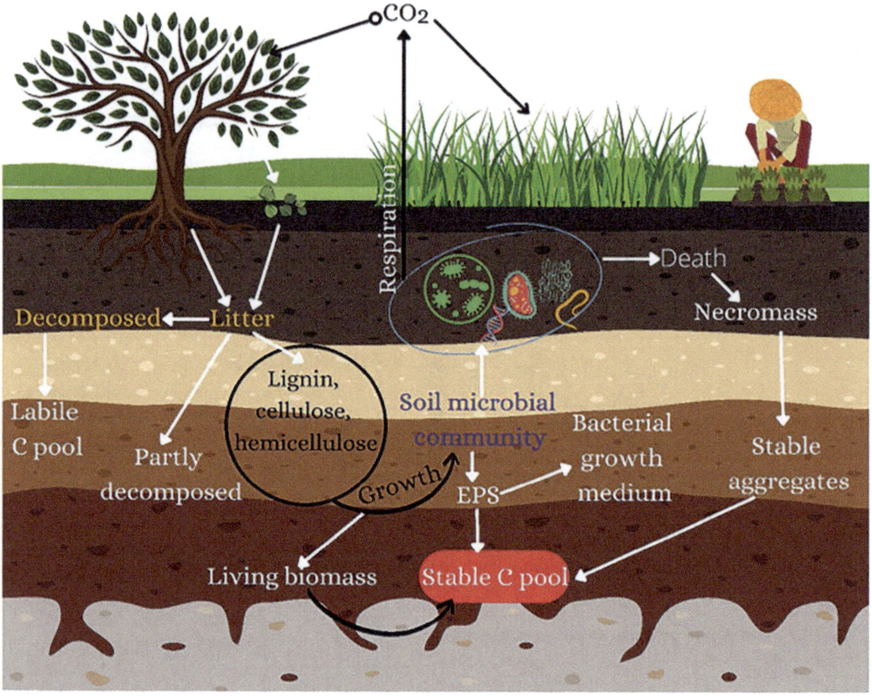

Fig. 10.2 Mechanism of bio carbon sequestration in agroforestry

through plants, animals, and microbes through aerobic respiration as carbon dioxide and anaerobic respiration as methane. A large amount of carbon dioxide and methane is released into the atmosphere during the burning of fossil fuels, forest fire, vehicular exhaust, land clearance for agriculture and other purposes. The plants store carbon both in above ground and belowground biomass. The above ground biomass encompasses leaves, twigs, stems, and branches and belowground as roots. The biomass on decomposition by the activity of microbes will transfer the carbon to a labile carbon pool. The lignin, hemicellulose, and cellulose content of the plant biomass is used for the growth and nourishment of soil microbes which in turn fixes carbon in the dead necromass to the soil stable carbon pool, and such carbon sequestration in the soil is called indirect soil carbon sequestration. Direct carbon sequestration occurs through direct chemical reaction of carbon dioxide with soil minerals and is converted into stable inorganic compounds such as calcium and magnesium carbonates and is stored in the soil pool. According to estimates, the soil and aboveground components of tree-based land-use systems carry the majority of the carbon (C), or around 60% and 30%, respectively.

10.3 Soil Carbon Sequestration

Soil carbon sequestration is one of the significant greenhouse gas (GHG) removal strategies and is estimated at about 4.8 Gt CO_2 eq/year. Soil carbon sequestration is also addressed as negative emission technology is through two ways, direct and indirect carbon sequestration (Fig. 10.3). Direct carbon sequestration occurs through direct chemical reaction of carbon dioxide with soil minerals and converted into inorganic compounds such as calcium and magnesium carbonates and is stored in the soil pool. The carbon thus formed is called as inorganic carbon or bound carbon which is in stable form and can be stored for a long period of time. The indirect carbon sequestration occurs when the dead or living plants biomass is acted upon by soil active microorganisms through the process of decomposition. The lignin, hemicellulose, and cellulose content of the plant biomass is used for the growth and nourishment of soil microbes which in turn fixes carbon in the dead necromass to the soil stable carbon pool (Nair et al. 2009a, b, c, d).

The pool of carbon in the soil is made up of 750 Pg of soil inorganic carbon and 1550 Pg of soil organic carbon, both at a depth of 1 m. According to an average calculation, the total soil carbon pool (2300 Pg) is more than both the atmospheric pool (770 Pg) and the vegetation pool (610 Pg) combined (Murthy et al. 2013; Lorenz and Lal 2014, b; Shi et al. 2018).

In an agroforestry system, the growth of the crop is a function of soil fertility, and soil organic carbon is an important factor determining the fertility of the soil. The litter from the tree decomposition is responsible for the enrichment of soil fertility for crop growth under an agroforestry system. The organic carbon content in the soil is a pool of atmospheric carbon dioxide that is sequestered indirectly contributing toward

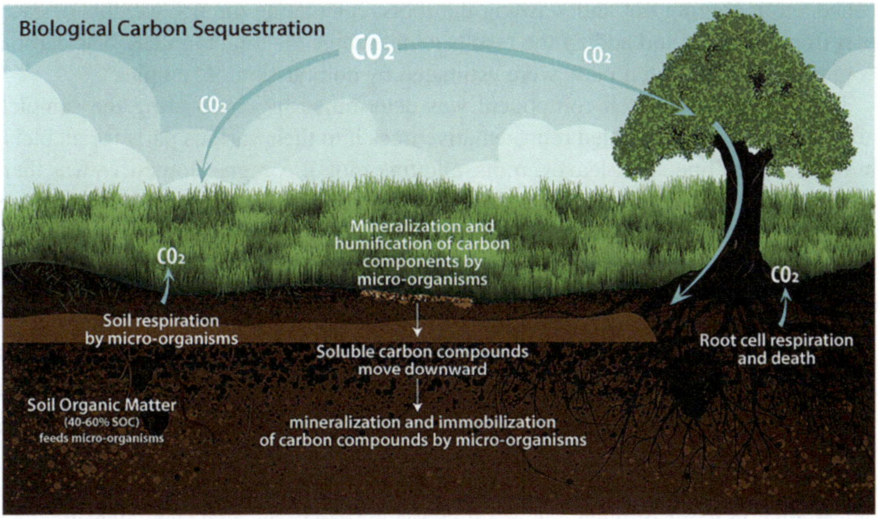

Fig. 10.3 Biological carbon sequestration in soils

the fertility of the soil and also plays a major role in determining the carbon storage in the ecosystem and regulating the concentration of carbon dioxide in the atmosphere. Hence, agroforestry can be adopted as the sustainable tool not only for the reduction in atmospheric carbon dioxide but also increases the crop productivity and enhances the sustainability of the system.

The protocols envisaged to quantify the soil carbon sequestration includes estimation of organic carbon dynamics in soil by wet digestion or dry combustion, measuring belowground living biomass, isotope measurements labelled by using either stable (^{13}C) or radio (^{14}C) isotopes and carbon dating. Apart from these various models such as CENTURY, RothC models, etc., were also developed to formulate the soil carbon pool. Recent spectroscopic methods, such as airborne spectroscopy, were also intended to assess the surface soil organic carbon utilizing multispectral and/or hyperspectral sensors mounted on aircraft, unmanned aerial vehicles (UAVs), or satellite platforms (Stefano and Jacobson 2018).

10.4 Carbon Stock Measurement

10.4.1 Aboveground

The above ground carbon sequestration is carried out by the trees' above ground biomass such as stems, leaves, branches, inflorescence, etc. The above ground carbon stock is measured in terms of harvested and standing biomass (Moraes et al. 1995; Guo and Gifford 2002). The traditional method of estimating the carbon stock is by harvesting the whole tree including roots. The method includes cutting down sample trees, separating different parts (such as the stems, leaves, inflorescence, etc.), digging out and washing the roots, calculating the dry weights of each part from samples, and adding the results to determine the total biomass. The carbon content of the harvested parts were estimated by combustion of samples.

The C content in each component was determined by combusting the samples after separating the collected representative trees into their various parts (branchlets, branches, dead branches, leaves, roots, and fine roots). A regression curve was then created using the calculated whole-tree biomass and carbon content. Such whole-tree harvesting processes need a lot of time and labour. In order to estimate whole-tree biomass, Dixon (1995) measured the volume of stem wood and multiplied it by the species-specific wood density. This result was then multiplied by 1.6. Root biomass was not included and it was assumed that 50% of the projected whole-tree biomass was made up of carbon. The worldwide forest biomass was then estimated in greater detail using this preliminary estimate. Variations in tree management can also be a problem; for example, trees in AFS may be pruned differently depending on management procedures, or they may grow in various ways because their spacing is different from that of natural (forest) systems. Agroforestry plots also differ from one another in terms of plant composition, planting patterns, and stand densities. As a result, estimating biomass output from local AFS is a tough undertaking that makes extrapolating results from one system to another exceedingly challenging.

10.4.2 Belowground Estimation

Understanding how belowground organic carbon dynamics in AFSs affect carbon stock depends on this determination, which is challenging. In addition to live root and hyphal biomass, microbial biomass, and Soil Organic Matter (SOM) in labile and more recalcitrant forms, organic C can take on a variety of distinct forms in soils. Measurement, estimation, and prediction of Soil Carbon Sequestration (SCS) are challenging tasks due to the complicated interplay of these several forms (Schulp et al. 2008). Carbon stock above ground and below ground has been presented in Fig. 10.4.

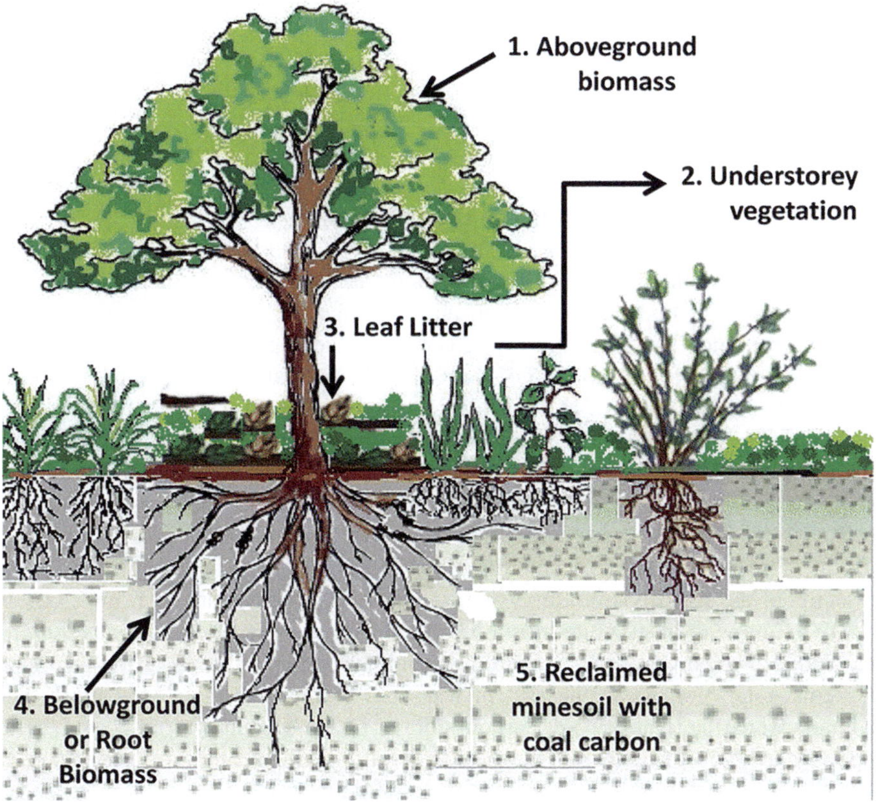

Fig. 10.4 Carbon stock above ground and below ground

10.5 Carbon Stocks in Agroforestry Systems in India

Through the enhancement of soil carbon and root biomass, carbon is stored in standing biomass above ground as well as below ground in diverse agroforestry systems. India's potential for sequestering carbon via agroforestry and other alternative land use systems was estimated to be 25 t C/ha over 96 M ha of land, or 68–228 Mg C/ha. However, this value varies by location according to biomass production. According to research by Jha et al. (2009), agroforestry can store 26% more carbon than farming in the Haryana plains, or about 83.6 t C/ha, up to a depth of 30 cm in the soil. However, the scale of the operation and the final use of the wood would determine the amount of carbon sequestration from forestry activities.

10.5.1 Agri-Silvicultural systems

Carbon sequestration in tree biomass: Maikhuri et al. (2001) projected that planted tree species on abandoned agricultural land may sequester 3.9 t/ha/year of carbon annually and 1.79 t/ha/year of carbon on degraded forest land. The intercropped *Alnus nepalensis* and *Dalbergia sissoo* plants with wheat and paddy had the maximum carbon sequestration rates of 0.256 t C/ha/year and 0.141 t C/ha/year, respectively.

10.5.1.1 Carbon Sequestration in Tree Biomass

According to Maikhuri et al. (2001), planted tree species on degraded forest land may absorb 1.79 t/ha/year of carbon and 3.9 t/ha/year of carbon on abandoned agricultural land. The highest rates of carbon sequestration were achieved by the intercropped *Dalbergia sissoo* and *Alnus nepaliensis* plants with wheat and rice, at 0.256 t C/ha/year and 0.141 t C/ha/year, respectively. Agri-silvicultural system based on *Gmelina arborea* that has been in place for 6 years sequestered 31.37 t C/ha. According to a different study, monocultures of trees and food crops sequestered 40% and 84% less carbon than agri-silviculture, showing that agroforestry systems have a greater capacity to sequester carbon. *Dalbergia sissoo*, at the age of 11 years, was able to accumulate 48–52 t/ha of biomass in an agri-silvicultural system. In an agri-silvicultural system where tree biomass ranged from 23.61 to 34.49 t C/ha with black gram-mustard, carbon dynamics involving various pruning techniques were investigated. According to studies on poplar-based agri-silvicultural systems, total biomass in the system was 25.2 t/ha, which is 113.6% more than solitary wheat cultivation. Net carbon storage in the system was 34.61 t C/ha as opposed to 18.74 t C/ha in single wheat cultivation. Albizia and mixed tree species, such as Mandarin, formed an agroforestry system that collected 1.3 t of biomass per hectare and stored 6939 kg of agricultural and tree biomass.

10.5.1.2 Enhancement of Soil Organic Carbon

Singh et al. (1989) found that *Populus deltoides* and *Eucalyptus hybrids* with Cymbopogon spp. increased SOC by 33.3 to 83.3% when planted alongside crops, with *Populus deltoides* showing the greatest increase in SOC. It has been found that agroforestry plantings with ages ranging from 6 to 20 years have boosted soil organic carbon. In an agroforestry system based on Poplars, trees were able to store more soil carbon in sandy clays than loamy sand during the first year of installation (6.07 t/ha/year) compared to the following years (1.95–2.63 t/ha/year). Traditional *Prosopis cineraria*-based systems cause SOC to rise by 50%, primarily as a result of leaf litter. After 5 years of planting, Samra and Singh (2000) noted increases in the status of soil organic carbon under *Acacia nilotica* + *Sacchram munja* of 0.39 to 0.52% and under *Acacia nilotica* + *Eulaliopsis binata* of 0.44 to 0.55%.

10.5.2 Silvipastoral Systems

10.5.2.1 Carbon Sequestration in Tree Biomass

The rate of biomass carbon storage in the silvipastoral system was 6.72 t C/ha/year in 8 years, which is two times more than the rate of 3.14 t C/ha/year from natural grassland, according to comparative studies conducted by National Research Centre for Agroforestry (NRCAF) in the year 2007 on biomass production from natural grassland and silvipastoral system composed of *Albizia amara*, *Dichrostachys cinerea*, and *Leucaena leucocephala*. Approximately 16,400 t/year of carbon is sequestered annually in farm forestry, which includes species like Eucalyptus sp., *Populus deltoides*, *Tectona grandis*, and *Anthocephalus chinensis* trees. In natural grassland in semi-arid Uttar Pradesh, species of *Eucalyptus tereticornis*, *Emblica officinalis*, *Albizia procera,* and *Albizia lebbeck* were introduced as part of a silvipastoral system.

10.5.2.2 Carbon Stored in Block and Boundary Plantations

In a study conducted by Kumar (2010) on four different agroforestry systems, including Eucalyptus hybrid boundary plantation + wheat, *Populus deltoides* block plantation + wheat, *Populus deltoides* block plantation + lemon grass and *Populus deltoides* boundary plantation + wheat it was estimated that total carbon sequestration rate [in trees] was 21.38, 70.59, 18.53, and 116.29 tonnes. For the Chirpine, Khair, mango, mixed plantations and Kino-based agricultural forestry systems in Uttaranchal, assessed a mitigation potential of 62.7, 48.5, 60.8, 61.7, and 37.6 t C/ha/year, respectively.

10.6 Estimation of Carbon Sequestration Potential for Agroforestry Systems

Under the Kyoto Protocol, agroforestry has been recognized as a viable global approach to reduce greenhouse gas emissions. And the reason for this is due to its potential in carbon sequestration. There are several agroforestry mechanisms with different carbon sequestration rates. In that aspect carbon sequestration can depend on type of climate, technology, time since land use change, and previous land use. In this regard, it is critical to understand carbon sequestration in various tree species in agroforestry technologies, as well as which agroforestry technologies offer the best value in terms of carbon sequestration. The carbon sequestration potential for any agroforestry system is estimated by the following methods:

1. Destructive method
2. Non-destructive algometric method

10.6.1 Destructive Method

The standard procedure for calculating biomass via destructive sampling is to cut down numerous sample trees and weigh their various components (e.g., branch, foliage, root, and stem). There are two methods employed for the estimation of carbon content using destructive sampling method

1. Destructive by weighing
2. Destructive with scaling

10.6.1.1 Destructive by Weighing

Carbon estimation in trees through destructive weighing methods involves measuring the biomass of the tree and converting it into an estimate of carbon content. Here is a general approach for conducting such measurements:

- Select a representative sample of trees: Choose a range of trees from the target populations that are representative of the species, age, and size distribution.
- Sample tree harvesting: Carefully select individual trees for destructive sampling. Ensure that the trees selected are healthy and not ecologically significant. Obtain necessary permissions and permits if required.
- Tree felling and sectioning: Cut down the selected trees and section them into different components, typically including the trunk, branches, and foliage.
- Weighing components: Weigh each component separately using a scale or balance with suitable precision. It is advisable to record weights in kilograms (kg) for accuracy.

- Moisture content determination: Measure and record the moisture content of each component, as this can affect the carbon content. This can be done by weighing a subsample of each component before and after drying in an oven.
- Carbon content determination: Convert the dry weight of each tree component to carbon content. The conversion factors differ for different tree components. For example, the carbon content of dry wood is usually assumed to be around 50% by weight.
- Summing carbon estimates: Sum up the carbon estimates of all the tree components to obtain the total carbon content for each tree.
- Extrapolation: Scale up the carbon estimates from the sampled trees to the entire population using appropriate statistical methods, considering the size and composition of the forest.

  Formula

$$W(f) = WW(f) \times DW(s)/WW(s)$$

  where,

- DW (f) = field dry weight in g
- WW (f) = field wet weight in g
- DW (s) = sample dry weight in g
- WW (s) = sample wet weight in g.

10.6.1.2 Destructive with Scaling

Estimating carbon content in trees through destructive sampling and scaling methods involves measuring the biomass of a tree and then converting it into carbon equivalents. Here is a step-by-step process for estimating carbon using destructive sampling and scaling:

- Select the trees: Choose a representative sample of trees from the target population. The sample size should be statistically significant to ensure accurate estimation.
- Destructive sampling: Cut down the selected trees and carefully measure the different components of the tree, including the stem, branches, leaves, and roots. Divide the tree components into sections or categories for easier measurement and analysis.
- Biomass measurement: Weigh each component of the tree using a scale or balance. It is important to separate the different components for accurate biomass determination. Measure the fresh weight of each section.
- Moisture content correction: Determine the moisture content of each tree component by collecting a subsample and drying it in an oven until it reaches a constant weight. Calculate the moisture content as a percentage of the fresh weight. Subtract the moisture content from the fresh weight to obtain the dry weight.

- Carbon content determination: Use conversion factors specific to the tree species to convert the dry weight biomass of each component into carbon equivalents. These conversion factors represent the average carbon content of different tree components.
- Scaling: Scale up the carbon content of the sample trees to estimate the carbon content of the entire population or a larger area. This involves applying appropriate statistical techniques to extrapolate the results from the sample to the population.
- Statistical analysis: Analyse the data collected from destructive sampling to estimate the mean carbon content per tree or per unit area, along with measures of uncertainty such as confidence intervals.
- Reporting: Present the estimated carbon content in a suitable format, such as tons of carbon per hectare or per individual tree, depending on the objectives of the study.

  Formula

$$V_b = (SA_1 + SA_2)/2 \times L$$

where,

- V_{cc} – volume with bark in m^3
- SA_1 – sectional area of the stem lower part in m^2
- SA_2 – sectional area of the upper stem in m^2
- L – stem section length in m

It is important to note that destructive sampling involves cutting down trees, which may not be feasible or desirable in certain situations. Alternative non-destructive methods, such as allometric equations based on tree measurements (e.g., diameter, height), can also be used to estimate carbon content without harming the trees.

10.6.2 Non-destructive Algometric Method

Non-destructive carbon estimation methods in trees allow for the assessment of carbon content without the need to cut down or harm the trees. These methods rely on various measurements and equations based on tree characteristics, such as diameter, height, and biomass allocation patterns. Non-destructive methods are widely used due to their efficiency, minimal ecological impact, and the ability to estimate carbon content in a non-invasive manner. They are particularly useful for large-scale assessments of carbon stocks in forests, ecological research, and monitoring efforts.

One commonly used non-destructive method for carbon estimation is the use of allometric equations. Allometry refers to the relationship between different tree parameters and biomass or carbon content. By measuring easily obtainable tree characteristics, such as diameter at breast height (DBH) and height, allometric equations can estimate the carbon content of the tree without the need for destructive sampling. These equations are developed using statistical analysis of data collected from destructive sampling and scaling methods. They provide a reliable and efficient means of estimating carbon content across different tree species and ecosystems.

Non-destructive methods can also utilize remote sensing techniques, such as LiDAR (Light Detection and Ranging) or aerial/satellite imagery, to estimate carbon content in trees. LiDAR uses laser pulses to measure the three-dimensional structure of the forest canopy, allowing for the estimation of tree height, canopy density, and aboveground biomass. Aerial or satellite imagery provides information about the spatial distribution and density of vegetation, which can be used to infer carbon content through statistical models and algorithms.

The advantage of non-destructive methods is their ability to estimate carbon content in a non-invasive manner, reducing the ecological impact on forests and preserving the integrity of the trees. These methods also allow for rapid and efficient carbon assessments across large areas, making them valuable for monitoring changes in carbon stocks over time and space. However, it is important to note that non-destructive methods rely on statistical models and equations that are developed based on specific tree species and ecosystems, and their accuracy may vary depending on the context and conditions in which they are applied.

10.6.2.1 Calculation of Above Ground Biomass (AGB)

Above ground biomass (AGB) is defined as "the aboveground standing dry mass of live or dead matter from tree or shrub (woody) life forms, expressed as a mass per unit area", typically Mg ha^{-1}.

The biometric values measured using tree biometry was utilized for calculating the above ground biomass. The volume arrived and the density measured was used to calculate the biomass content of the wood in metric tonnes per hectare as detailed below.

Above ground biomass of the plantation is calculated by using the following formula:

$$\text{Above ground biomass (AGB)} = \text{volume}\left(\text{m}^3/\text{tree}\right) \times \text{wood density}\left(\text{g/cm}^3\right)$$

The biomass expansion factor (BEF) is a ratio that quantifies the increase in aboveground biomass of a tree or plant as it grows from one stage or size to another:

$$\text{BEF} = (\text{total volume of trees/ha})/(\text{merchantable volume of trees/ha})$$

For calculating the above ground biomass, BEF was used to convert stem biomass to above ground biomass.

10.6.2.2 Calculation of Below Ground Biomass (BGB)

Belowground biomass in trees refers to the total mass of plant material present below the ground surface, including the roots and associated structures. It encompasses the root system, which plays a crucial role in nutrient and water uptake, anchoring the tree, and providing structural support.

BGB is calculated as per the standard procedure suggested by Pandya et al. (2013):

$$\text{Below ground biomass (BGB)} = 0.26 \times \text{AGB(ton)}$$

10.6.2.3 Estimation of Total Biomass (TB)

Total biomass comprises of both above ground and below ground biomass of individual trees in a plantation. Therefore by adding both above ground and below ground biomass of the plantation the total biomass was arrived.

Total biomass was estimated by using the following formula:

$$\text{Total biomass(TB)} = \text{Above ground biomass(AGB)} + \text{below ground biomass(BGB)}$$

10.6.2.4 Estimation of Weight of Carbon (C)

The average carbon content in trees is generally considered as 50% of the tree's total biomass. Therefore, carbon content in trees was calculated by multiplying the tree biomass by 50%.

Carbon content is estimated as follows:

$$\text{Carbon content} = \text{biomass} \times 0.50$$

10.6.2.5 Estimation of Total Quantity of Carbon Dioxide

Carbon dioxide equivalent is the ratio of the total weight of one molecule of carbon (44 g) to oxygen (12 g). Therefore, weight of carbon dioxide sequestered in the tree was calculated by multiplying the carbon content of the tree by 3.67:

Table 10.1 Estimated biomass of different agroforestry systems in different agroclimatic regions

Agroclimatic regions/states	Agroforestry systems	AGB (Mg ha^{-1})	BGB (Mg ha^{-1})	TB (Mg ha^{-1})
Northern Himalayas (Himachal Pradesh, Jammu and Kashmir, Uttarakhand)	Agrisilviculture	54.93	14.87	64.67
	Agrihorticulture	40.00	13.23	57.56
	Silvipasture	43.85	19.47	87.52
Indo-Gangetic region (Punjab, Haryana, Uttar Pradesh, and Bihar)	Agrisilviculture	33.82	3.76	23.85
	Silvipasture	38.41	9.32	50.72
Eastern and Northeastern India (West Bengal, Odisha, Assam, Sikkim, Meghalaya, Manipur)	Agrihorticulture	5.57	3.63	6.41
	Home garden	52.54	34.69	121.67
	Plantation crop-based agroforestry	40.46	13.36	87.16
	Boundary plantation	16.96	2.52	19.48
	Block plantation	186.20	25.33	220.20
Western and central India (Rajasthan, Gujarat, Maharashtra, and Madhya Pradesh)	Agrisilviculture	11.91	–	33.63
	Agrihorticulture	81.05	24.60	78.95
	Block plantation	79.24	21.84	120.09
Southern India (Karnataka, Andhra Pradesh, Tamil Nadu, and Kerala)	Agrisilviculture	37.37	11.87	35.96
	Plantation crop-based agroforestry	174.96	41.29	232.38
	Block plantation	170.9	69.49	239.8
	Coffee plantation	221.5	59.38	279.2

Source: Panwar et al., 2022

$$\text{Total } CO_2 \text{ equivalent} = \text{carbon content} \times 3.67$$

The biomass estimated under different agroforestry systems in different agroclimatic regions is given Table 10.1.

In an agricultural environment with larger net increases in carbon stocks, home gardens and block plantation agroforestry systems were observed to have higher carbon contents than other land uses. Agroforestry systems are now being adopted by developing nations as REDD+ strategic options to achieve climate change mitigation because they are financially viable, prevent deforestation, improve soil productivity, permanently sequester carbon in agricultural landscapes, and support growers.

As each agroforestry system differs based on site factors, tree species, the density and productivity of shade trees, as well as their longevity and subsequent use in processing systems, the production of litter, the rate of decomposition, and its incorporation in the soil matrix as soil carbon, nutrient cycling, and soil respiration, uncertainties in estimates of carbon stocks should be expected. Additionally, each system's management strategy plays a crucial role in determining how much carbon is added to and removed from each system. The system's resilience, or its capacity to tolerate climatic or other shocks and, so, retain carbon despite such disturbances may be more significant over the long run. The complexity and variety of the agroforestry management unit, as well as the characteristics of the landscape matrix in which agroforestry systems are located, influences all resilient mechanism in agroforestry systems. A functional landscape system must be viewed as an integrated landscape that includes flows of materials and services across system boundaries, from agroforests to natural forest patches, and more intensive land uses, such as plantations and annual crops. This is true from the perspectives of resilience and carbon storage. A detailed knowledge of the mechanisms and scales governing the allocation and partitioning of biomass in agroforestry plantings is necessary. Unfortunately, the exact nature of this driving force and its size are yet unknown. Due to a lack of data on changes in land use and land cover, there are also sizable uncertainties in the estimation of carbon fluxes into and out of systems.

10.7 Conclusion

Planting multipurpose tree species in non-forest land uses promotes biodiversity and carbon sequestration at the same time. When crops fail, trees provide an extra source of income. They also offer financial benefits from the non-carbon advantages. In order to produce valuable wood that is economically advantageous as well as wood for use as fuelwood and for construction purposes, it is useful to plant trees using a blend of fast- and slow-growing species. Agroforestry systems exhibit high soil and live biomass carbon accumulation, indicating their potential to provide the environmental service of carbon sequestration. Additionally, by preserving soil and preventing the burning of fuelwood derived from forests, agroforestry systems can aid in lowering CO_2 emissions. Agroforestry systems have the capacity to gather and store carbon, and they may develop into a technical alternative for reducing tropical deforestation rates while simultaneously providing rural populations with a wide range of goods and services.

Important knowledge gaps around C sequestration in AF include the following: (1) quantitative evaluation of carbon inputs and stocks in various AF systems, with special attention to deep soil carbon and its dependence on tree species and age; (2) optimization of the area allotted to trees and crops within each AF system to achieve maximum carbon sequestration, increase yield, maximize ecosystem services, and improve environmental conditions; (3) development of new remote sensing techniques to distinguish AF from the background of forests, plantations, and other agricultural areas.

It is important to note that the effectiveness of carbon sequestration in agroforestry systems can vary depending on factors such as site-specific conditions, management practices, and the longevity of the system. Additionally, carbon sequestration in trees and soil should be considered in the context of overall emissions reduction strategies and sustainable land management practices.

References

Chavan SB, Dhillon RS, Ajit, Rizvi RH, Sirohi C, Handa AK, Bharadwaj KK, Johar V, Kumar T, Singh P (2022) Estimating biomass production and carbon sequestration of poplar-based agroforestry systems in India. Environ Dev Sustain 24:13493–13521

De Stefano A, Jacobson MG (2018) Soil carbon sequestration in agroforestry systems: a meta-analysis. Agrofor Syst 92(2):285–299

Dixon RK (1995) Agroforestry systems: sources of sinks of greenhouse gases? Agrofor Syst 31(2): 99–116

FSI (2021) India State Forest Report 2021. Forest Survey of India: Dehradun, India

Guo LB, Gifford RM (2002) Soil carbon stocks and land use change: a meta analysis. Glob Chang Biol 8(4):345–360

IPCC (2007) Climate change 2007: synthesis report. In: Core Writing Team, Pachauri RK, Reisinger A (eds) Contribution of working groups I, II and III to the fourth assessment report of the intergovernmental panel on climate change. IPCC, Geneva

Jha B, Mohanty B, Tripathi A (2009) Drivers of agricultural diversification in India, Haryana and the greenbelt farms of India. Inst. of Economic Growth, Unvi. of Delhi Enclave, Delhi

Kumar S, Udawatta RP, Anderson SH (2010) Root length density and carbon content of agroforestry and grass buffers under grazed pasture systems in a Hapludalf. Agrofor Syst 80(1):85–96

Lorenz K, Lal R (2014) Soil organic carbon sequestration in agroforestry systems. A review. Agron Sustain Dev 34:443–454

Maikhuri RK, Nautiyal S, Rao KS, Saxena KG (2001) Conservation policy–people conflicts: a case study from Nanda Devi Biosphere Reserve (a world heritage site), India. For Policy Econ 2(3–4):355–365

Moraes JL, Cerri CC, Melillo JM, Kicklighter D, Neill C, Skole DL, Steudler PA (1995) Soil carbon stocks of the Brazilian Amazon basin. Soil Sci Soc Am J 59(1):244–247

Murthy IK, Gupta M, Tomar S, Munsi M, Tiwari R, Hegde GT, Ravindranath NH (2013) Carbon sequestration potential of agroforestry systems in India. J Earth Sci Climate Change 4(1):1–7

Nair PKR, Nair VD, Kumar BM, Haile SG, Dilly O, Pannell D (2009d) Soil carbon sequestration in tropical agroforestry systems: a feasibility appraisal. Environ Sci Policy 12:1099–1111

Nair PR, Kumar BM, Nair VD (2009a) Agroforestry as a strategy for carbon sequestration. J Plant Nutr Soil Sci 172(1):10–23

Nair PR, Nair VD, Kumar BM, Haile SG (2009b) Soil carbon sequestration in tropical agroforestry systems: a feasibility appraisal. Environ Sci Pol 12(8):1099–1111

Nair PR, Nair VD, Kumar BM, Showalter JM (2010) Carbon sequestration in agroforestry systems. Adv Agron 108:237–307

Nair PKR, Nair VD, Kumar BM, Haile SG, Dilly O, Pannell D (2009c) Soil carbon sequestration in tropical agroforestry systems: A feasibility appraisal. Environ Sci Pol 12:1099–1111

Pandya IY, Salvi H, Chahar O, Vaghela N (2013) Quantitative analysis on carbon storage of 25 valuable tree species of Gujarat, Incredible India. Indian Journal ofScientific Research, 4(1): 137–141.

Panwar P, Mahalingappa DG, Kaushal R, Bhardwaj DR, Chakravarty S, Shukla G, Gurung T (2022) Biomass production and carbon sequestration potential of different agroforestry systems in India: a critical review. Forests 13(8):1274

Samra JS, Singh SC (2000) Silvopasture systems for soil, water and nutrient conservation on degraded land of Shiwalik foothills (sub-tropical northern India). Indian J Soil Conserv 28(1): 35–42

Schulp CJ, Nabuurs GJ, Verburg PH, de Waal RW (2008) Effect of tree species on carbon stocks in forest floor and mineral soil and implications for soil carbon inventories. For Ecol Manag 256(3):482–490

Shi L, Feng W, Xu J, Kuzyakov Y (2018) Agroforestry systems: Meta-analysis of soil carbon stocks, sequestration processes, and future potentials. Land Degrad Dev 29(11):3886–3897

Singh K, Chauhan HS, Rajput DK, Singh DV (1989) Report of a 60 month study on fitter production, changes in soil chemical properties and productivity under Poplar (P.deltoides) and Eucalyptus (E. hybrid) interplanted with aromatic grasses. Agrofor Syst 9:37–45

Singh R, Shelar K, Duraisami M, Anderson W, Gautam RS (2021) Equitable and inclusive landscape restoration planning: learning from a restoration opportunity assessment in India. Ecological Restoration, 39(1–2):108–119

SSSA (2001). Carbon Sequestration: Position of the Soil Science Society of America (SSSA). Available from: www.soils.org/pdf/pos_paper_carb_seq.pdf (last accessed February 2010)

U.S. Department of Energy (2008) Carbon cycling and bio sequestration: integrating biology and climate through systems science, report from the March 2008 Workshop, DOE/SC-108. U.-S. Department of Energy Office of Science

Chapter 11
Sustainable Forest Management (SFM) for C Footprint and Climate Change Mitigation

Ruby Patel, Siddhartha Mukherjee, Biswabara Sahu, Bishnuprasad Dash, M. Jaison, Kodchire Avinash, and Puja Singh

11.1 Introduction

Nowadays government, stakeholders, and other common people are shifting from conventional forest management to sustainable forest management (SFM) to improve the productivity without affecting the environment. To implement the principles of sustainability, Sustainable Forest Management (SFM) requires adaptive measures, scrutiny of knowledge insights, and continuous monitoring of social, economical, and environmental factors. SFM is characterized by some identifiable features like trans-disciplinary, pluralistic, integrative nature, and heterogeneity that make it different from the conventional forest management approach. According to

R. Patel (✉) · K. Avinash
ICFRE-Institute of Forest Biodiversity, Hyderabad, Telangana, India
e-mail: patelr@icfre.org

S. Mukherjee
Faculty Centre of Agriculture, Rural and Tribal Development (ARTD), Ramakrishna Mission Vivekananda Educational and Research Institute, Ranchi, Jharkhand, India

B. Sahu
Department of Soil Science and Agricultural Chemistry, Faculty of Agriculture, Siksha O Anusandhan Deemed to be University, Bhubaneswar, Orissa, India

B. Dash
Department of Soil Science, M.S. Swaminathan School of Agriculture, Centurian University of Technology and Management, Paralakhemundi & Jatni, Odisha, India

M. Jaison
Department of Agricultural Chemistry and Soil Science, Bidhan Chandra Krishi Viswavidyalaya, Mohanpur, Nadia, West Bengal, India

P. Singh
Department of Soil Science and Agricultural Chemistry, Horticulture College, Khuntpani, Chaibasa, Birsa Agriculture University, Ranchi, Jharkhand, India

H. S. Jatav et al. (eds.), *Agroforestry to Combat Global Challenges*, Sustainable Development and Biodiversity 36, https://doi.org/10.1007/978-981-99-7282-1_11

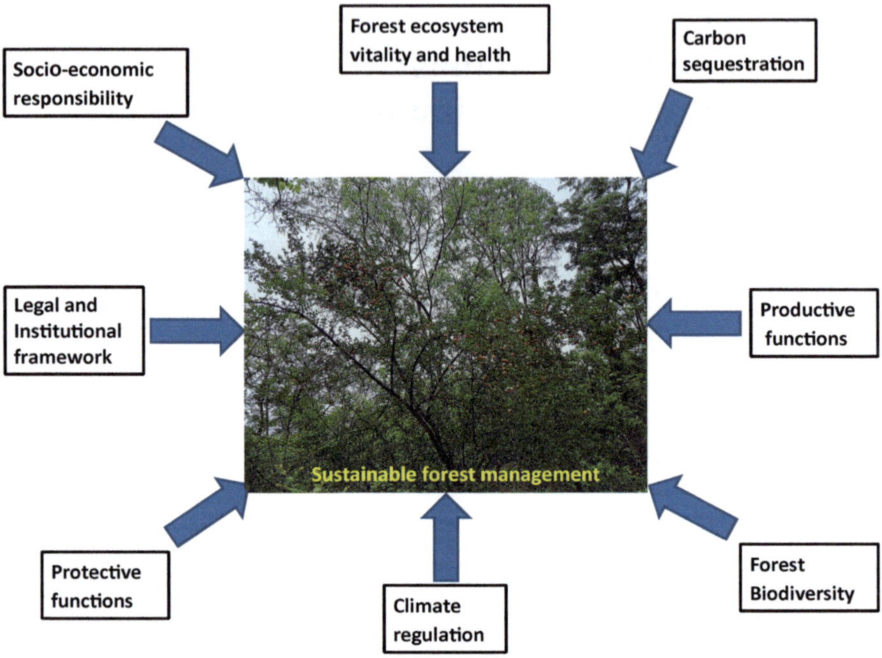

Fig. 11.1 Global-level thematic elements of SFM

Keenan (2015), approximately 30% of the Earth's land area is covered by forests, making it difficult to imagine a person who does not rely on forest products and services in their daily life (Köthke 2014). The economic contribution of forests is significant, providing various amenities that sustain livelihoods and safeguard our environment (MacDicken et al. 2015).

Managing the regenerative capacity of forests sustainably to ensure future benefits is a significant challenge, as emphasized by Nasi and Frost (2009) and Chow et al. (2013). This challenge gave rise to the concept of SFM, which recognizes the importance of sound management practices for both forest production and protection (Keeton and Crow 2009; Putz and Thompson 2020). The potential of SFM lies in two key aspects: the inherent ability of ecosystems to regenerate themselves and the adaptability of economic activities and social perceptions that influence human interaction with the ecosystem, ultimately safeguarding its long-term productivity and health, as stated by MacDicken et al. (2015). However, the planning of SFM must be tailored to climate change scenarios since climate plays a crucial role in adopting SFM strategies. Tropical, temperate, and boreal forests collectively provide diverse habitats for plants, animals, and micro-organisms, serving as homes to the majority of terrestrial species worldwide, according to the CBD (2009). Forests provide both productive and protective functions, supplying valuable resources and services while safeguarding the environment (Fig. 11.1). Sustainable management and conservation of forest resources are essential for ensuring their continued contribution to economic development, biodiversity conservation, climate regulation, and the well-being of present and future generations. The health of forests and the provision of various ecosystem services depend on species diversity, genetic diversity within species, and the diversity of forest types. The global-level thematic elements of SFM encompass key principles and components that are essential for the effective and holistic management of forests worldwide. These elements provide a framework for guiding policies, practices, and decision-making processes aimed at achieving sustainable outcomes for forests and the communities that depend on them. These elements include the productive and protective functions of forests, climate regulation, biodiversity conservation, forest health, socioeconomic considerations, and policy and institutional frameworks (Fig. 11.1).

11.2 Sustainable Forest Management (SFM) for C Footprint

SFM scheme assisted by forest certification with the green economy reduces the carbon footprint by C sequestration and also balances the economic, social responsibility, and environment to sustain the livelihood. SFM plays a key role in combating climate-related risk or climate variability and also ensures the environmental safety. SFM increases the carbon sequestration in both soil and plants which reduce the greenhouse gas emission to the atmosphere and lower the C footprint. SFM ensures the renewable carbon-neutral energy source from forest biomass and acts as a substitute for materials which are carbon intensive including cement and steel, thereby reducing C footprints and improving greenery economy.

11.3 Sustainable Forest Management (SFM) for Climate Change Mitigation in Tropical and Sub-tropical Regions

The implementation of SFM in tropical landscapes presents a unique challenge that necessitates disaggregated approaches for assessment, as highlighted by Köthke (2014). The definition of SFM in relation to tropical forests encompasses various categories, including managed, exploited, and unmanaged forests across landscapes, as well as protected areas, selectively logged natural forests, and logged forests subjected to additional silvicultural treatments (Putz and Thompson 2020). Many tropical forested areas face challenges such as weak administration, disputed land ownership, poverty, a high dependence on forests by local communities, intensified exploitation and deforestation, modest-to-high opportunity costs for forest conservation, and political conflicts.

The effectiveness of tropical protected areas within the framework of SFM relies heavily on governance, stakeholder consensus, adequate staff training and commitment, and sufficient funding (Bruner et al. 2001). In tropical and sub-tropical forests, SFM requires consideration of large enough landscapes that can sustain all values, including sustainable wood production and ecosystem services for people, without causing species losses. Individual protected areas often lack the size necessary to protect wide-ranging or rare species. However, proper management of buffer areas surrounding protected areas can establish connectivity, which is crucial for the persistence of certain species on a larger scale, as highlighted by Hodgson et al. (2011). Unmanaged or primary forests, which show no visible signs of human intrusion according to FAO (2018), are rapidly declining.

A recent article by Potapov et al. (2017) reported a global decline of 7.2% in the area of intact forests, defined as areas larger than 500 km^2 without roads, between 2000 and 2013. Intact forests are already absent in many tropical countries. Several large-bodied and heavily exploited animal species in tropical regions, such as the Asian elephant (*Elephas maximus*), African forest elephant (*Loxodonta cyclotis*), tiger (*Panthera tigris*), and harpy eagle (*Harpiaharpyja*), depend on intact forests for their survival (Kinnaird et al. 2003; Barlow et al. 2011; Birdlife International 2017; Roopsind et al. 2017). Furthermore, intact forests have high conservation value, as research shows that up to 94% of designated blocks of selectively logged forest remain intact due to factors such as absence of commercial timber, unfavourable conditions, poor planning, and insufficient supervision, as stated by Putz et al. (2019). The rate of intact forest loss has generally been higher compared to areas outside designated protected forests, although intact forest areas within parks also experience degradation.

In most tropical countries, timber stocks fail to recover to primary-forest levels within the officially designated minimum cutting cycle as per current regulations. For instance, in Amazonian Brazil, it takes over 60 years for timber volumes to recover after conventional timber harvest, as reported by Vidal et al. (2020). A meta-analysis based on numerous studies of yield recovery (over 100 publications) indicates significant variability in timber yield, with a decline of approximately 46% from the first to the second harvest (Putz et al. 2012). It is worth noting that while many tropical forests are logged multiple times, most of the reviewed studies focused on timber harvests from primary forests. Overall, the review suggests that despite the conservation potential of selective logging on a large scale, SFM is currently compromised in many tropical regions due to poor logging practices and premature re-entry logging after previous harvests (Sasaki et al. 2016; Ellis et al. 2019).

11.4 Sustainable Forest Management (SFM) for Climate Change Mitigation in Temperate Region

At the global level, policies and regulations pertaining to SFM have been reported to cover 97% of the world's forested areas. However, despite these efforts, sustainable forest operations and other factors such as fuelwood collection can contribute to forest degradation and negatively impact biodiversity within ecosystems. Disturbingly, more than 50% of the temperate broadleaf and mixed forest biome, as well as nearly 25% of the tropical rainforest biome, have experienced fragmentation or destruction due to various human activities, as documented by Secretariat of the Convention on Biological Diversity (SCBD) (2006).

To ensure the sustainable management of temperate forests, it is crucial to develop and implement appropriate and region-specific Forest Management Planning (FMP). MacDicken et al. (2015) conducted a global meta-analysis and observed that the proportion of land area under FMP was relatively high in temperate domains (63%), compared to tropical and subtropical domains, which had approximately 28% coverage (Table 11.1). However, the mere presence of an FMP does not guarantee its effective execution. Nevertheless, having an FMP in place is a positive step toward establishing conditions favourable for SFM. It is worth noting that successful long-term SFM can also be achieved without a written management plan, as evidenced by multi-generational family management of private forests. Monitoring of FMP implementation by governments plays a vital role in improving compliance with pre-determined strategies. The meta-analysis revealed that 40% of FMPs in tropical climates were monitored annually, followed by 38% in the boreal domain and 32% in the temperate domain (Fig. 11.2). In contrast, the subtropical domain showed a lower frequency of monitoring events, with only 22% of FMPs being monitored annually. On average, FMPs in the tropics were monitored and evaluated once every 2.5 years to ensure their smooth functioning (MacDicken et al. 2015). Globally, the adoption of forest management planning and the monitoring of plans has significantly increased, covering over 430 million hectares by 2014. However, it is worth noting that internationally verified certification is predominantly concentrated in the boreal and temperate climatic domains, accounting for

Table 11.1 Forest area with FMP by climatic domain. (Source: MacDicken et al. 2015)

	Forest under FMP		FMP for conservation	
Domain	Area (000 ha)	% of domain forest area	Area (000 ha)	% of domain forest area
Tropical	5,09,761	28.2	2,03,787	11.3
Temperate	4,24,971	63.1	2,09,428	31.1
Boreal	10,73,801	87.7	4,01,497	32.8
Boreal without Russia	2,58,656	63.1	7852	1.9
Sub-tropical	91,131	28.5	28,678	8.9
Total	20,99,664		8,43,391	

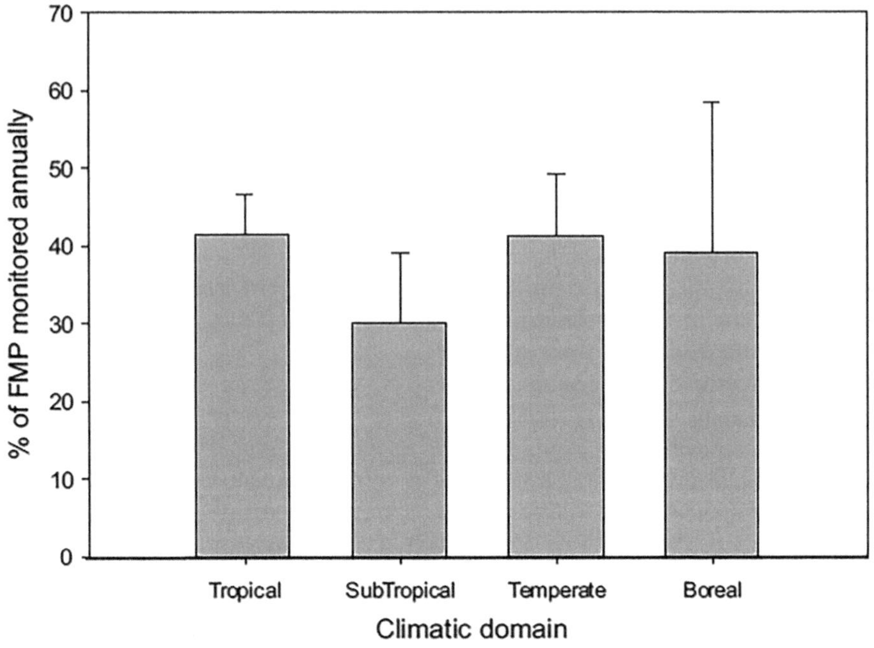

Fig. 11.2 Average proportion of forest management plans monitored annually by climatic domain (bars are the standard error of the mean) (Source: MacDicken et al. 2015)

90% of the total certifications. In contrast, only 6% of permanent forests in the tropical domain have been certified (MacDicken et al. 2015).

Over the next 30 years, it is projected that the consumption of primary timber products will increase, and the utilization of solid biofuels for electricity generation could be three times higher by 2030 compared to the present level (FAO 2018). Furthermore, it is anticipated that the demand for industrial round wood will rise by 50–75% by 2050, reflecting the overall growth in demand (Sedjo 2001). As a consequence of this escalating demand, the area of tropical forest plantations more than doubled between 1995 and 2005, reaching 67 million hectares, primarily concentrated in Asia. Plantations in boreal and temperate regions have also experienced some level of expansion, and this upward trend is expected to persist (ITTO 2006). However, the use of a limited number of tree species in these plantations and modified natural forests raises concerns regarding ecosystem resilience (Hagar 2007).

11.5 Forest Certification and REDD+ as New Approach for Healthy Forestry

Forest plays a very crucial role in fulfilling the material needs of humans as well as adding the aesthetic value to the society, protecting the environment from natural calamities, and also maintaining the quality of natural resources like soil, water, and environment by the process of waste water processing, ground water recharge, reducing effects of noise, erosion control, binding, and inactivation of toxic substances present in soil and water. Thus, along with the regeneration of new forests, it is of utmost importance to maintain the existing forests and to assess and monitor their functions through some scientific criteria. Forest certification is one such method which can promote better management of forests by providing environmental and financial assistance through forest and forest products.

11.5.1 What Is Forest Certification?

Forest certification is a program of judging the forest management practices by comparing with a series of pre-set standards based on environmental, social and economic perspectives for monitoring, and tracing and labelling the forest products, for example, timber and non-timber forest products and pulp, etc.

11.5.2 Aim of Forest Certification

1. Achieving SFM through market forces
2. Reducing greenhouse gas (GHG) emissions through healthy forests' sequestration potential for carbons

The government regulations are very unlikely to be imposed successfully at every corner of the world, improving the economic value of forest produce could be used as a way to motivate people for better forest management.

- Certification ensures better protection to the existing forests and adequate financial returns from sustainably managed forests.
- Secondly, a healthy forest can absorb a tremendous amount of carbon dioxide which can potentially reduce the carbon load in the environment.
- Along with climate change mitigation, the huge carbon sequestrating capacity of the forest trees offers a generation of carbon credits which can generate another source of income for the forest landowners.

11.5.3 Forest Certification Process

The two separate processes of forest certification deal with monitoring the forest management practices as well as movement of certified forest products from its origin to the point of sale (Fig. 11.3) (awsassets.wwfindia.org/). For example, the Forest Stewardship Council (FSC) is one of the globally established reliable certification schemes.

11.5.4 Forest Management Certification

In the forest certification method, the forestry operations are analysed to assess whether the operations are up to a predetermined set of standards or not. On meeting the desired standards, the landowner is issued with a certificate with the potential to market the products as certified products brought from the certified forests. Forest certification addresses the quality of forest management.

In order to attend SFM, some of the national and international criteria are being set. The indicators for forest management which are being set by the Montreal Process are as follows:

- Biodiversity conservation, maintenance of forest ecosystem and productivity of forests
- Conservation and maintenance of soil and water resources
- Observing the contribution of forest entities to carbon cycles
- Enhancement and maintenance of long-term socioeconomic benefits
- Construction of legal, institutional, and economic frameworks for sustainable management and conservation of forest (The Montreal Process 2015)

Broadly forest certification is done to identify and promote well-managed forest lands and to recognize the products of sustainably managed forests (Bettinger et al. 2016). Forest certification has certain benefits on the marketability of products obtained from such forests. For example, these products are eventually recognized as premium products and demand high price from the buyers. Also, the certified

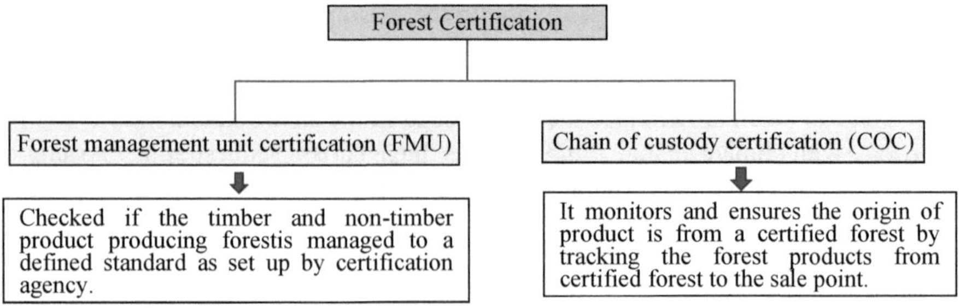

Fig. 11.3 Types of forest certification processes

products get access to certain high-quality markets, have new market penetration potential, and show increased sales too (Paluš et al. 2018; Yamamoto et al. 2014; Aguilar and Vlosky 2007). The forest certification scheme owing to its monetary benefit to the managers and land owners can better act as a way for healthy forest and for ensuring forest ecosystem services.

11.5.5 REDD+

Globally around 11% of global greenhouse gas emission is accounted to forest degradation and deforestation. The forest ecosystem can successfully uptake a huge amount of atmospheric CO_2 and store it in huge tree biomass and soils thereby carrying out an important role in climate change mitigation. Like forest certification, REDD+ also promotes SFM activities to reduce forest degradation by providing financial assistance to the countries promoting forest preservation and conservation activities.

REDD+ (also recognized in Article 5 of the Paris Agreement) is a framework to guide activities to reduce forest degradation and deforestation (Fig. 11.4). This aims to implement REDD+ activities at both the national and sub-national levels of government to reduce anthropogenic pressure on forests. The developing countries will receive payments in exchange for proof that they have reduced deforestation or for demonstrating their forest preservation activities.

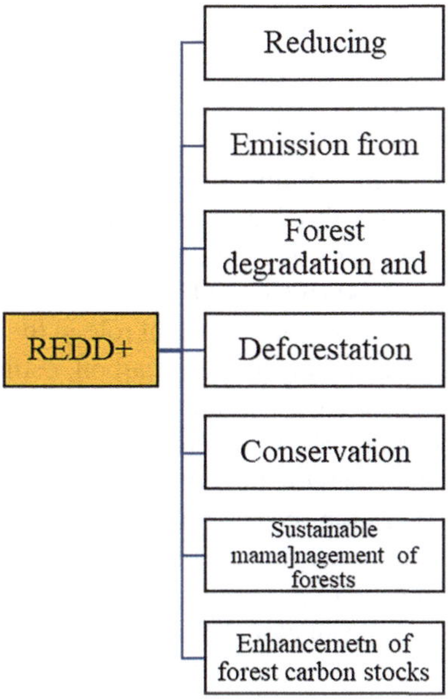

Fig. 11.4 Components of REDD+ (Cosslett 2013)

11.5.6 Phases of Working of REDD+

1. **Readiness phase**: Development of national strategies or action plans, policies and measures, and capacity-building.
2. **Implementation**: Implementation of the set policies and action plans. The results of such policy implementation are used as case studies for public demonstration. Also, the outcomes can be used as feedback for further capacity-building.
3. **Results-based report**: The results of the implemented actions are fully verified and reported and the countries are allowed to seek results-based payments (FAO.org).

11.6 Challenges for Sustainable Forest Management (SFM) to Mitigate Climate Change

Implementing SFM in a diverse country like India poses significant challenges. SFM, being aligned with sustainable development, carries crucial implications for the global economic landscape. The foremost challenge of the coming decades is climate change, which poses a threat to surpass the safe planetary boundaries for humanity (Rockstrom et al. 2009). Consequently, climate change has garnered considerable attention from the scientific community and policymakers, with a primary focus on mitigating human-induced interference with the climate system (IPCC 2007). Scientific research has provided irrefutable evidence that global warming is already occurring (IPCC 2007). The previously stable average global temperature is now rising at a rate of 0.2 °C per decade (Hansen et al. 2010), and the continued increase in greenhouse gas (GHG) concentrations will further elevate the temperature. The causes of this global climate change include the escalated release of greenhouse gases such as carbon dioxide (CO_2), methane (CH_4), nitrous oxide (N_2O), and ozone (O_3). Human activities, including land-use changes, deforestation, and the burning of fossil fuels, have been the primary contributors to increased carbon dioxide emissions. Without effective policies to restrict human-induced emissions, the global average temperature is projected to increase by 1.1–6.4 °C by 2100 (IPCC 2007).

Forests play a crucial role in the global carbon cycle. Forest ecosystems contain a substantial amount of carbon, estimated to be around 1200 gigatonnes (Gt), which represents a significant portion of terrestrial carbon and surpasses the amount of carbon (550 Gt) stored in the atmosphere (IPCC 2001). Climate, with its variability and changing patterns, exerts a significant influence on forest growth, development, migration, succession, and regeneration. While climate information has traditionally been incorporated into local or regional forest management decisions, the global impact of climate has often received limited attention. However, the projected changes in global climate pose threats to SFM. Therefore, it is imperative to develop

Table 11.2 Top ten countries in terms of forest area (2020)

Sl. No	Country	Forest Area (000 ha)	% of world forest area	% of country area
1	Russian federation	8,15,312	20	49.8
2	Brazil	4,96,620	12	59.4
3	Canada	3,46,928	9	38.7
4	USA	3,09,795	8	33.9
5	China	2,19,978	5	23.3
6	Australia	1,34,005	3	17.4
7	Democratic Republic of Congo	1,26,155	3	55.6
8	Indonesia	92,133	2	49.1
9	Peru	72,330	2	56.5
10	India	72,160	2	24.3
	Total	**4,85,438**	**66**	

Source: ISFR (2021)

a climate plan as an integral part of SFM, encompassing improved strategies for regeneration and protection to adapt to these changes.

Human-induced climate change presents potential risks to forests and poses future challenges for forest managers. Addressing climate change, through both mitigation and adaptation measures, necessitates transformative changes in forest management and research. Climate change is leading to increasing temperatures and alterations in precipitation patterns, including changes in snowfall and the timing, quantity, and variability of rainfall (IPCC 2013). Forests are long-lived ecosystems that are inherently complex to manage, both internally and externally. They are also susceptible to the impacts of long-term climatic changes, as are the societies and economies reliant on them. Climate change amplifies the significance of numerous existing challenges related to environmental, social, and economic changes.

In India, the total forest and tree cover spans 80.9 million hectares, accounting for 24.62% of the country's geographical area (ISFR 2021). Climate change hotspots within Indian forests, as identified by India State of Forest Report (ISFR) (2021), indicate that approximately 45–64% of forests in the country will experience adverse effects from climate change and rising temperatures by 2030. Among the states, except for Nagaland, Tripura, Meghalaya, and Assam, most forested regions in India are highly vulnerable to climate change. The Union Territory of Ladakh is expected to be particularly impacted by climate change and rising temperatures. The total forest cover in India spans 72,160 ha, equivalent to 24.3% of the country's geographical area, while the tree cover represents 2.91% of the geographical area (Table 11.2). According to ISFR (2021), the latest assessment indicates a combined increase of 0.38% in forest and tree cover at the national level, with forest cover increasing by 0.22% and tree cover by 0.76% (Table 11.3).

To achieve these goals, information, innovation, and implementation are the three essential factors that rely on human resources to make them a reality.

Table 11.3 Top ten countries for average net gain in forest area (2010–2020)

Sl. No	Country		Annual forest area gain
		Area (000 ha)	% 2010 forest area
1	China	1937	0.93
2	Australia	446	0.34
3	India	266	0.38
4	Chile	149	0.85
5	Vietnam	126	0.90
6	Turkey	114	0.53
7	USA	108	0.03
8	France	83	0.50
9	Italy	54	0.58
10	Romania	41	0.62

Source: IFSR (2021)

11.7 Challenges of SFM for Government, Research Scientists, and Institutions

Forests are a global resource, and effectively dealing with important issues related to their use and maintenance requires global participation. To fully understand and address the challenges in forest science, it is essential to establish an appropriate framework and enhance our capacity to generate knowledge for a sustainable future. A balanced approach is crucial for the successful management and development of forests, ensuring their existence at acceptable levels for the benefit of present and future generations.

1. Addressing the drivers of forest degradation and deforestation, and enabling SFM, requires greater innovation and better coordination in global forestry dialogue, national sector planning, and technical analysis. Demonstrating the potential of forests to reduce poverty, support economic growth, and provide environmental services at local and global levels necessitates close collaboration between donors and governments, linking forest sector activities with national strategies. Promoting forest ownership and access rights, as well as emphasizing stakeholder participation in policy formulation and implementation, are essential for poverty reduction and effective forest governance.
2. One of the main challenges with sustainability is its multidimensionality. Achieving a state of sustainable development requires progress in one dimension without compromising progress in other dimensions. Climate change, although initially perceived as an environmental issue, is closely interconnected with various sectors in society, such as energy. Climate policy cannot be confined solely to environmental policy as it encompasses multiple fields. National governments often struggle to determine how they will precisely achieve their goals, leading to a significant gap between expected emission reductions and global commitments.

3. The gap in forestry research capacity and the translation of practical results between developing and developed countries remains unacceptably wide. However, highly efficient and locally adapted low-cost technologies often contain a substantial amount of research-based knowledge. Unfortunately, the state of forestry research in many developing countries is characterized by a lack of political commitments, a shortage of scientists with diverse expertise, limited methodological and technological innovation, and inadequate funding for research programs. Consequently, forestry research in these places has not significantly evolved to address current and future global, national, or local issues, hindering the development of sustainable forest.
4. The implementation of climate policy administration patterns is not always clear-cut. In Italy, a significant number of responsibilities have been transferred to lower levels of government, while environmental issues remain predominantly centralized. In France, a central government policy is supplemented by the expectation that regions will develop their own plans incorporating climate change considerations, although only a few countries or regions have done so.
5. Challenges lie ahead for governments in the process of redefining the sharing of authority between different entities with regards to climate change. National governments grapple with the complexities of climate change and face difficulties in finding effective solutions within existing monitoring frameworks. Relations with lower levels of government are being revisited to foster greater unity in policymaking and explore new opportunities. Local governments play a crucial role in spatial planning, transportation, housing, and energy. SFM is a highly complex task that requires a comprehensive approach, including policy frameworks, strengthened governance, removal of market distortions, and engagement of market actors, full valuation and sharing of forest benefits through market mechanisms, capacity building, and mobilization of adequate financial resources.

References

Aguilar FX, Vlosky RP (2007) Consumer willingness to pay price premiums for environmentally certified wood products in the US. Forest Policy Econ 9:1100–1112

Barlow AC, Smith JL, Ahmad IU, Hossain AN, Rahman M, Howlader A (2011) Female tiger Panthera tigris home range size in the Bangladesh Sundarbans: the value of this mangrove ecosystem for the species' conservation. Oryx 45(1):125–128

Bettinger P, Boston K, Siry J, Grebner DL (2016) Forest management and planning. Academic Press

BirdLife International. (2017). Harpia harpyja (amended version of 2017 assessment). The IUCN Red List of Threatened Species 2017: e.T22695998A117357127. https://doi.org/10.2305/IUCN.UK.2017-3.RLTS.T22695998A117357127.en. [Accessed 6 September 2018].

Bruner AG, Gullison RE, Rice RE, Da Fonseca GA (2001) Effectiveness of parks in protecting tropical biodiversity. Science 291(5501):125–128

Chow J, Doria G, Kramer R, Schneider T, Stoike J (2013) Tropical forests under a changing climate and innovations in tropical forest management. Trop Conserv Sci 6(3):315–324

Convention on Biological Diversity (2009) A Good Practice Guide. Sustainable Forest Management, Biodiversity and Livelihoods. CBD, Montreal, Canada

Cosslett CE (2013) Understanding the potential impacts of REDD+ on the financing and achievement of sustainable forest management. In: A report prepared for the Secretariat of the United Nations Forum on Forests (UNFFS). fao.org/redd/en/ (Assessed on 13/04/2022)

Ellis PW, Gopalakrishna T, Goodman RC, Roopsind A, Griscom B, Umunay PM, Zalman J, Ellis E, Mo K, Gregoire TG, Putz FE (2019) Climate-effective reduced-impact logging (RIL-C) can halve selective logging carbon emissions in tropical forests. For Ecol Manag 438:255–266

FAO (2018) Terms and Definitions FRA 2020. In: Forest Resources Assessment Working Paper 188. Rome, Italy

Forest Survey of India (2021) India State of Forest Report. Ministry of Environment and Forests, Government of India, New Delhi

Hagar JC (2007) Wildlife species associated with non-coniferous vegetation in Pacific Northwest conifer forests: a review. For Ecol Manag 246(1):108–122

Hansen J, Ruedy R, Sato M, Lo K (2010) Global surface temperature change. Rev Geophys 48(4): RG4004

Hodgson JA, Moilanen A, Wintle BA, Thomas CD (2011) Habitat area, quality and connectivity: striking the balance for efficient conservation. J Appl Ecol 48(1):148–152

Intergovernmental Panel on Climate Change (IPCC) (2001) Climate Change 2001: the scientific basis. In: Houghton JT et al (eds) Contribution of Working Group I to the Third Assessment Report of the Intergovernmental Panel on Climate Change. Cambridge Univ. Press, New York, p 881

Intergovernmental Panel on Climate Change (IPCC) (2007) Climate Change 2007: the scientific basis. In: Solomon S et al (eds) Contribution of Working Group I to the Fourth Assessment Report of the Intergovernmental Panel on Climate Change. Cambridge Univ. Press, New York

IPCC (2013) In: Barros VR, Field CB, Dokken DJ, Mastrandrea MD, Mach KJ, Bilir TE, Chatterjee M, Ebi KL, Estrada YO, Genova RC, Girma B, Kissel ES, Levy AN, MacCracken S, Mastrandrea PR, White LL (eds) Climate change 2014: impacts, adaptation, and vulnerability. Part B: Regional aspects. Contribution of working group II to the fifth assessment report of the intergovernmental panel on climate change. Cambridge University Press, Cambridge, United Kingdom and New York, NY, p 688

India State of Forest Report (Forest Survey of India) (2021) New Delhi: Ministry of Environment and Forests, Government of India

International Tropical Timber Organization (ITTO) (2006) Global study on tropical forest plantations - Encouraging private sector investment in industrial plantation in the tropics – 1st Draft report. ITTO, Curitiba, Brazil

Keenan RJ (2015) Climate change impacts and adaptation in forest management: a review. Ann For Sci 72(2):145–167

Keeton WS, Crow SM (2009) Sustainable forest management alternatives for the Carpathian Mountain region: providing a broad array of ecosystem services. In: Ecological Economics and Sustainable Forest Management: Developing a Trans-disciplinary Approach for the Carpathian Mountains. Ukrainian National Forestry University Press, pp 109–126

Kinnaird MF, Sanderson EW, O'Brien TG, Wibisono HT, Woolmer G (2003) Deforestation trends in a tropical landscape and implications for endangered large mammals. Conserv Biol 17(1): 245–257

Köthke, M (2014) Costs of sustainable forest management in the tropics: State of knowledge. Thünen Working Papers (27).

MacDicken KG, Sola P, Hall JE, Sabogal C, Tadoum M, de Wasseige C (2015) Global progress toward sustainable forest management. For Ecol Manag 352:47–56

Nasi R, Frost PG (2009) Sustainable forest management in the tropics: is everything in order but the patient still dying? Ecol Soc 14(2):12

Paluš H, Parobek J, Vlosky RP, Motik D, Oblak L, Jošt M, Glavonji'c B, Dudík R, Wanat L (2018) The status of chain-of-custody certification in the countries of Central and South Europe. Eur J Wood Wood Prod 76:699–710

Potapov P, Hansen MC, Laestadius L, Turubanova S, Yaroshenko A, Thies C, Smith W, Zhuravleva I, Komarova A, Minnemeyer S, Esipova E (2017) The last frontiers of wilderness: Tracking loss of intact forest landscapes from 2000 to 2013. Sci Adv 3(1):e1600821

Putz FE, Baker T, Griscom BW, Gopalakrishna T, Roopsind A, Umunay PM, Zalman J, Ellis EA, Ellis PW (2019) Intact forest in selective logging landscapes in the tropics. Front Forest Global Change 2:30

Putz FE, Thompson ID (2020) Defining sustainable forest management (SFM) in the tropics. In book: Achieving sustainable management of tropical forests. 10.19103/as.2020.0074.19

Putz FE, Zuidema PA, Synnott T, Peña-Claros M, Pinard MA, Sheil D, Vanclay JK, Sist P, Gourlet-Fleury S, Griscom B, Palmer J (2012) Sustaining conservation values in selectively logged tropical forests: the attained and the attainable. Conserv Lett 5(4):296–303

Rockstrom J, Steffen W, Noone K, Persson A, Chapin FS III, Lambin EF, Foley JA (2009) A safe operating space for humanity: identifying and quantifying planetary boundaries that must not be transgressed could help prevent human activities from causing unacceptable environmental change, argue Johan Rockstrom and colleagues. Nature 461(7263):472–476

Roopsind A, Caughlin TT, Sambhu H, Fragoso JM, Putz FE (2017) Logging and indigenous hunting impacts on persistence of large Neotropical animals. Biotropica 49(4):565–575

Sasaki N, Asner GP, Pan Y, Knorr W, Durst PB, Ma HO, Abe I, Lowe AJ, Koh LP, Putz FE (2016) Sustainable management of tropical forests can reduce carbon emissions and stabilize timber production. Front Environ Sci 4:50

Secretariat of the Convention on Biological Diversity (SCBD). 2006. Global Biodiversity Outlook 2 . Accessed at: http:// www.cbd.int/GBO2

Sedjo RA (2001) From foraging to cropping: the transition to plantation forestry, and implications for wood supply and demand. Unasylva 204:24–32

The Montre'al Process, 2015. Criteria and Indicators for the Conservation and Sustainable Management of Temperate and Boreal Forests. 5th ed. Montre'al Process Liaison Office International Forestry Cooperation Office, Tokyo, September 2015. 31 p. Available from: http://www.montrealprocess.org/documents/publications/techreports/MontrealProcessSeptember2015.pdf (Accessed 5/28/2016)

Vidal, E., West, T. A. P., Lentini, M. W., de Souza, S. E. X. F., Klauberg, C. and Waldhoff, P. 2020. Sustainable Forest Management in the Brazilian Amazon

Yamamoto Y, Takeuchi K, Shinkuma T (2014) Is there a price premium for certified wood? Empirical evidence from log auction data in Japan. Forest Policy Econ 38:168–172

Chapter 12
Climate Change Mitigation Through Agro-Forestry Improves Natural Resource and Livelihood Security

Tanmoy Sarkar, Sudip Sengupta, Sahely Kanthal, and Suprabuddha Kundu

Abstract It is commonly known that human activity has a stronger impact on climate change than ever before. Deforestation and forest degradation lead to an increase in carbon emissions. However, this problem may be solved by using appropriate land and forest management. The improvement of forest C stocks by agroforestry is one method that significantly lowers atmospheric greenhouse gas emissions. In order to preserve natural resources, provide livelihood stability, advance the well-being of society and the economy, and sequester carbon, agroforestry is crucial, as this chapter critically investigates. Through the fusion of agricultural science and forestry studies, agroforestry is developing into a specialised area of study. Agroforestry is defined as an integrated agricultural system that includes woody trees, crops and herbaceous annual plants, and animals on a same land piece. The ability of agroforestry systems to sequester and store carbon, which may be fixed, absorbed, and stabilised in soils and plant structures (woody portions), is another benefit they provide. The establishment of databases for tracking tree and soil carbon stocks, however, necessitates the use of standardised procedures, which poses a substantial difficulty within this system. To improve their readiness in achieving national climate goals, regional countries must address additional obstacles like water scarcity, weak interactive governance, problems with farmer rights and land ownership, as well as insufficient financial support for small-scale farmers engaged in agroforestry. As a result, the supervision of agroforestry systems should place a high priority on maximising financial rewards for farmers while also attending to their needs for production (such as fuelwood, rich in nutrients fruits, food, and lumber) and protecting the environment (such as efficient nutrient cycling, watershed management, reducing soil erosion, and improving soil health and fertility).

Keywords Agro-forestry · Climate change · Livelihood · Mitigation · Sustainability

---

T. Sarkar (✉) · S. Sengupta · S. Kanthal · S. Kundu
School of Agriculture, Swami Vivekananda University, Kolkata, India
e-mail: tanmoys@svu.ac.in

H. S. Jatav et al. (eds.), *Agroforestry to Combat Global Challenges*, Sustainable Development and Biodiversity 36, https://doi.org/10.1007/978-981-99-7282-1_12

12.1 Introduction

Agroforestry is a practice that is used by more than 1.2 billion people globally and is a sustainable method of using land (Dawson et al. 2014a). It has received a lot of attention since it is a land-use approach that is both resource- and environment-friendly. Similar to this, the idea of agroforestry was developed in anticipation of the contribution that tree production on and off farms may make to the management of natural resources and the promotion of sustainable land use. The goal of agroforestry is to solve the economic, ecological, and social issues of the modern world by focusing on the function of trees in agricultural landscapes (Garrity 2004). In terms of ecosystem services, agroforestry provides a number of advantages, such as soil enrichment, biodiversity preservation, improved air and water quality, and carbon sequestration (Jose 2009). As it differs from other conservation strategies that call for stopping land use, agroforestry may be thought of as a type of "productive conservation" (Gold et al. 2004). Long planning horizons, uncertain costs and incomes, and the inclusion of permanent tree components alongside agricultural or animal components are characteristics of agroforestry systems from an economic perspective. Agroforestry systems are based on the core economic tenet that cooperative rather than individual output (monocropping) results in higher total net benefits. In the "climate-smart" environment, which incorporates policies and actions for both mitigation and adaptation, agroforestry systems play a significant role (Montagnini 2017). The potential of agroforestry, given its multiple environmental and economic benefits, may help the agricultural and forestry sectors and offer creative solutions to present difficulties including financial instability, environmental implications, and a bad public image. Farmers may maintain a decent standard of living while conserving land, water, and other natural resources by producing a sizable number of specialist items for markets (Gold et al. 2004). Agroforestry programmes help to the sustainability of diverse rural resources, which leads to the growth of more resilient agribusinesses and rural communities (Fig. 12.1). Agroforestry, according to Montagnini and Metzel (2017), can significantly help achieve a number of Sustainable Development Goals (SDGs), including reducing hunger (SDG 2), gender equality (SDG 5), clean water (SDG 6), affordable and clean energy (SDG 7), eradicating inequalities (SDG 10), addressing climate change (SDG 13), and sustainable forestry and restoration (SDG 15). The main goals of agroforestry systems are to improve human living standards, provide a sustainable basis for agricultural products, and make things better as they are now by increasing output and productivity. By looking at agroforestry systems as a potential alternative, it is possible to ensure sustainable and climate-smart agriculture (Somarriba et al. 2017). Agroforestry, which combines agriculture and forestry, maximises the advantages for economic, environmental, and social factors. Its environmentally friendly methodology is responsible for its rising acceptance as a tool for agricultural growth. As an alternative management method to conventional agricultural practices, agroforestry offers opportunities for farm economic success while satisfying the standards for decreased environmental impact. Agroforestry gives farmers in

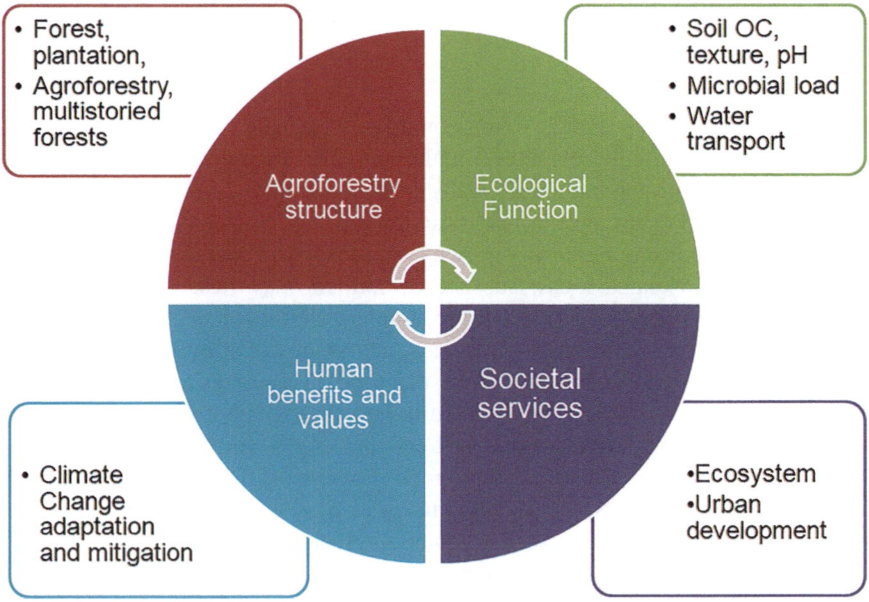

Fig. 12.1 The organisation and operation of ecosystems in interaction with societal systems serve as a representation of ecosystem services

underdeveloped countries access to business opportunities in regional and international markets. Although agroforestry-related products and services now have more access to the market, their value to rural and commercial growth is still overlooked. Lack of knowledge about existing agroforestry techniques and their potential for future agricultural growth may result in a lack of understanding of their critical function in local economies (Drew et al. 2004). Due to the scant academic attention devoted to agroforestry research, decision-makers are also left in the dark about how to create economic models that successfully address the SDGs (Rosenstock et al. 2020). To fully utilise and maximise the economic advantages of agroforestry as a climate-smart agricultural investment, a thorough grasp of its nature and features is necessary. The goal of this literature is to fill the knowledge gap and define the agroforestry sector. This chapter provides more information on the benefits of agroforestry systems, as well as the difficulties and possible rewards of making investments in them.

12.2 Agroforestry Adoption Characteristics in Developing Nations

The majority of people in developing countries, especially those who live in rural regions, rely on agriculture as their main source of income (Diao et al. 2010). Traditional agroforestry has traditionally been the primary source of income for rural inhabitants. For instance, forest communities often participate in activities like selling wood or consuming locally grown fruits and vegetables (Kalaba et al. 2010). Additionally, in isolated areas, subsistence farmers may raise animals, produce certain perennial plants, or grow crops to augment their income.

Around the world, agroforestry practices such as the use of multifunctional trees, riparian buffers, and better fallow are practised. These methods include agro-silvopastoral systems, which combine shrubs and trees with both crops and animals, silvopastoral systems, which combine trees and cattle, and silvoarable systems, which combine shrubs and trees with crops. Regional differences result from the adoption of modified versions of these strategies by several emerging nations. For instance, agroforestry is practised in many ways throughout Sub-Saharan Africa, including multilevel organised gardens on Mount Kilimanjaro in Tanzania and logical woodlots in Kenya (Mbow et al. 2014). Latin American nations exhibit comparable regional variances. For example, silvopastoral practices use eucalyptus, pine trees, and native grass species in Chile and Argentina to raise cattle (Browder et al. 2005). Silviculture is interplanted with coffee and native tree species in the Brazilian Cerrado and the Andean area, which includes Venezuela, Colombia, Peru, and Ecuador. Agro-silvopastoral practices combine the farming of animals, such as goats and cattle. While agroforestry systems with home gardens and agro-silvopastoral practices including cows, oil palm trees, and woodlands are more frequent in Southeast Asia, agroforestry systems with coffee or cacao and timber are not as common (Besar et al. 2020). According to Jara-Rojas et al. (2020), the existence of agroforestry systems is dependent on a number of variables, including the availability of resources, the sustainability of the economy, and the topographical, social and cultural, and ecological features of the region. Regression analysis and logistic analysis of models, for instance, were used in a research in Brazil's Atlantic rainforest to examine how the socioeconomic position affects farmers' aspirations for implementing agroforestry (McGinty et al. 2008). Farmers' adoption was found to be impacted by ideas about conservation, labour availability, and behavioural control perceptions. In a different research, Nguyen et al. (2020) found that the existence of coffee agroforestry systems in Northwest Vietnam was impacted by access to markets, ecological compatibility, and plot layout.

Agroforestry operations that are favourable to the community can be defined in a number of ways. To choose behaviours that are appropriate for adoption, it is necessary to identify all the salient differentiating characteristics, according to Mbow et al. (2014). The accessibility of natural resources and the farmer's location might affect the use of a certain approach or system (Jara-Rojas et al. 2020). For example, people who live close to forests and raise livestock may choose silvopastoral or agro-silvopastoral methods because they have access to resources (like cattle, goats, etc.) and can feed their animals by foraging shrubs or grasses that naturally occur in the forests, while they can also make additional money from crop or timber cultivation (Beyene et al. 2019). Furthermore, since they demonstrate greater rates of recycling, some forms of agroforestry that require little input and maintenance are more profitable and, as a result, are chosen by low-income farmers (Jezeer et al. 2018). Even yet, numerous agroforestry systems provide comparable functions, notably in regard to livelihoods and landscapes, and are appropriate for various agro-ecological zones. Consequently, it is impossible to choose the best agroforestry system for adoption using a single set of criteria. Depending on the resources at hand and the surrounding environment, agroforestry comprises a wide variety of practices and procedures. Understanding how agroforestry functions independently in various environmental, social, and political situations is crucial if one is to appreciate the motives for the adoption and use of agroforestry. When examining the selection criteria, it might be interesting to look at the trade-offs related to a farmer's choice to use a certain kind of land management, including non-agroforestry practices. Further study is required, especially in the creation of instruments for agroforestry intervention, as the aforementioned qualities may change across various temporal and geographical scales.

12.3 Role of Changing Climate on Agro-Forestry

12.3.1 Climate Change Exposure Risk for Smallholder Farmers

The production of agriculture and local lives are both threatened globally by the expected climate change. Farmers must modify their agricultural and land management practices in order to reduce the negative effects of a changing climate (Jarvis et al. 2011). Climate- or environment-driven adaptation can be a direct response to changing patterns in temperature and precipitation, but it can also come about as a consequence of actions taken to lessen the hazards associated with severe weather throughout the world when the change is not immediately obvious. Farmers' behaviour is impacted by environmental and social elements in addition to weather and temperature. According to Gourdji et al. (2013), it is clear that rising temperatures have a negative effect on agricultural productivity on a large scale. Warmer places, where the implications of temperature rises would be more obvious, are predicted to experience these negative effects more severely (Schlenker and Lobell 2010). Furthermore, because many of these warmer locations are predominately made up of poorer nations, it is anticipated that the effects of climate change would disproportionately harm those nations' poorest farming families (Schmidhuber and Tubiello 2007; Skoufias et al. 2011).

Numerous studies have used statistical techniques to analyse how climate change influences agriculture at the national or global level (Lobell and Field 2007; Gourdji et al. 2013), but they have only been able to address implications since household-level adjustments will be implicit in aggregated data. According to certain research (Schlenker and Roberts 2009; Schlenker and Lobell 2010), farmers either make long-term or short-term adjustments, but particular adaptation strategies have not been examined. Crop substitution, irrigation, and soil fertility improvements are a few examples of adaptations or changes that might aid in minimising production losses in the case of climatic shocks. The unique adaptation methods employed by households in diverse locations in response to climate changes are not adequately covered in the literature. Farmers may adjust their crop decisions consciously or unconsciously in response to changing and unpredictable environmental conditions. Given how difficult it is for someone to predict weather conditions that may have an effect on crop yields, producers may find it challenging to successfully adjust to these hazards on their own, even though variability in the weather has been demonstrated to have a major influence on yields of crops (Lobell et al. 2011). In order to overcome knowledge gaps, farmers need reliable climate information that provides them with anticipatory understanding of dangers to the environment (Rosenzweig and Udry 2013).

12.3.2 Climate Change Impact on Food Security

Due to the effects of climate change, it may become more difficult for people, communities, and nations to get appropriate amounts of high-quality food. Many emerging nations have seen a gain in buying power over the past thirty years as a result of falling real food prices and growing real earnings. However, recent market volatility has brought to light how susceptible to price changes marginalised and poor groups are. Given how challenging it is to predict and assess the effects of economic growth, the connection between the environment and food accessibility is extremely complex. Additionally, price rises may cancel out the advantages of income growth. Food insecurity trends are expected to deteriorate if income levels dramatically increase but stay low and the percentage of income dedicated to food remains high. According to empirical research, climate change may have an impact on people's capacity for self-sufficiency. One Ethiopian research, for instance, found a strong link between fluctuating rainfall patterns and economic activity. According to Conway and Schipper (2010), wetter years were related with larger SDP growth and drier years with less negative growth. Despite the fact that correlation does not indicate causation, this association shows that, in the absence of adaptive measures, climate change may have an influence on livelihoods. The possible effects of increased food costs and temperatures have been examined. According to Brinkman et al. (2009), rising food costs have resulted in decreased dietary variety and quality as well as a rise in malnutrition, notably stunting and deficits in macronutrients.

Researchers have also looked at how rising temperatures would affect food costs in the long run (Fischer et al. 2005; Nelson et al. 2009).

After the year 2050, temperatures are anticipated to increase gradually along with food costs, but pricing is also anticipated to rise substantially. According to a study by Nelson et al. (2010), real prices for wheat, rice, and maize could increase by 87–106%, 57–78%, and 54–58%, respectively, in contrast with the baseline year of 2010 by 2050 as a result of the adverse effects of climate change. According to research, food prices may rise sharply in the future, with particular Asian nations seeing increases in poverty of 20–50% as well as cost increases of 10–60% for essentials by 2030 (Hertel et al. 2010). Given how vulnerable agriculture is to climatic patterns, climate change may potentially have an impact on rural earnings. Rural income may suffer as a result of decreased agricultural output and rainfall (Morton 2007). Additional research shows how historical climatic data significantly influences the development of agriculture in particular places with moderate precipitation and rural incomes. It is predicted that rising temperatures that are beyond ideal levels and unpredictable rainfall patterns brought on by climate change would increase rural poverty levels and, as a result, lower the revenue needed to ensure food security.

12.3.3 Agro-Forestry Species Respond to Climate Change

By integrating tree species into cultivation as intercropping during two distinct cropping seasons as an enhanced fallow or on a short- and long-term basis as scattered intercropping, agroforestry practices are frequently used as a climate change solution (Hall et al. 2005). Increased carbon sinks, improved agricultural yields, and the maintenance of nutrients in the soil and biomass may result from tree species that are mostly leguminous. Agroforestry techniques may reduce evaporation by lowering wind speed and temperature (Lin 2010), providing shelter from radiation, and enhancing nutrient cycling by raising soil microorganism activity and organic carbon content. Deep-rooted trees may aid in the hydraulic lift process, which raises water and nutrients to the top layers, and primarily serve as "bioirrigators" (Barrios et al. 2012). Diverse crops and trees are generally more effective for natural resource use, create micro-climate for crop growth and diverse environmental stress conditions, namely, extreme weather, disease, and pest attack than the monocrops (Gaba et al. 2015). In the majority systems of agroforestry, the efficiency of nutrient and water consumption does not considerably differ between C_3 crops and perennial trees. Trees can grow to their full potential because they are perennial crops with better developed roots that can access bigger amounts of soil, nutrients, and water. Reduced soil understory evaporation, reduced excessive wind velocity, reduced light intensity, minimised solar radiation, and control of extremely high temperatures are all ways to improve gaseous exchange, the water table, and water use efficiency (Campi et al. 2009; Muthuri et al. 2009; Lasco et al. 2014). In a Mediterranean setting, the Cupressus arizonica tree canopy acts as a windbreak,

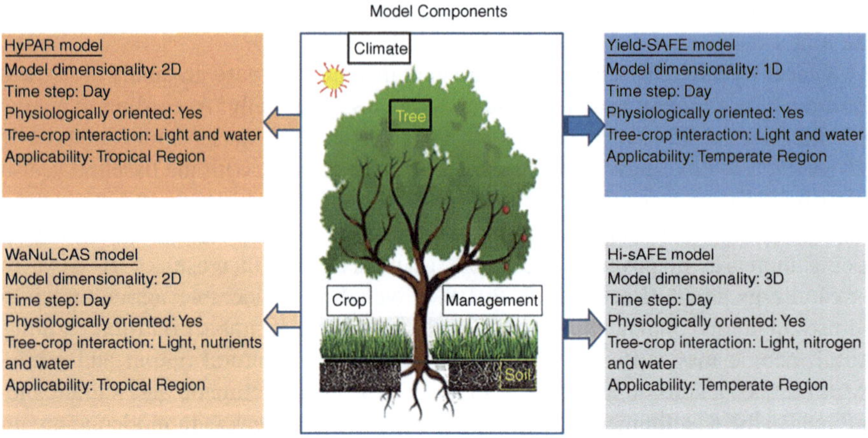

Fig. 12.2 Tree and crop interactions modelling overview used in agroforestry practices

which has a considerable impact on agricultural output and water usage efficiency, according to Campi et al. (2009). This decreases temperature and improves micro-climate. *Jugluans regia* plants with wheat crop (*Triticum aestivum*) increased crop yield production, this may be reduction of understory availability of light which decrease temperature and weed growth and minimum use of water. Selection of suitable trees at proper place depends on the selection of site, goals, and specific characters that increase positive interaction and decrease negative effects in aspects of climate change.

Crop modelling is a complex system in agro-forestry; its management depends on the tree–crop interaction and components (Fig. 12.2). In order to effectuate climate change interaction, tree-crop components must affect sustainability crop production and productivity. Therefore, crop modelling of agroforestry has a major role for increasing crop production and minimising environmental hazards into changing climate. The crop modelling systems needs to multiple objectives, researchers, village, private and government forest. However, this is challenging that crop grown together for agroforestry system mainly environmental condition, crop physiological interactions and management in terms of adequate placement of tree and crops. Different models are likely to be followed for diversified environmental conditions, agricultural crops, and trees, namely, agroforestry systems. These types of models may be organised into six categories (Burgess et al. 2019).

Estimating biomass and above-ground volume using allometric models
Models for canopy architecture
Models for soil status impact on agroforestry systems
Models for crop and tree growth (Yield-SAFE, WaNuLCAS, HyPAR)
Farm decision models (Foreage-SAFE, Farm-SAFE)
Land design models – mainly interpret the effect of agroforestry on land scale

To apply agroforestry practices, the WaNuLCAS and Hi-SAFE models have been used in a number of nations, especially in tropical and European locations (Dupraz et al. 2019). In the hedgerow cropping system used in agroforestry, the WaNuCLAS model, which models the growth of maize and sugarcane, is dynamically implemented. The WaNuLCAS model may also be used to tackle the issue of agricultural adaptation for mitigating climate change (Luedeling et al. 2014).

12.4 Agro-Forestry for Improving Livestock and Smallholder Livelihood Security

12.4.1 Agro-Forestry for Livestock and Livelihood Security

It is impossible to emphasise the importance of agroforestry systems in raising land productivity in order to support the growing human and animal populations. Agro-forestry has the ability to increase production and security while also generating jobs in rural regions. Dhyani et al. (2003, 2005) have emphasised the importance of agroforestry in conserving the native way of life and meeting the needs of marginalised groups. The economy has benefited greatly from the income and job creation brought forth by agroforestry. This is accomplished by fusing several commercial crops that provide a variety of goods and advantages, including wood agricultural products, food crops (such as vegetables, fruits, legumes, pulses, citrus fruit, and edible medicines), and timber agricultural products. In agricultural forests, trees provide a significant source of revenue and are essential for maintaining food security in times of adversity (Table 12.1). The multifunctional trees used in traditional agroforestry systems contribute significantly to social and cultural stability, food security, and revenue generation in rural communities. According to Dhyani et al. (2005), the Indian Himalayas may provide 5.763 million person days/year of labour and advantages for rural development through agroforestry.

The agricultural system founded on Acacia Senegal in some regions of the Barmer and Jodhpur regions of Rajasthan serves as the greatest example of the possibilities of agroforestry to improve lives. Acacia Senegal is the source of the highly valuable commercial product known as Arabic gum. The farmers also extracted significant amounts of Arabic gum using the gum inducer supplied by CAZRI in addition to growing grain for food and collecting crop leftovers for fodder (Roy et al. 2011). Methane (CH_4), nitrous oxide (N_2O), and 9% of the world's anthropogenic greenhouse gas emissions are all caused by the raising of cattle. The variety of production possibilities accessible to rural areas is increased by the raising of cattle, which are often acclimated to hard settings. Timber supplies produced from natural forests and woods are less accessible as a result of agricultural expansion through deforestation. The increasing importance of agroforestry as for the livestock industry is being highlighted by the difficulties that exist with human food and

Table 12.1 The five assets for sustaining livelihoods and ecosystem services are supported by trees

Ecosystem service		Livelihood resilience					References
		Societal	Natural	Physical	Economical	Human	
Supporting activities	Biodiversity	Diversely occupation based on diversity of tree species	Wide range of diversity supplies variation in tree species for crop improvement		Large diversity of tree species produces higher product and revenue earning	Enhanced health and nutrition	Leeuw et al. (2014)
	Fertility of soil		Tree species increase soil fertility		Direct or indirect impact on farmers' income	Increased health through livestock production	
	Soil moisture		Tree increase frequency of rain for crop production		Direct or indirect impact on farmers' income	Maintain better health by increasing livestock and crop production	

animal feed, the increase in consumer appetite for animal products globally, and its changing environmental repercussions (Dawson et al. 2014a, b).

12.4.2 Fodder Trees Increasing Income and Livestock Production

The addition of woody plants inside the system gives agroforestry a distinct advantage over other land use forms. The implementation of tree-based farming can increase economic resilience since it diversifies the types of goods produced (Amare et al. 2019). By offering rural inhabitants alternate sources of revenue, fodder, or food (such as edible wild fruits) during difficult times, the introduction of multifunctional trees, in particular, might boost the profitability of agroforestry (Gebru et al. 2019). Additionally, forests with higher economic value might provide the town with other cash streams to augment its regular agricultural earnings. Due to the slower growing phase, studies on teak-agroforestry structures in Indonesia have shown that despite their shorter recycling period, they may nevertheless provide up to 12% of a household's overall revenue. Damar (*Agathis dammara*) output accounted for up to fifty percent of the household's total income, according to research on damar cultivation in Pesisir, West Sumatra, Indonesia (Wollenberg and Nawir 2005). Additionally, compared to just 12% under the traditional non-agroforestry agricultural system, households' income increased by more than 50% after the establishment of a coffee agroforestry system in Lampung's Wey-Besay Watershed. The findings may be impacted by a variety of variables like the type of trees available, environmental conditions (such as the existence of pests, weather, etc.), and the price of commodities volatility, thus care must be given when evaluating the economic advantages of different systems. Agroforestry is yet another technique to improve the benefit-to-cost ratio. Growing woody plants with little need for input (chemical fertilisers, insecticides, etc.) is one strategy that can boost farmer revenue and lower production costs (Maia et al. 2021). The knowledge of the farmers, particularly in terms of how to successfully select appropriate plants and trees for their particular system, can have a considerable impact on the results. When grown alongside complementing crops, some trees can thrive. Contrarily, the choice of inappropriate tree or crop components may cause nutrient competition (Reynolds et al. 2007), resulting in lower yields and, as a result, lower earnings for the farmers. According to Iskandar et al. (2016), agroforestry has the potential to open up new job opportunities in rural areas in fields like furniture making, grain drying, and other off-farm pursuits. Because they may actively take part in industrial processes, women may also profit from having additional employment possibilities. Additionally, keeping jobs in rural regions can halt the departure of people from those areas, boosting the rural economy. However, caution must be taken when building industrial sites adjacent to primary forest areas due to the risk of human encroachment into these designated areas and the potential for ecological harm.

The application of agroforestry can provide food security for those living near woods in addition to producing revenue. Ickowitz et al. (2016) used geographic information to better understand the micronutrient intake of one- to five-year-old Indonesian children. They found a link between the national consumption of legumes and the use of agroforestry. Their specialised research showed a connection between agroforestry and more leafy greens and vitamin A-rich diets. Additionally, agroforestry practices were linked to an increase in meat intake, especially among silvopastoral farmers. Following its implementation, agroforestry training led to higher food production and greater diversity among low-income farmers, indicating enhanced food accessibility (Pratiwi and Suzuki 2019).

Agroforestry may also promote the societal participation of adopters. For instance, farming organisations may meet and discuss the method of cultivation, the choice of tree or crop species, the control of fertilisers, and other issues. A study by Roikhwanphut Mungmachon (2012) found that collecting was a traditional practice among Thailand's tiny forest villages. They had ongoing meetings to discuss their problems and work together to come up with solutions. The first phase involves examining their problems as a group, building on prior knowledge and traditional wisdom, then assimilating new information. This leads to a more informed and engaged community through peer-to-peer conversation and community participation.

12.4.3 Agro-Forestry Improving Ecosystem and Natural Resource

Agroforestry provides a range of environmentally conscious tactics that might enhance ecological services for rural residents. Rotation of crops, conserving soil (cover crop integration), enhanced fallows, and boundary plants are a few of these tactics. Crop diversification, which includes combining crops and trees, is another. As an example, adding pruning debris to the soil as an amendment can increase soil fertility and improve its physical structure (soil conservation) (Shrestha et al. 2018). The effectiveness of the pruning materials used in the system, however, can have an impact on the outcome of this practice. Plant wastes may decompose in soil, with the breakdown process varying depending on the C/N ratio of the soil. As a consequence, the quantity of minerals released into the soil may vary based on the kind of residues (Hossain et al. 2011), changing the chemistry of the soil and impacting crop development. Different decomposition rates, caused by changes in the C/N ratio, may have an impact on an agroforestry system's overall ability to sequester carbon. This might then affect the amount of carbon in the soil, either increasing it or decreasing it (Zhang et al. 2013). By creating homes for animals, the incorporation of several tree species in an agroforestry system enhances biodiversity (Assogbadjo et al. 2012). In addition, on steep slopes, the vast root networks of trees inside the soil matrix are essential for limiting soil erosion and reducing the risk of landslides

(De Souza et al. 2012). Through shadowing, the existence of trees in agroforestry systems may also change the microclimate, perhaps reducing the quantity of sunlight that acts as the temperature buffer around the farm. Due to the fact that extremely strong solar radiation can hinder crop physiology and growth, agroforestry can help boost crop growth and, therefore, agricultural production (Caron et al. 2018). Consideration must be taken when selecting a tree canopy, though, as excessive shadowing can severely limit penetration of light, which may stunt the development of co-cultivated crops and raise the risk of disease.

Agroforestry provides nearby communities with improved water conservation as one of its ecological advantages. Such ecological benefits are made possible by the greater water absorption of the integrated tree-crop system. A research done in Kenya that focused on an agroforestry system with maize and trees found that during the dry season, about 25% of the precipitation was absorbed inside the plant biomass. This demonstrates how effectively the system uses off-season precipitation, which makes up 15% to 20% of the yearly total. The system's capacity to hold onto moisture is demonstrated by the residual water that stays in the soil layers even after harvest (Lott et al. 2003). The organic carbon content of agroforestry soils can be further increased by the use of organic amendments. This can boost water retention, reducing excessive evaporation and water runoff. The choice of tree type is important, too, as different plant species may have different water uptake rates. The potential for water distinction between the soil and the surrounding environment leads plant roots to take in water while leaf stomata are open. This depends on a certain plant species' capacity to investigate the roots (Bayala and Prieto 2020). Additionally, an experiment was run to demonstrate the effect of vegetation decrease on rainfall, and the results showed a connection between greater precipitation rates and increasing vegetation density (which was achieved through more tree and shrub biomass). Reduced evapotranspiration and higher light reflection in the environment brought on by lower plant density may be to blame for the drop in rainfall seen in situations of decreasing vegetation density (Gonzalez et al. 2012). In addition, water cycle analysis emphasises the significance of controlling tree cover to increase rainfall (Ellison et al. 2012). So, one method for reducing dryness in some desert regions and boosting community resilience to climate change is agroforestry. These investigations, which rely on data correlation or modelling, were only carried out at the farm level, despite being promising. Therefore, additional research encompassing various geographic sites is required to corroborate such conclusions.

12.5 Adaptation and Mitigation Strategies

12.5.1 Climate Change Mitigation Via Sustainable Agroforestry

The capacity of farming systems in tropical and subtropical climates to adapt to shifting environmental circumstances is greatly improved by agroforestry systems. Additionally, these systems have a substantial capacity for carbon sequestration, which would lessen climate change. The goal of current agricultural research is to make agricultural systems more productive and resilient. While the resilience of these ecosystems mostly depends on their ability to adjust to unfavourable climatic changes, production is directly influenced by the capability to store and maintain carbon (Amare et al. 2019). While diligent observation, experimentation, and practice have allowed traditional civilisations to show their flexibility and resilience, dealing with the fast changes in the world's climate and other areas calls for more than a simple self-correcting mechanism. To reduce the effects of anticipated temperature rises and fulfil the requirements of families for money, food, and fuel, it asks for well-thought-out solutions, including supporting laws and regulations (Gebru et al. 2019). Stronger adaptation techniques are required to adequately manage the increased dangers offered by unanticipated or unexpected weather changes. Among the difficulties posed by adaptation are:

The use of crop varieties and effective management techniques appropriate for various soil and climate conditions is required to reduce heat stress on plants and animals.

Implementing strategies to reduce erosion and organic carbon loss in order to reverse land deterioration.

Increasing agricultural operations' innovative and sustainable diversification and promoting climate adaptation are two key strategies for managing climate-related hazards.

By utilising high-use efficiency technologies and techniques, applying irrigation practices, and adopting suitable soil and water conservation measures, rainfall may be captured, stored, and used efficiently.

Reducing nutrient mining and enhancing or sustaining soil fertility to maintain soil fertility and productivity.

Promoting tillage and fuel-saving management practices to cut emissions of greenhouse gases and encourage carbon sequestration.

Protecting against the burden of pests and diseases.

Increasing community resilience through more efficient use of resources and better investment targeting.

Making sure food and nutritional security is maintained.

12.5.2 Agroforestry's Role in Climate Change Adaptation

Farmers employ the well-known management method of agroforestry to grow trees and shrubs alongside annual crops. By using this strategy, you may reduce soil erosion, preserve soil fertility, diversify your revenue sources, increase and stabilise your income, and increase the effectiveness with which you utilise nutrients, water, and sunshine (Wu et al. 2020). Along with providing food and cash year-round, this approach also offers shade. Agroforestry also offers solid employment opportunities. The extensive research carried out by the World Agroforestry Centre and its associates over the past three decades has greatly increased our understanding of the various advantages, opportunities, and challenges associated with merging crop and tree production. The development of more trustworthy, scientifically based, and integrated systems that benefit from advancements in crop and tree management and breeding has also been made possible by this. For instance, planting swiftly growing, nitrogen-fixing shrubs and trees in agricultural fields is one of the several practices used in tropical areas to increase soil fertility and reduce erosion (Rao et al. 2007). The agricultural productivity of small-scale farms can be increased by choosing the appropriate tree and shrub species. Utilising agroforestry systems may boost output, encourage soil fertility, enhance microclimate, reduce soil erosion, and diversify income streams. The IPCC Third Assessment Report on Climate Change (IPCC 2001) notes that agroforestry also has a variety of economic, environmental, and social benefits. This demonstrates how flexible agroforestry systems are in solving different problems and offering a variety of benefits. For instance, the existence of trees on agroforestry farms improves soil fertility by reducing erosion, preserving soil organic matter and structure, increasing nitrogen content, taking in nutrients from deeper levels of soil, and fostering a closed nutrient cycle.

12.5.3 Agroforestry Systems as a Viable Tool for Mitigation to Climate Change

Agroforestry, which combines the production of food and/or animals with trees and shrubs, offers a significant chance to lessen the effects of climate change and promote adaptability. The acceptability of agroforestry will face a variety of obstacles, and it will not realise its full potential until the best and most durable solutions are made generally acknowledged. Therefore, this chapter's section is quite significant.

12.5.3.1 Agroforestry Systems for Moderating the Crop Microclimate

For many different kinds of plants and crops to reach their genetic potential, the environment must be almost ideal. Otherwise, they will not be able to realise their full potential. Any alteration of these elements, especially during the reproductive period, would immediately affect the yield and marketability of different crops. Even if the additional energy that has built up and been stored by the natural world cannot be released, a workable solution to lessen the impacts of local heat stress is to use agroforestry systems with the appropriate shade-giving plants.

On farms, trees have a considerable impact on the rate and duration of photosynthesis. This has an effect on soil water usage, transpiration, and plant development. The radiation flux, air temperature, wind speed, and crop saturation deficiency understorey are among these parameters, according to Monteith et al. (1991). The advantages of microclimatic changes are utilised in a wide variety of methods, such as the planting of shade trees, wind breaks, and shelter belts to lessen wind speed and prevent physical damage to crops, mulches that chill the soil, and diverse crop tree pairings to improve resource efficiency. However it is sensitive to hot herbs, such as cardamom, ginger, coffee, and chocolate.

The seasonal average for the surrounding temperature and solar radiation are often lower in microclimates caused by shadow. Beer et al. (1998) examined the literature on managing shade in cocoa and coffee farms and discovered that shade trees can minimise temperature extremes by up to 5 °C. According to Steffan-Dewenter et al. (2007), the removal of shade trees caused the soil's surface temperature to rise by 4.0 °C and the relative air humidity to drop by 12% at a height of 2 m. It was found that the soil temperature in semi-arid areas of Kenya was 6 °C lower at a depth of 5 to 10 cm than what was observed in an open environment (Belsky et al. 1993). In the Sahel, where soil temperatures often surpass 50 to 60 °C and are a significant barrier to farming a sustainable crop, Faidherbia trees lowered soil temperature at a depth of 2 cm by 5 to 10 °C depending on their movement of shadow (Vandenbeldt and Williams 1992). A popular method for enhancing microclimates is the creation of shelterbelts, which are horizontal rows of trees covering the ground. These trees are specifically used to control evapotranspiration and wind erosion, slowing down wind speed by making the surface rougher. The leeward side takes the brunt of the effects when shelterbelts are constructed properly, with impacts felt between 10 and 25 times the height of the belt below.

The production and economical benefits of these systems are still highly debated (Beer et al. 1998; Rao et al. 2007), despite the fact that the benefits of trees in shifting and warming microclimate conditions are well established. This is partly due to the intricate interactions seen in various agroforestry systems. The kind of crop and tree, the quantity and distribution of trees, the ages of the trees, the management of the crop and trees, and the changing temperature are the main biophysical elements affecting the effectiveness of mixed systems. Cereals like barley and millet are either moderately sensitive or less responsive to shade than other green horticultural crops like clover and alfalfa. When the annual crop is a C3 plant, which is regularly exposed to light outside, the net shadow impact is reportedly more favourable (Ong et al. 1996). Higher agricultural yields were seen in all climatic regimes, particularly in the humid and subhumid tropics, where the advantages of increased fertility outweigh the disadvantages of competition (Rao et al. 2007). Although maize was cultivated on Grevillea robusta, productivity was significantly decreased due to competition for water and nutrients (Ong and Swallow 2004).

12.5.3.2 Agroforestry Systems Provide Permanent Cover, which Is Very Successful in Preserving Soil and Water Resources

Soil deterioration is impacted by climate change in many different ways. Rao et al. (2007) found that higher temperatures and drier conditions reduce the amount of organic matter that may develop up in the soil, which results in a poorer soil structure, less precipitation absorption, increased runoff, and erosion. The severity, frequency, and extent of erosion were all predicted to be negatively impacted by the anticipated rise in the frequency of protracted rains (WMO 2005). The already horrible problem the continent is suffering will get worse as a result of these changes. Changes in the chemical, physical, and biological conditions are required to stop degradation and restore the soil's capacity for production. Systems that incorporate agroforestry are desirable because they may improve all three criteria. Better fallows, contour hedgerows, and other permanent covering agroforestry systems are essential to stop land deterioration. Permanent protection is provided by these systems, which also improve soil structure, increase infiltration, raise fertility, and stimulate biological activity.

In collaboration with the Institut de Recherche pour le Développement (IRD) and governmental agricultural research organisations in Kenya, the World Agroforestry Centre assessed the efficiency of expanded fallow for preventing soil erosion in western Kenya. In the experiment, *Tephrosia* species and *Crotalaria grahamiana* were utilised as fast-growing shrubs. The ability of these plants to reduce soil losses has been quite strong. During the fallow season, when trees reduce the influence of rainfall on the soil, the process of improving fallow begins. The growth of the soil's structure and biological activity, however, causes the soil to stay for a very long period after fallow clearing (Rao et al. 2007).

There are not many studies that examine the variety of soil life between planted fallows and natural or continuous cropping. Observations carried out in Muguga, Kenya, under natural forest, regularly harvested corn, one-year-old sesbania fallow, and grass fallow showed that sesbania fallows recovered the soil biological life to the same level as in natural forest. They were also much larger than in grass fallows or farmed regions. The World Agroforestry Centre has expanded the use of infrared spectroscopy to improve the diagnosis and monitoring of soil quality, allowing for quick evaluations of soils and many other organic resources (Shepherd et al. 2003). The technique aids in both a better knowledge of the complexity and a variety of local soils as well as soil quality monitoring for environmental protection. Large numbers of soil samples with georeferences may be described with ease using infrared spectroscopy. As a result, it is possible to extrapolate ground data from satellite imagery to cover quite large areas. Precision agriculture, farm advisory services, process studies, and large-area uses (soil survey, watershed management, pedo-transfer functions, soil quality indicators) all stand to gain from the use of IR spectroscopy to increase productivity and cut costs. In particular, it creates new possibilities for risk-based soil evaluation techniques that explicitly take into account the ambiguity of forecasts and interpretations of value assumptions for soil.

12.5.3.3 Agroforestry: An Important Route for the Sustainable Diversification of Agricultural Systems and Incomes

Farmers have long used diversification of their operations as a risk management strategy and as a way to take advantage of favourable weather conditions. This entails maximising complementarities and synergies between various farm operations while preserving their core characteristics. To decrease risk, take advantage of new market possibilities, investigate current market niches, increase output and on-farm processing, as well as other income-generating activities, diversification demands changing the makeup of farm operations (Dixon et al. 2001).

Diversification at the farm level refers to the adoption of multiple production activities including crops, animals, trees, and post-harvest processing that complement one another in terms of their ecological and/or economic properties. The most successful long-term strategy for diversifying agricultural systems is widely acknowledged to be integrated agroforestry systems. These techniques have been widely used, for instance, poplar trees, whose quick growth has made them popular on many South Asian farms. Home gardens with a variety of fruit and vegetable plants are widespread in Africa and considerably improve food security by supplying a year-round supply (Wezel and Bender 2003).

Home gardens may contribute up to 44% and 32%, respectively, of a household's protein and calorie needs, according to research (Torquebiau 1992). Home gardens are essential for supplying households with additional revenue in addition to meeting their fundamental necessities (Mendez et al. 2001). The success of agroforestry technology depends on increasing the market opportunities for small-scale farmers, particularly in niche markets and high-value items (Russell and Franzel 2004). The development of the small-holder tree product business in Africa is constrained by problems with outgrowing plans and contract farming, as well as by physical and social barriers to market access, forest regulations, and a lack of market knowledge regarding agroforestry items.

Nevertheless, a number of innovative initiatives have been implemented that show promise, including fuelwood programmes with contractual agreements, the development of small nurseries, adjustments to the policies governing the production of charcoal, the introduction of market information systems, and partnerships between the private sector, colleges and universities, and extension organisations (Russell and Franzel 2004). Traditional and non-traditional tree crops might be integrated into farming practices for growing fruits, nuts, short rotation woody agricultural products, biomass energy plantations, and medicinal herbs by setting up the proper market mechanisms (Rao et al. 2007; Hall and House 1993).

12.5.3.4 Agroforestry Systems Improve Rain Water Use Efficiency

Climate change is anticipated to make the existing situation worse because of the lack of water supplies. Water supply is impacted by climate change in both direct and indirect ways. Modifications in precipitation patterns are examples of direct consequences, whereas increased runoff and losses from evapotranspiration are examples of indirect effects. According to the Comprehensive Assessment of Water Management in Agriculture, if existing trends in food production and the environment are not reversed, there might be crises all over the world. This warning is based on in-depth investigations carried out as part of this project (CA 2007). Therefore, whether or not climate change happens, it is imperative to increase agricultural water output to manage the severe water shortages that are anticipated in the next 50 to 100 years.

Agroforestry has the capacity to utilise water resources more efficiently through a number of mechanisms as compared to annual crops. First, agroforestry systems with their permanent tree components, unlike annual systems where the land is exposed for extended periods, may use water retained in the soil after harvest and rainfall that arrives outside of the crop season. Second, by collecting a larger amount of the annual rainfall, reducing runoff, and utilising water stored in deep layers, agroforests boost the productivity of precipitation. Thirdly, there is more water available for transpiration as a result of climatic changes, including cooler air, stronger winds, and reduced agricultural water demands (Ong and Swallow 2004). However, due to the intricate interactions involved and the challenges of quantifying, we currently lack understanding regarding resource competition between the tree and crop components. The overwhelming bulk of the evidence is based on the analysis of information obtained from the system's above-ground components, which mostly demonstrate that trees have a negative influence on agricultural production. The competition for nutrients in moist, humid environments as well as the availability of water in semi-arid tropical regions are usually linked to this negative effect (Rao et al. 2007).

In a study conducted in Machakos, Kenya, soil water was measured over the course of three succeeding growing seasons in hedgerow intercropping systems with different *Senna* species, as well as an annual crop system with maize and cowpea. The study demonstrated the importance of water competition in semi-arid situations by showing that both hedgerow systems had lower soil water levels than the annual crop system, especially during times of water shortage. Compared to the slower-growing, lower-biomass *S. siamea*, *S. spectabilis* grew quickly and depleted soil water more quickly. The soil profile never entirely refilled even during the "short wet" season of 1994–1995, which received above-normal rainfall of 547 mm (50 percent greater than usual), because of significant water depletion in the previous season (Rao et al. 1998). Depending on tree density, rainfall volume, and distribution, the canopy of trees can drastically limit the quantity of water that reaches the soil by intercepting rain (Ong et al. 1996). Estimated losses can exceed 50%. The complementing nature of their vertical root systems is thought to be one advantage of agroforestry systems. According to studies on root growth, not all of the trees employed in agroforestry systems have deep, necessary roots; rather, they frequently have mixed and shallow root structures (vanNoordwijk et al. 1996). Plants prefer to develop surface roots in soils that lack nutrients and when the supply of water below the root zone is restricted, as it does during dry spells (Rao et al. 2004). Few plants have roots that can penetrate water tables that are rather deep. As a result, vertical root complementarity may not be as likely as initially believed (Ong and Swallow 2004), highlighting the need for management strategies to reduce competition. However, methods like side-trimming to reduce above-ground competition and periodic root pruning to reduce below-ground rivalry have been researched. It is yet unknown if root competition can be controlled on tropical farms.

12.5.3.5 Agroforestry Systems Offer an Affordable and Sustainable Way to Increase Soil Fertility

The fundamental barrier to higher productivity in many African countries is usually cited as the soil's nutrient depletion from continually growing crops without proper fertilisation or soil fallowing. According to studies, cultivated lands in 37 African nations have lost an average of 22 kg of nitrogen, 2.5 kg of phosphorus, and 15 kg of potassium per hectare year during the previous 30 years, which is equal to $4 billion in fertiliser costs (Sanchez 2002). However, it is extremely improbable that African farmers will be able to afford substantial investments in fertilisers given the current cost of fertilisers to resolve this issue. As a result, agroforestry systems have become a desirable and sustainable way to improve soil fertility.

The World Agroforestry Centre has made significant progress in identifying and promoting agroforestry systems intended to boost soil fertility. According to Sanchez et al. (1997), trees may help to deliver nutrients in four distinct ways: by increasing nutrient inputs into the soil, promoting internal nutrient cycling, reducing nutrient losses from the soil, and by providing environmental benefits. After carefully examining numerous soil fertility replenishment techniques, the World Agroforestry Centre created a system that consists of three components that can be used separately or together: (i) nitrogen-fixing leguminous tree fallows; (ii) the use of native rock phosphates in phosphorus-deficient soils; and (iii) biomass transfer of leaves from nutrient-accumulating shrubs. In the subhumid tropical regions of East and Southern Africa, leguminous trees from genera including *Sesbania*, *Tephrosia*, *Crotalaria*, *Glyricidia*, and *Cajanus* are interplanted with immature maize harvests. These trees are allowed to grow on fallow lands during dry seasons, accumulating 100 to 200 kg of nitrogen per hectare over a two- to six-year period. The nitrogen fixation achieved with this method is equivalent to the nitrogen fertiliser used by commercial farmers in industrialised countries to grow maize.

Biomass transfer utilising *Gliricidia* (mixed intercropping; *Gliricidia sepium*) or wild sunflower (*Tithonia diversifolia*) have also been investigated (Place et al. 2002). These methods have shown improvements in grain output of 50 to 200 kg N ha^{-1}. It should be understood, nonetheless, that not all agro-ecologies may be suited for these techniques. The usefulness of enhanced fallows in the semi-arid tropics of Africa has yet to be determined in shallow soils, poorly drained soils, and frost-prone locations since their growth and capacity to fix nitrogen are constrained by the longer dry seasons (Sanchez 2002).

12.5.3.6 Agroforestry Systems Limit Carbon Emissions and Sequester Carbon

Agroforestry has enormous potential to reduce CO_2 emissions and fight climate change by efficiently absorbing carbon from the atmosphere. The tree component of agroforestry systems may be a significant carbon sink on agricultural soils. Through three main mechanisms, trees help reduce carbon emissions: they sequester carbon through enhanced fallows and integration with trees, they conserve the existing carbon pools by preventing deforestation and implementing alternatives to slash-and-burn techniques, and they replace fossil fuels with biofuel and bioenergy plantations (Montagnini and Nair 2004). The effectiveness of agroforestry systems as carbon sinks has been examined in several research (IPCC 2001; Albrecht and Kandji 2003; Montagnini and Nair 2004). According to estimates, assuming a mean carbon content of 50% in aboveground biomass, agroforestry systems may store, on average, 9, 21, 50, and 63 Mg C ha^{-1} of carbon in semiarid, subhumid, humid, and temperate climates, respectively. With an estimated 630 x 106 hectares of land worldwide that are suitable for agroforestry, the widespread application of agroforestry greatly adds to its quantitative relevance (Palm et al. 2005). Improved fallows are without a doubt one of the most promising agroforestry strategies in the sub-humid tropics since they aim to restore nutrient-depleted soils. Recent studies have shown the potential for widespread adoption throughout southern and eastern Africa, including in drier regions like the Sudan-Sahel zone in West Africa. Short-rotation systems, which decrease aboveground carbon collection, are frequently used in improved fallows in place of the more permanent systems observed in the humid tropics. But over time, they store much more aboveground carbon than degraded land, croplands, or pastures (Albrecht and Kandji 2003). Although the capacity of agroforestry systems to store carbon is well acknowledged, it is important to consider any potential trade-offs between carbon storage and profitability when promoting these systems. One tonne of soil carbon can increase crop yields for agricultural soils by 20 to 40 kg ha^{-1} for wheat, 10 to 20 kg ha^{-1} for maize, and 0.5 to 1 kg ha^{-1} for cowpeas. The extent to which carbon sequestration schemes and carbon market activities will benefit smallholders is still unclear, according to Montagnini and Nair (2004). It is required to quantify carbon sequestration and its contribution to soil carbon pools more precisely in order to offer incentives and include these elements in carbon audits. However, agroforestry's potential for sequestering carbon has not yet been fully acknowledged or utilised. One notable challenge is the lack of empirical evidence for most of the proposed mechanisms explaining how agroforestry systems could effectively mitigate atmospheric CO_2 accumulation (Dhiman 2013).

12.6 Challenge and Future Outlook

Despite the large number of studies proving the advantages of agroforestry, many developing countries have not yet adopted it widely. One possible reason for this slow transition is the perception that agroforestry contradicts the prevailing notion of high-yield monoculture systems (Ollinaho and Kröger 2021). In contrast, agroforestry uses a more complex system with several elements, such as different tree species, crops, and/or animals, where the interaction between these elements is essential for getting the best outcomes in terms of the economy and environmental sustainability. As a result, whether the effects are beneficial or poor, a lot depends on how well-informed people are about efficient agricultural methods (Good Agricultural Practices, or GAP). For instance, lower yields or harvestable sections of crops or trees may result from competition for nutrients or light among the species included in the agroforestry system (Wu et al. 2020). The high nutrient and water needs of some tree and agricultural species may decrease soil resources and cause water loss (Mukhlis 2019). Therefore, agronomy becomes a vital skill for rural populations to successfully embrace agroforestry. Disseminating information within these communities necessitates additional personnel from NGOs, government organisations, and research institutes. Another barrier to adoption is the lack of explicit inclusion of agroforestry in national agendas promoting the transition to sustainable agriculture. Unlike more well-defined approaches like organic farming, the term "agroforestry" may be less familiar, contributing to this challenge. It is important to remember that different tree-based systems can exist under the agroforestry umbrella, but the results may vary depending on the particular components involved. Despite the fact that studies on the effects of agroforestry have been carried out for decades, the majority of these studies have concentrated on the farm level and have only addressed one aspect of the effects (social, economic, or environmental), with only a small number of thorough studies being carried out on larger scales like national or continental scopes (Ollinaho and Kröger 2021). Consequently, reaching a consensus on the effects of agroforestry remains challenging, partly due to the relatively limited political support it receives. Overcoming this obstacle requires future research to concentrate on examining the social, economic, and environmental impacts of agroforestry. Several attempts have been made to popularise agroforestry by collaborating with the business sector and emphasising the establishment of large-scale plantations. However, the practice of "industrial" agroforestry raises concerns, as it may lead to limited intercropping systems with a single tree species dominating, deviating from the more diverse components typically associated with agroforestry. Moreover, this "commercial" agroforestry approach may result in the conversion of virgin forests into mixed-commodity plantations (e.g., spices, palm oil, bananas), exacerbating biodiversity loss rather than safeguarding it. Clear lines must be drawn through legislation, particularly when defining the phrase "agroforestry", in order to stop "a new form" of deforestation. Outside of main forests, agroforestry can be used to improve biodiversity and restore soil quality in degraded regions. However, due to the fact that these resources are frequently under the jurisdiction of governments or enterprises, many rural areas, particularly those with smallholder farmers, have difficulty gaining access to the land, seedlings, and germplasm needed to create agroforestry systems (Gebru et al. 2019). Governments or NGOs can step in to remedy this issue by opening up degraded lands to anybody interested in putting agroforestry systems in place. To improve these communities' economic resilience, the government may also offer temporary aid in the form of market access, post-harvest equipment, or price stability.

12.7 Conclusion

At this time, the feeding of various stakeholders, including the general public and wood-based companies, is deemed necessary. It is acknowledged that the practice of agriculture alone will not be sufficient to fulfil the requirements of marginal and small-scale farmers because to climatic changes, growing demands on agriculture, declining land holdings, and the diversion of agricultural land to other uses. A solution to address these issues is found in agroforestry, which will contribute to the elevation of the standard of living for the agricultural community through the utilisation of a cluster strategy and value chain models, bringing all stakeholders together on a single platform. By incorporating animals and trees into fields, income can be increased, and job opportunities can be created in rural areas. Active efforts are being made by the research institutions of central and state governments to develop agroforestry and agricultural forestry on a broad scale, employing precise silvicultural methods and ensuring a structured pricing system. These initiatives not only meet the domestic and financial requirements of farmers but also yield positive environmental impacts. Ultimately, the implementation of agroforestry methods is necessary to preserve the agrarian character and natural (forest) resources of the nation. To fully unlock the potential of a tropical agroforestry system in a specific location, it is crucial that government action and the support of other private entities through the adoption of various policies are embraced.

Acknowledgement Authors are thankful to Honourable Chancellor, Vice-Chancellor, and Chief Operating Officer of Swami Vivekananda University for their constant technical support and guidance.

Conflict of Interest The authors declare that they have no known competing financial interests or personal relationships that could have appeared to influence the work.

References

Albrecht A, Kandji ST (2003) Carbon sequestration in tropical agroforestry systems. Agric Ecosyst Environ 99:15–27

Amare D, Wondie M, Mekuria W, Darr D (2019) Agroforestry of smallholder farmers in Ethiopia: practices and benefits. Small-Scale Forest 18:39–56

Assogbadjo AE, Kakaï RG, Vodouhê FG, Djagoun CAMS, Codjia JTC, Sinsin B (2012) Biodiversity and socioeconomic factors supporting farmers' choice of wild edible trees in the agroforestry systems of Benin (West Africa). Forest Policy Econ 14(1):41–49

Barrios E, Sileshi GW, Shepherd K, Sinclair F (2012) Agroforestry and soil health: linking trees, soil biota and ecosystem services. In: Wall DH (ed) Soil ecology and ecosystem services. Oxford University Press, Oxford, pp 315–330

Bayala J, Prieto I (2020) Water acquisition, sharing and redistribution by roots: applications to agroforestry systems. Plant Soil 453:17–28

Beer JW, Muschler RG, Somarriba E, Kass D (1998) Shade management in coffee and cacao plantations-a review. Agrofor Syst 38:139–164

Belsky AJ, Mwonga SM, Duxbury JM (1993) Effects of widely spaced trees and livestock grazing on understory environments in tropical savannas. Agrofor Syst 24:1–20

Besar NA, Suardi H, Phua MH, James D, Mokhtar MB, Ahmed MF (2020) Carbon stock and sequestration potential of an agroforestry system in Sabah, Malaysia. Forests 11(2):210

Beyene AD, Mekonnen A, Randall B, Deribe R (2019) Household level determinants of agroforestry practices adoption in rural Ethiopia. Forests Trees Livelihoods 28(3):194–213

Brinkman HJ, de Pee S, Sanogo I, Subran L, Bloem MW (2009) High food prices and the global financial crisis have reduced access to nutritious food and worsened nutritional status and health. J Nutr 140(1):153S–161S

Browder JO, Wynne RH, Pedlowski MA (2005) Agroforestry diffusion and secondary forest regeneration in the Brazilian Amazon: further findings from the Rondônia agroforestry pilot project (1992–2002). Agrofor Syst 65:99–111

Burgess, P., Graves, A., de Jalón, S. G., Palma, J., Dupraz, C., & van Noordwijk, M. (2019). Modelling agroforestry systems. In Agroforestry for sustainable agriculture (pp. 209-238). Burleigh Dodds Science Publishing. US.

CA (2007) Water for Food, Water for Life: A Comprehensive Assessment of Water Management in Agriculture. Earthscan, and Colombo: International Water Management Institute, London

Campi P, Palumbo AD, Mastrorilli M (2009) Effects of tree windbreak on microclimate and wheat productivity in a Mediterranean environment. Eur J Agron 30(3):220–227

Caron BO, Sgarbossa J, Schwerz F, Elli EF, Eloy E, Behling A (2018) Dynamics of solar radiation and soybean yield in agroforestry systems. An Acad Bras Cienc 90:3799–3812

Conway D, Schipper ELF (2010) Adaptation to climate change in Africa: challenges and opportunities identified from Ethiopia. Glob Environ Chang 21(1):227–237

Dawson IK, Carsan S, Franzel S, Kindt R, Van BP, Graudal L, Lilles JPB, Orwa C, Jamnadass R (2014a) Agroforestry, livestock, fodder production and climate change adaptation and mitigation in East Africa: issues and options. ICRAF working p 178. Nairobi, World Agroforestry Centre, p 18

Dawson, I. K., Carsan, S., Franzel, S., Kindt, R., van Breugel, P., Graudal, L., & Lillesø, J. P. B. (2014b). Agroforestry, livestock, fodder production and climate change adaptation and mitigation in East Africa: issues and options. World Agroforestry Center, Nairobi. Kenya.

De Souza HN, Goede RGM, Brussaard L, Cardoso IM, Duarte EMG, Fernandes RBA, Gomes LC, Pulleman MM (2012) Protective shade, tree diversity and soil properties in coffee agroforestry systems in the Atlantic rainforest biome. Agric Ecosyst Environ 146(1):179–196

Dhiman R (2013) Status and impact of commercial agroforestry in India. Indian J Agrofor 15(2): 55–67

Dhyani SK, Sharda V, Sharma AR (2005) Agroforestry for sustainable management of soil, water and environmental quality: looking back to think ahead. Range Manag Agrofor 26(1):71–83

Dhyani SK, Sharda VN, Sharma AR (2003) Agroforestry for water resources conservation: issues, challenges and strategies. In: Agroforestry: potentials and opportunities. Agribios. Agribios, Jodhpur, India

Diao X, Hazell P, Thurlow J (2010) The role of agriculture in African development. World Dev 38(10):1375–1383

Dixon JA, Gibbon DP, Gulliver A (2001) Farming systems and poverty: improving farmers' livelihoods in a changing world. FAO, Rome; Washington, D.C.: Word Bank

Drew WM, Alavalapati JRR, Nair PKR (2004) Determining agroforestry profitability using the policy analysis matrix. In: Valuing agroforestry systems, pp 59–78

Dupraz C, Wolz KJ, Lecomte I, Talbot G, Vincent G, Mulia R (2019) Hi-sAFe: a 3D agroforestry model for integrating dynamic tree–crop interactions. Sustainability 11(8):2293

Ellison D, Futter MN, Bishop K (2012) On the forest cover–water yield debate: from demand-to supply-side thinking. Glob Chang Biol 18(3):806–820

Fischer G, Shah M, Tubiello FN, van Velhuizen H (2005) Socio-economic and climate change impacts on agriculture: an integrated assessment, 1990–2080. Philos. Trans. R. Soc., B 360(1463):2067–2083

Gaba S, Lescourret F, Boudsocq S, Enjalbert J, Hinsinger P, Journet EP (2015) Multiple cropping systems as drivers for providing multiple ecosystem services: from concepts to design. Agron Sustain Dev 35:607–623

Garrity DP (2004) Agroforestry and the achievement of the millennium development goals. Agrofor Syst 61:5–17

Gebru BM, Wang SW, Kim SJ, Lee WK (2019) Socio-ecological niche and factors affecting agroforestry practice adoption in different Agroecologies of southern Tigray, Ethiopia. Sustainability 11:3729

Gold MA, Godsey LD, Josiah SJ (2004) Markets and marketing strategies for agroforestry specialty products in North America. Agrofor Syst 61:371–382

Gonzalez P, Tucker CJ, Sy H (2012) Tree density and species decline in the African Sahel attributable to climate. J Arid Environ 78:55–64

Gourdji SM, Mathews KL, Reynolds M, Crossa J, Lobell DB (2013) An assessment of wheat yield sensitivity and breeding gains in hot environments. Proc R Soc B Biol Sci 280(1752):20122190

Hall DO, House JI (1993) trees and biomass energy: carbon storage and/or fossil fuel substitution? Biomass Bioenergy 6(1–2):11–30

Hall NM, Kaya B, Dick J (2005) Effect of improved fallow on crop productivity, soil fertility and climate-forcing gas emissions in semi-arid conditions. Biol Fertil Soils 42:224–230

Hertel TW, Burke MB, Lobell DB (2010) The poverty implications of climate-induced crop yield changes by 2030. Glob Environ Chang 20(4):577–585

Hossain M, Siddique MRH, Rahman MS, Hossain MZ, Hasan MM (2011) Nutrient dynamics associated with leaf litter decomposition of three agroforestry tree species 1(Azadirachta indica, Dalbergia sissoo, and Melia azedarach) of Bangladesh. J For Res 22:577–582

Ickowitz A, Rowland D, Powell B, Salim MA, Sunderland T (2016) Forests, trees, and micronutrient-rich food consumption in Indonesia. PLoS One 11(5):e0154139

IPCC (2001) Climate change 2001: the scientific basis. Contribution of the working group 1 to the third assessment report of the IPCC. Cambridge University Press, Cambridge

Iskandar J, Iskandar BS, Partasasmita R (2016) Responses to environmental and socio-economic changes in the Karangwangi traditional agroforestry system, South Cianjur, West Java. Biodiver J Biol Diver 17(1):332–341

Jara-Rojas R, Russy S, Roco L, Fleming-Muñoz D, Engler A (2020) Factors affecting the adoption of agroforestry practices: insights from silvopastoral systems of Colombia. Forests 11(6):648

Jarvis A, Lau C, Cook S, Wollenberg E, Hansen J, Bonilla O, Challinor A (2011) An integrated adaptation and mitigation framework for developing agricultural research: synergies and trade-offs. Exp Agric 47(2):185–203

Jezeer RE, Santos MJ, Boot RGA, Junginger M, Verweij PA (2018) Effects of shade and input management on economic performance of small-scale Peruvian coffee systems. Agric Syst 162: 179–190

Jose S (2009) Agroforestry for ecosystem services and environmental benefits: an overview. Agrofor Syst 76:1–10

Kalaba KF, Chirwa P, Syampungani S, Ajayi CO (2010) Contribution of agroforestry to biodiversity and livelihoods improvement in rural communities of southern African regions. In: Tropical

rainforests and agroforests under global change: ecological and socio-economic valuations, pp 461–476. Environmental Science and Engineering. Springer, Berlin, Heidelberg.
Lasco RD, Delfino RJP, Espaldon MLO (2014) Agroforestry systems: helping smallholders adapt to climate risks while mitigating climate change. Wiley Interdiscip Rev Clim Chang 5(6): 825–833
Leeuw J, Njenga M, Wagner B, Liyama M (2014) An assessment of the resilience provided by trees in the drylands of eastern Africa. The world agroforestry Centre (ICRAF), Nairobi, Kenya, pp 1–182
Lin BB (2010) The role of agroforestry in reducing water loss through soil evaporation and crop transpiration in coffee agroecosystems. Agric For Meteorol 150(4):510–518
Lobell DB, Field CB (2007) Global scale climate–crop yield relationships and the impacts of recent warming. Environ Res Lett 2(1):014002
Lobell DB, Schlenker W, Costa-Roberts J (2011) Climate trends and global crop production since 1980. Science 333(6042):616–620
Lott JE, Khan AAH, Black CR, Ong CK (2003) Water use in a Grevillea robusta–maize overstorey agroforestry system in semi-arid Kenya. For Ecol Manag 180(1–3):45–59
Maia AG, Eusebio GDS, Fasiaben MDCR, Moraes AS, Assad ED, Pugliero VS (2021) The economic impacts of the diffusion of agroforestry in Brazil. Land Use Policy 108:105489
Mbow C, Van Noordwijk M, Luedeling E, Neufeldt H, Minang PA, Kowero G (2014) Agroforestry solutions to address food security and climate change challenges in Africa. Curr Opin Environ Sustain 6:61–67
McGinty MM, Swisher ME, Alavalapati J (2008) Agroforestry adoption and maintenance: self-efficacy, attitudes and socio-economic factors. Agrofor Syst 73:99–108
Mendez VE, Lok R, Somarriba E (2001) Interdisciplinary analysis of home gardens in Nicaragua: micro-zonation, plant use and socioeconomic importance. Agrofor Syst 51:85–96
Montagnini F (2017) Introduction: challenges for agroforestry in the new millennium. Agroforestry for Biodiversity Conservation and Food Sovereignty, Integrating Landscapes, pp 3–10
Montagnini F, Metzel R (2017) The Contribution of agroforestry to sustainable development goal 2: end hunger, achieve food security and improved nutrition, and promote sustainable agriculture. In: Montagnini F (ed) Integrating landscapes: agroforestry for biodiversity conservation and food sovereignty. pp 11–45. Advances in Agroforestry, vol 12. Springer, Cham.
Montagnini F, Nair PKR (2004) Carbon sequestration: an underexploited environmental benefit of agroforestry systems. Agrofor Syst 61:281–295
Monteith JL, Ong CK, Corlett JE (1991) Microclimatic interactions in agroforestry systems. For Ecol Manag 45(1–4):31–44
Morton JF (2007) The impact of climate change on smallholder and subsistence agriculture. Proc Natl Acad Sci U S A 104(50):19680–19685
Mukhlis I (2019) Food security for communities around the Forest in alleviating poverty. KnE Soc Sci 3:946–957
Muthuri C, Ong C, Craigon J, Mati B, Ngumi VW, Black CR (2009) Gas exchange and water use efficiency of trees and maize in agroforestry systems in semi-arid Kenya. Agric Ecosyst Environ 129(4):497–507
Nelson GC, Rosegrant MW, Koo J, Robertson R, Sulser T, Zhu T, Ringler C, Msangi S, Palazzo A, Batka M, Magalhaes M (2009) Climate change: impact on agriculture and costs of adaptation, vol 21. International Food Policy Research Institute
Nelson GC, Rosegrant MW, Palazzo A, Gray I, Ingersoll C, Robertson R, Tokgoz S, Zhu T, Sulser TB, Ringler C, Msangi S (2010) Food security, farming, and climate change to 2050: scenarios, results, policy options, vol 172. International Food Policy Research Institute
Nguyen MP, Vaast P, Pagella T, Sinclair F (2020) Local knowledge about ecosystem services provided by trees in coffee agroforestry practices in Northwest Vietnam. Land 9(12):486
Ollinaho OI, Kröger M (2021) Agroforestry transitions: the good, the bad and the ugly. J Rural Stud 82:210–221

Ong CK, Black CR, Marshall FM, Corlett JE (1996) Principles of resource capture and utilization of light and water

Ong CK, Swallow BM (2004) Water productivity in forestry and agroforestry. In: Water productivity in agriculture: limits and opportunities for improvement, pp 217–228

Palm CA, Vosti SA, Sanchez PA, Ericksen PJ (2005) Slash-and-burn agriculture: the search for alternatives. Columbia University Press

Place FS, Franzel S, De Wolf R, Rommelse R, Kwesiga FR, Nianf AI, Jama BA (2002) Agroforestry for soil-fertility replenishment: evidence on adoption processes in Kenya and Zambia. In: Natural resources management in African agriculture: understanding and improving current practices. CABI Publishing, Wallingford UK, pp 155–168

Pratiwi A, Suzuki A (2019) Reducing agricultural income vulnerabilities through agroforestry training: evidence from a randomised field experiment in Indonesia. Bull Indones Econ Stud 55(1):83–116

Rao KPC, Verchot LV, Laarman J (2007) Adaptation to climate change through sustainable management and development of agroforestry systems. J SAT Agric Res 4(1):1–30

Rao MR, Nair PKR, Ong CK (1998) Biophysical interactions in tropical agroforestry systems. Agrofor Syst 38:3–50

Rao MR, Palada MC, Becker BN (2004) Medicinal and aromatic plants in agroforestry systems. Agrofor Syst 61:107–122

Reynolds PE, Simpson JA, Thevathasan NV, Gordon AM (2007) Effects of tree competition on corn and soybean photosynthesis, growth, and yield in a temperate tree-based agroforestry intercropping system in southern Ontario, Canada. Ecol Eng 29:362–371

Roikhwanphut Mungmachon M (2012) Knowledge and local wisdom: community treasure. Int J Humanit Soc Sci 2(13):174–181

Rosenstock TS, Lubberink R, Gondwe S, Manyise T, Dentoni D (2020) Science direct inclusive and adaptive business models for climate-smart value creation $. Curr Opin Environ Sustain 42: 76–81

Rosenzweig M, Udry CR (2013) Forecasting probability. In: National Bureau of Economic Research NBER Working Paper No. 19334

Roy MM, Tewari JC, Ram R (2011) Agroforestry for climate change adaptations and livelihood improvement in Indian hot arid regions. Int J Agric Crop Sci 3:43–54

Russell D, Franzel S (2004) Trees of prosperity: agroforestry, markets and the African smallholder. Agrofor Syst 61:345–355

Sanchez PA, Izac AMN, Buresh RJ (1997) Soil fertility replenishment in Africa: an investment in natural resource capital. Replenishing Soil Fertility in Africa 51:1–46

Sanchez PA (2002) Soil fertility and hunger in Africa. Science 295:2019–2020

Schlenker W, Lobell DB (2010) Robust negative impacts of climate change on African agriculture. Environ Res Lett 5(1):014010

Schlenker W, Roberts MJ (2009) Nonlinear temperature effects indicate severe damages to US crop yields under climate change. Proc Natl Acad Sci U S A 106(37):15594–15598

Schmidhuber J, Tubiello FN (2007) Global food security under climate change. Proc Natl Acad Sci U S A 104(50):19703–19708

Shepherd KD, Palm CA, Gachengo CN, Vanlauwe B (2003) Rapid characterization of organic resource quality for soil and livestock management in tropical agroecosystems using near-infrared spectroscopy. Agron J 95(5):1314–1322

Shrestha BM, Chang SX, Bork EW, Carlyle CN (2018) Enrichment planting and soil amendments enhance carbon sequestration and reduce greenhouse gas emissions in agroforestry systems: a review. Forests 9(6):369

Skoufias E, Rabassa M, Olivieri S (2011) The poverty impacts of climate change: a review of the evidence. In: World Bank Policy Research Working Paper, (5622)

Somarriba E, Carreño-Rocabado G, Amores F, Caicedo W, Gillés O, de Pélichy S, Cerda R, Ordóñez JC (2017) Trees on farms for livelihoods, conservation of biodiversity and carbon storage: evidence from Nicaragua on this "invisible" resource. In: Integrating landscapes: Agroforestry for biodiversity conservation and food sovereignty. Springer, pp 369–393

Steffan-Dewenter I, Kessler M, Barkmann J, Bos M, Buchori D, Erasmi S, Faust H, Gerold G, Glenk K, Gradstein RS, Guhardja E, Harteveld M, Hertel D, Höhn P, Kappas M, Köhler S, Leuschner C, Maertens M, Marggraf R, Migge-Kleian S, Mogea J, Pitopang R, Schaefer M, Schwarze S, Sporn GS, Steingrebe A, Tjitrosoedirdjo SS, Tjitrosoemito S, Twele A, Weber R, Woltmann L, Zeller M, Tscharntke T (2007) Tradeoffs between income, biodiversity, and ecosystem functioning during tropical rainforest conversion and agroforestry intensification. Proc Natl Acad Sci U S A 104(12):4973–4978

Torquebiau E (1992) Are tropical agroforestry home gardens sustainable? Agric Ecosyst Environ 41(2):189–207

Vandenbeldt RJ, Williams JH (1992) The effect of soil surface temperature on the growth of millet in relation to the effect of Faidherbia albida trees. Agric For Meteorol 60(1–2):93–100

Wezel A, Bender S (2003) Plant species diversity of homegardens of Cuba and its significance for household food supply. Agrofor Syst 57:37–47

WMO (2005) Climate and Land Degradation WMO-No. 989 2005. World Meteorological Organization

Wollenberg E, Nawir AA (2005) Turning straw into gold: specialization among Damar agroforest farmers in pesisir, Sumatra. Forests Trees Livelihoods 15(4):317–336

Wu J, Zeng H, Zhao F, Chen C, Liu W, Yang B, Zhang W (2020) Recognizing the role of plant species composition in the modification of soil nutrients and water in rubber agroforestry systems. Sci Total Environ 723:138042

Zhang W, Hendrix PF, Dame LE, Burke RA, Wu J, Neher DA, Li J, Shao Y, Fu S (2013) Earthworms facilitate carbon sequestration through unequal amplification of carbon stabilization compared with mineralization. Nat Commun 4(1):2576

Luedeling, E., Kindt, R., Huth, N. I., & Koenig, K. (2014) Agroforestry systems in a changing climate—challenges in projecting future performance Current Opinion in Environmental Sustainability 6:1-7 https://doi.org/10.1016/j.cosust.2013.07.013

Part III
Ensuring Food Security Via Agro-Forestry System

Chapter 13
Interaction Between Belowground and Aboveground Resources in Tree-Crop Systems

Ankita Bhardwaj, Susmita Shil, M. N. Ashwath, Umakanta Dash, Vijay Kumar, and Preeti Vats

Abstract Agroforests are intricate systems whose management relies on copious interactions among their components. Complex interactions between trees, crops, soil, and climate have a significant impact on the productivity of agroforestry systems. The system's overall sustainability and yield are determined by its management and the harmony between positive and negative interactions. Intercropping of agricultural crops together with trees results in a number of changes through time and geography, mostly in response to environmental factors, biophysical interactions, and management choices (e.g., spatial-temporal selection). The essence and scale of interactions in agroforestry systems operate at the tree–crop interface (TCI) and soil–root interface (SRI). Ecologically, there exists complementarity when the relationship between the two is positive or synergistic, supplementary when the effect is neutral, and competitive when the interaction is negative or antagonistic. The selection of tree species is a crucial management action that affects the resource use between tree and crop components. Trees show either interference (shade effect, root competition, allelopathy, host for pests) or facilitation (nutrient pumps/safety nets, bio irrigators, mulch/litter production, amelioration of microclimate, erosion control, and reduction of weeds/pests). Several process-based models (WaNuLCAS, SCUAF, HyCAS, HyPAR, and APSIM) have been developed to simulate the intricate dynamics of tree–crop interactions. These models are known to simulate

A. Bhardwaj (✉) · S. Shil · V. Kumar
Department of Silviculture and Agroforestry, Kerala Agricultural University, Thrissur, Kerala, India

M. N. Ashwath
Department of Forest Biology and Tree Improvement, Kerala Agricultural University, Thrissur, Kerala, India

U. Dash
ICAR—Indian Agricultural Research Institute, Dirpai, Assam, India

P. Vats
Department of Forest Product and Wood Utilization, Kerala Agricultural University, Thrissur, Kerala, India

H. S. Jatav et al. (eds.), *Agroforestry to Combat Global Challenges*, Sustainable Development and Biodiversity 36, https://doi.org/10.1007/978-981-99-7282-1_13

physiological processes that regulate growth influenced by abiotic environmental factors such as soil, climate, and management. Investigations examining interactions among trees and crops provide insight into how different agroforestry components share and use resources, as well as how one component affects the growth and development of others. In agroforestry, interactions are measured under two main categories that are indirect in sequential systems as opposed to direct in simultaneous systems. All agroforestry systems must demonstrate interactions between trees and agricultural components in order to succeed. Therefore, to enhance both traditional and emerging systems, a thorough understanding of biophysical interactions, both above and belowground, is necessary.

Keywords Interactions · Agroforestry · Modelling · Productivity · Competition

13.1 Introduction

All the components of the biosphere are interlinked in such a way that ecological balance is maintained at equilibrium. Interlinkage provides ways for interaction, which happens either to benefit or harm the component at the interface. In simple terms, interaction is defined as sharing or exchanging of resources among the components. In the case of plants, it is not possible to see how this exchange of resources happens. This silent exchange we can only manifest by seeing the growth and developmental stages. Many determinants play an active role when two components undergo an interacting phase. Agroforestry, in terms of interaction, is simply described as the area where trees and agriculture coexist (Sinclair 2004). Agroforests which are envisaged for spatial and temporal complexities highlight the challenges to understand interactions. A comprehensive understanding is possible when all the practices, products, and services are simulated in a flexible model (Luedeling et al. 2016). The model simulates the production of a particular crop based on the interaction between crop features and environmental factors like weather and soil, which are in turn influenced by climate, inputs, and management. The models predict attainable yields which are far below the actual yields due to climatic uncertainties.

Agroforestry has been documented for providing numerous tangible and intangible benefits that vary with site-explicit responses of trees, crops/other elements with the intense spatial and temporal disparity in the farming perspective (Bayala et al. 2014; Coe et al. 2014). However, multispecies systems pose a significant challenge to systemic agronomy and contemporary agricultural research, because it is difficult to comprehend the effects of the different components that interact in these systems (Malézieux et al. 2009). Despite the benefits of mixing crops, it may be more challenging to comprehend how trees and crops interact ecologically in these intercropped systems (Moura et al. 2014). On the other hand, tree–crop interactions have been thought of a cause of harm. Most of the time, these factors are connected to how trees affect crop development by competing with them for light, moisture, and nutrients (Imo and Timmer 2000). The prime objective of writing this section is to synthesize conceptual and experimental information on interactional studies in tree–crop systems. The efforts are put forth to develop an understanding of the concept of interactions in the tree–crop system, their above and belowground nature, factors of influence, field scale mechanisms and models, the performance of component yield under various interactions in different agroforestry systems, and their economic implications. The knowledge gap is also discussed and prioritized for the need of future research in the aspect of interaction.

13.2 Resource use patterns at Tree-crop interface

The results, in general, of interactions between the woody and herbaceous components are believed to be promising over the long-term evaluation, which illustrates the rationales for the promotion of agroforestry (Smethurst et al. 2017). The secret to agroforestry success is to minimize the negative interactions and strengthen the positive ones to achieve the most beneficial results (Thevathasan and Gordon 2004). However, to maximize the beneficial interactions, it is vital to maintain soil cover and provide a sufficient release of nutrients, which will improve the soil's ability to hold water over the course of the crop cycle (Moura et al. 2010). Aguiar et al. (2010) suggested that a useful approach is to combine tree species that offer both low- and high-quality residues. Sharing of resources among the components under spatial configuration affects both plant structure as well as resource capture (Jones et al. 1998) (Fig. 13.1).

The effect of interaction (I) on crop yields under the two major agroforestry groups is denoted as follows (Rao et al. 1997):

$$\text{Simultaneous system}: \quad I = F + C + M + P + L + A$$
$$\text{Sequential systems}: \quad I = F + M + P + L + A$$

where, F is soil fertility (includes soil chemical (Sc), soil physical (Sp), soil biological (Sb) interactions), C is competition (competitive interactions for soil water (Sw), soil nutrients (Sn), and radiation (r)), M is microclimate, P is pest and diseases (interactions related to weeds (Pp), insects (Pi), and diseases (Pd)), L is soil conservation, and A is allelopathy.

The difference between the two types of systems is the absence of C in sequential systems, where interactions are indirect rather than direct as in simultaneous systems. The two counterparts for agroforestry were formulated in regard to classic crop production laws, that is, law of limiting factors and law of optimum. The converse of the law of limiting factors asserts that when a resource becomes more readily available in the environment of trees and crops, its proportion in the interaction between trees, environments, and crops decreases. The counterpart of the law of optimal asserts that as more of the other limiting factors in the tree–crop environment

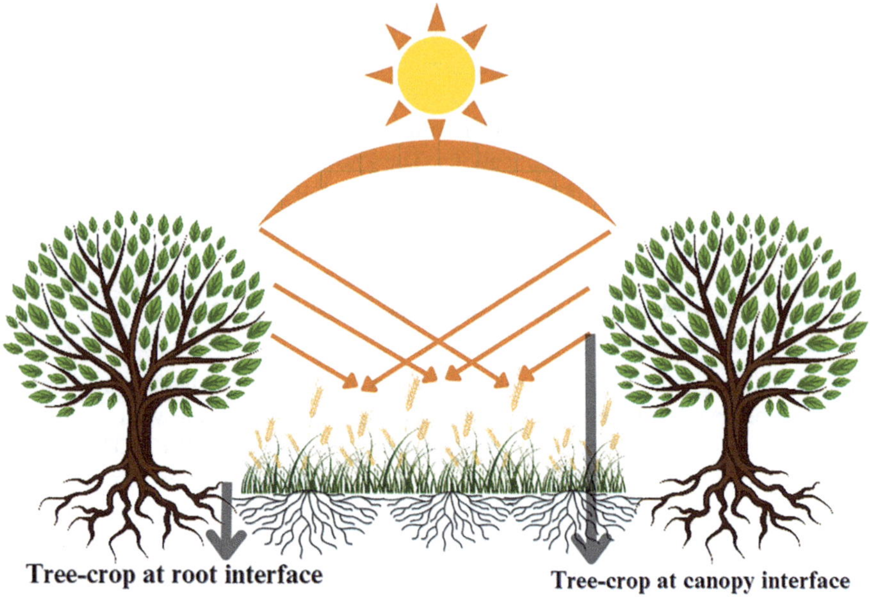

Fig. 13.1 Illustrations of resource use patterns among agroforestry components

become available, their proportion in the overall tree–environment–crop interaction increases. The rules govern synthesis and analysis of various agroforestry experiments. The analytical results of the alley cropping system demonstrated that nutrient capture by trees below the crop rooted zone results in net benefits for the crop, which is probably true for nitrogen but not for phosphorus. Additionally, the results indicated that the competitive nature of trees for water likely outweighs their water conservation effects (Kho 2000). The modelling effect of trees on crops started with the ecological field theory given by Wu et al. (1985). The theory states that a tree alters the accessibility of growth resources (light, nutrients, water) through its stem, crown, and roots (Fig. 13.2).

The effect of tree is depicted in terms of its distance and dimension. The three influential regions of trees (stem, root, and crown) or ecological fields behave differently to obtain relative availability of water, nutrients, and light around the tree. The tree's influence on light, water and nutrients creates three resource availability surfaces which together accounted with the resource requirement of agricultural crop give rise to the surface of ecological interference potential (EIP). The relative growth rate is 1-EIP (Pukkala 1998).

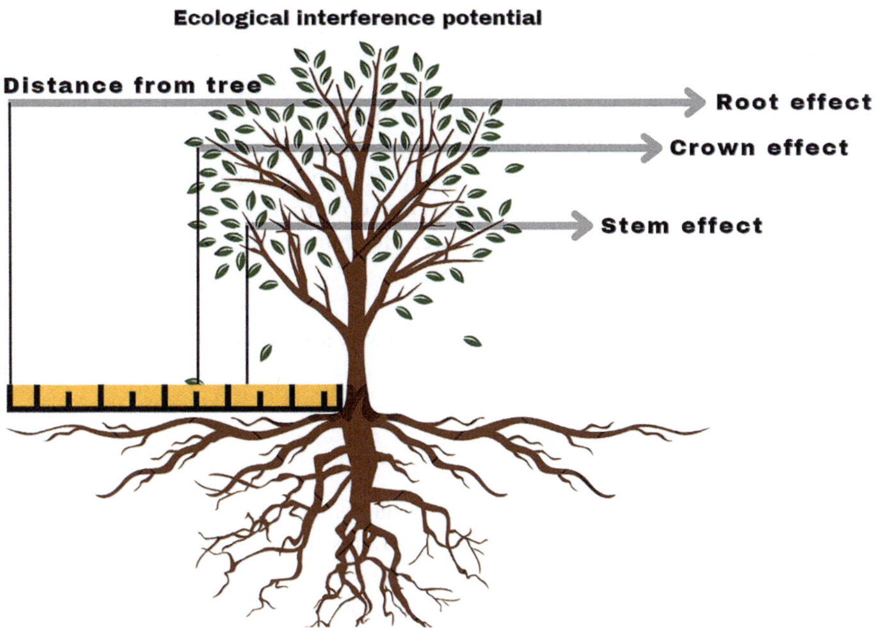

Fig. 13.2 Ecological interference of single isolated tree as a function of distance

13.3 Beneficial and Harmful Interactions

Umpteen factors influence tree–crop interactions. Trees offer shade, and their leaves may help with soil nutrient and nitrogen cycling. Crops can also offer trees with nutrients and litter for their betterment. However, crops and trees compete over water, nutrients, and light resources. It is challenging to measure changes and the availability of these elements directly for plants, rather, quantified indirectly by measuring the yield (Miina et al. 1999). In agroforestry systems, analysis of tree–crop interactions has uncovered equally beneficial and harmful interactions that take place above and below ground.

13.3.1 Beneficial Interaction

(a) Recycling of nutrients may be based on:

- Nutrients are taken while sinking to a deeper layer with the "safety net" of tree roots.
- Deep tree roots serve as a "nutrient pump", drawing nutrients up from weathered elements in a deeper stratum.

(b) Litter production: High-quality litter with a low C/N ratio, lignin concentration, and polyphenolic content will disintegrate quickly and provide nutrients.
(c) Mulch: Poor quality litter that has a high carbon to nitrogen ratio and high levels of lignin and polyphenols decomposes poorly and is suitable for mulching. Mulch helps keep the soil hydrated during the dry season.
(d) Due to roots turnover or direct transport from nodulated roots in adjacent proximity to crop roots, provid nitrogen to crop roots.
(e) The effects of trees and crops lower the burden of pests and diseases by enhancing potential biocontrol agents.
(f) Influence of trees on microclimate.
(g) Protracted effects on soil structure, soil organic matter management, and erosion.

13.3.2 Harmful Interaction

(a) Tree shade reduces the amount of light that reaches the crops.
(b) Root competition for moisture and/or nutrients in the topsoil between crop and trees. In this regard, the tree root architecture is crucial. Shallow tree root systems probably compete more with the crop for precious nutrients.
(c) Allelopathy refers to the chemical interactions between plants that can inhibit or promote the growth of other nearby plants. Some tree species release allelopathic chemicals that inhibit the growth, reducing yields and potentially damaging the overall health of the agroforestry system (*Eucalyptus*, *Acacia*, and *Robinia* spp.). In order to lessen allelopathy's detrimental impact on agroforestry, it is important to choose tree species that are less likely to release allelopathic chemicals. It is also possible to use agronomic practices, such as crop rotation or intercropping, to reduce the impact of allelopathic chemicals.
(d) Numerous pests and diseases that harm both trees and crops can be found in both.

Both above and below ground, a wide variety of intricate interactions including radiation exchange, water balance, food budget and cycling, shelter, and other microclimatic modifications take place (Singh et al. 2013). Given this, ICRAF researchers have developed an eq. ($T = F - C$) for quantifying tree–crop interaction that accounts for both the beneficial effects of tree and crop yield through the enrichment of soil fertility (F) and the detrimental effects resulting from crop competition for growth resources between tree and crop (C). If $F > C$, the interaction is positive; if $F < C$, it is negative; and if $F = C$, it is neutral. Interaction aids in knowing how the different agroforestry components use and share the environment's resources, as well as how the growth and development of one component may impact the others. The shifting trends toward agroecological, resilient, and sustainable farming systems has also modified the ways to manifest interactions in different ways. The 4 C approach (competition, complementarity, cooperation, and compensation) is the newly developed scientific and pedagogical way to represent the

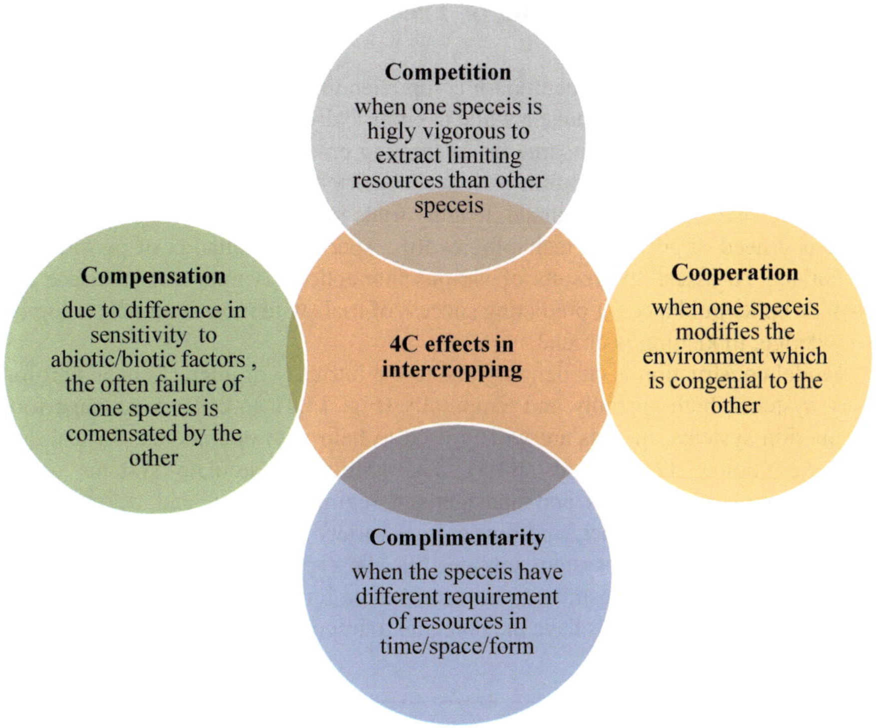

Fig. 13.3 The 4 C approach

processes and effects that occur simultaneously and dynamically between species during growing season (Fig. 13.3) (Justus et al. 2021).

13.4 Factors Affecting Tree–Crop Interactions

- **Species**: Appropriate tree–crop combination selection.
- **Sunlight**: Light-crowned trees; either choosing crops that can withstand shadow or managing tree growth to lessen shade on agricultural crops (Prunning).
- **Density**: Number of trees planted per hectare, growing as many trees as possible in a given area to minimize competition between crops and trees.
- **Age**: Competition is negligible in the early stages of the tree crop.
- **Site factors**: This has to do with site quality and carrying capacity.
- **Management**: The degree of tree crop management for the advantages of agricultural crops or raising system production overall.

13.5 Field Scale Modelling of Tree–Crop Interactions

Field testing has been main method for comprehending the complicated dynamics of agroforestry systems. It can take a while to empirically establish the attainability and correlative benefits of various tree-based farming practices in new environment due to factors like long tree gestation period, inconsistent funding, and other logistical issues. However, carefully planned, lengthy trials might provide data for years. But there is a need of advanced technologies for a speedier simulation of performance potentials. To assess the results of various interactions at tree–crop interface and provide better guidance for predicting success of trials with more in-depth ecological knowledge, modelling is crucial.

Model construction is challenging due to the intricacy of interactions in agroforestry systems, both spatially and temporally (Fig. 13.4). In order to manage food production systems, models are widely used to help with operational, tactical, and strategic choices (Zhu et al. 2003). The ability to simulate systems across pedoclimatic environments and management regimes, above- and below-ground interactions for light, water, and nutrients, a variety of potential yields, including food, fibre, fuel, and ecosystem services like the capture of excess nutrients, soil erosion, and carbon sequestration, are some of the foremost objectives for agroforestry models. These models have the potential to improve nutrition and food security

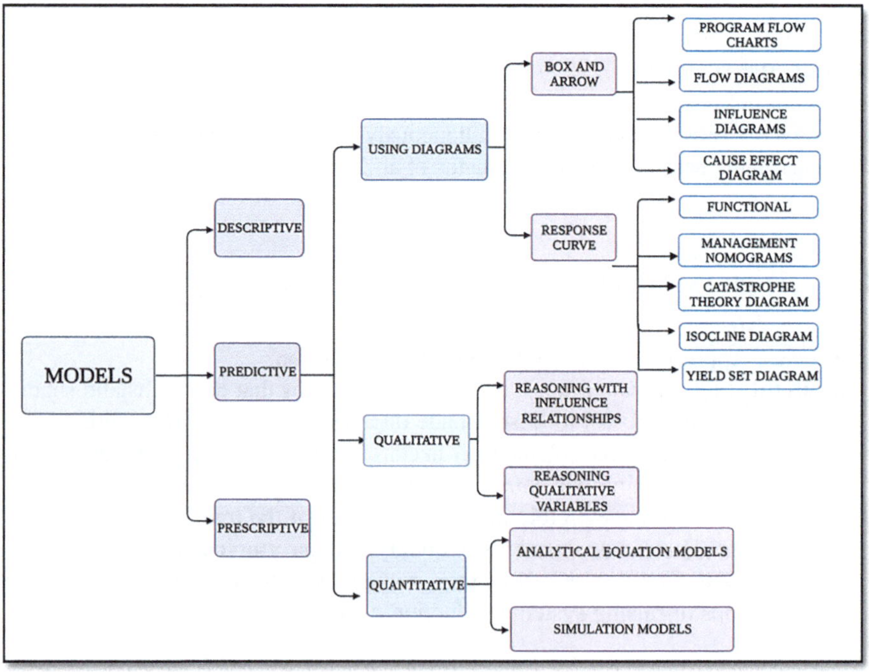

Fig. 13.4 Schematic representation of models classification

Table 13.1 Comparison between mechanistic and empirical models

Type	Mechanistic models	Empirical
Scope	Broad	Restricted
Model's approach to time	Variable	May be invariable
Connections to recorded data	Can be fragile	Normally good
Fidelity of simulations	Variable	Normally good
Normal usage	Testing and instructing	Governance
Normal intricacy	Intricate	Elementary

worldwide. Models are instruments created to improve the efficacy of research, management, or instruction (Burgess et al. 2019). Empirical or correlative models for agroforestry systems are inferior to process-based or mechanistic models because they are less accurate at capturing the complex dynamics of tree–crop interactions and are limited in the range of data that can be used for parameterization (Table 13.1).

The majority of conventional crop models are constrained to monocultures, where plant interactions are only allowed to occur when members of the same species exchange resources. Most of the time, models forecast the potential yield while also taking into account the actual yield, which is impacted by crop genotype, temperature, radiation, and management and is limited by limiting factors like water and nutrients. Simulating systems that are more complicated than a monoculture may be done using a few different methods. Examples include growing weeds in monocultures and utilizing intercropping systems (Deen et al. 2003). Since they presume all above and below ground stand components are horizontally homogenous, the majority of crop models are one-dimensional. Similar to tree stand models, which make the identical assumption, they are likewise 1D. On the other hand, for the majority of tree–crop systems, modelling discontinuous canopies and discontinuous root systems requires a 3D method. The 2D technique may be enough in some circumstances, but it is only applicable to some systems with established bilateral symmetry. Various agroforestry models are in vogue now but some main models which take into account both above and below ground interactions are briefly elucidated below.

13.5.1 WaNuLCAS

In order to offer a general pattern for agroforestry systems with significantly diverse geometries and temporal patterns, the Water, Nutrient, and Light Capture in Agroforestry Systems (WaNuLCAS) model was created (Fig. 13.5). It was created in an open-source environment so that users may edit it, albeit this is challenging in reality due to its complexity. The model is both broad enough to encompass a large range of characteristics and detailed enough to address particular requirements. The structure

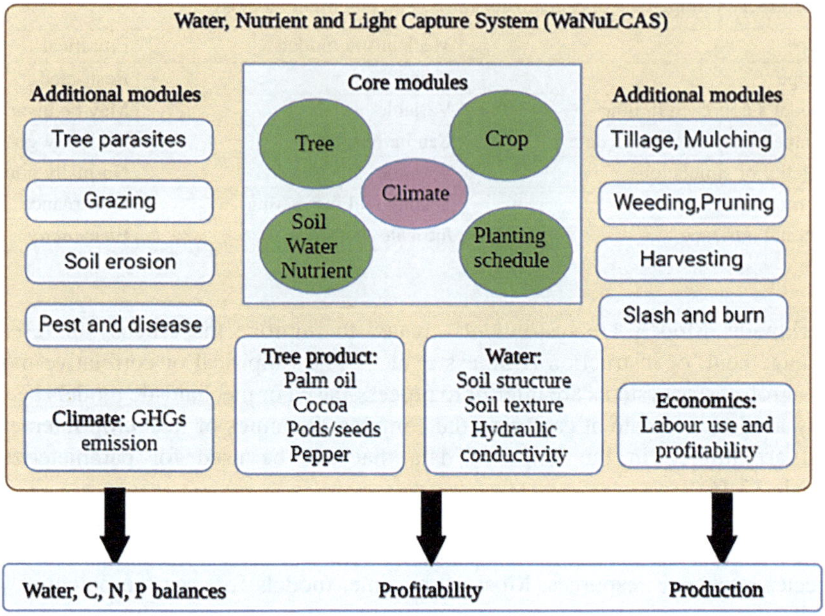

Fig. 13.5 Modules of WaNuLCAS model

of WaNuLCAS is based on a collection of inputs that specify the beginning conditions of the soil and tree attributes as well as dynamic inputs like rainfall.

Its main modules keep track of crop and tree growth as well as water, nitrogen, phosphorus, crop, and tree roots in four horizontal zones separated from the tree by vertical soil layers. The system's geographical definition and management scheduling calendar are also included. This generates a basic set of outputs (as net present value (NPV)) in terms of the vegetation standing stock, harvested goods, and profitability. To interact with the fundamental sub-models, numerous other processes, inputs, and outputs may be offered as optional sub-models (Noordwijk and Geijn 1996). With data on crop and tree root lengths that were captured in the soil profile, this model is often parameterized. Zones are used in an agroforestry setting to describe spatial patterns and model system activities. The STELLA environment, a popular and user-friendly ecological modelling tool, is used to code the model. The concept has been used to several issues. Utilized it, for instance, to examine water competition in agroforestry systems. WaNuLCAS was used to simulate the effects of evergreen and deciduous trees in semi-arid Central Kenya to assess the effects of tree leafing phenology on crop performance and soil water balance. Recent applications evaluated resource competition at the crop-soil-hedge interface in Thailand and looked at tree pruning and stand thinning options for teak + maize intercropping systems in Indonesia (Khasanah et al. 2015). Research on peat soils and the possibility for agroforestry systems to combine low greenhouse gas emissions with profitability are recent advances.

13.5.2 APSIM

The Agricultural Production Systems Simulator (APSIM) modelling framework was created to mimic biophysical processes in agricultural systems and forecast the economic and environmental effects of management practices and policy changes. Predicting the effects of climate risk and climate change has grown more crucial recently. APSIM has been extensively used in a variety of contexts, including on-farm decision-making, assessment of seasonal climate forecasts, farming system design, analysis of agribusiness supply chains, risk assessment for government policy, development of waste management guidelines, and as a research and exploration tool (Holzworth et al. 2010). Its modular architecture facilitates communication via a common protocol between different models (Keating et al. 2003). Its "plug in-pull out" design, which permits flexible routine recombination, makes it simple to add, remove, and swap out sub-models and subroutines within them. The APSIM framework currently includes more than thirty significant tree, pasture, and agricultural species (Probert et al. 1998). There are equivalent sub-models for all key soil processes that have an impact on agricultural systems, such as water, C, N, and P dynamics, and erosion (Paydar et al. 2005). According to Moore et al. (2014), APSIM can reproduce a variety of agricultural management options and allow the user to select complex crop rotations and land management regimes. The implementation of demanding scientific and software engineering methodologies to maintain integrity is a major strength of this architecture. Although it is not mainly based for an agroforestry modelling pattern, various forestry and agroforestry applications have been made possible by the framework's modular design.

13.5.3 SCUAF

In the 1980s, the Soil Changes Under Agroforestry (SCUAF) model was initially created with a focus on the effects of trees on soil carbon concentration and conservation (Fig. 13.6). In addition, it sought to forecast how land use will affect soil loss and medium-term production under certain climatic conditions (Young and Muraya 1990). SCUAF concentrates on trees' capacity to replenish soil properties in tropical humid regions (Young et al. 1998). Although tree and plant development rates are exogenous to the model, soil processes are nevertheless represented in detail.

However, the competition for light, nutrients, or water between trees and crops is not represented. It allows for slight extrapolation to longer time periods or other parameter conditions and gives interpolation across situations with clear bounds. The ability to simulate agricultural development processes with the level of accuracy and complexity found in models like APSIM and DSSAT (Decision Support System for Agrotechnology Transfer) is severely constrained by SCUAF's annual time scale. Although the model was developed more than 20 years ago, research has just lately used it (Lojka et al. 2008).

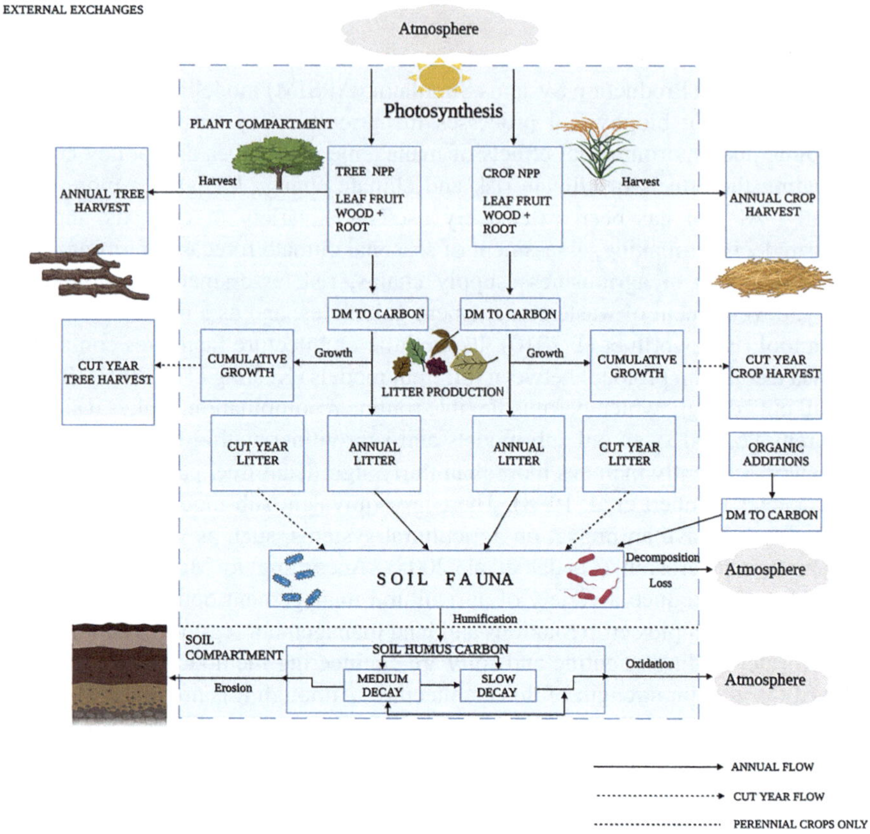

Fig. 13.6 Overview of SCUAF model

13.5.4 *HyPAR*

It is the amalgamation of Hybrid tree growth v 3.0 and tropical crop model PARCH (Predicting Arable Resource Capture in Hostile environments) model for cereal and pulse crops (Mobbs et al. 1997). The "gap model" used to represent trees in the hybrid model predicts nutrient fluxes for each tree at varying canopy heights and soil depths while simulating competition between individual trees with various physiological characteristics. A "big leaf" model, which simulates activities at several levels of the leaf canopy and then averages them horizontally across a sizable piece of land, is used to ascribe energy absorption, photosynthesis, stomatal conductance, and transpiration (Mobbs et al. 1998). Contrarily, the PARCH model simulates crop growth depending on the availability of several growth-promoting elements, including light, water, nitrogen, and phosphorus. Here, the negative effects of temperature extremes can be explained by the addition of additional stressors. The crop growth processes are not explicitly represented in this model; instead, they are

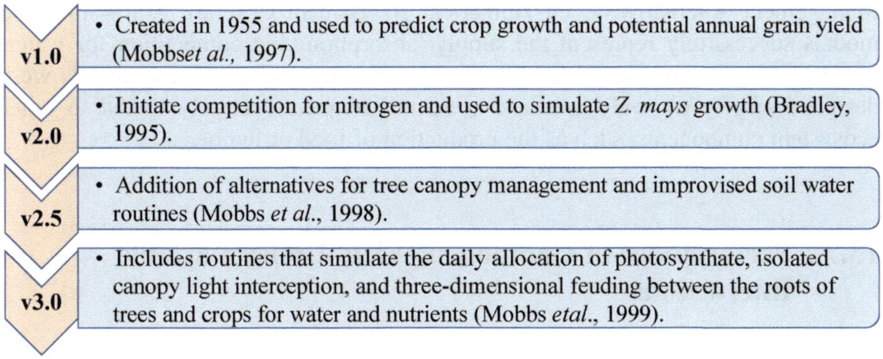

Fig. 13.7 Improvements to HyPAR from version 1.0 to 3.0

just recorded as process efficiency variables, making it less physiologically precise than its equal on the tree side (hybrid) or numerous other crop models now in use (such as DSSAT, APSIM). Bradley (1995) does not take into account the microclimatic interactions between trees and crops in this instance.

It provides a way to evaluate different agroforestry choices in light of a variety of soil types, climatic factors, and management techniques. HyPAR can represent up to 15 soil layers and, in addition to humus, takes into account five separate pools of soil organic matter (Das and Bauer 2012). The density of a plant's roots affects how well it can absorb water and nutrients (Khasanah et al. 2020). Despite being a major improvement over SCUAF, the HyPAR model does not seem to have seen much use once the project that served as its inspiration was finished (Fig. 13.7). Some of the reasons for this include the fact that the primary funding agencies have shifted their priorities away from a particular model in favour of the flexible modelling environment SIMILE as well as the realization that the "quick-fix" approach of allowing existing crop and tree models to interact with a single soil representation had limitations that could only be overcome by a fundamental reformulation (Matthews and Lawson 1997). WaNuLCAS and HyPAR were initially developed simultaneously, with the model ideas being cross-fertilized, until HyPAR as a separate model was largely placed on hold. In addition to HyPAR, another model called HyCAS (Hybrid tree growth model + GUMCAS) forecasts the performance of *Manihotesculenta* (Cassava) in agroforestry systems (Friend et al. 1997).

13.6 Below-Ground Interaction Models

In addition to the previously mentioned WaNuLCAS, SCUAF, HyPAR, HyCAS, and APSIM, Matthews et al. (2004) have studied a number of models for below-ground interactions between species, including CropSys, COMP8, GAPS (General-purpose Atmosphere-Plant-Soil Simulator), and Almanack (Agricultural Land

Management Alternatives with Numerical Assessment Criteria). A few of these models successfully represent the supply, absorption, and competition for water, carbon, nitrogen, and other nutrients for specific applications. These models were designed largely as study aids and were not frequently closely linked to other ecosystem components, such as the production of food or lumber.

13.7 Performance of Component Yield Under Variable Interactions

The performance of the agroforestry component and yield is determined by the nature and intensity of the interface among the components within the system. The influence of the tree component on the other component(s) frequently determines the net effect of these interactions. When trees are young, crop yields may not be influenced; nevertheless, crop yields beneath big trees may rise or decrease depending on the species. Given how slowly trees alter the soil environment, it would take many years before there would be any positive effects of trees on agricultural yields due to increased soil fertility. On the other side, it takes a few years to notice the detrimental impact of trees on competition for growing resources. Although fast-growing trees diminish agricultural yields quite rapidly, within two to three years, slow-growing trees may not have an impact on crop yields for many years.

13.7.1 Comparative Yield Performance—Monocrop v/s Intercrop

The potential of crops to produce sustainably is affected by the over-story components. Crops grown under shade and open may vary in growth performance, production, and quality. A study on soybean, tomato, and radish in association with *Xylia dolabiformis* reported that crop yield contributing parameters improved with a surge in planting distance from tree base. (Basak et al. 2009). Chauhan et al. (2012), stated that *Populus deltoides* trees grown at an 8 m × 3 m spacing had the maximum DBH (Diameter at Breast Height), crown diameter, and crown length with the highest wheat grain yield compared to lesser spacing. Singh and Bishnoi (2013) found that bean crops grew more vegetatively and produced more pods when grown under khejri trees as compared to sole bean cropping. On the contrary, Bhat (2015) purported declined plant height and yield of tomatoes under *Melia composita.* He explained it as a sign of severe rivalry between the tree and crop for scarce resources.

In an agrisilvicultural system of 20-year-old *Terminalia arjuna* and *Mitragyna parvifolia* with four varieties of green gram (*Vigna radiata*), all green grams fared better in open circumstances in terms of leaf count, branch count, and grain production (per plant and plot) (Kumar et al. 2015). However, it was shown that only plant height under Arjun and kalam trees was at its maximum. The shade effect, which can be controlled by routine branch pruning, may be the cause of the crop's comparatively poorer yield under tree cover. Likewise, Chavan and Dhillon (2019) reported a reduced yield of sorghum, berseem, cowpea, and wheat under the *P. deltoides* after the second year of planting due to competition. Similar observations were reported on turmeric, potato, and wheat under poplar plantations, where crop yield decreases with the oldness of poplar trees (Handa et al. 2019). Turmeric's growth and yield were enhanced under mango plantations (34.75 t ha^{-1}) compared to sole crops (Ali et al. 2018). Kumar et al. (2018) explained that plant growth was significantly higher under the shade of *Diospyrus embryophytes* and *Terminalia chebula.* The intercropping of medicinal plants (lemongrass, patchouli, citronella, palmarosa, and mango ginger) under coconut gardens positively influenced coconut yield. In contrast, the yield of all MAPs was reduced under tree canopy compared to sole cropping without altering the quality of produce (Padma et al. 2018). Similarly, green gram planted below Albizia, Grewia, and Subabulhas experienced less production compared to sole cropping (Gupta and Gupta 2017). Pandey et al. (2017) observed maximum growth and yield of ginger under the sapota + jatropha agroforestry system as compared to their sole crop. The yield of pineapple (9981 kg ha^{-1}) and aloe vera (8635 kg ha^{-1}) grown under *Acacia mangium* had shown maximum yield compared to the open sole cropping, whereas kalmegh (1239 kg ha^{-1}) and mango ginger (3300 kg ha^{-1}) performed well in the open as compared to *A. mangium* (Nayak et al. 2014). Turmeric and moong grown in the interspaces of poplar reported reduced yield with an increase in poplar's age (Chauhan et al. 2013). The cash crops like onion variety PRO-7 recorded 34.4% and 36.7% reductions in yield when planted with 4- and 5-year-old poplar (Bhardwaj et al. 2021).

13.7.2 Manifestation of Tree-Crop Interactions in Response to Spacing Regimes

The space between trees and the nature of the agricultural crop decides the density of the crop and the yield produced. The shade-loving crops may yield better under the canopy-developed tree compared to sun-loving crops. Several studies have shown negative and positive interactions in the final tree crop yield.

The bell pepper production under a silver oak-based agroforestry system showed maximum fruit yield under larger spacing compared to closer spacing of trees, as less space affected the light availability and competition for nutrient availability. The same system showed that the trees planted with larger spacing produced maximum leaf fodder, bast fibre, and torchwood (Kar et al. 2019). A similar increment trend was witnessed by Thakur et al. (2019) under the *Melia dubia* and *Cymbopogon flexuous*-based system. Systems volume (41.25 m ha^{-1}) and biomass (17.41 Mg ha^{-}

^{1}) increment was registered maximum at 3 × 3 m spacing, whereas minimal tree growth was recorded in 2 × 2 m.

Two baby corn varieties, that is, HIM-123 and DHM-107, under poplar-based system with different spacings, that is, 45 × 25, 55 × 25, and 65 × 25 cm, reported the maximum growth and yield under spacing of 55 × 25 cm for two years (Bhushan and Khare 2018). The okra intercropped *Melia composita* systems showed a negative effect of trees on okra production (Bhusara et al. 2018). Okra performed better in an open field than in the agrisilviculture system. But among the tree spacings, maximum growth and yield parameters were achieved in wider spacing in comparison to less space. Patil (2018) reported higher seed production of French bean and soybean (1042.9 and 875.8 kg ha^{-1}, respectively) planted under a young teak plantation with 4 × 2 m spacing. The same system in the same season showed a reduced yield of black gram and green bean, attributing to the nature of the agricultural crop used for developing the system, where black gram and green gram are well recommended for the open field than any agrisilvicultural system considering the final seed yield. The production of fennel and ajwain under different spacing of a 5-year-old poplar plantation (Rathee et al. 2017) showed that the seed yield declined up to 78.67% and 84.84% under 5 × 4 m spacing as compared to the sole fennel and ajwain plots.

Sharma and Pant (2017) experimented with maize cultivation in a poplar-based agroforestry system with tree spacing of 6 × 4 m and 4 × 4 m and open conditions. Maize's maximum plant height and grain weight were reported in open conditions (248.67 cm and 244.61 g, respectively) and minimum in the tree spacing of 4 × 4 m (210.19 cm and 236.06 g, respectively). An increase in wheat and paddy grain yield with an increase in distance from the base (6 m) of Poplar and Eucalyptus trees was observed by Gusain (2016).

13.7.3 Manifestation of Tree–Crop Interactions in Response to Nutrient Regimes

Farm nutrient management of a complex agroforestry system is a vital practice that directly influences the productivity of system components, ultimately defining the system's economics. Nutrient management in the earlier and later phase of the system influences the tree–crop interaction either by facilitation or competition. Additional application of nutrients directly increases the system's output; knowing the appropriate nutrient, application dosage, time of application, etc., plays a role in increasing or inhibiting growth and productivity.

Verma et al. (2019) reported better performance, namely, yield, oleoresin, and dry matter recovery, of two rhizomatous crops, turmeric and ginger, under organic practices than traditional prevalent methods, that is, without fertilizer or inoculations. Whereas Anuradha et al. (2018) reported that the combined application of 50% recommended dose of nitrogen fertilizer through an inorganic source and the remaining 50% through *Pongamia* cake resulted in higher fresh rhizome yield, that is, 43.94 t ha^{-1} in turmeric. A comparative analysis of different nutrient sources, namely, poultry manure (PM), NPK (Nitrogen, Phosphorus, Potassium) (200 kg per ha), Mg fertilizer (20 kg per ha), PM + Mg, NPK + Mg on the performance of turmeric showed that the yield of rhizome increased under application of NPK + Mg by 13.6% (Adekiya et al. 2019). Prajapati et al. (2018) reported the best results concerning plant growth height and maize cob yield, applied with NPK @ 100% + vermicompost @ 100% compared to other traditional practices. The highest grain yield, crude protein, and starch content in maize were reported with an application of vermicompost @ 5 t ha^{-1} with 75% RDF (Recommended Dose of Fertilizers) (Kumar 2014). The growth and yield of poplar and linseed were highest with the application of FYM 125% (Kaushal et al. 2019). The application of a 50% recommended dose of N P K and 100% Zn fertilizer resulted in a substantial surge in the number of pods per plant (19.53), seeds per pod (6.20), and pod yield (77.67 q ha^{-1}) in pea (Chethan et al. 2018). Kumar (2017) experimented the effect of INM (Integrated Nutrient Management) on wheat and paddy under the *Casuarina equisetifolia*-based agroforestry system resulting in higher grain yield in NPK (120:60:40 kg ha^{-1}) treatment. The brinjal planted in 3 × 2 m interspace of teak showed maximum growth and yield under 100% RDF (100:50:50 NPK ha^{-1}). Darjeeling tea production was highest under the application of 50% RDN (Recommended Dose of Nitrogen) through VC and urea respectively (Kumar et al. 2015).

13.8 Major Research Interests in Above and Below Ground Interactions

Contemporary developments and viewpoints regarding the functioning of above-ground and belowground agroforestry (AF) systems acknowledge the interaction between microclimate and soil water balance. The influence of trees in this interaction includes the equalization of water pressure and the significant role of tree roots in binding and anchoring soil. These combined effects help to mitigate the risk of landslides on sloping land, particularly during periods of heavy rainfall (Hairiah et al. 2020, Cardinael et al. 2020).

13.8.1 Amelioration of Microclimate

Tree canopy develops a microenvironment that modifies various meteorological parameters which in turn affects various physiological processes of intercrops. Several studies were carried out to study microclimatic effect under tree canopies. The influence of an agroforestry system based on poplar trees on the microenvironment was appraised by implementing various pruning treatments, including 50% and

75% pruning with and without topping, as well as lateral pruning with and without topping. Among microclimatic parameters, photosynthetically active radiation (PAR) and light intensity was found to increase with the increase of pruning intensity. During the agrisilviculture system, higher leaf area index and relative humidity were remarked in summers compared to winters. This could be due to the increased presence of leaf biomass during summers, resulting in more transpiration compared to pure crops (Singh et al. 2019). The PAR received by crops in intercropping systems is lower compared to crops grown in monoculture. On average, the daily PAR levels observed in intercropping systems based on jujube, apricot, and walnut were approximately 78.7%, 45.5%, and 20.1%, respectively, in contrast to the PAR levels in monoculture (Qiao et al. 2019). Plants adapt to different light regimes through changes in physiological demeanour. The maximum photosynthetic rate of understory crops typically occurs in the afternoon, depending on the prevailing weather conditions during their growth period. The growth and productivity of both rainy and winter season crops were significantly affected by the spacing configurations of the agroforestry system based on poplar trees, with the influence becoming more pronounced as the trees matured (Chavan and Dhillon 2019). Chauhan et al. (2013) studied the microclimatic interactions under agri-horti-silvicultural model involving poplar as timber tree, fruit trees, and agronomic crops (moong and turmeric). Bhardwaj et al. (2021) recorded higher relative humidity (42–72%) under poplar canopy while a lower range was recorded under open conditions (37–68%). Temperature variation showed that maximum temperature was lower under the canopy (14.6–36.8) and higher under the tree-less conditions (15.8–38). The value of PAR varied with phenological changes of poplar canopy. The PAR started to increase from January to the first fortnight of April and then it declined under full-fledged canopy while under open conditions it followed the increasing trend. The presence of PAR alters the microclimate in the area, subsequently impacting the eco-physiological parameters and ultimately influencing the crop yield (Sangwan et al. 2016).

13.8.2 Amelioration of Soil Physiochemical and Microbiological Properties

Greater significance was observed for saturated hydraulic conductivity in the agroforestry region compared to obvious regions (3 to 14 times) (Tshepiso et al. 2005). Baber et al. (2006) examined the physicochemical properties of soil at two depths in an agroforestry system (0–15 cm and 15–45 cm). Samples were taken 5, 10, 15, and 20 metres away from the eucalyptus trees. In the surface soil, pH, EC, OM, P, and K values declined with increasing distance from the trees. However, in the subsoil, OM and P values increased with distance, while PH, EC, and K values declined. It was recorded that soil bulk density, organic carbon, and dehydrogenase activity were improved under 1000 trees/ha, whereas available phosphorus, phosphatase activity, and exchangeable calcium were improved under 500 trees ha^{-1}. Therefore, to boost the physicochemical and biological properties of the soil, a tree density of 500 to 1000 trees per hectare has been found to be the optimal choice (Uthappa et al. 2015).

In 2009, Gupta et al. collected soil samples from both agroforestry and non-agroforestry sites containing poplar plants of diverse ages (1, 3, and 6 years) and varying soil textures (loamy sand and sandy clay). The purpose of the study was to investigate the soil organic carbon content to assess the amount of carbon sequestered in the soil. The average percentage of organic carbon in soil increased from 0.36 to 0.66% when an agroforestry system was used. Loamy sand had a higher organic carbon content than sandy clay soil. Pandey et al. (2010) conducted a study on *Acacia nilotica* trees (12 years old) to investigate the impact of three tree canopy positions, specifically mid-canopy, canopy edge, and canopy gap, on soil characteristics such as texture, organic carbon content, soil pH, as well as total mineral nitrogen and phosphorus. The results indicated that in comparison to the canopy gap, the area just underneath the mid-canopy experienced a 10% decrease in sand particles and a 9% increase in clay particles, respectively. With increasing soil depth, clay particles under any canopy location did not significantly drop. Mid-canopy and canopy edge positions had higher levels of soil organic C, total N, total P, mineral N (NO_3^- -N and NH_4^--N), and P compared to the canopy gap. Chauhan et al. (2012) conducted a study in an irrigated agro-ecosystem in India to assess the potential carbon sequestration of a common wheat intercropping system. They found that the presence of poplar blocks in the system increased organic carbon content in the top layer of soil (0–15 cm) due to the enrichment from litter and roots. Compared to open fields with wheat alone, the soil organic carbon content under poplar plantations was 0.42% as opposed to 0.32% under control conditions. In another study by Mao and Zeng (2013) in a semiarid temperate region of Northeast China, soil properties were investigated in croplands with 5-year-old poplar-based agroforestry systems. They found no significant differences in bulk density between croplands and agroforestry systems. However, they observed changes in total organic carbon, total nitrogen, nitrogen in microbial biomass, microbial metabolic quotient, and potential nitrogen mineralization rate. Tangjang et al. (2009) studied traditional agroforestry systems in northeast India and reported changes in microbial population and species composition attributed to factors such as plant residues, additional organic matter, vegetation, plant species composition, and soil mineral nutrients. Examining microbial biomass carbon (MBC) in semi-arid India, Benbi et al. (2012) investigated poplar-based agroforestry, rice-wheat, and maize-wheat cropping systems. Compared to maize-wheat systems (185 mg kg^{-1} soil) and rice-wheat systems (104 mg kg^{-1} soil), MBC was greater in agroforestry systems (203 mg kg^{-1} soil). According to Tian et al. (2013) microbial biomass values at various ginkgo spacings were substantially higher than in pure tea systems. The significant increases in soil microbial biomass C, N, and P under this AFS were mostly attributable to the gradual addition of various amounts of organic matter inputs through litter fall over time. As a result, microbial biomass, which depends on nutrient fluxes, has been utilized as a measure of soil fertility.

13.9 Economic Implications of Tree-Crop Interactions

Agroforestry systems are gaining more importance in the view of benefits to farmers and also by considering the ecological perspective. Incorporating diversified trees and crops are valued for better growth and/or household consumption. These shifts in farming system ensure the on-farm production and helps to manage "Hungry season". From an economic standpoint, agroforestry land-use systems will either have a higher production value at the same resource cost or the same output value at a lower resource cost than non-agroforestry land-use systems (Hoekstra 1987).

Khullar et al. (2010) conducted a study on the financial assessment of *Populus deltoides* based agroforestry models in Punjab and reported that the *P. deltoides*-based agroforestry system is more economically feasible and profit-making than the sole cultivation of paddy-wheat. The net returns from the agricultural model were 309,582 Rs ha^{-1}, whereas from the agroforestry model it was 1,086,389 Rs ha^{-1} (block plantation) and 540,679 Rs ha^{-1} (boundary plantation). Kareemulla et al. (2012) worked out the financial evaluation of *Populus deltoides*-based agroforestry system and reported that the main reason for adopting bund plantations and agrisilviculture system is the 70% additional income generation which provides nearly 20% of much-needed emergency sources of cash. In case of bund plantation of poplar with eight years of rotation, the NPV was Rs. 1,37,000/−, Rs. 1,27,000/−, and Rs. 1,18,000/− at discounted factors of 8%, 10%, and 12%, respectively, with benefit-cost ratio (B:C ratio) of 2.8 for all the three discount factors. Whereas, in the case of agrisilviculture with a rotation of seven years, NPV at the respected discount factors was Rs. 1,23,000/−, Rs. 1,11,000/−, and Rs. 1,01,000/−, respectively, while B:C ratios were 2.18, 2.15, and 2.12, respectively in comparison to the conventional crop rotation with a B:C ratio ranging from 1.34 to 1.42. On an average, both systems are socio-economically suitable and more profitable than monoculture. Devender et al. (2012) did a cost-benefit analysis of poplar-based agrisilviculture models adopted by farmer in N-W Indo-Gangetic plains during December 2000–06. They observed that productivity of the inter-sown crops reduced significantly after 2–3 years of poplar plantation which is compensated by the sale of green timber. Further they stated that the combination of poplar + sugarcane resulted in the highest net income of Rs. 64,355 ha^{-1} $annum^{-1}$ followed by poplar + turmeric (Rs. 59,543 $ha^{-1}annum^{-1}$) whereas the lowest net income was obtained from poplar + wheat/lentil (Rs. 18,719 $ha^{-1}annum^{-1}$) followed by sole poplar (Rs. 20,188 $ha^{-1}annum^{-1}$), which was higher than rice-wheat rotation (Rs. 22,970 $ha^{-1}annum^{-1}$) commonly practised in this region. Padma et al. (2018) noticed that coconut + patchouli intercropping system registered the highest net return (Rs. 1,43,705/−) and B:C ratio (2.84) followed by coconut + citronella (Rs. 1,08,870/−) with B:C ratio of 2.12, while monocropping of coconut gave net returns of Rs. 29,650/− with a B:C ratio of 1.60. Saresh et al. (2018) reviewed that in poplar-wheat and poplar-sugarcane intercropping systems the B:C ratio was 1.75 and 2.30, respectively, which was higher than their sole cultivation. Average net return ha^{-1} from poplar in combination with lemon grass, citronella, palmarosa, and

Japanese mint were Rs. 43,950, Rs. 40,160, Rs. 39,670, and Rs. 36,370, respectively. These agrisilvicultural systems offer an annual revenue of Rs. 70,000–80,000 per acre, which is three times more than what is primarily farmed in the central Punjab plains in a rice–wheat cycle. Likewise, Rana et al. (2017) studied the economic feasibility of growing sweet gourd alongside mango and guava in Gazipur, Bangladesh, and they claimed that the LER (Land Equivalent Ratio) of the intercropped mango and guava was 1.257 and 1.261, respectively. Pandey et al. (2016) reported that sapota + turmeric had the highest B:C ratio (4.85) whereas ginger sole cropping showed the lowest B:C ratio as 1.40. They stated that the cultivation of turmeric is more remunerative with sapota + jatropha as it provides a higher net income (Rs.3,55,350.60 ha^{-1}), B:C ratio (4.73 and lesser cost of cultivation (Rs.1,12,303.93 ha^{-1}) whereas in the case of ginger the cost of cultivation is higher (Rs. 1,59,343.88 ha^{-1}). Bhardwaj et al. (2021) reported higher returns from poplar+onion model with 60% IRR and B:C ratio of 2.94 in comparison to poplar +wheat model.

Apart from the products, ecosystem services also play a significant part in decision-making on land management (Ovando et al. 2016) and numerous ecological services are thought to be provided by agroforestry systems. Although there has been increasing interest in the Kyoto Protocol's clean development mechanism (CDM), this initiative holds promise for financial rewards for agroforestry systems carbon sequestration benefits and other ecosystem services. The sequestered carbon can be sold in carbon credit markets. Russell et al. (2010) calculated that transitioning from a continuous-cropping system to an agroforestry system that participated in carbon trading would cost \$109 ha^{-1}. Kay et al. (2019) conducted a study to assess the economic performance of marketable and non-marketable ecosystem services (ES) and dis-services in 11 contrasting landscapes of the European continent dominated by agroforestry land use. They found that the Mediterranean agroforestry systems tend to provide greater financial value compared to the agricultural system but in continental and Atlantic regions the agricultural system tends to be more profitable. However, when the related ES's economic values were taken into account, agroforestry has become more profitable in comparison. So, integrating trees with field crop in the same area are far more productive in comparison to sole cultivation of trees and agricultural crops.

13.10 Conclusion

The productivity of intercropping mixture consisting of a tree–crop combination is regulated by numerous growth determinants. More information on interactional studies between trees and crops is required to move towards low-input and diversified systems and also to manifest the beneficial relationship between trees and crops. In order to promote efficient and productive tree crop combinations across a wide range of environmental conditions, it is very important that reliable predictions of agroforestry practices be made with flexible models. Technological alternatives is the area where future research needs to be focused, which would promote better resource use to harness the benefits of each component (trees/crops/livestock). Moreover, combined efforts with indigenous knowledge and scientific methods will provide better solutions.

References

Adekiya AO, Alori ET, Aboyeji CM, Dunsin O, Adegbite KA, Aremu CO, Bamiro O, Ejue WS, Okunlola FO, Adesola OO (2019) MgO fertilizer sole and combined with organic and inorganic fertilizers: effect on soil chemical properties, turmeric performance, and quality in a tropical alfisol. Sci World J 2019:8

Aguiar ACF, Bicudo SJ, Costa Sobrinho JRS, ALS M, Coelho KP, Moura EG (2010) Nutrient recycling and physical indicators of an alley cropping system in a sandy loam soil in the pre – Amazon region of Brazil. Nutr Cycl Agroecosyst 86:189–198

Ali MM, Rahman MM, Islam S, Islam MA, Alam MR, Bari MS, Nahar MN (2018) Varietal performance of turmeric under mango based agroforestry system. An J Plant Sci 9:995–1003

Anuradha UB, Patil UB, Kurubar AR, Ramesh G, Hiregoudarv S (2018) Effect of integrated nutrient management on growth and yield of turmeric (Curcuma longa L.) cv. Salem. Int J Curr Microbiol Appl Sci 7:3196–3203

Baber S, Ahmad MF, Bhatti A (2006) The effect of eucalypts camaldulensis on soil properties and fertility. J Agri Biol Sci 1(3):47–50

Basak S, Hasan MK, Islam MS, Wadud MA (2009) Performance of radish, tomato and soybean during the first year of Lohakat (Zylia dolabriformis) plantation. J Environ Sci Nat Resour 4(1): 185–189

Bayala J, Sanou J, Teklehaimanot Z, Kalinganire A, Ouédraogo SJ (2014) Parklands for length and soil water content in semi-arid agroforestry. Plant Soil 201(2):197–207

Benbi DK, Brar K, Toor AS, Singh P, Singh H (2012) Soil carbon pools under poplar-based agroforestry, rice-wheat, and maize-wheat cropping systems in semi-arid India. Nutr Cycl Agroecosyst 92(1):107–118. https://doi.org/10.1007/s10705-011-9475-8

Bhardwaj A et al (2021) Optimization of onion planting time and variety under Populus deltoides-based agroforestry system in North-Western India. Agrofor Syst 95:533–546. https://doi.org/10.1007/s10457-021-00600-2

Bhat SA (2015) Effect of tree spacing and organic manures on growth and yield of vegetable crops under Melia composita Willd. based agroforestry system. Ph.D. Thesis, Department of Silviculture and Agroforestry, Dr. YS Parmar University of Horticulture and Forestry, Solan (HP) p 251

Bhusara JB, Dobriyal MJ, Thakur NS, Gunaga RP, Tandel MB (2018) Performance of okra (Abelmoschus esculentus L. Moench) under different spatial arrangements of Melia composita based agroforestry system. Int J Curr Microbiol Appl Sci 7:3533–3542

Bhushan A, Khare N (2018) Effect of organic manure and spacing on the growth and yield performance of baby corn under poplar based agroforestry system. Int J Curr Microbiol Appl Sci 7:3015–3019

Bradley RG (1995) Modelling the growth and water use of tropical cereals in semi-arid environments. Dissertation, University of Nottingham buffering climate risk and sustaining agricultural production in the Sahel of West Africa. Curr Opin Environ Sustain 6:28–34

Burgess P, Graves A, de Jalón SG, Palma J, Dupraz C, van Noordwijk M (2019) Modelling agroforestry systems. In Agroforestry for sustainable agriculture; Burleigh Dodds Science Publishing Limited, Cambridge, UK, pp 209–238

Cardinael R, Mao Z, Chenu C et al (2020) Belowground functioning of agroforestry systems: recent advances and perspectives. Plant Soil 453:1–13. https://doi.org/10.1007/s11104-020-04633-x

Chauhan SK, Dhillon WS, Singh N, Sharma R (2013) Physiological behaviour and yield evaluation of agronomic crops under Agri-horti-silviculture system. Int J Plant Res 3:1–8

Chauhan, S., Singh Brar, M. and Sharma, R. (2012) 'Performance of poplar (Populus deltoides bartr.) and its effect on wheat yield under agroforestry system in irrigated agro-ecosystem, India tree crop interaction for sustainability view project defining growth, quality and biomass production of different bamboo species in central plains of Punjab view project', (October). Available at: http://research.guilan.ac.ir/cjes

Chavan SB, Dhillon RS (2019) Doubling farmers income through Populus deltoides-based agroforestry systems in northwestern India: an economic analysis. Curr Sci 117:219–226

Chethan KVR, David AA, Thomas T, Swaroop N, Rao S, Hassan A (2018) Effect of different levels of N P K and Zn on physico-chemical properties of soil growth parameters and yield by pea (*Pisum sativum* L.). J Pharmacogn Phytochem 7:2212–2215

Coe R, Sinclair F, Barrios E (2014) Scaling up agroforestry requires research 'in' rather than 'for' development. Curr Opin Environ Sustain 6:73–77

Das R, Bauer S (2012) Bio-economic analysis of soil conservation technologies in the mid-hill region of Nepal. Soil Tillage Res 121:38–48

Deen W, Cousens R, Warringa J, Bastiaans L, Carberry P, Rebel K, Riha S, Murphy C, Benjamin LR, Cloughley R, Cussans J, Forcella F, Hunt T, Jamieson P, Lindquist L, Wang E (2003) An evaluation of four crop: weed competition models using a common data set. Weed Res 43(2): 116–129

Devender C, Afzal A, Bhatia JN (2012) Assessment of agroforestry based two-tier-cropping system in Ambala district of Haryana. Agric Update 7(3/4):210–213

Friend A, Stevens A, Knox R, Cannell M (1997) A process-based, terrestrial biosphere model of ecosystem dynamics (Hybrid v3. 0). Ecol Model 95(2):249–287

Gupta RK, Gupta SK (2017) Growth and performance of trees and green gram (Vigna radiata L. wilczek) in agrisilvicultural system in the dryland conditions of Jammu region. Indian J Ecol 44:326–329

Gusain D (2016) Evaluation of wheat-paddy cropping system under poplar and eucalyptus-based agroforestry system. . M.Sc. thesis. College of Forestry, Uttarakhand University of Horticulture and Forestry, p 207

Hairiah K, Widianto W, Suprayogo D, Van Noordwijk M (2020) Tree roots anchoring and binding soil: reducing landslide risk in Indonesian agroforestry. Land 9(8):256. https://doi.org/10.3390/land9080256

Handa AK, Dev I, Rizvi RH, Kumar N, Ram A, Kumar D, Kumar A, Bhaskar S, Dhyani SK, Rizvi J (2019) Successful agroforestry models for different agro-ecological regions in India. Central Agroforestry Research Institute (CAFRI), Jhansi and South Asia Regional Programme of World Agroforestry (ICRAF), New Delhi, p 203

Hoekstra DA (1987) Economics of agroforestry systems. Agrofor Syst 5(3):293–300

Holzworth DP, Huth NI, de Voil PG (2010) Simplifying environmental model reuse. Environ Model Softw 25(2):269–275

Imo M, Timmer VR (2000) Vector competition analysis of a Leucaena-maize alley cropping system in western Kenya. For Ecol Manag 126:255–268

Jones M, Sinclair FL, Grime VL (1998) Effect of tree species and crown pruning on root length and soil water content in semi-arid agroforestry. Plant Soil 201:197–207

Justes E, Bedoussac L, Dordas C, Frak E, Louarn G, Boudsocq S, Li L (2021) The 4c approach as A way to understand species interactions determining intercropping productivity. Front Agric Sci Eng 8(3):387–399. https://doi.org/10.15302/J-FASE-2021414

Kar S, Chandel A, Pant KS (2019) Effect of different tree spacings of Grewia optiva and combined doses of poultry manure and vermicompost on growth and yield of bell pepper under agrisilvi culture system. Int J Curr Microbiol Appl Sci 8:2213–2223

Kareemulla K, Rizvi RH, Kumar K, Dwivedi RP, Singh R (2012) Poplar agroforestry systems of western Uttar Pradesh in northern India: a socioeconomic analysis. For Trees Livelihoods 15(4): 375–381

Kaushal GS, Rajiv U, Vijaykumar R (2019) Effect of organic and inorganic fertilizer on growth of linseed (*Linum usitatissimum* L.) under poplar based agroforestry system. J Tree Sci 38:48–51

Kay S, Graves A, Palma JH, Moreno G, Roces-Díaz JV, Aviron S, Chouvardas D, Crous-Duran J, Ferreiro-Domínguez N, de Jalón SG, Măcicăşan V (2019) Agroforestry is paying off–economic evaluation of ecosystem services in European landscapes with and without agroforestry systems. Ecosyst Serv 36:100896

Keating BA, Carberry PS, Hammer GS, Probert ME, Robertson MJ, Holzworth D, Huth NI, Hargreaves ING, Meinke H, Hochman G, McLean G, Verburg K, Snow V, Dimes J, Silburn M (2003) An overview of APSIM, a model designed for farming systems simulation. Eur J Agron 18(3–4):267–288

Khasanah N, van Noordwijk M, Slingerland M, Sofiyudin M, Stomph D, Migeon AF, Hairiah K (2020) Oil palm agroforestry can achieve economic and environmental gains as indicated by multifunctional land equivalent ratios. Front Sustain Food Syst 3:122. https://doi.org/10.3389/fsufs.2019.00122

Khasanah NM, Perdana A, Rahmanullah A, Manurung G, Roshetko JM, van Noordwijk M (2015) Intercropping teak (Tectona grandis) and maize (Zea mays): bioeconomic trade-off analysis of agroforestry management practices in Gunungkidul. West Java Agrofor Syst 89(6):1019–1033

Kho R (2000) A general tree- environment- crop interaction equation for predictive understanding of agroforestry systems. Agric Ecosyst Environ 80(87–100)

Khullar V, Gill RIS, Singh B, Kaur N (2010) Economic evaluation of poplar (Populus deltoides) based forestry and agroforestry models in Punjab, India. Indian J Soc Res 51:51–67

Kumar A, Anand R, Singh I, Rawat P, Tewari SK (2018) Performance of turmeric (Curcuma longa L.) and soil characteristics under different agroforestry tree species in Uttarakhand. India J Pharmacogn Phytochem 7:2150–2154

Kumar N (2017) Effect of fertilizer and organic manures on growth and yield attributes of wheat and paddy variety under casuarina (Casuarina equisetifolia) based agrisilviculture system. Int J Pure Appl Biosci 5:879–887

Kumar R, Singh M, Bera B (2015) Influence of organic, inorganic and combined based fertilizers on bush physiology of Darjeeling tea (Camellia sinensis L.). Int J Basic Appl Biol 2:265–271

Kumar VS (2014) Effect of integrated nutrient management on soil fertility and yield of maize crop (Zea mays) in entichaplustart in Tamil Nadu, India. J Appl Nat Sci 6:294–297

Lojka B, Lojkova J, Banout J, Polesny Z, Preininger D (2008) Performance of an improved fallow system in the Peruvian Amazon—modelling approach. Agrofor Syst 72(1):27–39

Luedeling E et al (2016) Field-scale modeling of tree-crop interactions: challenges and development needs. Agric Syst 142:51–69

Malézieux E, Crozat Y, Dupraz C et al (2009) Mixing plant species in cropping systems: concepts, tools and models: a review. In: Lichtfouse E, Navarrete M, Debaeke P, Véronique S, Alberola C (eds) Sustainable agriculture. Springer, Dordrecht, pp 43–62. https://doi.org/10.1051/agro:2007057

Mao R, Zeng DH (2013) Effect of land-use change from cropland to poplar-based agroforestry on soil properties in a semiarid region of Northeast China. Fresenius Environ Bull 22(4a):1077–1084

Matthews R, Lawson G (1997) Structure and applications of the HyCAS model. Agrofor Forum 8: 14–17

Matthews R, Van Noordwijk M, Gijsman AJ, Cadisch G (2004) Belowground interactions in tropical agroecosystems. CAB International, Wallingford, UK, pp 41–60

Miina J, Saramaki J, Mwihomeke S (1999) In: Temu AB, Lund G, Malimbwi RE, Kowero GS, Kleinn C, Malende Y, Kone I (eds) 'Describing tree/crop interaction in agroforestry', Off-forest tree resources of Africa. A proceedings of a workshop held at Arusha, Tanzania. The African Academy of Sciences, pp 220–226

Mobbs D, Crout N, Lawson G, Cannell M (1997) Structure and applications of the HyPAR model. Agrofor Forum 8(2):10–14

Mobbs DC, Cannell MGR, Crout NMJ, Lawson GJ, Friend AD, Arah J (1998) Complementarity of light and water use in tropical agroforests: I. Theoretical model outline, performance and sensitivity. For Ecol Manag 102(2–3):259–274
Moore AD, Holzworth DP, Herrmann NI, Brown HE, de Voil PG, Snow VO, Zurcher EJ, Huth NI (2014) Modelling the manager: representing rule-based management in farming systems simulation models. Environ Model Softw 62:399–410
Moura EG et al (2014) Interactions among leguminous trees, crops and weeds in a no-till alley cropping system. Int J Plant Prod 8(4):441–456
Moura EG, Serpa SS, Santos JGD, Costa Sobrinho JRS, Aguiar ACF (2010) Nutrient use efficiency in alley cropping systems in the Amazonian periphery. Plant Soil 335:363–371
Nayak MR, Behera LK, Mishra PJ, Bhola N (2014) Economics and yield performance of some short duration fruit and medicinal crops under agrisilvicultural system in rainfed uplands of Odisha. J Appl Nat Sci 6:274–278
Ovando P, Campos P, Oviedo JL, Caparrós A (2016) Ecosystem accounting for measuring total income in private and public agroforestry farms. For Policy Econ 71:43–51
Padma E, Ramanandam G, Dorajee A, Rao M, Kalpana H, Maheswarappa HP (2018) Performance of medicinal and aromatic crops as intercrops in coconut garden under east coast of Andhra Pradesh. Int J Pure Appl Bio 6:421–426
Pandey SBS, Jadeja DB, Manohar NS, Tandel MB (2016) Economic comparison of intercropping of ginger and turmeric under sapota-jatropha based agro-forestry systems in south Gujarat. Int J Sci Environ Tech 5(5):3635–3642
Pandey SBS, Pandey M, Jadeja DB, Tandel MB, Nayak D (2017) Growth and yield of ginger (Zingiber officinale L.) under sapota-jatropha based agroforestry systems in south Gujarat. J Pharmacogn Phytochem 6:247–251
Pandey CB, Singh AK, Sharma DK (2000) Soil properties under Acacia nilotica trees in a traditional agroforestry system in central India. Agrofor Syst 49:53–61
Patil HY (2018) Physiological investigations on legumes in teak based agroforestry system. Ph.D. thesis. Department of Crop Physiology, College of Agriculture, University of Agriculture Sciences, Dharwad, Karnataka, p 320
Paydar Z, Huth N, Ringrose-Voase A, Young R, Bernardi T, Keating B, Cresswell H (2005) Deep drainage and land use systems. Model verification and systems comparison. Aust J Agric Res 56(9):995–1007
Prajapati VK, Swaroop N, Masih A, Lakra R (2018) Effect of different dose of NPK and vermicompost on growth and yield attributes of maize (Zea Mays L.) cv. MM2255. J Pharmacogn Phytochem 7:2830–2283
Probert ME, Dimes JP, Keating BA, Dalal RC, Strong WM (1998) APSIM's water and nitrogen modules and simulation of the dynamics of water and nitrogen in fallow systems. Agric Syst 56(1):1–2
Pukkala T (1998) Modelling tree-crop interactions in agroforestry systems. Modelling the Growth of Tree Plantations and Agroforestry Systems in Southern and East Africa. University of Joensuu, Faculty of Forestry, pp 117–130
Qiao X, Sai L, Chen X, Xue L, Lei J (2019) Impact of fruit-tree shade intensity on the growth, yield and quality of intercropped wheat. PLoS One 14:203–238
Rana S, Khatun MA, Mahboob MG, Wadud MA, Rahman GMM (2017) Performance of sweet gourd under fruit tree based agroforestry practices in char land ecosystem. J Agrofor Environ 11(1–2):165–170
Rao M, Nair P, Ong C (1997) Biophysical interactions in tropical agroforestry systems. Agrofor Syst 38:3–50
Rathee P, Kaushik N, Khajuria S, Singh P and Manjeet (2017) Performance of fennel and ajwan as intercrops under different spacings of poplar plantations in north-western, India. Int J Pure Appl Biosci 5:901–907

Russell AE, Raich JW, Arrieta RB, Valverde-Barrantes O, González E (2010) Impacts of individual tree species on carbon dynamics in a moist tropical forest environment. Ecol Appli 20(4):1087–1100

Sangwan AK, Dhillon WS, Chauhan SK, Singh NP, Batth MK (2016) Performance of garlic under Agri-horti-silvicultural system in relation to physiological behaviour and yield. Indian J Ecol 43: 724–729

Saresh NV, Verma A, Sankanur D (2018) Poplar (Populus deltoides) based agroforestry systems: an economically viable livelihood option for the farmers of North India. In: Pandey CB, Gaur MK, Goyal RK (eds) Climate change and agroforestry adaptation, Mitigation and Livelihood Security. New India Publishing Agency

Sharma G, Pant KS (2017) Effect of integrated nutrient management and tree spacing on production potential of maize (Zea mays) under poplar-based agroforestry system. Int J Curr Microbiol Appl Sci 6:2692–2697

Sinclair F (2004) Agroforestry. In: Burley J, Evans J, Youngquist JA (eds) Encycl For Sci. Elsevier, Amsterdam, pp 27–32

Singh J, Bishnoi M (2013) Performance of bean crops under Khejri (Prosopis cineraria) canopy in agroforestry systems of arid zone cineraria canopy in agroforestry systems of arid zone. Int J Agric Sci Res 2(12):318–321

Singh J, Singh A, Chauhan SK (2019) Impact of manipulating poplar canopy on microenvironment and marigold yield under Agri-silviculture system. Indian J Agrofor 21:48–52

Singh R, Jhariya N, Kumar M, Abhishek R (2013) Tree crop interaction in agroforestry system, vol 10, pp 15–16

Smethurst PJ (2017) Accurate crop yield predictions from modelling tree-crop interactions in gliricidia-maize agroforestry. Agric Syst 155:70–77. https://doi.org/10.1016/j.agsy.2017.04.008

Tangjang S, Arunachalam K, Arunachalam A, Shukla AK (2009) Microbial Population Dynamics of Soil Under Traditional Agroforestry Systems in Northeast India. Res J Soil Biol 1:1–7. https://doi.org/10.3923/rjsb.2009.1.7

Thakur NS, Mohanty S, Hegde HT, Chauhan RS, Gunaga RP, Bhuva DC (2019) Performance of Melia dubia under Cymbopogon spp. based agroforestry systems. J Tree Sci 38:28–34

Thevathasan NV, Gordon AM (2004) Ecology of tree intercropping systems in the north temperate region: experiences from southern Ontario, Canada. Agrofor Syst 61:257–268

Tian Y, Cao F, Wang G (2013) Soil microbiological properties and enzyme activity in Ginkgo-tea agroforestry compared with monoculture. Agrofor Syst 87(5):1201–1210

Tshepiso, Seobi SH, Anderson RP, Gantzer CJ (2005) Influence of grass and agroforestry buffer strips on soil hydraulic properties for an Albaqualf. Am J Soil Sci Society:893–901

Uthappa AR, Bana OPS, Kumar M, Kanwal M (2015) Soil physico-biochemical properties as influenced by varying tree densities in poplar (poplar deltoides Bartr. Ex Marsh) based agroforestry system. Ind J Agrofor 17(1):81–90

Van Noordwijk M, Van de Geijn SC (1996) Root, shoot and soil parameters required for process-oriented models of crop growth limited by water or nutrients. Plant Soil 183(1):1–25

Verma V, Patel R, Deshmukh N, Jha AK, Ngachan S, Singha A, Deka BC (2019) Response of ginger and turmeric to organic versus traditional production practices at different elevations under humid subtropics of North-Eastern India. Ind Crop Prod 136:21–27

Wu HI, Sharpe PJ, Walker J, Penridge LK (1985) Ecological field theory: a spatial analysis of resource interference among plants. Ecol Model 29(1–4):215–243

Young A, Muraya P (1990) SCUAF: soil changes under agroforestry. ICRAF, Nairobi

Young A, Menz KM, Muraya P, Smith C (1998) SCUAF version 4: A model to estimate soil changes under agriculture, agroforestry and forestry. Australian Centre for International Agricultural Research, Canberra, Australia

Zhu Z, Arp ZA, Meng F, Bourque CP-A, Foster NW (2003) Forest nutrient cycling and biomass model (ForNBM) based on year-round, monthly weather conditions: part II: calibration, verification, and application. Ecol Model 170:(1)13-27

Chapter 14
Restoration of Degraded Soils for Food Production Through Agroforestry

Vinita Partel, Rajesh Kumar Meena, Vishnu K. Solanki, and Tulika Kumari

Abstract The effect of land degradation is mostly seen on the economically weak peoples because economically weak peoples are more dependent on natural resources. Almost all agroecologies and terrestrial biomes face the problem of land degradation, not only low-income countries but high-income countries as well. In the world, approximately 2 billion ha land is affected by land degradation in many forms of natural occurring and human generated. The main contribution in land degradation is done by water erosion. The world priority is restoration of ecology of ecosystems which are degraded. Agroforestry approaches like several species of fruit trees and forest, arable crops, medicinal crops of high value and forages are used for the rehab of degraded land from eroded soil, mining, deforestation, degradation of rangeland and intensive agriculture. To diversify and intensify farming system, agroforestry can be used through indigenous tree species integration which helps to maintain the sustainability also. The involvement of various integrated farming systems by including horticultural trees and various multipurpose trees helps to build up the soil fertility.

Keywords Arable · Erosion · Agroforestry · Ecosystem · Deforestation

14.1 Introduction

Soil degradation explains the consequences when soil quality declines and capacity to support plants and animals gets also declined. Certain chemical, physical or biological qualities also go towards downfall due to soil degradation. It is a global problem that has large impacts upon everybody through food insecurity and higher food prices, through environmental hazards and climate change and through loss in

V. Partel · V. K. Solanki
Jawaharlal Nehru Krishi Vishwa Vidyalaya, Jabalpur, MP, India

R. K. Meena (✉) · T. Kumari
Dr. Rajendra Prasad Central Agricultural University, Samastipur, India
e-mail: rkmeena@rpcau.ac.in

H. S. Jatav et al. (eds.), *Agroforestry to Combat Global Challenges*, Sustainable Development and Biodiversity 36, https://doi.org/10.1007/978-981-99-7282-1_14

biodiversity and services of ecosystem. Land degradation chiefly describes the life support land resource loss by salinization, desertification, soil erosion, acidification, etc. Deforestation reports severe degradation problems because it may cause serious soil erosion, fertile top soil removal and flood. Land degradation causes loss of economic productivity or biological activity and complication of rainfed cropland, rangeland or irrigated cropland, woodlands, pasture and forest arising from uses of land or from a process or process combinations which also include the processes arising from the patterns of habitation and activities of human beings like (i) erosion of soil occurring by water or wind; (ii) degradation of many properties of soil like biological, chemical and physical as well as economic; and (iii) depriving the natural vegetation on a long-term basis. Land degradation will hold on the important factor which will frighten the village livelihoods, increasing the dispute over restricted natural resources and also bringing about forced migration. In advance instance of land deterioration, drought and desertification, the whole village, social and community are enforced to move from own original inherited place to areas having already competitiveness over scanty resources and in this way come up with conflict at excess risk. The major role played by sustainable use of land is soil protection for human security, food and climate (Lal 2014; Amundson et al. 2015). After that also, land degradation is a worldwide occurrence which is influenced by natural and social factors; it occurred commonly in agroecologies and terrestrial ecosystem, in advanced industrialized countries as well as in low-income countries (Nkonya et al. 2016).

The primary effects of land degradation are reduced production of food, storage of water, biodiversity losses, organic carbon removal and diminished ecosystem functions (IUCN et al. 2015; Gilbey 2019). Land degradation has several primary causes, including incorrect utilization of land, soil erosion, soil carbon losses, water logging, mining, desertification, imbalance of nutrients and reduced soil biodiversity (FAO and ITPS 2015).

Understanding the scope of land degradation at global level by knowing both its impacts and causes, its imperative on environment as well as socio-economic could impose a basic threat to security and peace from communities of the local people to the whole continents. This chapter will explain in a nutshell these potential and linkages by considering the human securities. The basic purpose is to convince the decision-makers about the urgent need for action against the land degradation and meanwhile highlighting the economical and realistic solutions which also include the embracing and step-up of sustainable land management practices and ecosystem re-establishment activities. The unchecked soil degradation can entirely lose its productive capacity for human use, and this will again reduce if the steps are not taken to stop the further degradation and not restoring the productivity. The productive capacity of land is lowered temporarily or permanently by land degradation (Hurni 1993).

When negative activities of humans become supplemental to the factors which is natural, then degradation occurs, and the negative human activities are system of irrigation which is inefficient and not corresponding with requirement of soil and water; overgrazing, deforestation over-cultivation and industrial pollution also increase, and population is also another factor for land degradation (UNCCD 2003). The significant land degradation consequences are commonly well known which have negative impacts on human life and the environment. This involves non-timber forest products (NTFP) shortages and firewood scarcity, and other woods become scarce also (EFAP 1994), the water of springs and water bodies dry, sediments are deposited in dams, landslides and floods cause the diseases which are waterborne; climate change, biodiversity loss and desertification all these affect negatively land efficiency and food safety (Robert et al. 2008).

14.2 Agroforestry

Agroforestry is a system in which perennial woody shrubs and trees intentionally grow by mixing on the similar land use system with crops or animals in some forms of time sequence or in spatial arrangement. Alternate land use systems on agroforestry lessen the erosivity caused by runoff and eroding of soil through reducing the rainfall intensity by low height coverage, litter on surface, hindering the overland flow, binding the roots and ameliorating the health of the soil. When there are species which are compatible and availability of sufficient soil-water resources fulfills the water demand of both the species, then the satisfactory crop yield can be achieved. Agroforestry plays a significant role for providing food security as well as nutritional security and reducing land degradation (Pretty and Bharucha 2014; Dagar and Tewari 2017; Lal 2004; Nair 2007).

The three basic components of agroforestry are trees, crops and animals which are classified as follows:

1.	Agri-silvicultural	Agriculture crop, trees, shrubs
2.	Silvi-pastoral	Trees, pastures/animals/grass
3.	Agri-silvi-pastoral	Crops/pasture, animals and trees
4.	Agri-horticulture	Agriculture crop/fruit trees
5.	Silvi-horticulture	Forest trees and fruits
6.	Silvi-hortipastoral	Trees and fruits, pastures and animal

However, some component trees like subabul have prolific habit of seeding which results in higher growth rate of weeds and reduces the production of primary crops. The impact of rooting and shading of *Acacia nilotica* on cultivated crops is as much as 20 m, and the main drawback of the land management system of agroforestry is harbouring pest and diseases and birds; reduction in the scope of mechanization makes the system more requiring labour and having tree allelopathy effects on cultivated crops. Agroforestry also includes a high diversification on the degraded and waste lands for sustaining several ecological functions (Schoeneberger et al. 2012).

14.3 Soil Degradation Types

14.3.1 Biological Soil Degradation

The microbial and biological activity of soil is adversely affected by the factors like soil micro flora and fauna. It also has an impact on yield. It is also studied that cultivating the similar agriculture crop on similar unit of land every year, i.e. monocropping, causes the increase on the pest and disease attack. In the Nilgris, cultivation of potato is threatened by fatal nematodes, and if not checked on time, it can cause risk in potato cultivation in that area. Microbial activity and biomass get reduced by the excessive use of pesticide. Nitrification is inhibited by the incorporation of many pesticide chemicals (like bromacil, amitrole, picloram, atrazine, etc.). Different pesticides also inhibit some leguminous crop growth, its nitrogen fixation and their nodulation. Oil shales and heavy metal disposal and spillage of crude oils contaminate the soil and have bad impact on micro flora, as a result affecting the productivity of soil and causing soil demeaning. It is reported that soil biota is a pointer for soil richness and affecting structure of soil (Nebiyou and Muluneh 2016).

A huge number of diverse living organisms in a complex system and varied communities are possessed by the soil. The presence of organisms which are billions in number like actinomycetes, bacteria, algae, nematodes, protozoa, fungi and cyanobacteria is revealed by the microscopic examination of a soil sample, and a complicated web of biological activity is formed in the ecosystem by the interaction of these diverse organisms. The functions and activities of soil biological communities are influenced by many factors of the environment like acidity, moisture and temperature and also by the activities of human like management practices which include agricultural and forestry. Physico-chemical soil deterioration is the direct result of unsuitable soil management practices. Covering the land with the required shrubs and trees is the key strategy to control the biological degradation. In the sustainability of ecosystem, the vital role is played by the soil organisms. Additionally, they serve as the primary driving force behind the management of soil organic matter dynamics, the release of greenhouse gases, carbon sequestration in the soil, the modification of soil physical structure and water regimes and the impact on plant health. Biological degradation mostly manifests as a reduction of the organic matter in the soil, reduction of the vegetation cover and decrease in biological activity (Nebiyou and Muluneh 2016).

14.3.2 Physical Soil Degradation

Physical soil qualities including texture, structure, aggregate stability, porosity, permeability or compaction and crusting are fundamentally affected negatively by physical degradation (Mitiku et al. 2006). The change in the size distribution of the pores and the total volume of the soil are mostly reflected by the soil degradation which brought about the solid phase reorganization. Compaction means reorganization in the subsoil, and if it is called sealing or crusting, it means reorganization occurs in the surface. If we see worldwide, then compaction, sealing and crusting are the major soil physical management problems. Compaction of soil also mostly happened in farming systems which are mechanized, in which the soil has to tolerate the continuously heavy loads of machineries. By the use of machineries for the clearance of forest and in agricultural-based industries in the tropics, soil suffers mostly compaction problems (Nebiyou and Muluneh 2016).

14.3.3 Chemical Soil Degradation

The chemical fertility of the soil is negatively impacted both directly and indirectly by changes in one or more of its chemical qualities (Suraj et al. 2001). The activities of soil life process get interfered by the large amount of toxic chemicals which are present in chemically degraded soils. Nutrient element mobility, nutrient availability and nutrient uptake are also interfered with these toxic elements, as organic matter is the primary factor for the productivity and the development of the plants which is declined by the chemical degradation of the land due to repeated tillage in cropping system. Therefore, chemical degradation reduction is one of the sustainable management, which can be achieved by the tree-based or shrub-based farming systems like by adopting agroforestry practices. Chemical degradation involves the collection of chemicals which influence the soil biological activity of soil (Logan 1990).

14.4 Possibilities to Execute the Rehabilitation of Degraded Land

There should be involvement of some short-term advantages, either material or monetary, including the foreseen future beneficial impacts:

- Behaviour of rural people, behaviour and perceptions should have interest on the principle of the programme regarding rehabilitation—the public should believe that their advantage is based upon the long-run change only.
- Complete knowledge of the connectivity, peculiarities and challenges of dry land ecosystems, especially for dry land forests and woodlands.
- It should conduct the soil productiveness improvement, hydrological processes, etc.
- Accessible uses of land should be studied, and the attributes of land should be matched with land uses so as to find out the causes of degradation.

14.5 Soil Degradation Causes (Gupta et al. 2020)

The major factors for degradation or unproductiveness of soils are the following:

1. Using higher pesticides
2. Waterlogging conditions
3. Soil salinity
4. Soil erosion
5. Other factors

14.5.1 *Pesticides*

In modern agriculture there is large contribution of pesticides which endures the food security. It means the chances of incorporated pesticides going inside the soil and degrading some aspects of soil property should not be ignored (Fig. 14.1). Such effect chances are more when pesticides are incorporated at a higher dose from long durations (Hance et al. 2001); higher dose of pesticide leads to toxicity. According to a study, using less than 0.1% of pesticides added to crops will really reach the intended insect. The remaining chemicals will find their way into the environment and may contaminate the soil, air and water (Pimentel and Levitan 1986).

14.6 Waterlogging

The equilibrium of water of any area is distressed due to surplus recharge; soil becomes waterlogged. Overland water flow towards the basin; leakage from canals and supply system, heavy rains and tide flooding are the major sources of water.

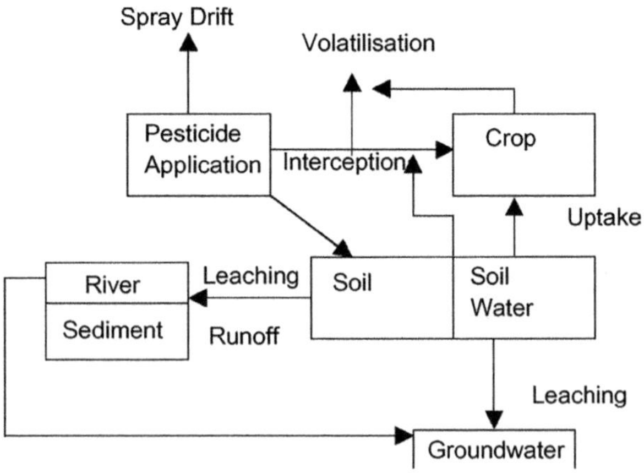

Fig. 14.1 Pesticide incorporated pathways for a crop (Arias-Estevez et al. 2008)

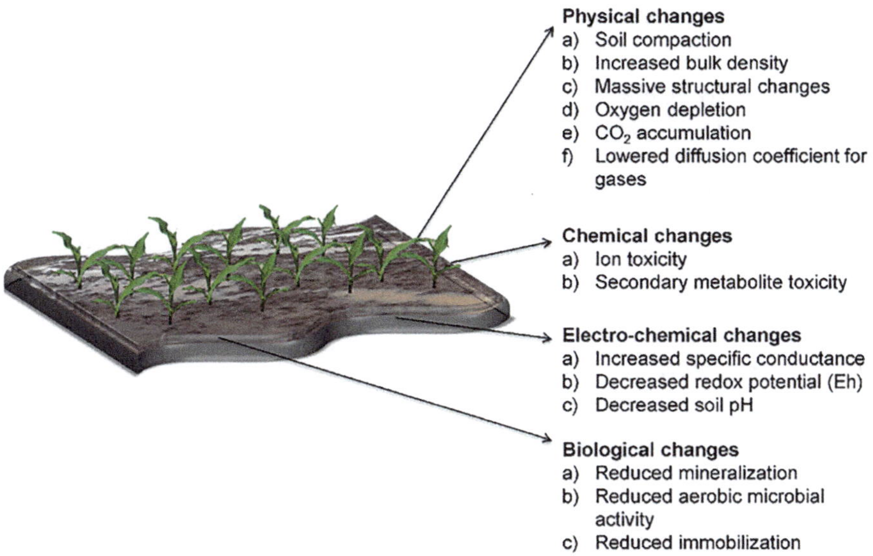

Fig. 14.2 Soil properties affected by waterlogging (Nuruzzaman Manik et al. 2019)

There are some conditions which cause waterlogging like low porosity of subsurface horizons, interior drainage, natural basins without outlet for water, less absorbing capacity of surface soils and hindrances to natural run of rain water. The speedy augmentation in the water table is due to irrigation by canal in highly productive areas. In arid and semiarid areas, irrigation by canal expansion is also straight relevant by the waterlogging and salinity problems. The disturbance in hydrological cycle is caused by the surface irrigation water inefficient use, deprivation of land development, poor drainage and leakage resulting in high water table. Canals of arid and semiarid regions are mostly affluent in soluble salts. When the same water are used as irrigation water, the salts rise to the surface by capillary action, and these salts are deposited on the upper surface of the soil as a coating or crust after drying up the water. Soil having more organic matter content, because of less temperature and waterlogging, can be surely affected due to both factor changes. Waterlogging is a global problem distressing 16% of the soils in the United States, 10% in Russia and agricultural land and crop production of India, Pakistan and China which depends on irrigation (Yaduvanshi et al. 2014). Soil properties affected by waterlogging have been presented in Fig. 14.2.

14.7 Salinity (Saline and Alkali Soils)

The productivity is affected directly by salinity, making the soil unsuitable for crop growth, and also indirectly by lowering the productivity by having the serious impacts over the nutrient availability. The alkalinity's undesirable impact on nutrient accessibility is because of deflocculating effect of sodium ions. Extreme irrigation in agriculture largely contributes to the increase in troubles of secondary salinization and waterlogging (Qadir et al. 2007).

14.7.1 Erosion

The major reason behind the soil degradation is the soil erosion. In the erosion of the soil, the top layer having more fertility and containing essential nutrients is lost due to which soil becomes essential mineral deficient resulting in loss of productivity (Fig. 14.3). Forest destruction or deforestation reducing the rainfall frequency and leading the soil erosion and damages the agriculture property. When the soil is abrupt, sloppy or simply erodible, then deforestation causes the degradation of the soil fast. Erosion of soil by wind and water is the chief factor to destruct the natural vegetation cover. In India soil erosion is extensive and a severe problem for its continued existence. It occurs in arid and semiarid lands, forest lands, agricultural lands and in areas where disturbances take place geologically or naturally (Saroha 2017).

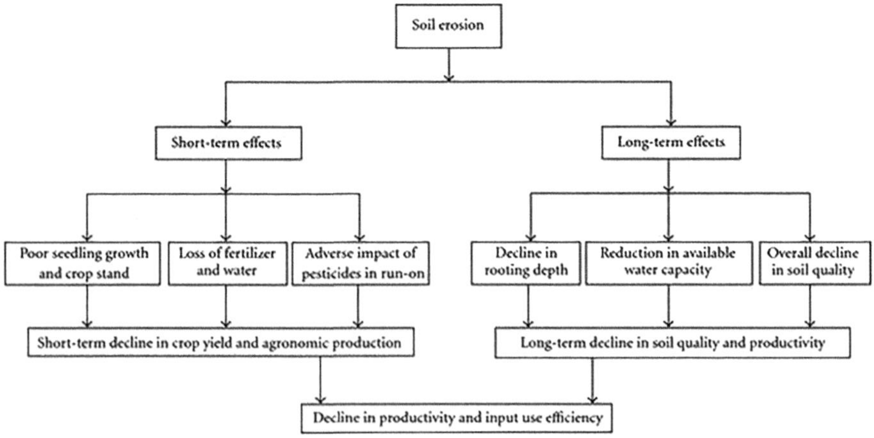

Fig. 14.3 Impacts of soil erosion (Lal et al. 2003)

14.8 Shifting/Jhum Cultivation

This kind of cultivation is primarily practiced in the northeastern states of India. It is a kind of slash and burn cultivation method. Mostly the forest land is slashed and burned after reaping of crops. The subsequent cultivation will be done on a different piece of land, and the burned land will be left without cultivation for some period; the length of gap years between two cultivations in a land was 10–20 years in the early periods. As the population increases due to which land availability is compact, the gap is also getting reduced to merely 2–3 years. This causes severe deforestation, removal of wild animal habitation, environmental pollution, etc. The firing of forest causes gradual degradation and soil erosion. Jhum cultivation is a significant trade partner for the conservative communal orders as their essential sources of food, shelter, medicine, shelter and particular goods and services (Bhattacharjee et al. 2020).

14.9 Extension of Cultivation to Marginal Land

The land use increases tremendously day by day due to high growth in population. Although marginal lands are viable for cultivation, they become low in fertility and susceptible to degradation. Marginal lands are the lands in dry and semi-dry areas, abrupt sloppy lands and sandy soils (Honson et al. 2015).

14.10 Improper Crop Rotation

In place of more balanced cereal-legume rotations, intensive cropping methods of profit-making crops are adopted by farmers due to land shortage, population increase and economic pressure. During the last two decades, cooking crop area decreased, and non-consumable crops area enlarged. Soil fertility is lost due to elimination of huge quantity of nutrients because of intensive cultivation (Yaduvanshi et al. 2014).

14.11 Fertilizer Misuse

Fertility of soil decreased due to long-time intensive farming. The efficiency of soil is maintained by the farmers by incorporating chemical fertilizers which make them take away by the use of organic manures. Even though using fertilizers production can be maintained at the same level that provides undersupplied minerals, their usage mostly causes deficiencies of other nutrients. In several parts of the world, restoration at the landscape scale is being used to repair the harm that anthropogenic ecosystem degradation has done to biodiversity and human welfare (Honson et al. 2015).

14.12 Overgrazing

In India due to agricultural land expansion, pasture ground area is shrinking as time passes. Recent satellite data clears that pasture land area is extremely degraded. Overgrazing is also the reason for this deprived situation of pasture lands. Forest soils have degraded because of indiscriminate and unchecked graze on forest land. Overgrazing directly results to vegetation disappearance, making it as a primary cause of water and wind erosion in dry lands (Dass et al. 2011).

14.13 Mining

The physical, chemical and biological properties of soil are affected through mining. The intensity of impact of mining depends upon the chemical and physical properties of the waste generated. The layers of the soil change as the uppermost soil is changed into interior within the dumps. Nearly all organic materials and plant and mineral nutrients are missing from the eroded material. There is a need of mining for diverse uses which causes degradation of many locations and leads to no biomass production in the affected area (Gupta et al. 2020).

Soil conservation called the management of land which depends upon the land capacity includes best applied management practices, leading the commercial crop production with no degraded land (Nebiyou and Muluneh 2016). Soil degradation types have been depicted in Fig. 14.4.

14.14 Agrostological Methods

Agrostological method refers to utilize the grasses to check the erosion of soil, to reduce the runoff and to improve the moisture storage. Grasses having dense canopy cover on the surface of the soil and the rooting which is profuse and holds the soil provide the best protection against erosion and runoff. Various in situ techniques for the conservation of moisture are given below (Dagar and Tewari 2017).

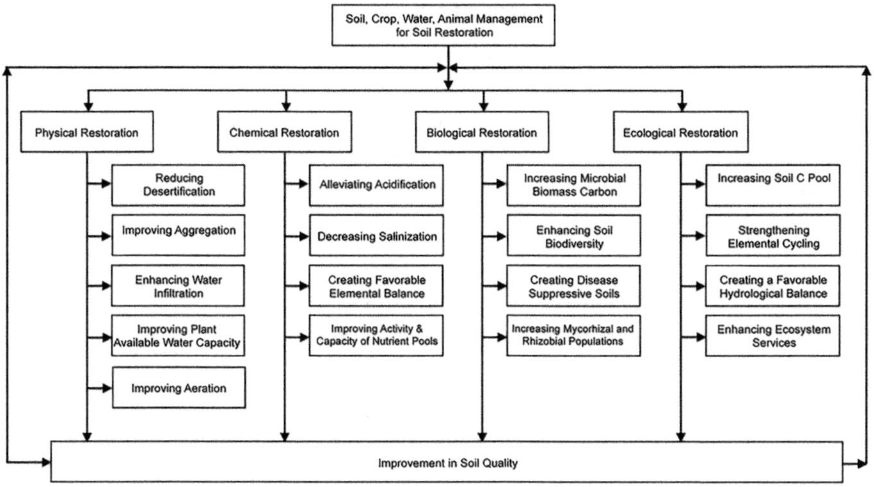

Fig. 14.4 Soil degradation types

14.14.1 Restoration by Rangelands

The recommendation for eroded, degraded, shallow gravelly soils is raising the perennial grasses to establish the grassland or pastures. Canopy of grasses catches the rainfall, lessens the splash erosion, controls the runoff and increases the soil moisture storage from rainfall. It is also an internationally well-known, ecologically exclusive system that sustains affluent biodiversity and intellectual and recreational value (White et al. 2000).

14.14.2 Restoration by Ley Farming

In the restoration by ley farming, grasses, legumes trees, shrubs and annual crops are cultivated in rotation. Legumes and grasses, e.g. *Stylosanthes* and *Cenchrus*, are cultivated for the duration of 4–5 years, then followed by yearly crops, e.g. sorghum for the period of 2 years. The coverage of soil with legumes and grasses improves the moisture conservation. It is advised that grassy weed attack was lessened by rotating graze legume pasture with wheat in comparison to the cropping systems which did not contain grazed pasture (Martin 1996).

14.14.3 Restoration Follows the Strip Cropping Pattern with Grasses

Grasses and fodder crops are cultivated alternately and across the slope, helping to control the runoff and erosion and increasing moisture infiltration capacity of the soil. Grasslands store most of the carbon in the soil in comparison to the vegetation unlike the forest (White et al. 2000).

14.14.4 Restoration by Vegetative Barriers

Vegetative barriers comprise the one or two rows of perennial grasses across the slope and along the contour at suitable intervals. They create a hindrance for unhindered runoff and transportation of the soil. Plantation of vetiver should be done in rows at 40 m spacing in the slope of 0.5%. Before the start of the monsoon, plough furrows are exposed by using disc plough. The holes at a depth of 5–8 cm at the interval of 20 cm are formed, and the planting of two slips in one hole should be done in the commencement of monsoon. The surrounding soil of root zone should be compressed. Soil erosion and runoff are prevented by the barriers of vetiver. Vetiver holds the soil by allowing the surplus runoff to run by their canopy with no loss of the soil. The effects of various vegetation barriers growing in trench-cum-bund arrangements were seen in runoff, moisture, loss of nutrients, soil fertility and crop production in the rainfed uplands in the southern Orissa watershed (Dass et al. 2011).

Vegetative barriers need less maintenance and have adapted to drought also. It also doesn't show any border effect over the adjacent row crops. It hikes up the yield of a crop by 10–15% by allowing the uniform spreading of water to sloppy portion in the field as a result making the uniform plant stand. It stores the moisture in the soil. The harvesting of fodder can also be taken for the animals if any fodder grasses, e.g. marvel grass or *Cenchrus glaucus*, are used. Barriers of vegetation are more useful for black soil. It does not permit any operations in black soil like contour bunding which develops the crack in summers and gives the way for water loss. Therefore, the maintenance of vegetation barrier is able to successfully apply in the black soil up to 4–5 years. The replanting material can be taken from the previous barriers after 4–5 years by 'quartering'. Vegetative barriers are the biological procedure which is substitute, which efficiently preserve the soil and water from the surface runoff moderation and allow the extra infiltration time (Krishnagowda et al. 1990).

14.15 Wind Breaks and Shelterbelts Uses

Wind breaks are the structure which hinders the passing of the wind and decreases the wind speed; at the same time, shelterbelts are tree rows planted for the crop protection from the winds which desiccate, prevent from soil erosion and provide a favourable micro-climate. Mostly, shelterbelt provides the protection from the wind to 2–5 times of its height in the windward side and 30 times in the leeward side. The excellent protection from the winds will be provided by the conical cross section of wind breaks. Windward side is the direction from which the wind blows, and leeward side is the direction to which the wind is blowing. The planting of shelterbelts must be one across the wind direction. Wind flow is not obstructed completely by them. Some amount of wind pass the shelterbelts, and the remaining are diverted and deflect from the shelterbelts depending upon their porosity. In this way without turbulence wind speed can be reduced (Bird et al. 1992).

If the speed of the wind reduces, ultimately losses done by evaporation get reduced, and therefore, availability of water is more for the plants. The good impact of shelterbelts is further visible during drought period, and it also reduces the wind erosion. Wind breaks which are correctly distributed can decrease the wind speed (30–50%) on 5% of the area and reduce the loss of soil by 80% (Bird et al. 1992). Shelterbelt and advantages of windbreak have been depicted in Fig. 14.5.

14.16 Tree Farming

In place where arable crops are not profitable, trees flourish well and yield sufficiently. Farmers are interested on tree farming due to labour cost, farm operation scarcity at peak time and constant crop failure because of drought. Generally on undersized farms, MPT_S means a single tree provides multiple products, like timber, fuel, fodder, nitrogen, food, resins, fibre, medicines, shade, etc. *Prosopis cineraria* is the best MPT_S for Gujarat plains and hills; after that *Leucaena*, *Dalbergia sissoo*, *Ailanthus excelsa* and *Eucalyptus hybrid* are the best trees. Multipurpose tree species are intentionally planted to give many considerable produce and service. Multipurpose trees planted on the bunds of soil reduce the surplus runoff, drain the concentrated runoff, decrease the soil erosion and finally check the land degradation (Abebe and Tolera 2021).

14.17 Alternate Land Use Systems (ALUS)

Marginal and sub-marginal lands must be used effectively for planting to fulfill the growing demand of fruit, fibre, food, fuel and fodder. However, these types of lands are not capable of maintaining production, which leads to an imbalance in the

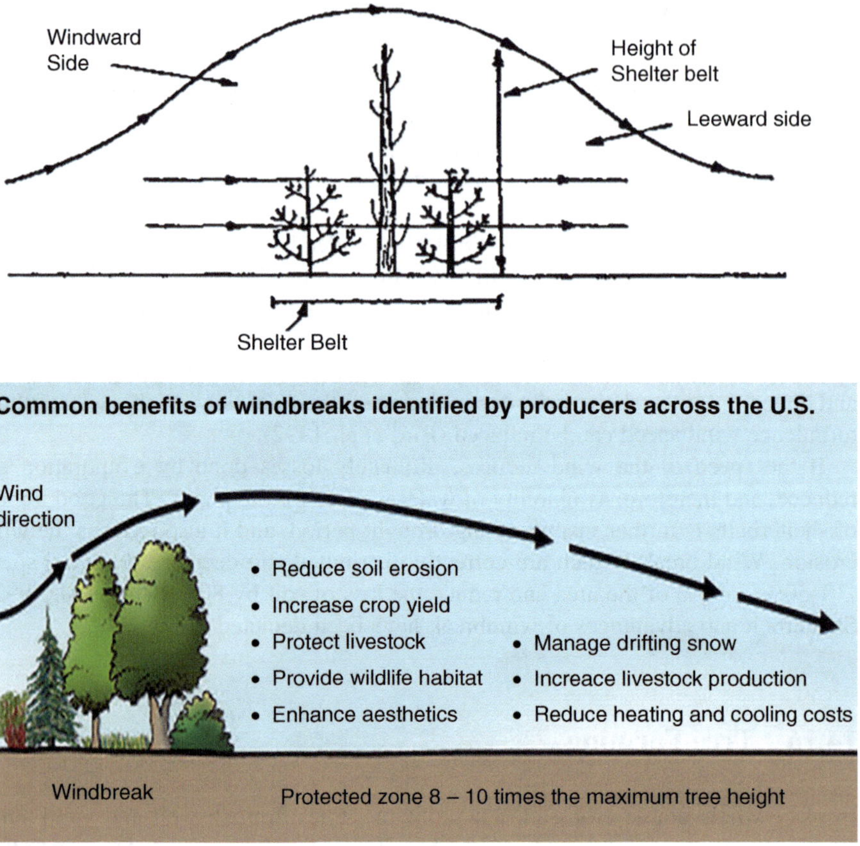

Fig. 14.5 Shelterbelt and advantages of windbreak (Mathew et al. 2021)

ecosystem. Therefore, an alternative land use system introduces a system of land use which is unusual from the conservative systems. Alternative land use is the system in which land are used for another production system to equate its capacity closely to the latest land use system and target extra economical and biological productivity. Alternative land use is the practice of using land in a different way than it was originally intended to produce goods or services in order to better fit the land's capabilities to new land uses and to achieve longer-lasting economic, biological and biological productivity (Reddy 2011).

14.17.1 Alternative Land Use Systems Advantages

1. By enhancing profitability and biological productivity optimizes resource use.
2. Resource base quality is enhanced and conserved.

3. Integrate the arable crops, pastoral crops and livestock.
4. To make the agriculture which is not completely independent with off-farm inputs.
5. Generated the employment opportunities.
6. Improve the living standards of farmers.

Many kinds of land use are incorporated on watershed by depending upon the component of farming system. Alternate land use systems available for all land categories target the assured income with least risk by using utilizing available resources efficiently. Agroforestry, ley farming and tree farming are commonly known alternate land use system. The potential to reduce nutrient loss through soil conservation is an important management factor in agroforestry (Abhishek et al. 2017).

14.18 Conclusion

Pressure of increasing population, use of natural resources unsustainably, farm land use for non-agricultural relevant purposes continuously and the constant ruin of the ecosystem which is rich in biological diversification are all the facts which cause the situation in which food security is the major challenge. Overexploitation of naturally available resources, industrialization and mining for overutilization of earth assets are removing the soil organic matter, and in return biodiversity of soil and fertility of soil and agricultural productivity are also getting eroded on a long-term basis. Agroforestry techniques are used to restore damaged soil from mining, eroded soil, deforestation, degradation of rangeland and intensive agriculture. These techniques include a variety of species of forest, arable crops, fodder crops, fruit crops and medicinal and aromatic crops. To diversify and intensify farming system, agroforestry can be used through indigenous tree species integration which helps to maintain the sustainability also.

References

Abebe BB, Tolera AG (2021) Farmland trees for the improvement of crop yield, soil fertilities, soil and water conservation, and carbon sequestration: a review. Agric Sci Res J 11:154–168

Abhishek R, Jhariya MK, Toppo P, Oraon PR (2017) Role of agroforestry in nutrient cycling, Kishan World. J Agric Rural Dev 44(7):38–40

Amundson R, Berhe A, Hopmans J, Olson C, Sztein AE, Sparks D (2015) Soil and human security in the 21st century. Science 348(6235):1261071

Arias-Estevez M, López-Periago E, Martìnez-Carballo E, Simal-Gandara J, Juan-Carlos M, Garcıa-Rıo L (2008) The mobility and degradation of pesticides in soils and the pollution of groundwater resources. Agric Ecosyst Environ. 123:247–260

Bhattacharjee D, Barau AA, Haque ME, Afrad MSI (2020) Knowledge of Chakma women on shifting cultivation: a comparative study between Bangladesh and India. Asian J Agric Ext Economics Soc 28:57–66

Bird PR, Bicknell D, Bulman PA, Burke SJA, Leys JF, Parker JN, van der Sommen FJ, Voller P (1992) The role of shelter in Australia for protecting soils, plants and livestock. Agrofor Syst 18: 59–86

Dagar JC, Tewari VP (eds) (2017) Agroforestry: anecdotal to modern science. Springer, Singapore, p 879

Dass A, Sudhishri S, Lenka NK, Patnaik US (2011) Runoff capture through vegetative barriers and planting methodologies to reduce erosion, and improve soil moisture, fertility and crop productivity in southern Orissa, India. Nutr Cycl Agroecosyst 89:45–57

EFAP (1994) Ethiopian Forestry Action Program. Final Report, Ministry of Natural Resources Development and Environmental Protection, Addis Ababa, Ethiopia

FAO and ITPS (2015) Status of the World's Soil Resources (SWSR) – main report. Food and agriculture organization of the united nations and intergovernmental technical panel on soils, Rome

Gilbey B (2019) Reviving land and restoring landscapes: policy convergence between forest landscape restoration and land degradation neutrality. IUCN, Gland, viii, p 34

Gupta SR, Dagar JC, Demel T (2020) Agroforestry for rehabilitation of degraded landscapes: achieving livelihood and environmental security. Ed. Agroforestry for degraded landscapes. Springer, Singapore

Hance RJ, Hussain M, Ferris IG (2001) Impact of long-term pesticide usage on soil properties using radiotracer techniques. IAEA, Vienna. IAEA-TECDOC, p 1248

Honson C, Buckingham K, Dewitt S, Laestactus L (2015) The restoration diagnostic method for developing forest landscape restoration strategies by rapidly assessing the status of key success factors. World Resources Institute, Washington

Hurni H (1993) Land degradation, famines and resource scenarios in Ethiopia. In: Pimentel D (ed) World soil erosion and. Conservation Cambridge University Press, Cambridge, pp 27–62

IUCN, Davies J, Gudka M, Laban P, Metternicht G, Alexander S, Hannam I (2015) Land degradation neutrality: implications and opportunities for conservation. IUCN, Gland

Krishnagowda KT, Krinappa AM, Panduranga P, Hegde BR (1990) Live bunds for soil and moisture conservation under dryland conditions. In: International symposium on water erosion, sedimentation and resource conservation. October, 9–13, CSWCRTI, Dehradun

Lal R (2004) Soil carbon sequestration impacts on global climate change and food security. Science 304:1623–1627

Lal R (2014) Societal value of soil carbon. J Soil Water Conserv 69(6):186A–192A

Lal R, den Biggelaar C, Wiebe KD (2003) Measuring on-site and off-site effects of erosion on productivity and environmental quality. In: Proceedings of the OECD Expert Meeting on Soil Erosion and Soil Biodiversity Indicators, Rome, Italy, March 2003

Logan TJ (1990) Chemical degradation of soil. Adv Soils Sci. 11:187–221. Springer

Martin CC (1996) Weed control in tropical ley farming systems: a review. Aust J Exp Agric 36: 1013–1023

Mathew MS, Bentrup G, Kellerman T, Katherine M, Straight R, Ameyaw L (2021) Windbreaks in the United States; A systematic review of producer- reported benefits, challenges, management activities and drivers of adoption. Agric Syst. 187:103032

Mitiku H, Herweg K, Stillhardt B (2006) Sustainable Land Management – A New Approach to Soil and Water Conservation in Ethiopia

Nair PKR (2007) The coming age of agroforestry. J Sci Food Agric 87:1613–1619

Nebiyou M, Muluneh M (2016) A review paper on: the role of agroforestry for rehabilitation of degraded soil. J Biol Agric Healthcare 6:128–135

Nkonya E, Mirzabaev A, Von Braun J (2016) Economics of land degradation and improvement: an introduction and overview. In: Kwon H-Y, Nkonya E, Johson T, Graw V, Kato E, Kihiu E (eds)

Economics of land degradation and improvement – a global assessment for sustainable development. Springer, Cham, pp 1–14
Nuruzzaman Manik SM, Pengilley G, Dean G, Field B, Shabala S, Zhou M (2019) Soil and crop management practices to minimize the impact of waterlogging on crop productivity. Front Plant Sci 10:1–23
Pimentel D, Levitan L (1986) Pesticides: amount applied and amount reaching pest. Bioscience 27: 178–185
Pretty J, Bharucha ZP (2014) Sustainable intensification in agricultural systems. Ann Bot 114: 1571–1596
Qadir M, Oster JD, Schubert S, Noble AD, Sahrawat KL (2007) Phytoremediation of sodic and saline-sodic soils. Adv Agron 96:197–247
Reddy S(2011) Alternate land based farming system in rainfed agriculture. In: Compendium of winter school organized by CRIDA, Hyderabad w.e.f. 10 Feb-02 March, 2011
Robert J, Zomer A, Antonio Trabucco BC, Deborah A, Bossio C, Louis V, Verchot A (2008) Climate change mitigation: a spatial analysis of global land suitability for clean development mechanism afforestation and reforestation. Agriculture, Ecosystems and Environment, World Agroforestry Centre, 126, pp 67–80, Nairobi, Kenya
Saroha J (2017) Soil erosion: causes, extent and management in India. Int J Creative Res Thoughts 5:1321–1330
Schoeneberger MM, Bentrup G, de Gooijer H, Soolanayakamahally R, Sauer T, Brandle J, Zhou X, Current D (2012) Branching out: agroforestry as a climate change mitigation and adaptation for agriculture. J Soil Water Conserv 67:128–136
Suraj B, Reddy CP, Manda V (2001) Scientific management and development of saline land. Addressing the challenges of land degradation: an overview and Indian perspective. In: Proceedings of the national seminar Land resource management for food and environmental security. Soil Conservation Society India, pp 73–83
UNCCD (2003) Community level workshop on best practices in Agroforestry and Soil Conservation, Palapye, Botswana, 2–4 December 2003; Available from Secretariat
White RP, Murray S, Rohweder M (2000) Pilot analysis of global ecosystems: grassland ecosystems. World Resources Institute, Washington, DC
Yaduvanshi N, Setter T, Sharma S, Singh K, Kulshreshtha N (2014) Influence of waterlogging on yield of wheat (Triticum aestivum), redox potentials, and concentrations of microelements in different soils in India and Australia. Soil Res 50:489–499

Chapter 15
Nanofertilizers: A Novel Technology for Enhancing Nutrient Use Efficiency of Crops and a Relevance to Agroforestry

Sarbasree Goswami, Satish Kumar Singh, Sayon Mukherjee, Astha Pandey, Adyasha Priyadarshini, Abhik Patra, Surendra Singh Jatav, Ayush Bahuguna, Asik Dutta, and Gorantla Prathap Reddy

Abstract Nanofertilizers are important in agriculture and agroforestry to boost nutrient use efficiency, lower fertilizer waste, and lower cultivation costs while also enhancing crop growth and yield. Nanofertilizers provide more surface area for different metabolic reactions in the plant which accelerates photosynthesis and increases the amount of dry matter and productivity of crops and trees. Nanofertilizers are particularly useful for precise nutrient management in precision agriculture. According to studies, applying nanofertilizers minimizes soil toxicity, reduces the risk of adverse side effects from overdosing, and increases the nutrient use efficiency. Due to the scarcity of arable land and water, the development of the agricultural sector can only be achieved by improving resource use efficiency through efficient utilization of modern technologies. One of these technologies is nanotechnology, which has the potential to completely transform agricultural systems. Consequently, nanotechnology has a great potential to promote sustainable agriculture, particularly in underdeveloped nations.

Keywords Nanofertilizer · Agriculture · Environment sustainability · Agroforestry

15.1 Introduction

Newer technological advancements are highly entailed to enhance the ever-increasing food demands of human being with a reducing cultivable area making agricultural productivity vulnerable. Sometimes we depend on the use of synthetic resources which may make it easy (Dwivedi et al. 2016). Miscellaneous problems related to agriculture can be addressed using nanotechnology. The demand of using

S. Goswami · S. K. Singh · S. Mukherjee (✉) · A. Pandey · A. Priyadarshini · A. Patra · S. S. Jatav · A. Bahuguna · G. P. Reddy
Institute of Agricultural Sciences, Banaras Hindu University, Varanasi, India

A. Dutta
ICAR-Indian Institute of Pulses Research, Kanpur, India

H. S. Jatav et al. (eds.), *Agroforestry to Combat Global Challenges*, Sustainable Development and Biodiversity 36, https://doi.org/10.1007/978-981-99-7282-1_15

nanotechnology in agriculture field is increasing in trend motivating researchers to engage in quality research work in the nanotechnology field (Shang et al. 2019). Norio Taniguchi first time explained nanotechnology in 1974 at Tokyo Science University. It is basically the technology of reducing or increasing the size of matter in nanoscale. Nanoparticle can be defined as a particle of at least one dimension lower than 100 nm. Size range of the particle is between 1 and 100 nm. There are mainly two approaches for making nanoparticles like bottom-up and top-down approach. Among these bottom-up approaches is making of nanoparticles from atomic structures which is regulated by thermodynamic means such as self-assembly (Ferrari 2005). But top-down approach is making of nanoparticles from macromolecular structures including photolithography, nanomoulding, dip-pen lithography, and nanofluidic (Peppas 2004). Bottom-up approach of making nanoparticle is an easier approach than top-down approach. A nanoparticle has totally different and unique physiochemical properties than its bulk component. These changes can be attributed to altered interactions between particles at reduced size. A nanoparticle easily penetrates the cell wall of plants and animals which is used by nanotechnologists to conduct research at cellular level with a greater effectiveness than conventional methodologies.

Fertilizer is a prime input in agriculture; but overuse of it leads to crucial environmental problems related to chemical pollution. This may lead to reduction of soil fertility and development of barren lands over the long run (Vermeulen et al. 2012). Some of biological species are getting vulnerable and even endangered due to over-application of chemicals in agricultural fields (He et al. 2019). So sustainable use of agricultural inputs with advanced technology is on demand. Introduction of nanotechnology in agricultural sectors expedited site-specific restricted delivery of plant nutrients which facilitates nominal use of fertilizers (Xiao et al. 2013). Beside this, agrochemicals and other plant products used for plant protection might be improved with the use of nanotechnology leading to enhancement of crop yield. Nanosensors are the newer headway in nanotechnology research (Chen et al. 2016) that promote crop productivity by improving the measurements related to fertilizer release patterns depending on crop requirements at peak demand or growth period of crop (Prasad et al. 2017). Thus, nanotechnology as an emerging science may secure the sustainability of intensive agricultural movements towards a "green world".

Nanotechnology can be used for enhancing fertilizer use efficiency besides making it environmentally safe for use (Janmohammadi et al. 2016). Nanofertilizers have higher reactivity due to increased solubility than bulk fertilizers (Naderi and Danesh-Shahraki 2013). Dispersibility of insoluble nutrients by application of nanofertilizers enhances their plant availability by reducing fixation at clay minerals or other adsorption sites. Fertilizers formulated with nanomaterials can be readily bioavailable and prolongs the nutrient supply period in soil facilitating a greater bioavailability (Rameshaiah et al. 2015). Conventional nitrogenous fertilizers only release nitrogen for 500 hours, but nanofertilizers release it till 1000 h (Rahale 2011). Slow release or controlled release fertilizers made by coating with nanomaterials are called as nanocomposites. Application of these nanocomposites is reported safer for germination of wheat seeds, growth, and seedling emergence and assists delivery of nutrients when a plant requires it more (Zhang et al. 2006).

Around the world, nanotechnology is used in the agriculture field, health, and environment studies, among many others. Researchers have found that using nanotechnology to address climate change issues may be advantageous for agroforestry. By creating new products and features in plants using nano-engineered catalysts, nanotechnology has also been found to be more effective by strengthening the energy efficiency of agroforestry. In the agroforest-based paper industries, its potential has been demonstrated. The development of wood and agriculturally based products using nanotechnology may surprise the agricultural economy (McCrank 2009). The applications to which nanotechnology is put greatly influence the likelihood of disputes over it. However, reviews in terms of applications of nanotechnology in agroforestry are still limited despite its importance and potential in agroforestry management.

Nanotechnology-enabled sensors have been identified by researchers as a way for the agroforestry sector to deal with climate change issues (Yadav et al. 2014). Water resource management has shown the potential of systems that are nanotechnology enabled. By creating new products and features in the plants using nano-engineered catalysts, nanotechnology has also been found to be more effective by strengthening the energy efficiency of agroforests. In the agroforest-based paper industries, nanotechnology has demonstrated its potential. Consider the paper industry, which at presently is concentrating on the development of nanomaterials with new paper features enabled by nanotechnology, such as surface texture, optical resistance, electronic properties, and barriers.

Agroforestry is a crop-plant and tree-dominated community that offers a variety of goods and ecosystem services, including food, fiber, wood, oxygen, water, and the regulation of the earth's chemistry. However, the ratio of supply (forest and resource production) to demand (human demands) is very low, necessitating the use of nanotechnology to maintain the ratio that is fundamentally needed to achieve sustainability. Numerous databases have been created, and numerous studies and investigations have been conducted globally regarding the nanotechnology applications in the agroforestry sector. As a result, the current paper also discussed specifics of opportunities and nanotechnology applications in agroforestry.

15.2 Benefits of Using Nanofertilizers

- Nutrient use efficiency has been enhanced.
- Reduction of overuse of fertilizers and wastage.
- Reduction in cost of cultivation.
- Meet up the crop nutrient demand throughout the crop growth periods.
- Enhancement of crop growth up to a certain level after that reduction of crop growth occurs due to toxicity.
- Requirement of amounts of fertilizers is very less.

- Increase photosynthesis rate, make available more surface area for metabolic reactions, and higher biomass production.
- Crop yield has increased.
- Biotic stress and abiotic stress protection.
- Precise release of nutrients in response to biological and environmental demands.
- Use of nanofertilizers reduces the frequency of fertilizer application.
- High penetration power leads to higher efficiency of fertilizer applications that ultimately leads to increase in crop yield.
- High produce quality.
- Generally, the produce has no side effects.

15.3 Classification of NFs

The classification of nanofertilizers has been presented in Table 15.1.

15.3.1 Zeolite-Based Nanofertilizers

Zeolite has been used long since as a growing medium because of its good physiochemical characters which contains nearly 50 different minerals (Markovich et al., 1994). It has three-dimensional crystalline rigid structures with high pore spaces and CEC due to isomorphous substitution of Si with Al in tetrahedral sheet (Ayan 2001). Nano-zeolite, on the other hand, has hexagonal symmetry with an aperture of about 0.71 nm size leading to cavities of 0.48 × 124 × 1.07 nm size and a Si/Al ratio of around 3.0 (Bruhwiler 2005). Nano-zeolite crystal is made up of cages connected by double six-membered rings that form columns in the C-direction. When these columns are connected, 12-membered rings with free diameters ranging from 0.71 nm to 1.26 nm are formed (Agger et al., 2005). After suitable partial modification, zeolite with a more surface area holds a wider range of positive and negative nutrient ions.

Table 15.1 Classification of nanofertilizers

Classification	Comments
Nanoscale fertilizers	These are made up of nutrient-rich nanoparticles
Nanoscale additives	Traditional fertilizers containing nanoscale additives are referred to as nano-scale additive fertilizers
Nanoscale coatings	Traditional fertilizers, coated or loaded with nanoparticles, are known as nanoscale coating fertilizers

15.3.2 Naturally Occurring Nanoparticles in Soils

Since the existence of life on earth, nanoparticles are there. During evolution of life, exposure to nanoparticles has been encountered by all, and adoption of tolerance mechanism is a part of this (Buffle 2006). Soil organic and inorganic colloids including nano-sized silt and clay particles, dissolved organic carbon, and sesquioxides all are commonly occurring nanoparticles (Nowack and Bucheli 2007). Incomplete or anaerobic combustion of coal, petroleum products, and even plant parts and plant-derived substances present in soils is of nano-sized substances (Maurice and Hochella 2008). Nanoparticles such as fulvic acids, humic acids, proteins, sugars, particulate organic carbon, and even some viruses are diligently entangled in biological activity. Fullerene, carbon nanoparticles are detected entrapped under the ice core for 10,000 years (Murr et al. 2004). Discrete nanoparticles are scanty to find in soils. Organic nanoparticles occur as coatings on inorganic colloids and minerals (Chorover et al. 2007). Thus, separation and subsequent collection of nanoparticles from soil are generally impractical (Banfield and Zhang 2001). Nanoparticles have large surface to volume ratio which makes it very reactive and can be used as an effective material for carbon sequestration purpose (Khedr et al. 2006). Nanoparticles can affect nutrient transport, pollutant adsorption, and fixation of elements and organic molecules (Mani and Mondal 2016). Weathering of primary minerals also produces nanoparticles like allophane, silicate clays, and amorphous oxides (Nowack and Bucheli 2007). Production of nanoparticles by microorganisms through the use of redox metals by metabolic pathways is another promising source of natural nanoparticles. Several bacteria and fungi produce nanoparticles of iron, Zn, ZnS, Si, Al, and oxy-hydroxides of these elements (Maurice and Hochella 2008). Moreover, presence of montmorillonite nano-clay in soil alters the shrinking and swelling behavior than its bulk counterparts (Haack et al. 2008). One of the most promising facts about natural nanoparticles is that when the particle size reduced from 6 nm to 2 nm, distortion of some particle sites has been noticed (Michel et al. 2007).

15.4 Synthesis/Production of Nanoparticles

15.4.1 Top-Down Methods

The bulk material is often broken down into its corresponding nanosized structures or particles using the top-down technique. These methods are an extension of those that have been utilized to create particles with a diameter of less than a micron. This technique needs substrates, like zeolites or other nanomaterials, that have been ball-grounded for hours of time in order to acquire the nanodimension. In order to prevent agglomeration of the product's heterogeneous nanoparticles, stabilizing agents like polymers or surfactants must be added. NPs have an affinity for anions, which can be used to effectively load the anionic nutrients in the produced nanoparticles for using them as slow delivery/released fertilizers.

15.4.2 Bottom-Up Methods

The term "bottom-up approach" describes the process of building up material from the bottom up, such as at atom level, molecule level, or cluster level (Fig. 15.1). It suggests that some chemical processes start with molecules in a solution and progress through molecular association to produce nanoparticles (NPs). The size of the particle is regulated by chemically controlled process. Emulsion, co-precipitation, micelle formation, and reverse micelle formation are examples of regulated synthesis methods for NP that place a priority on minimizing aggregation or clotting or coagulation and producing homogeneous nanoparticles. To ascertain their functionality, such as solubility, dispersion, and stability, they must first be synthesized and then physiochemically and mechanically evaluated.

15.4.3 Hybrid Nanofibers

Hybrid nanofibers are composed of an organic matrix (typically a polymer) and a scattered inorganic phase in the form of uniformly dispersed nanoparticles (NFs). In *Abelmoschus esculentus*, Tarafder et al. (2020) showed that hybrid NFs might release slowly up to 14 days. They created hydroxyapatite that has been urea-modified since it is a source of calcium, phosphate, and nitrogen. The modified hydroxyapatite might potentially be combined with nanoparticles of Cu, Zn, and Fe, so that they may be able to enhance the absorption amount of these mineral elements in plant significantly.

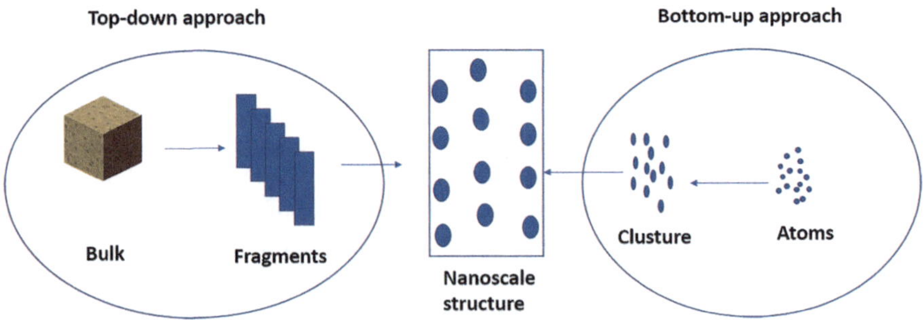

Fig. 15.1 Top-down and bottom-up approach of synthesis of nanoscaled particles

15.4.4 *Biogenic Synthesis*

In fact, the term "Biogenic synthesis or green synthesis" refers to the creation of nanoparticles (NPs) through sustainable, non-toxic, and environmentally friendly processes. The goal of this strategy is to reduce the use of potentially harmful chemicals and energy-intensive processes frequently connected to conventional nanoparticle synthesis. In biogenic synthesis or green synthesis, metal and metal oxide nanoparticles are produced using natural resources like plant extracts, actinobacteria, fungi, bacteria, and algae. These biological agents contain a variety of bioactive substances that can function as reducing agents, stabilizers, and capping agents for nanoparticle formation, including phenolic compounds, enzymes, proteins, and polysaccharides. Biological synthesis and application of nanofertilizer in agriculture have been presented in Fig. 15.2.

15.5 Methods of Application of Nanofertilizers

15.5.1 *Foliar Application*

In foliar application of nanofertilizers, trichomes, stomata, stigma, and hydathodes absorb NPs, which are then distributed throughout the plant by the phloem and xylem. Preparation of different types of nanofertilizers has been presented in Fig. 15.3.

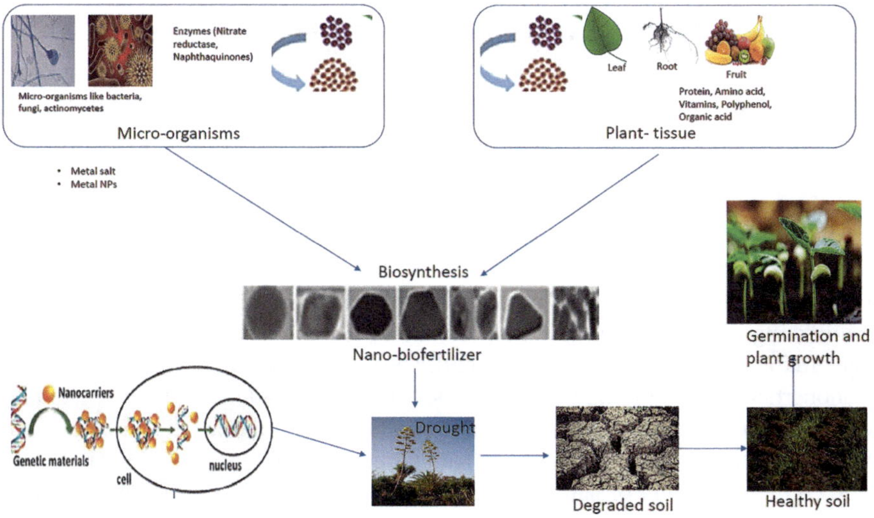

Fig. 15.2 Biological synthesis and application of nanofertilizer in agriculture

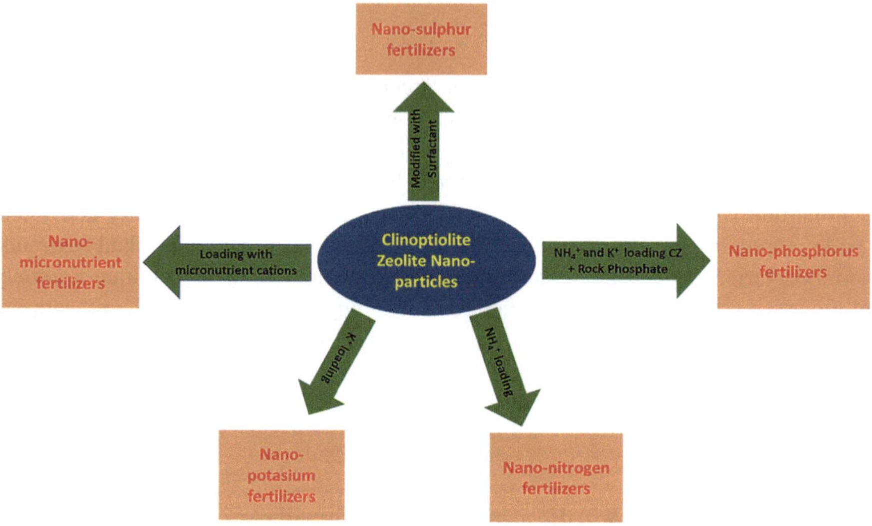

Fig. 15.3 Preparation of different types of nanofertilizers

15.5.2 Soil Application

NPs enter the root's epidermis, traverse its endodermis, and then enter the xylem, where they are carried to the plant's aerial portion. When they are between 3 and 8 nm in size, NPs infiltrate the cell wall through pores.

15.6 Benefits of Foliar Application of Nanofertilizers

After green revolution intensive agricultural practices encountered boost-up in crop productivity which enhanced the demand of fertilizers. With time a need was felt to save the environment from pollution from excess use of fertilizers by enhancing their use efficiency. In this respect, foliar application reveals better performance than soil application of fertilizers due to restriction of nutrient availability by fixation in soil, runoff, leaching, transformation to insoluble forms, and microbial immobilization. Foliar application is used for correcting the deficiency of nutrients in a fastest way by minimizing environmental pollution (Romheld 1999). Nano-formulation improves penetration of substances through cuticular pores and stomata (Eichert et al. 2008). Nanoparticles move through phloem vessels from application site to the inner parts of plants. It may also be transported through aquaporin, ion channels, or by complex formation with transporter proteins (Kurepa et al. 2010). Many researchers have studied the movement of nanoparticles in plant system. For example, movement of calcium oxide nanoparticle occurred through phloem tissue in groundnut plant documented by Deepa et al. (2015). Phloem movement of nanoparticles has also

Table 15.2 Impact of foliar application of nanofertilizers on different crop plants and trees

S. No.	Crop	Nanofertilizer	Impact on crop	References
1	Wheat	Nano-chitosan NPK	Enhancement of leaf growth, yield	Abdel-Aziz et al. (2018)
2	Peppermint	Nano-iron	Height of plant, dry weight, yield increment	Rostami et al. (2017)
3	Pearl millet	Nano-zinc	Root and shoot length, dry biomass, yield increment	Tarafder et al. (2020)
4	Cotton	Nano-zinc	Plant height, fresh and dry weight, chlorophyll content enhancement, and higher activity of antioxidants	Rezaei and Abbasi (2014)
5	*Satureja hortensis* L.	Nano-zinc	Content of chlorophyll and essential oil increases	Vafa et al. (2015)
6	Sunflower	Nano-ZnS	Oil content of seeds, plant growth, and yield increased	Singh & Kumar (2017)
7	Groundnut	Nano-zinc oxide	Enhancement of pod yield	Prasad et al. (2012)
8	Rice	Nano-zinc	Relative leaf water content, shoot and root growth, yield enhancement	Upadhyaya et al. (2015)
9	Wheat	Nano Zn-Fe oxide	Enhancement of total chlorophyll content, soluble sugars, enzymatic activities, and yield	Babaei et al. (2017)
10	Maize	Nano-Zn	Increment of test weight, harvest index, and chlorophyll content	Farnia et al. (2015)
11	*Moringa peregrina*	Nano-ZnO and Nano-Fe_3O_4	Number of leaves per plant, dry weight, sugars, proteins, enzyme activity enhancement	Amira et al. (2015)
12	*Cucurbita pepo L.*	Nano-SiO_2	Reduction of chlorophyll degradation, H_2O_2 level, and enhancement of photosynthesis, water use efficiency, and stomatal conductance	Siddiqui et al. (2014)
13	Cucumber	Nano-SiO_2	Maintenance of turgidity, elasticity, strength of cell wall	Yassen et al. (2017)

been established by Abdel-Aziz et al. (2018) in wheat and Raliya et al. (2015a) in watermelon. Impact of foliar application of nanofertilizers on different crop plants and trees has been presented in Table 15.2.

15.7 Nano-wastes

Nanoparticles produced by engineering methods create waste which are extremely toxic to biota and are commonly called as nano-waste (Table 15.3). They enter the aquatic environment from human use and ultimately come to food chain and lead to

Table 15.3 Classification of nano-waste

Class	Toxicity	Characteristics
I	Very low toxic	No specific requirements for disposal
II	Harmful – toxic	Waste management at its best during the handling, transportation, and disposal processes
III	Toxic – very toxic	Proper waste management protocols
IV	Toxic – very toxic	Should be disposed of at specialized, approved sites for hazardous waste
V	Very toxic – extremely toxic	Only at authorized areas for specific hazardous waste streams

bioaccumulation of these wastes which is very harmful to human being. So, management of these is highly needed to reduce human health risk (Musee 2011). These can be divided into five classes based on toxicity level (Mishra et al. 2020).

15.8 Different Types of Nanofertilizers and Their Uses in Agricultural Crop Production

15.8.1 Nitrogen Nanofertilizer

Nitrogen is the key ingredient of chlorophyll pigment, which is essential for the photosynthesis process, and N is also a vital component of different plant enzymatic proteins which regulate plant metabolic activities (Sinfield et al. 2010). As the use efficiency of nitrogen by the crops is very low due to different losses of nitrogen, it is very important to supply the nutrient when the crop needs it to minimize the loss. Application of nanotechnology can release N according to the crops' need and can increase N use efficiency (Naderi and Danesh-Shahraki 2013; Suman et al. 2010). Zeolite chips loaded with urea (Millan et al. 2008) and nanocomposite bearing N (Jinghua 2004) can be used as a slow N releaser and efficiently increase uptake by the plants. Application of clinoptilolite zeolite (CZ) can reduce the NO_3^- and NH_4^+ concentration in the leachate as it increases the surface area of soil and cation exchange capacity (Huang and Petrovic 1994), and the retained ammonium is generally helpful to slow release (Kithome et al. 1998). It also decreases mission of NH_3 from manures (Amon et al. 1997). It is observed that application of ammonium sulfate $(NH_4)_2SO_4$ when loaded into CZ can minimize nitrate leaching and increase N use efficiency of crops in sandy soils as it inhibits nitrification of ammonium to nitrate resulting in reduced nitrate leaching (Perrin et al. 1998). In addition to CZ, zeolite can decrease ammonia volatilization by securing ammonium-nitrogen upon exchanging cites (Lefcourt and Meisinger 2001a, b). Ammonia volatilization was reduced by 50% with the addition of 6.25% zeolite. As it possessed extensive surface area, mixing it with conventional nitrogenous fertilizers can

Fig. 15.4 Urea nanofertilizer introduced by IFFCO

reduce the loss of nitrogen. Urea nanofertilizer introduced by IFFCO is presented in Fig. 15.4.

15.8.2 Phosphorus Nanofertilizer

Several studies revealed that as compared to conventional fertilizer, P supplied through nanofertilizer remains available in soil for the plants for a long time. The release of P in the soil is not significantly influenced by the nanofertilizer, according to regression analysis between the treatments. Bansiwal et al. (2006) outlined the release of P from surface-modified zeolite (SMZ) continuing even after 1080 h of percolation study which shows its potential application as a slow-release P fertilizer. Rahale (2011) examined the release pattern of PO_4^{3-} from surface-modified zeolite in a percolation reactor and discovered that while traditional fertilizer releases nutrients for only up to 10–12 days, nano-formulations release phosphate for a longer duration of 40–50 days.

15.8.3 Potassium Nanofertilizer

K was released from nano-zeolite gradually and steadily, according to Zhou and Huang (2007). It might be as a result of zeolites' ability to exchange some nutritional cations for other ions. Rezaei and Movahedinaeini (2009) found that as the equilibrium K concentration increases, more potassium becomes adsorbed on the zeolite. Rahale (2011) noted that although soil potassium fixation and dynamic equilibrium together maintain soil potassium availability, nanotechnology can further enhance the availability and controlled release of nutrients. Compared to commercial fertilizer, which immediately released a significant amount of nutrients followed by a release of low and uneven quantities until day 30, the nanofertilizer demonstrated an early burst and a subsequent sluggish release even after 60 days (Fujinuma and Balster 2010). Potassium nanofertilizer enhances the absorption of nutrients like nitrogen, phosphorus, potassium, calcium, and magnesium by plants. Nano-K fertilizer can also have an immediate positive impact on plant development. Although K nanofertilizer treatment stimulates both shoot and root growth, the effect is more pronounced on the roots, leading to greater root system effectiveness. Particularly in alkaline-calcareous soils and low soils, K nanofertilizer efficiently increases soil fertility and crop yield (Rajaei 2010).

15.8.4 Sulfur Nanofertilizer

Studies conducted by Li and Zhang (2010) indicated that using surfactant-modified zeolite (SMZ) as fertilizer additives can function well as a sulfate carrier. Sulfate can be released slowly and leaching can be reduced as a result. Release rate of SO_4^{2-} can be decreased by 5–7 folds when SMZ is used. Different studies showed that application of nano-based sulfur fertilizer increased agricultural yield factors and crop growth characteristics. The controlled or gradual release of nutrients is regulated by nano-sulfur, which also improves plant uptake and nutrient utilization (Hochella et al. 2008). The nano-based S is a powerful fungicide that can be used to treat powdery mildew on okra by blocking the germination of conidial spores (*Erysiphe cichoracearum*). According to Subramanian et al. (2022), whereas the sulfur release from gypsum, a common form of sulfur fertilizer, stopped after 35 days, the release from nano-sulfur has continued for 42 days. Plants fed with nano-sulfur produced considerably more dry matter (11–12%), seed yield (15%), and oil (14.7%) than plants given regular gypsum fertilizer.

15.8.5 Calcium Nanofertilizer

In comparison to applying calcium alone, according to research by Xiumei et al. (2005), adding nano-$CaCO_3$ to organic manures and humic acid significantly accelerated groundnut crop growth and its development. In comparison to nitric acid-calcium, the combination of nano-$CaCO_3$ and humus acid improved the absorption of nutritional elements like calcium, nitrogen, and phosphorus. The nutrient content in the shoot and root of groundnut increased by 0.72% and 0.32%, 1.3% and 0.43%,

0.08% and 0.04%, and 0.49% and 0.01%, respectively. Different applications of calcium and potassium nano-chelate fertilizer increased the production of sweet basil as compared to the control as reported by Ghahramani et al. (2013).

15.8.6 Magnesium Nanofertilizer

Magnesium plays a pivotal role in different plant enzymatic activities like RNA polymerases, ATPases, phosphatases, and protein kinases (Shaul 2002). As it is central in the chlorophyll molecule, it plays an important role in photosynthesis (Scott and Robson 1990). According to Khordadi Varamin et al. (2020), foliar application of nano-Mg and chitosan fertilizers enhances the total chlorophyll production, grain yield, oil content, and also the production of sugars and proline content while reducing the activity of the enzymes like catalase, peroxidase, and ascorbate peroxidase. This may increase sesame yield under water stressed conditions. The nitrate reductase enzyme's activity is increased by the foliar application of Mg nanoparticles. According to Salcido-MartíNez et al. (2020), the foliar application of nano-magnesium improves the size, structure, and function of chloroplasts, along with the transmission of electrons in photosystem II in addition to a favorable light absorption and photosynthetic activities and an increased release of enzymes to achieve the mobilization of nutrients, resulting in a better uptake of these and a consequently higher yield. The number of opening bolls per plant, plant yield (58 g/ plant), and cotton seed yield (1729 kg/ha) was all significantly higher after foliar treatment of 60 ppm concentration of 50 nm size MgO nanoparticles (NPs) (Kanjana 2020). This also increased the uptake and accumulation of other macronutrients in cotton plants.

15.8.7 Zinc Nanofertilizer

According to Slaton et al. (2005), plants absorb more zinc when ZnO particles are smaller. This is due to nano-particulate ZnO's small size, high specific surface area, and greater reactivity when compared to bulk ZnO. According to several researches, plant grains treated with ZnO NP had significantly greater total N contents and decreased crop water stress indices. Nano-slow-releasing fertilizer's qualities enhance several physiological traits of plants and grain nutritional characteristics; as a result, crops and plants benefit from its application. ZnO NP application has been shown by Rizwan et al. (2019a) to have favorable effects on the physiological, qualitative, and quantitative parameters of *Zea mays* L. (maize) and *Triticum aestivum* L. (wheat). Plants' carbohydrate, oil, and protein content has also been seen to rise (Schmidt et al. 2016; Matzen et al. 2019). To boost the pearl millet's resistance to plant fungal infection, the ZnO NPs increased the defense enzyme activity like polyphenol oxidase, phenylalanine ammonia-lyase, lipoxygenase, and peroxidase. Additionally, Rizwan et al. (2019a, b) hypothesized that maize and wheat treated with ZnO NP had larger chlorophyll contents, which increase photosynthetic efficiency and can raise starch, oil, total protein, and dry mass constituents (Bellesi et al. 2019; Schmidt et al. 2016; Matzen et al. 2019). García-López et al. (2018) reported that foliar fertilization with ZnO NPs dramatically boosted the capsaicin content of pepper fruit.

15.8.8 Boron Nanofertilizer

The use of nano-boron resulted in the production of the greatest number of perfect flowers, as shown by Abbasi et al. (2012) and Perica et al. (2001). This resulted from more boron being available for various metabolic processes taking place in plant cells. Plants can more effectively use nutrients when using a nano-chelated mix. Additionally, it has been claimed that nanofertilizers have greater solubility and greater reactivity than their bulk equivalents (Naderi and Danesh-Shahraki 2013). With the use of nano-B, vegetative and reproductive growths were balanced. This might be as a result of boron's function in cell elongation and division, nitrogen and carbohydrate metabolism, sugar transport, and indole acetic acid production. The quality of fruit crops is improved better by applying identical concentrations of boron in nano-chelated form as compared to boric acid. The slow-release capability of B nanofertilizers during the growth stage of fruit crops can facilitate the flow of nutrients into the fruit mesocarp, thereby satisfying the needs for cell division and cell enlargement.

15.8.9 Copper Nanofertilizer

Treatments with copper nanoparticles have been shown to greatly raise the flavonoid content of basil plants. It can be inferred that Cu NPs must be applied topically to basil in order to improve its quantity and quality (Abbasifar et al. 2020). Different studies showed that copper nanoparticle fertilization significantly increases chlorophyll content in leaf. Application of nano-Cu results in enhanced nutrient consumption, less soil toxicity, and minimal adverse consequences from overfertilization. Shah and Belozerova (2009) demonstrated that application of copper nanoparticles at 130 and 600 mg/kg significantly increased the growth of lettuce seedlings by 40 and 91%, respectively. However, at greater concentrations (1000 mg/L), it has been reported that the Cu nanoparticles were hazardous and stunted the growth of bean (*Phaseolus vulgaris* L.), wheat (*Triticum aestivum*), and yellow zucchini seedlings (Lee et al. 2008) and Cucurbita pepo (Musante and White 2012).

15.8.10 Iron Nanofertilizer

A study reported by Liu et al. (2005) suggests that nano-Fe_2O_3 improves the growth and photosynthesis of groundnut. According to Sheykhbaglou et al. (2010), soybean yield and pod and leaf dry weight rose due to nano-iron oxide. Liu et al. (2005) also reported that application of nano-Fe_2O_3 significantly enhanced chlorophyll pigment content in groundnut leaves. Activity of the enzyme catalase was observed to be the highest by application of nano-iron oxide (Ghafari and Razmjoo 2013). A study by Ghodsi et al. (2012) reported that nano-Fe_2O_3 increased plant height and seed yield of sunflower. Iron chelate nanofertilizers have been recognized as a beneficial source of bivalent iron for crops due to their excellent stability and ability to release iron gradually over a wide pH range. These fertilizers enhance the ratio of ferrous iron to ferric iron on the chelate surface, leading to an increase in chlorophyll production in plants (Hokmabadi et al. 2006). In a study by Burger et al. (2007), the effects of nano-iron chelate fertilizer on the qualitative and quantitative characteristics of cut flowers were investigated. The researchers found that treatments with 1 and 1.5 g/L of nanofertilizer had a positive and significant impact. This suggests that the application of nano-iron chelate fertilizers can enhance the growth and development of cut flowers, potentially leading to improved quality and yield.

15.8.11 Manganese Nanofertilizer

According to Pradhan et al. (2013), nano-Mn treatment increased mung bean root growth at about 52% and shoot growth by 38%. It also increased eggplant yield by 22% (Elmer and White 2016). A soil study reported that foliar nano-manganese (Mn) (0.1–1 mg/L) spray boosted tomato fruit output by 6.2% under Fusarium disease stress. However, the nano-formulation was 6% less effective than the bulk form at increasing biomass yield in disease-infested conditions (Elmer et al. 2018). Studies have indicated that nano-manganese (Mn) is superior to other Mn forms for enhancing wheat grain yield. Nano-manganese (Mn) foliar application increased soil and shoot P, decreased soil nitrate N, and increased shoot and grain Mn concentrations. Thus, foliar application of nano-manganese (Mn) may be a tactics to control its nebulous effects in soil (Dimkpa et al. 2018).

15.8.12 Molybdenum Nanofertilizer

According to Gad and Kandil (2013), applying molybdenum nanofertilizer can improve yield by up to 39.8% while reducing the amount of nitrogenous fertilizer used by up to 25%. According to Preetha and Balakrishnan (2017), both the nitrogen cycle and metabolism are closely tied to Mo. This nutrient's availability must be adequate to improve yield.

15.9 Nanotechnology Applications in Agroforestry

The invasion of alien species, mostly in the form of weeds, puts agroforestry at risk. It is difficult to control these invasives at the agroforest ecosystem level. The use of nanoherbicides to manage these on a large scale has only been suggested by a small number of studies (Chinnamuthu and Boopathi 2009). The risk of forest fires is rising due to the thickness of the population, the slow spread of cities, attacks on the wildland-urban interface (WUI), and changes in land use plans that conflict with societal and environmental assurance. These problems are getting worse as a result of rising temperatures and shifting climatic patterns. The Fire Mash, a proven nanotechnology solution, emerges in wildlands destroyed by forest fires with quick flame concealment, complete smothering, and a quick biological rebuilding impact. By holding tight vertical surfaces, it enables the surface in direct contact with the fire front to be secured against the fire. The fire mash can put out a fire by eliminating both the chemical reaction and oxygen delivery to the fire. It can reduce the oxygen content on the contact surface to 8%. It consists of recyclable ultra-fine grade fiber made from lignocellulosic pulp that was removed from trees using environmentally friendly and sustainable tree cultivation techniques.

15.9.1 Nanotechnology Applications Regarding Stress Management in Agroforestry

Global changes, such as global warming, abiotic factors, edaphic factors, or pathological factors, can all be stressors for agroforestry (Sharma et al. 2017). Nanosensors can detect minute quantities of microbes, humidity, and toxic pollutants. These methods have wide applications in agriculture field to increase crop production, productivity, and disease resistance, among other things (Baruah and Dutta 2009). Utilization of nanotechnology to detect signal stress factors in plants, including nitric oxide (NO), reactive oxygen species (ROS), calcium (Ca^{2+}), methyl salicylate, sucrose, glucose, and abscisic acid (ABA) might be one of the promising approaches. For the purpose of pathogen control or disease detection, these studies can be expanded to agroforest ecosystems. Additionally, site-specific gene transfer and expression for desired products/characters are made easier by nanotechnology, which shortens the time needed to transfer genes from alien organisms. The ability of engineered nanomaterials (ENM), particularly nanopolymer-coated seeds, to withstand water stress and germinate under favorable conditions has been reported by researchers. The process of "greening" desert regions with very little rainfall and water availability may be helpful (Giraldo et al. 2019).

Nanobiosensors based on nanotechnology have the capacity to detect early signs of stress due to plant disease, soil moisture, stress from a lack of nutrient resources, etc. Researchers are working to create a microsystem based on nanobiosensors that will make it easier to provide individual trees with the right amount and timing of watering. This kind of system is useful for protecting against pathogens like nematodes and conserving water resources. Handheld instruments can provide a variety of data that can be combined to produce hyperspectral measurements that can reveal details about the levels of chlorophyll, nitrogen, and plant diseases

(Singh and Singh 2018).

15.9.2 Nanotechnology Applications in Monitoring Agroforestry

For the management of agroforests, agroforestry monitoring is important and essential in terms of coverage, health, and ecological services. Quartz crystal microbalance (QCM) devices, carbon nanotubes (CNTs), surface plasmon resonance (SPR) sensors, ion-sensitive sensors (ISEs), and other nanotechnology-enabled biosensor systems may offer the opportunity to measure and estimate parameters such as gaseous exchange, water requirement, and other factors that may be used for accounting ecological services provided by agroforestry at the micro-level (Kaushik et al. 2015). These micro-level data are useful for researching how climate change affects vegetation and the agroforest system (Yadav et al. 2014). The technology might also be useful for creating sensors for satellite and space technology applications that allow for remote agroforest monitoring. In order to monitor forests and risks associated with them, such as the effect of air pollution on forests, pathogen attack, etc., rapidly and on a large scale, it is now essential to use satellite and space technology. Planners can create scientific management plans for the agroforests using current information about the health of the forest (Singh and Singh 2019). Materials created using nanotechnology are used to create sensors that measure how the earth's features react to air pollutants, light, moisture, etc. (Wendt and Potkonjak 2011). These sensors could be installed in satellite platforms or spacecraft or stations, giving them the ability to estimate trace gases and pollutants and the impact they have on the health of the environment and the crew. Additionally, these can be used to determine which species in the forest are most tolerant of high levels of air pollution. The species that were screened in the study mentioned can also be utilized for creating green spaces and mitigating urban air pollution (Sharma and Sharma 2018). These plants can contribute to improving air quality by absorbing pollutants and releasing oxygen through processes such as photosynthesis. In addition, hyperspectral data obtained from instruments like Hyperion-EO1, Advanced Visible Infra-Red Sensor (AVIRS), and their newer generations have proven valuable in assessing the health of forests. These data provide information on various indicators such as plant diseases, nitrogen content, and chlorophyll content, which are crucial for monitoring the overall health and vitality of forest ecosystems (Singh and Singh 2018). Additionally, these sensors can provide species distribution over a sizable area, which can be used to manage the forest at the species level (Singh and Singh 2019). Nanomaterials may be used to enhance these sensors for more precise measurements, but much more research is needed in this area. The forest ecology, its evaluation, potential risks to forest health, and related studies are also included in this application area for nanotechnology in forestry.

15.9.3 Nanotechnology Applications in Agroforestry-Based Wood Products (Paper and Pulp) Industry

As we all know that the main product of the forest is paper and pulp. Since a very long time ago, humans have made extensive use of paper in their daily lives. Important agreements, the Vedas, and scientific manuscripts are regarded as valuable resources for the advancement of humanity. The storage and preservation of paper from moisture, dust, and biological agents, however, are a serious problem because of its hydrophilic nature, fibrous architecture, highly porous nature, ultraviolet radiation-based degradation, microbial assault, and high-water vapor transmission rate (Richardson and Grubb 2013). In this regard, nanotechnology holds significant promise and may offer a special chance to preserve and safeguard paper from deterioration caused by factors like moisture, light, temperature, dust, and biological agents. Researchers and scientists have used nanostructured materials to create durable paper. Each type of nanomaterial has a unique set of mechanisms. Some main and important uses of nanotechnology in the paper manufacturing sector include (a) the creation of new materials, (b) the use of nanofiltration to stop water circulation, (c) the creation of coating materials, and (d) the production of nanoscale assemblers (Mohieldin et al. 2011). It has been discovered that nano-engineered fiber compounds and materials hold promise for producing paper materials with exceptional strength. There are four different processes for creating nanoscale cellulose fibers: (1) bacterial biosynthesis, (2) microfibrillated cellulose, (3) electrospinning, and (4) synthesis of nanorods, also known as cellulose whiskers (Mohieldin et al. 2011). Due to their abundance, inexhaustibility, nanofibrillar makeup, self-assemble ability into fine-tuned architectures, and ability to be prepared multifunctionally, cellulose and lignocellulose have extraordinary potential as nanoengineered materials (Moon 2008).

15.9.4 Nanomaterials on Growth and Development of Crop Plants and Trees

Nanofertilizers are economical and environmentally responsible agents that support highly effective crop plant nutrition and boost crop plant production. Crop plants receive nutrients from nanofertilizers in three different ways: The nutrient can be protected in one of three ways:

1. Nanoparticles made of nanotubes or nanoporous materials
2. Emulsion or nanoscale particles
3. Thin protective polymer film

Nanofertilizers make nutrients available to plants by releasing them gradually, effectively, and precisely. Zinc oxide (ZnO) nanoparticles have been found to increase peanut yield (*Arachis hypogaea*). Similarly, the use of silicon dioxide (SiO_2) nanoparticles has been shown to enhance plant biomass and the levels of biomolecules such as proteins, phenols, and chlorophyll in maize grains. Additionally, the application of low concentrations of these nanoparticles has been found to promote various growth aspects in different plant species. For hexaploidy wheat, the root growth is improved when exposed to low concentrations of ZnO nanoparticles. In mustard (*Brassica juncea*), black gram (*Phaseolus mungo*), rice (*Oryza sativa*), and tobacco (*Nicotiana tabacum*), the seed germination and seedling growth are enhanced when treated with these nanoparticles. Furthermore, tobacco cell growth has been observed to increase by 16% when exposed to ZnO nanoparticles at low concentrations (USDA 2013).

The usage of nanomaterials (NMs) fosters growth of horticultural crops and their development, much like it does for field crops. Spraying nano-boron on the mango canopy improves overall yield production and improves the chemical characteristics of fruits, which is probably related to the increase in chlorophyll and other vital nutrient elements in the leaves. Also, mango trees that have been sprayed with nano-zinc produce fruits that are heavier, more plentiful, contain more chlorophyll and carotene, and have higher amounts of various nutrient elements like N, P, K, and Zn (Alabdallah et al. 2020). Similar to this, using fertilizers containing nanocontent of boron and zinc raises fruit yields, fruit quality, and raises the proportion of total soluble sugars (TSS) to maturity index, total sugars, and also total phenols in the pomegranate fruits (Ismail et al. 2021).

15.9.5 Nanosensors in Precision Agroforestry

The research and development of nanoscale delivery systems for agricultural chemicals, such as pesticides and fertilizers, are made possible by nanotechnology. The effectiveness of active ingredients is increased, while negative effects on the environment are reduced thanks to nanoencapsulation techniques (Khot et al. 2012).

Agrochemicals are shielded from deterioration, their off-target effects are diminished, and their bioavailability is increased by nanocarrier systems like liposomes or polymeric nanoparticles (Raliya et al. 2018). By enhancing the availability and uptake of nutrients in agroforestry systems, nanotechnology contributes to precision nutrient management. According to Khodakovskaya et al. (2013), controlled release mechanisms for nutrients are provided by nanofertilizers, such as nanoparticle-based formulations, ensuring their availability to plants for an extended period of time. Real-time monitoring of soil nutrient levels by nanosensors integrated into precision agriculture systems enables precise and targeted nutrient applications (Singh et al. 2017). Farmers and forestry professionals can manage microclimate conditions, optimize irrigation practices, and evaluate environmental impacts by using real-time data from nanoscale sensors on these variables (Das et al. 2019).

15.9.6 Agroforestry Systems Promote Food Security Through Several Mechanisms

With the daunting task of nourishing an increasing population while contending with climate change and depleting natural resources, food security is a crucial global issue. By combining trees with the production of agricultural crops and livestock, agroforestry systems provide a sustainable way to improve food security. Multiple advantages of this strategy include increased biodiversity, healthier soil, water conservation, and improved climate resilience. Agroforestry systems can improve agricultural output, earning capacity, and nutrition by utilizing the synergies with trees and crops.

1. **Diversification of Production:** Agroforestry systems enable the growing of a variety of plants, animals, and trees. By increasing the variety of food sources available, this diversification lowers the chance of crop failure and improves dietary diversity. Tree crops, like fruit and nut trees, can increase nutritional value and provide opportunities for income.
2. **Nutrient Cycling and Soil Fertility:** In agroforestry systems, trees support soil fertility and nutrient cycling. In order to reduce the demand for synthetic fertilizers, nitrogen-fixing trees, for instance, can capture nitrogen from the atmosphere and make it accessible to crops. Tree organic matter and leaf litter improve soil fertility, which increases the yield and hardiness of agricultural crops.
3. **Water Management and Conservation:** By lowering soil erosion, enhancing water infiltration, and lowering runoff, agroforestry systems aid in water management. Tree canopies lessen soil surface evaporation, maintaining moisture in the soil for crop growth. Because of their extensive root systems, trees are better able to withstand droughts and have access to more water than crops.
4. **Climate Resilience:** By providing a buffer against extreme weather and improving agricultural systems' adaptability, agroforestry systems support climate resilience. In agroforestry systems, trees act as windbreaks, reduce temperature

extremes, and shield crops from too much sunlight. Additionally, trees absorb carbon, which slows down global warming.

5. **Livelihoods and Income Generation:** Agroforestry systems can give farmers access to additional revenue sources. High-value tree crops like cacao, coffee, or timber species can be added to increase income diversity and stability. Additionally, agroforestry systems can open doors for the processing and selling of tree products with added value, resulting in jobs and bettering rural livelihoods.

The promise of agroforestry to strengthen food security with increased crop productivity, increase soil fertility, and diversify income sources has been highlighted in numerous studies. For instance, a study by Nair et al. (2009) emphasized the significance of agroforestry in strengthening food security in sub-Saharan Africa through the integration of high-value tree crops with staple crops. Another research investigation by Garrity et al. (2010) showed the role of agroforestry in strengthening food security in sub-Saharan Africa.

15.9.7 Role of Nanotechnology in Strengthening Food Security Through its Application in Agroforestry Systems

Through its use in agroforestry systems, nanotechnology has the potential to significantly improve food security. Researchers and practitioners hope to increase crop productivity, lower resource inputs, and lessen environmental impacts by utilizing nanotechnology-based solutions. Here, we will talk about how agroforestry systems can use nanotechnology to promote food security.

1. **Enhancing Nutrient Management:** Innovative methods for effective nutrient management in agroforestry systems are provided by nanotechnology. For instance, nutrients can be formulated into nanofertilizers to release gradually, ensuring that plants have access to them for a long time. According to Khodakovskaya et al. (2013), these nanofertilizers can decrease nutrient losses, increase nutrient use efficiency, and lessen environmental contamination. Additionally, the targeted delivery of bioactive substances like micronutrients and growth regulators to plants can be improved by nanoencapsulation, which enhances the uptake of nutrients and overall performance of crops (Singh et al. 2017).
2. **Precision Delivery of Pesticides and Agrochemicals:** Pesticides and agricultural chemicals can be delivered precisely in agroforestry systems thanks to nanoencapsulation and nanocarrier systems. By increasing the active ingredients' stability, solubility, and controlled release, nano-formulations can increase their effectiveness while using fewer chemicals overall (Khot et al. 2012). By increasing target specificity, minimizing off-target effects, and lowering environmental

contamination, nanoscale delivery systems can also improve pest and disease management (Gogos et al. 2017).
3. **Efficient Water Management:** In order to maximize water use in agroforestry systems, sensors and irrigation systems based on nanotechnology are essential. Nanosensors can measure soil moisture, which allows for precise irrigation planning and decreases water waste (Das et al. 2019). Additionally, soil can retain more water thanks to the use of nanomaterials like hydrogels, which increases drought resistance and water use effectiveness (Khan et al. 2020). Regarding the problems associated with water scarcity in arid regions, nanotechnology also provides potential remedies for water purification and desalination (Kang et al. 2018).
4. **Improving Crop Protection and Stress Tolerance:** Crop protection and tolerance to stress in agroforestry systems could be improved with nanotechnology-based solutions. Plant resistance to a variety of abiotic stresses, such as drought, high temperatures, and salinity, can be improved with the use of nano-formulations of growth regulators for plants (Raliya et al. 2015a, b). In order to prevent pathogens and pests from entering, nanomaterials like nanoparticles and nanocoating can act as physical barriers (Pandey et al. 2019).
5. **Soil Health and Restoration:** In agroforestry systems, nanotechnology aids in managing and restoring soil health. By providing the degradation and disposal of pollutants, nanoremediation techniques can be used to remediate polluted soil (Mukherjee et al. 2018). Additionally, soil parameters like organic matter content, pH, and microbial growth can be monitored by nanosensors, which can help with management decisions and the assessment of soil health (Zhang et al. 2019).

15.10 Nanomaterials for Farmland Restoration

Due to overcultivation, water scarcity, and climate change, the arable and fertile farmlands will become arid. Farmland restoration may benefit from NMs like hydrogels (potassium polyacrylate), nanoclays, nano-zeolites, which can increase soil water retention at about 50–70%, also reduce soil compactness/hardiness at about 8–10% (Sekhon 2014). Application of non-water-soluble polymers like hydrogels in drylands increased crop yield while preventing leaching and improving soil texture, evaporation, and microbial activity. NMs are safe for the environment, do not harm plants, and break down quickly into CO_2, nitrogen (N), and water. In the same way, heat-resistant fertilizers like nano-zeolites improve soil aeration, microbial growth, water-holding capacity, nutrient use efficiency, and avoiding soil contamination by taking and absorbing heavy toxic metals from their parent compounds (Saponaro et al. 2016). The effects of drought stress can raise oxidative stress and lipid peroxidation in plants by causing the production of oxygen radicals. Plants with narrow leaves, stunted growth, affected foliar matrix, lower biomass contents, etc. are effects visible to our naked eyes. Nanoparticles like hydrogel lower the effect of drought on plants, resulting in less stress and oxygen radical production.

15.11 Nanomaterials as Growth Enhancers

Studies suggest that hydrogel usage in agriculture can significantly decrease the requirement of synthetic chemical fertilizers while maintaining crop growth, yield, and nutritional status and its value. In areas with comparable ecological constraints, such as arid and semi-arid climates, it would indeed a more suitable practice when it comes to sustainable agriculture. Additionally, potassium polyacrylate usage is secure and less toxic, protecting agroecosystems from contamination.

15.12 Nanomaterials Induced Biomass Accumulation

Carbon dots range in size from 1 to 10 nm and have fluorescent characteristics. When pH is neutral, carbon dots become excited, absorb a variety of UV light, and typically express a bluish color. These carbon dots absorb sunlight in order to increase photosynthesis, which results in higher biomass synthesis. For instance, single-walled carbon dots increase the rate of electron transfer by 49% by boosting the near-infrared (NIR) fluorescence light-harvesting rate and reduces reactive oxygen species (ROS) in chloroplasts, enhancing photosynthetic efficiency of plants, crop yield, and biomass production (Giraldo et al. 2014). When nano-TiO_2 was used, spinach's photosynthesis rate improved by 3.13 times more (Zheng et al. 2005).

15.13 Conclusion

In the case of conventionally using fertilizers, the efficiency is too low – for nitrogenous, phosphatic, and potassium fertilizers, the efficiency ranges from 20 to 50%, 10 to 25%, and 35 to 40%, respectively. Nanofertilizers benefit the agricultural sector by lowering the amount of conventionally using fertilizers that are currently using and increasing crop yields. In addition to being eco-friendly solutions, minimizing the leaching and volatilization loss, it has a significant economic benefit for growers. In comparison to traditional nutrients, nanonutrients are more effective and affordable. The various kinds of nanofertilizers have a significant effect on crop productivity, the preservation of natural resources, and lowering the expenditure on fertilizers for crop production. By using the correct dosage and concentration, nanofertilizers encourage healthy crop growth and yield. This clearly demonstrates that although nanotechnology has been used extensively in the agricultural sector, there is still a wider spectrum of potential applications in forest sector and agroforestry industry. The nanoparticles mentioned above can have both beneficial and harmful effects. Thus, these could be used in accordance with rules and guidelines from science. Through nanotoxicity, engineered nanoparticles (ENPs) can affect molecular and physiological characteristics in agriculture and agroforestry systems. In order for research to be effective, awareness must be raised in both the nanotechnology and nanotoxicology fields.

15.14 Future Research Perspectives

Further research can be conducted on how NFs are maintained in soil, their behavior in the environment, and their transport pathway. Through metagenomics, the potential consequences of usage of NF on soil micro-organisms can be investigated with reference to soil applications of NF. Evaluating how plants interact with NFs is another intriguing area to research. For a sustainable agriculture, it is important to investigate the biological and biochemical interactions of the NF in the soil as well as the degradation of the NFs. To secure and sustain the expanding research opportunities in the field of applications of nanotechnology in agroforest management, future opportunities in this area lie in significant R&D grants, an increase in the number of technical institutions, and effective research collaborations with well-established labs. The production, use, and effects of various nanoparticles are being tracked by scientists. This is done to strike a balance between the advantages of the technique and any potential negative effects.

References

Abbasi Y, Bakhshi D, Forghani A, Sabouri A, Porghauomy M (2012) Effect of macro and micronutrients sprays on fruit quality and quantity of Zard and Rowghani olive (*Olea europaea* L.) cultivars in northern Iran. Am Eurasia J Agric Environ Sci 12:1548–1552

Abbasifar A, Shahrabadi F, ValizadehKaji B (2020) Effects of green synthesized zinc and copper nano-fertilizers on the morphological and biochemical attributes of basil plant. J Plant Nutr 43(8):1104–1118

Abdel-Aziz HMM, Hasaneen MNA, Omer AM (2018) Foliar application of nano chitosan NPK fertilizer improves the yield of wheat plants grown on two different soils. Egypt J Exp Biol (Bot) 14(1):63–72

Agger JR, Meza LI, Cundy CS, Plaisted RJ, Anderson MW (2005) Control of zeolite surface nucleation. Stud Surf Sci Catal. 158:35–42. Elsevier

Alabdallah NM, Al-Ajlouni MM, Ismail A (2020) Effect of nano-boron foliar application on mango trees. Jordan Journal of Agricultural Sciences 16(4):749–757

Amira SS, Souad AEF, Essam D (2015) Alleviation of salt stress on Moringa peregrina using foliar application of nanofertilizers. J Hortic For 7(2):36–47

Amon M, Dobeic M, Sneath RW, Philips VR, Misselbrook TH, Pain BF (1997) A farm-scale study on the use of clinoptilolite zeolite and deBodorase for reducing odor and ammonia emissions from broiler houses. Bioresour Technol 6:229–237

Ayan S (2001) Utilization of zeolite as plant growing media. East Mediterr For Res Inst J 7:97–111

Babaei K, Seyed Sharifi R, Pirzad A, Khalilzadeh R (2017) Effects of bio fertilizer and nano Zn-Fe oxide on physiological traits, antioxidant enzymes activity and yield of wheat (Triticum aestivum L.) under salinity stress. J Plant Interact 12(1):381–389

Banfield JF, Zhang H (2001) Nanoparticles in the environment. Rev Mineral Geochem 44(1):1–58

Bansiwal AK, Rayalu SS, Labhasetwar NK, Juwarkar AA, Devotta S (2006) Surfactant-modified zeolite as a slow-release fertilizer for phosphorus. J Agric Food Chem 54(13):4773–4779

Baruah S, Dutta J (2009) Nanotechnology applications in pollution sensing and degradation in agriculture: a review. Environ Chem Lett 7(3):191–204

Bellesi FJ, Arata AF, Martínez M, Arrigoni AC, Stenglein SA, Dinolfo MI (2019) Degradation of gluten proteins by Fusarium species and their impact on the grain quality of bread wheat. J Stored Prod Res 83:1–8

Bruhwiler L (2005) Controlling size and morphology of zeolite. Mater Synth 31:9–19

Buffle J (2006) The key role of environmental colloids/nanoparticles for the sustainability of life. Environ Chem 3(3):155–158

Burger J, Gochfeld M, Jeitner C, Gray M, Shukla T, Shukla M, Burke S (2007) Kelp as a bioindicator: does it matter which part of 5 M long plant is used for metal analysis. Environ Monit Assess 128(1–3):311–321

Chen YW, Lee HV, Juan JC, Phang SM (2016) Production of new cellulose nanomaterial from red algae marine biomass *Gelidium elegans*. Carbohydr Polym 151:1210–1219

Chinnamuthu CR, Boopathi PM (2009) Nanotechnology and agroecosystem. Madras Agric J 96(1/6):17–31

Chorover J, Kretzschmar R, Garcia-Pichel F, Sparks DL (2007) Soil biogeochemical processes within the critical zone. Elements 3(5):321–326

Das R, Kanmani P, Rhim JW (2019) Nano-biosensors for sensing and monitoring of pesticide residues in food and environment. Trends Food Sci Technol 83:91–106

Deepa M, Sudhakar P, Nagamadhuri KV, Balakrishna Reddy K, Giridhara Krishna T, Prasad TNVKV (2015) First evidence on phloem transport of nanoscale calcium oxide in groundnut using solution culture technique. Appl Nanosci 5(5):545–551

Dimkpa CO, Singh U, Adisa IO, Bindraban PS, Elmer WH, Gardea-Torresdey JL, White JC (2018) Effects of manganese nanoparticle exposure on nutrient acquisition in wheat (*Triticum aestivum* L). Agronomy 8:158

Dwivedi S, Saquib Q, Al-Khedhairy AA, Musarrat J (2016) Understanding the role of nanomaterials in agriculture. In: Microbial inoculants in sustainable agricultural productivity. Springer, New Delhi, pp 271–288

Eichert T, Kurtz A, Steiner U, Goldbach HE (2008) Size exclusion limits and lateral heterogeneity of the stomatal foliar uptake pathway for aqueous solutes and water-suspended nanoparticles. Physiol Plant 134(1):151–160

Elmer W, White JC (2016) The use of metallic oxide nanoparticles to enhance growth of tomatoes and eggplants in disease infested soil or soilless medium. Environ Sci Nano 3:1072–1079

Elmer W, De La Torre-Roche R, Pagano L, Majumdar S, Zuverza-Mena N, Dimkpa C, Gardea-Torresdey J, White JC (2018) Effect of metalloid and metallic oxide nanoparticles on Fusarium wilt of watermelon. Plant Dis 102:1394–1401

Farnia A, Omidi MM, Farnia A (2015) Effect of nano-zinc chelate and nano-biofertilizer on yield and yield components of maize (*Zea mays* L.), under water stress condition. Indian J Nat Sci. 5(29):4614

Ferrari M (2005) Cancer nanotechnology: opportunities and challenges. Nat Rev Cancer 5(3): 161–171

Fujinuma R, Balster NJ (2010) Controlled-release nitrogen in tree nurseries. Res Commun 2:123–126

Gad N, Kandil H (2013) Evaluate the effect of molybdenum and different nitrogen levels on cowpea (*Vigna unguiculata*). Int J Appl Sci Res Rev 9(3):1490–1497

García-López JI, Zavala-García F, Olivares-Sáenz E, Lira-Saldívar R, Barriga-Castro E, Ruiz-Torres NA, Ramos-Cortez E, Vázquez-Alvarado R, Niño-Medina G (2018) Zinc oxide nanoparticles boosts phenolic compounds and antioxidant activity of Capsicum annuum L. during germination. Agronomy 8:215

Garrity D et al (2010) Evergreen agriculture: a robust approach to sustainable food security in Africa. Food Secur 2(3):197–214

Ghafari H, Razmjoo J (2013) Effect of foliar application of nano-iron oxidase, iron chelate and iron sulphate rates on yield and quality of wheat. Int J Agron Plant Prod. 4(11):2997–3003

Ghahramani A, Akbari K, Yousofpour M (2013) Evaluation the effect of application calcium and potassium nano chelate fertilizers on quantitative and qualitative traits of sweet basil. The First National Conference of Nanotechnology Applications in Industry, Mine, Agriculture and Medicine, Karaj, Research Institute of Materials and Energy. (In Farsi)

Ghodsi A, Astaraei AL, Emami H, Mirzapour MH (2012) Effect of nano iron oxide powder and urban solid waste compost coated sulfur on sunflower yield and yield components in saline-sodic soil. Environ Sci 9(3):111–118

Giraldo JP, Landry MP, Faltermeier SM, McNicholas TP, Iverson NM, Boghossian AA et al (2014) Plant nanobionics approach to augment photosynthesis and biochemical sensing. Nat Mater 13(4):400–408. https://doi.org/10.1038/nmat3890

Giraldo JP, Wu H, Newkirk GM, Kruss S (2019) Nanobiotechnology approaches for engineering smart plant sensors. Nat Nanotechnol 14(6):541–553

Gogos A, Knauer K, Bucheli TD (2017) Nanomaterials in plant protection and fertilization: current state, foreseen applications, and research priorities. J Agric Food Chem 65(40):8725–8740

Haack EA, Johnston C, Maurice PA (2008) Siderophore sorption to montmorillonite. Geochim Cosmochim Acta 72:3381–3397

He X, Deng H, Hwang HM (2019) The current application of nanotechnology in food and agriculture. J Food Drug Anal 27(1):1–21

Hochella MF, Lower SK, Maurice PA, Penn RL, Sahai N, Sparks DL, Twining BS (2008) Nano minerals, mineral nanoparticles, and earth systems. Science 319:1631–1635

Hokmabadi H, Haidarinezad A, Barfeie R, Nazaran M, Ashtian M, Abotalebi A (2006) A New Iron chelate Introduction and Their Effects on Photosynthesis activity, Chlorophyll Content and nutrients Uptake of Pistachio (*Pistacia vera* L.). 27th International Horticultural Congress and Exhibitions Seoul, Korea, August 13–19

Huang ZT, Petrovic AM (1994) Clinoptilolite zeolite influence on nitrate leaching and nitrogen use efficiency in simulated sand-based golf greens. J Environ Qual 23:1190–1194

Ismail A, Alabdallah NM, Al-Ajlouni MM (2021) Effect of nano-boron and nano-zinc foliar spraying on yield and quality of pomegranate (Punica granatum L.) Fruits. Horticulturae 7(3):53

Janmohammadi M, Amanzadeh T, Sabaghnia N, Dashti S (2016) Impact of foliar application of nano micronutrient fertilizers and titanium dioxide nanoparticles on the growth and yield components of barley under supplemental irrigation. Acta Agric Slov 107(2):265–276

Jinghua G (2004) Synchrotron radiation, soft X-ray spectroscopy and nano-materials. Int J Nanotechnol 1:193–225

Kang Y, Liu L, Liu J, Guo H, Shi L (2018) Nanotechnology approaches for water purification: a review. J Environ Manag 222:324–340

Kanjana D (2020) Foliar application of magnesium oxide nanoparticles on nutrient element concentrations, growth, physiological, and yield parameters of cotton. J Plant Nutr 43(20): 3035–3049

Kaushik A, Kumar R, Jayant RD, Nair M (2015) Nanostructured gas sensors for health care: an overview. J Pers Nanomed 1(1):10–23

Khan MMA, Mobin M, Abbas ZK, Al-Shaeri M (2020) Nanotechnology: emerging tool for agriculture. Environ Sci Pollut Res 27(10):10099–10119

Khedr MH, Omar AA, Abdel-Moaty SA (2006) Reduction of carbon dioxide into carbon by freshly reduced CoFe2O4 nanoparticles. Mater Sci Eng 432(1–2):26–33

Khodakovskaya M, Dervishi E, Mahmood M, Xu Y, Li Z, Watanabe F, Biris AS (2013) Carbon nanotubes are able to penetrate plant seed coat and dramatically affect seed germination and plant growth. ACS Nano 7(11):10654–10662

Khot LR, Sankaran S, Maja JM, Ehsani R, Schuster EW (2012) Applications of nanomaterials in agricultural production and crop protection: a review. Crop Prot 35:64–70

Kithome M, Paul J, Lavkulich W (1998) Kinetics of ammonium absorption and desorption by the natural zeolite clinoptilolite. Soil Sci Soc Am J 62:622–623

Kurepa J, Paunesku T, Vogt S, Arora H, Rabatic BM, Lu J, Wanzer MB, Woloschak GE, Smalle JA (2010) Uptake and distribution of ultrasmall anatase TiO_2 Alizarin red S nanoconjugates in Arabidopsis thaliana. Nano Lett 10(7):2296–2302

Lee WM, An YJ, Yoon H, Kweon HS (2008) Toxicity and bioavailability of copper nanoparticles to the terrestrial plants mung bean (*Phaseolus radiatus*) and wheat (*Triticum aestivum*): plant agar test for water-insoluble nanoparticles. Environ Toxicol Chem 27:1915–1921

Lefcourt AM, Meisinger JJ (2001a) Effect of adding alum or zeolite to dairy slurry on ammonia volatilization and chemical composition. J Dairy Sci 84(8):1814–1821

Lefcourt AM, Meisinger JJ (2001b) Effect of adding alum or zeolite to diary slurry on ammonia volatilization and chemical composition. J Dairy Sci 84:1814–1821

Li Z, Zhang Y (2010) Use of surfactant-modified zeolite to carry and slowly release sulfate. Desalin Water Treat 21(1–3):73–78

Liu XM, Zhang FD, Zhang SQ, He XS, Fang R, Feng Z, Wang Y (2005) Effects of nano-ferric oxide on the growth and nutrients absorption of peanut. Plant Nutr Fertil Sci 11:14–18

Mani PK, Mondal S (2016) Agri-nanotechniques for plant availability of nutrients. In: Plant nanotechnology. Springer, Cham, pp 263–303

Markovic V, Takac A, Ilin Z (1994) Enriched zeolite as a substrate component in the production of pepper and tomato seedlings. Hydroponics Transplant Prod 396:321–328

Matzen N, RavnJørgensen J, Holst N, NistrupJørgensen L (2019) Grain quality in wheat-Impact of disease management. Eur J Agron 103:152–164

Maurice PA, Hochella MF (2008) Nanoscale particles and processes: a new dimension in soil science. Adv Agron 100:123–153

McCrank J (2009) Nanotechnology applications in the forest sector. Natural Resources Canada, ISBN: 978-1-100-12065-2. Available on https://cfs.nrcan.gc.ca/pubwarehouse/pdfs/29382.pdf

Michel FM, Ehm L, Antao SM, Lee PL, Chupas PJ, Liu G, Strongin DR, Schoonen MA, Phillips BL, Parise JB (2007) The structure of ferrihydrite, a nanocrystalline material. Science 316(5832):1726–1729

Millan G, Agosto F, Vazquez M (2008) Use of clinoptilolite as a carrier for nitrogen fertilizers in soils of the Pampean regions of Argentina. Ciencia Investig Agrar 35:293–302

Mishra AK, Tiwari A, Arya R, Panchal D (2020) Environmental nanotechnology: global framework and integrative strategies of nanowaste management. In: Handbook of environmental materials management. Springer, Cham, pp 1–31

Mohieldin SD, Zainudin ES, Paridah MT, Ainun ZM (2011) Nanotechnology in pulp and paper industries: a review. In: Key engineering materials, vol 471. Trans Tech Publications, Switzerland, pp 251–256

Moon RJ (2008) Nanomaterials in the forest products industry. In: Mc Grpubnameaw-Hill yearbook of science and technology, 2008. McGraw-Hill, New York, pp 225–228

Mukherjee A, Majumdar S, Servin AD, Pagano L, Dhankher OP, White JC, Peralta-Videa JR (2018) Carbon nanomaterials in agriculture: a critical review. Front Plant Sci 9:145

Murr LE, Soto KF, Esquivel EV, Bang JJ, Guerrero PA, Lopez DA, Ramirez DA (2004) Carbon nanotubes and other fullerene-related nanocrystals in the environment: a TEM study. JOM 56(6):28–31

Musante C, White JC (2012) Toxicity of silver and copper to *Cucurbita pepo*: differential effects of nano and bulk-size particles. Environ Toxicol 27(9):510–517

Musee N (2011) Nanowastes and the environment: Potential new waste management paradigm. Environ Int 37(1):112–128

Naderi MR, Danesh-Shahraki A (2013) Nanofertilizers and their roles in sustainable agriculture. Int J Agric Crop Sci 5(19):2229–2232

Nair PKR et al (2009) Agroforestry as a strategy for carbon sequestration. J Plant Nutr Soil Sci 172(1):10–23

Nowack B, Bucheli TD (2007) Occurrence, behavior and effects of nanoparticles in the environment. Environ Pollut 150(1):5–22
Pandey P, Singh S, Bhatt R, Sharma PK (2019) Role of nanotechnology in plant disease management. In: Nanoscience in food and agriculture, vol 6. Springer, Cham, pp 283–306
Peppas, N. A. (2004). Intelligent therapeutics: biomimetic systems and nanotechnology in drug delivery. Advanced drug delivery reviews, 56(11), 1529-1531.
Perica S, Brown PH, Connell JH, Nyomora AM, Dordas C, Hu H, Stangoulis J (2001) Foliar boron application improves flower fertility and fruit set of olive. Hortic Sci 36:714–716
Perrin TS, Dros DT, Boettinger JL, Norton JM (1998) Ammonium-loaded clinoptilolite: a slow-release nitrogen fertilizer for sweet corn. J Plant Nutr 21:515–530
Pradhan S, Patra P, Das S, Chandra S, Mitra S, Dey KK, Akbar S, Palit P, Goswami A (2013) Photochemical modulation of biosafe manganese nanoparticles on Vigna radiata: a detailed molecular, biochemical, and biophysical study. Environ Sci Technol 47:13122–13131
Prasad TNVKV, Sudhakar P, Sreenivasulu Y, Latha P, Munaswamy V, Reddy KR, Sreeprasad TS, Sajanlal PR, Pradeep T (2012) Effect of nanoscale zinc oxide particles on the germination, growth and yield of peanut. J Plant Nutr 35(6):905–927
Prasad R, Bhattacharyya A, Nguyen QD (2017) Nanotechnology in sustainable agriculture: recent developments, challenges, and perspectives. Front Microbiol 8:1014
Preetha PS, Balakrishnan N (2017) A review of nano fertilizers and their use and functions in soil. Int J Curr Microbiol App Sci 6(12):3117–3133
Rahale S (2011) Nutrient release pattern of nanofertilizer formulation. PhD (Agri.) Thesis. Tamil Nadu Agricultural University, Coimbatore
Rajaei M (2010) Plant nutrition. Text booklet of Islamic Azad University of Jahrom. Khayam Press, West Palm Beach, FL
Raliya R, Nair R, Chavalmane S, Wang WN, Biswas P (2015a) Mechanistic evaluation of translocation and physiological impact of titanium dioxide and zinc oxide nanoparticles on the tomato (Solanum lycopersicum L.) plant. Metallomics 7(12):1584–1594
Raliya R, Tarafdar JC, Biswas P (2015b) Enhancing the mobilization of native phosphorus in the mung bean rhizosphere using ZnO nanoparticles synthesized by soil fungi. Appl Environ Microbiol 81(10):3506–3513
Raliya R, Tarafdar JC, Biswas P (2018) Enhancing the efficiency of uptake and transport of nutrients in plants using nanomaterials. In: Advances in agronomy, vol 148. Academic Press, Cambridge, MA, pp 199–247
Rameshaiah GN, Pallavi J, Shabnam S (2015) Nano fertilizers and nano sensors–an attempt for developing smart agriculture. Int J Eng Res Generic Sci 3(1):314–320
Rezaei M, Abbasi H (2014) Foliar application of nanochelate and non-nanochelate of zinc on plant resistance physiological processes in cotton (*Gossypium hirsutum* L.). Iran J Plant Physiol 4(4): 1137–1144
Rezaei M, Movahedinaeini SAR (2009) Effects of ammonium and Iranian natural zeolite on potassium adsorption and desorption kinetics in the loess soil. Int J Soil Sci 4:1–19
Richardson DE, Grubb MJ (2013) Evaluation of mineral filler interaction with wood extractive colloids in pulp and paper mill process waters using flow cytometry. J For 3(2):16–21
Rizwan M, Ali S, Ali B, Adrees M, Arshad M, Hussain A, Zia ur Rehman M, Waris AA (2019a) Zinc and iron oxide nanoparticles improved the plant growth and reduced the oxidative stress and cadmium concentration in wheat. Chemosphere 214:269–277
Rizwan M, Ali S, Zia ur Rehman M, Adrees M, Arshad M, Qayyum MF, Ali L, Hussain A, Chatha SAS, Imran M (2019b) Alleviation of cadmium accumulation in maize (*Zea mays* L.) by foliar spray of zinc oxide nanoparticles and biochar to contaminated soil. Environ Pollut 248:358–367
Romheld V (1999) Foliar nutrient application: challenge and limits in crop production. In: Proc. of the 2nd International Workshop on Foliar Fertilization, April 4–10, 1999. Bangkok, Thailand, pp 1–34

Rostami M, Movahedi Z, Davari MR, Siahpoosh S (2017) Effect of foliar application of biofertilizer and nano-fertilizers on morpho-physiological characteristics of peppermint (Mentha piperita L.). Tropentag 2017-Future Agriculture: Socialecological Transitions and Bio-cultural shifts
Salcido-MartíNez A, Sanchez E, Licon-Trillo LP, Perez-Alvarez S, Palacio-MáRquez A, Amaya-Olivas NI, Preciado-Rangel P (2020) Impact of the foliar application of magnesium nanofertilizer on physiological and biochemical parameters and yield in green beans. Not Bot Horti Agrobot Cluj-Napoca 48(4):2167
Saponaro S, Sezenna E, Mastorgio A (2016). Pollutants in groundwater. Filtration Mater Groundwater: A Guide Good Pract 1:1–44
Schmidt M, Horstmann S, De Colli L, Danaher M, Speer K, Zannini E, Arendt EK (2016) Impact of fungal contamination of wheat on grain quality criteria. J Cereal Sci 69:95–103
Scott BJ, Robson AD (1990) Distribution of magnesium in subterranean clover (*Trifolium subterranean* L.) in relation to supply. Aust J Agric Res 41:499–510
Sekhon B (2014) Nanotechnology in agri-food production: an overview. Nanotechnol Sci Appl 7: 31. https://doi.org/10.2147/NSA.S39406
Shah V, Belozerova I (2009) Influence of metal nanoparticles on the soil microbial community and germination of lettuce seeds. Water Air Soil Pollut 197:143–148
Shang Y, Hasan MK, Ahammed GJ, Li M, Yin H, Zhou J (2019) Applications of nanotechnology in plant growth and crop protection: a review. Molecules 24(14):2558
Sharma B, Sharma S (2018) Air pollution tolerance of temperate woody vegetation growing along the national highway no 5 from Solan to Shimla Himachal Pradesh. Ph.D. Thesis, submitted to FRI deemed to be University, Dehradun, pp 1–9
Sharma J, Upgupta S, Jayaraman M, Chaturvedi RK, Bala G, Ravindranath NH (2017) Vulnerability of forests in India: a national scale assessment. Environ Manag 60(3):544–553
Shaul O (2002) Magnesium transport and function in plants: the tip of the iceberg. Biometals 15: 309–323
Sheykhbaglou R, Sedghi M, Tajbakhshshishevan M, Sharifi SR (2010) Effects of nano-iron oxide particles on agronomic traits of soybean. Not Sci Biol 2:112–113
Siddiqui MH, Al-Whaibi MH, Faisal M, Al Sahli AA (2014) Nano-silicon dioxide mitigates the adverse effects of salt stress on Cucurbita pepo L. Environ Toxicol Chem 33(11):2429–2437
Sinfield JV, Fagerman D, Colic O (2010) Evaluation of sensing technologies for on-the-go detection of macro-nutrients in cultivated soils. Comput Electron Agric 70:1–18
Singh MD, Kumar BA (2017) Bio efficacy of nano zinc sulphide (ZnS) on growth and yield of sunflower (Helianthus annuus L.) and nutrient status in the soil. Int J Agric Sci. 9:3795
Singh D, Singh S (2018) Geospatial modeling of canopy chlorophyll content using high spectral resolution satellite data in Himalayan Forests. Clim Change Environ Sustain 6(1):20–34
Singh D, Singh S (2019) Mapping of Dominant Tree Species and their Biophysicochemical Characterization using Geospatial Techniques in part of Doon Valley. Ph.D. Thesis, submitted to FRI deemed university, Dehradun, India
Singh S, Mishra S, Singh VP, Tiwari DD (2017) Nanotechnology in agroecosystem: implications on plant biology. In: Nanoscience in food and agriculture, vol 3. Springer, Cham, pp 287–316
Slaton NA, Gbur EE, Wilson CE, Richard NJ (2005) Rice response to granular zinc sources varying in water-soluble zinc. Soil Sci Soc Am J 69:443–452
Subramanian KS, Rajeswari R, Yuvaraj M, Pradeep D, Guna M, Yoganathan G (2022) Synthesis and characterization of nano-sulfur and its impact on growth, yield, and quality of sunflower (*Helianthus annuus* L.). Commun Soil Sci Plant Anal. https://doi.org/10.1080/00103624.2022.2072867
Suman PR, Jain VK, Varma A (2010) Role of nanomaterials in symbiotic fungus growth enhancement. Curr Sci 99:1189–1191
Tarafder C, Daizy M, Alam MM, Ali MR, Islam MJ, Islam R, Khan MZH (2020) Formulation of a hybrid nanofertilizer for slow and sustainable release of micronutrients. ACS Omega 5(37): 23960–23966

Upadhyaya H, Shome S, Tewari S, Bhattacharya MK, Panda SK (2015) Effect of Zn nano-particles on growth responses of rice. In: Singh B, Kaushik A, Mehta SK, Tripathi SK (eds) Nanotechnology: Novel perspectives and prospects, Studium LLC: Houston, TX, pp 508–512

USDA (2013) Joint research done by Southern University, Louisiana State University and the US forest Department resulted in to the development of a new nano-particle Copper Carbon Core–Shell Nanoparticle (CCCSNs)

Vafa ZN, Sirousmehr AR, Ghanbari A, Khammari I, Falahi N (2015) Effects of nano zinc and humic acid on quantitative and qualitative characteristics of savory (*Satureja hortensis* L.). Int J Biosci 6(3):124–136

Varamin JK, Fanoodi F, Sinaki JM, Rezvan S, Damavandi A (2020) Foliar application of chitosan and nano-magnesium fertilizers influence on seed yield, oil content, photosynthetic pigments, antioxidant enzyme activities of sesame (*Sesamum indicum* L.) under water-limited conditions. Not Bot Horti Agrobot Cluj-Napoca 48(4):2228–2243. https://doi.org/10.15835/nbha48411852

Vermeulen SJ, Aggarwal PK, Ainslie A, Angelone C, Campbell BM, Challinor AJ, Hansen JW, Ingram JS, Jarvis A, Kristjanson P, Lau C, Wollenberg E (2012) Options for support to agriculture and food security under climate change. Environ Sci Pol 15(1):136–144

Wendt JB, Potkonjak M (2011) Nanotechnology-based trusted remote sensing. Sensors 2011: 1213–1216. IEEE

Xiao M, Song F, Jiao J, Wang X, Xu H, Li H (2013) Identification of the gene Pm47 on chromosome 7BS conferring resistance to powdery mildew in the Chinese wheat landrace Hongyanglazi. Theor Appl Genet 126(5):1397–1403

Xiumei L, Fudao Z, Shuqing Z (2005) Responses of peanut to Nano-calcium carbonate. Plant Nutr Fertil Sci 11(3):385–389

Yadav RC, Patra AK, Purakayastha TJ, Singh R, Kumar C (2014) Effect of engineered nanoparticles of Fe and Zn oxides on enzyme activity and bacterial abundance in soil at ambient and elevated atmospheric CO_2. Proc Natl Acad Sci India Sect B Biol Sci 84(3):649–656

Yassen A, Abdallah E, Gaballah M, Zaghloul S (2017) Role of silicon dioxide nano fertilizer in mitigating salt stress on growth, yield and chemical composition of cucumber (Cucumis sativus L.). Int J Agric Res 22:130–135

Zhang F, Wang R, Xiao Q, Wang Y, Zhang J (2006) Effects of slow/controlled-release fertilizer cemented and coated by nano-materials on biology. II. Effects of slow/controlled-release fertilizer cemented and coated by nano-materials on plants. Nanoscience 11:18–26

Zhang M, Han J, Jiang F, Jiang F, Liu H (2019) Recent advances and perspectives of nanosensors applied in soil management for precision agriculture. J Sens 2019:4351809

Zheng L, Hong F, Lu S, Liu C (2005) Effect of nano-TiO2 on strength of naturally aged seeds and growth of spinach. Bter 104(1):083–092. https://doi.org/10.1385/BTER:104:1:083

Zhou JM, Huang PM (2007) Kinetics of potassium release from illite as influenced by different phosphates. Geoderma 138:221–228

Chapter 16
Sustainable Agroforestry-Based Approach to Achieve Food Security Through Soil Health

Shubhashree Sahu and Hitesh Gupta

Abstract Earth is one of the unique planets of the solar system containing life in the presence of land, air, and water, of which the land is the most important platform for growth and development of plants. The land and water contribute to carry it. The upper crust of land, i.e., solum/soil, bears productive potential. It is a critical resource for food production. Since the twentieth century, intensive farming has substantially deteriorated the soil. Around 33% of soil is already degraded and 90% could become degraded by 2050. The ever-increasing food demand would not allow for lowering intensity or coverage of cultivation; therefore the "Sustainable management" of land resources seems to be the most suitable option. This division of the book evaluated how soil and agroforestry are interconnected concepts that relate to sustainable land use and food production. Agroforestry practices vary depending on the type of environment and needs of the community; however, the characteristic feature is the deliberate management of trees, crops, and animals to achieve multiple benefits. These benefits include increased soil fertility, reduced erosion, improved water quality, increased biodiversity, and increased carbon sequestration. By maintaining year-round surface cover that shields the soil from water and wind erosion, the perennial woody flora used in agroforestry techniques helps to conserve soil. Enhancement of soil organic carbon, available nitrogen (N), phosphorus (P), potassium (K), calcium (Ca), and magnesium (Mg) (macronutrients) under different conditions is observed. Even improvement in soil bio-physical condition is also reported in many studies. Soil health is a critical component of agroforestry, as the soil provides the foundation for the entire ecosystem. In agroforestry, soil health is often improved through the use of organic matter, such as compost or cover crops, and the reduction of tillage. Both soil and agroforestry land use systems are interlinked and share a complementary relation, i.e., one being responsible in elevating and maintaining the status of another.

Keywords Agroforestry · Ecosystem · Sustainable · Organic carbon

---

S. Sahu (✉) · H. Gupta
College of Forestry, Ranichauri, Uttarakhand, India

V.C.S.G. Uttarakhand University of Horticulture and Forestry, Bharsar, Uttarakhand, India

H. S. Jatav et al. (eds.), *Agroforestry to Combat Global Challenges*, Sustainable Development and Biodiversity 36, https://doi.org/10.1007/978-981-99-7282-1_16

16.1 Introduction

Earth is the third planet in our solar system and the only one to support life. It has a diameter of about 12,742 km and is estimated to be around 4.54 billion years old. The Earth is made up of various layers, including the crust, mantle, and core, and is composed of a variety of elements such as oxygen, silicon, and iron. One of the most important components of the Earth is soil. Soil is the thin uppermost crust on the earth's surface that supports plant growth and is essential for agriculture and forestry. It is composed of a mixture of organic matter, minerals, and water and provides a habitat for a wide range of microorganisms.

Soil formation involves a complex interaction of various physical, chemical, and biological factors. These factors include climate, parent material, topography, organisms, and time. Climate influences soil formation by affecting the rate of weathering and erosion, while parent material provides the mineral components of soil. Topography influences soil formation through its effect on soil depth and the amount of water that it receives. Organisms, (plants, animals, and microorganisms) play an important role in soil formation by contributing to organic matter and nutrients. Finally, time is a critical factor in soil formation, as it takes hundreds to thousands of years for soil to develop. Soil is not a homogenous material but varies in composition and properties depending on its location and the processes that formed it. The physical properties of soil determine its ability to retain water, air, and nutrients and affect plant growth. The chemical properties of soil influence the availability of nutrients and biota that flourishes in it.

Soil is not only important from the viewpoint of plant growth and agriculture, but it also plays a crucial role in climate change mitigation. Soil can store large amounts of carbon, with estimates suggesting that global soil carbon stocks are about 3.3 times greater than the carbon stored in the atmosphere. However, human activities such as deforestation, intensive agriculture, and land use change have led to significant losses of soil carbon. To address the issue of soil carbon loss, there has been a growing interest in soil carbon sequestration practices. Soil carbon sequestration refers to the process of capturing atmospheric carbon dioxide and storing it in the soil in the form of organic matter. This can be achieved through various practices, such as conservation tillage, cover cropping, crop rotation, agroforestry, and the use of organic amendments.

Research has shown that soil carbon sequestration practices cannot only reduce greenhouse gas emissions but also improve soil health and fertility, thereby boosting crop yields and biodiversity. For instance, a study published in the journal *Nature Climate Change* found that the adoption of conservation agriculture practices, such as reduced tillage and crop rotations, could lead to an increase in global soil carbon stocks by 0.25 to 0.8 gigatons per year, equivalent to removing 0.5 to 1.6 billion metric tons of carbon dioxide from the atmosphere annually.

Another important aspect of soil is its microbiome. The soil microbiome is critical for soil health and fertility, as it plays a vital role in nutrient cycling, carbon sequestration, and disease regulation. Recent research has shown that soil microbiota is highly diverse and complex, with the potential to influence plant growth and productivity. Furthermore, it is also sensitive to variations in land use, management practices, and environmental conditions. Human activities such as intensive agriculture, deforestation, and pollution alter the soil microbiome's composition and function, leading to negative impacts on soil health and ecosystem services. Therefore, understanding the soil microbiota and its response to different management practices is essential for developing sustainable land management strategies.

Soil degradation is a significant environmental concern, as it leads to crop yield reduction, loss of biodiversity, and accelerated soil erosion. Some of the factors that contribute to soil degradation include erosion, compaction, salinization, acidification, and pollution. Sustainable land management practices, such as conservation tillage, crop rotation, and the use of cover crops, are recommended to protect and preserve soil. Earth and soil are intimately connected, with soil playing a critical role in supporting life on our planet. Understanding the formation, properties, and degradation of soil is essential for sustainable land management and to ensure the progressive utilization of agricultural and forestry systems.

16.2 Agroforestry as a Land Use

Land use refers to how land is utilized, managed, and modified by humans. It encompasses a variety of activities and practices, such as agriculture, forestry, urban development, and conservation. Agroforestry is a land use system that involves the integration of trees with crops and/or livestock in a way that provides ecological, economic, and social benefits. The trees in an agroforestry system may be grown for timber, fruit, fuelwood, or other products and may provide shade, wind protection, erosion control, and other ecosystem services. The crops and livestock in an agroforestry system may benefit from the shade, nutrients, and other benefits provided by the trees and may also contribute to the overall productivity of the system. Agroforestry systems can range from simple arrangements of trees in crop fields to complex, multi-layered agroforests that mimic natural ecosystems.

Several different types of land use can be integrated with agroforestry practices, depending on the local context and ecological conditions.

- Agriculture: Agroforestry can be used to enhance agricultural productivity by integrating trees with crops or livestock. For example, shade-grown coffee is a popular agroforestry system that combines coffee production with the cultivation of shade trees, which provide a habitat for birds and other wildlife.
- Forestry: Agroforestry can also be used to manage forested landscapes sustainably. For example, silvopastoral systems combine trees with grazing livestock, providing economic benefits for farmers while also protecting soil, water, and biodiversity.
- Urban development: Agroforestry can be integrated into urban landscapes in the form of urban forestry, green roofs, and other green infrastructure. These

practices can provide a range of benefits like improved air and water quality, reduced urban heat effects, and enhanced biodiversity.

Overall, integrating agroforestry practices into different land use systems can help to promote sustainable land use, conserve natural resources, and support the livelihoods of rural communities.

16.3 Basic Concepts of Agroforestry

Agroforestry is a sustainable land use system that involves the integration of trees, crops, and/or livestock in a single land use unit. Here are some basic concepts related to agroforestry:

- Agroforestry systems: There are different types of agroforestry systems, including alley cropping, silvopasture, forest farming, and home gardens.
- Tree-crop interactions: In agroforestry systems, trees and crops can interact in a variety of ways. For example, trees can provide shade and reduce soil erosion, while crops can provide additional income and improve soil fertility.
- Environmental benefits: Agroforestry can have a range of environmental benefits, such as improving soil health, reducing greenhouse gas emissions, and providing habitat for wildlife.
- Economic benefits: Agroforestry systems can also provide economic benefits, such as diversifying income streams for farmers and increasing the productivity of the land.
- Social benefits: Agroforestry systems can have social benefits, such as improving food security, enhancing cultural practices, and promoting community development.
- Management practices: Agroforestry systems require specific management practices, such as selecting appropriate tree and crop species, managing competition between trees and crops, and controlling pests and diseases.
- Sustainability: Agroforestry is a sustainable land management system that can help to address environmental, economic, and social challenges in agriculture. It can contribute to food security, poverty reduction, and environmental conservation.

16.4 Advantages and Disadvantages of Agroforestry over Other Land Uses Presented in Tables 16.1 and 16.2

Table 16.1 Advantages of agroforestry over other land uses

Agroforestry systems	Other land use
Diverse benefits	Limited benefits
Agroforestry systems provide multiple benefits such as food production, timber and non-timber products, soil conservation, water management, biodiversity conservation, carbon sequestration, and climate change mitigation	Other land uses such as monoculture agriculture or grazing systems typically focus on a single output and may not provide the same range of benefits
Improved soil health	Degraded soil health
Agroforestry systems promote healthy soils by increasing organic matter content, improving soil structure, reducing soil erosion, and improving nutrient cycling	Other land uses can degrade soil health by removing nutrients, compacting soil, and reducing soil organic matter content
Increased resilience	Vulnerability to environmental changes
Agroforestry systems are more resilient to environmental changes such as droughts, floods, and climate variability due to their diverse cropping systems and tree cover	Other land uses may be more vulnerable to environmental changes as they often rely on single crops or livestock and may not have the same level of soil protection or water retention
Cost-effective	Expensive
Agroforestry systems can be more cost-effective than other land uses due to their ability to produce multiple products on the same piece of land, reducing the need for additional inputs or land purchases	Other land uses may require more inputs such as fertilizer, pesticides, and additional land purchases to achieve the same level of productivity
Improved livelihoods	Limited livelihood opportunities
Agroforestry systems can improve livelihoods by providing diverse income sources and employment opportunities for farmers and rural communities	Other land uses may provide limited livelihood opportunities, particularly for small-scale farmers, and may contribute to rural poverty

Table 16.2 Disadvantages of agroforestry over other land uses

Agroforestry systems	Other land use systems
Time and labor-intensive to set up and maintain	Monoculture management is less labor and time intensive
A long-term investment is required for benefits to be realized	The short-term focus of many farmers and policy-makers
Reduced yields in initial years while trees are being established	Lower short-term returns than monoculture cropping
The complexity of management can be daunting for some farmers	Simpler land use systems may be easier to manage
Limited market demand for some agroforestry products	Higher market demand for some monoculture crops
Limited access to technical and financial support for farmers	More support available for conventional monoculture farming
Can be challenging to integrate into existing land tenure systems	Existing land tenure systems are already well adapted by the society

16.5 Soil-Related Problems

Soil-related issues such as soil erosion, soil contamination, and soil depletion can have significant impacts on agricultural productivity, food security, and the environment. Here are some examples of these issues and their management:

Agriculture is one of the primary activities that depend on soil, and the practices associated with agriculture can have significant impacts on soil health. Some of the soil problems related to agriculture include the following:

1. Soil erosion: Soil erosion occurs when soil is removed from the land surface due to factors like wind, water, and tillage practices. It can lead to a severe reduction in soil fertility and crop productivity. For example, in the United States, the erosional loss of topsoil is estimated to result in a reduction of crop yields by 1.5% per year.
2. Soil compaction: Soil compaction occurs when the soil becomes densely packed, reducing pore space and thereby restricting the movement of air, water, and nutrients. It can lead to reduced crop productivity and increased susceptibility to drought and flooding (Brouder and Hofman 2019). For example, in India, soil compaction has been identified as a major constraint to rice production.
3. Soil salinization: Soil salinization occurs when the salt content of the soil increases, leading to reduced soil fertility and plant growth. Soil salinization can occur due to factors such as irrigation with saline water and the use of salt-based fertilizers. For example, in Egypt, soil salinization is a major problem in the Nile Delta region, where the use of irrigation has led to salt accumulation in the soil.
4. Soil acidification: Soil acidification occurs when the soil pH decreases, leading to reduced soil fertility and plant growth. Soil acidification can occur due to factors such as acid rain, soil leaching, and the use of acidifying fertilizers. For example, in Brazil, the use of nitrogen fertilizers has led to soil acidification, which has negatively impacted crop productivity.
5. Soil nutrient depletion: Soil nutrient depletion occurs when the soil becomes depleted of essential nutrients such as nitrogen, phosphorus, and potassium, leading to reduced crop productivity. Soil nutrient depletion can occur due to factors such as intensive agriculture, the use of chemical fertilizers, and soil erosion. For example, in China, the overuse of chemical fertilizers has led to soil nutrient depletion and decreased crop yields.
6. Soil pollution: Soil pollution occurs when contaminants such as pesticides, heavy metals, and industrial chemicals are introduced into the soil, leading to reduced soil fertility and potential health risks for humans and wildlife. Soil pollution can occur due to factors such as improper disposal of hazardous waste, use of chemical pesticides and fertilizers, and industrial activities. For example, in China, the use of pesticides and fertilizers has led to widespread soil pollution, particularly in rural areas.
7. Soil biodiversity loss: Soil biodiversity loss occurs when the diversity of organisms in the soil, such as bacteria, fungi, and earthworms, decreases, leading to

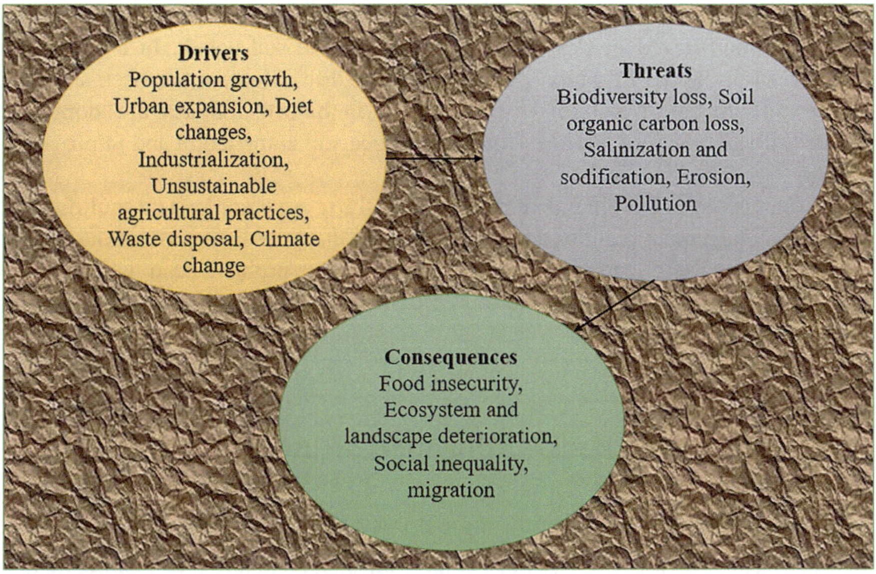

Fig. 16.1 The entire process of soil degradation

reduced soil fertility and ecosystem functioning. Soil biodiversity loss can occur due to factors such as intensive agriculture, land use change, and soil disturbance. For example, in Europe, intensive agricultural practices have led to a decline in soil biodiversity, which has been linked to reduced soil fertility and crop productivity.

8. Soil degradation: Soil degradation is a general term that refers to the reduction of soil quality and productivity due to various factors, such as soil erosion, soil compaction, nutrient depletion, and pollution (Fig. 16.1). Soil degradation can lead to reduced crop yields, increased soil erosion, and loss of biodiversity. For example, in Africa, soil degradation has been identified as a major challenge to agricultural development, with an estimated 65% of agricultural land affected by degradation.
9. Desertification: Desertification is a process by which fertile land becomes degraded and desert-like due to factors such as soil erosion, nutrient depletion, and climate change. Desertification can lead to reduced crop productivity, loss of biodiversity, and increased soil erosion. For example, in China, desertification is a major problem in the north-western regions, where overgrazing and intensive agricultural practices have led to soil degradation and desertification.

Management practices such as crop rotation, reduced tillage, improved irrigation, agroforestry, soil conservation, and land restoration can help to address these soil problems and promote soil health and productivity. For example, in Africa, the adoption of agroforestry practices has been shown to improve soil fertility and crop yields while also providing additional benefits such as carbon sequestration and biodiversity conservation. In China, the implementation of land restoration programs has been found to reduce desertification and improve soil health. In Pakistan, the adoption of improved irrigation practices such as drip irrigation has been shown to reduce soil salinization and improve crop yields. In the United States, the adoption of reduced tillage practices has been found to reduce soil compaction and improve soil health.

In conclusion, soil problems related to agriculture are complex and multifaceted, and their management requires a holistic approach that takes into account the specific causes and impacts of each problem. Addressing these problems can help to ensure the sustainability of agriculture and food production while also protecting soil health and the environment.

16.6 Soil-Related Issues: Mitigation Measures

Here are some mitigation measures for soil-related issues:

1. Soil erosion control: Soil erosion can be controlled by planting vegetation on bare land or steep slopes, using mulch or erosion control blankets, and establishing vegetative buffer strips along streams or rivers. Contour farming and terracing can also help prevent soil erosion.
2. Soil conservation: Soil conservation practices include crop rotation, conservation tillage, cover cropping, and organic farming. These practices help maintain soil health, reduce soil erosion, and improve soil fertility.
3. Soil remediation: Soil remediation involves removing contaminants from soil and restoring the soil's natural properties. Common methods of soil remediation include bioremediation, phytoremediation, and soil washing.
4. Soil management: Proper soil management practices can help prevent soil degradation and improve soil health. These practices include minimizing soil disturbance, avoiding overgrazing, and limiting the use of chemical fertilizers and pesticides.
5. Soil testing: Soil testing can help identify soil nutrient deficiencies and soil pH imbalances. This information can be used to develop a fertilizer and nutrient management plan that promotes soil health and reduces the risk of soil contamination.
6. Land use planning: Land use planning can help reduce soil-related issues by identifying areas suitable for agricultural activities and areas that need to be preserved for conservation purposes. This can help prevent soil degradation, erosion, and contamination.

In a broader sense, soil-related issues can have significant impacts on the environment, human health, and the economy. Mitigation measures such as soil erosion control, soil conservation, soil remediation, soil management, soil testing, and land use planning can help preserve soil health and maintain the productivity of the land.

It is essential to adopt these measures to ensure the sustainable use of soil resources for future generations.

16.7 Contribution of Agroforestry Toward Soil Profile

16.7.1 Soil Enhancement and Microenvironment Alleviation Through Agroforestry

The diverse implications of agroforestry on ecosystem functions and amenities due to the immediate and long-term impacts of trees vary by crop type, temperature, and geography. As a "security curtain" against the loss of nutrients from the process of nutrient turnover, trees play a significant role in the replenishing of nutrients by recapturing and pumping back drained nutrients through deep roots. Agroforestry offers an exceptional opportunity to absorb and store carbon in the soil that is lost as a result of increased agricultural productivity, intensive farming, and fertilizer application (Chatterjee et al. 2018). Agroforestry encourages more efficient utilization of assets than single-crop farming due to the different functional and architectural characteristics of the various elements accumulated in a multipurpose canopy. The proliferation of trees on farms improves the landscape's drainage capacity, organic matter (OM), readily accessible potassium, accessible phosphorus, soil carbon stocks, and lower bulk density (BD). These factors increase the soil's ability to retain water by increasing its water-holding capacity (WHC), which in turn progressively releases it to plants like a sponge.

The incorporation of organic matter helps to reduce bulk soil density and aids in the consolidation of soil. In the desert and semi-arid areas, this lower BD of the soil improves subterranean recharging, air circulation, water dispersion in the rhizosphere, and soil nutritional quality. The most significant source of essential nutrients and biological carbon in agroforestry systems is an overabundance of litter precipitated by the withering of leaves and twigs. The efficiency of nutrient use in agribusiness is influenced either directly or indirectly by soil organic carbon (SOC). The improved accessibility and assimilation of soil with high OM and a vibrant deep root structure will improve the efficiency of nitrogen usage. Additionally, mycorrhizae are presumably provided by the increased microbial variety brought on by the addition of OM, which releases P and makes it available to plants. The total organic carbon comprised of the liable pool and a non-liable pool (Fig. 16.2). According to the IPCC report (IPCC 2000), the area presently under agroforestry worldwide is 400 million hectares with an expected increase in carbon stock by 0.72 Mg C ha/year, with an estimated potential for sequestering 45 Tg C/year by 2040.

As a result of the nitrogen fixation in the trees, the amount of soil and nitrogen recirculating through the decomposing leaf litter is increased, and the long-term stability of soil nitrogen is improved by organic matter additions (Montagnini and

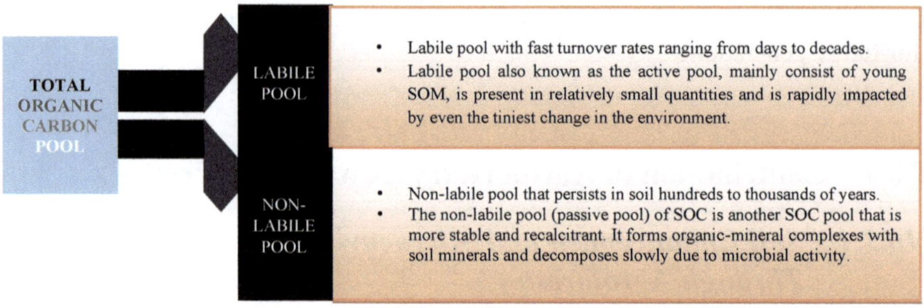

Fig. 16.2 The total organic carbon comprised of the liable pool and non-liable pool

Nair 2004). Plants with a reputation for N2 fixation and green farming include *Gliricidia*, *Leucaena*, and *Sesbania*. In the context of agroforestry, substantial nutrient collection by tree roots is seen as a supplementary nutrient input because such nutrients would be otherwise drained out from the crop at depths where crop roots are absent. In agroforestry, phosphorus is frequently a crucial nutrient. More effective utilization of nutrients might be achieved by combining organic and inorganic sources of phosphorus. Due to the extremely low levels of accessible phosphorus in the subsoil, it is anticipated that the deep capture of P will be insignificant. Though many agroforestry systems do store P in their biomass and recycle it into the soil through litter breakdown, this process does not count as an external input. However, some of the soil's less accessible inorganic forms of phosphorus may be changed into accessible organic forms through cycling. The addition of both leguminous and non-leguminous species, such as durian and rambutan, to cocoa-based agroforestry has had an advantageous effect on the soil's nutritional qualities (Wartenberg et al. 2020). According to a study by Riyadh et al. 2018 in Bangladesh, the soils of various crops about jackfruit-based agroforestry had lower soil temperatures (3.37–9.25%), higher soil moisture (10–20%), and higher total nitrogen (9–19%) levels than in fields that were open for the growing season.

Shade plantings change the micro-ecosystem underneath coffee plants by minimizing forthcoming radiation, moderating maximum temperatures, reducing the temperature intensity, and raising minimum temperatures. Coffee plants are protected from extreme heat and radiation by shade trees in coffee gardens, which also lessen seasonal changes in the coffee leaf area (Lin 2007).

16.7.2 Agroforestry to Boost Soil Health and Productivity

Agroforestry systems play a crucial role in boosting soil health and productivity. Through the integration of trees, crops, and livestock, agroforestry enhances various aspects of soil health. It is always argued that the presence of woody perennials in agroforestry systems affects several bio-physical and bio-chemical processes that determine the health of soil substrate. The most obvious effects of trees on soil include amelioration of erosion primarily through surface litter cover and understory vegetation; maintenance or increase of organic matter and diversity through continuous degeneration of roots and decomposition of litter; nitrogen fixation; enhancement of physical properties such as soil structure, porosity, and moisture retention due to the extensive root system and the canopy cover; and absorb and recycle nutrients in the soil that would otherwise be lost through leaching. The increased organic matter content improves soil structure, enhancing its ability to retain water and nutrients (Jose 2009). Furthermore, the extensive root systems of trees help to break up compacted soil layers, promoting better aeration and root penetration. It also facilitates nutrient cycling within the system. Trees, particularly nitrogen-fixing species, capture atmospheric nitrogen and convert it into a form that is available to plants. This process improves soil fertility and provides a sustainable source of nutrients for crops. In addition, the integration of livestock in agroforestry systems contributes to nutrient cycling. The manure produced by livestock serves as a valuable organic fertilizer that enriches the soil with nutrients, benefiting both trees and crops. By improving soil health, agroforestry enhances soil productivity. The presence of diverse tree species and associated vegetation provides a range of ecosystem services. Trees act as windbreaks, reducing the impact of strong winds on crops and preventing soil erosion. Moreover, the complex root systems of trees facilitate efficient water uptake, improving water availability for crops and enhancing drought resistance. The combination of trees, crops, and livestock in agroforestry systems creates a diverse and resilient agricultural landscape. This diversity reduces the risk of crop failure and pest outbreaks associated with monoculture systems. The presence of diverse vegetation attracts beneficial insects and birds, contributing to natural pest control and pollination services. Overall, agroforestry boosts soil health and productivity through the addition of organic matter, nutrient cycling, improved soil structure, and enhanced water retention. The diverse and integrated nature of agroforestry systems creates a resilient environment that promotes natural processes, reduces reliance on external inputs, and fosters sustainable agricultural practices (Young 1997). A schematic depiction illustrating nutrient relationships and the benefits of "optimal" agroforestry systems in contrast to typical agricultural and forestry systems is displayed in Fig. 16.3.

16.7.3 Restoration of Degraded Land

In India, approximately 120.72 million hectares of land, which accounts for 37 % of the total geographical area, is affected by various forms of soil degradation (e.g., water erosion, 93 million hectares; wind erosion, 11 million hectares; salt-affected soils, 6.74 million hectares; and 16.53 million hectares of open forest area; ICAR 2010). There are various processes leading to land degradation (Fig. 16.4). Agroforestry has played a significant role in recent times in the reclamation of wastelands, including desert areas and lands that have undergone degradation due to salinization,

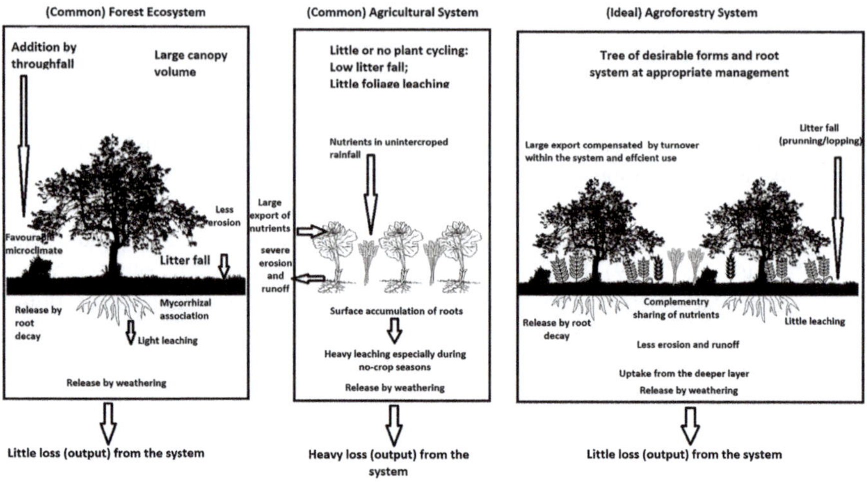

Fig. 16.3 Schematic representation of nutrient relations and advantages of ideal agroforestry systems in comparison with common agriculture and forestry

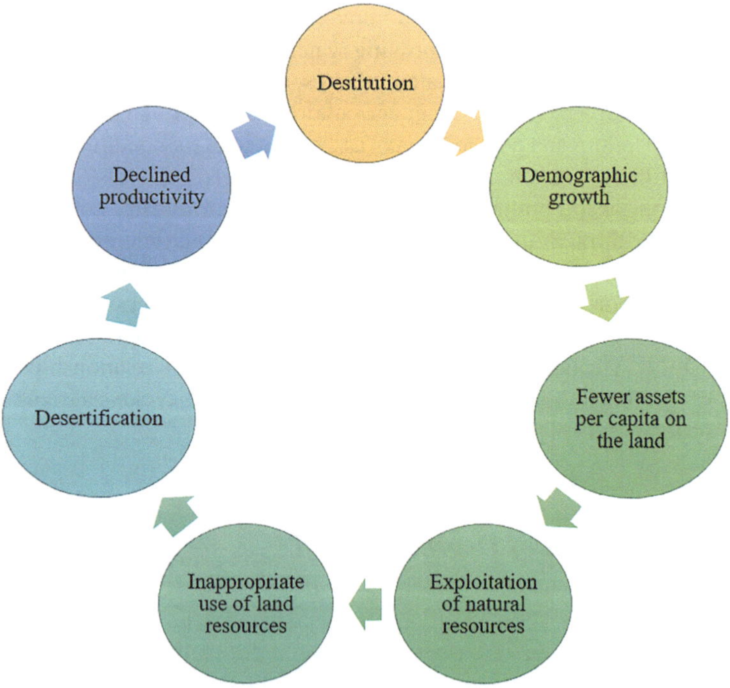

Fig. 16.4 The series of events that trigger and perpetuate land deterioration

as well as those affected by ravines, gullies, and other forms of water and wind erosion hazards.

Agroforestry research has demonstrated the potential of numerous salt-tolerant trees and shrubs to contribute to the biological improvement and rehabilitation of salt-affected lands (CSSRI 2010–2012). The restoration of degraded saline and sodic soils through agroforestry can be attributed to changes in various soil parameters. In many cases, the presence of trees gradually enhances the soil fertility of degraded lands, leading to higher levels of soil organic carbon, total nitrogen, available phosphorus, and exchangeable potassium, calcium, and magnesium. Simultaneously, there is a reduction in soil salinity and sodicity, indicated by decreases in exchangeable sodium, pH, and electrical conductivity, which progressively diminish as the trees mature (Dhyani et al. 1994). These combined processes contribute to the improvement of productivity in saline and sodic soils by enhancing nutrient availability and alleviating the negative effects of sodicity.

Long-term cultivation of tree plantations has been demonstrated to enhance the physical, chemical, and biological characteristics of soil in agroforestry systems (SAS). The presence of trees decreases the bulk density of SAS, leading to increased soil porosity, water retention capacity, field capacity, permeability, and rate of water infiltration (Mishra et al. 2004). Evaluations of soil chemical properties in conventional agroforestry systems in the northeastern region have revealed significant enhancements within a timeframe of 10–15 years. These improvements encompass a notable elevation in soil pH, organic carbon (C) content, exchangeable calcium (Ca), magnesium (Mg), potassium (K), and available phosphorus (P) levels, as measured using the Bray's P test. Among various agroforestry practices (AFP), the combination of areca nut, jackfruit, black pepper, and *Cinnamomum* (tejpata) exhibited the highest accumulation of organic carbon (2.91%). This was followed by the combination of areca nut, betel vine, and assorted trees, which demonstrated an organic carbon increase of 1.85%. Conversely, degraded land exhibited a mere 0.78% rise in organic carbon within the same period. Furthermore, all agroforestry interventions resulted in a significant elevation in exchangeable Ca, Mg, K, and Na compared to adjacent degraded lands. The presence of exchangeable aluminum (Al), a potential cause of soil infertility, was eliminated within 10–15 years of implementing agroforestry practices. This elimination of Al was attributed to the addition of fresh organic matter, which complexed the exchangeable Al during decomposition, possibly forming Al-humate complexes, and the accumulation of Ca, Mg, K, and Na cations. Consequently, these processes led to an increase in soil pH ranging from 0.6 to 1.7 units under the various AFPs (Singh et al. 1994). The research conducted at the National Research Centre for Agroforestry in Jhansi (NRCAF) over a span of two decades has successfully developed suitable agroforestry technologies for the restoration of degraded lands in the semi-arid Bundelkhand region. Extensive afforestation endeavors have been undertaken in the arid zones of India, particularly in 11 districts of western Rajasthan, to combat desertification and wind erosion. These initiatives involve the establishment of shelterbelts and windbreaks, as well as the stabilization of dunes. The shelterbelts serve to mitigate the adverse effects of wind and create a favorable micro-environment for crops (NRCAF, 2012)

16.7.4 Nutrient Recycling and Intercrop Yield

Agroforestry plays a crucial role in facilitating efficient nutrient cycling, which in turn benefits intercrops. In agroforestry systems, the nitrogen (N) requirement of crops can be partially fulfilled through various mechanisms, including the decomposition rate of organic mulches, biological nitrogen fixation (BNF), and residue management. Trees contribute to N inputs in agroforestry systems through BNF and deep nutrient capture. The presence of active nodules in the roots of leguminous species indicates that BNF can supply significant amounts of N to crops through litter decomposition in the soil. Additionally, non-fixing trees such as cassia accumulate more N in their leaves compared to nitrogen-fixing legumes, likely due to their larger root volume and nutrient-capturing ability, which can be incorporated into the soil as green leaf manure. Tree species, such as *Gliricidia*, *Leucaena*, and *Sesbania*, are known for their N_2 fixation capacity and their potential for green manuring. Deep nutrient capture by tree roots, particularly at depths beyond the reach of crop roots, represents an additional source of nutrients in agroforestry systems. Without the presence of trees, these nutrients would otherwise be leached beyond the crop's reach. However, when tree litter decomposes, these captured nutrients are transferred to the soil, acting as inputs. By incorporating legumes into agroforestry systems, the nitrogen self-sufficiency of the system can be achieved. Introducing suitable legumes, such as *Dolichos lablab*, *Clitoria ternatea*, *Atylosia scarabaeoides*, *Macroptilium atropurpureum*, and *Stylosanthes* species, in rangelands, pastures, silvipastures, and agroforestry practices holds significant importance. In agroforestry systems, even non-leguminous trees like *Alnus*, *Myrica*, and *Casuarina*, which form associations with *Frankia*, are widely recommended due to their nitrogen-fixing capabilities. For instance, *Casuarina equisetifolia* has a nitrogen-fixing potential of 50–80 kg N/ha/year, while *Alnus nepalensis* can fix 29–117 kg N/ha/year (Sharma and Kapoor 2005). Phosphorus (P) is often a critical nutrient in agroforestry. Combining organic and inorganic sources of P can result in more efficient nutrient utilization. Deep phosphorus capture is typically negligible due to the low concentrations of available phosphorus in the subsoil. While many agroforestry systems accumulate P in their biomass and return it to the soil through litter decomposition, this cycling does not constitute an external input to the system. However, through cycling, some less available inorganic forms of phosphorus in the soil can be converted into more readily available organic forms. Furthermore, the beneficial interaction with mycorrhizal fungi associated with trees enhances nutrient uptake from deeper soil layers, particularly for less mobile nutrients like phosphorus. Incorporating tree crops into farming systems promotes improved nutrient cycling and availability within the system.

16.7.5 Impact of Agroforestry on Soil Biota

The integration of trees with crops has a positive impact on the physical and chemical properties of soil, as well as on the soil microbiota. This results in increased biological soil fertility, which indirectly promotes plant growth by enhancing nutrient cycling. Soil organisms, particularly microorganisms, play a significant role in plant productivity and health. They are involved in a variety of processes, including C-transformation, nutrient cycling, and aggregate formation. Nematodes, collembola, acari, diplopoda, earthworms, fungi, and various insects are all important soil organisms that influence C-transformation and nutrient cycling. Soil engineers, such as ants, termites, and earthworms, play important roles in aggregate formation and maintaining the soil structure. Centipedes, ground or rove beetles, predatory mites, collembola, and carnivorous nematodes are important for biological control. Agroforestry systems have more soil microbes than soil cropping systems, and these microbes are more diverse and functional. This is expected to result in increased biological soil fertility in these systems. The integration of trees with crops can improve soil health by increasing the diversity and activity of soil organisms. This, in turn, can improve soil fertility and productivity, which can benefit plant growth (Altieri 1999).

Agroforestry practices contribute to the enrichment of soil biodiversity compared to monocropping by providing habitat, microclimate heterogeneity, moisture regulation, and buffering effects, which serve as refugia for various soil organisms (Brussaard et al. 2007). The litter and root exudates from trees supply microbial communities with essential energy sources like amino acids, sugars, and organic acids, among other substances. The presence of trees creates favorable conditions for soil microflora to thrive, leading to higher microbial diversity rates in agroforestry systems relative to monoculture. The diversity and activity of soil fauna significantly influence soil health, with the presence of *P. reticulatum*, for example, boosting microbial activity and nematode diversity. This, in turn, enhances the breakdown of soil organic matter, nutrient mineralization, and nutrient enrichment. Agroforestry systems, particularly alley cropping with rows of trees integrated into farms, promote the abundance of row-associated soil bacteria, thereby contributing to the overall diversity and functional diversity. In terms of nitrogen fixation, trees in agroforestry systems promote the growth of N-fixing bacteria such as *Bradyrhizobium* and *Mesorhizobium*. Conversely, the abundance of nitrifying bacteria like *Nitrosospira* and *Nitrospira* tends to be lower in monoculture tree systems compared to agroforestry systems. The adoption of land use practices like agroforestry, which involve consistent deposition of plant residues (such as litter and roots), plays a crucial role in ecosystem functioning. It contributes significantly to glomalin production, protection of soil carbon, and enhanced activity of arbuscular mycorrhizal fungi (AMF). Overall, agroforestry systems provide a beneficial environment that supports diverse microbial communities, enhances soil functions, and contributes to important ecosystem services through factors such as glomalin production, soil carbon protection, and increased AMF activity.

16.7.6 Food Security: A Major Challenge

A major challenge to global food security is the need to approximately double food production over the next few decades, especially due to rapidly growing demand from the developing world (Food and Agriculture Organization of the United Nations (FAO), International Fund for Agricultural Development, and World Food Programme, 2015). Traditional methods of increasing yields, such as the utilization of chemical inputs, genetic advancements, and mechanization, have become customary. However, these conventional agricultural practices have also contributed significantly to various social and environmental challenges, including the alteration of climate patterns, the decline in biodiversity and ecosystem stability, degradation of land, water insecurity, and disruption of social systems. Soil degradation affects human nutrition and health through its adverse impacts on quantity and quality of food production. The decline in crop yield and agronomic production exacerbates food insecurity that currently affects 854 million people globally (Lal 2020).

There is now a widespread consensus that a shift is needed from the current narrow focus on yield toward a more "multifunctional" approach in agriculture that not only prioritizes but also enhances broader societal and environmental objectives within the framework of sustainable intensification (United Nations General Assembly 2015). This international agreement provides a comprehensive and cohesive structure for multifunctional agriculture by integrating food security (Sustainable Development Goal 2) with environmental, climate, and social goals, emphasizing the importance of a multigoal approach. A multigoal approach acknowledges that each agricultural option will have varying effects on individual goals, and different stakeholders will assign varying degrees of importance to each goal. Food being a fundamental need in human life, it is imperative to prioritize achieving sufficient yield to meet the SDG of food security within a multigoal agriculture approach. In the realm of physical and human geography, agroforestry emerges as one of the most multifunctional agricultural systems, involving the simultaneous cultivation of trees and crops on the same land. While agroforestry has been practiced for ages and has often been studied for its ecological benefits and associations with peasant farmers, scientific evidence now demonstrates that the adoption of agroforestry can lead to yield increases of up to twofold, depending on the crop type, local conditions, and level of expertise. These yield improvements are attributable to the various ecosystem services provided by trees, including enhanced soil nutrient status through processes like nitrogen fixation, reduction of crop stress by mitigating temperature and rainfall extremes, prevention of soil erosion through root binding, and regulation of water supply via hydraulic uplift facilitated by tree roots. Furthermore, agroforestry offers highly sustainable outcomes by maintaining soil fertility and even restoring degraded lands (Nair 1993). However, it is important to note that food security (SDG2) is not solely dependent on yield (FAO 2008; World Bank 2015). Resilience to climate change and shocks, which can give rise to severe food crises, represents a critical additional component. Agroforestry enhances crop resilience to various anticipated climate change effects, such as drought or higher temperatures, by improving water infiltration and storage while mitigating evaporation and temperature extremes (Pandey 2007). It also bolsters livelihood resilience since trees provide free ecosystem services, reducing dependence on volatile external commodity markets. During poor harvests, trees serve as alternative sources of both income and food, such as fruits, fodder, or fuel. Apart from advancing food security, agroforestry has the potential to enhance multiple social dimensions of the Sustainable Development Goals (SDGs) (Van Noordwijk et al. 2011). It can serve as a pathway out of poverty, a significant driver of hunger (World Bank 2015), as the combination of increased yield, low costs, and additional tree-based farm products can substantially augment net farm income. In agroforestry systems, reduced reliance on external chemical inputs and greater resilience to market fluctuations contribute to a sense of control, equity, and dignity in agricultural work. Therefore, agroforestry techniques are likely to be widely applicable across a significant proportion of global farmland.

16.8 Sustainable Agroforestry

Sustainable agroforestry is an integrated land management approach that combines the principles of agriculture and forestry to create a productive and resilient system while promoting ecological balance and sustainability. It involves the intentional integration of trees, crops, and livestock on the same land, utilizing their interactions to maximize benefits and minimize negative impacts.

16.8.1 Key Features of Sustainable Agroforestry

- Soil Improvement: Agroforestry systems contribute to soil health and fertility. Trees help to fix nitrogen, enhance nutrient cycling, and increase organic matter content in the soil. Their deep root systems improve soil structure, prevent erosion, and enhance water retention capacity.
- Water Management: Agroforestry practices can help regulate water availability and reduce water-related risks. Tree canopies intercept rainfall, reducing soil erosion and runoff. Their root systems act as natural filters, promoting water infiltration and reducing the risk of water pollution.
- Food Security and Nutrition: Sustainable agroforestry systems can enhance food security by diversifying food sources and improving nutrition. The inclusion of fruit and nut-bearing trees provides additional food options, while diverse crops and livestock contribute to a more balanced diet. In Africa, agroforestry has been shown to provide a number of benefits to farmers. For instance, it can enhance soil fertility in many situations and improve farm household resilience through the provision of additional products for sale or home consumption (Mbow et al. 2014). The agri-horticulture system (paddy + areca catechu) that ensures the

Fig. 16.5 Agri-horticulture system (paddy + areca catechu) on the foothills of the Velliangiri Mountains, adjacent to the Nilgiri Biosphere Reserve, Coimbatore

state's food security is shown in Fig. 16.5 at the Isha Foundation's headquarters, Coimbatore.

- Biodiversity and Ecological Balance: Agroforestry systems promote biodiversity by incorporating a variety of tree species, crops, and livestock. The diverse plant and animal life contributes to ecosystem health, including enhanced soil fertility, pest control, and nutrient cycling.
- Conservation of Natural Resources: Agroforestry practices emphasize the conservation and efficient use of natural resources. Trees help prevent soil erosion, improve water quality by reducing runoff, and enhance water infiltration. Additionally, they provide shade, reduce evaporation, and improve microclimatic conditions.
- Increased Resilience and Productivity: By diversifying the farm landscape, agroforestry systems are more resilient to climate variability and extreme weather events. Trees act as windbreaks, protect crops and livestock, and provide a source of renewable energy. The intercropping of trees and crops can increase overall productivity through improved nutrient cycling and enhanced soil fertility.
- Carbon Sequestration and Climate Mitigation: Trees play a vital role in carbon sequestration, helping to mitigate climate change. Agroforestry systems contribute to carbon storage in both above-ground biomass and in the soil. They also reduce greenhouse gas emissions associated with conventional agricultural practices, such as synthetic fertilizer use (Albrecht and Kandji 2000).
- Climate Adaptation: Agroforestry plays a crucial role in climate adaptation by providing farmers with resilient systems that can withstand climate change impacts. The diverse structure of agroforestry landscapes buffers against extreme temperatures, reduces heat stress on crops and livestock, and provides shade and shelter.

Table 16.3 Role of sustainable agroforestry in soil health and food security

Role of sustainable agroforestry	Impact on soil health	Impact on food security
Soil health improvement	Enriches soil with organic matter Improves soil structure and fertility Enhances nutrient availability Reduces erosion and compaction	Diversifies food sources
Nutrient cycling and soil fertility	Facilitates nutrient cycling Fixes nitrogen from the atmosphere Provides organic fertilizer through livestock manure	Increased food production to lessen the gap between supply and demand
Diversification of food sources	Provides a variety of food products from trees, crops, and livestock	Reduces dependency on single crops or livestock breeds
Enhanced crop and livestock resilience	Acts as windbreaks, reducing wind damage to crops and livestock Provides shading, reducing heat stress on crops and livestock	Increases resilience to pests, diseases, and market fluctuations
Conservation of genetic diversity	Preserves genetic diversity of tree species, crops, and livestock	Enhances resilience of agricultural systems and promotes adaptation
Improved water management	Reduces soil erosion and surface runoff Enhances water infiltration and groundwater recharge	Promotes better water retention to benefit the crops
Economic and social benefits	Diversifies income sources for farmers and rural communities Promotes community engagement and knowledge sharing	Provides employment opportunities Contributes to rural development

- Economic and Social Benefits: Agroforestry systems offer economic benefits by diversifying income streams. Farmers can generate revenue from multiple sources, including timber, fruits, nuts, and agricultural products. Additionally, agroforestry can provide employment opportunities and contribute to rural development.
- Landscape Restoration and Conservation: Agroforestry can contribute to landscape restoration efforts, particularly in degraded or deforested areas. Planting trees and integrating them with agriculture help restore ecosystem functions, enhance wildlife habitat, and reconnect fragmented landscapes (Table 16.3).

16.9 Conclusion

Agroforestry is a sustainable land use system that combines the cultivation of crops, livestock, and trees in a single integrated system. It has the potential to improve soil conditions and mitigate problems such as soil erosion, nutrient depletion, and climate change. Incorporation of trees into agricultural landscapes, agroforestry systems can increase soil organic matter, improve soil structure, and enhance soil fertility. Trees can also protect soil from erosion by reducing wind, intercepting rainfall and enhancing soil stability. Additionally, trees can sequester carbon from the atmosphere, thereby mitigating climate change. Agroforestry has the potential to improve soil conditions and mitigate a range of environmental problems. It is a promising approach to sustainable agriculture that can benefit farmers, ecosystems, and society as a whole.

References

Albrecht A, Kandji ST (2000) Carbon sequestration in tropical agroforestry systems. Agric Ecosyst Environ 99:15–27

Altieri MA (1999) The ecological role of biodiversity in agroecosystems. Agric Ecosyst Environ 74(1–3):19–31

Brouder SM, Hofman VL (2019) Soil compaction in cropping systems: a review of the nature, causes, effects, and strategies for improvement. Soil Tillage Res 194:104292

Brussaard L, De Ruiter PC, Brown GG (2007) Soil biodiversity for agricultural sustainability. Agric Ecosyst Environ 121(3):233–244

Chatterjee N, Nair PR, Chakraborty S, Nair VD (2018) Changes in soil carbon stocks across the Forest-Agroforest-Agriculture/Pasture continuum in various agroecological regions: a meta-analysis. Agric Ecosyst Environ 266:55–67

CSSRI (2010–2012) Annual Reports 2010-2011, 2011-12, and 2012-13. Central Soil Salinity Research Institute, Karnal, India

Dhyani SK, Singh BP, Chauhan DS, Prasad RN (1994) Evaluation of MPTS for agroforestry systems to ameliorate fertility of degraded acid alfisols on sloppy lands. In: FSI.2009. India State of Forest Report 2009. Forest Survey of India, Dehradun

Food and Agriculture Organization of the United Nations (2008) An introduction to the basic concepts of food security. Rome, Italy

ICAR (2010) Degraded and wastelands of India: status and spatial distribution. Indian Council of Agricultural Research, New Delhi, p 158

IPCC (2000) Special report on land use. Land Use Change and Forestry. Summary for Policy Makers (eds.) Land use, land-use change, and forestry. A special report of the IPCC. Cambridge University Press, Cambridge

Jose S (2009) Agroforestry for ecosystem services and environmental benefits: an overview. Agrofor Syst 76(1):1–10. https://doi.org/10.1007/s10457-009-9229-7

Lal R (2020) Soil degradation as a reason for inadequate human nutrition. Food Secur 12(2): 259–278

Lin BB (2007) Agroforestry management as an adaptive strategy against potential microclimate extremes in coffee agriculture. Agric For Meteorol 144(1–2):85–94

Mbow C, Noordwijk MV, Luedeling E, Neufeldt H, Minang PA, Kowero G (2014) Agroforestry solutions to address food security and climate change challenges in Africa. Curr Opin Environ Sustain 6:61–67

Mishra A, Sharma SD, Pandey R, Mishra L (2004) Amelioration of a highly alkaline soil by trees in northern India. Soil Use Manag 20:325–332

Montagnini F, Nair PKR (2004) Carbon Sequestration: an under exploited environmental benefit of agroforestry systems. Agrofor Syst 61:281–295

Nair P (1993) An introduction to agroforestry. Springer, Dordrecht

NRCAF (2012) Garhkundar-Dabar watershed project- a successful model for Bundelkhand. Technical Bulletin 2/2012

Pandey DN (2007) Multifunctional agroforestry systems in India. Curr Sci 92(4):455–463

Riyadh ZA, Rahman MA, Saha SR, Hossain MI (2018) Soil properties under jackfruit based agroforestry systems in Madhupur tract of Narsingdi district. J Sylhet Agric Univ 5:173–179

Sharma PK, Kapoor KK (2005) Problems and prospective of nitrogen fixation in agroforestry systems. Proc Indian Natl Sci Acad B71:145–161

Singh BP, Dhyani SK, Prasad RN (1994) Traditional agroforestry systems and their soil productivity on degraded alfisols/ultisols in hilly terrain. In: Singh P, Pathak PS, Roy MM (eds.). Agroforestry systems for degraded lands, pp 205–214

United Nations (2015) Sustainable Development Goals: Goal 15 - Life on land. https://www.un.org/sustainabledevelopment/conservation-and-sustainable-use-of-ocean-resources/

Van Noordwijk M, Hoang MH, Neufeldt H, Öborn I, Yatich T (eds) (2011) How trees and people can co-adapt to climate change: reducing vulnerability through multi- functional agroforestry landscapes. Nairobi, World Agroforestry Centre (ICRAF)

Wartenberg AC, Blaser WJ, Roshetko J, Van'Noordwijk M, Six J (2020) Soil fertility and *Theobroma cacao* growth and productivity under commonly intercropped shade-tree species in Sulawesi, Indonesia. Plant Soil 453:87–104

World Bank (2015) Ending poverty and hunger by 2030. An agenda for the global food system. Author, Washington, DC

Young A (1997) Agroforestry for soil management, 2nd edn. CAB International, Wallingford

Chapter 17
Soil, Water, and Biodiversity Conservation Through Agroforestry for Crop Production

R. K. Meena, Tulika Kumari, V. K. Solanki, Vinita Partel, S. L. Singh, and Rashmi Sinha

Abstract Agroforestry is integrated with agriculture crops, perennial trees, and livestock or pasture. Agroforestry has the potential for sustainable development and increases the overall income of farmers though intervention of trees with agriculture crops. Nowadays the population of the world is increasing day by day, and total agriculture of land is decreasing; simultaneously demand of food also increases; to fulfill the demand of food, farmers are using high quantity of different chemical fertilizer, pesticides, insecticides, and herbicides for increasing the yield; it results to soil degradation and also affects environmental conditions. Agroforestry is a suitable tool to mitigate the soil degradation problems and mitigate climate change. Agroforestry is competent to protect natural resources like soil, water, nutrients of soil, ecological biodiversity, mainline the temperature and air quality by various agroforestry practices under different regions and conditions. It is estimated that the requirement of water will be increased up to 19% by 2050 and 5% of productivity is decreased due to soil degradation. Agroforestry is able to improve the infiltration rate, water-holding capacity of soil, improve the productivity of soil, and increase the nutrient status of soil and control of soil erosion. Agroforestry system gives good strength to conserve the soil and water. Different agroforestry systems are very diverse in nature, cultivating different food crops, fruit crops, and fodder crops and integrated with livestock. Due to the diverse nature of agroforestry, it is able to provide food security and health for the farmer. It is estimated that 75% foods are obtained from 12% of plants and animal that we are cultivating on a large scale. In this chapter the main focus is on the role of agroforestry in the conservation of soil, water, and environmental and ecological biodiversity and its role on the crop production.

R. K. Meena (✉) · T. Kumari · S. L. Singh · R. Sinha
Dr. Rajendra Prasad Central Agricultural University, Samstipur, Bihar, India
e-mail: rkmeena@rpcau.ac.in

V. K. Solanki · V. Partel
Jawaharlal Nehru Krishi Vishwa Vidyalaya, Jabalpur, MP, India

H. S. Jatav et al. (eds.), *Agroforestry to Combat Global Challenges*, Sustainable Development and Biodiversity 36, https://doi.org/10.1007/978-981-99-7282-1_17

Keywords Agroforestry · Sustainability · Soil and water conservation · Biodiversity conservation · Crop diversity and crop production

17.1 Introduction

Agroforestry is an old traditional farming practice, which is integrated with agriculture crops, trees, and livestock with pasture (Saravanan and Berry 2021). Agroforestry provides multiple benefits that include soil conservation, enhance nutrient status of soil, increase soil productivity, enhance biodiversity, increase the water-holding capacity of soil, improve microclimate, improve water quality, improve the nutrient cycle, improve the infiltration rate, increase the groundwater, etc. (Razafindratsima et al. 2021). Agroforestry also provides as a platform for sustainable way to improve the wetlands and soil and water conservation (Waldron et al. 2017). Agroforestry has potential to provide long-term agricultural benefits, and it provides an alternative for strengthening climate change resilience while contributing to food security, income, social health, and environmental sustainability. Agroforestry practices have capacity to maintain soil fertility in agroforestry systems through accumulation of biomass and organic carbon in the form of leaf litter, twigs, and dead branches of trees (Hinsinger et al. 2011). Agroforestry is also helpful to control soil erosion and reduce the losses of soil from the primary upper surface, and it contributes to maintain the soil productivity by improving soil conservation practices. Approximately 1.2 billion of the world populations are actively involved in agroforestry (Zomer et al. 2016; Garrity 2004). Agroforestry is capable to balance between demands, consumption, and natural degradation.

Different agroforestry practices like alley cropping, windbreak, shelterbelt, home garden, protein bank, taungya cultivation, etc. maintain good diversity and secure the foods for the human being and manage the nutrient cycle, agriculture sustainability, and environmental ecology. Soil and water are the major input for agricultural production. However, due to the development activity, soil properties are affected and start soil degradation. Degradation of soil is a serious problem in present time (Gardner 1996). Researcher believes that soil degradation and soil erosion have calamitous impact on agricultural production capacity (Scherr and Yadav 1996). It is estimated 5% productivity of loss is due to soil degradation (Crosson 1995). This chapter acknowledges the role of agroforestry for sustainable development, climate change, and mitigation, how agroforestry maintains the soil productivity, and how to conserve the ecological biodiversity and determine the effects on crop production.

17.2 Common Agroforestry Systems Practiced in India

India is a wide country; and it has 328.7 million hectares total geographical area; out of total area, 139.4 million hectares area is reported under the net sown area. Climatic conditions of India are differing throughout the country. Due to the variation in climatic conditions throughout the country, demand of fodder, food, fuel wood, and timber varies by region to region. Variation in demand leads to growing of different crops, trees, shrubs, and grasses to fulfill the demands under the different agroforestry system. Various agroforestry systems practiced in India are as follows in Table 17.1.

17.3 Need of Soil and Water Conservation

The world's population continues to increase, and demand of food and water is increasing day by day. It is estimated that the agriculture water consumption is increasing by 19% up to the 2050. The need of soil conservation and water management is more essential than ever, particularly as all world suffer from climate change. There is only one way to conserve the soil and water through decreasing the soil hampering from runoff and increase the rate of infiltration. Only agriculture crops are not capable to control the runoff rate. Agroforestry system gives good strength to soil and is capable to increase the infiltration rate of water into the soil. Deep tap root system of trees also improves the groundwater table and quality of water confining nutrient and metals which are accumulated on surface of soil and also in subsurface of soil. As an outcome, fertility of soil enhanced (Dury 1991). Therefore, agroforestry plays an important role to the resolution the problem of conserving the soil and water. Plantation of trees on agricultural land and farmland can diminish water requirement; it also helps in keeping hold of water for field crop and gives insurance to watersheds.

17.4 Role of Agroforestry for Soil and Water and Crop Production

Soil conservation is understood here generally to control the erosion of soil and manage the fertility of soil (Young 1989a, b). Agroforestry practices have a direct positive impact on soil fertility improvement and soil erosion control. Agroforestry systems are capable to control soil erosion through providing mulching of the upper surface of soil by tree canopy and leaf litter; trees also reduce the water erosion of soil through creating the barrier runoff and reduce the current of the water. In arid region trees are planted in agroforestry system; trees not only control the soil erosion; it also helps in water management through reducing the temperature and decreasing the rate of transpiration. Silvipastoral agroforestry systems should be incorporated when measuring possibility for control of erosion and water management in arid region. Legume trees and shrubs are preferred in all agroforestry system to fix the atmospheric nitrogen, improve the soil fertility, and improve crop productivity. Forest tree species have the ability to fix the atmospheric nitrogen; in this

Table 17.1 Common agroforestry systems practiced in India

Sr. no.	Name of system	Description
1.	Taungya system	Agricultural crops are interplanted with trees in the same area for 2–3 years. Agricultural crops are grown up to the closing of tree canopy. Taungya system has provided food for the forest workers and forage for cattle. This system is classified into three categories: (1) village taungya, (2) departmental taungya, (3) leased taungya system
2.	Integrated taungya	Integrated taungya is similar to taungya system; however here, once tree canopies are close, then grazing animals are allowed in the place of agricultural crops
3.	Improved fallow in shifting cultivation	Fellow tree species are introduced in degraded land to improve the soil health and minimize the soil and soil nutrient losses. Keep in mind that fellow species must be legume
4.	Alley cropping	In alley cropping system, agriculture crops are growing in alley between the rows of trees. Trees preferably legumes are regularly prune to prevent the shade effects on agriculture crop
5.	Trees on farmland	Farmers planted trees on their field to get additional income, food security, soil improvement, and amelioration of environment. Trees are planted to protect farm from adverse climatic conditions
6.	Scattered trees on farm	In this system different legume trees are planted in scattered pattern. Scattered trees are planted when crop cultivation on land becomes permanent. Trees are planted in a scattered pattern so that they do not compete for the light, water, and nutrient with the main crop. Planted trees must have deep root system, have capacity to fix the atmospheric nitrogen, and produce high leaf litter
7.	Boundary line plantation	Fast-growing tree species are planted on the boundary of farm land, road side, river side, live fences for farm, and control of soil erosion
8.	Shelterbelts	In shelterbelt trees, shrubs and grasses are planted at the right angle to the prevailing wind. Main function of the shelterbelt is protection of main crop from the strong wind
9.	Windbreaks	In windbreak a strip of trees are planted; generally one or two row trees are planted around the food crop. Main function of windbreak is protect the food crop from warm and cold wind and demarcation of farm boundary
10.	Home garden	Home garden is a multilayer farming system. It is practiced in high rainfall area in south India in which different trees, food crops, medicinal plants, and fruit trees are grown in different layers
11.	Multipurpose trees on farm land	Farmers planted few multipurpose tree species in farm with the agriculture crop. The trees normally planted there have economic importance
12.	Soil conservation hedges	Trees and shrubs are planted for the soil conservation purpose
13.	Aqua-forestry	In this system some flowering trees (*Moringa oleifera*, *Leucaena leucocephala*) are planted on the edge of the pond. Flowers and pods of the trees are good food for the fish

(continued)

Table 17.1 (continued)

Sr. no.	Name of system	Description
14.	Apisilviculture	Apisilviculture system different nectar producing species are growing to produce honey
15.	Protein bank	Perennial trees are growing for the production of fodder for the livestock. These practices mainly follow in dry areas where food is scarce in the summer season

Brown et al. (2018)

contest few examples are *Leucaena* species which fix the atmospheric nitrogen at the rate 400–500 kg/ha, *Acacia* species which fix the atmospheric nitrogen at the rate 270 kg/ha and *Alnus* species which fix the nitrogen at the rate 100–300 kg/ha (Misra 2011).

17.4.1 Effect of Agroforestry on Soil Fertility and Soil Health

In agroforestry systems plant nutrient status improves through decomposition of leaf litter, self and artificial pruning of trees, and atmospheric nitrogen fixation. Few nutrients are not available to the annual crops; generally they are known as unavailable nutrients, because they are out of reach of the rooting zone of agricultural crop; agroforestry trees have capacity to be brought under the system from deeper layers to upper layer of soil through deep root system of trees.

Agroforestry helps to increasing soil organic carbon into above- and below-ground by adding the leaf litter from trees which is incorporated in agroforestry system. Soil organic carbon is the important parameter to assume about soil health. Soil properties, viz., physical, chemical, and structural, are affected by the stored soil organic matter. Soil organic carbon also affects the biological properties like bulk density, water-holding capacity, infiltration rate, aggregate stability, cation exchange capacity, and availability of nitrogen. Agroforestry helps to increase the availability of plant nutrients. The incidence of competition and facilitation relation is noticed when plants are mixed in agroforestry system as functional groups (Jose et al. 2004a, b). Hence, trees function in agroforestry model as a facilitative well understood by proficient nutrient cycling. Yengwe et al. (2018) reported that the nutrient value of *Faidherbia albida* is incorporated with maize. It is also examined that *F. albida* is able to supply nitrogen 18 kg/ha/year and capable to improve soil microbe diversity and population. The microorganisms are measured significantly for ecosystem sustainability as well as soil health because microorganisms are important for the decomposition of leaf litter, soil organic matter, and nutrient recycling, thereby improving the chemical, physical, and biological status of soil, which will finally improve the fertility of soil and long-term sustainability. The soil microorganisms which includes both primary and secondary decomposers, performs nutrient cycling, break the organic matter and the availability of essential nutrients specially N mineralization. In agroforestry practices micro- and macroorganisms are generally found on the top soil surface. Dense network trees' fine roots are beneficial for plentiful mycorrhizal association, which affects the availability of nutrients in a positive manner to the participating crops. Density of population and soil fauna diversity are good tools to know about the soil condition and rehabilitation of ecosystem quality.

Health of soil can be also affected by impurity and poisonous martial. Agroforestry has the potential for the remediation of the pollutant site; some trees can be utilized for phytoremediation of the polluted area. Kaur et al. (2018) reported the tolerance capacity against the cadmium (Cd) of different four trees species, i.e., *Leucaena leucocephala*, *Melia azedarach*, *Dalbergia sissoo*, and *Eucalyptus tereticornis*.

17.4.2 Control of Soil Erosion Through Using Agroforestry Practices

Trees, shrubs, and herbs have been always planted to prevent the soil erosion. There are various techniques used to reduce soil erosion, strategies used to direct plant growth to slow down soil erosion, and soil stabilization. Direct use means to trees incorporated with the annual crop that improves the soil cover, decomposition rate, soil organic matter, and carbon content of soil. In the tropic region, agroforestry practices are extensively utilized to minimize the soil erosion; various agroforestry practices are used in different regions. Suitable agroforestry practices to control soil erosion for different regions are given in Table 17.2.

Grewal et al. (1994) studied the effect of two agroforestry systems, *viz.* (i) *Leucaena*-based agroforestry system and (ii) *Leucaena* and napier grass *(Pennisetum purpureum)*-based agroforestry system on degraded land of foot-hill ecosystem of the sub-tropical northern India and concluded that agroforestry system was a more effective conservation as compared to the traditional system. Grewal et al. (1992) studied the different agroforestry systems of Shiwalik region of the sub-Himalayas of northern India, viz., agri-silvi-horticulture system (*Leucaena* + lemon + papaya + turmeric) on class I, agri-silvicultural system (cluster beans + *Leucaena* gave) adopted on class II land, and silvipasture system on sandy loam soil under class III incorporated with (*Eucalyptus tereticornis* + bhabar grass) eucalyptus in most upper story and bhabar grass in the understorey. Sloping (25-30%) gravelly under the land capability class IV were best suited to grow Bhabbar and *Acacia* species. The results show the superiority of agroforestry systems over the conventional agriculture system on all land capability classes and losses of soil and soil nutrients approximately removed by bhabar grass. Investigated the sub-humid climate region of the western Himalayas and suggested that agroforestry systems with *Leucaena* hedgerows or tree-row barriers of *Leucaena* or *Eucalyptus* are a viable alternative to the conventional maize-wheat rotation when soil erosion control

Table 17.2 Common agroforestry practices for control of soil erosion (Young 1989a, b)

Agroforestry practice	Suitable condition	Remarks
Combinations of plantation crop	Humid region to moist subhumid climatic conditions	Agricultural plantation crops are planted densely with multipurpose tree species planted to prevent the soil erosion on least moderate slopes
Home garden and multistory tree gardens	Home gardens are developed mostly in humid and subhumid region	Home garden have great potential to control soil erosion through combination of herbs, shrubs, and trees with abundant litter
Alley cropping	Followed in humid, subhumid, and maybe semi-arid region	Alley cropping huge potential to minimize the soil erosion on gentle to moderate slopes
Shelterbelts and windbreaks	Practices mainly followed in dry, arid, and semi-arid region	Deflect the current of wind and reduce the wind velocity and control the wind erosion, highly effective for sandy soil
Silvipasture practices	Followed mainly in arid region and semi-arid area, sometime followed in subhumid climates	This system acts as an opportunity for pasture improvement by including trees and shrubs

is the main goal. Agricultural crops and trees must be grown in separate blocks if grain production is the main goal in order to provide timber, fuel, and other materials.

The use of agroforestry techniques prevents soil erosion. *Senna siamea* mulch applied to the upper surface reduces soil loss to just 13% of the usual loss, and alley cropping reduces soil loss by 2% in semi-arid Kenya (Kiepe 1996).

Leucaena leucocephala and maize plots were combined in an agroforestry system in Malawi in 1994 by Banda et al. They found that soil loss was 2 tonnes/ha/year, which was lower than the usual loss of 80 tonnes/ha/year. Indian Himalayan valley with less steep slope (4%), where countercultivation on maize plot was practiced and reported to have reduced runoff by 27% and soil loss by 45%. In a similar manner, *Leucaena* hedges and contour tree rows reduced soil loss by 48% and additional runoff by 40%, respectively, as compared to maize fields. In regions that weren't planted, soil loss dropped from 39 tonnes/ha/year to 12.5 tonnes (Narain et al. 1998). Roose and Ndayizigiye (1997) noticed that the threat of soil erosion is reduced in Rwanda by adopting agroforestry models. Angima et al. (2003) studied the tropical highland area of Kenya and reported the harshness of soil erosion, loss of soil 2.2 to 10¯tons/ha/year. Soil erosion rate was observed at 100–200 tons/ha/year in the Philippines in agricultural filed, similar rate of soil erosion in alley cropping was decreased by 51 tons/ha/year (Paningbatan et al. 1995). Sarminah et al. (2018) studied the effects of *Arachis hypogaea* and *Falcataria moluccana*-based agroforestry system for rehabilitation and soil conservation of degraded land and reported that land having slightly steep slope 15–25%, where 90% *Falcataria moluccana* survival was recorded and 70–80% ground was covered with the *Arachis hypogaea* and rate of soil erosion was 20.05 ton/ha/year reported, with an erosion hazard index of 0.80 (low); the survival percent of *Falcataria moluccana* was noticed to be 90%, ground cover with *Arachis hypogaea* was 50–60%, and soil erosion was noticed 45.50 ton/ha/year on steeper ground 25–40% with an erosion hazard index of 3.25 (moderate) and a low hazard level.

17.4.3 Windbreak Use for Erosion Control

Windbreak is one of the important agroforestry practices which is used for the control of soil erosion. It is mainly practiced in the arid and semi-arid region and where value of land is very high. The primary objective of windbreak is provide shelter for crops and animals. Windbreaks reduce wind direction, the mean wind speed, and airflow. As a result of windbreak, aerial environment, plant, and soil environmental conditions are changed because of changes to the following processes (Fig. 17.1). The design of windbreak is special and must be depending on the local climatic condition, spot condition, and objectives (Wright and Stuhr 2002). For the

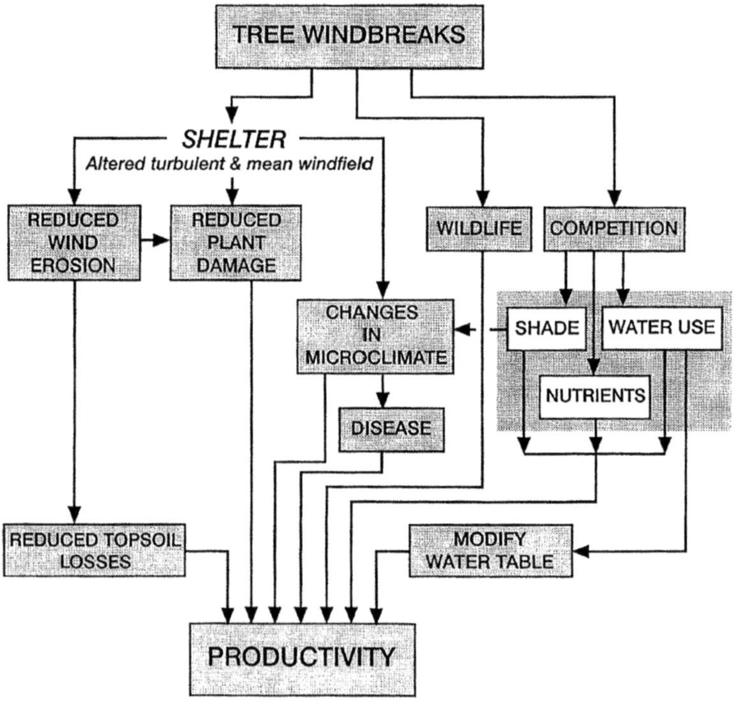

Fig. 17.1 Mechanisms of a windbreak affecting microclimate and plant productivity (Cleugh 1998)

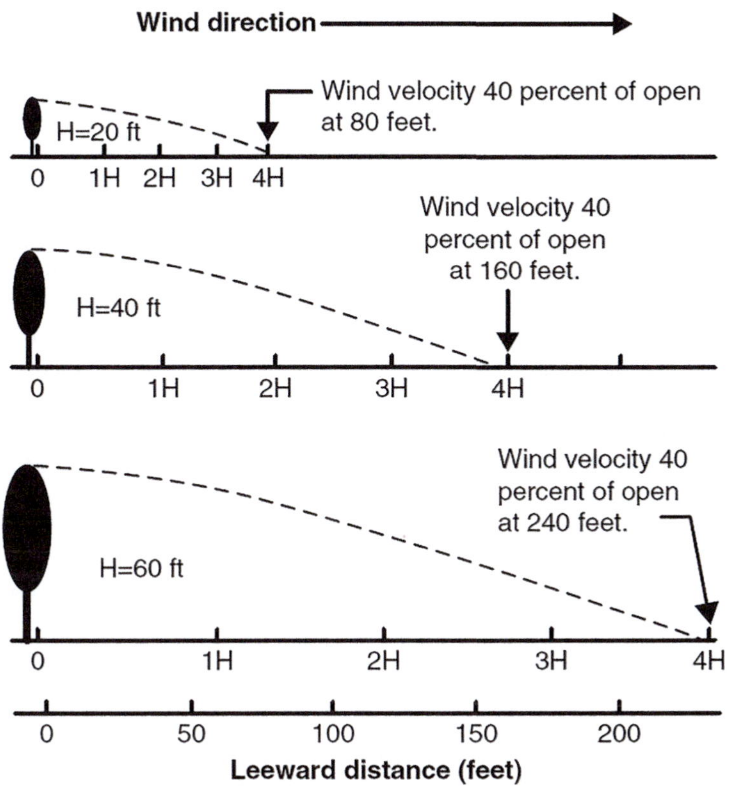

Fig. 17.2 The relationship between protected area and height of windbreak in leeward side (Wright and Stuhr 2002)

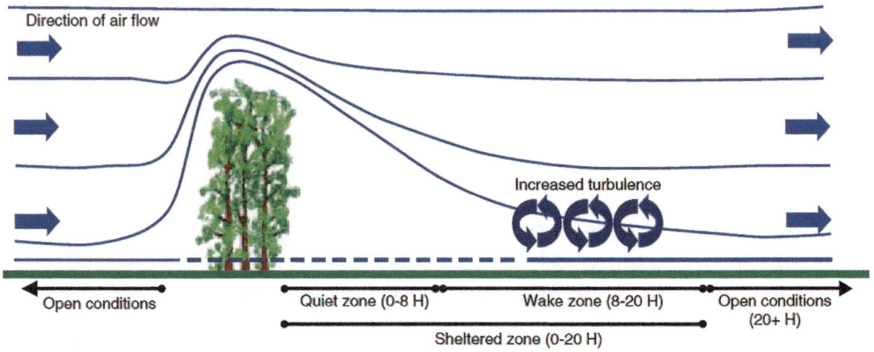

Fig. 17.3 Microclimate zone of windbreak in leeward side (H = tree height; Sudmeyer et al. 2007)

purpose of designing, the height of the windbreak must be considered in relation to the required wind protection's required leeward distance (Figs. 17.2 and 17.3).

An ideal windbreak would consist with two to four row of trees and shrubs; in a perfect windbreak, double row of fast-growing tall tree species are planted at central core, and next two rows of shrubs and small tree species which have muscular root are planted on both sides of the core. However, the form of trees changes throughout the growing period. It is essential to plant a number of species with different growth rates, form, architecture, shapes, and sizes in several lines. Few species of trees that develop quickly must be used to set up the needed impact as earlier possible. Suitable trees species for the windbreaks are *Andropogon gayanus*, *Faidherbia albida*, *Acacia nilotica*, *Acacia holosericea A.Cun.*, *Azadirachta indica, Bauhinia rufescens*, *Acacia senegal, Eucalyptus cladocalyx*, *Eucalyptus gomphocephala*, *Casuarina* spp., *Acacia torta, Pinus pinaster*, and *Anacardium occidentale* (Sudmeyer and Scott 2002). Legume plant species like *Senna siamea*, *Setaria*, *Leucaena*, and *Calliandra* are utilized as live fence in hilly region of Malawi and Rwanda with alley cropping (Banda et al. 1994; Roose and Ndayizigiye 1997; Kiepe 1996).

Windbreak density also plays an important role to check wind erosion. Utility of windbreak is increasing with increasing density of windbreak and it is decreasing with decreasing the density of windbreak. However, high-density windbreaks cause more harm than a perfect windbreak because they erode the soil on the windward side and destroy the crops on the leeward side. The wind speed is reduced by 20% in the leeward and windward side in the protected area of windbreak. The effect of windbreak is calculated as expressed as multiple height of windbreak tallest tree in the row (Fig. 17.2); theoretically, the practical impacts of windbreaks extend to 15–20 times of tallest tree height in leeward side and 2–5 times in the windward side. However, Sudmeyer and Scott (2002) conducted a research in Australia to assess the effect of windbreak on wind speed and microclimate on 450 m thick bay between two *Pinus pinster* over 4 years and reported that the wind speed decreases by 20% when direction of wind is perpendicular to the windbreak; it extended 18 times height of windbreak, and overall the growing season wind current reductions are greater than 20% up to the 3–6 times of height of windbreak. Windbreak affects the rate of soil erosion and also affects the composition of species within the windbreak. Conducted experiment in southwest Niger, to determine the effects of windbreak on soil erosion, and reported that, within 20 m distance strips of the perennial grass *Andropogon gayanus* reduced 6–55% soil flux annually and 2 m height hedges of *Bauhinia rufescens* reduced 47–77% soil flux annually as compared with unprotected plots. Puri et al. (1992) Studies the beneficial effects of *Dalbergia sissoo* shelterbelts and windbreaks in arid and semiarid region of Haryana under the social forestry program and reported that windbreak reduces the wind speed 15–45% and increases the growth of trees and productivity up to 4 times distance of the tree height in leeward side. Plant growth and yield were recorded high in protected area. Thevs et al. (2019) also reported the effect of a single row windbreak which is capable to reduce the speed by 60% compared to the open field condition.

17.5 Role of Agroforestry for Water Conservation and Crop Production

Continually rising food demand is accompanied by an increase in global population. It is an assumption that the requirement of water for agricultural will increase by 19% up to 2050. So there are essential needs to improve water management in agriculture, especially as we face already the climate change challenges.

Kort (1988) examined the impact of windbreak on crop and observed that yield of crop is higher in protected field by windbreak as compared to crops cultivated in open field conditions. Positive changes in yield percentage were observed in dry and semidry areas. Wheat yield was increased in the Chuy Valley, Kyrgyzstan, by 28% after planted trees as windbreak (Bulychev and Onishenko 1979). In India poplar-based agroforestry systems were recorded cost-effective compared to monocropping cultivation (Dwivedi et al. 2007); extra income from the poplar-based agroforestry is the main inspiration to engage in agroforestry (Dwivedi et al. 2007). In an interview campaign (Smith et al. 2021) and survey across the United States, it was found out that the majority of people and farmers maintain the windbreak for the indirect benefits from the windbreak, like protection of crop from the strong wind and control of soil erosion, efficiently increasing the yield and quality of production and efficiently utilizing the water.

17.5.1 Agroforestry Can Increase the Water Use Efficiency

Annual crops can only utilize few amount of available water. In India, sorghum share in transpiration at 41% of rainfall, and millet share in transpiration at 6 to 16% of annual rainfall in Nigeria; the rest water run off, drain and evaporate. Introducing some agroforestry perennial tree species in this agriculture system can capture a large amount of rainfall water. Study has shown that 70% annual rainfall water can be used by combinations of *Grevillea* trees with maize annual crop. Trees planted in agro-forestry field while regularly practicing the pruning of roots and shoots can increase the water utilization efficiency and facilitate new economic opportunities. Thevs et al. (2021) investigated the effects of single row windbreak with cotton (*Gossypium hirsutum* L.), corn (*Zea mays* L.), and rice (*Oryza sativa* L.) on water consumption on field condition and reported that windbreak helps in saving the water and reducing the consumption of water in irrigated condition. Windbreak trees regulate the water utilization of crops, improve the total yields of crop, and provide extra profit-making sources. Windbreak also controls the speed of wind, which is the main reason beyond to decrease crop water consumption Alemu (2016). Study conducted across Australia and New Zeeland to examine the effects of windbreak by Baker et al. (2018) reported that windbreak increases the crop yield and reduces the water requirement. The total protected area is depending upon the structure and design and height of windbreak (Peri and Bloomberg 2002). This explains the sharp decrease in water use and the lower water requirement during the early stage of tree development (up to 5 years) compared to a later time. Crop water consumption is affected by windbreak, and water consumption reduced by 6.6% compared with traditional cropping system in Northern China (Liu et al. 2018a, b). Baker et al. (2018) reported that tree windbreaks increase the yield and minimize the damage from irrigation.

17.5.2 Agroforestry Can Help Retain Water and Provide Watershed Protection

Agroforestry can improve crop productivity in several ways: improving soil organic matter, infiltration rate, and water storage, improving the physico-chemical properties of soil and enhancing the soil biological activity, and improving nutrient supplies through atmospheric nitrogen fixation and reduced leaching and soil erosion. Agroforestry improves all soil properties that are complementary to retain water and increases the soil moisture. Another important function of trees on agriculture field is providing shade, helping to maintain soil moisture. Peat land soil is managed with the forest plant as combining agricultural crop; because of forest plant species and agricultural crops, CO_2 absorption can increase, which increases the photosynthetic efficiency and production of oxygen. Total biomass production also increases due to the increasing photosynthetic efficiency and oxygen production. It can be beneficial for maintaining soil organic matter and preventing soil erosion. The effectiveness of rehydrating and water-holding capacity is improved by organic matter in the soil (Lestari and Mukhlis 2021). Intercropping rubber and pineapple (IRP) system makes a microclimate which plays a significant role to soil and water conservation. Organic matter in soil also will increase through adding leaf litter and its decomposition, which increases the water retention ability of water (Lestari and Mukhlis 2021). High densities of hedgerow plantations and plant wastes used as mulch helped in increasing infiltration rates and decreasing runoff in semi-arid Kenya and mountainous areas of the Philippines (Paningbatan et al. 1995; Kiepe 1996). Wu et al. (2019) studied the effect of agroforestry practices in karst rocky desertification (KRD) problematic areas and observed that effects of agroforestry practice help in the conservation of soil and water and increase the carbon sequestration, and agroforestry practices are also helpful for increasing soil fertility.

17.6 Biodiversity Conservation Through Agroforestry and Crop Production

Biodiversity means variability in living flora and fauna present in an ecosystem (Hamilton 2005; Carnus et al. 2006; Gugerli et al. 2008). Biodiversity is categorized as species diversity, genetic diversity, and ecological diversity (Mace et al. 2012;

Convention on Biological Diversity United Nations 1992; Srivastava and Vellend 2005; Swift et al. 2004). Flora and fauna present in an ecosystem affect each other in different manner, such as microorganisms, plants, animals, etc. (Vandermeer and Perfecto 1995). They play an important role to support the different ecological function to sustain the existence on earth and also provide other ecological services which are essential to support the life on earth, i.e., human economies, fauna, water quality, qualitative planting material and seeds, pollination, development of soil, nutrient management, control of pest and disease in plants and humans, carbon sequestration, and regulation of climate change and cultural services (Carnus et al. 2006; Balvanera et al. 2013; Mace et al. 2012; Mergeay and Santamaria 2012).

One of the main factors contributing to the loss of biodiversity is the expansion of agricultural land use (Udawatta et al. 2019). Thus, intensified land use for agricultural activities needs the search of new technologies for the biodiversity conservation while increasing the agricultural productions. Biodiversity conservation in agricultural field can be maximized by decreasing few management intensity, like application of chemical fertilizer, herbicide, insecticide, and pesticide. Another important way to conserve the biodiversity is add in natural or semi-natural areas landscapes nearby agricultural areas (Gonthier et al. 2014).

17.6.1 Need of Biodiversity Conservation for the Crop Production

All over the world facing the problem of losses of biodiversity, this is the major issue for the worldwide community, and it is very essential to reduce the rate of global biodiversity defeat and destruction (Sala et al. 2000a, b; Tilman et al. 2001). At the present time, a number of effects are observed in ecosystem function due to the losses of the species (Cardinale et al. 2012), such as increase of the population of insects due to the losses of some bird species from the ecosystem, and a number of ecological services provided by a number of species have economic importance (Daily 1997; Perrings et al. 2006). There are a number of potential species present on earth which are used for the food purpose. FAO reported that the 75% of food is currently obtained from 12% of plants and animal that we are cultivating on a large scale. However, they've extended a number of varieties of crops, fruits, vegetables, and animal. Diversity is very important for the cultivated crops and animals because it helps in the development of new improved varieties due to the large gene pool for traits like disease and pest resistance. Cultivation of only a few number of varieties of different crops makes sure to fulfil the demand of food, and biodiversity becomes vulnerable to threats (Vigouroux et al. 2011). Because of cultivating one or two varieties of crops continuously in a single field, it dangers the local biodiversity, food nutritional value, human livelihood security, and other ecological services, and also crops become susceptible to the different diseases and pests (Thrupp 2000; Sistla et al. 2016). Conservation of wild relative crop plants is also essential from the industrial agriculture; it is also important for the point of development of new improved varieties of different crops: wild populations are kept the valuable genes.

17.6.2 Biodiversity and Climate Change

Modern agricultural industry is a danger for the biodiversity at global level being one of the most important reasons for the climate change due to emissions like poisonous chemicals from the chemical fertilizers and other factories of modern agriculture. It is estimated that 20% greenhouse gas is released from the agricultural industry annually.[2] Climate change is the most important serious hazard to biodiversity now. Even effects of climate change are seen on remote location area of biodiversity where human scarcely touched. Impacts of increasing in temperature already have an example on migratory birds. Due to climate change, weather cycle can also completely change, and effects are seen on the plant and soil community.[3]

17.6.3 How Monocropping Destroys Biodiversity

Various researchers observed that the single-species cultivation or plantations is evidently having a number of negative impact on social and environmental conditions which is not good economic benefit (Alem et al. 2015). Monocropping cultivation practiced by farmers provides simplicity and less effort and a regular supply of feed to industry, but it is a danger to biodiversity (Jacques and Jessica 2012). Monocrop cultivation requires large number of chemical in high quantity that is responsible for the decrease of the presence of wild species on agriculture field and off the agriculture filed. The chemical composition of pesticides, insecticide, and herbicides is designed in such manner to remove the pests that can damage or fight with main crop crops; however, at the same time, these chemicals are also harmful to plants and animals, which is outside the farm fields. Monoculture cultivation is not suitable for soil; it is a major cause of soil erosion and degradation (Baltodano 2000; Bowyer 2006). Single-species cultivation or plantations are not capable for the efficient utilization of available soil nutrient, because all roots are present in same surface and uptake same nutrients from same surface, which significantly leads to soil degradation and nutrient losses (Liu et al. 2018a, b). Agroforestry systems always have shown great floristic and structure diversity as compared to the single cropping system; agroforestry system has shown great faunal diversity because faunal diversity is always related with floristic diversity (Jose 2012).

17.6.4 Role of Agroforestry for Biodiversity Conservation and Crop Production

Biodiversity has great importance and a number of benefits and losses and changes at the global level at an alarming rate (Pimm et al. 1995; Jose 2012). Population growth rate is continuously increasing at the global level; due to the increasing population, the consumption rate per capita also increases, and better dietary requirement with enhancing income has resulted in overutilization of earth's ecological diversity. It is estimated that total population will reach 9.5 billion up to 2050 if it is not controlled timely (Udawatta et al. 2019). Another important factor is climate change which is also responsible for the decline of the biodiversity (Pimm and Biodiversity 2008; Sala et al. 2000a, b). Due to increasing population demand of food, to fulfill the demand of food, agricultural industry and agriculture activity continuously increase which are responsible for the decline of biodiversity, and deforestation is also a major reason for the losses of ecological and biological diversity and function (Steffan-Dewenter et al. 2007; Culman et al. 2010). Continuous cultivation of only one or two crops or varieties on the similar agriculture land leads to the loss of agriculture biodiversity and affects livelihood, food security, food nutrition, and other ecological services (Thrupp 2000; Sistla et al. 2016). Largely modern agriculture is responsible for the decline in agriculture biodiversity and ecosystem biodiversity; agricultural lands need better management plans to provide better service to biodiversity survival (Kleijn and Sutherland 2003). For example, in the countries of Europe, agriculture-related habitats and land support 50% of the population of both plants and animals (Kristensen 2003). Therefore, agricultural practices are used for the support to improving biodiversity and its conservation. Bring to a focus from the traditional agroforestry to more advanced agroforestry systems and approach to establish a better connection among agroforestry and biodiversity preservation (McNeely and Schroth 2006). Agroforestry received great attention from the last two decades for the conservation approaches of biodiversity (McNeely and Schroth 2006; Buck et al. 2004). Agroforestry has been a recognized measure to conserve rich species biodiversity throughout the world (Mendez et al. 2001; Borkhataria et al. 2012). Agroforestry plays five major roles to conserve the biodiversity (Jose 2012). (1) Agroforestry offers habitat for plants and animals that can withstand little disturbance. (2) Agroforestry aids in protecting the genetic diversity of threatened plant and animal species. (3) A more effective, sustainable alternative to conventional agricultural methods that may be responsible for destroying natural habitats is provided by agroforestry, which aids in reducing the rate at which natural habitats are being converted. (4) Agroforestry creates pathways between habitat remnants, which may increase their dependability and aid in the preservation of species of plants and animals that are dependent on particular geographic regions. (5) Agroforestry supports other ecological services such as soil erosion control, groundwater recharge, increasing soil nutrient status, maintaining the nutrient cycle, soil health, air quality, pollination, pest management, fire retardation, and cultural services such as improving aesthetic, recreational, and cultural values to preserve biological and ecological diversity (Torralba et al. 2016; Schulze et al. 2004; Udawatta et al. 2011). Agroforestry system can help to improve the variety of pollinator, which is necessary for the production of food. Agroforestry also helps in maintaining the population of wild plant diversity (Varah et al. 2013). Pollinator service is valuable for the flowering plants and crops. It is estimated that 75% of the world's key crops and 90% of blooming plants are pollinated by insects, and animal pollinators impact an estimated 35% of food production (Klein et al. 2003). Agroforestry also helps conserve the adjoining habitat. Agroforestry can be utilized as a tool in association with suitable conservation practices to buffer biodiversity loss because a number of agroforestry practices have 50–80% diversity of similar natural habitat diversity; it is helpful for the further conservation of biodiversity. It is reported that agroforestry helps in biodiversity improvements in both temperate and tropical regions (Huang et al. 2002; Noble and Dirzo 1997; Dollinger and Jose 2018). Few findings have reported that biological and ecological diversity is higher in agroforestry systems compared to that in forests (Steffan-Dewenter et al. 2007; Sistla et al. 2016; Huang et al. 2002). Agroforestry systems have a 60% higher diversity of species than forests, according to a meta-analysis. Another meta-analysis in Europe noticed an overall better impact of agroforestry on biodiversity (Torralba et al. 2016). Sistla et al. (2016) studied 17 agroforestry sites, 8 sites adjacent to secondary forest, and 7 pasture sites in the Pearl Lagoon Basin. They found that agroforestry sites have higher surface soil carbon percent, nitrogen content, and pH relative than secondary forest. The agroforestry system is found to have high biodiversity and to be very beneficial for crop production because species richness, phylogenetic variety, and natural resource biodiversity are all quite low in grassland.

Transformation from agroforestry to monoculture has reduced biodiversity richness of agriculture field (Perfecto et al. 1996; Schroth et al. 2004). Despite having more biodiversity than nearby agricultural and forest systems, agroforestry systems typically contain fewer native species due to extensive management (Noble and Dirzo 1997). Recent research has shown a strong positive correlation between biodiversity and ecosystem services and has strongly advocated biodiversity protection to boost ecological services (Rands et al. 2010; Hooper et al. 2005; Gallai et al. 2009). Humans receive a variety of essential services from biodiversity, including direct and indirect advantages and the regulation of environmental processes (Rands et al. 2010; Alkemade et al. 2009). Biodiversity contributes to maintain the healthy physical and mental health of human (Barton and Pretty 2010). Biodiversity contributes to pest management agriculture and pollination and provides long-term flexibility to conflict and climate change (Hooper et al. 2005). Biodiversity contributes significantly to economic and social growth (Gallai et al. 2009). At the genetic, species, and farming system levels, biodiversity provides essential ecological and biological services and functions for the generation of agricultural products (Thrupp 2000). Ecosystem services, food production, animal and plant revenue, as well as increased health risks and malnutrition were all seen to be negatively impacted by a decline in biodiversity (Leakey et al. 2006). A significant number of researchers have found out the positive bound between agroforestry and ecosystem services of agroforestry (Jose 2012; Steffan-Dewenter et al. 2007; Perfecto et al. 1996).

Agroforestry is able to improve the land productivity as combinations of trees, crops, and pasture animal together. Combination of all components (trees, crops, pasture, and animals) could use natural resources more efficiently than monoculture cropping pattern or trees (values) (Torralba et al. 2016; Jose et al. 2004a, b). In Kyoto Protocol, it is accepted that agroforestry is a measure for carbon sequestration for mitigation of the climate change. All these benefits can be directly related with diversity of agroforestry systems.

17.7 Conclusion

The utilization of various tree species and other techniques used in agroforestry systems might offer alternate ways to improve soil fertility and preserve agricultural productivity, with significant real-world implications for the sustainability of agriculture. A sustainable ecosystem depends on healthy soil and water as its primary resources. Agroforestry, as a sustainable land management practice, has shown positive effects on its role in improving soil quality, water management, nutrient cycling, efficient utilization of water, improving and maintaining soil microorganisms and their activity, and maintaining soil organic carbon. Conservation and protection practices in agroforestry are helpful for crop production, sustainable development, and climate change mitigation.

References

Alemu MM (2016) Ecological benefits of trees as windbreaks and shelterbelts. Int J Ecosyst 6:10–13

Alem H, Kibebew K, Heluf G (2015) Characterization and classification of soils of Kabe Subwatershed in South Wollo Zone, Northeastern Ethiopia. Afr J Soil Sci 3(7):134–146

Alkemade R, Van Oorschot M, Miles L, Nellemann C, Bakkenes M, Brink B (2009) GLOBIO3: A framework to investigate options for reducing global terrestrial biodiversity loss. Ecosystems 12:374–390

Angima SD, Stott DE, O'Neill MK, Ong CK, Weesies GA (2003) Soil erosion prediction using RUSLE for central Kenyan highland conditions. Agric Ecosyst Environ 97(1–3):295–308

Baker TP, Moroni MT, Mendham DS, Smith R, Hunt MA (2018) Impacts of windbreak shelter on crop and livestock production. Crop Pasture Sci 69:785–796. https://doi.org/10.1071/CP17242

Baltodano J (2000) Monoculture forestry: a critique from an ecological perspective. In: Tree Trouble: A Compilation of Testimonies on the Negative Impact of Large-scale Monoculture Tree Plantations Prepared for the 6th COP of the FCCC. Friends of the Earth International, p 2e10

Balvanera P, Siddique I, Dee L, Paquette A, Isbell F, Gonzalez A, Byrnes J, O'Connor MI, Hungate BA, Griffin JN (2013) Linking biodiversity and ecosystem services: current uncertainties and the necessary next steps. Bioscience 64:49e57. https://doi.org/10.1093/biosci/bit003

Banda AZ, Maghembe JA, Ngugi DN, Chome VA (1994) Effect of intercropping maize and closely spaced *Leucaena* hedgerows on soil conservation and maize yield on a steep slope at Ntcheu, Malawi. Agrofor Syst 27:17–22

Barton J, Pretty J (2010) What is the best dose of nature and green exercise for improving mental health? A multi-study analysis. Environ Sci Technol 44:947–3955

Borkhataria RR, Collazo JA, Groom MJ (2012) Species abundance and potential biological control services in shade vs. sun coffee in Puerto Rico. Agric Ecosyst Environ 151:1–5

Bowyer JL (2006) Forest plantations threatening or saving natural forests? Arborvitae (IUCN/ WWF Forest Conservation Newsletter) 31:8e9

Brown SE, Miller DC, Ordonez PJ, Baylis K (2018) Evidence for the impacts of agroforestry on agricultural productivity, ecosystem services, and human well-being in high-income countries: a systematic map protocol. Environ Evid 24. https://doi.org/10.1186/s13750-018-0136-0

Buck LE, Gavin TA, Lee DR, Upho NT, Behr DC, Drinkwater LE, Hively WD, Werner F, Ecoagriculture R (2004) A review and assessment of its scientific foundations. Cornell University, Ithaca, NY

Bulychev AS, Onischenko LP (1979) Field protection wind-breaks in Chuy valley and their efficiency. Frunze

Cardinale BJ, Duffy JE, Gonzalez A, Hooper DU, Perrings C, Venail P, Narwani A, Mace GM, Tilman D, Wardle DA, Kinzig AP, Daily GC, Loreau M, Grace JB, Larigauderie A, Srivastava DS, Naeem S (2012) Biodiversity loss and its impact on humanity. Nature 486(7401):59–67. https://doi.org/10.1038/nature11148

Carnus JM, Parrotta J, Brockerhoff E, Arbez M, Jactel H, Kremer A, Ponette Q, Walters B (2006) Planted forests and biodiversity. J For 104(2):65e77

Cleugh HA (1998) Effects of windbreaks on airflow, microclimates, and crop yields. Agrofor Syst 41:55–84

Convention on Biological Diversity, United Nations (1992). cbd.int/doc/legal/cbd-en.pdf

Crosson P (1995) Soil erosion and its on-farm consequences: what do we know? Discussion Paper 95-29, Resources for the Future, Washington, D.C.

Culman SW, Young MA, Hollander AD, Ferris H, Sanchez-Moreno S, O'Green AT, Jackson LE (2010) Biodiversity is associated with indicators of soil ecosystem functions over a landscape gradient of agricultural intensification. Landsc Ecol 25:1333–1348

Daily GC (1997) Nature's services: societal dependence on natural ecosystems. Island Press, Washington, DC, p 185. Destruction. Oakland: Food First Books; 1995

Dollinger J, Jose S (2018) Agroforestry for soil health. Agrofor Syst 92:213–219

Dury SJ (1991) Agroforestry for soil conservation: Anthony Young. CAB International, Wallingford, UK, in cooperation with ICRAF; 1989, p 276

Dwivedi RP, Singh KKR, Rizvi RH, Chauhan J (2007) Socio-economic analysis of agroforestry systems in western Uttar Pradesh. Indian Res J Ext Educ 7:18–22

Gallai N, Salles JM, Settele J, Vaissiere BE (2009) Economic valuation of the vulnerability of world agriculture confronted with pollinator decline. Ecol Econ 68:810–821

Gardner G (1996) Shrinking fields: cropland loss in a world of 8 billion, p 55. Worldwatch Paper 131, Washington, DC

Garrity DP (2004) Agroforestry and the achievement of the Millennium Development Goals. Agrofor Syst 61(1):5–17

Gonthier DJ, Ennis KK, Farinas S, Hsieh HY, Iverson AL, Batary P, Rudolphi J, Tscharntke T, Cardinale BJ, Perfecto I (2014) Biodiversity conservation in agriculture requires a multi-scale approach. Proc R Soc B 281:20141358. https://doi.org/10.1098/rspb.2014.1358. rspb.royalsocietypublishing.org

Grewal SS, Mittal SP, Dyal S, Agnihotri Y (1992) Agroforestry systems for soil and water conservation and sustainable production from foothill areas of north India. Agrofor Syst 17: 183–191

Grewal SS, Juneja ML, Singh K, Singh S (1994) A comparison of two agroforestry systems for soil, water and nutrient conservation on degraded land. Soil Technol 7:145–153

Gugerli F, Englisch T, Niklfeld H, Tribsch A, Mirek Z, Ronikier M, Zimmermann NE, Holderegger R, Taberlet P, IntraBioDiv Consortium (2008) Relationships among levels of biodiversity and the relevance of intraspecific diversity in conservation a project synopsis. Perspect Plant Ecol Evol Syst 10(4):259e281. https://doi.org/10.1016/j.ppees.2008.07.001

Hamilton AJ (2005) Species diversity or biodiversity? J Environ Manag 75(1):89e92. https://doi.org/10.1016/j.jenvman.2004.11.012

Hinsinger P, Betencourt E, Bernard L, Brauman A, Plassard C, Shen J, Tang X, Zhang F (2011) P for two, sharing a scarce resource: soil phosphorus acquisition in the rhizosphere of intercropped species. Plant Physiol 156:1078–1086

Hooper DU, Chapin FS, Ewel JJ, Hector A, Inchausti P, Lavorel S, Lawton JH, Lodge DM, Loreau M, Naeem S, Schmid B, Setala H, Symstad AJ, Vandermeer J, Wardl DA (2005) Effects of biodiversity on ecosystem functioning: a consensus of current knowledge. Ecol Monogr 75: 3–35

Huang W, Luukkanen O, Johanson S, Kaarakka V, Raisanen S, Vihemaki H (2002) Agroforestry for biodiversity conservation of nature reserves: functional group identification and analysis. Agrofor Syst 55:65–72

Jacques PJ, Jessica JR (2012) Monocropping cultures into ruin: the loss of food varieties and cultural diversity. Sustainability 4:2970–2997. https://doi.org/10.3390/su4112970

Jose S (2012) Agroforestry for conserving and enhancing biodiversity. Agrofor Syst 85:1–8

Jose S, Gillespie AR, Pallardy SG (2004a) Interspecific interactions in temperate. Agroforestry. Agrofor Syst 61–62:237–255. https://doi.org/10.1023/B:AGFO.0000029002.85273.9b

Jose S, Gillespie AR, Pallardy SG (2004b) Interspecific interactions in temperate agroforestry. Agrofor Syst 61:237–255

Kaur B, Singh B, Kaur N, Singh D (2018) Phytoremediation of cadmium-contaminated soil through multipurpose tree species. Agrofor Syst. https://doi.org/10.1007/s10457-017-0141-2

Kiepe P (1996) Cover and barrier effects of Cassia siamea hedgerows on soil conservation in semiarid Kenya. Soil Technol 9:161–171

Kleijn D, Sutherland WJ (2003) How effective are European agri-environment schemes in conserving and promoting biodiversity? J Appl Ecol 40:947–969

Klein AM, Stean-Dewenter I, Tscharntke T (2003) Fruit set of highland coffee increases with the diversity of pollinating bees. Proc R Soc Lond Ser B Biol Sci 270:5–961

Kort J (1988) Benefits ofwindbreaks to field and forage crops. Agric Ecosyst Environ 22123:165–190

Kristensen P (2003) EE Core Set of Indicators; Revised Version April 2003; European Environment Agency: Copenhagen, Denmark, 2003

Leakey R, Tchoundjeu Z, Schreckenberg K, Simons T, Shackleton S, Mander M, Wynberg R, Shackleton C, Sullivan C (2006) Trees and markets for agroforestry tree products: targeting poverty reduction and enhanced livelihoods. In: Garrity D, Okono A, Grayson M, Parrott S (eds) World agroforestry into the future. World Agroforestry Centre, Nairobi, Kenya, pp 11–22

Lestari Y, Mukhlis (2021) Peatland water conservation by agroforestry system. E3S Web Conf 305: 03004. https://doi.org/10.1051/e3sconf/202130503004

Liu CLC, Kuchma O, Krutovsky KV (2018a) Mixed-species versus monocultures in plantation forestry: development, benefits, ecosystem services and perspectives for the future. Global Ecol Conserv. 15:e00419. https://doi.org/10.1016/j.gecco.2018.e00419

Liu Z, Yu X, Jia G, Zhang J, Zhang Z (2018b) Water consumption by an agro ecosystem with shelter forests of corn and Populus in the North China Plain. Agric Ecosyst Environ 265:178–189. https://doi.org/10.1016/j.agee.2018.05.027

Mace GM, Norris K, Fitter AH (2012) Biodiversity and ecosystem services: a multilayered relationship. Trends Ecol Evol 27(1):19e26. https://doi.org/10.1016/j.tree.2011.08.006

McNeely JA, Schroth G (2006) Agroforestry and Biodiversity conservation-traditional practices, present dynamics, and lessons for the future. Biodivers Conserv 15:549–554

Mendez VE, Lok R, Somarriba E (2001) Interdisciplinary analysis of home gardens in Nicaragua: micro-zonation, plant use and socioeconomic importance. Agrofor Syst 51:85–96

Mergeay J, Santamaria L (2012) Evolution and biodiversity: the evolutionary basis of biodiversity and its potential for adaptation to global change. Evol Appl 5(2):103e106. https://doi.org/10.1111/j.1752-4571.2011.00232.x

Misra PK (2011) Soil fertility management in agroforestry system. Int J Biotechnol Biochem 7(5): 637–644

Narain P, Singh RK, Sindhwal RS, Joshie P (1998) Agroforestry for soil and water conservation in the western Himalayan Valley Region of India. 1. Runoff, soil, and nutrient losses. Agrofor Syst 39:175–189

Noble IR, Dirzo R (1997) Forests as human-dominated ecosystems. Science 277:522–525

Paningbatan EP, Ciesiolka CA, Coughlan KJ, Rose CW (1995) Alley cropping for managing soil erosion of hilly lands in the Philippines. Soil Technol 8(3):193–204

Perfecto I, Rice RA, Greenberg R, Van DVME (1996) Shade coffee: a disappearing refuge for biodiversity. Bioscience 46:598–608

Peri PL, Bloomberg M (2002) Windbreaks in southern Patagonia, Argentina: a review of research on growth models, windspeed reduction, and effects on crops. Agrofor Syst 56:129–144

Perrings C, Jackson L, Bawa K, Brussaard L, Brush S, Gavin T, Papa R, Pascual U, De Ruiter P (2006) Biodiversity in agricultural landscapes: saving natural capital without losing interest. Conserv Biol 20:263–264. https://doi.org/10.1111/j.1523-1739.2006.00390

Pimm S, Biodiversity L (2008) Climate change or habitat loss-which will kill more species? Curr Biol 18:R117–R119. https://doi.org/10.1016/j.cub.2007.11.054

Pimm SL, Russell GJ, Gittleman JL, Brooks TM (1995) The future of biodiversity. Science 269: 269–350

Puri S, Singh S, Khara A (1992) Effect of windbreak on the yield of cotton crop in semiarid regions of Haryana. Agrofor Syst 18:183–195

Rands MR, Adams WM, Bennun L, Butchart SH, Clements A, Coomes D, Entwistle A, Hodge I, Kapos V, Scharlemann JPW (2010) Biodiversity conservation: challenges beyond 2010. Science 329:1298–1303. https://doi.org/10.1126/science.1189138

Razafindratsima OH, JFM K, Sills EO, Mutta DN, Song C, Kabwe G, Castle SE, Kristjanson PM, Ryan CM, Brockhaus M, Abd Terry S (2021) Reviewing the evidence on the roles of forests and tree-based systems in poverty dynamics. For Policy Economics 131:102576. https://doi.org/10.1016/j.forpol.2021.102576

Roose E, Ndayizigiye F (1997) Agroforestry, water and soil fertility management to fight erosion in tropical mountains of Rwanda. Soil Technol 11(1):109–119

Sala OE, Chapin FS, Armesto JJ, Berlow E, Bloomfield J, Dirzo R, Huber-Sanwald E, Huenneke LF, Jackson RB, Kinzig A, Leemans R, Lodge DM, Mooney HA, Oesterheld M, Poff NL, Sykes MT, Walker BH, Walker M, Wall DH (2000a) Global biodiversity scenarios for the year 2100. Am Assoc Adv Sci 287:1770–1774

Sala OE, Chapin FS, Armesto JJ, Berlow E, Bloomfield J, Dirzo R, Huber-Sanwald E, Huenneke LF, Jackson RB, Kinzig A, Rik L, Lodge DM, Mooney HA, Oesterheld M, LeRoy Poff N, Sykes MT, Walker BH, Walker M, Wall DH (2000b) Global biodiversity scenarios for the year 2100. Science 287:1770–1774

Saravanan S, Berry N (2021) Agroforestry practices in Tamil Nadu, India-a boon for farmers for livelihood security. Curr Sci 120(4):644–653

Sarminah S, Karyati K, Simbolon J, Tambunan E (2018) Rehabilitation and soil conservation of degraded land using sengon (*Falcataria moluccana*) and peanut (*Arachis hypogaea*) agroforestry system. Biodiversitas 19(1):222–228

Scherr SJ, Yadav S (1996) Land degradation in the developing world: implications for food, agriculture, and the environment to 2020. IFPRI Discussion Paper No. 14, Washington, DC, p 36

Schroth G, Da Fonseca GAB, Harvey CA (2004) Agroforestry and biodiversity conservation in tropical landscapes. Island Press, Washington, DC, p 523

Schulze CH, Waltert M, Kessler PJ, Pitopang R, Veddeler D, Mühlenberg M, Gradstein SR, Leuschner C, Stean-Dewenter I, Tscharntke T (2004) Biodiversity indicator groups of tropical land-use systems: comparing plants, birds, and insects. Ecol Appl 14:1321–1333

Sistla SA, Roddy AB, Williams NE, Kramer DB, Stevens K, Allison SD (2016) Agroforestry practices promote biodiversity and natural resource diversity in Atlantic Nicaragua. PLoS One 11(9):e0162529. https://doi.org/10.1371/journal.pone.016252

Smith MM, Bentrup G, Kellermann T, MacFarland K, Straight R, Ameyaw L (2021) Windbreaks in the United States: a systematic review of producer-reported benefits, challenges, management activities and drivers of adoption. Agric Syst 187:103032

Srivastava DS, Vellend M (2005) Biodiversity-ecosystem function research: is it relevant to conservation? Annu Rev Ecol Evol Syst 36:267–294. http://www.jstor.org/stable/30033805

Steffan-Dewenter I, Kessler M, Barkmann J, Bos MM, Buchori D, Erasmi S, Faust H, Gerold G, Glenk K, Gradstein SR (2007) Tradeoffs between income, biodiversity, and ecosystem functioning during tropical rainforest conversion and agroforestry intensification. Proc Natl Acad Sci 104:4973–4978. https://doi.org/10.1073/pnas.0608409104

Sudmeyer RA, Scott PR (2002) Characterisation of a windbreak system on the south coast of Western Australia. 1. Microclimate and wind erosion. Aust J Exp Agric 42:703–715

Sudmeyer RA, Bicknell D, Coles N (2007) Tree windbreaks in the wheatbelt. Department of Agriculture and Food, Government of Western Australia, Bulletin 4723. ISSN 1833-7236

Swift MJ, Izac AM, Van Noordwijk M (2004) Biodiversity and ecosystem services in agricultural landscapes are we asking the right questions? Agric Ecosyst Environ 104(1):113e134. https://doi.org/10.1016/j.agee.2004.01.013

Thevs N, Gombert AJ, Strenge E, Lleshi R, Aliev K, Emileva B (2019) Tree wind breaks in Central Asia and their effects on agricultural water consumption. Land 8:167–183. https://doi.org/10.3390/land8110167

Thevs N, Kumar A, Roland L (2021) Water productivity of tree wind break agroforestry systems in irrigated agriculture-An example from Ferghana Valley, Kyrgyzstan. Trees For People 4:100085. https://doi.org/10.1016/j.tfp.2021.100085

Thrupp LA (2000) Linking agricultural biodiversity and food security: the valuable role of agrobiodiversity for sustainable agriculture. Int Aff 76(2):283–297. https://doi.org/10.1111/1468-2346.00133

Tilman D, Fargione J, Wolff B, Antonio CD, Dobson A, Howarth R, Schindler D, Schlesinger WH, Simberloff D, Swackhamer D (2001) Forecasting agriculturally driven global environmental change. Science 292:281–284. https://doi.org/10.1126/science.1057544

Torralba M, Fagerholm N, Burgess PJ, Moreno G, Plieninger T (2016) Do European agroforestry systems enhance biodiversity and ecosystem services? A meta-analysis. Agric Ecosyst Environ 230:150–161

Udawatta RP, Garrett HE, Kallenbach RL (2011) Agroforestry buffers for nonpoint source pollution reductions from agricultural watersheds. J Environ Qual 40:800–806

Udawatta RP, Rankoth LM, Jose S (2019) Agroforestry and biodiversity. Sustainability 11:2879. https://doi.org/10.3390/su11102879

Vandermeer J, Perfecto I (1995) Breakfast of biodiversity: the truth about rainforest. Institute for Food and Development Policy, - Nature – 185

Varah A, Jones H, Smith J, Potts SG (2013) Enhanced biodiversity and pollination in UK agroforestry systems. J Sci Food Agric 93:2073–2075. https://doi.org/10.1002/jsfa.6148

Vigouroux Y, Barnaud A, Scarcelli N, Thuillet AC (2011) Biodiversity, evolution and adaptation of cultivated crops. C R Biol 334:450–457

Waldron A, Garrity D, Malhi Y, Girardin C, Miller DC, Seddon N (2017) Agroforestry can enhance food security while meeting other sustainable development goals. Trop Conserv Sci 10:1–6. https://doi.org/10.1177/1940082917720667

Wright B, Stuhr K (2002) Windbreaks: an agroforestry practice. USDA Agroforestry Note 25, USDA Forest Service NAC, University of Nebraska, Lincoln, p 4

Wu Q, Liang H, Xiong, K. and. Li, R. (2019) Eco-benefits coupling of agroforestry and soil and water conservation under KRD environment: frontier theories and outlook. Agrofor Syst 93: 1927–1938. https://doi.org/10.1007/s10457-018-0301-z

Yengwe J, Gebremikae MT, Buchan D, Lungu O, Neve SD (2018) Effects of *Faidherbia albida* canopy and leaf litter on soil microbial communities and nitrogen mineralization in selected Zambian soils. Agrofor Syst. https://doi.org/10.1007/s10457-016-0063-4

Young A (1989a) Agroforestry for soil conservation. CAB International, Wallingford, p 276. (with index)

Young A (1989b) Agroforestry for Soil Conservation. Science and Practice of Agroforestry No. 4. International Council for Research in Agroforestry, Nairobi, Kenya

Zomer RJ, Neufeldt H, Xu J, Ahrends A, Bossio D, Trabucco A, Noordwijk MV, Wang M (2016) Global tree cover and biomass carbon on agricultural land: the contribution of agroforestry to global and national carbon budgets. Sci Rep 6:29987. https://doi.org/10.1038/srep29987

Chapter 18
Breeding of Jatropha For Oil, Phorbol and Quantitative Traits for Sustainable Yield Under Agroforestry System

Hausila Prasad Singh, Vinay Rojaria, Noopur Singh, Saroj Chauhan, and Om Prakash Raigar

Abstract Continuous emission of carbon from fossil fuels into the environment causes an adverse effect in the form of climate change and global warming. To overcome these issues, the world is moving towards zero carbon emission using alternative renewable resources of energy such as biodiesel. In this context, different biofuel potential plants are screened, and *Jatropha curcas* has been suggested as a promising biofuel crop. But because of toxic nature with low oil percentage in most of the *Jatropha* germplasm, it requires a precise breeding programme to increase oil percentage and other desirable features. After rigorous breeding programme, the first cultivar Chhatrapati of *Jatropha* was released, but low seed yield was the major drawback. This indicates a more precise breeding programme required for the development of high yielding desirable cultivars. In the current scenario, the cultivar JPNT 1 is the first commercial non-toxic high yielding *Jatropha* with 62–64% kernel oil. Similarly, cultivar JO S2 was released from the National University of Singapore having 45.90% seed oil content. Various researchers from different institutes are focusing on breeding of *Jatropha* for high seed yield, oil content and non-toxicity for the development of promising cultivars/hybrids. Therefore, this book chapter has critically reviewed on profound information of breeding outcomes for high oil, low phorbol, quantitative traits and promising developed and released genotypes yet.

Keywords Biofuel · Breeding · Inheritance · Mapping · Phorbol

H. P. Singh (✉) · V. Rojaria
Department of Genetics and Plant Breeding, BRD PG College, Deoria, India

N. Singh
Department of Genetics and Plant Breeding, Institute of Agriculture and Natural Sciences, Deen Dayal Upadhyaya Gorakhpur University, Gorakhpur, India

S. Chauhan
Department of Agricultural Entomology, Institute of Agriculture and Natural Sciences, Deen Dayal Upadhyaya Gorakhpur University, Gorakhpur, India

O. P. Raigar
Department of Plant Breeding and Genetics, Punjab Agricultural University, Ludhiana, India

H. S. Jatav et al. (eds.), *Agroforestry to Combat Global Challenges*, Sustainable Development and Biodiversity 36, https://doi.org/10.1007/978-981-99-7282-1_18

18.1 Introduction

Rising prices, depleting natural resources, increasing environmental pollution and global warming are major drawbacks of natural oil. To address these constraints, researchers reviewed plant biomass as a source of renewable energy which resulted in production of biofuel from damaged foodgrains, sugarcane, algae, etc. India has started the National Biofuel Mission in 2003 which aims at cultivation of shrubs and trees containing non-edible oil that can be easily cultivated on non-arable land (Das 2020; Saravanan et al. 2020). Under this mission, five tree-borne oilseeds (TBOs), viz. *Pongamia pinnata*, *Azadirachta indica*, *Simmondsia chinensis*, *Vernicia fordii* and *Jatropha curcas*, have been identified based on their oil content and agronomic performance (Dhyani et al. 2015). Amongst TBOs *Jatropha curcas* has a short gestation period (Singh et al. 2013) with high seed oil content determined to be as an alternate source of fossil fuel (Banerji et al. 1985). Crop stores a large quantity of hydrocarbons, thus prevailing environmental concerns; hence, it can be used as a renewable energy resource for the commercial production of biodiesel (Mandpe et al. 2005; Ghosh et al. 2007; Alherbawi et al. 2021a).

Breeding of *Jatropha* focuses on high kernel yield and kernel oil content which was started in Egypt and India in 2005, respectively (Gour 2015; Hegazy 2017). The crop is a large shrub that belongs to the family Euphorbiaceae and is native of tropical America. It is a monoecious crop with protandrous nature; thus geitonogamy and xenogamy favour pollination (Gour 2015). Seed contains 20–40% oil (Samra et al. 2014), and the variation in oil content is subjected to allogamous nature of the crop (Kaushik et al. 2007). *Jatropha* oil has the potential to be converted into biodiesel/jet fuel substitute or extender by the process of transesterification (Foidl et al. 1996; Tapanes et al. 2008; Deng et al. 2010; Alherbawi et al. 2021b). Geographical distribution of *Jatropha* exists into toxic and non-toxic form; toxic nature is due to the presence of phorbol-12-myristate-13-acetate (PMA) (Makkar et al. 2011; Francis et al. 2013; Salazar-Villa et al. 2020; Alherbawi et al. 2021a). The PMA content in non-toxic *J. curcas* is below 0.1 mg/g in kernel (Makkar et al. 1998a; Rodrigues et al. 2021). Defatted oilcake obtained from non-toxic varieties can be utilized as cattle feed; in addition to this, it also protects human as well as animal health from the adverse effect of phorbol in the environment released due to incomplete combustion of oil from the vehicles.

Major constraints for its commercial adoption are the lack of high yielding varieties, inconsistent yield over years, presence of toxic phorbol esters and absence of key traits for domestication (Achten et al. 2010; Montes and Melchinger 2016; Jonas et al. 2021). Several varieties have been acclaimed in the crop (Gour 2015), but only a few are utilized commercially due to the presence of high yield and key traits of domestication (Achten et al. 2010; Montes and Melchinger 2016; Vargas-Carpintero et al. 2021). Makkar and Becker (2009) reviewed the potential of non-toxic *Jatropha* for the production of biofuels that could enhance the economic viability of *Jatropha* seed oil-based biodiesel production. This book chapter provides a single platform to discuss the current conclusive outcome for oil, phorbol and quantitative traits through breeding from various organizations involved in biofuel programmes.

18.2 Breeding History

Naturally, the available genotype is in its wild form; thus breeding is required for increasing seed oil content and reducing the toxic level of PMA. *Jatropha* improvement through breeding has been started with the work of Makkar et al. (1998a, b), and evidence from literature indicated that very few breeding efforts have been made (Francis et al. 2018, 2019) to improve *Jatropha* genotypes in the toxic or non-toxic background with high oil and stabilized seed yield. High yielding development programme for Jatropha has been started since 2005 (Alfredo and Quintero 2017). Achten et al. (2010) recommended that domestication of *Jatropha* to a level requires at least 15 years of conventional breeding, but this period was shortened by using molecular breeding. Planned use of wide crosses in Jatropha has boosted the conventional breeding programme for introgression of desirable traits (Basha and Sujatha 2007; Popluechai et al. 2009). Crossing of *J. curcas* with *J. integerrima* resulted in high seed yield as reported by Maghuly and Laimer (2013). Yue et al. (2013) suggested that genetic improvement made through conventional breeding provides higher seed yield; however expected seed oil percentage was unanticipated (Yue et al. 2013). To overcome this concern, speed breeding is a remedial solution that can be achieved through marker-assisted selection and genomic selection. Currently, breeders are using combinatorial breeding approach to improve quantitative traits (Peixoto et al. 2021). The major objectives for *Jatropha* breeding programmes are high yield with improved oil content along with development of non-toxic plants with resistant to biotic and abiotic stresses (Rao et al. 2008; Mishra 2009; Maghuly and Laimer 2013; Singh et al. 2021).

18.3 Breeding for Oil, Phorbol Esters and Quantitative Traits

Maghuly and Laimer (2013) defined plant breeding as "identifying and selecting desirable genetic variations of useful traits for crop improvement". In the absence of desirable variation, plant breeder proceeds towards a complementary mutation tool to create genetic variation or to alter the genotype (Jiang et al. 2012). Dhakshanamoorthy et al. (2011) achieved mutants which flowered early and had increased number of fruits/plant and kernel yield/plant. In order to lower the toxicity level in plants, traditional, chemical and biochemical methods have been used extensively. In this regard, conventional genetic tools have been utilized to shorten breeding cycle, developing better yielding genotypes with greater oil yield in toxic

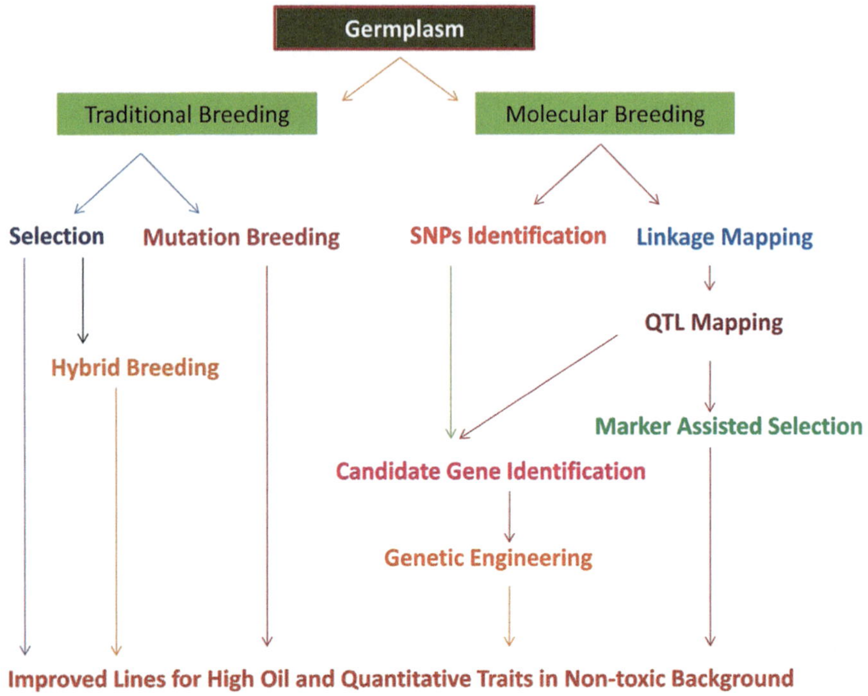

Fig. 18.1 Application of various breeding tools for improving existing germplasm in *J. curcas* for high oil, low phorbol and quantitative traits

and non-toxic backgrounds (Fig. 18.1). Large-scale application of molecular markers has been used substantially to estimate genetic diversity with the help of RAPD markers (Ganesh et al. 2008), RAPD and AFLP markers (Sudheer Pamidimarri et al. 2009a, b), RAPD and ISSR markers (Basha and Sujatha 2007) and SNP markers (de Souza et al. 2021) for identification of diverse lines, characterization, polymorphism and transferability across *Jatropha* (Kumar et al. 2011), but none of these studies emerged with its practical utilization.

18.3.1 Breeding for Oil

In Indian accession, the oil content has been reported from 28 to 37% with seed index variation from 49 to 79 g (Kaushik et al. 2007; Achten et al. 2010). Although *Jatropha* starts bearing fruits after the third month of transplanting, however it becomes economically viable only after the third year onwards. On the contrary, Rao et al. (2008) reported more than 2 kg of dry seed/plant after the fifth year. In 2008, Virgin Airlines first demonstrated use of *Jatropha* oil with 20% blending as jet fuel (Alherbawi et al. 2021b). India ranks first among the developing nations to use partially blended *Jatropha* biofuel in aviation industries (The Hindu 2018). In order to achieve zero carbon emission by 2070, the Indian government has directed to use 20% blended ethanol in petrol by 2025 (NITI Aayog 2020). After these successful attempts, breeders are emphasizing to enhance the oil quality for utilization of *Jatropha* as jet fuel (Alherbawi et al. 2021b).

Ewunie et al. (2021) identified variation in seed and kernel oil content at different altitudes in Ethiopian genotypes of *Jatropha curcas*. The outcome of this research represented that 35.84% and 64.15% of fruit's dry weight is contributed by shell and seed, respectively, whereas seed index varied from 50.53 to 68.97 g, kernel (% in dry seeds) from 49.07 to 66.74% and kernel oil content from 47.10 to 59.32%, respectively, in Ethiopian genotypes. Similarly, Pant et al. (2006) used composite seed samples of six trees to estimate seed oil variation in *Jatropha* plantations at different conditions, viz. arable (T1) and non-arable (T2), and at three different altitudes E1, E2 and E3 ranging 400–600 m, 600–800 m and 800–1000 m, respectively, at Himachal Pradesh, India. G x E interaction was significant for oil yield. The highest oil yield was recovered in non-arable (42.34%) in E1 (43.19%) and lowest in arable conditions (34.97%) in E3 (30.66%). Thus, it indicated that the variability in kernel characteristics and oil yield varied with geographical location. Later, Kaushik et al. (2007) described that the variability in oil yield is due to the allogamous nature of the crop. The study on variations in the vertical split of matured capsules and kernel oil content (Ghatak and Gour 2014) was recorded highest in capsules with four vertical splits (52.58%). Singh (2016) described plants with higher number of seeds with low seed index indicate the presence of seeds without kernel or kernel with low density, thus producing low oil content.

The oil estimations performed in kernel generate reliable data due to absence of shell. The kernel oil yield (%) varied from 47.08 to 58.12% (Ginwal et al. 2004), 13 to 58.20% (Gohil and Pandya 2009), 35.50 to 51.10% (Rao et al. 2009), 57.40 to 57.50% (Makkar et al. 2011), 23.44 to 52.58% (Ghatak and Gour 2014), 35.60 to 66.08% (Francis et al. 2018), 33.80 to 44.20% (Andrianirina et al. 2019), 50.60 to 60.30% (Salazar-Villa et al. 2020) and 47.10 to 59.32% (Ewunie et al. 2021).

However, reports on oil content (%) in whole seeds ranged from 33.02 to 39.12% (Ginwal et al. 2004), 28 to 38.8% (Kaushik et al. 2007), 29.80 to 37.05% (Rao et al. 2008), 8.10 to 30.6% (Gohil and Pandya 2009), 27.40 to 35.5% (Shabanimofrad et al. 2011), 17.50 to 36.70% (Sunil et al. 2011) and 25.16 to 30.13% (Jonas et al. 2021).

18.3.2 Breeding for Phorbol Esters

The genotypes which carry high level of anti-nutritional and toxic factors are of limited use (Devappa et al. 2010; Francis et al. 2013; Alherbawi et al. 2021a). Makkar et al. (1998a) collected non-toxic and toxic genotypes from Nicaragua, Nigeria and Mexico. Physical, chemical, biological and biochemical methods were devised and used to reduce these antinutrient levels, whereas genetic method is the best suited to utilize non-toxic sources (Abou-Arab et al. 2019; Salazar-Villa et al. 2020; Zhang et al. 2021). Phorbol content in non-toxic *J. curcas* does not exceed the threshold toxicity level, namely, 0.1 mg/g (Makkar et al. 1998a; Jonas et al. 2021; Rodrigues et al. 2021). The variation for phorbol content (Makkar and Becker 1997) was studied in 18 provenances from Western and Eastern Africa, Northern and Central America and Asia. They identified non-toxic lines from Mexico, whereas the level of phorbol varied from 0.87 to 3.32 mg/g of kernel. Identification of non-toxic lines in *J. curcas* from Mexico and a new species *J. platyphylla* (Makkar et al. 2011) with 60% oil in kernel opened a new avenue to transform *Jatropha* potential in non-toxic background. Identification and classification of toxic and non-toxic *Jatropha* have been successfully demonstrated by molecular techniques (Basha and Sujatha 2007; Basha et al. 2009; Tanya et al. 2011; King et al. 2013; Trebbi et al. 2019; de Souza et al. 2021) which is evident from practical breeding for genetic improvement.

18.3.3 Breeding for Quantitative Traits

Interspecific hybridization was resorted to create genetic variability and introgress desirable genes (Basha and Sujatha 2009; Sudheer Pamidimarri et al. 2009a, b; Senthil et al. 2009) for incorporating yield contributing traits for high kernel yield and kernel oil content due to low to moderate genetic variation in Indian collections ascertained by molecular diversity analysis (Basha and Sujatha 2007; Sudheer Pamidimarri et al. 2009a, b). The derivatives of such hybrids were utilized by Singh (2016) in backcross breeding programme to develop population for identification and selection with few important traits of domestication. Chromosomal locations of various qualitative and quantitative traits are represented below (Fig. 18.2). Landmark approach citing a combination of conventional breeding, molecular genetics and introgression of pleiotropic QTLs for plant growth and kernel yield has been described (Sun et al. 2012; Xia et al. 2018; Arockiasamy et al. 2021; dos Santos et al. 2021) for *Jatropha* improvement. QTL mapping to decipher the genetic basis of kernel yield using *J. curcas* x *J. integerrima* crosses indicated that (1) two diverse parents contribute for favourable alleles, (2) bidirectional nature of transgressive segregants and (3) association of chromosomal regions with more than two parameters indicating their complex nature and involvement of pleiotropy or linkage. These findings emphasized on use of elite *Jatropha* varieties as recurrent parents to incorporate favourable alleles for growth and kernel yield by backcross breeding.

Through molecular breeding various quantitative trait loci (QTLs) have been identified for fatty acid and total oil content (Liu et al. 2011); plant height, number of branches, female flowers and fruits (Sun et al. 2012); seed length (Ye et al. 2014); plant height, stem diameter, number of branches, seeds/plant, 100 seed weight and oil content (King et al. 2015); fruit yield (Xia et al. 2018); and *Jatropha* mosaic virus (Kancharla et al. 2019). Work on linkage and QTL mapping (Liu et al. 2011; Sun

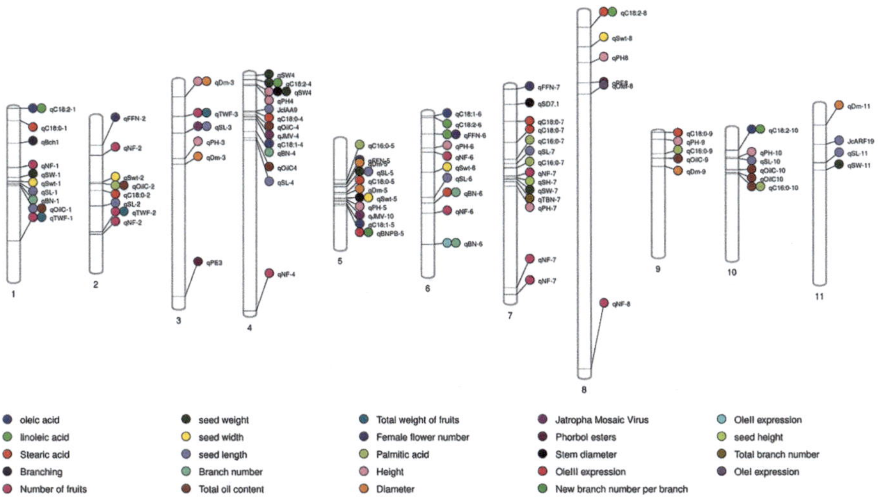

Fig. 18.2 Chromosomal locations of various qualitative and quantitative traits in *Jatropha*

et al. 2012) at TLL and National University of Singapore is among the pioneer institutes which led to the development of *Jatropha* variety JO S2 through mass selection (Yi et al. 2014). The transgenic approach also helps in the improvement of *Jatropha* by *Agrobacterium tumefaciens* infection (Pamidimarri et al. 2009; Li et al. 2007). Also the achievement was obtained for the biosynthesis of fatty acid (Li et al. 2008).

Screening of all available literatures reveals the presence of QTLs for various qualitative and quantitative traits from chromosome number 1 to 11, in which QTLs for oil content were located on chromosomes 1, 2, 4, 5, 6, 7, 8, 9 and 10; for phorbol it was located on chromosomes 3 and 8, while for various quantitative traits, it was present on all chromosomes. Yepuri et al. (2022) used 411 SNP markers to identify 83 QTLs for yields and oil content by using 11 linkage groups. Identification and utilization of QTLs are required to develop new cultivar of *Jatropha* with improved agronomic performance for seed and oil content (King et al. 2015). The detailed summarized information of QTLs for oil, phorbol and various quantitative traits have been represented in Table 18.1 and Fig. 18.2, respectively.

18.4 Inheritance of Phorbol

The F_1 hybrid developed between toxic and non-toxic plants and their first-generation backcross could be used to detect inheritance of toxicity (Achten et al. 2010). Estimation of PMA in seeds and its various parts, viz. shell, kernel, tegmen, endosperm and cotyledon, revealed its presence in both endosperm and tegmen (more concentrated in tegmen) (He et al. 2011; Kumar et al. 2018). The presence of

Table 18.1 A summarized information on identified QTLs for oil, phorbol and quantitative traits

Sl. No.	QTLs	Traits	Chr.	Marker interval	Start position	End position	Position	Lod	R^2/PVE	References
1	qC18:1-1	Oleic acid C18:1 (%)	1	Jcuint057	0.000	0.000	0.000	18.400	36.000	Liu et al. (2011)
2	qC18:2-1	Linoleic acid C18:2 (%)	1	Jcuint057	0.000	0.000	0.000	16.500	34.100	Liu et al. (2011)
3	qC18:0-1.1	Stearic acid C18:0 (%)	1	1,398,420–12,336,456	2.000	25.100	9.900	3.570	6.400	King et al. (2015)
4	qBch1.1	Branching	1	–	0.000	25.090	25.000	3.680	11.200	King et al. (2015)
5	qNF-1.1	Number of fruits	1	Jatr749	42.400	42.400	42.400	2.470	5.500	Sun et al. (2012)
6	qSW-1	Seed weight	1	Jatr749	42.400	57.200	49.400	5.170	10.000	Ye et al. (2014)
7	qSwt-1	Seed width	1	Jatr749	42.400	55.900	52.400	3.020	3.600	Ye et al. (2014)
8	qSL-1	Seed length	1	Jatr722	47.600	58.230	53.100	2.660	3.800	Ye et al. (2014)
9	qBN-1	Branch number	1	Jatr722	54.100	54.100	54.100	3.440	5.600	Sun et al. (2012)
10	qSL-1	Seed length	1	Jatr722	53.100	58.300	55.100	10.680	17.500	Ye et al. (2014)
11	qOilC-1	Total oil content (%)	1	Jatr722	55.100	55.100	55.100	2.300	4.600	Liu et al. (2011)
12	qNF-1.2	Number of fruits	1	KK914295.1_173222 Jatropha666	93.880	95.930	94.670	5.273	18.967	Xia et al. (2018)
13	qNF-1.2	Number of fruits	1	KK914295.1_173222 Jatropha666	93.880	93.930	94.670	4.893	19.633	Xia et al. (2018)
14	qTWF-1	Total weight of fruits	1	KK914295.1_173222 Jatropha666	93.880	93.930	94.670	4.893	19.633	Xia et al. (2018)
15	qFFN-2	Female flower number	2	Jatr691	20.700	20.700	20.700	2.000	4.500	Sun et al. (2012)
16	qNF-2.1	Number of fruits	2	KK914286.1_1650230 Jatropha580	26.950	27.920	27.920	6.567	19.367	Xia et al. (2018)
17	qSwt-2	Seed width	2	Jcuint143	43.300	55.300	46.400	13.990	19.900	Ye et al. (2014)
18	qC16:0–2	Palmitic acid C16:0 (%)	2	Jcuint143	47.400	47.400	47.400	2.600	0.100	Liu et al. (2011)

19	qOilC-2	Total oil content (%)	2	Jcuint143	47.400	47.400	47.400	2.500	4.900	Liu et al. (2011)
20	qC18:0–2	Stearic acid C18:0 (%)	2	curcin2	52.600	52.600	52.600	2.600	5.300	Liu et al. (2011)
21	qSL-2	Seed length	2	Jatr964	52.600	69.300	63.300	2.920	4.900	Ye et al. (2014)
22	qNF-2.2	Number of fruits	2	KK914708.1_14953 Jatropha3259	81.400	81.550	81.510	6.733	19.333	Xia et al. (2018)
23	qNF-2.3	Number of fruits	2	KK914708.1_14953 Jatropha3259	81.400	81.510	81.510	5.257	17.400	Xia et al. (2018)
24	qTWF-2	Total weight of fruits	2	KK914708.1_14953 Jatropha3259	81.400	81.510	81.510	5.257	17.400	Xia et al. (2018)
25	qNF-2.4	Number of fruits	2	KK914399.1_1189397 Jatropha1302	83.230	83.580	83.500	5.367	18.000	Xia et al. (2018)
26	qPH-3	Height	3	Jatr1054	11.900	11.900	11.900	3.550	6.300	Sun et al. (2012)
27	qDm-3	Diameter	3	Jatr1054	11.900	11.900	11.900	8.110	12.700	Sun et al. (2012)
28	qNF-3.1	Number of fruits	3	KK914370.1_355263 Jatropha1134	25.580	26.050	25.760	5.520	18.033	Xia et al. (2018)
29	qNF-3.2	Number of fruits	3	KK914370.1_355263 Jatropha1134	25.760	25.940	25.760	4.340	16.067	Xia et al. (2018)
30	qNF-3.3	Number of fruits	3	KK914370.1_355263 Jatropha1134	25.580	26.050	25.760	5.520	18.033	Xia et al. (2018)
31	qTWF-3	Total weight of fruits	3	KK914370.1_355263 Jatropha1134	25.760	25.940	25.760	4.340	16.067	Xia et al. (2018)
32	qJMV-3	Jatropha mosaic virus	3	RJM613 - RJM1472	21.900	58.900	37.000	4.500	–	Kancharla et al. (2019)
33	qSL-3	Seed length	3	Jcuint221	27.800	50.600	37.100	2.200	3.000	Ye et al. (2014)
34	qPH-3	Height	3	Jcuint179	55.800	55.800	55.800	3.500	5.000	Sun et al. (2012)
35	qDm-3	Diameter	3	Jcuint048	59.900	59.900	59.900	3.850	4.900	Sun et al. (2012)
36	qPE3.1	Phorbol esters	3	CN_SSR326–CN_SSR32	152.500	156.500	153.000	3.820	14.100	Amkul et al. (2017)

(continued)

Table 18.1 (continued)

Sl. No.	QTLs	Traits	Chr.	Marker interval	Start position	End position	Position	Lod	R^2/PVE	References
37	qSW4.2	Seed weight	4	1,407,326–12,327,601	0.000	19.000	1.340	5.040	16.100	King et al. (2015)
38	qSW4.3	Seed weight	4	–	0.000	52.000	4.000	3.440	13.200	King et al. (2015)
39	qC18:2–4.1	Linoleic acid C18:2 (%)	4	–	0.000	36.000	4.000	4.750	11.100	King et al. (2015)
40	qPH4.1	Height	4	G37	1.000	13.000	7.050	3.030	9.200	King et al. (2015)
41	qSD4.1	Stem diameter	4	G37	5.000	11.210	7.050	4.350	14.900	King et al. (2015)
42	qSD4.2	Stem diameter	4	G37	0.670	10.000	7.050	3.700	8.900	King et al. (2015)
43	qSW4.1	Seed weight	4	G37	1.300	15.000	7.050	7.900	22.600	King et al. (2015)
44	qPH4.2	Height	4	–	3.340	25.730	8.000	3.190	7.000	King et al. (2015)
45	JcIAA9	Seed length	4	IAA9	26.200	26.200	26.200	4.000	5.500	Ye et al. (2014)
46	qC18:0–4.1	Stearic acid C18:0 (%)	4	–	23.000	39.000	27.000	6.010	12.300	King et al. (2015)
47	qOilC-4.1	Total oil content (%)	4	Jatr872	29.600	29.600	29.600	5.000	11.100	Liu et al. (2011)
48	qJMV-4	Jatropha mosaic virus	4	RJM236 - RJM1157	58.000	88.600	30.600	3.100	–	Kancharla et al. (2019)
49	qC18:1-4.1	Oleic acid C18:1 (%)	4	–	2.000	34.300	32.000	4.730	13.300	King et al. (2015)
50	qBN-4	Branch number	4	Jatr854	41.000	41.000	41.000	3.510	5.500	Sun et al. (2012)
51	qOilC4.2	Total oil content (%)	4	–	0.000	57.100	45.500	3.270	10.800	King et al. (2015)

52	qSL-4	Seed length	4	Jcuint028	45.600	52.300	47.100	4.000	5.500	Ye et al. (2014)
53	qNF-4.1	Number of fruits	4	KK915534.1_3177 Jatropha5670	167.290	168.110	168.110	5.093	18.033	Xia et al. (2018)
54	qNF-4.2	Number of fruits	4	KK915534.1_3177 Jatropha5670	167.290	168.110	168.110	5.093	18.033	Xia et al. (2018)
55	qC16: 0-5.1	Palmitic acid C16: 0 (%)	5	–	19.200	41.600	28.000	5.480	13.200	King et al. (2015)
56	qFFN-5	Female flower number	5	OleI	30.300	30.300	30.300	3.380	7.000	Sun et al. (2012)
57	qDm-5.1	Diameter	5	OleI	31.300	31.300	31.300	3.880	5.600	Sun et al. (2012)
58	qSW-5.1	Seed weight	5	Jcuint002	32.200	36.020	33.700	9.330	17.700	Ye et al. (2014)
59	qSL-5.1	Seed length	5	Jcuint002	32.020	36.500	33.700	2.850	3.800	Ye et al. (2014)
60	qSL-5.2	Seed length	5	Jcuint002	31.300	34.000	33.700	7.760	11.600	Ye et al. (2014)
61	qSW-5.2	Seed weight	5	Jcuint002	33.700	33.700	33.700	2.240	5.200	Sun et al. (2012)
62	qC18:0-5	Stearic acid C18:0 (%)	5	Jatr746	37.300	37.300	37.300	6.900	13.000	Liu et al. (2011)
63	qDm-5.2	Diameter	5	Jatr746	38.300	38.300	38.300	15.030	21.100	Sun et al. (2012)
64	qSD5.1	Stem diameter	5	G123	26.000	44.020	41.100	3.230	8.500	King et al. (2015)
65	qSwt-5.3	Seed width	5	Jatr735	40.900	42.800	41.900	14.440	19.700	Ye et al. (2014)
66	qPH-5	Height	5	Jatr945	42.100	42.100	42.100	5.810	8.500	Sun et al. (2012)
67	qJMV-10	Jatropha mosaic virus	5	RJM1836 - RJM2234	143.700	187.300	43.600	3.300	–	Kancharla et al. (2019)
68	qC18:1-5	Oleic acid C18:1 (%)	5	Jatr739	45.100	45.100	45.100	2.300	3.400	Liu et al. (2011)
69	qOleIII-5	OleIII expression ($\Delta\Delta C_T$)	5	Jatr739	46.200	46.200	46.200	3.100	11.700	Liu et al. (2011)
70	qBNPB-5	New branch number per branch	5	Jatr739	46.200	46.200	46.200	2.980	7.900	Sun et al. (2012)
71	qC18: 1-6.1	Oleic acid C18:1 (%)	6	–	2.000	11.000	2.000	3.470	10.800	King et al. (2015)

(continued)

Table 18.1 (continued)

Sl. No.	QTLs	Traits	Chr.	Marker interval	Start position	End position	Position	Lod	R^2/ PVE	References
72	qC18: 2-6.1	Linoleic acid C18:2 (%)	6	–	0.000	7.000	3.000	5.050	11.900	King et al. (2015)
73	qC18: 2-6.2	Linoleic acid C18:2 (%)	6	Jatr301	15.000	15.000	15.000	2.400	3.800	Liu et al. (2011)
74	qFFN-6	Female flower number	6	Jatr301	15.800	15.800	15.800	6.410	13.600	Sun et al. (2012)
75	qPH-6	Height	6	Jcuint312	25.300	25.300	25.300	2.560	4.000	Sun et al. (2012)
76	qNF-6.1	Number of fruits	6	Jatr839	28.300	28.300	28.300	4.970	12.400	Sun et al. (2012)
77	qSwt-6	Seed width	6	Jatr798	40.600	44.600	42.000	6.220	7.400	Ye et al. (2014)
78	qSL-6	Seed length	6	Jatr798	42.600	62.000	60.000	3.380	5.000	Ye et al. (2014)
79	qC18:0-6	Stearic acid C18:0 (%)	6	Jcuint036	64.000	64.000	64.000	3.900	7.100	Liu et al. (2011)
80	qBN-6	Branch number	6	Jcuint036	64.300	64.300	64.300	4.040	6.900	Sun et al. (2012)
81	qNF-6.2	Number of fruits	6	KK914342.1_53519 Jatropha915	71.850	72.170	71.940	5.837	18.067	Xia et al. (2018)
82	qNF-6.3	Number of fruits	6	KK914342.1_53519 Jatropha915	71.850	72.170	71.940	5.837	18.067	Xia et al. (2018)
83	qOleII-6	OleII expression ($\Delta\Delta C_T$)	6	Jatr152	93.400	93.400	93.400	2.600	6.400	Liu et al. (2011)
84	qBN-6	Branch number	6	Jcuint111	93.400	93.400	93.400	3.580	5.900	Sun et al. (2012)
85	qFFN-7	Female flower number	7	Jcuint151	2.000	2.000	2.000	4.160	9.600	Sun et al. (2012)
86	qSD7.1	Stem diameter	7	–	6.000	22.000	13.000	4.310	10.200	King et al. (2015)
87	qC18: 0-7.1	Stearic acid C18:0 (%)	7	–	13.000	31.000	25.000	8.340	16.100	King et al. (2015)
88	qC18: 0-7.2	Stearic acid C18:0 (%)	7	Jatr883	40.300	40.300	40.300	2.300	4.000	Liu et al. (2011)

89	qC16: 0-7.1	Palmitic acid C16: 0 (%)	7	Jatr802	52.100	52.100	52.100	3.100	7.400	Liu et al. (2011)
90	qSL-7	Seed length	7	Jatr802	48.600	58.200	55.100	5.510	8.600	Ye et al. (2014)
91	qC16: 0-7.2	Palmitic acid C16: 0 (%)	7	–	45.000	73.500	58.000	3.360	7.800	King et al. (2015)
92	qNF-7.1	Number of fruits	7	Jatr866	64.800	64.800	64.800	3.150	7.300	Sun et al. (2012)
93	qSH-7	Seed height	7	Jatr866	63.800	68.300	65.800	5.060	7.400	Ye et al. (2014)
94	qSW-7	Seed weight	7	Jatr866	66.800	66.800	66.800	2.700	4.900	Sun et al. (2012)
95	qTBN-7	Total branch number	7	Jatr610	68.900	68.900	68.900	3.400	8.400	Sun et al. (2012)
96	qPH-7	Height	7	Jatr610	71.900	71.900	71.900	2.970	4.400	Sun et al. (2012)
97	qNF-7.2	Number of fruits	7	KK914970.1_774896 Jatropha4408	143.590	143.780	143.620	5.010	20.067	Xia et al. (2018)
98	qNF-7.3	Number of fruits	7	KK914970.1_774896 Jatropha4408	143.590	143.780	143.620	5.010	20.067	Xia et al. (2018)
99	qNF-7.4	Number of fruits	7	KK914352.1_426180 Jatropha984	148.290	148.310	148.290	5.110	20.967	Xia et al. (2018)
100	qNF-7.5	Number of fruits	7	KK914352.1_426180 Jatropha984	148.290	148.310	148.290	5.110	20.967	Xia et al. (2018)
101	qC18: 0-8.1	Stearic acid C18:0 (%)	8	–	2.000	21.000	11.000	5.340	10.900	King et al. (2015)
102	qC18: 2-8.1	Linoleic acid C18:2 (%)	8	JCT23	2.000	27.000	11.500	4.260	9.900	King et al. (2015)
103	qSwt-8	Seed width	8	Jatr892	12.000	35.200	25.000	4.850	8.000	Ye et al. (2014)
104	qPH8.1	Height	8	–	0.000	53.000	36.000	3.180	7.000	King et al. (2015)
105	qPE8.1	Phorbol esters	8	NG288C–NG286A	48.500	51.500	50.000	4.060	15.490	Amkul et al. (2017)
106	qOleI-8	OleI expression ($\Delta\Delta C_T$)	8	Jcuint277	58.200	58.200	58.200	1.900	5.300	Liu et al. (2011)

(continued)

Table 18.1 (continued)

Sl. No.	QTLs	Traits	Chr.	Marker interval	Start position	End position	Position	Lod	R^2/ PVE	References
107	qNF-8	Number of fruits	8	KK914383.1_100448 Jatropha1196	253.750	253.780	253.780	5.413	18.367	Xia et al. (2018)
108	qNF-8	Number of fruits	8	KK914383.1_100448 Jatropha1196	253.750	253.780	253.780	5.413	18.367	Xia et al. (2018)
109	qC18:0-9	Stearic acid C18:0 (%)	9	Jatr859	0.000	0.000	0.000	9.200	17.900	Liu et al. (2011)
110	qPH-9	Height	9	Jatr859	5.000	5.000	5.000	3.090	4.500	Sun et al. (2012)
111	qC16:0-9	Palmitic acid C16: 0 (%)	9	Jatr859	15.000	15.000	15.000	2.600	7.200	Liu et al. (2011)
112	qOilC-9	Total oil content (%)	9	Jatr698	18.600	18.600	18.600	2.500	5.200	Liu et al. (2011)
113	qDm-9	Diameter	9	Jatr698	24.600	24.600	24.600	2.700	5.200	Sun et al. (2012)
114	qC18:1-10	Oleic acid C18:1 (%)	10	Jcuint180	15.200	15.200	15.200	4.000	5.900	Liu et al. (2011)
115	qC18:2-10	Linoleic acid C18:2 (%)	10	Jcuint180	15.200	15.200	15.200	3.000	4.600	Liu et al. (2011)
116	qPH-10	Height	10	Jcuint081	20.600	20.600	20.600	4.630	6.700	Sun et al. (2012)
117	qSL-10	Seed length	10	Jcuint081	20.600	22.300	21.600	3.170	4.000	Ye et al. (2014)
118	qOilC-10.1	Total oil content (%)	10	–	0.000	32.200	29.000	3.810	11.700	King et al. (2015)
119	qOilC10.2	Total oil content (%)	10	JCT27	4.000	32.200	31.000	4.310	12.100	King et al. (2015)
120	qOilC10.3	Total oil content (%)	10	–	1.000	32.200	32.000	3.050	11.800	King et al. (2015)
121	qC16: 0-10.1	Palmitic acid C16: 0 (%)	10	–	0.000	32.200	32.000	3.120	7.300	King et al. (2015)
122	qDm-11	Diameter	11	Jatr684	14.300	14.300	14.300	2.980	4.600	Sun et al. (2012)
123	JcARF19	Seed length	11	ARF19	23.500	30.050	28.500	16.690	29.600	Ye et al. (2014)
124	qSL-11	Seed length	11	Jatr684	39.600	45.600	41.800	15.000	27.900	Ye et al. (2014)
125	qSW-11	Seed weight	11	Jatr684	40.600	55.600	46.500	4.750	8.600	Ye et al. (2014)

PMA in endosperm during analysis may be due to the non-excision of the complete layer of about 26–30 cell thickness of tegmen (Corner 1976). Sujatha et al. (2005) reported maternal inheritance of PMA; later it was confirmed by Basha and Sujatha (2009) and Kumar et al. (2018). Sujatha et al. (2005) reported the toxic trait with monogenic control. Later, King et al. (2013) explained maternally controlled monogenic dominant trait of PMA which was later mapped by Trebbi et al. (2019) on linkage group 8 using SNP markers. However, based on literature available, now it is confirmed that it is controlled by QTLs that were mapped on chromosome numbers 3 and 8, respectively (Amkul et al. 2017).

18.5 Status of Released Varieties

Several cultivars of *Jatropha* have been released in public and private sectors, but only a few cultivated commercially due to the presence of superior yield and other agronomic performance. The first released variety of *Jatropha* was Chhatrapatithat which has been released from Sardarkrushinagar Dantiwada Agricultural University in 2006 with seed yield ranging from 1000 to 1100 kg/ha with 49.2% seed oil content (Gour 2015). Three genotypes, viz. JJH34-6, JJH 9-1 and JJH H1-5 (non-toxic), have been identified from Jawaharlal Nehru Krishi Vishwa Vidyalaya, seed yield ranging from 1500 to 2000 kg/ha (Gour 2015). Seed yield in three Mexican varieties, viz. Grand Victoria, Doña Aurelia and Don Rafael, has been reported to be 0.9–1.98 tonnes/ha in the fourth year and 1.9–3.6 tonnes/ha in the fifth year onwards (López-Guillén et al. 2019). Alfredo and Quintero (2017) developed high yielding genotypes that yielded 3.5–4 tonnes seeds/ha. The first high yielding variety acclaimed was JO S2 (Yi et al. 2014) from the National University of Singapore, developed through mass selection by focusing on traits viz. kernel yield, kernel oil yield, fatty acid composition, phosphorus and phorbol content. Seed yield of this variety has been reported to be 2.95 tonnes/ha in the first year and 4.25 tonnes/ha in the second year. This variety is characterized by early flowering, better self-branching, high flowers/bunch, high fruits/bunch and, more importantly, better uniformity amongst plants. Don Rafael is being used as pollinator for Latin American gynoecious cultivars, viz. Gran Victoria and Doña Aurelia.

On the other hand, private sectors have also undertaken *Jatropha* breeding programme. The first commercial non-toxic high-yielding *Jatropha* variety JPNT 1 has been released from Jatropower company with seed yield 2–2.5 tonnes/ha and oil content 40%, respectively. The company also released the world's first commercial hybrid of *Jatropha* JPH 1 with a yield of 4 tonnes dry seed/ha and 37% oil content. Several other cultivars released by private organizations are JP 1010, JP 47, JP 40, JP 1003 and JP 1064. Similarly, Jatrosolutions has released cultivars Greenfuels, Desertgreen, Ediblenut and Proteinfeed. A brief summary of seed yield and oil content in released cultivars has been represented in Fig. 18.3. Promising genotypes that emerged through combinatorial breeding may be evaluated to ascertain the value of derived plants as "Value of Varieties" and compared with the data

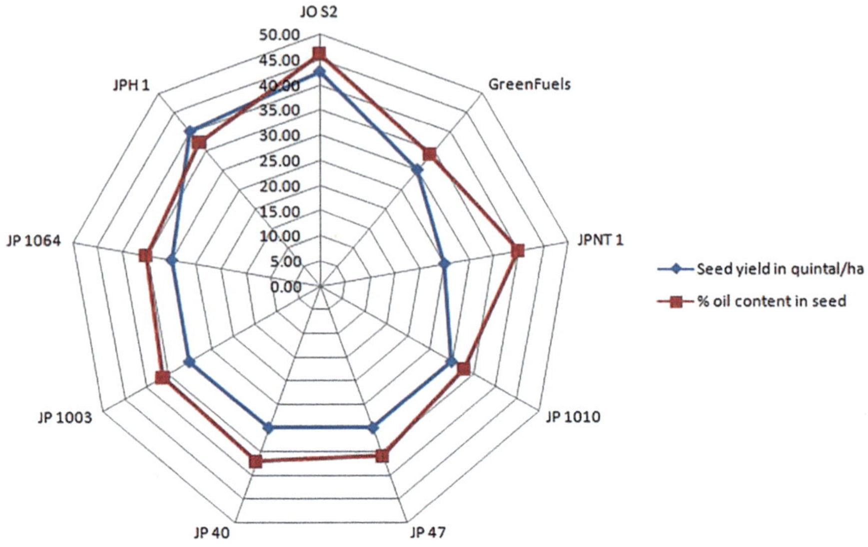

Fig. 18.3 A brief summary of seed yield and oil content in released varieties of *Jatropha*

on quantitative traits focused on varieties/hybrids JO S2, JPNT 1, JP 1010 and JPH 1 for validation, identification and utilization in future *Jatropha* breeding programmes. Development of high-yielding varieties with clonal propagation provides identical plants to mother plants that help in quick multiplication of promising developed genotypes (Gressel 2008). Recently, Olloqui et al. (2021) have reported edible non-toxic variety Sevangel which is a rich source of protein, Ca, Mg, K and dietary fiber.

18.6 Economic Feasibility

Jatropha can be grown on marginal fertile soil which are poorly managed, thus suitable to grow on wasteland. The crop has a recommended 3 × 3 m spacing under normal and 3 × 1.5 m under high density planting; thus an additional income can be generated using intercropping with chickpea, vegetables, etc. The crop has low gestation period, yet the economic yield is viable after 3–4 years of transplanting. The investment capital required for cultivation is substantially low. It could be a worrisome situation for the farmers due to differences in the gestation period and economic yield, future price, etc. under commercial cultivation. As per literature available, the economic feasibility of acclaimed high-yielding variety JO S2 and hybrid JPH 1 can be estimated based on dry kernel yield and oil percentage. The production of dry seed yield/plant for variety JO S2 and hybrid JPH1 is 4.25 and 4.00 tonnes/ha along with seed oil content 45.90 and 37%, respectively. Using the above data, estimated oil yield/ha obtained for variety JO S2 is 1950.75 kg/ha and for hybrid JPH 1 is 1480 kg/ha. Thus estimated profits obtained from JO S2 and JPH 1 are very less while considering price as 1$/lit.

Although the initial investment (viz. land preparation, transplanting and irrigation) for *Jatropha* cultivation is low, still profit generated is very less; thus, it is advisable not to promote large-scale commercial cultivation of *Jatropha*. Net present value (NPV) and benefit cost (BC) ratio analysis suggested about economic non-feasibility of *Jatropha* cultivation (Ntaribi and Paul 2019). However, a study conducted by Baral et al. (2020) suggested that the economic viability of *Jatropha* could lead to positive signs when the yield exceeds 5 tonnes/ha which might be achievable after 2030. Environmental constituents act as the key parameter in the selection of location for profit-oriented cultivation (Najafi et al. 2021). On the other hand, it might be economically favourable if the cultivation started onto wasteland with minimum initial investment and least annual maintenance charges.

18.7 Conclusion

Most of the *Jatropha* breeding programmes are initiated with interspecific hybridization due to limited variability in indigenous collection of *J. curcas*. The derived crosses between *J. curcas* × *J. integerrima* could be repeatedly backcrossed with elite lines to combine and develop plants with high kernel yield and high oil yield in non-toxic background. Since, inheritance of phorbol is controlled maternally; thus, selection of non-toxic female parent is the key step in the breeding programmes of non-toxic cultivars/hybrids. As compared to toxic genotypes, non-toxic genotypes express high number of kernels/plant, seed index and oil yield. The plant derived from combinatorial breeding should be evaluated for qualitative and quantitative data with the acclaimed varieties/hybrids for validation, identification, development and utilization in future breeding programs.

Acknowledgements Authors are thankful to all reviewers whose critical comments and suggestions helped to improve the quality of this book chapter. The authors receive no specific funding for this work.

Declaration of Competing Interest The authors declare that they have no known competing interest or personal relationships that could have appeared to influence the work reported in this book chapter.

CRediT Authorship Contribution Statement **Hausila Prasad Singh:** conceptualization; data curation; investigation; methodology; supervision; roles/writing – original draft; writing – review and editing; validation. **Vinay Rojaria:** data curation; investigation; formal analysis; writing – review and editing. **Noopur Singh:** data curation; methodology; supervision; roles/writing – original draft; writing – review and editing; validation. **Saroj Chauhan:** conceptualization; data curation; methodology; roles/writing – original draft; writing – review and editing. **Om Prakash Raigar:** software; formal analysis; visualization.

References

Abou-Arab AA, Mahmoud MH, Ahmed DMM, Abu-Salem FM (2019) Comparative study between chemical, physical and enzymatic methods for *Jatropha curcas* kernel meal phorbol ester detoxification. Heliyon 5:e01689. https://doi.org/10.1016/j.heliyon.2019.e01689

Achten WMJ, Nielsen LR, Aerts R, Lengkeek AG, Kjær ED et al (2010) Towards domestication of *Jatropha curcas*. Biofuels 1:91–107. https://doi.org/10.4155/bfs.09.4

Alfredo ZC, Quintero PV (2017) New clonal varieties of jatropha. In: Tsuchimoto S (ed) The jatropha genome. Compedium of plant genomes. Springer, Cham, pp 275–288

Alherbawi M, AlNouss A, McKay G, Al-Ansari T (2021a) Optimum sustainable utilisation of the whole fruit of *Jatropha curcas*: an energy, water and food nexus approach. Renew Sust Energ Rev 137:110605. https://doi.org/10.1016/j.rser.2020.110605

Alherbawi M, McKay G, Mackey HR, Al-Ansari T (2021b) *Jatropha curcas* for jet biofuel production: current status and future prospects. Renew Sust Energ Rev 135:110396. https://doi.org/10.1016/j.rser.2020.110396

Amkul K, Laosatit K, Somta P, Shim S, Lee SH, Tanya P, Srinives P (2017) Mapping of QTLs for seed phorbol esters, a toxic chemical in *Jatropha curcas* (L.). Genes 8:205. https://doi.org/10.3390/genes8080205

Andrianirina ZT, Martin M, Dongmeza E, Senger E (2019) Effects of genotype, direct sowing and plant spacing on field performance of *Jatropha curcas* L. Agronomy 9:465. https://doi.org/10.3390/agronomy9080465

Arockiasamy S, Kumpatla J, Hadole S, Yepuri V, Patil M, Shrivastava V et al (2021) Breeding and biotechnological efforts in *Jatropha curcas* L. for sustainable yields. Oil Crop Sci 6:180–191. https://doi.org/10.1016/J.OCSCI.2021.10.004

Banerji R, Chowdhury AR, Misra G, Sudarsanan G, Verma SC, Srivastava GS (1985) Jatropha seed oils for energy. Biomass 8:277–282. https://doi.org/10.1016/0144-4565(85)90060-5

Baral NR, Neupane P, Ale BB, Quiroz-Arita C, Manandhar S, Bradley TH (2020) Stochastic economic and environmental footprints of biodiesel production from *Jatropha curcas* Linnaeus in the different federal states of Nepal. Renew Sust Energ Rev 120:109619. https://doi.org/10.1016/J.RSER.2019.109619

Basha SD, Sujatha M (2007) Inter and intra-population variability of *Jatropha curcas* (L.) characterized by RAPD and ISSR markers and development of population-specific SCAR markers. Euphytica 156:375–386. https://doi.org/10.1007/s10681-007-9387-5

Basha SD, Sujatha M (2009) Genetic analysis of *Jatropha* species and interspecific hybrids of *Jatropha curcas* using nuclear and organelle specific markers. Euphytica 168:197–214. https://doi.org/10.1007/s10681-009-9900-0

Basha SD, Francis G, Makkar HPS, Becker K, Sujatha M (2009) A comparative study of biochemical traits and molecular markers for assessment of genetic relationships between

Jatropha curcas L. germplasm from different countries. Plant Sci 176:812–823. https://doi.org/10.1016/j.plantsci.2009.03.008

Corner EJH (1976) The seeds of dicotyledons. Cambridge University Press, Cambridge

Das S (2020) The National Policy of biofuels of India - A perspective. Energy Policy 143:111595. https://doi.org/10.1016/J.ENPOL.2020.111595

de Souza CA, dos Santos A, Laviola BG, Teodoro LPR, Teodoro PE, Rodrigues EV (2021) Genetic diversity and population structure in Jatropha (*Jatropha curcas* L.) based on molecular markers. Genet Resour Crop Evol 1:1–10. https://doi.org/10.1007/S10722-021-01224-2

Deng X, Fang Z, Liu YH (2010) Ultrasonic transesterification of *Jatropha curcas* L. oil to biodiesel by a two-step process. Energy Convers Manag 51:2802–2807. https://doi.org/10.1016/J.ENCONMAN.2010.06.017

Devappa RK, Makkar HPS, Becker K (2010) Nutritional, biochemical, and pharmaceutical potential of proteins and peptides from jatropha: review. J Agric Food Chem 85:6543–6555. https://doi.org/10.1021/jf100003z

Dhakshanamoorthy D, Selvaraj R, Chidambaram ALA (2011) Induced mutagenesis in *Jatropha curcas* L. using gamma rays and detection of DNA polymorphism through RAPD marker. Comptes Rendus Biol 334:24–30. https://doi.org/10.1016/j.crvi.2010.11.004

Dhyani SK, Devi VS, Handa AK (2015) Tree borne oilseeds for oil and biofuel. Jhansi- India: Technical Bulletin 2/2015. ICAR-CAFRI, Jhansi, p 50

dos Santos A, Rodrigues EV, Laviola BG, Teodoro LPR, Teodoro PE, Bhering LL (2021) Increasing selection gain and accuracy of harvest prediction models in Jatropha through genome-wide selection. Sci Rep 111:1–9. https://doi.org/10.1038/s41598-021-93022-0

Ewunie GA, Lekang OI, Morken J, Yigezu ZD (2021) Characterizing the potential and suitability of Ethiopian variety *Jatropha curcas* for biodiesel production: variation in yield and physicochemical properties of oil across different growing areas. Energy Rep 7:439–452. https://doi.org/10.1016/j.egyr.2021.01.007

Foidl N, Foidl G, Sanchez M, Hackel S (1996) *Jatropha curcas* L. as a source for the production of biofuel in Nicaragua. Bioresour Technol 58:77–82. https://doi.org/10.1016/S0960-8524(96)00111-3

Francis G, Oliver J, Sujatha M (2013) Non-toxic jatropha plants as a potential multipurpose multiuse oilseed crop. Ind Crop Prod 42:397–401. https://doi.org/10.1016/j.indcrop.2012.06.015

Francis G, Oliver J, Mulpuri S (2018) High yielding and trait specific genotypes and genetic associations among yield and yield contributing traits in *Jatropha curcas* L. Agrofor Syst 92: 1417–1436. https://doi.org/10.1007/s10457-017-0089-2

Francis G, John O, Piergiorgio S, Mulpuri S (2019) Apomixis as a tool for development of high yielding clones and selections in *Jatropha curcas* L. Genet Resour Crop Evol 67:727–743. https://doi.org/10.1007/s10722-019-00851-0

Ganesh RS, Parthiban KT, Senthil KR, Thiruvengadam V, Paramathma M (2008) Genetic diversity among Jatropha species as revealed by RAPD markers. Genet Resour Crop Evol 55:803–809. https://doi.org/10.1007/s10722-007-9285-7

Ghatak J, Gour VK (2014) Oil content variation in different parts and across different species of Jatropha. Biosci Trends 7:2926–2927

Ghosh A, Chaudhary DR, Reddy MP, Rao SN, Chikara J, Pandya JB et al (2007) Prospects for jatropha methyl ester (biodiesel) in India. Int J Environ Stud 64:659–674. https://doi.org/10.1080/00207230701766499

Ginwal HS, Rawat PS, Srivastava RL (2004) Seed source variation in growth performance and oil yield of *Jatropha curcas* Linn. in Central India. Silvae Genet 53:186–192. https://doi.org/10.1515/sg-2004-0034

Gohil RH, Pandya JB (2009) Genetic evaluation of Jatropha (*Jatropha curcas* Linn.) genotypes. J Agric Res 47:221–228

Gour VK (2015) Research and Development Status of *Jatropha curcas* L. Potential feedstock for Biofuels- Analysis & Way Forward. National Workshop on TBOs- Way Ahead. October 15–16, p 56

Gressel J (2008) Transgenics are imperative for biofuel crops. Plant Sci 174:246–263. https://doi.org/10.1016/j.plantsci.2007.11.009

He W, King AJ, Khan MA, Cuevas JA, Ramiaramanana D, Graham IA (2011) Analysis of seed phorbol-ester and curcin content together with genetic diversity in multiple provenances of *Jatropha curcas* L. from Madagascar and Mexico. Plant Physiol Biochem 49:1183–1190. https://doi.org/10.1016/j.plaphy.2011.07.006

Hegazy A (2017) Utilization of wastewater for cultivation of Jatropha in Egypt. In: Tsuchimoto S (ed) The Jatropha genome. Springer, Cham, pp 219–257. https://doi.org/10.1007/978-3-319-49653-5_14

Jiang Y, Cai Z, Xie W, Long T, Yu H, Zhang Q (2012) Rice functional genomics research: progress and implications for crop genetic improvement. Biotechnol Adv 30:1059–1070. https://doi.org/10.1016/j.biotechadv.2011.08.013

Jonas M, Ketlogetswe C, Gandure J (2021) Quantification of Phorbol-12-myristate 13-acetate in Jatropha seed oil and cake at different stages of fruit maturity. Int J Environ Stud 1:1–10. https://doi.org/10.1080/00207233.2021.1893103

Kancharla N, Jalali S, Narasimham JV, Nair V, Yepuri V et al (2019) De novo sequencing and hybrid assembly of the biofuel crop *Jatropha curcas* L. identification of quantitative trait loci for geminivirus resistance. Genes. https://doi.org/10.3390/genes10010069

Kaushik N, Kumar K, Kumar S, Kaushik RS (2007) Genetic variability and divergence studies in seed traits and oil content of Jatropha (*Jatropha curcas* L.) accessions. Biomass Bioenergy 31: 497–502. https://doi.org/10.1016/j.biombioe.2007.01.021

King AJ, Montes LR, Clarke JG, Affleck J, Li Y et al (2013) Linkage mapping in the oilseed crop *Jatropha curcas* L. reveals a locus controlling the biosynthesis of phorbol esters which cause seed toxicity. Plant Biotechnol J 11:986–996. https://doi.org/10.1111/pbi.12092

King AJ, Montes LR, Clarke JG, Itzep J, Perez CAA, Jongschaap REE, Visser RGF, Van Loo EN, Graham IA (2015) Identification of QTL markers contributing to plant growth, oil yield and fatty acid composition in the oilseed crop *Jatropha curcas* L. Biotechnol Biofuels 8:1–17. https://doi.org/10.1186/s13068-015-0326-8

Kumar YH, Ranjan A, Asif MH, Mantri S, Sawant SV, Tuli R (2011) EST-derived SSR markers in *Jatropha curcas* L.: development, characterization, polymorphism, and transferability across the species/genera. Tree Genet Genomes 7:207–219. https://doi.org/10.1007/s11295-010-0326-6

Kumar S, Singh HP, Saha R, Gour VK (2018) Identification of toxic and non-toxic lines in Jatropha. Int J Curr Microbiol Appl Sci 7:3167–3177. https://doi.org/10.20546/ijcmas.2018.706.372

Li M, Li H, Jiang H, Pan X, Wu G (2007) Establishment of an Agrobacterium-mediated cotyledon disc transformation method for Jatropha curcas. Plant Cell Tissue Organ Cult 92:173–181. https://doi.org/10.1007/s11240-007-9320-6

Li J, Li MR, Wu PZ, Tian CE, Jiang HW, Wu GJ (2008) Molecular cloning and expression analysis of a gene encoding a putative β-ketoacyl-acyl carrier protein (ACP) synthase III (KASIII) from *Jatropha curcas*. Tree Physiol 28:921–927. https://doi.org/10.1093/treephys/28.6.921

Liu P, Wang CM, Li L, Sun F, Liu P, Yue GH (2011) Mapping QTLs for oil traits and eQTLs for oleosin genes in Jatropha. BMC Plant Biol 11:132. https://doi.org/10.1186/1471-2229-11-132

López-Guillén G, Solís Bonilla JL, Martínez Valencia BB, Herrera Parra E, ZamarripaColmenero A (2019) Agronomy of Jatropha curcas in Mexico. In: Mulpuri S, Carels N, Bahadur B (eds) Jatropha, challenges for a new energy crop. Springer, Singapore, pp 255–272

Maghuly F, Laimer M (2013) *Jatropha curcas*, a biofuel crop: functional genomics for understanding metabolic pathways and genetic improvement. Biotechnol J 8:1172–1182. https://doi.org/10.1002/biot.201300231

Makkar HPS, Becker K (1997) Potential of *J. curcas* seed meal as a protein supplement to livestock feed; constraints to its utilization and possible strategies to overcome constraints. In: Gübitz GM, Mittelbach M, Trabi M (eds.) Biofuels and Industrial Products from *Jatropha curcas*. DBV Graz. pp 190–205

Makkar HPS, Becker K (2009) *Jatropha curcas*, a promising crop for the generation of biodiesel and value-added coproducts. Eur J Lipid Sci Technol 111:773–787. https://doi.org/10.1002/ejlt.200800244

Makkar HPS, Aderibigbe AO, Becker K (1998a) Comparative evaluation of non-toxic and toxic varieties of *Jatropha curcas* for chemical composition, digestibility, protein degradability and toxic factors. Food Chem 62:207–215. https://doi.org/10.1016/S0308-8146(97)00183-0

Makkar HPS, Becker K, Schmook B (1998b) Edible provenances of *Jatropha curcas* from Quintana Roo state of Mexico and effect of roasting on antinutrient and toxic factors in seeds. Plant Foods Hum Nutr 52:31–36. https://doi.org/10.1023/A:1008054010079

Makkar HPS, Kumar V, Oyeleye OO, Akinleye AO, Angulo-Escalante MA, Becker K (2011) *Jatropha platyphylla*, a new non-toxic Jatropha species: physical properties and chemical constituents including toxic and antinutritional factors of seeds. Food Chem 125:63–71. https://doi.org/10.1016/j.foodchem.2010.08.037

Mandpe S, Kadlaskar S, Degen W, Keppeler S (2005) On road testing of advanced common rail diesel vehicles with biodiesel from the *Jatropha curcas* plant. Soc Auto Eng Inc 26:356–364. https://doi.org/10.4271/2005-26-356

Mishra DK (2009) Selection of candidate plus phenotypes of *Jatropha curcas* L. using method of paired comparisons. Biomass Bioenergy 33:542–545. https://doi.org/10.1016/j.biombioe.2008.08.004

Montes JM, Melchinger AE (2016) Domestication and breeding of *Jatropha curcas* L. Trends Plant Sci 21:1045–1057. https://doi.org/10.1016/j.tplants.2016.08.008

Najafi F, Sedaghat A, Mostafaeipour A, Issakhov A (2021) Location assessment for producing biodiesel fuel from *Jatropha Curcas* in Iran. Energy 236. https://doi.org/10.1016/j.energy.2021.121446

NITI Aayog (2020) Expert Committee on Roadmap for Ethanol Blending in India; by 2025. pp 1–85. https://www.niti.gov.in/expert-committee-roadmap-ethanol-blending-india-2025. (Accessed 30 Dec 2021)

Ntaribi T, Paul DI (2019) The economic feasibility of Jatropha cultivation for biodiesel production in Rwanda: a case study of Kirehe district. Energy Sustain Dev 50:27–37. https://doi.org/10.1016/j.esd.2019.03.001

Olloqui EJ, Castaneda-Ovando A, Evangelista-Lozano S, Alanis-Garcia E, Ramirez-Moreno E, Valadez-Vega C, Anorve-Morga J (2021) Measurement of nutrients and minor components of a non-toxic variety of *Jatropha curcas*. J Food Meas Charact:1–9. https://doi.org/10.1007/s11694-021-01229-6

Pamidimarri DVNS, Sinha R, Kothari P, Reddy MP (2009) Isolation of novel microsatellites from *Jatropha curcas* L. and their cross-species amplification. Mol Ecol Resour 9:431–433. https://doi.org/10.1111/j.1755-0998.2008.02471.x

Pant KS, Khosla V, Kumar D, Gairola S (2006) Seed oil content variation in *Jatropha curcas* Linn. in different altitudinal ranges and site conditions in H.P. India. Lyonia 11:31–34

Peixoto MA, Evangelista JSPC, Coelho IF, Alves RS, Laviola BG, Fonseca e Silva F, de Resende MDV, Bhering LL (2021) Multiple-trait model through Bayesian inference applied to *Jatropha curcas* breeding for bioenergy. PLoS One 16:1–11. https://doi.org/10.1371/journal.pone.0247775

Popluechai S, Breviario D, Mulpuri S, Makkar HPS, Raorane M (2009) Narrow genetic and apparent phenetic diversity in Jatropha curcas: initial success with generating low phorbol ester interspecific hybrids. Nat Prec. https://doi.org/10.1038/npre.2009.2782.1

Rao GR, Korwar GR, Shanker AK, Ramakrishna YS (2008) Genetic associations, variability and diversity in seed characters, growth, reproductive phenology and yield in *Jatropha curcas* (L.) accessions. Trees 22:697–709. https://doi.org/10.1007/s00468-008-0229-4

Rao MRG, Ramesh S, Rao AM, Gangappa E (2009) Exploratory studies on components of variability for economic traits in jatropha (*J. curcas* L.). Karnataka J Agric Sci 22:967–970

Rodrigues DA, Demuner AJ, Barbosa LCA, Pereira GAM, Fabris JD et al (2021) Removing phorbol esters from the biomass to add extra value to the byproduct from deoiling seeds of

Jatropha curcas in the biodiesel industry. Biomass Convers Biorefinery:1–13. https://doi.org/10.1007/s13399-020-01188-w

Salazar-Villa E, Alcaraz-Melendez L, Leon-Felix J, Heredia JB, Soto-Landeros F, Angulo-Escalante MA (2020) Morphological variability and oil content of *Jatropha platyphylla* Müll. Arg. germplasm as determined using multivariate analysis. Sci Hortic 261:108968. https://doi.org/10.1016/j.scienta.2019.108968

Samra S, Samocha Y, Eisikowitch D, Vaknin Y (2014) Can ants equal honeybees as effective pollinators of the energy crop *Jatropha curcas* L. under Mediterranean conditions? GCB Bioenergy 6:756–767. https://doi.org/10.1111/gcbb.12105

Saravanan AP, Pugazhendhi A, Mathimani T (2020) A comprehensive assessment of biofuel policies in the BRICS nations: implementation, blending target and gaps. Fuel 272:117635. https://doi.org/10.1016/J.FUEL.2020.117635

Senthil KR, Parthiban KT, Govinda RM (2009) Molecular characterization of Jatropha genetic resources through inter-simple sequence repeat (ISSR) markers. Mol Biol Rep 36:1951–1956. https://doi.org/10.1007/s11033-008-9404-3

Shabanimofrad M, Rafii YM, Said Saad M, Edaroyati MWP, Biabanikhanehkahdani A, Latif M (2011) Diversity of physic nut (*Jatropha curcas*) in Malaysia: application of DIVA-geographic information system and cluster analysis. Aust J Crop Sci 5:361–368

Singh HP (2016) Identification of non toxic lines in new crosses and backcross population of Jatropha curcas L. Ph.D. thesis, Jawaharlal Nehru Krishi Vishwa Vidyalaya, Jabalpur. https://krishikosh.egranth.ac.in/handle/1/5810002528

Singh B, Singh K, Shukla G, Goel VL, Pathre UV, Rahi TS, Tuli R (2013) The field performance of some accessions of *Jatropha curcas* L. (Biodiesel Plant) on degraded sodic land in North India. Int J Green Energy 10:1026–1040. https://doi.org/10.1080/15435075.2012.738336

Singh HP, Raigar OP, Chahota RK (2021) Estimation of genetic diversity and its exploitation in plant breeding. Bot Rev. https://doi.org/10.1007/s12229-021-09274-y

Sudheer Pamidimarri DVN, Pandya N, Reddy MP, Radhakrishnan T (2009a) Comparative study of interspecific genetic divergence and phylogenic analysis of genus Jatropha by RAPD and AFLP: genetic divergence and phylogenic analysis of genus Jatropha. Mol Biol Rep 36:901–907. https://doi.org/10.1007/s11033-008-9261-0

Sudheer Pamidimarri DVN, Singh S, Mastan SG, Patel J, Reddy MP (2009b) Molecular characterization and identification of markers for toxic and non-toxic varieties of *Jatropha curcas* L. using RAPD, AFLP and SSR markers. Mol Biol Rep 36:1357–1364. https://doi.org/10.1007/s11033-008-9320-6

Sujatha M, Makkar HPS, Becker K (2005) Shoot bud proliferation from axillary nodes and leaf sections of non-toxic *Jatropha curcas* L. Plant Growth Regul 47:83–90. https://doi.org/10.1007/s10725-005-0859-0

Sun F, Liu P, Ye J, Lo LC, Cao S, Li L, Yue GH, Wang CM (2012) An approach for jatropha improvement using pleiotropic QTLs regulating plant growth and seed yield. Biotechnol Biofuels 5:1–10. https://doi.org/10.1186/1754-6834-5-42

Sunil N, Sujatha M, Kumar V, Vanaja M, Basha SD, Varaprasad KS (2011) Correlating the phenotypic and molecular diversity in *Jatropha curcas* L. Biomass Bioenergy 35:1085–1096. https://doi.org/10.1016/j.biombioe.2010.11.030

Tanya P, Dachapak S, Tar MM, Srinives P (2011) New microsatellite markers classifying nontoxic and toxic *Jatropha curcas*. J Genet 92:76–78. https://doi.org/10.1007/s12041-011-0093-7

Tapanes NCO, Gomes ADA, de Mesquita CJW, Ceva Antunes OA (2008) Transesterification of *Jatropha curcas* oil glycerides: theoretical and experimental studies of biodiesel reaction. Fuel 87:2286–2295. https://doi.org/10.1016/J.FUEL.2007.12.006

The Hindu (2018). https://www.thehindu.com/business/Industry/spicejet-operates-indias-first-biojet-fuel-flight/article24790919.ece

Trebbi D, Ravi S, Broccanello C, Chiodi C, Francis G, Oliver J, Mulpuri S, Srinivasan S, Stevanato P (2019) Identification and validation of SNP markers linked to seed toxicity in *Jatropha curcas* L. Sci Rep 9:10220. https://doi.org/10.1038/s41598-019-46698-4

Vargas-Carpintero R, Hilger T, Mossinger J, Souza RF, Barroso AJC, Tiede K, Lewandowski I (2021) Acrocomia spp.: neglected crop, ballyhooed multipurpose palm or fit for the bioeconomy? A review. Agron Sustain Dev 416:1–35. https://doi.org/10.1007/S13593-021-00729-5

Xia Z, Zhang S, Wen M, Lu C, Sun Y, Zou M, Wang W (2018) Construction of an ultrahigh-density genetic linkage map for *Jatropha curcas* L. and identification of QTL for fruit yield. Biotechnol Biofuels 11:1–10. https://doi.org/10.1186/S13068-017-1004-9

Ye J, Liu P, Zhu C, Qu J, Wang X, Sun Y, Sun F, Jiang Y, Yue G, Wang C (2014) Identification of candidate genes JcARF19 and JcIAA9 associated with seed size traits in Jatropha. Funct Integr Genomics 14:757–766. https://doi.org/10.1007/s10142-014-0400-5

Yepuri V, Jalali S, Mudunuri V, Pothakani S, Kancharla N, Arockiasamy S (2022) Genotyping by sequencing-based linkage map construction and identification of quantitative trait loci for yield-related traits and oil content in Jatropha (*Jatropha curcas* L.). Mol Biol Rep. https://doi.org/10.1007/S11033-022-07264-W

Yi C, Reddy C, Varghese K, Ngoc T, Bui TNH, Zhang S, Kallath M, Kunjachen B, Ramachandran S, Hong Y (2014) A new *Jatropha curcas* variety (JO S2) with improved seed productivity. Sustainability 6:4355–4368. https://doi.org/10.3390/su6074355

Yue GH, Sun F, Liu P (2013) Status of molecular breeding for improving *Jatropha curcas* and biodiesel. Renew Sust Energ Rev 26:332–343. https://doi.org/10.1016/j.rser.2013.05.055

Zhang Z, Chang Y, Tang H, Zhao H, Chen X, Tian G, Liu G, Cai J, Jia G (2021) Bio-detoxification of *Jatropha curcas* L. cake by a soil-borne *Mucor circinelloides* strain using a zebrafish survival model and solid-state fermentation. J Appl Microbiol 130:852–864. https://doi.org/10.1111/jam.14825

Part IV
Socio-Economic Impact of Agro-Forestry System

Chapter 19
Breeding Potentials of Wild Forest Rattans Palms to Ensure Food Security

Dimjaneng Haokip, S. Senjam Jinus, Ningthaipuilu Longmei, Sanjukta Samurailatpam Sharma, Th. Bhaigyabati, Yendrembam K. Devi, Akoijam Ranjita Devi, B. Lalramhlimi, Vandana Mutum, Ningombam Swapana, Athokpam Kalpana, and J. Vanitha

Abstract Nontimber forest products (NTFPs) are often regarded to be a "Silver bullet." Attention on vital conservation and sustainable development of these assets are highly appreciated to boost the indigenous livelihoods by many researchers. And as such, rattan canes have been recognized as one of the world's most valuable NTFPs. Apart from commercial cane-ware products, rattan resources are yet to be explored in terms of their food value, therapeutic, medicinal properties and most pivotal is its breeding aspects of traits, therefore, research aiming at crop

D. Haokip · N. Longmei
College of Forestry and College of Agricultural Sciences, FEEDS Group of Institution, Hengbung, Kangpokpi, Manipur, India

S. Senjam Jinus (✉) · V. Mutum · A. Kalpana
School of Horticulture, and School of Agricultural Sciences, Pandit Deen Dayal Upadhyay Institute of Agricultural Sciences, Utlou, Bishnupur and Kameng Khunjao Campus, Lamlai, Manipur, India
e-mail: singhsenjam@gmail.com

S. S. Sharma
Institute of Bioresources and Sustainable Development, Regional Centre,, Gangtok, Sikkim, India

T. Bhaigyabati
Department of Botany, Pravabati College, Mayang Imphal, Manipur, India

Y. K. Devi
Department of Entomology, Lovely Professional University, Phagwara, Punjab, India

A. R. Devi · J. Vanitha
SRM College of Agricultural Sciences, SRM Institute of Science and Technology, Chengalpattu (Dist.), Tamil Nadu, India

B. Lalramhlimi
Department of Vegetable Science, Bidhan Chandra Krishi Viswavidyalaya, Nadia, West Bengal, India

N. Swapana
Department of Chemistry, Manipur Technical University, Takyelpat Imphal, Manipur, India

H. S. Jatav et al. (eds.), *Agroforestry to Combat Global Challenges*, Sustainable Development and Biodiversity 36, https://doi.org/10.1007/978-981-99-7282-1_19

improvement program through conventional and nonconventional approaches like tissue culture for mass propagation, application of various breeding methods for further plant traits selection, bio-fortification for new food and medicinal plants through selective breeding and systematic molecular investigation of the genomic information of potential wild rattan resources will provide a foothold for further studies and investigations of its food and biochemical composition. This chapter sketches the recent findings of rattans palms on morphological and taxonomical aspects, economical importance, pollination behavior, medicinal and food values, practices of agriculture, and nontimber forest foods management under an outline of breeding approaches which can be useful in future traits improvement program in rattans for food security and mass commercialization.

Keywords Nontimber forest products · Conventional and nonconventional rattans breeding · Food values and medicinal properties

19.1 Introduction

Rattans are an important group of nontimber forest products which are often described as "green gold" because of their versatility and multitude of uses. The word "Rattan" is originated from the Malay word "rotan" because of its unusual climbing nature. Rattan palms are prevalent exclusively confined to moist evergreen forest of African (Ariane et al. 2017), South Asian region, and southern part of China, the Malay Archipelago, Australia, and the western Pacific. However, Southeast Asian region has the highest diversity of rattan genera and species (Hong et al. 2002).

Being an ecological and economical asset of tropical and subtropical rainforests, rattans are a component of forest under-storey in tropical and subtropical forest ecosystems. Each species has its own ecological and altitudinal preferences and may occur from sea level to 2000 m above sea level. In order to reach sunlight, they climb up on other plants with recurved hooks, an organs specially adapted. They are found to grow in broad range of soil types. Several species have been found in gap vegetation, swamps, and seasonally inundated forest or on dry ridge tops (Sunderland and Dransfield 2002). Several rattan species (*Calamus, Daemonorops, Eremospatha, Korthalsia*, and *Laccosperma*) have evolved morphological transformation providing sites for ant's nesting (Sunderland and Dransfield 2002; Chan et al. 2012). Sunderland (2004) also reported the association between ants and rattans in the forest of tropical Africa.

Economically, rattans are a potential source for commercial utilization and trade (Singh et al. 2004). Stalks without the sheath of rattans can be considered as "Cane" and is used in wide range of productions, either as whole, round-form or in split forms in cottage industries. They are also utilized by local communities for various purposes such as basketry, mats and containers and thus, providing a significant way of livelihood for several local people (Sunderland and Dransfield 2002; Renuka 2007a, b).

Importance of NTFPS for the local forest dwellers and being a vital component of under-storey in tropical and subtropical for forest ecosystems are well recognized by the scientific communities for decades of years. Such considerations are thus resulted in recommenced research program for NTFPs into a viable approach for the conservation and commercial cultivation adopting the modern breeding tools and agroforest management (Belcher 2002).

And as such, rattan canes have been recognized as one of the world's most valuable NTFPs (Ros-Tonen 2000). In fact, rattan canes is a high valued export earner, only second to timber in the world ranking of primary forest product exports (Nuruzzaman 2001). The prominent suppliers of rattan in the world market are Indonesia (Van Valkenburg 2002), Myanmar and Vietnam (Meijaard et al. 2014) with Indonesia itself exporting 80% of the world's demand of rattans in 2008 (Hirschberger 2011). In terms of profitability, large cane enterprises profit more than the small and medium cane enterprises (Alamgir et al. 2006). Due to multiple usages, evidences of unscientific hunting of rattan from the wild forest of Asian and Malaysian where most of the rattans exploited are from the wild stocks (Dransfield et al. 2002a, b). In fact, about 90% of commercial products in global markets are the main sources of forest's extraction (Hunter and Lou 2000).

Rattans also form a part of traditional medicines and therapeutic potential for the treatment of common ailments like stomach disorder and intestinal worms in Bangladesh. Indigenous people of Bangladesh used the various parts of rattans like roots, shoot tips and young leaves are consumed as vegetables (Renuka 2001). Despite their ecological, commercial, and medicinal importance, knowledge regarding diversity, taxonomy, economic, medicinal, and breeding of rattan resource is still in infancy and poorly known. In order to distinguished rattan species of commercial importance and breeding strategies from others, it is important to have essential taxonomic and genetic diversity knowledge. This will also help in understanding conservation, sustainable management and breeding objectives as well as in the development of rattan resource. Meanwhile, studies conducted so far on medicinal properties and food values of rattans also reveal that investigations have been done only on few rattan species. Other rattan species remain unexplored. It is very probable that the unexplored rattan resources may also possess phytochemical constituents which may show medicinal properties. Traditional knowledge documented so far has also revealed that certain species of rattans do possess medicinal value and have been used by the local people traditionally for curing various ailments. Among which the claims of antihyperglycemic activity of the plant is intriguing. However, these claims need scientific validation. In this paper, we highlighted the recent findings on morphological and taxonomical aspects, economical importance, pollination behavior, medicinal and food values, practices of agriculture and nontimber forest foods management under an envelope of breeding approaches which can be useful in future crops improvement program in rattans.

19.2 Distribution of Rattans

Globally, there are 17 genera and about 650 species of rattans namely, *Eugeissona* Griff., *Calamu, Eleiodoxa, Eremospatha, Korthalsi, Laccosperma, Lepidocaryum, Mauritia, Mauritiella, Metroxylon, Myrialepis, Oncocalamus, Pigafetta, Plectocomia, Plectocomiopsis, Raphia,* and *Salacca* (Baker and Dransfield 2016). Out of the 17 genera, the genus *Calamus* is predominantly an Asian genus. It is the largest genus comprising of 520 species (Baker and Dransfield 2016; Uhl and Dransfield 1988). The highest concentration of rattan species are considered to be found in tropical regions of Asia particularly in the Borneo, Malay Peninsula, and Sumatra. High concentration of rattan species are also confined to secondary centers of diversity of Indochina and New Guinea (Sreekumar and Henderson 2014). The three endemic genuses of African rattans are *Laccosperma, Eremospatha, and Oncocalamus* and the genus *Calamus*. The remaining rattan genera are mostly distributed in Southeast Asian countries spreading further eastwards and northwards (Sunderland and Dransfield 2002).

Under 4 genera viz. *Calamus, Laccosperma, Eremospatha* and *Oncocalamus*, Sunderland (2012) reported 22 species of rattans from African regions. South-East Asia accounts for the largest rattan diversity in the world. The rattans of this region are reported and studied by various researchers. According to Dransfield (1979), there are 104 species comprises of 8 genera of rattans namely; *Korthalsia, Plectocomia, Plectocomiopsis, Myrialepis, Calospatha, Daemonorops, Calamus*, and *Ceratolobus* with descriptions and anatomical drawings of each species from Peninsular Malaysia, 23 species and 1 variety from Sarawak (Dransfield 1992), 82 species which include 8 species and 2 varieties that are endemic in Sabah (Dransfield 1984). Reported 57 species under 6 genera, namely, *Calamus, Ceratolobus, Daemonorops, Korthalsia, Plectocomia, and Plectocomiopsis* from East Kalimantan, Indonesia. Khou (2008) reported 23species belonging to 6 genera from Cambodia consisting of *Calamus, Korthalsia, Daemonorops, Plectocomia, Plectocomiopsis,* and *Myrialepsis*. Henderson and Dung (2014) reported 6 genera of rattans from Vietnam which consists of genus *Daemonorops, Calamus, Plectocomia, Plectocomiopsis, Myrialepsis,* and *Korthalsia*. Tesoro (2007) reported 64 species belonging to 4 genera of rattans viz. *Calamus, Daemonorops, Korthalsia,* and *Plectocomia* from the Philippines. In Thailand, 83 species of rattans belonging to 7 genera viz. *Calamus, Ceratobolus, Daemonorops, Korthalsia, Plectocomia, Plectocomiopsis,* and *Myrialepsis* reported by Vongkaluang (2007). Xu et al. (2000) reported three genera of rattans which included 40 species and 21 varieties namely *Calamus, Daemonorops,* and *Plectocomia* from China. Six genera viz. *Calamus, Plectocomia, Plectocomiopsis, Myrialepsis, Korthalsia,* and *Daemonorops* and 31 species are reported from Myanmar (Renuka 2007a, b).

The rattans of Indian subcontinent are not excluded from research. Several authors reported on the diversity and distribution of rattans from different parts of the Indian subcontinent. Alam (1990) reported 6 *Calamus* species and 1 *Daemonorops* species from Bangladesh including a key based on vegetative characters, a key to genera, and their descriptions. Paudel and Chowdhary (2005) reported two genera viz. *Calamus* and *Plectocomia* and 7 species of rattans from Nepal. The rattans of Sri Lanka are studied by De Zoysa and Vivekanandan (1991) and reported 7 species of rattans under two genera, *Calamus* and *Daemonorops*. In our country India, according to Uma Shaanker et al. 2004, rattans are represented by 5genera *viz. Calamus, Daemonorops, Korthalsia, Plectocomia,* and *Salacca* with 61 species. To date, contribution of Beccari (1911) is still accepted and remains fresh as the standard for rattans of Asia. He described 164 *Calamus* species and 77 *Daemonorops* in his series of volumes "Systematic enumeration of the species *Calamus* and *Daemonorops*."

19.3 Morphological Studies of Rattans

The subfamily Calamoideae possesses the most morphologically diverse of the six subfamilies which is being recognized best among the family of Arecaceae. Uhl and Dransfield (1988) recognize 2 tribes and 22 genera within the subfamily. Evolutionary strategies such as selection and massive genetic drift, and phenotypic variation have led to morphological differences (Abdelkrim et al. 2005; Serebryanaya and Shipunov 2009). The parts of a rattan can be divided into two groups: the vegetative part which includes roots, stems, sheaths, leaves, climbing organs, and spines and the reproductive parts which includes inflorescence, flowers, and fruits. The vegetative parts are very useful to identify rattan at the genus level, and could be a useful index to identify at extend of species level, while the reproductive parts are often required for species identification.

Since, morphological markers are cheap and fast, they are routinely carried out for the identification of genetic diversity of plants. Characterization based on morphological data plays an important role in the analysis of genetic resources for proper understanding and utilization of diverse genepool among taxa and their characters. For morphological characterization, commonly employed traits include plant organs exhibiting phenotypic variability such as leaves, flowers and stems. Although, environmental conditions influence morphological markers because of which true genetic differences or similarities may not be represented through observations; the concept of plant morphological studies still forms an integral part in molecular genetic analysis, evolutionary biology and plant systematic studies (Sattler and Rutishauser 1997a, b).

The phenetic or morphometric study which is used to express a correlation of overall similarities and relationships among taxa, can also be extended to give phylogenetic or diagnostic systems and can be applied to many other fields of endeavor (Sneath and Sokal 1973). It involves the application of multivariate techniques in systematics. Recently, phenetic taxonomy is also defined as taxometrics (Rogers 1963a, b) or multivariate morphometrics (Blackith and Reyment 1971). In applied research, the adoption of recognized multivariate statistical algorithms, in particular, Cluster analysis and Principle Component Analysis

(PCA) are significant strategy for the classification of germplasm and analyzing genetic relationships.

Each individual under investigation are simultaneously analyzed by these statistical algorithms. Such measurement methods are widely used in the genetic diversity analysis (Mohammadi and Prasanna 2003). Also, similarities between variables can be uncovered using PCA. In other words, independent impact of a particular character to the total variance can be measured using PCA. In spite of the indication, the greater the coefficients, the greater will they be discriminated between the accessions (genotypes). Cluster analysis on the other hand, is concerned with classifying previously unclassified materials. Cluster analysis measures the distance between any two genotypes on the basis of character or trait value. This way, cluster analysis classifies the whole population panel, whereas PCA identifies which variable or trait is responsible for classifying the population into cluster. The character that contributes maximum variation is usually considered for selection. The use of PCA for morphological characterization enables us to identify minimum descriptors that effectively account for the majority of the diversity saving time and effort for future characterization efforts. For measuring plant genetic diversity, PCA is the simplest among the formal, standardized and repeatable methods (Hoogendijk and Williams 2002).

Considering the vital tools mentioned above in rattans science, multivariate analysis approach has been used for characterizing germplasm of rattans. The study conducted by Sarmah et al. (2007a, b) uses morphological markers to evaluate genetic relationships among different rattan genera and identifies 24 sets of morphological descriptors for characterizing rattan germplasm which are listed here; plant height (m), plant girth (with leaf sheath) (cm), nature of stem, stem type, stem diameter (cm), leaf length (Petiole base to apex) (m), leaf sheath texture, leaf sheath auricle, leaf sheath spines, leaflet pairs on rachis, leaflet pattern, leaflet length (cm), leaflet breadth (cm), leaflet area (cm^2), leaflet shape, leaflet color, leaflet lamina, rachis claws, petiole spine length (cm), spine shape, spine clusters, fruit color, fruit length (cm), and fruit diameter (cm).

According to the Cladistic systematics, classification and ranking of organisms are based exclusively on the "recency of common descent." Recognition of species membership in taxa is done by the joint possession of derived ("apomorphous") characters. Branching points simultaneously giving grouping and ranking (Mayr 1974). The relationships between taxa are represented by a cladogram.

Baker et al. (1999) conducted a study on subfamily Calamoideae using morphology and cladistic analysis to draw phylogenetic relationship between 22 genera of the subfamily Calamoideae. The study was based on 66 morphological characters on 31 taxa on Calamoideae and related subfamilies. The study revealed that the subfamily Calamoideae is monophyletic and the tribes Calameae and Lepidocaryeae and the subtribe Plectocomiinae are monophyletic as well. However, the subtribes Metroxylinae and Calaminae were found to be nonmonophyletic. The study also revealed the inadequacies of using morphometric analysis for establishing phylogenetic relationship and suggests alternative approach such as analysis of DNA data for reliable estimate of calamoid phylogeny.

19.3.1 Molecular Diversity Study of Calamoid Palms

Knowing the science of genetic diversity of a region may help in the creation of strategies which are effective in their preservation and future usages. For developing optimum strategies for the conservation of plant species an understanding of the patterns of genetic variation within and among populations of the plant species is needed. For the development of a strategy which is scientifically efficient, encompassing and resource effective for gene pool conservation, knowledge of the pattern of genetic diversity is necessary. Since the richness of genetic diversity in the population to be preserved will allow the rise of new evolutionary genetic combinations, and thus characterization of rattan genetic diversity is important. This genetic diversity in turn will present greater ability for evolution and adaptation to changes in environmental conditions. For genetic improvement and preservation genetic diversity is an important characteristic. Further, rattans being dioecious in nature and as such identification of sex of the plant is difficult at a very young age. This poses a problem in plant improvement breeding. Therefore, characterization based on molecular markers of rattan (Fagen et al. 2004) species may contribute to explaining genetic diversity and sex determination of rattan plants. Further, it may also help in planning effective management strategies and conservation (Singh et al. 2004) for future utilization of rattan resources.

For investigating genetic variation (Geleta and Bryngelsson 2009a, b; Li et al. 2011; Patel et al. 2015), clonal diversity and population genetic structure (Ramesha et al. 2007, Sikdar et al. 2010, Li et al. 2011, Jaisankar et al. 2020), and phylogenetic analysis (Ray et al. 2010; Haider et al. 2012; Priya et al. 2016), ISSR markers (Inter Simple Sequence Repeat) are one of the most useful molecular markers. ISSR markers have also been adopted for the evaluation of genetic diversity within rattans (Ramesha et al. 2007, Ambida et al. 2012, Asra et al. 2014, Jaisankar et al. 2020) as well as in sex determination in *Calamus tenius* (Sarmah and Sarma 2011) and *Calamus guruba* (Sinha et al. 2017a, b). RAPD markers have also been used to study genetic diversity among rattan genotypes (Sreekumar and Renuka 2006, Jaisankar et al. 2020), to provide technological basis for future molecular studies in rattan and the related species (Li et al. 2004), sex determination in *Calamus simplicifolius* (Yang et al. 2005).

19.3.2 Taxonomic and Phylogenetic Studies of Calamoid Palms

Rattans belong to the subfamily Calamoideae which was previously known as "Lepidocaryoideae" whose stems are harvested for the production of cane furniture and many other products. The subfamily Calamoideae was established by Griffith (1844) in his "Palms of British India." It belongs to the family Arecaceae (Palmeae) and is one of the most species rich and heterogeneous plants showing much variation in their habits and natural habitats, with 17 genera and two tribes, of which the genus *Calamus* is the largest genus comprising of 520 species (Baker and Dransfield 2016).

Rattans are taxonomically one of the most difficult groups and identification of rattans in the field is very difficult, sometimes even for a taxonomist. They flower annually and hence mostly these plants are seen in the vegetative condition, which makes the problem of identification more complicated since inflorescence plays a crucial role in identification to the species level. An incorrect identification and naming may lead to confusion and also contribute to misleading information (Renuka 2000).

Earlier, the studies on calamoid palms were restricted to specific countries and regions. Books monographs and field guides which are based primarily on vegetative and reproductive morphological characters such as plant habit, stem nature, leaf morphology, spine characters, inflorescence type and fruit shape were produced from different regions of the world (Griffith 1844; Hooker 1879; (Beccari 1908, 1911, 1918; Blatter 1926; Dransfield and Uhl 1986; Basu 1992; Renuka 1992; Renuka and Vijayakumaran 1994a, b; Thomas and Haridasan 1997; Fisher and Dransfield 1977; Khou 2008). A sound knowledge of rattan taxonomy is needed for the development of rattans and such taxonomic information provides a chance to solve problems regarding identification of rattans in the field which is the first step in any development and conservation-related activities.

The first classification of the family Arecaceae was conducted by Moore Jr (1973). Moore's classification recognized five major lines of evolution in 15 groups. This classification was based on precise interpretations of morpho-anatomical specialization. However, due to lack of sufficient information to construct a formal hierarchy the major groups had no taxonomic status. Following Moore's preliminary work; Uhl and Dransfield (1988) formulated a new system based on the 15 major groups of palms defined by Moore. In this new system of classification, rattans were placed in the subfamily II; "Calamoideae" consisting of 22 genera (*Calamus, Calospatha, Ceratobolus, Calaminae*, *Daemonorops, Eleidoxa, Eremospatha, Eugeissona, Korthalsi, Laccosperma, Lepidocaryum, Metroxylon, Myrialepsis, Mauritia, Mauritiella, Oncocalamus, Pigafetta, Pogonotium, Plectocomiopsis, Plectocomia, Retispatha, Raphia,* and *Salacca*) which are divided into 8 subtribes (Ancistrophyllinae, Eugeissoninae, Metroxylinae, Oncocalaminae, Plectocomiinae, Pigafettinae, and Raphiinae) and 2 tribes (Lepidocaryeae and Calameae). They described members of the Calamoideae as bearing scales which represent one of the several derived characters that members of the subfamily have in common and having a unique gynoecium consisting of three lateral connate carpels with open ventral sutures which are unique to the subfamily Calamoideae and distinguished them from other subfamilies of the palm family Arecaceae. The two tribes were separated on the basis of leaf structures. Lepidocaryeae possess palmate leaf-type while the tribe Calameae has pinnate leaf.

The taxonomic classification of rattans underwent some changes with the advance of molecular taxonomy. Several researchers started resorting to molecular data for plant systematics studies. Molecular markers such as, DNA sequences from nuclear ribosomal ITS region and the rps16 intron of chloroplast has been used to determine the phylogenetic relationship among different genera of rattans (Baker et al. 2000a), 5S nrDNA Spacer Sequence was used to investigate phylogenetic relationships among the rattan palm genera *Calamus, Ceratolobus, Calospatha, Daemonorops, Pogonotium,* and *Retispatha* (Baker et al. 2000b), nuclear and plastid DNA markers were used to resolve generic relationships between subtribe Ancistrophyllinae of African rattans (Faye et al. 2016). Systemic and phylogenetic research has been revolutionized by molecular data. And such data are now extensively used in systematic laboratories (Pleijel et al. 2008). The application of molecular markers for phylogenetic methods gave an impetus to the understanding of evolutionary trends in plants. Phylogenetic reconstructions also aid in the discovery of greater plant diversity and assists biologists in choosing areas or species to prioritize in their conservation efforts, identifying scientific importance of the plants including their economic value to humans and creating floras, monographs and inventories (Cameron 2010).

When morphological data for subfamily Calamoideae was analyzed, it failed to produce a well-supported phylogeny (Baker et al. 1999). And since there is a broad spectrum of morphological diversity that is encompassed by the subfamily, relationships within the Calamoideae are not easy to deduce (Baker et al. 2000a; b; c; d). Previous phylogenetic analyses by various researchers in palm family (Arecaceae) revealed that the subfamily Calamoideae is monophyletic (Asmussen 2000; Baker et al. 1999). Baker et al. (2000a) had carried out phylogenetic analysis of subfamily Calamoideae. The analysis was based on data of nrDNA ITS and cpDNA rps16 Intron sequence. The analysis found that the subfamily Calamoideae is monophyletic and the tribe Calameae is paraphyletic with Lepidocaryeae nested within it. The result also shows three major clades: the African rattan clade consisting of the common African rattan genera, namely, *Eremospatha*, *Laccosperma,* and *Oncocalamus*; the Lepidocaryeae-Raphia clade, which comprise of the fan-leaved New World tribe Lepidocaryeae and also the African genus *Raphia*; and the Asian clade which encompasses all Asian genera except *Eugeissona*. The *Eugeissona's* position was found to be ambiguous and could not be resolved inside in none of the three major clades. The genus *Calamus* was also found to be paraphyletic.

The nonmonophyly of genus *Calamus* was also reported by Baker et al. (1999), Baker et al. (2000a) and Baker et al. (2000b). This led to a process in the reduction of genera of Calaminae. Since, the goal of phylogenetic systematics is to identify and discard the nonmonophyletic taxa. First, the genus *Calospatha* was sank into *Calamus* (Baker and Dransfield 2008); followed by sinking of genus *Retispatha* into *Calamus* (Henderson and Floda 2015) and finally the genera *Ceratolobus*, *Daemonorops,* and *Pogonotium* were sank into *Calamus* (Baker 2015). The genus *Calamus* was thus, expanded to include *Ceratolobus*, *Daemonorops*, *Pogonotium,* and *Retispatha*. Anatomical evidence supports the re-delimitation of the Calamusgenus (Tomlinson et al. 2011). Changes were also made in the classification of tribes and subtribes within the family Calamoideae. Subtribes Raphiinae, Korthalsiinae, and Salaccinae were added into the new classification system and the subtribe Oncocalaminae was subsumed with subtribe Ancistrophyllinae which comprises mainly of African rattans. This led to taxonomic classification of rattans into 3 tribes (previously 2) (Eugeissoneae, Calameae, and Lepidocaryeae), 9 subtribes (previously 8) (Ancistrophyllinae, Metroxylinae, Calaminae, Plectocomiinae, Pigafettinae, Raphiinae, Raphiinae, Korthalsiinae, and Salaccinae) at the suprageneric level and at the genus level, into 17 genera (previously 22) (*Calamus, Eugeissona, Eremospatha, Eleiodoxa, Korthalsia, Laccosperma, Lepidocaryum, Mauritia, Mauritiella, Myrialepis, Metroxylon, Oncocalamus, Pigafetta, Plectocomia,, Plectocomiopsis, Raphia,* and *Salacca*) (Baker and Dransfield 2016). Not much research on calamoid phylogenetics has been published after Genera Palmarum - the Evolution and Classification by Dransfield et al. (Dransfield et al. 2008a, b), although several studies may be currently in progress. This also implies that modification in the systematics of calamoid palms cannot be ruled out in the future.

19.4 Economic Valuation of Rattans

Rattans are an important component of the forest which is well known for their utility since time immemorial. The rattan stem without sheath, known as "Canes," always applauded of highly valuable forest assets in the world. For utility purposes, canes' position is only next to timber and possibly equal to bamboos (Basu 1992). In the rural economy, canes play a major role for their daily breads by engaging large number of people who resides in far flung places by extracting the canes from the forests and again processing them for using it all in the small-scale and cottage industries. These urban people are being engaged in the manufacture of cane products like baskets, mats, furniture, tables, etc.

Either in terms of utilization or their market importance, potential economic value of NTFPs is often underestimated or unknown. To asses and quantify the value of the canes products is therefore a challenge. And also the transformation of the usage of many of these canes products to be socially and ecologically viable for subsistence and development is another challenge (Saulei and Aruga 1994). Indonesia accounts for the largest export of rattan canes in the global trade while China is the most important canes importing country. Global imports of rattan canes amounts to 62,000 tonnes while the value of import amounts to US$59.6 million in 2008 (Hirschberger 2011). According to the finding of INBAR (2014), bamboo and rattans trades mostly occur within and between Asia and Europe. The most important source of bamboo and rattan products is Asia. And the important importing markets are Europe, Asia, and North America. According to the UN Comtrade database, the international export of bamboo and rattan products in 2017 was estimated to be USD 1.7 billion (INBAR 2017).

On the other hand, although rattans are a potential economic resource, reports on the rapid decline of rattan resources, particularly commercially important species and large diameter species are being reported. Decline of rattan resources is attributed to natural constraint such as vanishing of forest which led to reducing habitats which are suitable for rattan plants and also overutilization (Sastry 2002; Sunderland

2001a, b; Dransfield 2002a, b). Extinction of the commercial species and no plantation to secure rattan supply causes lack of sufficient supply in both quality and quantity. Moreover, since most rattan producing countries do not have rattan resources inventories and even if they have the inventories they are generally not up to date or simply approximate estimations. As the true volume and rate of growth of rattan resources are not known, the allowable annual cut cannot be determined in terms of sustainability but by the demand current in the rattan industry. This results in the overexploitation of commercially important rattan (Hirschberger 2011). Therefore, preparing an inventory of the rattans in terms of demand and supply chain, volume extracted by locals and estimating the income derived from rattan gathering and rattan industry becomes crucial.

19.5 Medicinal Properties of Rattans

Ethnobotanical studies conducted so far have confirmed the used of rattan plants in traditional medicine by the indigenous people around the world. Apart from being one of the most important NTFPs usually used by local people for various handicrafts works. Reports on other traditional uses of these plants indicate that the shoots, fruits, and leaves are consumed as a delicacy (Renuka 2001; Sarmah 2010a, b) and as such encourage scientific validation for its consumption as functional food (Thakur and Sheth 2015). According to Wangyal 2012 and Borah et al. 2013) this plant is traditionally known to have several therapeutic potential against stomach disorder, intestinal worms, healing nausea. In the traditional Chinese medicines, red resin obtained from the immature fruit of *Daemonorops draco* was used for wound healing and also believed to possess hemostatic, antiseptic, antibacterial, antiviral properties, and it was also valued as medicine in Europe due to its astringent property, externally used for wound healing and internally used to alleviate internal traumas, chest pains, postpartum bleeding and menstrual irregularities (Baja-Lapis 2009). Therapeutic potential and medicinal uses (health tonic) of various species of *Calamus* was also reported (Islam et al. 2015; Sunderland and Dransfield 2002; Jin 2005). It has been reported that *Calamus tenius* is use as herbal medicine in treating Diabetes mellitus (Mitali and Palash 2013) by the local communities of Assam (Tag et al. 2012).

Calamus sp. is reported to possess antibacterial activity, antiseptic and antidiabetic properties and in ayurvedic preparations for treating fever, piles, dyspepsia, antihyperglycemic activity, etc. (Thakur and Sheth 2015; Sarkar et al. 2018; Salusu et al. 2021). Palmitic acid, iso-eugenol, calamine, calamol, etc. which were present in *Calamus* oil were extracted from the roots and is used in perfumery and flavoring of liquors (Anon 1992). Huangcan et al. (1991) conducted an analysis to evaluate the nutrient content of rattan shoots of two commercial species, viz., *Daemonarops marga-ritae* Hance and *Calamus simplicifolius* Wei and found that rattan shoots contain a lot of protein, fat, carbohydrate, eight kinds of amino acids and various nutrient elements and vitamins, which are essential nutrient substance for human being and concluded that the shoots can be exploited as a valuable vegetable with low sugar and high protein.

19.5.1 Rattan as a Source of Food

Apart from its various uses say furniture and other canes items, some species of rattan are also edible. Traditionally, edible shoots of rattans are considered as nontimber forest products. In South-East Asian countries, tender shoots are consumed either as raw or as vegetable. In Philippines (Durst et al. 1994a, b), France, United States, Lao PDR (People's Democratic Republic) and Thailand young shoot of various species of rattan are also consumed which has considerably rich amount of starch (Manohara 2013). Several shoots bearing rattan have been adopted as crop plants in Thailand and Lao PDR (Laos). *Calamus tenuis* Roxb., is one of the major species planted in Laos (Dransfield et al. 2002a, b). Report on nutritional profile says that rattans are rich in proteins, carbohydrates, minerals and fiber content. As a result of which, in various countries across South-East Asia rattan preferred to be the most popular dietary supplement for rural populace (Manohara 2013). According to Saikia and Khan 2011 shoots of rattans have therapeutic potential against stomach disorder and intestinal worms.

In India, particularly in Assam, *Calamus tenuis* Roxb. shoots are consumed either by frying, roasting or boiling. Delicious shoots curry are made in combination with other food items fish, meat, red ant eggs, elephant apple, mustard flakes in a form of traditional style by wrapping with edible leaves and with black gram pulses. March to May has highest consumption of the shoot when compared to other months of the year (Thakur and Sheth 2015). Low socio-economic group and people living in villages near forest, those who are unemployed or unskilled worker, consumed *Calamus tenuis* Roxb. shoots as a mouth-watering food items rather than therapeutic purpose. This lesser known plant, *Calamus tenuis* Roxb., still remains as a forest crop which is meagerly available in the market even though they confined as traditional delicacy of the region (Thakur and Sheth 2015). In Manipur, rattan fruit are used as obligatory fruit items during the important religious festival "*Cheiraoba.*" Despites all these, still therapeutic potentials of rattans is known by few people in this region (Durst et al. 1994a, b).

There is a species called, *Calamus rotang* which possesses an edible fruits. Matured fruits are roundish, similar in size with hazelnut and covered with small, shining scales, laid like shingles, one upon the other. The kernel is surrounded by subacid pulp which can be sucked out and eaten. The fruit is pickled with salt and eaten at tea-time. In other types of species, a gelatinous pulp which is either sweet or sour, surrounds the seeds. This pulp is usually eaten raw and the taste is similar to citrus. Renuka 2001 and Sarma, 2010 mentioned its traditional uses as desirable food item. Thakur and Sheth 2015 encourage consumption of rattans as functional food with proper scientific investigation on nutritional values.

19.6 Breeding Strategies and Improvement in Rattan

Rattans are mostly collected from the forest. This increases in unhindered rattan harvesting results in overexploitations of natural resources and forest biodiversity. This in turn makes it important to develop commercial rattan cultivations strategy which is based on scientific breeding approach. Adoption of eco-friendly and sustainable commercial practices will help in overcoming lost natural resources due to overexploitation.

To satisfy the growing market for good canes varieties, economical rattan species should possessed commercially acceptable agronomical and yield attribute traits.

Identification of dominant strains, selection then further development of quality seed orchards of indigenous and exotic species, so that improved commercial rattan species can be produced. According to Zhao et al. (2017), as there has been much less systematic development to dates, improvement this plant using genetics tools has become a vital option for yields and yield attributes. It is obvious that any rattan improvement plan should take into account the fact that selection has to be done based on stability in diverge agro climatic zones and plant's end use. However, most species are able to grown at different climatic zones and adapt to different soil conditions. On the contrary, monsoonal rattans may enjoy some advantages as they can thrived water rich climate and a wide adaptability to areas with similar climatic characteristics both in the northern and southern hemisphere (Shim 1995).

19.6.1 Breeding Objectives of Rattan Species Consist of the Following Traits

Production of suckers: Sucker production increases with clump dimension in clustering rattans species: Traits like long stolons and rhizomes which are found in *C. trachycoleus* should be selected to facilitate production of new stems. Species having the single stemmed that exhibits growth of poly-suckers traits should be avoided because the dominant growth is hampered, resulting in reduction of quality and weakened the growth of canes.

Rate of Growth: Microclimatic and nutrient factor affects the growth rate. However, in environments where the factors are similar, growth rate depends both on production rate of internodes and its length.

Internodes length: Of all the species, the intermodal length varies; *C. subinermis*, *C. manan*, and *C. trachycoleus* are 31 cm, 22 cm, and 22 cm, respectively. This trait is one of the most important morphological parameter for the selection of rattan species both for the cane production and for the purposes of manufacturing furniture products.

Differences in nodal diameter: Better quality will have possessed more in fewer differences in nodal diameter, so uniformity of the cane formation will be determined based on the changes in the nodal diameter.

Diameter of cane: Climatic factors and species determined the diameter of canes.

Skin's color and blemish: Selection of species and treatment method determines color and blemish of skin.

Inflorescence: The inflorescence present in the rattan species results in the reduction of the internode length in the immediate distance. This also affects the shape of the cane. Generally, late flowering canes or even those ones that do not flower annually are preferred.

Fruits: This horticultural trait of importance is lesser known and neglected most and should concentrate on food value and yield attributes based on: improve pulp size, large fruit size, small size seeds, and number of fruit per cluster, higher juice content, higher TSS, and nutritional value of the fruits.

Possible strategies for rattans breeding program are mentioned below:

First, for the short term methods, indigenous plants can be selected and grown immediately having traits of good quality seeds stock selected from the diverse species, then the high viable seeds and good germination can be chosen from them as high quality seed stocks for commercial plantations. However, this may result in a heterogeneous population. Second, for a long term strategy for obtaining optimum results, recurrent selection of quality seeds through provenance trials methods by selection wide range of collection of seeds, sites selection, growing of seeds, replication system on different soil, and climatic condition for the measurement of traits selection and interpretation of best rattans plants.

Some issues in preserving the vigor seeds for a lengthy period of season and it is very difficult to reproduced seed which is selected from the aged-old stock for the uniform fruit bearing. Besides, immediate bearing of seed and prediction of vigor plants in a very quick glance is a difficult job for breeder, agronomist, and horticulturist, as rattans take normally 7–10 years to reach its maturity stage. Other alternative to being adopting seedling from seeds is vegetative propagation, however, it is tedious expensive and difficult to carry out on a large scale, due to inadequate production.

Correct species identification, their tools are being mentioned above, has been prioritized in recent research because: First, it is an effective approach to transmit information and predict the traits of rattan. Second, it helps to finding the best methods for conservation and development also needs assessment and inventory.

Third, it makes the identifications of the genetic variation in the species and suitable sites easier. Although few extrinsic parameters influence the exploration extraction rates and natural re-establishment of rattans species, whichever itself is determined by environmental circumstances, and have a practical results on adopting bona fide in-situ methods and feasible use of the genetic resource for immense economic well-being.

There is a shortage of fundamental knowledge regarding the taxonomy, phylogeny, genetic variation, and geographic variation of rattan species. The information that is now known is dispersed throughout numerous publications. These hindered or delayed the systematic improvement programs for breeding rattan plants. Rattan is propagated through seed, but due to their spiny nature and the absence of flowers and fruits most of the year, it made difficult for crossing and hybridization. Most rattans and bamboos can produce seeds, which allows for selfing and makes selfed lines valuable. According to Zhao et al. (2017), evaluation studies using some plantation species of unimproved wild-type germplasm produced significant progress, and these trials are relevant to rattan species. For a rigorously managed, well-designed, and sufficient scale evaluation, the initial survey, determination of target regions for sampling, as well as the proper sample processes play a crucial role. Additional data on cytology, pollination, fruit growth, and other ecological and biological processes must be gathered throughout evaluation trials. Through evaluation, it is possible to identify superior genotypes that will serve as the foundation for systematic improvement programs involving crossing and selection. Although difficult, crossing operations are possible in rattan. It yields twice as much oil annually as the finest commercial material, according to an evaluation trial of oil palm utilizing germplasm obtained in Nigeria in the 1970s.

Each population in an assessment trial should be characterized by a number of naturally occurring progenies who share the same maternal parent; as a result, half-sibs will be assumed for each population. The progeny produced through a trial of controlled pair-wise crossing allowed to estimates the heritability of quantitative characters of interest which may response to selection, like any other cross-pollinating species. In their separate reviews of bamboo and rattan, Banik (1997) and Shim (1995), respectively, reported species hybridization. Natural hybrids, in the case of rattan, are unknown. Shim recognizes the challenges associated with using species hybrids in a seed-propagated crop until standardized vegetative propagation techniques are available. The multiplication rate through seed is likely to raise sufficient material for commercial uses, but it should be more helpful when it accomplishes this by clonal means. Germplasm exploration and evaluation are taken into account as a component of a long-term plan for crop improvement. It is intended to increase a crop's genetic diversity in terms of the genes controlling polygenic traits and to supply the primary gene sources (pest and disease resistance) and other desired features that are not present in commercial cultivars. These significant genes could be introgressed from unimproved germplasm to cultivated variety through the recurrent backcross method.

With the advance of molecular markers and genome sequencing technologies, it will enhance the genetic improvement of rattan. Teulat et al. (2000) identified six coconut microsatellite markers and used for cross-amplification in four genera (*Korthalsia, Zalacca, Daemonorops*, and *Calamus*) of rattan by Rao et al. (2007). Sreekumar and Renuka (2006) reported DNA analysis of *C. thwaitesii* population at Goa (India) shown the high genetic diversity. With the studies of molecular markers, Lyngdoh et al. (2005) identified four diversity hot spot sites at north-eastern Himalayas (India) for conserving *C. flagellum*. Baker et al. (1999) successfully classified the palm genus Phoenix using nuclear 5S nontranscribed spacer region and this study inspired the rattan researchers to use this region of DNA for classification and relationship analysis of *Calamus* and five related genera (Baker et al. 2000b). From this study of the 5S nrDNA data, Baker et al. (2000b) found that the genus *Calamus* belongs to paraphyletic group with four major lineages. DNA barcoding and molecular phylogeny could provide insight understanding of systematics taxa. The slow rate of evolution of palm DNA restricts the use of plastid as well as nuclear gene regions in molecular systematics of palms. With the introduction of low copy nuclear regions have gained to use plastid as well as nuclear gene regions in molecular systematics of palms (Kurian et al. 2017). Molecular phylogeny together with biogeography could contributes broaden insight about the distribution pattern of extant species as well as their origin of ancestral area. With the advance of super barcodes along with whole genome sequencing could provide promising platform to strengthen the classification of palms species, in the near future.

Sex of dioeciously palm genera was usually studies from floral characteristics however; it does not provide convenient throughout the year due to seasonal flowering and fruiting in rattan. Development of gender specific markers would enhance to screening rattan plants at the early seedling stage. Yang et al. (2005) generated a male specific RAPD molecular marker nearly 500 bp length for determination of sex on *C. simplicifolius*. Similarly, Sarmah and Sarma (2011) reported ISSR4_600 is a putative sex-linked marker for *C. tenuis*. The introduction of high-throughput techniques like genotyping-by-sequencing (GBS) and restriction site associated DNA sequencing (RAD-Seq) can accelerate the early sexing procedures in rattans.

Zhao et al. (2017) analyzed in-depth transcriptomic sequencing on *Daemonorops jenkinsiana* to characterize the cirrus development at different developmental stages. They assembled 404,875 transcripts and 61,569 high-quality unigenes were identified, of which nearly 76.16% were annotated and classified by seven authorized databases. Additionally, 14,693 microsatellites markers of transcriptome-based were identified. Out of it, 168 designed SSR primer pairs, 153 were validated and 16 pairs were used for the polymorphic analysis of 25 rattan accessions. Zhao et al. (2018a, b) developed two chromosome-level genome assemblies of *C. simplicifolius* and *D. jenkinsiana* using Illumina, Pacific Biosciences, and Hi-C sequencing data and their study shown that four Arecaceae plants clustered together while the divergence time between *C. simplicifolius* and *D. jenkinsiana* was approximately 19.3 million years ago. It provides a fundamental resource for functional genomics that would enhance germplasm utilization for breeding and also act as reference genomes for comparative studies between and among different species.

19.6.2 Pollination in Rattan

Both wind and insect pollinate rattan flower. Rattan, *Calamus* inflorescence morphology is thought to be anemophily. Considering the fact that the adaptable flower arrangment can provide an easy pathway for wind to carry over the pollen. Though wind pollination is not a main pollination in Rattan. In Lee and Jong (1995) and his coworker have provided the information on wind pollinated rattan *Calamus* species which only 88% of the pollen can be dispersed in 3.5 m Lee and Jong (1995).

Basically, rattan flowers which are entomophily emitted various scents that attracted the pollinating agents. Rattans palms are spiny dioecious plants that flower once a year and also pollinated by different insect species and so considered as entomophilous flower. In the insect-plant mutualistic relationship, insect gets the nectar and introduced pollen as rewards for the pollination service. Ants help rattan palm with seed harvesting and dissemination (Berg 1975) as well as serving as a good pollinator (Peakall and Beattie 1991; Liu et al. 2019). Ants although seem to pollinate certain type of rattan species where the flowers are unisexual. For flowers which are spatially separated they hardly transfer pollen as reported by Renuka (1998). Other pollinating agents are stingless bees (Trigona) and paper wasps (Vespidae) pollinate *Ceratolobus castaneus*. Kidyoo and McKey (2012) also found two Trigona bee species which are good pollinators of *C. castaneus* flowers, visiting both male and female flowers. The unisexual inflorescence of the *Ceratolobus* is enclosed, allowing the pollinator to enter only for short time from the apical split. According to Dransfield (1979) *Ceratolobus* produces a musty odor that attracts small staphylinid and some other small beetle for the pollination.

In other rattan species *Daemonorops lewissiana* is pollinated by ants that crawl in female and male flower, however *D. didymophylla* (Kiew and Muid 1989) likely to visit by *Trigona melina* to male but not the female flowers. Similarly, in 1979, Dransfield had also reported many Hymenopterans such as *Trigona* species, honey bees visiting the male flowers of *Plectocomia* sp. and other Coleopteran beetles have visited attracting its musty odor emitting from the inflorescences. Calamus inflorescences produces sour odor which is attracted by several wasps and flies (Dransfield 1979a, b, c). Lee et al. (1995) was also reported several nocturnal insects are active pollinators belonging to pyralid, noctuid, and moths.

19.7 Agroforestry Management

Rattan plants are propagated through seed, wildings, suckers, rhizomes, or by cuttings. But stem cutting method of propagation are not used in North Eastern Indian canes species as they do not branch out aerially. As rattans are spiny climbing palms, they must be interplanted with trees to give them support and shade during the seedling stage of the rattans. The yield and quality of rattan are also influenced by the supporting trees as reported by Weidelt, 1990.

Rattan plants prefer areas with abundant and well-distributed rainfall, where the soil is fine clay and rich in humus Campbell et al. (2017). Goswami et al. (2000) reported that rattans prefer strong to medium acidic to dark colored loamy soil with moderate water holding capacity by analyzing the rattan growing soils of Arunachal Pradesh. The spacing for the plantation of rattan depends on their species as different species of rattan plant have different growth habit as for clump forming species, 5 m and more spacing are require while single stemmed species require 2–5 m spacing.

Rattan plants have a great potential as agroforestry crop. Other forest tree and fruit trees required 25–35 years for one rotation but rattan plants need 6–7 years for bearing of fruits and 11–15 years for the first pole to be harvested. As rattan plants need support trees for their proper growth, other forest tree and fruit and plantation trees can be served as the support tree so integration of rattan plant in a community forest or any tree farm can add to productivity of land as well as protection and conservation of watershed areas while waiting for the harvest of rattan. Not only this, as rattan plant need at least approximately 3–4 years to develop the climbing organs (flagellum or cirrus) so during the early years for the establishment of the rattan plantation we can go for the cultivation of arable crop between the rows of the rattan plants or rattan plants can be planted along the boundaries and stream margins of a newly established agroforestry system. As for example- the indigenous rice-rattan swidden agroforestry which is called as the *qaiya-aneya* system practice in south-west China in which the farmers interplants the rattan seeds in open new swidden field for upland rice, particularly near remaining stumps. After harvesting rice for several time, the land are left as fallow for the rattan canes which can be harvested after 7–10 years (Xu et al. 2000).

If rattan plants are integrated in an agroforestry system then there is a scope for getting a perennial source of income to the farmers. As during the early stage of establishment farmer can get income from the fruit tree or other plantation crop like coconut, arecanut palm, jack, custard apple which are used as support for the rattan plant and also from the arable crop growth between the rows of the rattan plants.

19.8 Climate Change—Rattan

Human civilization has a negative impact on climate change. It is severely affecting agricultural production in terms of yield loss, decreasing natural resources, and increasing pests and diseases. To increase agricultural production, support elevated carbon (C) sink, and combat climate change, the agroforestry system provides a potential method. As India is rich in biodiversity, agroforestry can provide climate-resilient agriculture. Agroforestry consists of multispecies that can produce micro-climate and conserve soil properties. Soil is a plant growth medium developed from the combined effect of climate and living matter, it can change over a period of time. Soil is an important source of cultivation that determines growth; provides water and nutrients for plants. Large-scale adaption of agroforestry can create additional national carbon sinks for CO_2, increase the use of renewable energy and increase soil fertility (Kay et al. 2019). Societies are responsible for the effects of climate change (Billi et al. 2019). It can be reduced by the sustainable management of forests and reducing dependence on fossil fuels.

Rattan forest provides replacements for higher carbon emission material for products made from plastics and steel. It is an important land use system. Rattan trees are having solid stems with a group of spiny (formed from modified leaf tips or inflorescences) climbing palms or canes. The canes of rattans are harvested for their solidity, durability, lightweight, flexibility, and strong nature (Meijaard et al. 2014). More products we can get from rattan like baskets, handicraft items, furniture, and important raw material in the cottage industry (Sun and Liu 2022). In Asian countries like India, China, and Malaysia it is an important traditional agroforestry system. Rattan agroforestry includes heritage values of traditional ecological knowledge. For the climate-smart agriculture with bamboo and Rattan agroforest system, it is a powerful tool to maintain the stability of the slop, and protect soil from erosion and degraded lands can be restored. Due to climate change, the ecosystem is severely affected and it can affect traditional livelihood activities. Both rattan and bamboo can provide benefits in socio-economic and environmental dimensions to rural people to cope with potential "climate shocks." As it grows easily in difficult climatic conditions, it can be recommended as the most suitable strategy to work on under extreme climatic events.

19.9 Future Perspectives

Apart from commercial cane-ware products, rattan resources are yet to be explored in terms of their food value, therapeutic, medicinal properties and most pivotal is its breeding aspects of traits, therefore, research aiming at crop improvement program through conventional and nonconventional approaches like tissue culture for mass propagation, application of various breeding methods for further traits selection, molecular approaches to design new varieties and documentation of bioactive potential as well as its potential against the new-normal diseases along with their systematic molecular investigation of the genomic information of potential rattan resources will provide a foothold for further studies and investigations of its biochemical composition. This may also open a new gateway to various possibilities of discovery and may boost towards the current scientific field of improving human health care system and food security.

Breeding technologies have been utilized to improve the quality and quantity characters of rattan. Despite, the knowledge of genetic structure underlying the important characters of rattan have not been known clearly. This critically hindrance the overall understanding of its molecular biology for scientific research, actual production and in-depth study of comparative genome analyses between and among related species. Development of chromosome-level reference genomes in rattan made feasible for comparative genome analyses and other downstream applications viz., the development of biomarkers, the identification of functional genes, and molecular design breeding. Similarly, genomic, transcriptomic, and metabolomic analyses of rattan traits will facilitate through the development of high-quality genome assemblies of rattan. These investigations lay a framework for future research on the employment of these genes to improve rattan quality and diversity within rattan germplasm.

References

Abdelkrim J, Pascal M, Samadi S (2005) Island colonization and founder effects: the invasion of the Guadeloupe islands by ship rats (*Rattus rattus*). Mol Ecol 14(10):2923–2931. https://doi.org/10.1111/j.1365-294X.2005.02604.x

Alam MK (1990) Rattans of Bangladesh. Bulletin 7. *Bulletin-Plant Taxonomy Series, Bangladesh Forest Research Institute*, No. 7. Plant taxonomy series

Alamgir M, Bhuiyan MA, Jashimuddin M, Alam MS (2006) Economic profitability of cane based furniture enterprises of Chittagong City Corporation Area, Bangladesh. J For Res 17(2): 153–156. https://doi.org/10.1007/s11676-006-0036-y

Ambida RC, Gabor CR, Panes VA (2012) Preliminary molecular analysis of Philippine rattan germplasm collection. Ateneo De Manila University, pp 1–24

Anon (1992) *Calamus*; The wealth of India, CSIR, New Delhi, vol. 3

Ariane C, Gardiner L, Ouattara D, Stauffer F, Sonké B, Couvreur T (2017) An endangered west African rattan palm: Eremospatha dransfieldii. Biodiversity Data Journal 5:e11176. https://doi.org/10.3897/BDJ.5.e11176

Asmussen CB (2000). Phylogeny of the palm family (Arecaceae) based on rps16 intron and trnL-trnF plastid DNA sequences. Monocots: systematics and evolution. Monocuts: systematics and evolution: proceedings of the 2. international conference on the comparative biology of the monocuts, Sydney, pp 525–535

Asra R, Syamsuardi S, Mansyurdin M, Witono JR (2014) The study of genetic diversity of *Daemonoropsdraco* (Palmae) using ISSR markers. Biodivers J Biol Divers 15(2). https://doi.org/10.13057/biodiv/d150201

Baja-Lapis AC (2009) Specialty rattans of the ASEAN. Blumea-Biodivers Evol Biogeogr Plants 54(1-2):39–43. https://doi.org/10.3767/000651909X476003

Baker WJ (2015) A revised delimitation of the rattan genus Calamus (Arecaceae). Phytotaxa. 197(2):139–152. https://doi.org/10.11646/phytotaxa.197.2.7

Baker WJ, Dransfield J (2008) Calospatha subsumed in Calamus (Arecaceae: Calamoideae). Kew Bull 63(1):161–162. https://doi.org/10.1007/s12225-007-9007-5

Baker WJ, Dransfield J (2016) Beyond Genera Palmarum: progress and prospects in palm systematics. Bot J Linn Soc 182(2):207–233. https://doi.org/10.1111/boj.12401

Baker WJ, Dransfield J, Harley MM, Bruneau A (1999) Morphology and cladistic analysis of subfamily Calamoideae (Palmae). In: Henderson A, Borchsenius F (eds) Evolution, variation and classification of plants. Memoirs-New York Botanical Garden. The New York Botanical Garden Press, Southern Boulevard, NY, pp 307–326

Baker WJ, Asmussen CB, Barrow SC, Dransfield J, Hedderson TA (1999) A phylogenetic study of the palm family (Palmae) based on chloroplast DNA sequences from the trnL—trnF region. Plant Syst Evol 219(1–2):111–126. https://doi.org/10.1007/BF01090303

Baker WJ, Dransfield J, Hedderson TA (2000b) Phylogeny, character evolution, and a new classification of the *calamoid palms*. Syst Bot 1:297–322. https://doi.org/10.2307/2666644

Baker WJ, Hedderson TA, Dransfield J (2000c) Molecular phylogenetics of subfamily Calamoideae (Palmae) based on nrDNA ITS and cpDNA rps16 intron sequence data. Mol Phylogenet Evol 14(2):195–217. https://doi.org/10.1006/mpev.1999.0696

Baker WJ, Hedderson TA, Dransfield J (2000d) Molecular phylogenetics of Calamus (Palmae) and related rattan genera based on 5S nrDNA spacer sequence data. Mol Phylogenet Evol 14(2): 218–231

Banik RL (1997) Domestication and improvement of bamboo. INBAR working paper/International Network for Bamboo and Rattan; no. 10

Basu SK (1992) Rattans (Canes) in India: A monographic Revision. Rattan Information Centre. Forest Research Institute Malaysia, Kepong, p 141

Beccari O (1908) Asiatic palms-Lepidocaryeae part I: The species of *Calamus*. Ann R Bot Gard Calcutta. 11:1–518

Beccari O (1911) Asiatic palms–Lepidocaryeae. Part II. The species of Daemonorops. Ann R Bot Gard Calcutta 12(1):1–237

Beccari O (1918) Asiatic Palms-Lepidocaryeae: part 3: The species of the genera *Ceratolobus*, *Calospatha*, *Plectocomia*, *Plectocomiopsis*, *Myrialepis*, *Zalacca*, *Pigafettia*, *Korthalsia*, *Metroxylon* and *Eugeissona*. Ann R Bot Gard Calcutta 12:156–195

Belcher B (2002) CIFOR research: forest products and people rattan issues. In: Dransfield J, Tesoro FO, Manokaran N (eds) Rattan. Current research issues and prospects for conservation and sustainable development. FAO, Roma, IT. (Serie Non-wood forest products No. 14); pp 49–62

Berg RY (1975) Myrmecochorous plants in Australia and their dispersal by ants. Aust J Bot 23(3): 475–508. https://doi.org/10.1071/BT9750475

Billi M, Blanco G, Urquiza A (2019) What is the 'social' in climate change research? A case study on scientific representations from Chile. Minerva 57:293–315. https://doi.org/10.1007/s11024-019-09369-2

Blackith RE, Reyment RA (1971) Multivariate morphometrics, p 412

Blatter E (1926) The palms of British India and Ceylon. H. Milford; Oxford University Press, Oxford, pp 246–255

Borah S, Kakoti BB, Mahato K, Kumar M (2013) Investigation of in-vitro anthelmintic activity of *Calamus leptospadix* Griff. shoot in Indian adult earthworm (*Pheretimaposthuma*). J Appl Pharm Sci 3(6):156–159. https://doi.org/10.7324/JAPS.2013.3626

Cameron KM (2010) On the value of taxonomy, phylogeny, and systematics to orchid conservation: implications for China's Yachang Orchid Reserve. Bot Rev 76(2):165–173. https://doi.org/10.1007/s12229-010-9052-x

Campbell MJ, Edwards W, Magrach A et al (2017) Forest edge disturbance increases rattan abundance in tropical rain forest fragments. Sci Rep 7:6071. https://doi.org/10.1038/s41598-017-06590-5

Chan CS, Cheng J, Loh JY, Tan E, Loo AH, Tang JH (2012) Observations of ants and aphids in the rattan species Korthalsiaechinometrabecc. and Korthalsia rostrata blume. Raffles B Zoo 25:133–139

De Zoysa N, Vivekanandan K (1991) The bamboo and rattan cottage industry in Sri Lanka: livelihoods in danger. Forestry Information Service, Forestry Research and Information Project, Forest Dept., Battaramulla, Sri Lanka

Dransfield J (1979a) A manual of the rattans of the Malay Peninsula. A manual of the rattans of the Malay Peninsula, p 29

Dransfield J (1979b) A monograph of Ceratolobus (Palmae). Kew Bull 34:1–10. https://doi.org/10.2307/4117966

Dransfield J (1979c) A manual of the rattans of the Malay Peninsula. Forest Department, Ministry of Primary Industries, Malaysia, p 29

Dransfield J (1984) Rattans of Sabah. Sabah Forest Department

Dransfield J (1992) The rattans of Sarawak. Royal Botanic Gardens, p 230

Dransfield J (2002a) General introduction to rattan: The biological background to exploitation and the history of rattan research. Rattan: Current research issues and prospects for conservation and sustainable development, pp 23–34

Dransfield J (2002b) General introduction to rattan-The biological background to exploitation and the history of rattan research. In: Dransfield J, Tesoro FO, Manokaran N (eds) Rattan current research issues and prospects for conservation and sustainable development. FAO, Rome, pp 23–34

Dransfield J, Uhl NW (1986) An outline of a classification of palms. Principes 30:3–11

Dransfield J, Tesoro FO, Manokaran N (2002a) Rattan current research issues and prospects for conservation and sustainable development. Based on the outcome of the FAO expert consultation on Rattan development held at FAO, Rome, Italy, pp 23–34

Dransfield J, Tesoro FO, Manokaran N (2002b) Rattan: Current research issues and prospects for conservation and sustainable development. Rome, FAO

Dransfield J, Uhl NW, Asmussen CB, Baker WJ, Harley MM, Lewis CE (2008a). Genera Palmarum-The Evolution and Classification of the Palms. https://doi.org/10.34885/92

Dransfield J, Uhl NW, Lange CBA, Baker WJ, Harley MM, Lewis CE (2008b) Genera Palmarum: the evolution and classification of palms. Kew Publishing, Richmond

Durst PB, Ulrich W, Kashio M, Khan SA, Kunshan S, Gupta BN, Silitonga T, Yong PL, Khatri DB, Iqbal Sial M, Neri BS (1994a) Non-wood forest products in Asia. RAPA Publication (FAO), Rome

Durst PB, Ulrich W, Kashio M (1994b) Non-wood forest products in Asia. Rapa Publication, Bangkok

Fagen L, Hua Y, Guangtian Y, Huangcan X, Siming G (2004) Genomic DNA extraction and RAPD protocols for rattan. For Res 17(6):824–828

Faye A, Pintaud JC, Baker WJ, Vigouroux Y, Sonke B, Couvreur TL (2016) Phylogenetics and diversification history of African rattans (Calamoideae, Ancistrophyllinae). Bot J Linn Soc 182(2):256–271. https://doi.org/10.1111/boj.12454

Fisher JB, Dransfield J (1977) Comparative morphology and development of inflorescence adnation in rattan palms. Bot J Linn Soc 75(2):119–140. https://doi.org/10.1111/j.1095-8339.1977.tb01481.x

Geleta M, Bryngelsson T (2009a) Inter simple sequence repeat (ISSR) based analysis of genetic diversity of *Lobelia rhynchopetalum* (Campanulaceae). Hereditas 146(3):122–130. https://doi.org/10.1111/j.1601-5223.2009.02111.x

Geleta M, Bryngelsson T (2009b) Inter simple sequence repeat (ISSR) based analysis of geneticdiversity of *Lobelia rhynchopetalum* (Campanulaceae). Hereditas 146(3):122–130

Goswami M, Haridasan K, Barik SK (2000) Soil characteristics of some cane (rattan) bearing areas in Arunachal Pradesh-a preliminary investigation. Arunachal For News. 18(1&2):107–109

Griffith W (1844) Palms of British East India. Calcutta J Nat Hist 5:1–103

Haider N, Nabulsi I, MirAli N (2012) Phylogenetic relationships among date palm (*Phoenix dactylifera* L.) cultivars in Syria using RAPD and ISSR markers. J Plant Biol Res 1(2):12–24

Henderson A, Dung NQ (2014) Notes on rattans (Arecaceae) from Vietnam. Phytotaxa 8(1): 25–33. 2010

Henderson A, Floda D (2015) Retispatha subsumed in Calamus (Arecaceae). Phytotaxa. 192(1): 58–60. https://doi.org/10.11646/phytotaxa.192.1.8

Hirschberger P (2011) Global rattan trade: pressure on forest resources, analysis and challenges. WWF-Austria, Vienna

Hong LT, Ramanatha Rao V, Amaral W (2002) Rattan genetic resources conservation and use IPGRI's perspective and strategy. Non-Wood Forest Products (FAO), Rome

Hoogendijk M, Williams DE (2002) Characterizing the genetic diversity of home garden crops: some examples from the Americas. In: Watson JW and Eyzaguirre PB (eds) Home gardens and in situ conservation of plant genetic resources in farming systems. IPGRI: Proceedings of the second international home gardens workshop, Witzenhausen, Germany, pp 34–40, p 34

Hooker JD (1879) Flora of British India, Vol. II

Huangcan X, Zaizhi Z, Guangtian Y (1991) Nutrition evaluation of rattan shoots [J]. For Res 4 (4):459–462. http://lykxyj.xml-journal.net/en/article/id/19910422

Hunter IR, Lou Y (2000) Presentation of INBAR's programme on rattan. In: Dransfield J, Tesoro FO, Manokaran N (eds) Rattan Current Research Issues and Prospects for Conservation and Sustainable Development. FAO, Rome, pp 43–18

INBAR (2014) Bamboo value chains and inclusive green growth. In: International Market of Bamboo and Rattan

INBAR (2017) Bamboo and rattan commodities in the international market. In: Trade overview, pp 1–29

Islam SA, Miah MA, Habib MA, Rasul MG (2015) Growth performance of Calamus tenuis Roxb. (Jali bet) in the coastal homesteads of Bangladesh. J Biosci Agric Res 4(02):74–79

Jaisankar I, Subramani T, Velmurugan A, Singh AK (2020) Genetic diversity revealed among rattan genotypes from Andaman and Nicobar Islands based on RAPD and ISSR markers (Assessed 14 Sept 2020)

Jin CB (2005) Medicinal properties and common usages of some palm species in the Kampung Peta community of Endau-Rompin National Park. Johor. J Trop Med Plants 6(1):79

Kay S, Rega C, Moreno G, Herder M, Palma J, Borek R, Crous-Duran J, Freese D, Giannitsopoulos M, Graves AR, Jäger M, Lamersdorf N, Memedemin D, Mosquera-Losada MR, Pantera A, Paracchini M-L, Paris P, Roces-Díaz JV, Rolo V, Herzog F (2019) Agroforestry creates carbon sinks whilst enhancing the environment in agricultural landscapes in Europe. Land Use Policy 83:581–593. https://doi.org/10.1016/j.landusepol.2019.02.025

Khou EH (2008) A field guide of the rattan of Cambodia. WWF greater Mekong-Cambodia country programme, Cambodia

Kidyoo AM, McKey D (2012) Flowering phenology and mimicry of the rattan *Calamus castaneus* (Arecaceae) in southern Thailand. Botany 90(9):856–865. https://doi.org/10.1139/b2012-05

Kiew R, Muid M (1989) Bees and palms in Peninsular Malaysia. Principles 33:74–77

Kurian A, Sreekumar VB, Dev SA, Muralidharan EM (2017) A review on molecular studies of rattans, with special attention to the Genus Calamus (Arecaceae). J Bamboo Rattan 16(3): 97–114

Lee YF, Jong K, Swaine MD, Chey VK, Chung AY (1995) Pollination in the rattans *Calamus subinermis* and *C. caesius*. Sandakiana 6:15–39. sandakania-6-015-030

Lee YF, Jong K (1995) Pollination in the rattans Calamus subinermis and C. Caesius (Palmae: Calamoideae). Sandakania 6(1):15–39

Li FG, Yang H, Yin GT, Xu HC, Gan SM (2004) Genomic DNA extraction and RAPD protocols for Rattan. For Res 17(6):824–828

Li GT, Ai CX, Zhang LS, Wei HR, Liu QZ (2011) ISSR analysis of genetic diversity among seedling walnut (*Juglans* spp.) populations. J Plant Genet Resour 12(4):640–645

Liu K, Mansor A, Ruppert N, Lee CY, Azman NM, Fadzly N (2019) Rattan litter-collecting structures attract nest-building and defending ants. Plant Signal Behave 14(8):1621245. https://doi.org/10.1080/15592324.2019.1621245

Lyngdoh N, Santosh SH, Ramesha BT, Rao MN, Ravikanth G, Narayani B, Ganeshaiah KN, Uma Ganeshaiah R (2005) Rattan species richness and population genetic structure of *Calamus flagellum* in North-Eastern Himalaya, India. J Bamboo Rattan. 4(3):293–307. https://doi.org/10.1163/156915905774310016

Manohara TN (2013) Nutritional Evaluation of Shoots of Two Rattans of Northeast India—*Calamus flagellum* Griff. ex Mart. and *C. floribundus* Griff. (Arecaceae) 1. Economic Bot 67(3):263–268. https://doi.org/10.1007/s12231-013-9237-4

Mayr E (1974) Cladistic analysis or cladistic classification? J Zoolog Syst Evol Res 12(1):94–128

Meijaard E, Achdiawan R, Wan M, Taber A (2014) Rattan: The decline of a once-important non-timber forest product in Indonesia. CIFOR

Mitali G, Palash M (2013) In vitro antidiabetic and antioxidant activity of *Calamus erectus*Roxb. Fruit: a wild plant of Darjeeling Himalaya. Int J Pharm Bio Sci 4(2):671–684

Mohammadi SA, Prasanna BM (2003) Analysis of genetic diversity in crop plants—salient statistical tools and considerations. Crop Sci 43(4):1235–1248. https://doi.org/10.2135/cropsci2003.1235

Moore HE Jr (1973) The major groups of palms and their distribution. Gentes Herbarium 11:27–140

Nuruzzaman M (2001) National report on the state of bamboo and rattan development in Bangladesh. Forest Department, Ban Bhavan, Dhaka, Bangladesh

Patel HK, Fougat RS, Kumar S, Mistry JG, Kumar M (2015) Detection of genetic variation in Ocimum species using RAPD and ISSR markers. 3 Biotech 5:697–707. https://doi.org/10.1007/s13205-014-0269-y

Paudel SK, Chowdhary CL (2005) Managing rattan as a common property: a case study of community rattan management in Nepal. J Bamboo Rattan 4(1):81–91

Peakall R, Beattie AJ (1991) The genetic consequences of worker ant pollination in a self-compatible, clonal orchid. Evolution 45(8):1837–1848. https://doi.org/10.1111/j.1558-5646.1991.tb02691.x

Pleijel F, Jondelius U, Norlinder E, Nygren A, Oxelman B, Schander C, Sundberg P, Thollesson M (2008) Phylogenies without roots? A plea for the use of vouchers in molecular phylogenetic studies. Mol Phylogenet Evol 48(1):369–371. https://doi.org/10.1016/j.ympev.2008.03.024

Priya K, Indira EP, Sreekumar VB, Renuka C (2016) Assessment of genetic diversity in *Calamus vattayila*Renuka (Arecaceae) using ISSR markers. J Bamboo Rattan 15(1/4):61–69

Ramesha BT, Ravikanth G, Nageswara Rao M, Ganeshaiah KN, Uma Shaanker R (2007) Genetic structure of the rattan Calamus thwaitesii in core, buffer and peripheral regions of three protected areas in central Western Ghats, India: do protected areas serve as refugia for genetic resources of economically important plants? J Genet 86(1):9–18. https://doi.org/10.1007/s12041-007-0002-2

Rao NM, Ramesha BT, Ravikanth G, Ganeshaiah KN, Umashankar R (2007) Cross-species amplification of coconut microsatellite markers in rattans. Silvae Genet. 56:282–286

Ray S, Madhumita J, Sandip M (2010) Phylogenetic relationship among six economically important species of *Asparagus* utilizing RAPD, ISSR and isozyme polymorphism. Biores Bull 3:153

Renuka C (1992) Rattans of the western Ghats: A taxonomic manual (No. 584.5 R423). Kerala Forest Research Institute, Kerala (India) IDRC, Ottawa, Ont. (Canadá)

Renuka C (1998) Genetic Diversity and Conservation of Certain Species of Rattans in Andaman and Nicobar Islands and Southern India: Final Report of the Project KFRI 243/95, May 1995 to April 1998 (No. 157). Kerala Forest Research Institute

Renuka C (2000) Field identification key for Rattans of Kerala. In: KFRI Research Report 188, p 34

Renuka C (2001) Uses of rattan in South Asia. UNASYLVA-FAO, pp 7–8

Renuka C (2007a) Status of rattan resources and uses in South Asia. In: Dransfield J, Tesoro FO, Manokaran N (eds) Rattan: Current Research Issues and Prospects for Conservation and Sustainable Development. FAO, Rome, p 101

Renuka C (2007b) Status of rattan resources and uses in South Asia. In: Dransfield J, Tesoro FO, Manokaran N (eds) Rattan: Current research issues and prospects for conservation and sustainable development. FAO, Rome, pp 101–112

Renuka C, Vijayakumaran TT (1994a) Some new species of rattans from Andaman and Nicobar Islands. Rheedea 4(2):120–128

Renuka C, Vijayakumaran TT (1994b) Some new species of rattans from Andaman and NicobarIslands. Rheedea 4(2):120–128

Rogers DJ (1963a) Taximetrics-new name, old concept. Brittonia. pp 285–290

Rogers DJ (1963b) Taximetrics-new name, old concept. *Brittonia*, pp 285–290

Ros-Tonen MA (2000) The role of non-timber forest products in sustainable tropical forest management. Holzalsroh-und Werkstoff 58(3):196–201. https://doi.org/10.1007/s001070050413

Saikia P, Khan ML (2011) Diversity of medicinal plants and their uses in homegardens of upper Assam, Northeast India. Asian J Pharm Biol Res 1(3):296–309

Salusu HD, Ariani F, Budiarso E, Kusuma IW, Arung ET (2021) Increased Benefits of *Calamus manan*Miq. Fruit by its potential bioactivity. In: Joint Symposium on Tropical Studies (JSTS-19). Atlantis Press, Amsterdam, pp 180–185. https://doi.org/10.2991/absr.k.210408.030

Sarkar H, Sharin T, Al Mamun MS (2018) Assessment of *Calamus tenuis* fruits extract on blood glucose level elevation and its antibacterial potency. J Appl Pharm Sci. 6(3):01–06

Sarmah R (2010a) Commonly used non-timber forest products (NTFPs) by the Lisu tribe in Changlang district of Arunachal Pradesh, India. J Front Res 5:68–77

Sarmah R (2010b) Commonly used non-timber forest products (NTFPs) by the Lisu tribe in Changlang district of Arunachal Pradesh, India. Sibsagar Coll Teachers Res J 5:68–77

Sarmah P, Sarma RN (2011) Identification of a DNA marker linked to sex determination in Calamus tenuis Roxb., an economically important rattan species in northeast India. Mol Breed. 27(1):115–118. https://doi.org/10.1007/s11032-010-9510-5

Sarmah P, Barua PK, Sarma RN (2007a) Morphological characterization of some Calamus species of north-east India. Plant Genetic Resources Newsletter. Bioversity International. Rome. Italy
Sarmah P, Barua PK, Sarma RN (2007b) Morphological characterization of some Calamus species of north-east India. Plant Genetic Resources Newsletter. Bioversity International. Rome. Italy, pp 21–29
Sastry CB (2002) Rattan in the twenty-first century: an outlook. In: Dransfield J, Tesoro FO, Manokaran N (eds) Rattan current research issues and prospects for conservation and sustainable development. FAO, Rome, pp 237–244
Sattler R, Rutishauser R (1997a) The fundamental relevance of morphology and morphogenesis to plant research. Ann Bot 80(5):571–582. https://doi.org/10.1006/anbo.1997.0474
Sattler R, Rutishauser R (1997b) The fundamental relevance of morphology and morphogenesis toplant research. Ann Bot 80(5):571–582
Saulei SM, Aruga JA (1994) The status and prospects of non-timber forest products development in Papua New Guinea. The Commonwealth Forestry Review, pp 97–105
Serebryanaya A, Shipunov A (2009) Morphological variation of plants on the uprising islands of northern Russia. In: Annales BotaniciFennici Finnish Zoological and Botanical Publishing Board, pp 81–89
Shaanker UR, Ganeshaiah KN, Srinivasan K, Rao VR, Hong LT (2004) Bamboos and rattans of the Western Ghats: population biology, socio-economics and conservation strategies. ATREE, Bangalore, IPGRI, Malaysia and UAS, Bangalore
Shim PS (1995) Domestication and improvement of rattan. INBAR Working Paper, New Delhi
Sikdar B, Bhattacharya M, Mukherjee A, Banerjee A, Ghosh E, Ghosh B, Roy SC (2010) Genetic diversity in important members of Cucurbitaceae using isozyme, RAPD and ISSR markers. Biol Plant 54(1):135–140. https://doi.org/10.1007/s10535-010-0021-3
Singh HB, Puni L, Jain A, Singh RS, Rao PG (2004) Status, utility, threats and conservation options for rattan resources in Manipur. Curr Sci 10:90–94
Sinha P, Nanda S, Joshi RK, Panda PC (2017a) Development of a sequence-tagged site (STS) marker for sex identification in the dioecious rattan species Calamus gurubaBuch.-Ham. Mol Breed 37(3):1–9. https://doi.org/10.1007/s11032-017-0630-z
Sinha P, Nanda S, Joshi RK, Panda PC (2017b) Development of a sequence-tagged site (STS) marker for sex identification in the dioecious rattan species *Calamus guruba*Buch.-Ham. Mol. Breed. 37(3):22
Sneath PH, Sokal RR (1973) Numerical taxonomy. The principles and practice of numerical classification, p 573
Sreekumar VB, Henderson A (2014) Nomenclatural notes on Indian Calamus (Arecaceae). Phytotaxa 166(2):145–149. https://doi.org/10.11646/phytotaxa.166.2.6
Sreekumar VB, Renuka C (2006) Assessment of genetic diversity in Calamus thwaitesii BECC. (Arecaceae) using RAPD markers. Biochem Syst Ecol 34(5):397–405. https://doi.org/10.1016/j.bse.2005.12.002
Sunderland TC (2001a) Rattan resources and use in West and Central Africa. *UNASYLVA-FAO*, pp.18–23
Sunderland TH (2001b) Rattan resources and use in West and Central Africa. Unasylva (FAO)
Sunderland TC (2004) Ant and rattan associations in forest of tropical Africa. Ghana J For. 15:13–19. https://doi.org/10.4314/gjf.v15i1.36897
Sunderland TC (2012) A taxonomic revision of the rattans of Africa (Arecaceae: Calamoideae). Phytotaxa. 51(1):1–76. https://doi.org/10.11646/PHYTOTAXA.51.1.1
Sunderland TC, Dransfield J (2002) Species profiles of rattans. In: Dransfield J, Tesoro FO, Manokaran N (eds) Rattan–current research, issues and prospects for conservation and sustainable development. FAO, Rome, pp 9–22
Sun Y, Liu X (2022) How design technology improves the sustainability of intangible cultural heritage products: a practical study on bamboo basketry craft. Sustainability 14(19):12058. https://doi.org/10.3390/su141912058

Tag H, Kalita P, Dwivedi P, Das AK, Namsa ND (2012) Herbal medicines used in the treatment of diabetes mellitus in Arunachal Himalaya, northeast, India. J Ethnopharmacol 141(3):786–795. https://doi.org/10.1016/j.jep.2012.03.007

Tesoro FO (2007) Rattan resources of The Philippines, their extent, production, utilization and issues on resource development. In: Dransfield J, Tesoro FO, Manokaran N (eds) Rattan: current research issues and prospects for conservation and sustainable development. Food and Agriculture Organisation of the United Nations (FAO), Rome, pp 167–182

Teulat B, Aldam C, Trehin R, Lebrun P, Barker JH, Arnold GM, Karp A, Baudouin L, Rognon F (2000) An analysis of genetic diversity in coconut (Cocos nucifera) populations from across the geographic range using sequence-tagged microsatellites (SSRs) and AFLPs. Theor Appl Genet. 100(5):764–771. https://doi.org/10.1007/s001220051350

Thakur PK, Sheth M (2015) Traditional consumption and therapeutic beliefs of Calamus tenuis Roxb. edible shoots of forest village natives of Dibrugarh district of Assam, India. Int J Adv Res 3(12):1530–1538

Thomas S, Haridasan K (1997) *Calamus nambariensis*Becc. - an interesting rattan palm from Arunachal Pradesh. Arunachal For News 15:29–30

Tomlinson PB, Horn JW, Fisher JB (2011) The anatomy of palms: Arecaceae-Palmae. p 276. https://doi.org/10.1093/acprof:osobl/9780199558926.001.0001

Uhl NW, Dransfield J (1988) "Genera Palmarum", a new classification of palms and its implications. Adv Econ Bot 1:1–9

Van Valkenburg J (2002) Rattan in East Kalimantan, Indonesia: species composition, abundance, distribution and growth in some selected sites. In: Dransfield J, Tesoro FO, Manokaran N (eds) Rattan: current research issues and prospects for conservation and sustainable development. Non-wood forest products 14. FAO, pp 199–223

Vongkaluang I (2007) Thai rattan in the early 2000. In: Dransfield J, Tesoro FO, Manokaran N (eds) Rattan: current research issues and prospects for conservation and sustainable development. FAO, Rome, pp 183–189

Wangyal JT (2012) Ethnobotanical knowledge of local communities of Bumdeling wildlife sanctuary, Trashiyangtse, Bhutan. Indian J Tradit Knowl 11(3):447–452

Xu HC, Yin GT, Li YD, Fu JG, Zhang WL (2000) Distribution and utilization of rattans in China. In: Xu HC, Rao AN, Zeng BS, Yin GT (eds) Research on rattans in China: conservation, cultivation, distribution, ecology, growth, phenology, silviculture, systematic anatomy and tissue culture. IPGRI-APO, Serdang, Malaysia, p 148. http://agris.upm.edu.my:8080/dspace/handle/0/9562

Yang H, Gan SM, Yin HT, Xu HC (2005) Identification of random amplified polymorphic DNA markers linked to sex determination in *Calamus simplicifolius* CF Wei. J Integr Plant Biol 47(10):1249–1253. https://doi.org/10.1111/j.1744-7909.2005.00113.x

Zhao H, Sun H, Li L, Lou Y, Li R, Qi L, Gao Z (2017) Transcriptome-based investigation of cirrus development and identifying microsatellite markers in rattan (*Daemonoropsjenkinsiana*). Sci Rep 7(1):1–3. https://doi.org/10.1038/srep46107

Zhao H, Zhao S, International Network for Bamboo and Rattan, Fei B, Liu H, Yang H, Dai H, Wang D, Jin W, Tang F, Gao Q, Xun H, Wang Y, Qi L, Yue X, Lin S, Gu L, Li L, Zhu T, Wei Q, Su Z, TBWA W, Ofori DA, Muthike GM, Mengesha YM, de Castro E Silva RM, Beraldo AL, Gao Z, Liu X, Jiang Z (2017) Announcing the genome atlas of bamboo and rattan (GABR) project: promoting research in evolution and in economically and ecologically beneficial plants. Gigascience 6(7):1–7. https://doi.org/10.1093/gigascience/gix046. PMID: 28637269; PMCID: PMC5570132

Zhao H, Wang S, Wang J, Chen C, Hao S, Chen L, Fei B, Han K, Li R, Shi C, Sun H (2018a) The chromosome-level genome assemblies of two rattans (*Calamus simplicifolius* and *Daemonoropsjenkinsiana*). Giga Sci 7(9):1–11. https://doi.org/10.1093/gigascience/giy097

Zhao H, Wang S, Wang J, Chen C, Hao S et al (2018b) The chromosome-level genome assemblies of two rattans (*Calamus simplicifolius* and *Daemonoropsjenkinsiana*). Giga Sci 7:1–11

Chapter 20
An Insight into Prevalent Agroforestry Land Use Systems of North Western Himalayan Region, India: Challenges and Future Prospects

Harish Sharma, K. S. Pant, Rohit Bishist, Prem Prakash, and Krishan Lal Gautam

Abstract The north-western Himalayan region of India is comprised of Jammu & Kashmir, Ladakh Himachal Pradesh and Uttarakhand, covering about 10% geographical area of the country. The region is ecologically as well as biologically rich in diversity and source of livelihood to large no. of people constituting 89.90%, 69.40%, 72.62% and 61.33% of rural people in Himachal Pradesh, Uttarakhand, J&K and Ladakh, respectively. India has just 2% of the world land resources yet it supports about 18% of the human population and 12% of the livestock population throughout world. Exploitive resource use due to fast growing human and livestock population coupled with natural and anthropogenic disturbances cause degradation of the land and bio-resources thereby affecting the fragile ecosystem. Changing climatic conditions and the increasing land-use conflicts call for the development of such sustainable land use systems that reconcile the production from the agriculture along with the provision of multiple ecosystem services, including climate change mitigation. Estimates suggest that about 30% of the emission reductions and carbon sequestration can be contributed by the sustainable land use interventions to meet the target set by Paris agreement. Agroforestry is practiced traditionally in north-western Himalayan region as is evident from the various multipurpose tree species deliberately retained by farmers on their farmland. The various traditional land uses are the outcome of the topographical features, socio-economic conditions, cultural and aesthetic values in the region. Besides providing multiple benefits, such as food, fodder, fuelwood, fibre etc., agroforestry systems act as a cushion against the several

H. Sharma (✉)
Faculty of Agricultural Sciences, DAV University, Jalandhar, Punjab, India

K. S. Pant
Department of Silviculture and Agroforestry, College of Horticulture and Forestry, Dr. YS Parmar University of Horticulture and Forestry, Neri, Hamirpur, India

R. Bishist · P. Prakash · K. L. Gautam
Department of Silviculture and Agroforestry, College of Forestry, Dr. YS Parmar University of Horticulture and Forestry Nauni, Solan, HP, India

H. S. Jatav et al. (eds.), *Agroforestry to Combat Global Challenges*, Sustainable Development and Biodiversity 36, https://doi.org/10.1007/978-981-99-7282-1_20

ecological hazards associated with the developmental activities and helps in conservation of resources in a sustainable manner. Various traditional agroforestry practices reported in the north-western Himalayan region are agrisilviculture, agrihorticulture, agrisilvihorticulture, agrihortisilviculture, hortisilviculture, silvopastoral, pastoralsilviculture, agrisilvopastoral, pastoralsilvihorticulture etc. varying in structural and functional composition as per the needs and preferences of the farmers well adapted to the ecological conditions. This article is an overview of the various agroforestry practices prevalent, their compositional variation, bio-economic productivity and carbon stock potential in north-western Indian Himalayan region.

Keywords Agroforestry · Himalayas · Composition · Productivity · Carbon

20.1 Introduction

Himalaya forms the northern boundary of the India and is geographically vast ranging from Nanga Parbat in the west to Namcha Barwa in the east, having complex and diverse ecosystems (Rawat and Vishvakarma, 2011; Kumar et al., 2018a, b). Indian Himalayan region, covering about 12% of the geographical area of the country (ISFR, 2013), is inhabited by about 51 million people practicing hill agriculture in fragile and diverse ecosystems. Owing to richness in biological as well as socio-cultural diversity, the region has been identified as one of the 34 biological hotspots (Tiwari et al., 2017). Western Himalayan region, constituting 10% of Indian geographical area (ISFR, 2013), comprised of J&K and Ladakh, Himachal Pradesh and Uttarakhand is agro-biodiversity rich region with large number of species under cultivation (Singh, 2009). With variation in site factors such as altitude, slope, temperature, humidity, rainfall, edaphic factors and distance from snowline or plains, have led to the diversified farming landscapes. In Himalayan states of the country, indigenous agroforestry systems form an integral part of the communities and planting trees on farms helps farmers to satisfy their multifarious needs, which leads to an increase in tree cover and thereby reducing the burden on existing forests (Phondani et al., 2020). Further, agroforestry being an integrated farming system plays a key role in sustaining the fragile ecosystems of the region (Kaler et al., 2017) and investment risk of farmers' because they diversify their crop and income source, which reduces economic and social risks (Lefroy, 2009). The knowledge of agroforestry has been continuously used as a way to tackle problems of rural livelihood in India traditionally. The area under agroforestry during next four decades is expected to increase to 53 million ha from 25.32 million ha presently; therefore, agroforestry land use will be having substantial contribution in meeting the societal requirements through increase in production and provision of environmental benefits as well (Dhyani et al., 2013). In India, agroforestry practices are mostly traditional and practiced in a variety of ways (Solanki, 1998; Sharma, 1996) subjected to multiple factors like demographic, socio-economic, cultural factors, as well as farmers' experiences. Agroforestry systems in India have a lot of component diversification both structurally and functionally, which mainly depends upon the temperature, topography, elevation, aspect, edaphic properties and rainfall pattern (Combe, 1982; Nair and Dagar, 1991; Tiwari, 1995). Several agroforestry systems, their floristic diversity, biological productivity, carbon sequestration potential, amelioration of soil physico-chemical properties etc. in north-western Himalayan region have been delineated by Toky et al., 1989; Khosla and Toky (1996); Thakur et al. (2004). Various traditional agroforestry practices reported in the north-western Himalayan region are agrisilviculture, agrihorticulture, agrisilvihorticulture, agrihortisilviculture, hortisilviculture, silvopastoral, pastoralsilviculture, agrisilvopastoral, pastoralsilvihorticulture etc. with structural and functional composition varying in accordance with day-to-day needs and preferences of the farmers well suited with ecological conditions. This article gives an overview of the various agroforestry practices prevalent, their compositional variation, bio-economic productivity and carbon stock potential in north-western Indian Himalayan region.

20.2 Agroforestry Systems in the North-Western Himalayas

Agroforestry in tropical, sub-tropical and temperate region is being practiced traditionally (Kumar et al., 2018a, b). Agroforestry, incorporating tree, crop and livestock component, is a multidisciplinary land use system satisfying productive as well as protective objectives (Singh et al., 2015). In Indian Himalayan region also, agroforestry has been recognized as the productive land use; however, regional causes of adoption, factors causing changes in traditional practices and socio-economic development associated with agroforestry need to be studied thoroughly. Agroforestry helps in satisfying the diverse and multifarious needs of the humans along with providing economic benefits as well as environmental services in the form of carbon sequestration, watershed protection and climate change mitigation and adaptation. Tree-based systems affect local economy by economic stabilization, product diversification, food and fuel security, improvement of natural environment (Dar et al., 2018). With dramatic changes in the altitudinal ranges in the western Himalayan region, vegetation pattern also changes (Tewari et al., 2017) and so is the composition of the agroforestry systems. Over the years, farmers have accustomed several multipurpose tree species on their farmlands which have evolved into extant agroforestry practices. The traditional tree-based systems prevalent in the western Himalayan region are generally location specific regarding relevance, performance and adoption (Dar et al., 2018) and depends mainly on the topography, altitude, climate, edaphic factors etc. Traditionally prevalent as well as adopted agroforestry systems in any area provides much needed information for the extension and further improvement in the systems as it is time tested regarding its potential and possible constraints under specific conditions prevailing in the area. The major agroforestry systems in the Indian north western Himalayan region have been summarized in the Table 20.1. Major agroforestry systems practiced in the region comprised of

Table 20.1 Composition of prevalent agroforestry systems in north-western Himalayan region

State/Union Territory	Agroforestry systems	Forest Trees	Fruit Trees	Agricultural crops	Grasses	References
J&K	Agrisilviculture, Agrihorticulture, Hortiagriculture, Hortisilviculture, Hortisilvopastoral, Hortiagrisilviculture, Silvopastoral, Hortipastoral, Agrisilvipastoral, Agrihortisilviculture	*Populus deltoids, Populus nigra, Ulmus wallichiana, Aesculus indica, Salix alba, Robinia pseudoacacia, Morus alba, Pinus wallichiana, Ailanthus altissima, Cedrus deodara, Albizzia lebbeck, Acacia catechu, Dalbergia sissoo, Abies pindrow, Picea smithiana*	*Malus* spp., *Prunus amygdalus, Juglans regia, Prunus persica, Prunus avium, Cydonia oblonga, Punica granatum, Vitis vinifera*	*Zea mays, Brassica juncea, Daucas carota, Raphanus sativus, Brassica rapa, Oryza sativa, Solanum lycopersicum, Abelmoschus esculentus, Glycine max, Triticum aestivum, Pisum sativum, Brassica oleracea var. capitata, Brassica oleracea var. botrytis, Solanum tuberosum, Allium sativum, Allium cepa, Cucumis sativus, Lagenaria siceraria, Momordica charantia, Phaseolus vulgaris, Vigna radiata*	*Avena sativa, Trifolium repens., Dactylis glomerata, Festuca pretense, Aegilops tauschii, Echinochola crusgalli, Lolium perenne, Lolium multiflorum, Bromus japonicas, Poa* spp., *Bromus inermis, Cynodon dactylon, Chrysopogon fulvus, Dicanthium* spp.	Choudhary et al. (2012); Kashyap et al. (2014); Gupta and Arora, (2015); Banyal et al. (2016); Khaki et al. (2016); Ahmad et al. (2017); Islam et al. (2017); Dar et al. (2018); Zahnoor et al. (2022)
Ladakh	Agrisilviculture, Agrihortisilviculture, Agrihorticulture, Silvopastoral,	*Populus* spp., *Salix* spp., *Ulmus wallichiana, Juniperus* spp.	*Malus domestica., Prunus armeniaca., Juglans regia, Morus alba*	*Triticum aestivum, Hordeum vulgare, Avena sativa, Fagopyrum*	*Avena sativa, Trifolium* spp., *Medicago sativa, Iris lactea*	Pateria et al. (2003); Butola et al. (2012); Banyal et al. (2016); Handa, (2019);

	Hortisilvopastoral, Hortipasture			*esculentum, Fagopyrum tataricum, Panicum miliaceum, Solanum tuberosum, Brassica* spp., *Pisum sativum, Allium cepa, Coriandrum sativum, Carum carvii, Chenopodium album, Setaria italic, Brassica oleracea var. capitata, Brassica oleracea var. botrytis, Phaseolus vulgaris*		Namgial et al. (2020)
Himachal Pradesh	Agrisilviculture, Agrihorticulture, Agrisilvihorticulture, Agrihortisilviculture, Hortiagriculture, Silvopastoral, Hortipastoral, Pastoralsilviculture, Agrisilvopastoral	*Grewia optiva, Celtis australis, Dalbergia sissoo, Toona ciliata, Morus alba, Bauhinia variegata, Melia composita, Albizia chinensis, Acacia catechu, Quercus spp, Cedrus deodara, Pinus roxburghii, Robinia pseudoacacia, Ulmus villosa, Salix alba, Pinus wallichiana, Abies pindrow, Picea*	*Mangifera indica, Citrus* spp., *Prunus domestica, Litchi chinensis, Psidium guajava, Carica papaya, Phyllanthus emblica, Malus domestica, Prunus persica, Pyrus communis, Prunus armeniaca, Diospyros kaki, Punica granatum, Juglans regia, Prunus amygdalus, Pistacia vera*	*Zea mays, Oryza sativa, Vigna mungo, Solanum lycopersicum, Abelmoschus esculentus, Glycine max, Triticum aestivum, Hordeum vulgare, Brassica juncea, Cicer arietinum, Pisum sativum, Brassica oleracea var. capitata, Brassica oleracea var. botrytis, Solanum*	*Apluda mutica, Imperata cylindrica, Chrysopogon montanus, Seteria glauca, Cymbopogon martinii, Heteropogon contortus, Dicanthium annulatum, Themeda anathera, Paspalum notatum, Apluda mutica, Andropogon nardus, Pennisetum*	Kumari et al. (2008); Chisanga et al. (2013); Goswami et al. (2014); Kashyap et al. (2014); Gupta et al. (2017); Rajput et al. (2017); Tiwari et al. (2018); Kumar et al. (2018a, b); Salve and Bhardwaj, (2020)

(continued)

Table 20.1 (continued)

State/Union Territory	Agroforestry systems	Forest Trees	Fruit Trees	Agricultural crops	Grasses	References
		smithiana, Populus ciliata		*tuberosum, Allium sativum, Allium cepa, Avena sativa, Amaranthus hypochondriacus*	*clandestinum, Dactylis gloemerata, Arundinella nepalensis Agrostis* spp., *Poa annua, Trifolium repens, Cynodon dactylon*	
Uttarakhand	Agrisilviculture, Agrisilvihorticulture, Agrisilvopastural, Silvipastural, Agrihorticulture, Agrihortisilviculture, Silvihorticulture, Hortipastoral	*Quercus leucotrichophora, Grewia oppositifolia, Celtis australis, Pinus roxburghii, Prunus cerasoides, Rhododendron arboretum, Pyrus pashia, Pinus wallichiana, Cedrus deodara, Ficus auriculata, Ficus palmata, Morus serrata, Melia azedarach, Myrica esculenta, Toona ciliata*	*Juglans regia, Prunus armeniaca, Prunus domestica, Malus domestica, Pyrus pyrifolia, Musa paradisiacal, Punica granatum, Carica papaya, Psidium guajava, Prunus* spp., *Emblica officinalis, Citrus* spp., *Mangifera indica*	*Zea mays, Triticum aestivum, Oryza sativa, Phaseolus vulgaris, Eleusine coracana, Amaranthus caudatus, Amaranthus spinulosa, Vigna umbellate, Vigna mungo, Cajanus cajan, Glycine max, Hordeum vulgare, Lens culinaris, Brassica campestris, Pisum sativum, Lycopersicon esculentum, Abelmoschus esculentus, Spinacea oleracea, Brassica juncea, Brassica*	*Cynodon dactylon, Andropogon munroi, Apluda mutica, Euphorbia hirta, Avena fetua, Setaria* spp., *Paspalum* spp.	Kashyap et al. (2014); Mahato et al. (2016); Vikrant et al. (2018); Yadav et al. (2018); Yadav et al. (2019); Bhatt and Parihaar. (2020); Gariya et al. (2020); Himshikha et al. (2020); Kumar et al. (2021a, b)

				oleracea var. capitata, Brassica oleracea var. botrytis, Coriandrum sativum, Curcuma longa, Zingiber officinale, Capsicum annuum, Allium cepa, Allium sativum		

agrisilviculture, agrisilvihorticulture, agrihorticulture, agrihortisilviculture, hortiagriculture, hortisilviculture, hortisilvopastoral, hortiagrisilviculture, silvopastoral, pastoralsilviculture, hortipastoral, agrisilvipastoral and silvihorticulture, as reported in literature. In J&K, important tree species are *Populus* spp., *Salix* spp., *Ulmus wallichiana*, *Ailanthus altissima*, *Morus alba*, *Aesculus indica* etc. while, in Ladakh region *Populus* spp., *Ulmus wallichiana*, *Salix* spp., *Juniperus* spp. are major tree species. In Himachal Pradesh and Uttarakhand, most of the tree species prevalent are similar such as *Grewia* spp., *Celtis australis*, *Quercus leucotrichophora*, *Toona ciliata*, *Cedrus deodara*, *Morus* spp., *Melia* spp. etc. Being in hilly terrain, local people depend on the forests for several day-to-day needs. Mountain farming systems are generally characterized by presence of livestock component which provide milk, meat, manure and draught power (Nautiyal et al., 2018) in the areas where farm mechanization is having limited scopes. The basic requirement for livestock rearing is fodder availability which is generally fulfilled from fodder grown in community land, forest land and crop residues. India is having about 11% of the world livestock population that is supported on the land area constituting about 2% globally (Roy et al., 2019) creating challenges for fulfilment of the fodder requirement. The issues of fodder availability need to be addressed as feed constitutes about 70% cost of milk production alone, which, in turn is responsible for the 20–60% lower productivity of livestock in Indian conditions. According to report (ICAR-IGFRI, 2021) there is 49.17% shortage of fodder in Jammu and Kashmir, 40–45% in Ladakh (Tewari et al., 2016) and about 33% in Himachal Pradesh (NITI Aayog, 2018). Through adoption of alternate land use systems such as silvopastoral, hortipasture etc. it is possible to increase the productivity of the land along with fulfilment of the fodder requirement, reduction of grazing pressure as well as positive environmental implications (Roy et al., 2019). Further, with wide altitudinal variations in the Himalaya region, the climatic conditions also vary significantly with some regions being covered under snow during winters. Under such conditions, fuelwood serves as an important source of energy for which people mostly depend on the forest resources (Kumar et al., 2020). Studies reported that 93% of the population in Himachal Pradesh uses fuelwood as the source of energy (Parikh, 2011; TERI, 2015) out of which 94% of the fuelwood users depends on the forests for this. Fuelwood consumption per capita per day (in kg) in Jammu and Kashmir varies from 0.05–5.50, in Himachal Pradesh varies from 0.91–5.13 kg, while, in Uttarakhand varies from 1.13–8.75 (Kumar et al., 2020) showing the dependence of the inhabitants on the fuelwood. Govt. initiatives such as Pradhan Mantri Ujjwala Yojana are helpful in meeting the objectives of the clean energy and simultaneously integrated farming practices are also having key role in meeting the demand of fuelwood to certain extent and also to reduce the pressure as well as exploitive utilization of the natural resources. The annual availability of the fuelwood (in million tones) from the tree outside forests (TOF) in J & K including Ladakh, Himachal Pradesh and Uttarakhand is 0.365, 0.290 and 0.297 respectively, in comparison to 0.02 million tones, 50 tones and 0.05 million tones fuelwood available from forests in the respective UTs/ states (Dar and Ahmad 2016). Further, availability of the fuelwood on the farmland will also facilitate the utilization of the cow dung as organic manure in the farm instead of burning it as energy source. In addition to the fodder and fuelwood requirement farmers are also dependent on the natural resources for their timber and small wood needs which generally results into exploitation of the resources when the need turns into greed. Agroforestry not only provides ecological services but also economic benefits as 65% of the timber requirement in the country is met from TOF (GoI, 2016). In the current scenario when there are lack of data for demand as well as supply of tree-based products and natural forests are closed for the protection and conservation purpose, there is greater scope for the promotion of the agroforestry practices (Parthiban et al., 2021). Further, a dedicated agroforestry policy facilitating the selection of suitable species for the specific region, provision of providing quality planting material, permissive felling and transit regulations as well as marketing facilities may encourage the mass towards adoption of scientific agroforestry interventions.

20.3 Biomass Production of Agroforestry Systems

Photosynthesis is the process involved in the manufacturing of the food by the primary producers through transformation of the light energy in chemical energy and the product formed is either used or is stored. The energy is stored in the plants in the form of biomass and is having great importance to other individuals present on other tropic levels as well as humans as the stored energy can be harvested to be used as food, fuel, fibre and several other uses (Roberts et al., 1985). Plant biomass is the weight of the biological material contained in aboveground and belowground portion of plant and is generally expressed as plant dry matter dried to constant weight. Biomass served as the primary source of the fuel anciently since humanity became familiar with fire (Fekete, 2013). In the current scenario, fossil fuels have become common source of energy but still biomass energy is an important and preferable source of energy for the poor people that may due to its cheapness and easy availability from the forest area. Global concern towards the woody biomass is increasing due to increased fossil fuel prices, emissions resulting from burning of fossil fuels as well as threat resulting from catastrophic wildfires (Proto et al., 2014). Agroforestry practices having deliberate incorporation of the woody perennials into the land use therefore has immense potential for the production as well as storage of biomass. Biomass production of trees in agroforestry is generally estimated on the basis of region specific allometric equations developed for specific tree species. Biomass production of agroforestry systems depends on several factors such as physiography, structural and functional composition, age and density of trees, specific management practices, environmental, socio-economic, interaction of components affecting efficiency of resource use etc. (Goswami et al., 2014; Rajput et al., 2017; Chisanga et al., 2018; Singh et al., 2020; Panwar et al., 2022). The biological production potential of the prevalent agroforestry practices in the north western Himalayas based on literature review has been summarized in Table 20.2. A lot of work regarding the biomass production potential of the agroforestry systems has

Table 20.2 Biomass production potential of agroforestry systems in north-western Himalayan region

State/Union Territory	Agroforestry systems	Aboveground biomass (Mg/ha)	Belowground biomass (Mg/ha)	Total biomass (Mg/ha)	References
J&K	Agrisilviculture	6.70–159.41	1.58–71.55	15.94–202.59	Ajit et al. (2017); Panwar et al. (2022)
	Agrihorticulture	15.79–137.56	2.40–34.39	18.19–171.95	Zahnoor et al. (2021); Panwar et al. (2022)
	Silvopastoral	34.49–53.20	9.01–34.42	43.51–136.42	Panwar et al. (2022)
Ladakh	Agrisilviculture	17.11	6.03	23.14	Namgial (2018)
	Agrihortisilviculture	19.11	8.05	27.16	Namgial, (2018)
	Agrihorticulture	16.15	6.97	23.12	Namgial, (2018)
	Silvopastoral	16.91	9.51	26.43	Namgial, (2018)
	Hortisilvopastoral	19.95	10.93	30.88	Namgial, (2018)
Himachal Pradesh	Agrisilviculture	6.70–159.41	1.58–71.55	13.47–202.59	Goswami et al. (2014); Singh et al. (2015); Gupta et al. (2017); Panwar et al. (2022)
	Agrihorticulture	9.58–137.56	2.40–34.39	12.29–171.95	Goswami et al. (2014); Singh et al. (2015); Gupta et al. (2017); Rajput et al. (2017); Chisanga et al. (2018); Singh et al. (2020); Panwar et al. (2022)
	Agrisilvihorticulture	15.15–67.97	4.30–20.20	16.31–88.17	Goswami et al. (2014); Gupta et al. (2017); Thakur, (2020)
	Agrihortisilviculture	13.26–85.49	3.38–23.08	18.40–108.60	Goswami et al. (2014); Bammanahalli, (2016); Gupta et al. (2017); Chisanga et al. (2018); Thakur, (2020); Janju, (2021); Sharma et al. (2021)
	Hortiagriculture	14.17–26.42	3.99–7.10	19.26–33.26	Janju, (2021); Singh et al. (2020)
	Silvopastoral	4.58–162.80	1.33–35.70	5.92–198.20	Goswami et al. (2014); Singh et al. (2015); Gupta et al. (2017); Rajput et al. (2017); Chisanga et al. (2018); Singh et al. (2019); Sharma et al. (2021); Panwar et al. (2022)
	Hortipastoral	11.24–24.97	3.23–6.33	14.47–31.30	Thakur, (2020); Singh et al. (2020); Janju, (2021)

	Pastoralsilviculture	3.77–10.58	0.59–3.27	5.13–13.85	Bammanahalli, (2016); Singh et al. (2019); Thakur, (2020); Janju, (2021)
Uttarakhand	Agrisilviculture	6.7–159.41	1.58–71.55	0.52–202.59	Newaj et al. (2016); Vikrant et al. (2018); Kumar et al. (2021a, b); Panwar et al. (2022)
	Silvopasture	34.49–53.20	9.01–34.42	43.51–136.42	Panwar et al. (2022)
	Agrihorticulture	15.79–137.56	2.40–34.39	0.33–171.95	Yadav et al. (2016); Yadav et al. (2017); Vikrant et al. (2018); Adhikari et al. (2019); Panwar et al. (2022)
	Agrihortisilviculture	–	–	0.13–1.37	Vikrant et al. (2018)

been carried out in the states of Himachal Pradesh and Uttarakhand, while in union territories of Jammu & Kashmir and Ladakh, work has been carried out regarding the identification of the agroforestry, which needs to be further elaborated to assess the productivity of the prevalent systems. From Table 20.2, it can be seen that the total biomass production potential of agrisilviculture is highest (202.59 Mg/ha) among prevalent systems in Jammu and Kashmir. Also, lowest total biomass production (15.94 Mg/ha) is also reported in agrisilviculture system in the region. The preponderance of fast growing tree species such as *Populus* spp., *Salix* spp., *Robinia pseudoacacia* etc. may be the reason for more accumulation of the biomass as contributed by perennial component, while varying tree densities retained as per farming practices may affect the overall productivity of the system. Aboveground and belowground biomass is also more in agrisilviculture system with overall range varying between 6.70–159.41 Mg/ha and 1.58–71.55 Mg/ha, respectively. In Ladakh, hortisilvopastoral resulted in maximum aboveground, belowground and total biomass viz., 19.95 Mg/ha, 10.93 Mg/ha and 30.88 Mg/ha, respectively attributed to the diverse components, more tree density and specific practices adopted for the management of the system. In Himachal Pradesh, aboveground biomass is reported maximum under silvopastoral system (162.80 Mg/ha), while belowground biomass (71.55 Mg/ha) under agrisilviculture. Dominance of forest trees in silvopastoral system may be the factor for the higher aboveground biomass as contributed through tree component, however, management practices in agrisilviculture system as well as withdrawal of nutrition by components from different zones in the soil may have resulted in better belowground biomass in agrisilviculture system. In total biomass production is highest (202.59 Mg/ha) in agrisilviculture that may be due to higher tree density as well as differences in management practices. In Uttarakhand also, agrisilviculture system is reported most productive among all the systems with aboveground biomass production potential to a tune of 159.41 Mg/ha, belowground production potential of 71.55 Mg/ha with total biomass production to the tune of 202.59 Mg/ha. Tree density along with the type of species incorporated plays significant role in influencing the productivity of the system.

20.4 Carbon Stock Potential of Agroforestry Land Uses

Currently, climate change is among one of the most important topic of discussion world over that bring up unique challenges directly or indirectly. Concentration of the GHGs (greenhouse gases) in the atmosphere shows the equilibrium between the source (natural and anthropogenic activities) and sink (biosphere and ocean). The concentration of CO_2 in earth's atmosphere is 413.20 $\pm$ 0.2 ppm, methane 1889 $\pm$ 2 ppb and nitrous oxide 333.20 $\pm$ 0.1 ppb that is 149%, 262% and 123% above the pre-industrial level, respectively, and considered main cause behind this global warming (WMO, 2021). It is believed that through alternate cultivation practices of the agricultural and forest crops this increase in the concentration of the CO_2 can be checked and can be partially mitigated through biomass production (Jose and Bardhan, 2012). International concern about the changing climatic conditions resulted in the Kyoto protocol in 1997 and ever since this protocol, agroforestry has been highlighted as a sustainable strategy for the mitigation of the increasing concentration of CO_2 throughout the world. Agroforestry being the deliberate incorporation of woody perennial on the farmland helps in storage of higher amount of biomass carbon through carbon sequestration as compared to monocropping and thus plays an important role in mitigation as well as adaptation of climate change. In addition to the carbon stored in the form of biomass aboveground, agroforestry also helps in the storage of considerable amount of carbon belowground. However, for the adoption of agroforestry in the carbon sequestration, projects under the schemes such as clean development mechanism exact information of the carbon stored aboveground, belowground and in soil are needed. Carbon stock potential of the agroforestry practices in the north western Himalayan region has been collected from literature of the area and highlighted through Table 20.3. In J&K, highest (71.78 Mg C/ha) vegetation carbon stock is reported under agrisilviculture having the carbon range 32.61–71.78 Mg C/ha, while soil carbon stock range is reported equal to 25.99–58.07 Mg C/ha. Range of carbon stored is more for vegetation in agrisilviculture and agrihorticulture land use systems, while silvopastoral system has more carbon stored in soil as that of vegetation which may be due to more litter addition along with root decay material in the soil as contributed by fine roots of the grasses (Goswami et al., 2014). In Ladakh region, maximum vegetation carbon (44.59 Mg C/ha) is reported to have stored under silvopastoral system that may be due to more tree density, while, soil carbon is reported to have stored more (64.34 Mg C/ha) under agrihorticulture system that may be due to management practices adopted for agriculture as well as horticulture components as both the components hold economic values. In Himachal Pradesh, maximum total carbon (109.93 Mg C/ha) is reported to have stored under silvopastoral land use ascribed to continuous carbon accumulation by the perennial component which is present in more number under silvopastoral system and is the major cause for the higher vegetation carbon (71.61 Mg C/ha) stored in silvopastoral land use system. Agrisilvihorticulture system is reported to have stored maximum soil carbon (56.70 Mg C/ha) which is quite identical to the soil carbon stored under agrihorticulture, agrihortisilviculture, silvopastoral and agrisilviculture system. Diverse composition of the land use system may be responsible for the more soil carbon as facilitated by the more addition of litter as well as better decomposition. In Uttarakhand also, silvopastoral system was reported to have stored more vegetation carbon to a tune of 51.14 Mg C/ha, while soil carbon was more (64.34 Mg C/ha) in agrihorticulture system. Overall, maximum carbon storage (79.92 Mg C/ha) is found under agrihorticulture system attributed to more biomass stored by the fruit tree component as compare to sole cropping. The biomass production is subjected to the composition of the system as affected by the factors of the locality (Yadav et al., 2017; Adhikari et al. 2019).

Table 20.3 Carbon storage (vegetation + soil) in agroforestry systems of north-western Himalayas

State/Union Territory	Agroforestry systems	Vegetation carbon (Mg C/ha)	Soil carbon (Mg C/ ha)	Total carbon (Mg C/ ha)	References
J&K	Agrisilviculture	32.61–71.78	25.99–58.07	97.77	Ajit et al. (2017); Panwar et al. (2022)
	Agrihorticulture	29.61	64.34	–	Zahnoor et al. (2021); Panwar et al. (2022)
	Silvopastoral	44.59	47.63	–	Panwar et al. (2022)
Ladakh	Agrisilviculture	11.57–32.61	11.78–58.07	–	Namgial, (2018); Panwar et al. (2022)
	Agrihortisilviculture	13.58	11.71	–	Namgial, (2018)
	Agrihorticulture	11.56–29.61	10.65–64.34	–	Namgial, (2018); Panwar et al. (2022)
	Silvopastoral	13.21–44.59	11.34–47.63	–	Namgial, (2018); Panwar et al. (2022)
	Hortisilvopastoral	15.44	11.63	–	Namgial, (2018)
Himachal Pradesh	Agrisilviculture	8.44–52.95	9.37–51.19	35.11–87.99	Singh et al. (2015); Bammanahalli, (2016); Gupta et al. (2017); Singh et al. (2019); Panwar et al. (2022)
	Agrihorticulture	8.64–51.65	17.05–55.64	36.58–96.67	Singh et al. (2015); Bammanahalli, (2016); Gupta et al. (2017); Rajput et al. (2017); Singh et al. (2019); Panwar et al. (2022)
	Agrisilvihorticulture	11.17–44.08	19.80–56.70	49.97–100.78	Bammanahalli, (2016); Gupta et al. (2017)
	Agrihortisilviculture	12.10–46.65	12.40–54.06	32.12–100.71	Bammanahalli, (2016); Gupta et al. (2017); Singh et al. (2019)
	Silvopastoral	15.34–71.61	17.96–53.12	46.13–109.93	Gupta et al. (2017); Rajput et al. (2017); Singh et al. (2019)
	Pastoralsilviculture	1.19–4.94	20.18–32.62	29.72–38.32	Bammanahalli, (2016)
Uttarakhand	Agrisilviculture	7.00–38.84	10.35–15.50	18.39–25.17	Newaj et al. (2016); Bhattacharjya et al. (2017); Kumar et al. (2021a, b); Panwar et al. (2022)
	Silvopasture	42.34–51.14	40.69–49.75	–	Kumar et al. (2021a, b); Panwar et al. (2022)
	Agrihorticulture	21.93–44.14	35.78–64.34	79.92	Yadav et al. (2017); Vikrant et al. (2018); Adhikari et al. (2019); Rathore et al. (2020); Panwar et al. (2022)

20.5 Socio-Economic Impact of Agroforestry Systems

The combined measure of the social and economic position with respect to others in the society represents the socio-economic condition of the society. It is having general influence on the resource accessibility, societal livelihood pattern, food security etc. (Roy et al., 2013) and greatly influences the farm-based enterprises by affecting the organization, management, production and marketing of the enterprise. The understanding of socio-economic factors holds great importance in farming systems and helps in formulating the policies for the well-being of the society as ignorance of socio-economic aspects results in the suffering of the various developmental programs (Sood et al., 2008). Agroforestry and socio-economic considerations act as two phases of the same coin as improved socio-economy affects the integration of trees on the farm land on one hand, while, adoption of the agroforestry helps in improvement of the socio-economy of the farming families. Agroforestry is having vast potential for the improvement of the society as can be realized through its benefits to the vulnerable sections mainly marginal and small farmers, women and children (Murthy et al., 2016). Throughout the country various studies confirm the positive impact of the agroforestry land use on farmer socio-economic in terms of women welfare, upliftment of the marginal sections, food security, improved financial resilience, reduced crop failure, regular employment and income, increased land productivity, annual and periodic economic benefits from multiple outputs. Generation of more than 5.76 million mandays per year from agroforestry if implemented on an area of 75,500 ha in Indian Himalayan region shows the employment potential of this sustainable land use in the region and as an option for rural development in the challenging terrains of the Himalayas (Arunachalam et al., 2020). Agroforestry in the Himalayan region plays an important role owing to the topographical factors which on interacting with different socio-economic parameters get modified in various location-specific systems. Although much of the research has been carried out on the identification, productivity, carbon sequestration potential, yet there is dearth of research work highlighting the impact of adoption these systems on the socio-economic condition of the farmers. This poses constraint in framing the suitable policies for the betterment of the farming community but on the other hand offers a scope that can be addressed in the future research projects.

20.6 Challenges Associated with Farming Communities in North-Western Himalayas

India has become the most populous country surpassing China, and agriculture is the important sector providing employment to about half of the population; however, the share of agriculture towards GDP has declined since independence to about 17.8% (Sharma and Raina, 2021). Western Himalayan region is generally characterized by the variations in topography, edaphic factors, climate and land use practices. Being hilly and mountainous terrain, the ecosystems in the north western Himalayan region are fragile with respect to topography, geological hazards, land degradation, land use and land cover, biodiversity etc. (Saha and Kumar, 2019). Several anthropogenic activities including deforestation, indiscriminate and over utilization of resources, faulty agricultural practices etc. along with challenging and unstable terrain has resulted in soil erosion, depletion of land resources, lower productivity etc. Keeping in mind the vulnerability of the bio-physical characteristics of Himalayan region necessary actions are required in order to maintain the sustainability of the ecosystem. Sustainable land use practices as well as their management can help in acting as sink to the carbon along with providing livelihood opportunities to the rural population and help in reducing the vulnerability of the natural resources towards changing climatic conditions. Agroforestry can help in the stabilization of the fragile landscapes through the addition of litter, binding of soil by extensive root network thereby preventing the soil erosion, provision of multiple products improving the socio-economic conditions, preventing the pressure on natural resources such as forests, pastures etc. Although agroforestry seems the most suitable land use facilitating the fulfilment of the needs in a conservative way but the limited land resource seems hindering its true potential. The average land holding size in the western Himalayan region has declined for all the categories and has come down to about 1 ha on an average. The condition is even worse by the continuous fragmentation of this limited asset making farming non-viable from food as well as income point of view. Land fragmentation is one of the major causes for the reduced agricultural productivity in the Himalayan region (Shukla et al., 2018). As hill farming is mainly done manually and is dependent on draught animal, the land fragmentation leads to increase of input costs involved in agriculture thus turning the asset into liability. Farmers having limited land area have less scope of incorporating trees on the farmland as over agricultural component. But, diversified farming can help getting better benefits along with natural security towards total crop failure. In addition to the fragility of the ecosystems in the north western Himalayas, presence of cold desert region also make the region susceptible to vagaries of climate and livelihood more difficult. Cold desert in the western Himalayas exists in Leh and Kargil districts of Ladakh, Lahul & Spiti as well as some pockets of Chamba district and some areas in Janvi valley of Uttarkashi district (Tewari and Kapoor, 2013). Herbaceous plants of annual and perennial nature along with few bushes dominate the vegetation of the cold desert region which is generally xerophytic or mesophytic in nature. The area under cultivation in the cold desert region is very less which is generally flatter portion of valleys, but, with increase in population people are cultivating sloppy area also which has resulted in ecosystem degradation. Integrated land use as well as management techniques are necessary for the ecological restoration of the area which includes management of pastures, plantations, livestock component in harmonic association. Agroforestry seems the answer to all the problems concerned which along with the fulfilment of the basic need of the agricultural crop helps in the provision of the fodder, fuelwood, fruit, fibre, timber etc.

20.7 Future Perspective

- Strengthening of the research and extension activities towards the land use systems having better potential from ecological as well as economical point of view such as horticulture-based and pasture-based systems.
- Development of the fodder tree-based systems for checking the fodder scarcity, enhancing carbon stock potential of the land use and facilitating the rearing of the livestock.
- Holistic approach towards the estimation of biological productivity and carbon sequestration potential of location specific agroforestry systems so that degraded and wastelands can be reclaimed with the system having high production potentials.
- Tree breeding techniques for the exploitation of the quality planting material for mass propagation, distribution of planting material to the farmers and the socio-economic development of the society.
- Tree-based farming systems should be popularized among farmers residing in fragile areas based on suitable models developed and tested regarding their feasibility in terms of checking natural hazards as well as act as a source of livelihood.
- As choice of tree components for incorporation in farming systems is limited in cold desert region and generally includes *Salix* spp. and *Populus* spp. So, research needs to be focused on the genetic improvement of the species and development of superior clones having better productivity and adoptability by the people.

References

Adhikari B, Lodhiyal N, Lodhiyal LS (2019) Carbon sequestration efficiency of mango-based agri-horticulture system in Shiwalik plain of Central Himalaya, India. Indian J Agrofor 21(1):75–80

Ahmad S, Khan PA, Verma DK, Mir N, Singh JP, Dev I, Roshetko JM (2017) Scope and potential of hortipastoral systems for enhancing livestock productivity in Jammu and Kashmir. Indian J Agrofor 19(1):48–56

Ajit HAK, Dhyani SK, Bhat GM, Malik AR, Dutt V, Masoodi TH, Uma JA (2017) Quantification of carbon stocks and sequestration potential through existing agroforestry systems in the hilly Kupwara district of Kashmir valley in India. Curr Sci 113(4):782–785

Arunachalam A, Handa AK, Dev I, Kaushal R, Panwar P, Dhyani SK (2020) Alternate land use and agroforestry systems for resource conservation and enhanced productivity in the hills. Indian J Hill Farming 33(1):176–193

Bammanahalli S (2016) Productivity and carbon storage in prevalent agroforestry systems in subtropical region of Himachal Pradesh. Ph.D. Thesis. Dr YS Parmar University of Horticulture and Forestry, Nauni Solan (HP), P 209

Banyal R, Qaiser KN, Abidi RA (2016) Temperate agroforestry systems in Jammu and Kashmir. In: Agroforestry for increased production and livelihood security, pp 41–59

Bhatt GD, Parihaar RS (2020) Use of remote sensing and geographic information system on agroforestry ecosystem in Himalayan Region of Uttarakhand. In: Seema S, Singh RB, Kumar M (eds) Geoecology of landscape dynamics. Springer, Singapore, pp 167–185

Bhattacharjya S, Bhaduri D, Chauhan S, Chandra R, Raverkar KP, Pareek N (2017) Comparative evaluation of three contrasting land use systems for soilcarbon, microbial and biochemical indicators in North-Western Himalaya. Ecol Eng 103:21–30

Butola JS, Malik AR, Baba JA (2012) Livelihood diversification in cold arid desert of Indian Himalaya: urgent need of transforming traditional agroforestry system with emphasis on adoption of herbal farming. Int J Med Aromatic Plants 2(3):544–548

Chisanga K, Bhardwaj DR, Sharma S (2013) Bio-economic appraisal of agroforestry systems in Dry Temperate Western Himalayas. J Tree Sci 32(1&2):1–7

Chisanga K, Bhardwaj DR, Pala NA, Thakur CL (2018) Biomass production and carbon stock inventory of high-altitude dry temperate land use systems in North Western Himalayas. Ecol Process 7(22):1–13

Choudhary P, Tandon V, Sharma R (2012) Traditional agroforestry practices for sustainable livelihood in Rajouri district of Jammu and Kashmir. J Tree Sci 31(1&2):102–107

Combe J (1982) Agroforestry techniques in tropical countries: potential and limitations. Agrofor Syst 1:13–28

Dar MD, Ahmad S (2016) Current status and prospects of fuel wood species in North-western Himalayan Region–a review. Ann Agri-Bio Res 21(2):164–167

Dar M, Qaisar KN, Ahmad S, Wani AA (2018) Inventory and composition of prevalent agroforestry systems of Kashmir Himalayas. Adv Res 14(1):1–9

Dhyani SK, Handa AK, Uma (2013) Area under agroforestry in India: an assessment for present status and future perspective. Indian J Agrofor 15:1–11

Fekete BM (2013) Biomass. Elsevier, Alpharetta, GA, pp 83–84

Gariya K, Dwivedi GK, Kumar V, Tewari SK (2020) Agrobiodiversity, composition and functional characteristics of Homegardens in Bhimtal Block of Nainital District, Uttarakhand, India. Int J Agric Environ Biotechnol 13(2):1–11

GoI (2016) Sub mission on agroforestry operational guidelines. Department of Agriculture, Cooperation and Farmer Welfare, Government of India, p 30

Goswami S, Verma KS, Kaushal R (2014) Biomass and carbon sequestration in different agroforestry systems of a Western Himalayan watershed. Biol Agric Hortic 30(2):88–96

Gupta RD, Arora S (2015) Agroforestry as alternate landuse system for sustaining rural livelihoods in Himalayan ecosystem. In: Bhan S, Arora S (eds) Advances in soil and water resource management for food and livelihood security in changing climate. Soil Conservation Society of India, New Delhi, pp 656–669

Gupta B, Sarvade S, Singh M (2017) Species composition, biomass production, and carbon storage potential of agroforestry systems in Himachal Pradesh. In: Gupta SK, Panwar P, Kaushal R (eds) Agroforestry for increased production and livelihood security. New India Publishing Agency, New Delhi, pp 242–270

Handa AK, Dhyani SK, Rizvi J (2019) Guidelines to produce quality planting material of agroforestry species. ICAR (CAFRI) & WAC, p 42

Himshikha NAK, Poonia PK, Kumar M, Daneva V (2020) Evaluation of existing agroforestry systems in plain region of Uttarakhand, India. Indian J Agrofor 22(2):44–50

ICAR-IGFRI (2021) Fodder resources development plan for Jammu & Kashmir. ICAR-Indian Grassland and Fodder Research Institute, Jhansi, p 56

ISFR (2013) Indian State of Forest Report 2013. FSI, Dehradun, p 252

Islam MA, Qaisar KN, Bhat GM (2017) Indigenous knowledge in traditional agroforestry systems of Kashmir valley: current challenges and future opportunities. Int J For Crop Improv 8(1): 68–77

Janju S (2021) Appraisal of existing agroforestry systems in Seraj valley of District Mandi, HP. M. Sc. Thesis. Dr YS Parmar University of Horticulture and Forestry Nauni-Solan (HP), p 79

Jose S, Bardhan S (2012) Agroforestry for biomass production and carbon sequestration: an overview. Agrofor Syst 86:105–111

Kaler NS, Gupta B, Negi V (2017) Traditional agroforestry systems in cold desert region of Himachal Pradesh, India. Popular Kheti 5(2):35–36

Kashyap SD, Dagar JC, Pant KS, Yewale AG (2014) Soil conservation and ecosystem stability: natural resource management through agroforestry in Northwestern Himalayan Region. In: Dagar JC et al (eds) Agroforestry systems in India: livelihood security & ecosystem services, advances in agroforestry. Springer, India, pp 21–55

Khaki BA, Wani AA, Bhardwaj DR, Singh VRR (2016) Soil carbon sequestration under different agroforestry land use systems. Indian Forester 142(8):734–738

Khosla PK, Toky OP (1996) Status of agroforestry in the western Himalaya. Current researches in plant sciences. pp 367–377

Kumar BM, Handa AK, Dhyani SK, Arunachalam A (2018a) Agroforestry in the India Himalayan region. In: Gordon AM, Newman SM, Coleman BRW (eds) Temperate agroforestry systems, 2nd edn. CAB International, Wallingford, pp 153–171

Kumar P, Thakur CL, Rai P, Attri K (2018b) Identification of existing agroforestry systems and socio-economic assessment in Kandaghat block of Solan district, HP, India. Int J Curr Microbiol App Sci 7(4):3815–3826

Kumar B, Singh K, Sharma J, Gairola S (2020) A comprehensive review of fuelwood resources and their use pattern in rural villages of western Himalaya, India. Plant Arch 20(2):1949–1958

Kumar A, Tewari S, Singh H, Kumar P, Kumar N, Bisht S, Devi S, Nidhi KR (2021a) Biomass accumulation and carbon stock in different agroforestry systems prevalent in the Himalayan foothills, India. Curr Sci 120(6):1083–1088

Kumar S, Bijalwan A, Singh B, Rawat D, Yewale AG, Riyal MK, Thakur TK (2021b) Comparison of carbon sequestration potential of *Quercus leucotrichophora*–based agroforestry systems and natural forest in Central Himalaya, India. Water Air Soil Pollut 232:350

Kumari A, Sehgal RN, Kumar S (2008) Traditional agroforestry systems practiced in Lahul (Lahul & Spiti) and Kinnaur Districts of Himachal Pradesh. Indian Forester 8:1003–1010

Lefroy E (2009) Agroforestry and the functional mimicry of natural ecosystems. In: Nuberg I, George B, Reid R (eds) Agroforestry for natural resource management. CSIRO Publishing, Collingwood, pp 23–36

Mahato S, Dasgupta S, Todaria NP, Singh VP (2016) Agroforestry mapping and characterization in four districts of Garhwal Himalaya. Energy Ecol Environ 1(2):86–97

Murthy IK, Dutta S, Varghese V, Joshi PP, Kumar P (2016) Impact of agroforestry systems on ecological and socio-economic systems: a review. Global J Sci Front Res Environ Earth Sci 16(5):15–27

Nair PKR, Dagar JC (1991) An approach to developing methodologies for evaluating agroforestry systems in India. Agrofor Syst 16:55–81

Namgial J (2018) Floral spectrum and productivity of different land uses in Leh region of Himalayan cold desert. M.Sc. Thesis. Department of Silviculture and Agroforestry, Dr YS Parmar University of Horticulture and Forestry, Nauni, Solan, HP, p 66

Namgial J, Prabhakar M, Bishist R, Gautam KL, Sharma H, Sharma A (2020) Identification of different land use systems in Leh region of Himalayan Cold Desert. Int J Curr Microbiol App Sci 9(11):1–11

Nautiyal M, Tiwari P, Tiwari JK, Rawat DS (2018) Fodder diversity, availability and utilization pattern in Garhwal Himalaya, Uttarakhand. Plant Arch 18(1):279–287

Newaj R, Dhyani SK, Ajit PR, Rizvi RH, Alam B, Chavan SB, Handa AK, Jha A (2016) Carbon sequestration potential of agroforestry systems in Indo-gangetic plains of India. India Farming 66(3):41–43

NITI Aayog (2018) Demand and supply projections towards 2033. The Working Group Report, p 220

Panwar P et al (2022) Biomass production and carbon sequestration potential of different agroforestry systems in India: a critical review. Forests 13(1274):1–15

Parikh J (2011) Hardships and health impacts on women due to traditional cooking fuels: a case study of Himachal Pradesh, India. Energy Policy 39:7587–7594

Parthiban KT, Fernandaz CC, Sudhagar RJ, Sekar I, Kanna SU, Rajendran P, Devanaand PS, Vennila S, Kumar NK (2021) Industrial agroforestry- a sustainable value chain innovation through a consortium approach. Sustainability 13:1–14
Pateria HM, Sharma OP, Bhatt BP (2003) Sustainability through agroforestry in Himalayas: an overview. Proceeding. In: Bhatt BP, Bujarbaruah KM, Sharma YP, Patiram (eds) Approaches for increasing agricultural productivity in hill and mountain ecosystem. ICAR Research Complex for NEH Region, Umiam, Meghalaya, pp 129–138
Phondani PC, Maikhuri RK, Rawat LS, Negi VS (2020) Assessing farmers' perception on criteria and indicators for sustainable management of indigenous agroforestry systems in Uttarakhand, India. Environ Sustain Indicators 5:1–8
Proto AR, Zimbalatti G, Abenavoli L, Bernardi B, Benalia S (2014) Biomass production in agroforestry systems: V.E.Ri.For project. Adv Eng Forum 11:58–63
Rajput BS, Bhardwaj DR, Pala NA (2017) Factors influencing biomass and carbon storage potential of different landuse systems along an altitudinal gradient in temperate northwestern Himalayas. Agrofor Syst 91:479–486
Rathore AC, Mehta H, Sharma NK, Gupat AK, Alam NM, Islam S, Dogra P (2020) Performance of litchi (Litchi chinensis Sonn.) based agrihorticultural land uses in rainfed condition on degraded lands in North Western Himalayas, India. Agrofor Syst 94:2225–2236
Rawat YS, Vishwakarma SCR (2011) Pattern of fodder utilization in relation to sustainability under indigenous agroforestry systems, north-western Himalayas, India. Environ Int J Sci Technol 6: 1–13
Roberts MJ, Lomg SP, Tieszen LL, Beadle CL (1985) Measurement of plant biomass and net primary production. In: Techniques in nioproductivity and photosynthesis, pp 1–19
Roy ML, Chandra N, Kharbikar HL, Joshi P, Jethi R (2013) Socio-economic status of hill farmers: an exploration from Almora District in Uttarakhand. Int J Agric Food Sci Technol 4(4):353–358
Roy AK, Agrawal RK, Bhardwaj NR, Mishra AK, Mahanta SK (2019) Revisiting national forage demand and availability scenario. In: Roy AK, Agrawal RK, Bhardwaj NR (eds) Indian fodder scenario: redefining state wise status. ICAR-AICRP on Forage Crops and Utilization, Jhansi, pp 1–21
Saha SK, Kumar AS (2019) Northwest Himalayan ecosystems:issues, challenges and role of geospatial techniques. In: Navalgund RR et al (eds) Remote Sensing of Northwest Himalayan Ecosystems. Springer, Singapore, pp 3–14
Salve A, Bhardwaj DR (2020) Soil carbon stock and nutrient study in different agroforestry systems at Kinnaur district, Himachal Pradesh. Plant Arch 20(2):4251–4260
Sharma KK (1996) Agroforestry in farming system development. Indian Forester 122:547–559
Sharma V, Raina A (2021) Problems and prospects of Himalayan farmers and farming: a case study of District Kishtwar, Jammu and Kashmir. Reg Econ Dev Res 2(1):82–95
Sharma J, Sharma SD, Sharma K, Sharma D (2021) Appraisal of biological yield and economic returns of existing afroforestry systems in Tehsil Bangana of Unda District of Himachal Pradesh, India. Biol Forum Int J 13(2):413–419
Shukla R, Chakraborty A, Sachdeva K, Joshi PK (2018) Agriculture in the western Himalayas – an asset turning into a liability. Dev Pract 28(2):318–324
Singh AK (2009) Probable agricultural biodiversity heritage sites in India: II. The Western Himalayan Region. Asian Agri Hist 13(3):197–214
Singh R (2019) Studies on identification and socio-economic status of existing agroforestry systems in Shimla District of Himachal Pradesh. Ph.D. Thesis. Dr YS Parmar University of Horticulture and Forestry Nauni-Solan (HP), p 144
Singh M, Gupta B, Sarvade S, Awasthe RK (2015) Biomass and carbon sequestration potential in different agroforestry systems in Giri catchment of North Western Indian Himalayas. India J Agrofor 17(2):42–48
Singh R, Bhardwaj DR, Pala NA, Rajput BS (2019) Biomass production and carbon stock potential of natural vegetation, agroforestry and cultivated land use systems along altitudinal gradient in north western Himalayas. Range Manag Agrofor 40(1):94–103

Singh R, Pant KS, Tiwari P, Dhiman R (2020) Biomass productivity of different horticulture based agroforestry systems in North-Western Himalayas. J Pharmacogn Phytochem 9(5):862–866
Solanki KR (1998) Agroforestry research in India. Special issue. Indian J Agric Sci 68:559–566
Sood KK, Najiar C, Singh KA, Handique P, Singh B, Rethy P (2008) Association between socio-economic parameters and agroforestry uptake: evidences from eastern Himalaya. Indian J For 31(4):559–564
TERI (2015) Green growth and forestry in Himachal Pradesh. The Energy and Research Institute, New Delhi, p 16
Tewari VP, Kapoor KS (2013) Western Himalayan cold deserts: biodiversity, eco-restoration, ecological securities and concerns. Ann Arid Zone 52(3&4):223–230
Tewari JC, Pareek K, Raghuvanshi MS, Kumar P, Roy MM (2016) Fodder production system- a major challenge in cold arid region of Ladakh, India. MOJ Ecol Environ Sci 1(1):00005
Tewari VP, Verma RK, Gadow KV (2017) Climate change effects in the western Himalayan ecosystems of India: evidence and strategies. For Ecosyst 4(13):1–9
Thakur A (2020) Appraisal of existing agroforestry systems in Chuhar valley of district Mandi, H.P. M.Sc. Thesis. College of Horticulture and Forestry, Neri, Dr YSPUHF Nauni, Solan (HP), p 113
Thakur NS, Gupta NK, Gupta B (2004) Phytosociology analysis of woody and non woody components under some agroforestry systems in Western Himalaya: case study. Indian J Agrofor 6:65–71
Tiwari DN (1995) Agroforestry Systems. International Book Distributors. Dehradun, pp 31–107
Tiwari P, Pant KS, Singh R (2018) System units under prevalent agroforestry systems in North-Western Himalayas and their constraints. J Pharmacogn Phytochem 7(6):2758–2764
Toky OP, Kumar P, Khosla PK (1989) "Structure and function of traditional agroforestry systems in Garhwal Himalaya" Biomass and productivity. Agrofor Syst 9:47–70
Vikrant KK, Chauhan DS, Rizvi R (2018) Biomass and carbon potential in agroforestry systems along altitudes in Tehri District of Uttarakhand. In: Agroforestry for climate resilience and rural livelihood, pp 171–178
WMO (2021) State of the Global Climate 2021- WMO Provisional Report. World Meteorological Organization, Geneva, Switzerland, 47
Yadav RP, Bisht JK, Gupta B, Mondal T (2016) Toward the carbon sequestration potential of agroforestry practices to combat climate change in Kumaon Himalaya, India. In: Bisht JK et al (eds) Conservation agriculture. Springer, Singapore, pp 293–313
Yadav RP, Bisht JK, Bhatt JC (2017) Biomass, carbon stock under different production systems in the mid hills of Indian Himalaya. Trop Ecol 58(1):15–21
Yadav RP, Bisht JK, Meena VS, Choudhary M (2018) Sustainable agroecosystems for livelihood security in Indian Himalayas. In: Sustainability of agroecosystems, pp 63–77
Yadav RP, Gupta B, Bhutia PL, Bisht JK, Pattanayak A (2019) Sustainable agroforestry systems and their structural components as livelihood options along elevation gradient in central Himalaya. Biol Agric Hortic 35(2):73–95
Zahnoor S, Dutt V, Mughal AH, Pala NA, Rashid M, Khan PA (2022) Influence of intercropping on soil properties and nutrient status of fruit based agroforestry systems in north western region of Indian Himalaya. Indian J Agrofor 24(1):85–92
Zahoor S, Dutt V, Mughal AH, Pala NA, Qaisar KN, Khan PA (2021) Apple-based agroforestry systems for biomass production and carbon sequestration: implication for food security and climate change contemplates in temperate region of Northern Himalaya, India. Agrofor Syst 95: 367–382

Chapter 21
Assessment of Impact of Land Uses on Soil Carbon Stock and Quality

I. A. Jimoh and J. Aliyu

Abstract Deforestation due to urbanization depletes soil quality and increases greenhouse gas emission into the atmosphere leading to climate change. Hence, agroforestry in terms of plantation is an appropriate system for mitigating atmospheric gasses and increasing soil quality. Forest ecology is estimated to store about 70% of above and below organic carbon. This study aimed to assess the impact of land uses on soil carbon stock and quality in an ecosystem. Among all the land uses, natural forest have high porosity, exchangeable bases, organic carbon, total nitrogen, carbon stock, micronutrients, and soil quality than other land uses. Mixed plantation was reported to have low bulk density than other monoculture plantation, such as teak, gmelina, eucalyptus, and acacia and were attributed to high quantity of litter falls under mixed plantation as compared to low litter fall under monoculture plantation sites. Mixed plantations were significantly higher in organic carbon, litter fall, and exchangeable bases and soil quality than soils under monoculture plantations. Soil under teak, eucalyptus, and acacia were characterized with low moisture content, high bulk density, low organic carbon, and potassium content. The low organic carbon in teak was attributed to the high rate of mineralization of litter. Teak was reported to have higher litter fall and faster rate litter decomposition than eucalyptus, thus contributing higher nitrogen and phosphorus to soil. Teak contributes higher extractable Fe and Cu relative to other forest trees. Gmelina was reported to contribute higher organic carbon to soil than teak, while acacia have the lower nitrogen than other tree species. Higher bulk density and low nutrient reserve, which result to poor quality predominates under elephant grass and farmland, were ascribed to continuous disturbance leading to low organic carbon in the land uses. Sustainable land use management is required to increase soil carbon stock and quality for sustainability.

I. A. Jimoh (✉)
Department of Geography, Gombe State University, Gombe, Nigeria
e-mail: iajimoh@gsu.edu.ng

J. Aliyu
Department of Soil Science, Ahmadu Bello University, Kaduna, Nigeria

H. S. Jatav et al. (eds.), *Agroforestry to Combat Global Challenges*, Sustainable Development and Biodiversity 36, https://doi.org/10.1007/978-981-99-7282-1_21

Keywords Agroforestry · Land uses · Soil quality · Carbon stock · Climate change

21.1 Introduction

Soil organic carbon content and sequestration is influenced by the nature and type of land use of an area (Albaladejo et al. 2013). Vegetation influences soil properties through addition, removal, transformation, and redistribution of matter through a series of processes, which brings about biocycling of mineral elements and a change in the environment (Shukla 2009). The decomposition of leaf litter releases humic and fulvic acids, which play an important role in biocycling of metals and transformation of clay minerals in the soil system (Samndi 2012). Changes in land use and management of agroecosystems have the potential to release a considerable amount of soil organic carbon (SOC) stored in the soil or lead to increased carbon sequestration, thereby reducing atmospheric carbon dioxide (CO_2) concentration (Lal 2005); making the soil a source or sink for greenhouse gasses. Soil organic carbon (SOC) is sensitive to changes in land use and a change from natural or semi-natural forested ecosystems to other land uses, often leads to significant changes in SOC stock (Wilson et al. 2008). The conversion of forest land use to permanent farmland due to urbanization can reduce carbon stock by an average of 20% over a period leading to climate change (Powers et al. 2011). Carbon sequestration refers to the removal of carbon from the atmosphere through photosynthesis, dissolution, and storage in the soil as organic matter (OM) or secondary carbonates (Lal 2005). Through this process, carbon storage in soil is enhanced and its loss minimized, thereby reducing the chance of global warming by the reduction of atmospheric concentration of CO_2. There are five carbon pools; these include: oceanic pool which is the largest; followed by geologic, pedologic (soil), biotic, and atmospheric pool (Lal 2005). Among all these pools, the pedological pool is the pool that can easily be used to sequester carbon through recommended management practices (Lal 2005). Soil organic carbon is one of the major determinants and indicators of soil fertility, quality, and productivity in an ecosystem (Reeves 1997).

Soil quality is the ability of a soil to function within ecosystem boundaries in order to sustain biological productivity (plant and animal), maintain environmental quality (air and water), and support/promote plant and animal health. As a result, soil quality has a significant impact on health and productivity of a specific ecosystem as well as the environment that surrounds it (Doran and Parkin 1994). Bouma and Mc-Bratney (2013) listed soil functions to include biomass production, climate regulation, heritage, hydrologic storage, and pollution control. Land use changes have direct relationship with soil organic carbon, soil carbon sequestration, and soil quality. Severe depletion of vegetal cover degrades the soil organic carbon pool, soil quality, reduces biomass productivity, adversely impacts water quality, and also increases climate change severity (Lal 2005). The protection of soil quality under intensive land use system and fast economic development is becoming a major challenge for sustainable resource use in many developing countries of the world (Karlen et al. 1997).

To adequately maintained soil quality, basic assessment of soil health/quality is necessary to evaluate the degradation status and changing trends resulting from different land use and land cover (Odunze et al. 2017). Nigeria is currently undergoing a wide range of changes in its land use because of human activities (anthropogenic); bush burning, shifting cultivation, fuelwood harvesting, urbanization, industrial, and infrastructural development (Odunze 2017). Forest land use types influence carbon fluxes in an ecosystem through litter quality, biomass deposition, and turnover rates. Forest reserves were gazetted for the purpose of biodiversity conservation in Nigeria. These forest reserves were established for the purpose of increasing productivity, conserving biodiversity, and arresting desertification in northern savannahs of Nigeria (Nwadialor 2001). Anthropogenic factors such as continuous cultivation, bush burning, deforestation, and urbanization have been militating against realization of this objective. Forests, grasslands, and shrublands have been reported to be high in soil carbon due to high litter input and controlled soil temperature. With increase in man's activities, such as land conversion, soil carbon has significantly reduced; this was attributable to low litter deposition, high mineralization rates due to exposure to surface temperature, and intensive erosion (Cao et al. 2013). Studies investigating the effect of land use land cover change on soil carbon in Afaka forest show that thick forest decrease from 4.1% in 1986 to 3.8% in 2018, while light forest and built up generally increases from 11.8% and 4.8% from 1986 to 22.0% and 15.9% in 2018, respectively. The decrease in thick forest and increase in light forest and built up were clear evidence of forest degradation due to anthropogenic and natural factors.

Further, information on the effects of land use on soil quality and carbon sequestration are limited. For sustainable forest management and to keep up to date with changes in the forest reserve, there is need to compile land use effects on soil quality and carbon stock as it impacts the ecosystem. By accelerating the breakdown of organic matter, decreasing replenishment through litter input, and accelerating soil erosion, deforestation depletes soil organic matter and soil nutrient stores, leading to soil deterioration (Odunze 2015). Degradation throws the equilibrium off, which causes harm to the soil, water, and flora. Food security will be threatened since such degraded soils won't be able to sustain the growth of plants that can absorb CO_2 to slow down climate change (global warming). There is a need for intensive review of the relationship between forest land uses and soil quality for appropriate recommendations to be made for sustainable forest management. This will reduce the rate of soil and forest degradation and the amount of carbon released into the atmosphere to address climate change. This study intends to fill the gap in knowledge as a measure for a sustainably managed soil under forest condition. The ability of soil to sustain ecosystem services and to operate better and be maintained is largely dependent on increasing and preserving soil (Soil Carbon Initiative 2011). This can be accomplished by using carbon-sequestering agricultural land management techniques (Minasny et al. 2012; Odunze et al. 2017). There is need to understand which land use will sequester carbon and enhanced soil quality for sustainable development; hence, carbon stock accretion under some forest trees was reviewed. This will aid policy decisions on tree species to be used for afforestation and reforestation programs.

21.2 Effects of Forest Land Use on Soil Physical Properties

The addition of organic materials through litter fall and remains of vegetal materials into the soil control processes in the soil system. Vegetation cover favors the maintenance of good soil structure regardless of texture. Plant root systems and organic matter accumulation exert fundamental control over the creation and stabilization of soil structure (Russell, 1971 in Samndi 2012). Vegetation cover helps to reduce the rate of runoff, favor infiltration, as well as help in binding particles together, thus supporting soil aggregation (Lawal et al. 2009).

21.2.1 Effects of Forest Land Use on Particle Size Distribution

The most stable physical property of soil is soil texture, which also influences other soil properties like soil structure, consistency, moisture regime, infiltration rate, runoff rate, erodibility, workability, permeability, root penetrability, and soil fertility (Landon 1991). Soil texture has previously been described as a nearly permanent property of the soils that rarely changes with land use, vegetation, age and management, or conservation (Ahukaemere et al. 2016). According to an investigation of effect of forest trees on particle size distribution conducted by Mohd et al. (2018), it is indicated that soils under teak, eucalyptus, and acacia had the same sandy loam texture while soils under mixed plantation had clay loam texture. Studies by Egbuchua and Bosah (2011) observed that soils under teak plantation in four different locations were loam sand in texture. Ahukaemere et al. (2016) noted similar loamy sand texture of soils under 7- and 23-year-old vegetation. Kadeba and Adeuayi (1985) in their study of soils in Nimbia forest reported no evidence of textural changes because of afforestation with pine. Contrarily, Samndi and Jibrin (2012) observed a change of texture from very gravelly clay loam under younger teak plantation to gravelly clay texture under older Teck plantation and attributed it to soil aging under older plantations. Jaiyeoba (1995) observed a significant coarsening of texture under eucalyptus and mango plantation compared to a natural vegetation. Bargali et al. (1993) in their study on the effect of replacing natural vegetation with eucalyptus species reported a decrease in the proportion of fine soil particles especially at the early stage of eucalyptus development which was attributed to the open canopy of eucalyptus which favors erosive influence of water penetrating faster than natural forest with closed canopy. Generally, the influence of vegetation on texture is most evident within the profile; that is, between the surface and subsoils. Soils under plantation are dominated by high infiltration than runoff when compared to farmland or open land use. This supports argilluviation as water moves downward into the profile and thus easily influences soil texture, impacting clayey texture in the subsoils relative to surface soils (Samndi and Jibrin 2012). Jimoh (2021) in study observed that eucalyptus and mixed plantation were significantly higher in sand content than soils under teak, natural forest, and gmelina plantation and sand amount decreases with soil depth.

21.2.2 Effects of Forest Land Use on Bulk Density

Bulk density (BD) is an important factor that determines soil quality and ecosystem functions (Munishi 2012). Variation in land uses had been reported to influence soil bulk density. Mongia and Bandyopadhyay (1994) observed that replacing virgin forest in Andaman with plantation species like Pine dalbergiodes, Tectona grandis, and Eucalyptus **guineensis** plantations led to rapid deterioration in soil physical characteristic particularly bulk density of soil surface increased to a range of 1.28 to 1.49 M/g^3 compared to bulk density of 1.05 M/g^3 in the virgin forest. Mohd et al. (2018) also noted that bulk density of soils under mixed plantation were significantly lower BD (1.29 M/g^3) compared to the soils under teak (1.37 M/g^3), eucalyptus (1.50 M/g^3), and acacia plantation (1.42 M/g^3). The mixed plantation was characterized with low bulk density and was ascribed to the soil's high organic carbon content which raises the soil's overall quality. The high BD values in monocultural plantation was attributed to low amount of litter falls, that adds less amount of organic carbon into the soil system. Odunze et al. (2019) also observed a significantly high BD on cultivated land than forested soils in northern Nigeria. Similar research conducted by Are et al. (2018) also observed significantly higher BD under elephant grass and cultivated land than soils under secondary forest and Leucaena plantation in southwestern Nigeria. Higher BD was attributed to higher silt content in cultivated areas which causes crusting/surface sealing, trampling of livestock during grazing, continuous cultivation, and intensive mechanization that have resulted in the higher bulk density (Odunze et al. 2012; Are et al. 2018). Generally, low BD value favors root growth and development while high BD value inhibits root development.

21.2.3 Effects of Forest Land Use on Soil Porosity

Soils of Nigeria savanna vary in porosity due to land use, soil type, and management practices. Bargali et al. (1993) reported soils under natural vegetation have a higher porosity than soils under eucalyptus because of the open canopy of eucalyptus which supports direct raindrop impact blocking pore space. Loss of fine soil particle under eucalyptus was also attributed to soil compaction which leads to a reduction in available pore space for microbial activities and lower soil moisture retention capacity. Available pore space influences the rate of microbial decomposition and mineralization (Elliot et al., 1980 in Bargali et al. 1993).

21.2.4 Effects of Forest Land Use on Soil Moisture Content

Differences in forest land use types influenced soil moisture content. Mohd et al. (2018) noted that soils under mixed plantation had significantly higher moisture content than soils under eucalyptus and teak, while soils under acacia recorded the least moisture content. They attributed it to high litter layer on the surface and clay loam texture of the soil which helps to retain moisture and protects it from evaporation. Further, Srivastava (1993) estimated that the eucalyptus species had higher water-holding capacity than a nearby open area even after three consecutive drought years. Haghnazari et al. (2015) reported that the differences in soil moisture content was due to variation in particle size distribution, organic matter content, and soil type. Soils with higher organic carbon, clay content, and rainfall amount will have high moisture content. Imadojemu et al. (2018) observed that high moisture contents in soils of Ekpoma, southern Nigeria attributed it to higher organic carbon due to dense vegetation of the ecology.

21.3 Effects of Forest Land Use on Soil Chemical Properties

There are number of variables, including nutrient uptake, leachates from tree bark, leaves and roots, and organic acids from decomposing organic matter affect how land uses impact soil chemical properties. According to Kodama and Schnitzer (1976), decomposing litters produces water soluble chemicals that are crucial to the cycling of metals, weathering, and alteration of clay minerals in the soil. Humic and fulvic acids produced from litter decomposition are classified as naturally occurring poly electrolytes capable of transforming soil minerals (Samndi 2012). The humic acid favors humification and acid leaching. Compounds that leave the soil through solution are Ca, Mg, and K, and are referred as leaching losses.

21.3.1 Effects of Forest Land Use on Soil Reaction (Soil pH)

Soil pH is a function of the types of land use in an area because of differences in inherent base content of their litter (William 1979). The decomposition of litter from vegetation produces organic compounds which decrease soil pH in the ecosystem (Killham 1994). Different forest land uses release varying types of organic acid because of their chemistry. Organic acids produced from the decomposition of forest litter play an important role in soil acidity. Mohd et al. (2018) observed that soils under mixed plantation had the lowest pH value and were significantly lower than soil under teak and acacia plantation because of the large litter deposited by different tree species when compared to the other sites. Soils under eucalyptus trees were also reported to be low in pH and statistically like mixed plantation. Eucalyptus is an invasive species which process acids to its surrounding to inhibit the survival of other plants species. According to Haan (1977), decrease in soil pH under eucalyptus was ascribed to accumulation and slow decomposition of litters that produces acids in the forest soil. Richard (1995) in Samndi (2006) compared soil pH under pines and gmelina and reported that pH decreased significantly beneath pines and increased significantly beneath gmelina arborea. The author further stressed that the increase beneath gmelina was possible probably due to gmelina having high affinity for extractable Ca. This confirms the report of Salifu and Meyer (1998) who also stated a significant positive correlation between soil Ca and pH under teak plantation which they attributed to ability of teak plantation to function as cation pump. Samndi and Jibrin (2012) noted an increase in soil pH with plantation age under teak.

Samndi and Jibrin (2012) attributed a slight increase in pH with plantation age under teak. They also reported a decrease in pH in the subsurface soil which was ascribed to reduction in organic carbon, basic elements uptake, and leaching of nutrients. Similarly, Lawal et al. (2014) noted higher pH in soil associated with teak plantation and surface soil pH were higher than subsoils pH. Offiong et al. (2009) asserted that variations in biomass returned to the soil, both in terms of quantity and quality, influence soil reactivity. Punyisa et al. (2012) also reported lower pH under perennial crops than annual crops in oxisols of Thailand and attributed it to higher organic matter content of the soils under perennial crops. Jimoh (2021) reported significantly higher soil pH under eucalyptus and farmland than other forest land use, and surface soils were higher in soil pH than subsoils and were attributed to higher organic carbon in surface soils.

21.3.2 Effects of Forest Land Use on Exchangeable Bases

Exchangeable calcium (Ca), potassium (k), magnesium (Mg), and sodium (Na) constitute the exchangeable bases. According to Ohta (1990), exchangeable K and Mg are intensively used by soil microorganism and plantations for growth and development. These nutrients are then gradually released into the soil as the trees mature, thus increasing these exchangeable cations in the soil. Okoro et al. (2000) noted that the content of Ca varied with plantation species and decrease with depth, while Mg did differ with both plantation species and soil depth. These nutrient concentrations also varied with plantation age as observed by Braise et al. (1995) who stated that exchangeable Ca and Mg concentrations decrease linearly with plantation age. Tan (1980) in Samndi (2006) observed that humic and fulvic acids from organic matter decomposition are capable of mobilizing dissolve K from potassium feldspar, biotite, and muscovite. Mohd et al. (2018) also ascribed exchangeable potassium variation with plantation type. They reported that soils under mixed plantation were significantly higher in potassium than soils under eucalyptus, teak, and acacia. Similarly, Okoro et al. (2000) studied soil properties under 28-year-old plantations of teak, Terminalia, Nauclea, gmelina, and natural forest in southern Nigeria and opined that natural forest recorded the highest exchangeable Ca, Mg, and Na, while soil under *Nauclea diderrichii* recorded the highest K content indicating species preference for K. Adeboye et al. (2011) in their research observed that soils under plantations (teak, gmelina, and Cashew) had significantly higher exchangeable bases (Ca, Mg, and K) compared to arable soils. Similarly, Odunze et al. (2019) noted significantly higher (Ca, Mg, and K) under forested soils than cultivated land. Higher amounts of nutrients in forested soils were ascribed to less leaching losses and recycling of nutrients by plantation deep roots to the surface. Bush burning of litter under plantations was attributed to causing liming effect of ashes which increases the exchangeable bases under the vegetation (Nounamo et al. 2002).

21.3.3 Effects of Forest Land Use on Cation Exchange Capacity

The ability of soil to bind, hold, and exchange cations against leaching is determined by the cation exchange capacity (CEC), which is affected by the type of land use (Olorunfemi et al. 2016). Forested land use differs in the amount of litter and organic matter supply. Soil organic carbon and clay are the two major determinants of soil CEC (Raji 2011). Punyisa et al. (2012) observed significant higher CEC values in perennial crops over annual crops in Thai Oxisols. Samndi and Jibrin (2012) reported that soil CEC varies with plantation ages. They submitted that soils under older plantations recorded lowest CEC as compared to soils under younger plantations. Ashesh and Ramarkrishnan (1987) noted that low organic matter content was found under the older stand of tree, could be attributed to the fact that dry matter production in such older trees is mostly directed to the boles with very little allocation to the leaves. Additionally, they discovered reduced surface soil effective cation exchange capacity (ECEC) beneath the older plantation, which is an indication of older trees' utilize higher cations. Similar findings were made by Braise et al. (1995), who found that the effective cation exchange capacity (ECEC) of the forest floor of boreal species in northwest Quebec declined linearly with stand age. Lawal et al. (2014) also showed an increase in ECEC with soil depth, which they attributed to the eluviation of exchangeable bases to the subsoil in teak plantation. According to Ahukaemere et al. (2016), ECEC over older vegetation decreased with soil depth, whereas ECEC increased over younger vegetation. The ability of the older plantation to recycle nutrients back to the soil surface may be responsible for the observed drop in ECEC with depth under older vegetation, while leaching under younger vegetation may be the cause of the observed increase in ECEC with depth.

21.3.4 Effects of Forest Land Use on Exchangeable Acidity

The age of a tree plantation, the forms of land use, the parent material of the soil, and fertilizer applied all affect soil acidity (Alekseeva et al. 2011; Abe et al. 2006). Most of the aluminum (Al) absorbed by tea plants, which is known to be a typical accumulator, accumulates in the leaves. It has been suggested that soil acidification in tea plantations is caused by the biogeochemical cycling of Al in tea litter. According to Okoro et al. (2000), the exchangeable acidity of soil under Terminalia had the highest value exchangeable acidity while soils under teak had the least value. Samndi and Jibrin (2012) reported that plantation age influences exchangeable acidity where soils under older plantation were higher in acidity than soils under younger plantation. The trees spices in older plantations are characterized by high nitrogen usage, which depleted the soil's exchangeable bases, was attributed for their higher acidity. The above observation was corroborated by the findings of Braise et al. (1995); they observed that exchangeable acidity increased linearly with plantation age. Egbuchua and Bosah (2011); Ahukaemere et al. (2016) reported high exchangeable acidity values in soils of humid forest than those reported by Samndi and Jibrin (2012) in soils of savanna forest. The higher acidity value in soils of humid forest was attributed to higher rainfall of the environment which is associated with high leaching effects.

21.3.5 Effects of Forest Land Use on Base Saturation Percentage

Varying tree species and plantation ages had a significant influence on soil base saturation percentage (BSP). Samndi and Jibrin (2012) reported that BSP was generally low to medium in soils of northern guinea savanna, and soils under younger plantations had higher BSP values than soils under older plantations. Further, they also noted a decrease in BSP with soil depth reflecting organic carbon distribution. This contradicts the findings of Ahukaemere et al. (2016) who observed that older plantation had higher BSP than the younger plantation and attributed it higher organic carbon. Lawal et al. (2014) also reported high BSP under teak plantation in the southern guinea savanna zone as compared to values reported by Samndi and Jibrin (2012) in northern guinea savanna.

21.3.6 Effects of Forest Land Use on Soil Organic Carbon

The amount of organic material contributed to the soil by various land uses influences the qualities of the soil. The organic materials are characterized by a different chemical composition which in turn influences the soil processes and characteristics differently. Samndi (2006) noted that organic carbon content in soils varies with vegetation or tree types. They observed that organic carbon under pine with grass was between 15% and 19% lower under pine only. Okoro et al. (2000) also reported natural vegetation had higher soil organic carbon than soils under Terminalia, Nauclea, and gmelina with teak recording the least amount, which they attributed to the slower rate of mineralization of litter fall under teak. They also reported that organic carbon was significantly different with soil depth over the entire plantation. Similarly, Alekseeva et al. (2011) observed higher organic carbon on soils under tea plantation relative to natural vegetation and a significant decrease in SOC with depth under the two land uses in eastern China. Similarly, Rezaei et al. (2012) observed same in their study in soils under tea plantations in Iran. Ahukaemere et al. (2016) noted that soils under 23 years vegetation had higher organic carbon contents than soils under 7 years plantation, which was attributed to higher litter accumulations and lesser rate of mineralization. Mohd et al. (2018) corroborated the contribution of varying tree species to different rates of soil carbon. They observed that soils under mixed plantation had significantly higher organic carbon than soils under teak, eucalyptus, and acacia trees and attributed it to massive amounts of litter falling into the ground, decomposition, and mineralization of these litters to organic carbon in the soil. A study conducted by Singh et al. (1993) observed that teak has a higher yearly leaf litter fall than eucalyptus and degrade more quickly. Jimoh (2021) also observed high organic carbon under eucalyptus and mixed plantation than other land uses (mixed plantation, gmelina, teak, and cultivated land), and surface soils were higher in carbon than subsoils in savanna region of Nigeria. Studies on variation in soil carbon content in relation to land uses are show in Fig. 21.1.

21.3.7 Effects of Forest Land Use on Soil Total Nitrogen

Forest land use types and ages have different effects on soil quality and characteristics through exudate release, plant component decomposition, and rooting activities (Quideau et al. 2001). The quantity of organic matter and the rate of decomposition affect how much nitrogen is present in the soil at any given time. According to Mohd et al. (2018), organic matter is the main source of nitrogen in forest stands. They also noted a significant positive association between organic matter and total nitrogen, which highlights how significant organic matter is as a source of nitrogen. The humus layers of the forest floor and the surface horizon contain most of the total nitrogen in forest soils. Nitrogen mineralization in soils differs with land use and leaf quality. Species with high foliar concentrations of

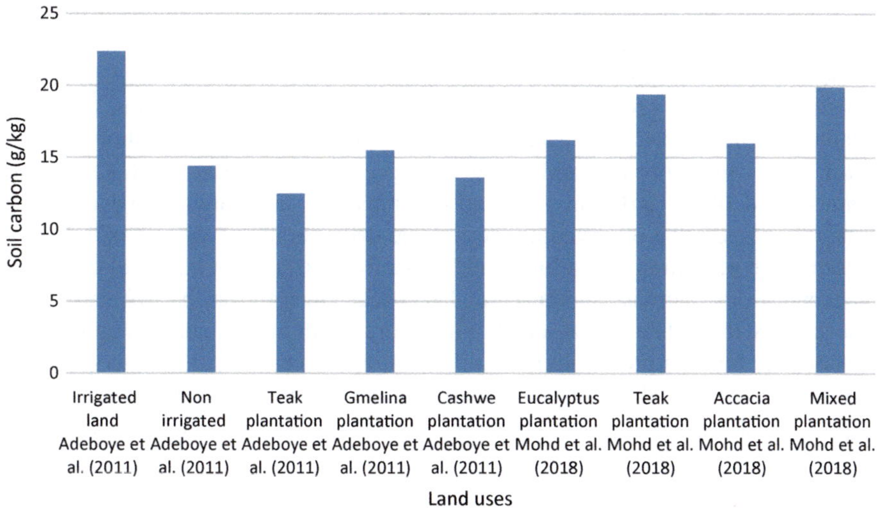

Fig. 21.1 Variation in soil carbon content with land uses

nitrogen and recycling of accumulated nutrient contribute to higher soil nitrogen content (Richard, 1995 in Samndi 2006). Other possible reasons for higher N content under trees are large nitrogen uptake by roots from the soil and thus concentrating it below the trees through litter falls as hypothesized by Browaldh (1995). Okoro et al. (2000) in their study on effects of monoculture plantation on soil of the tropic observed that natural forest had significantly higher total nitrogen while teak recorded the least value. Punyisa et al. (2012) also reported significantly higher total nitrogen on perennial crops than annual crops. According to Mohd et al. (2018), mixed plantations had the highest levels of total nitrogen, followed by teak plantations, while eucalyptus and acacia plantations had the lowest levels. Contrary to Prescott (1995) who observed that the mixed plantations had higher nitrogen level in the soil and attributed it to rapid rate of litter decomposition. Rezaei et al. (2012) noted that soils under natural forests have higher nitrogen contents than soils under tea plantations. Samndi and Jibrin (2012) and Ahukaemere et al. (2016) observed a reduction in total nitrogen with increase in tree age, and they attributed it to synergistic interaction between the trees effective nitrate absorption and an increased nitrogen mineralization.

21.3.8 Effects of Forest Land Use on Available Phosphorus

Forest land use influences the distribution of phosphorus, as Richard (1995) in Samndi (2006) reported that species with high foliar nutrient concentration and recycling property accumulate more nutrients beneath them, thus increasing soil nutrient content. The content and distribution pattern of soil phosphorus strongly depends upon the biological activity of the soil as well as its chemical behavior in each condition. According to Nwoboshi (1970), 80%–90% of the teak annual minerals uptake is immobilized, which over time depletes the soil's nutritional levels. Mohd et al. (2018) reported higher available phosphorus under the mixed plantation site followed by teak, while eucalyptus and acacia plantation recorded least value. In Venezuela and Nigeria, respectively, Marquez et al. (1993) and Samndi and Jibrin (2012) investigated the impact of teak chronosequence on soil characteristics. They observed a considerable decrease in the amount of soil phosphorus with plantation age. Aluko and Fagbernro (2001) noted that trees depend on phosphorus for biomass synthesis, and that teak immobilizes phosphorus, which causes teak to deplete soil available phosphorus with age plantations and may be responsible for their greater nutrient utilization. Rezaei et al. (2012) observed higher phosphorus value under forest soils than Tea plantations. Pande and Sharma (1993) reported that teak and sal saved more nutrients than pine and eucalyptus plantation.

21.3.9 Effects of Forest Land Use on Soil Carbon to Nitrogen Ratio (C/N)

The soil C/N ratio is an index of soil quality and is frequently interpreted as a measure of the soil's ability to mineralize nitrogen. Low soil C/N ratio will speed up the process of microbial decomposition of organic matter and nitrogen, which is not favorable for carbon sequestration. Conversely, high soil C/N ratio can slow down the decomposition of organic matter and organic nitrogen by restricting the activities of soil microbes. Lower C/N ratios are a sign of rapid organic N mineralization and high levels of humification (Bai et al. 2005). High soil C/N ratios may hinder decomposition from occurring (Aerts 1997).

Lorenzo and Welington (2018) in their research on the impacts of eucalyptus and Pinus forest management on soil organic carbon in Brazilian forested savanna found that the C/N ratios at the forest floor and surface soil were greater in eucalyptus and Pinus stands than in natural forests. Increased C/N ratio, especially in tropical soils where biotic factors have great effect on decomposition, is frequently associated with decreased SOC quality and may have an impact on nitrogen cycling and fertility (Aerts 1997).

21.4 Effects of Forest Land Use on Carbon Sequestration and Stock

Soil carbon sequestration involves the conversion of atmospheric carbon dioxide into plant materials through photosynthesis, and the release of organic carbon into soil organic carbon pools because of their death, decay, and decomposition (Powlson et al. 2011). When proper management practices are employed, soil organic matter plays a significant role in the global carbon cycle by functioning as a sink for atmospheric CO_2 (Paustian et al. 2000). Depending on how the land is used and managed, higher SOC resulted from soils under natural forest, gmelina/teak plantation, and managed artificial grassland at the 0–30 cm depth, with values of 9510.9, 8987.8, and 7906.6 gC/m^2, respectively, while soils under conventionally tilled and continuous cropped had lower SOC stocks of 1978.5 and 2768.7 gC/m^2 in 0–30 cm depth, respectively. Thus, a 45-year-old gmelina forest had a soil carbon storage of 8987 gC/m^2, whereas areas of this forest that were constantly cleared and farmed for 15 years had a carbon stock that was 75% lower (1978 gC/m^2). Different management practices affect SOC content if the soil and according to Johnston et al. (2009), the conversion of acropland to grasslands or forests tends to increase soil carbon stocks and vice versa. Jimoh (2021) observed that the carbon stock of soils that were continually farmed and conventionally tilled were 25% lower than the carbon stock of soils under conservation tillage. Anikwe (2010) observed that the conversion of forests to croplands in southeast Nigeria resulted in a 50%–75% loss of the region's soil carbon pool. Land removal and ongoing agriculture have been blamed for these decreases in SOC, and this was corroborated by the findings of Bationo et al. (2007). Other factors that affect the amounts and rates of carbon sequestration include the climate (temperature and rainfall), soil texture, bulk density, clay mineralogy, soil depth, and agricultural practices. In addition, Onweremadu et al. (2011) reported significantly higher SOC stock under fallow soils followed by soils under pineapple plantation, while soils under cassava plantation recorded the least SOC stock. These variations have also been influenced by tillage and soil management practices (Denef et al. 2004). In a study conducted by Mbah and Idike (2011) on soils under natural forest, gmelina forest, alley crop farming, and sewage sludge dumpsite land uses in southern Nigeria demonstrate higher amount of carbon stored in natural forests. They observed that natural forests contained more soil carbon stock than alley cropping, sewage sludge dumps, and gmelina forests by 27%, 37%, and 62%, respectively. Furthermore, Odunze et al. (2019) claimed that the carbon stock in the Afaka forest was much higher than that of the nearby cultivated area because of the accumulation and decomposition of forest biomass. According to Odunze et al. (2017), the low carbon stock found under farmed land were caused by a rapid rate of organic matter mineralization brought on by agricultural operations and the high diurnal temperature of the savanna environment. According to Are et al. (2018), soils in southwest Nigeria have significantly greater soil carbon stocks under elephant grass, Leucaene, and guinea grass than soils under cultivated land and secondary forest. This was due to the annual contribution of substantial amounts of biomass to the soil system. Additionally, Jimoh (2021) observed that soils in mixed plantations had the highest carbon stock levels (1846.6 t C) while soils under gmelina plantation recorded the least (1080.3 t C/ha) among the six land system in northern Nigeria. Studies on variation in soil carbon stock in relation to land uses are shown in Fig. 21.2.

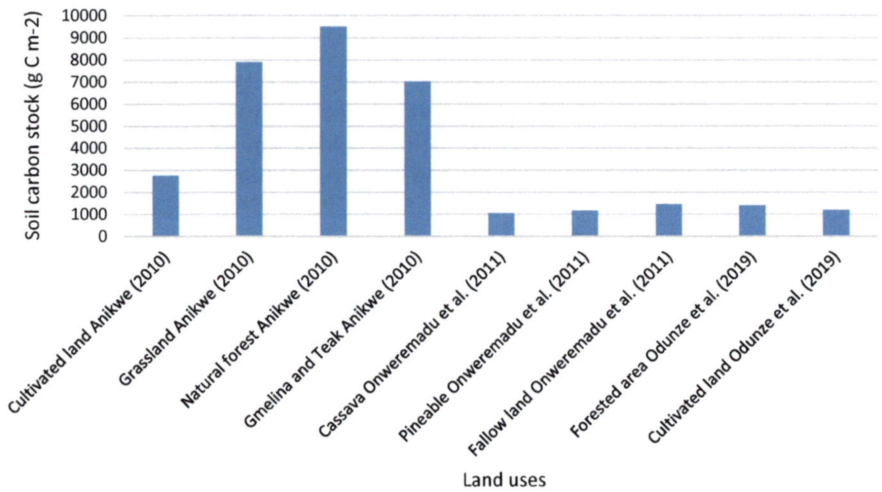

Fig. 21.2 Variation in soil carbon stock with land uses

21.5 Effects of Forest Land Use on Soil Micronutrients

Macronutrients as compared to micronutrients are nutrients that plants need in considerably lesser amounts. According to Tisdale et al. (2003), micronutrients form stable complexes with soil organic matter-components and are more readily available to plants in their organically bound forms than in their inorganic forms in the pools, insoluble inorganic precipitates, and those held in primary minerals. Tree species vary with respect to micronutrient cycling and accumulation. Some species accumulate large amounts of micronutrients in their litter while other species accumulate very little. Gideon et al. (2016) reported that soils under Cacao plantation accumulate significantly higher Cu content and least amount of Fe. Contrarily to secondary forest where significantly higher Fe and least amount of Cu were accumulated, soils under gmelina plantation accumulated significantly higher Fe than soils under cocoa plantation. Further, Albert (2016) reported significantly higher Mn concentration on soils under eucalyptus compared to soils under natural forest. The higher Mn recorded under eucalyptus was ascribed to the strongly acidic nature and reduction in soil pH under eucalyptus. The mean concentration of Zn was significantly higher under natural forest than in eucalyptus. This confirms the reports of Tererai et al. (2014) who also reports the significantly higher amount of Zn in eucalyptus spices than soils in natural forest. In terms of Fe, soils under eucalyptus were higher than natural forest, though significantly similar. Similarly, the concentration of copper in natural forest was higher than in eucalyptus plantation. Similarly, Tererai et al. (2014) reported that the soil accessible Cu was lower in the severely invaded site with Eucalyptus spp. compared to the uninvaded site in the Berg watershed in South Africa, but that the difference between them was not statistically significant. Geetha (2008) also reported that soils under teak had significantly higher

Fe and Cu content, while Zn and Mn were not significant when compared to soils under natural forest and eucalyptus. Soils under natural forests were also higher in all the micronutrients than eucalyptus except extractable Fe which was higher in eucalyptus than natural forest. Generally, all extractable micronutrients were higher in younger teak than older teak. Replanted plantations were reported to have poorer micronutrient availability than coppiced eucalyptus plantations. Age-related declines in Zn and Mn availability were seen in eucalyptus plantings (Sangha and Jalota 2005). Under eucalyptus and teak plantations, Jimoh (2021) found a strong and adverse association between organic carbon and extractable iron which according to Stevenson (1991) attributed it to complexation with soil organic carbon, especially when the metal–organic complex has a low solubility.

21.6 Effects of Forestland Use on Soil Quality

Understanding the functionality of soil under various land uses and management approaches regarding aggrading, maintaining, or degrading soil is largely dependent on soil quality (Karlen et al. 2003). The measurement of the soil quality is crucial in developing an early warning system for negative effects from changing the kind of land use since soil quality differs with different land uses. In a study conducted by Demessie et al. (2011) on Ethiopia's natural forest, coniferous forest, and eucalyptus plantation, they observed a significantly difference regarding soil quality index (SQIs). They noted that low SQIs under eucalyptus were due to rapid growth, capacity to absorb nutrients from the soil quickly, frequent harvesting and transportation of woods from the eucalyptus plantation. They also observed that Coniferous species were harvested during a longer rotation cycle, which may have allowed for a relative increase in soil quality through nutrient recycling via root absorption from the subsurface and by litter fall and the decomposition on the soil's surface. Demessie et al. (2011) observed that the quality of soil under undisturbed forest and those under longer rotation were significantly higher than those under previously cultivated land. Odunze et al. (2019) showed that the best quality soils were found under forest-covered lower slope terrain (SQ1), followed by farmed middle slope and forest land use type (SQ2). The quality of cultivated upper slope soils for use in crop production was ranked least (SQ6), primarily because of higher bulk density that could cause soil compaction and lower organic carbon as a result of continuous cultivation. Cultivated lower slope and cultivated land use types soils were ranked SQ5 and SQ6, respectively.

21.7 Conclusion

Forest land use influences soil properties through innumerable processes, which brings about biocycling of mineral elements within the soil. Loss of fine soil particles occurred more under eucalyptus was also attributed to soil compaction which leads to a reduction in soil quality. Higher BD predominates under elephant grass and cultivates land than soils under secondary forest and Leucaena plantation. Similarly, soils under natural vegetation have a higher porosity than soils under eucalyptus because of the open canopy of eucalyptus which supports direct raindrop impact blocking pore space. Soils under mixed plantation had the lowest pH value and were significantly lower than teak and acacia plantation because of the large litter when compared to the other sites. Also soils under mixed plantation had higher organic carbon than soils under teak, eucalyptus, and acacia trees. Low total nitrogen while teak forest was due to rapid decomposition of its litter while mixed plantation had the highest amount. High nitrogen in mixed plantation was due to higher plant litter production. Soils under natural forest, has the highest SOC stock than gmelina/teak plantation and managed artificial grassland. Soil under eucalyptus had the highest concentration of micronutrient relative to other forest land use types.

References

Abe SS, Hashi I, Masunaga T, Yamamoto S, Honna T, Wakatsuki T (2006) Soil profile alteration in a brown forest soil under high-input tea cultivation. Plant Prod Sci 9:457–461

Adeboye MKA, Bala A, Osunde AO, Uzoma AO, Odofin AJ, Lawal BA (2011) Assessment of soil quality using soil organic carbon and total nitrogen and microbial properties in tropical agroecosystems. Agric Sci 2:34–40

Aerts R (1997) Climate, leaf litter chemistry and leaf litter decomposition in terrestrial ecosystems: a triangular relationship. Oikos 80:353–361

Ahukaemere CM, Onweremadu EU, Ndukwu BN, Okoli NH (2016) Pedogenesis of two lithologically similar soils under vegetation of contrasting features in Ohaji, South-Eastern Nigeria. Agro-Sci J Trop Agric Food Environ Ext. 15(3):34–40

Albaladejo J, Ortiz R, Garcia-Franco N, Navarro AR, Almagro M, Pintado JG, Martínez-Mena M (2013) Land use and climate change impacts on soil organic carbon stocks in semi-arid Spain. J Soils Sediments 13(2):265–277

Albert KM (2016) Effects of Eucalyptus Plantation on Soil Physico-Chemical Properties in Thiririka Sub-Catchment, Kiambu County, Kenya. MSc. Thesis in the School of Pure and Applied Sciences of Kenyatta University

Alekseeva A, Xu R-K, Zhao A-Z, Kalinin P (2011) Effect of soil acidification induced by a tea plantation on chemical and mineralogical properties of Alfisols in eastern China. Environ Geo Chem Health 33:137–148. https://doi.org/10.1007/s10653-010-9327-5

Aluko AP, Fagbemro JA (2001) The role of tree species on land use systems in organic matter and nutrient availability in degraded Ultisols of Onne, south-south-Nigeria. Paper Presented at 26th Annual Conference of Soil Science Society of Nigeria, Ibadan

Anikwe MAN (2010) Carbon storage in soils of Southeastern Nigeria under different management practices. Carbon Balance Manag 5(5):1–7. http://www.cbmjournal.com/content/5/1/5

Are KS, Ande OT, Fademi IO (2018) Carbon sequestration under different land cover in soils derived from sandstones in Southwestern Nigeria. In: Oluwatosin GA, Ande OT, Adediran JA, Anikewe MAN (eds.) Sustainable Management of Soil and Water Resources for Food Security, Climate Change Adaptation and Mitigation. Proceedings of the 42nd Annual Conference of Soil Science Society of Nigeria (SSSN) held at Institute of Agricultural Research and Training, Obafemi Awolowo University, Moor Plantation, Ibadan, March 12–16, 2018, pp 494–503

Ashesh KD, Ramarkrishnan PS (1987) Above-ground biomass and nutrient contents in an age series khasi pine Pinus kesiya. For Ecol Manag 18:61–72

Bai JH, Ouyang H, Deng W, Zhu YM, Zhang XL, Wang QG (2005) Spatial distribution characteristics of organic matter and total nitrogen of marsh soils in river marginal wetlands. Geoderma 124:181–192

Bargali SS, Sighh RP, Joshi M (1993) Changes in soil characteristics in Eucalyptus plantation replacing natural broad-leaved forest. J Veg Sci 4:25–28

Bationo A, Kihara J, Vanlauwe B, Waswa B, Kimetu J (2007) Soil organic carbon dynamics, functions and management in West African agro-ecosystems. Agric Syst 94:13–25

Bouma J, Mc-Bratney A (2013) Framing soils as an actor when dealing with wicked environmental problems. Geoderma 200–201:130–139

Braise S, Bergeron CC, Pare D (1995) Changes in nutrient availability and forest floor characteristics in relation to stand age and forest composition in Northwestern Quebec. For Ecol Manag 76:181–189

Browaldh M (1995) The influence of trees on nitrogen dynamics in an agrisilvicultural system in Sweden. Agrofor Syst 30(3):301–313

Cao J, Wang X, Sun X, Zhang L, Tian Y (2013) Effects of grazing intensity on soil labile organic carbon fractions in a desert steppe area in Inner Mongolia. Springerplus 2(1):S1

Demessie A, Singh B, Lal R (2011) Soil carbon and nitrogen stocks under plantations in Gambo District, Southern Ethiopia. J Sustain For 30(6):496–517. https://doi.org/10.1080/10549811.2010.550547

Denef K, Six J, Merckx R, Paustian K (2004) Carbon sequestration in microaggregates of no-tillage soils with different clay mineralogy. Soil Sci Soc Drn J 68:1935–1944

Doran JW, Parkin TB (1994) Defining and assessing soil quality. In: Doran JW, et al (eds.) Defining Soil Quality for a Sustainable Environment, Soil Science Society of America Special Publication No. 35, Madison, WI, 3–21

Egbuchua CN, Bosah BO (2011) Characterization of some forest soil under teak (Tectona Grandis) plantation in a Humid Forest Region of Nigeria. Int J Sci Nat. 2(3):821–882

Geetha T (2008) Impact of teak and eucalypt monoculture on soils in the highlands of Kerala Forest Research Institute. PhD Thesis in Environmental Chemistry, University of Science and Technology, India

Gideon IK, Ogbonna AN, Nzegbule EC (2016) Soil Physico-chemical properties in three land use systems (Theobroma Cacao, Gmelina Arborea and secondary Forest) in Umuahia north of Abia state. Nigeria Canadian Journal of Agriculture and Crops 1(1):19–29. www.onlinesciencepublishing.com

Haan SD (1977) Humus, its formation, its relation with the mineral part of the soil and its significance for soil productivity. Soil Organic Matter Studies, IAEA Vienna 1:21–30

Haghnazari F, Shahgholi H, Feizi M (2015) Factor affecting the infiltration of soil: review. International Journal of Agronomy and Agricultural Research (IJAAR) 6(5):21–35

Imadojemu PE Osujieke DN, Obasi SN, Mbe JO, Dibofori EG (2018) Characterization and classification of some soils of Edo State formed under different parent materials. FUW Trends Sci Technol J. 3(1):201–206. www.ftstjournal.com e-ISSN: 24085162; p-ISSN: 20485170

Jaiyeoba IA (1995) Changes in Soil Properties related to different Land uses in part of Nigerian Semi-arid Savanna. Soil Use Manag 11:84–89

Jimoh AI (2021) Land use effect on soil development, quality and carbon sequestration in Afaka forest reserve, Kaduna State, Nigeria. Unpublished Ph.D., Thesis, Department of Soil Science, Ahmadu Bello University Zaria, p 216

Johnston AE, Poulton PR, Coleman K (2009) Soil organic matter: its importance in sustainable agriculture and carbon dioxide fluxes. Adv Agron 101:1–57
Kadeba O, Adeuayi EA (1985) Impact on soils of plantation of Pinus caribaea stands in natural tropical savanna. For Ecol Manag 13:27–39
Karlen DL, Mausbach MJ, Doran JW, Cline RG, Harris RF, Schuman GE (1997) Environmental quality. Advances in soil science. CRC Press, Boca Raton, FL
Karlen DL, Doran JW, Weinhold BJ, Andrews SS (2003) Soil quality: Humankind's foundation for survival. J Soil Water Conserv 58:171–179
Killham K (1994) Soil ecology. Cambridge University Press, Cambridge
Kodama H, Schnitzer M (1976) The dissolution of micas by fulvic acid. Geoderma 15:381–391
Lal R (2005) Forest soils and carbon sequestration. For Ecol Manag 220(1):242–258
Landon JR (1991) Booker tropical soil manual: a handbook for soil survey and agricultural land evaluation in the tropics and subtropics. Longman, London, p 474
Lawal HM, Ogunwole JO, Uyovbisere EO (2009) Changes in soil aggregate stability and carbon sequestration mediated by land use practices in a degraded dry savanna Alfisol. Trop Subtrop Agroecosyst 10:423–429
Lawal BA, Tsado PA, Eze PC, Idefoh KK, Zaki AA, Kolawole S (2014) Effects of Slope position on some properties of Soils under Tectona Grandis Plantation in Minna, Southern Guinea Savanna of Nigeria. Int J Res Agric For 1(2):37–43
Lorenzo ML, Welington D (2018) Effects of eucalyptus and pinus forest management on soil organic carbon in Brazilian wooded-savanna. In: Forest biomass and carbon. IntechOpen, London, pp 61–81. https://doi.org/10.5772/intechopen.72684. Chapter 24
Marquez O, Hernandez R, Torres A, Franco W (1993) Changes in the physiochemical properties of soils in chronosequence of Tectona grandis. Turrialba 43:37–41
Mbah CN, Idike FI (2011) Carbon storage in tropical agricultural soils of Southeastern Nigeria under different management practices. Int Res J Agric Sci Soil Sci 1(2):53–57
Minasny B, McBratney AB, Hong SY, Sulaeman Y, Kim MS, Zhang YS (2012) Continuous rice cropping has been sequestering carbon in soils in Java and South Korea for the past 30 years. Global Biogeochem Cycles 26:GB3027. https://doi.org/10.1029/2012GB004406
Mohd S, Kumar S, Kumar P, Gupta MK (2018) A comparative study of soil physicochemical properties between eucalyptus, teak, acacia and mixed plantation of Jhilmil Jheel wetland, Haridwar-Uttrakhand. Int J Sci Res Publ 8(1):378–338. www.ijsrp.org
Mongia AD, Bandyopadhyay AK (1994) Soil nutrients under natural and planted forest in Island ecosystem. J Indian Soc Soil Sci 42:43–46
Munishi JA (2012) "Pedological characterisation of soils and assessment of their agricultural potential in selected Districts of Mbeya Region, Tanzania," MSc (Soil Science and Land Management) dissertation. Department of Soil Science, Sokoine University of Agriculture, Morogoro, Tanzania, p 136
Nounamo L, Yemefack M, Tchienkoua M, Njo-mgang R (2002) Impacts of natural fallow duration on topsoil's characteristics of a Ferralsol in southern Cameroon. Niger J Soil Res. 3:52–57
Nwadialor IJ (2001) An assessment of spatio-temporal variabilities of deforestation for sustainable forestry development: a case study of Afaka Forest Reserve, Nigeria. International Conference on Spatial Information for Sustainable Development Nairobi, Kenya 2–5 October 2001
Nwoboshi LC (1970) Studies on nutrient cycle in forest plantation: preliminary observations in litter fall and macronutrient returns in teak *Tectona grandis* L. F. plantation. Niger J Sci 4(231):239
Odunze AC (2015) Soil conservation for mitigation and adaptation to a changing climate: sustainable solutions in the Nigerian savanna ecology. Int J Plant Soil Sci 8(4):1–12. ISSN: 2320-7035 www.sciencedomain.org
Odunze AC (2017) Soil Health: The Key to Agricultural Revolution in Nigeria. An Inaugural Lecture. Ahmadu Bello University, Series N0. 07/17, p 54
Odunze AC, Jinshui W, Liu S, Zhu H, Ge T, Wang Y, Luo Q (2012) Soil quality changes and quality status: a case study of the Subtropical China Region Ultisols. Br J Environ Clim Change Sciencedomain Int 2(1):37–57

Odunze AC, Musa YD, Abdulkadir A (2017) Soil organic carbon concentrations and stocks under maize/legume cropping system in Alfisols of a Savanna Zone, Nigeria. Br J Appl Sci Technol. 21(1):1–12

Odunze AC, Hamza H, Oyinlola EY (2019) Slope position and land use effect on select soil properties, quality and carbon stock in surface soils at Afaka forest area, northern guinea savanna of Nigeria. Curr J Appl Sci Technol 32(4):1–13

Offiong RA, Atu JE, Njar GN, Iwara AI (2009) Effects of land use change on soil physicochemical properties in South-south Nigeria. Afr J Environ Pollut Health 7(2):47–51

Ohta S (1990) Initial Soil changes associated with afforestation with Acacia auriculiforms and Pinu kesiye on denuded grassland of the Pantabangan area Central Luzon, Philippines. Soil Sci Plant Nutr 36(4):633–643

Okoro SP, Aighewi A, Osagie CO (2000) Effects of selected monoculture plantation species on humid tropical soils of southern Nigeria. Indian J Agric Sci 70(2):105–109

Olorunfemi IE, Fasinmirin JT, Ojo AS (2016) Modeling cation exchange capacity and soil water holding capacity from basic soil properties. Eurasian J Soil Sci 5(4):266–274

Onweremadu EU, Osuaku SK, Udebuani AC, Lekwa M (2011) Carbon sequestration in soils of three land use types in Owerri Area, Southeastern Nigeria. Niger J Agric Food Environ 7(3): 20–24

Pande PK, Sharma SC (1993) Biochemical cycling and nutrient conservation strategy in some plantations. Indian Forester 1(19):299–305

Paustian K, Six J, Elliott ET, Hunt HW (2000) Management options for reducing CO2 emissions from agricultural soils. Biogeochem 48:147–163

Powers JS, Corre MD, Twine TE, Veldkamp E (2011) Geographic biasoffield observations of soil carbon stocks with tropical land-use changes precludes spatial extrapolation. Proc Natl Acad Sci 108(15):6318–6322

Powlson DS, Whitmore AP, Goulding KWT (2011) Soil carbon sequestration to mitigate climate change: a critical re-examination to identify the true and the false. Eur J Soil Sci 62:42–55

Prescott CE (1995) Does nitrogen availability control rates of litter decomposition in forests? Plant Soil 168–169:83–88

Punyisa T, Robert JG, Kumut S (2012) Effects of land use on some physical, chemical, and mineralogical characteristics of Thai Oxisols. Sci Asia 38:82–89. https://doi.org/10.2306/scienceasia1513-1874.2012.38.082

Quideau SA, Chadwick OA, Trumbore SE, Johnson-Maynard JLG, R.C and Anderson, M.A. (2001) Vegetation control on soil organic matter dynamics. Org Geochem 32:247252

Raji BA (2011) Predicting the cation exchange capacity of soils of the Nigerian Savanna. Niger J Soil Sci 21(2):22–23

Reeves DW (1997) The role of soil organic manure in maintaining soil quality in continuous cropping systems. Soil Tillage Res 43:131–167. https://doi.org/10.1016/S0167-1987(97)00038-X

Rezaei N, Roozitalab MH, Ramezanpour H (2012) Effect of land use change on soil properties and clay mineralogy of forest soils developed in the caspian sea region of Iran. J Agric Sci Technol 14:1617–1624

Salifu KF, Meyer WL (1998) Physiochemical properties of soil associated with logged forest area converted to Teak (Tectona grandis) in Ghana Commonwealth. For Rev. 77, 91–99.157, 159–160

Samndi AM (2006) An Evaluation of Soil Properties and Development under Teak (Tectona grandis) Plantation of Various ages in Southern Guinea Savanna of Nigeria. Unpublished PhD dissertation, Ahmadu Bello University Zaria Nigeria

Samndi AM (2012) Vegetation effects on pedogenetic forms of iron and aluminium and mineralogical properties of basaltic soils in the southern guinea savanna of Nigeria. Bayero J Pure Appl Sci. 5(1):139–148

Samndi AM, Jibrin JM (2012) Pedogenesis and classification of soils under Teak (Tectona grandis Linn. f) plantation of various ages in the Southern Guinea Savanna of Nigeria. Asian J Agric Sci 4(1):16–25

Sangha KK, Jalota KR (2005) Value of ecological services of Exotic Eucalyptus tereticornis and native Dalbergia sisos tree plantations of north-western India. Conserv Soc 3:92–109

Shukla PK (2009) Nutrient Dynamics of teak plantations and their impact on soil productivity. A case study from India XIII Wold Forest Congress. Buenos Aires Argentina, pp 18–23

Singh O, Sharma DC, Rawat JK (1993) Production and decomposition of leaf litter in sal, teak, eucalypts and poplar forests in Uttar Pradesh. Indian Forester 119:112–121

Soil Carbon Initiative (2011) Soil Carbon Summit Outcomes Document [online]. Available from: http://ussc.edu.au/events-special/soilcarbon-initiative

Srivastava AK (1993) Change in physical and chemical properties of soil in irrigated eucalyptus plantation in Gujarat state. Indian Forester 119(3):226–231

Stevenson FJ (1991) Organic Matter Micronutrients Reactions in Soils. In: Mortverdt JJ, Cox FR, Shuman LM, Welch RM (eds) Micronutrients in agriculture. SSSA, Madison, pp 145–186

Tererai F, Gaertner M, Jacobs SM, Richardson DM (2014) Eucalyptus camaldulensis invasion in riparian zones reveals few significant effects on soil physico-chemical properties. River Res Appl 7(4):1–12. https://doi.org/10.1002/Rra

Tisdale SL, Nelson WL, Beaton JD, Havlin JL (2003) Soil fertility and fertilizers, 5th edn. Prentice-Hall, New Delhi

William LP (1979) Properties and management of forest soils. John Wiley and sons, New York, p 500

Wilson BR, Growns I, Lemon J (2008) Land-use effects on soil properties on the north-eastern slopes of New South Wales: implications for soil condition assessment. Aust J Soil Res 46:359–367

Chapter 22
Litter Fall Decomposition and Its Effects on Nutrient Accretion to Soil Under Agroforestry Systems

Sudip Sarkar and Trisha Sinha

Abstract Litter fall is an essential component of the nutrient cycle in agroforestry systems, which is directly relevant to nutrient accretion and restoration, soil organic matter buildup, biodiversity stability, and other ecological functions. In agroforestry, the production of litter fall and the rate of its decomposition are key determining variables for soil nutrient accumulation. Litter deposition plays a key role in the interaction between plants and soil because it aids in incorporating nutrients and carbon from plants into the soil. According to the morphological characteristics of the vegetation, such as tree size, species variety, and density, litter generation provides essential information about the performance of the soil system and is related to soil nutrient dynamics. Therefore, for sustainable soil management, a better understanding of the crucial processes in the nutrient cycle is essential. This chapter discussed the role of plant litter in agroforestry systems with the aim of evaluating the significance of plant litter in agroforestry systems for soil nutrient availability. Then, the major factors that affect tree litter production and decomposition were recognized. Further, the potential of plant litter to enhance the accumulation of soil nutrients in agroforestry systems is discussed.

Keywords Agroforestry · Litter fall · Nutrient cycle · Decomposition

22.1 Introduction

Litter fall plays a vital role for nutrient retrieve in soil system. Agroforestry system is well known for its large litter fall producing capability. Tree leaves can comprise up to 80% of surface litter fall in agroforestry systems, remaining includes of young twigs and stems (Kotowska et al. 2016). The loss of litter mass through decomposition is the total of carbon dioxide (CO_2) emission and release of plant nutrients.

S. Sarkar (✉)
ICAR-Research Complex for Eastern Region, Patna, India

T. Sinha
Department of Plant Physiology, Gopal Narayan Singh University, Sasaram, Bihar, India

H. S. Jatav et al. (eds.), *Agroforestry to Combat Global Challenges*, Sustainable Development and Biodiversity 36, https://doi.org/10.1007/978-981-99-7282-1_22

Litter decomposition goes through many processes, particularly heterotrophic use of organic compounds in litter. Surface water leaching and the actions of soil living organisms not directly responsible for CO_2 emission to the atmosphere, although they accelerate litter decaying. The CO_2 released throughout the process of microbial decomposition can up to 25% of total soil surface CO_2 release also called as soil respiration. In agroforestry system, soil nitrogen (N), phosphorus (P), and calcium (Ca) produced from tree litter fall through decomposition are available for uptake of crops. This chapter summarizes the role of litter fall in nutrient accretion process in agroforestry systems.

22.2 Litter

In the ecological perspective, the term "litter" has two different meanings: (i) the layer of plant materials available on surface soil and (ii) residues (leaves, twigs, stems, etc.) separated from a plant. A layer of litter is different from the mineral layer as the former comprises of dead plant materials. It is hard to identify the accurate starting point of litter decomposition that is separated from a living plant (Cornelissen et al. 2017). A dead stem of a tree may have led to partial decaying before it falls on the soil. Similarly, sometimes whole tree may die and decompose entirely before it falls to ground. A significant amount of litter that falls on the forest floor has a major impact on the dynamics of the forest ecosystem (Chandra and Bhardwaj 2015).

22.3 Decomposition of Litter

The role of litter fall decomposition is very important in the nutrient balance of an agroforestry system, where vegetation is highly subjected by plant nutrient recycling from plant litter fall (Nonghuloo et al. 2020). The decomposition of litter fall goes through breakdown of organic matter (OM) and released as CO_2 in the atmosphere via. Faunal respiration. Plant nutrients pass through physical, chemical, and biological pathways during the formation of complex compounds to simple forms. The rate of litter fall decomposition is crucial to buildup of carbon and nutrient stocks in any ecosystem. Slow decomposition promotes higher accumulation of OM and nutrients in soil, while a high rate of decomposition accelerates the uptake of plant nutrients, as the availability of nutrients is increased. Climatic factors, including precipitation, temperature, and seasonal changes, may affect the presence of soil organisms those considerably influence decomposition rate. The type of litter directly affects the presence and actions of soil living organisms throughout the decomposition process (Wagg et al. 2014). Litter decomposition has direct impact on soil health which is critical for plant nutrient availability, as listed in Table 22.1.

Table 22.1 Direct effect of litter decomposition on soil health

Effect	Mechanism
Reduction in excess temperature of soil	Litter capture light which decreases soil temperature (Udawatta et al. 2021).
Reduction in evapotranspiration from the soil	Forms a blockade to moisture diffusion and reduce temperature (Liu et al. 2015).
Reduce water requirement	Litter fall hold a significant portion of precipitation (Zagyvai-Kiss et al. 2019).
Increase of CO_2 in soil	Litter fall increase microbial decomposition, which adds more CO_2 in soil (Hsieh et al. 2016).

22.4 Major Factors Affecting Litter Fall Decomposition

The process of litter fall decomposition comprises of two concurrent steps: first, the mineralization as well as humification of organic compounds through the activities of soil microbes, and second, the carbon and nitrogen of organic compounds are slowly mineralized (Yan et al. 2018). These processes largely depend on factors such as moisture, temperature, contents of litter fall, and soil living organisms (García-Palacios et al. 2016). The physio-chemical status of ecosystem, litter fall content, and microbial population are the three major factors leading litter fall decomposition (Yue et al. 2018). Temperature is regarded as one of the most vital factors influencing all the biological processes involving the litter fall degradation. Thus, the rate of litter fall decomposition can effect by a slight chance of temperature. It is well known that the rise of soil microbial actions is parallel with temperature increase up to a certain limit in a specific environment. Along with climate, the role of chemical composition of litter fall in decomposition was reported by some studies. The freshly detached litters from plants are a readily accessible content for soil living fauna. The quality of litter fall has direct influence on the decomposition, as decomposition rate usually decreases during the decaying process due to the elimination of readily available carbon and the accumulated nutrients. According to Liu et al. (2010), different forms of leaf litter have influence on the rate of decomposition and soil-dwelling microorganisms. The major factors that effect on litter decomposition are drawn in Fig. 22.1.

22.5 Role of Trees and Litter Quality

In the agroforestry system, plant parts fall on the ground as organic deposits called litter, which makes a major share of OM in the soils (Bargali et al. 2018). Litter fall comprises several groups of organic compounds. Among these organic compound groups, there are four most important soluble classes present in litter: sugars, glycerides, phenolics, and hydrocarbons. The soluble sugars are hard to metabolize, which include mostly mono and oligosaccharides. The composition of these organic

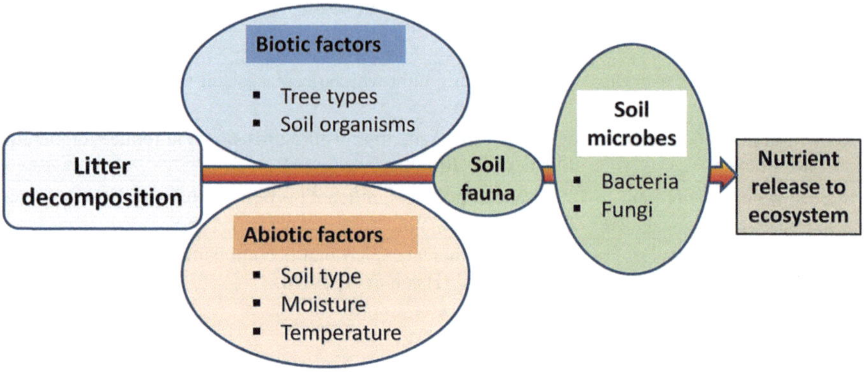

Fig. 22.1 Major factors influence litter decomposition

compounds depends on their origin of plant species as well as part (leaf, bark, and stem). The quality of litter fall is determined by the presence of nutrients and cell wall contents that affect the degradation rate and nutrient release. The primary nutrients consider for evaluating quality of liter are nitrogen, phosphorous, and potassium. Lignin, cellulose, and hemicelluloses are major constituents of cell wall influence decaying of litter. Lignin constitutes for around 20%–45% of the total litter production, but can exceeds 60% or below 5% depending on type and growth stage of plants in the system (Le Floch et al. 2015). Lignins are more flexible molecules than cellulose and their structure varies with the kind of plants. Deciduous plants have irregular shares of guaiacyl and syringyl lignin, whereas conifers species commonly have guaiacyl lignin. Cellulose and hemicellulose are two other important carbohydrates present in litter in terms of amount. Celluloses are polymers of sugars comprising about 10%–50% of the total litter volume which largely differ and depends on litter origin. Hemicelluloses quantities also vary in litter to litter related to plant species (Schäfer et al. 2019). Variations in the litter properties, such as leaf hardiness, lignin, cellulose, nitrogen contents, and the C/N ratio, are the main causes of differences in decaying rate. Out of all these characters, lignin and nitrogen concentration of litter fall are the most considerable factors responsible for decomposition. Along with the quantity of nitrogen and lignin, the lignin/nitrogen ration is another important deciding factor of decaying rate. The low C/N ratio and high lignin contents slower the decaying rate, whereas high nitrogen, phosphorous, and potassium concentration accelerates decomposition. Through the evaluation of the relationship between litter volume and decomposition, litter characters can be used to predict the decomposition rate of specific plant species and can be helpful in developing biogeochemical models. The degradation of coniferous tree leaves are slower than deciduous tree leaves, as the later have broad leaves rich in phosphorus and potassium, more cellulose and less lignin presence. The decaying of *Simaroubha* litter fall was quicker over *Litchi* or *Kadamb* litter fall; furthermore, leaf litter fall disappeared much faster than branches or twigs (Sarkar and Das 2020). Litter fall under agroforestry canopy is generally softer and vanishes faster over litter fall exposed to sun. This is mainly due to the environment developed under agroforestry canopy, which is highly suitable for the activities of decomposers. The seasonal variations of litter fall decomposition of same species in a specific location were reported (Strickland et al. 2015). Climatic differences can be a foremost factor effecting litter fall degradation on a large regional scale. The contents of nutrients greatly differ from species to species (Zukswert and Prescott 2017). The litter fall of *Simaroubha* has more contents of N (1.6%) over the leaf litter of *Litchi* (0.7%). Therefore, the type of plants present in agroforestry systems has direct link with the nutrient accretion of litter fall (Sarkar and Das 2020).

22.6 Role of Soil Flora and Fauna

The number and proportion of soil flora and fauna are recognized to have direct effect on the litter breakdown at different stages of degradation. Microbial degradation of detached plant parts on the agroforestry floor has important influence on soil carbon as well as energy pathway in the environment. Some species of microbes are foremost decomposer resulting nutrient release in soil. Species diversity of soil fungi is lower over the bacteria in agroforestry system, because of fast bacterial growth and multiplication. In 1 g of soil, the number of bacterial species can be thousands, while the total number of species is higher than four million (Lu et al. 2017). Fungi are the principal decomposer with 80% higher potentiality to vanish plant debris compared to other microbes (Pausch et al. 2016). Although, fungal activities vary from season to season. In addition to fungi, bacteria also contribute significantly to the mineralization of organic matter and make up roughly 27% of the soil's total microbial biomass (Van Leeuwen et al. 2017). Litter degradation by bacteria and fungi is generally faster at nutrient-rich environment and could react negatively in response to stress conditions. The multiplication of microbes can be stunt by moisture scarcity. The rise of temperature makes the role of soil moisture more crucial to retain higher microbial activities. Thus, the litter decaying increases with rising of temperature as well as moisture. Microbial activities, especially fungi, on the plant litter, may start before detached from plants, but the actual decomposition initiates only after the litter reaches the ground. Their actions in relation to litter are greatly influenced by the features of the litter, the kind of soil, and variations in these properties over the period of time. Litter decaying is further affected by the amount and nutrient content of the litter input, which mainly depends on species of the plants. Apart from these microbes, soil biota are also contents of invertebrates. Arthropods stay alive in the litter mass and surface soil, acting as a vital role in the ecosystems by helping plant residue decomposition and mineralization of nutrients. Soil faunal actions mostly assist to acclimatize the litter and stimulate microbial activities. The labile parts of litter may be taken by soil microorganisms, hence it is susceptible to fast degradation. The structural labile compounds, such as cellulose has goes through fast split by enzymes into sugars, which further readily taken by microorganisms. Though other structural compounds, such as lignin is too huge to pass via cell membranes, thus stay unaffected to cellular enzymes because of their rough chemical structure and complex formation. Earthworms have huge ability to decompose litter fall, which may be two to three times quicker over soil living small invertebrates (Ulyshen 2016).

22.7 Decomposition Processes of Polymers in Litter

22.7.1 *Cellulose*

Cellulose is structured in a crystal form in the plant fiber, making it hard to breakdown by microbes. Cellulose is degraded by microbes using cellular enzymes. At first, it is decomposed to oligomers or monomers of glucose, which further involved into the microbial metabolization. Several soil organisms have capability of decomposing the more amorphous type of cellulose. The wood decaying fungi and white-rot basidiomycete have been used to deteriorate lignocellulosic materials (Miyauchi et al. 2020). The enzymes responsible for degradation are different in nature as well as have specific characteristics. The glucanases have ability to degrade crystal like cellulose. Cellobiose dehydrogenase is present in a range of fungi and plays role in lignin and cellulose decomposition. The bacterial decomposition of cellulose is generally hydrolytic, though the process is different from fungal decomposition. In case of bacteria, decomposition takes place by a group of cellulolytic enzymes through collective mechanism. The foremost bacterial species those are able of use cellulose are *Bacillus*, *Clostridium*, *Achromobacter*, *Cytophaga*, *Pseudomonas*, *Cellulomonas*, *Cellvibrio*, and *Sporocytophaga* (Mazzoli et al. 2018; Chukwuma et al. 2021). The process use by actinomycetes to degrade the cellulose is similar to fungi and they can decompose the crystal-like form of cellulose, although many bacterial species have the capability to decompose the lignocelluloses-like complex compounds. The presence of cellulose can act like a best stimulator; however, glucose suppresses the cellulose production.

22.7.2 *Hemicelluloses*

The total assimilation of hemicelluloses ranges from 25% to 30% in wood (Ulyshen 2016). There are huge structural differences between the hemicelluloses present in litters of softwood and hardwood. The hemicelluloses are comprised of both linear and branched heteropolymers. Decomposition of hemicelluloses also needs supplementary complex enzyme systems for cellulose hydrolysis. The decaying of these molecules needs the collective actions of different hydrolytic enzymes. Important bacteria responsible in the use of hemicellulose are *Achromobacter*, *Pseudomonas*, *Bacillus*, *Vibrio*, *Streptomyces*, and *Lactobacillus* (Chukwuma et al. 2021).

22.7.3 Lignin

The decomposition of lignin differs among the three main groups of decomposers, that is, white, soft, and brown-rot fungi. So far, most of the different enzymatic processes of lignin decomposition are not well defined, except white-rot fungi (Kuuskeri et al. 2016). White-rot fungi can completely mineralize lignin to CO_2 and water, resulting in the faster decomposition of lignin by lignocellulosic complex. On the other hand, brown-rot fungi generally degrade the cellulose and hemicelluloses compositions in wood, but have the ability to generate lignin decomposition. The soft-rot fungi unable to degrade lignin, but they soften wood through break down of the cell wall.

22.8 The Effect of Soil Nutrient on Litter Decomposition Rate

It has been extensively accepted that the influence of soil nutrients status to the litter degradation and majority of the findings of soil nutrient accessibility related to litter degradation are indirectly studied, such as litter fall generally immobilizes nutrients at the time of initial decomposition process, signifying that fresh litter comprises inadequate amount of nutrients required for the growth, and development of microbes responsible for decomposition (Zhou et al. 2015). Soil nutrient restricting litter degradation suggested that the rate of litter decomposition is positively linked with its nutrient presence. In comparison to crop land, agroforestry system has faster decomposition rate, resulting in easy accumulation of nutrients (particularly nitrogen and phosphorous) in initial stages of decomposition. Litter fall decay rate in bare land is slow mainly due to nutrient poor soil with high C/N ratio (Werner and Homeier 2015). The observation by Zhou et al. (2015) recommended that the loss of tree residue and nutrients release were more from agroforestry system over farmland when considering on several litter fall features and decomposition rate at several sites in southern China. Though, nutrient pattern do limit degradation in some forest systems suggested by findings that have mainly evaluated nutrient limitation directly by analyzing the results of degrading litter to increase nutrient release. The application of fertilizer can enhance litter decomposition or shows no influence on the decomposition rate or even reduces decomposition (Tongkaemkaew et al. 2018). The application of nitrogen may stimulate the degradation of lignin by accelerating the production of ligninolytic enzymes. Nitrogen use is more effective when lignin holds a large part of the litter, but, in general, agroforestry system litter predominantly comprised of leaves, twigs, and young stems which contain more labile fractions of litter. In general, an increase of nutrient supply reduces amounts of immobilized nitrogen and phosphorous, while increase their release from litter fall. To understand the influence of nutrient addition through litter degradation on carbon emission is crucial, which also helps for a better understanding of the local carbon cycle. It is not necessary that addition of nutrients will enhance fast litter decay rate; indeed, it can increase the release of nutrients. Therefore, the litter production in an ecosystem is positively correlated with nutrient supply and the sustainability of nutrient cycles. Recent findings suggested that there is significant influence of nutrient use efficiency of forest by the leaf seasonal litter fall dynamics.

22.9 Litter Nutrient Release Influenced by Soil Nutrient

The content of nitrogen, phosphorus, and mineralization of soil nutrients were linked with nutrient discharge of litter, though, the magnitude of quality is hard to predict, as the nutrients concentration in different ageing leaves varies, nutrient leaching, and plant species diversity (Del Giudice and Lindo 2017). The availability of nutrients in the soil, especially inadequate amounts of nitrogen and phosphorus has an impact on the production of agroforestry litter. There is evidence that nitrogen mineralization and litter decomposition are unrelated, despite the fact that nitrogen mineralization was influenced by litter deterioration (Zhang et al. 2010). The net nitrogen mineralization is considerably linked with the tree litter fall deterioration, as nitrogen mineralization may influence the litter composition or decomposer population at the period of decomposition. Nitrogen fertilization inspired more degradation of the low lignin content litters over the high lignin content litters. Nitrogen addition through human activities may raise litter decaying process more in agroforestry systems with low-lignin litter over those with high-lignin litter. The chemical properties of upper soil are influenced by the nitrogen and phosphorous release pattern. The role of fertilization in litter nutrient release is stronger over its influence in degradation. Hence, the supply of soil nutrients has direct influence on the quality of litter. Though, the influence of the anthropogenic soil nutrient addition to litter decay process may stimulate differently in different decomposer groups. A healthy soil may generate faster litter decomposition process over a degraded soil. The quantity of litter input and its direct association with the soil has more impact of humus production over only litter quantity.

22.10 Litter Decomposition and Nutrient Cycling

In terrestrial ecosystems, organic matter degradation in soil is a vital process of plant nutrient cycling. The soil nutrients of the litter fall are uptake by plants, continuing nutrient cycling and providing nutrients to the ecosystem. The total quantity and composition of litter fall materials, as well as the chemical and physical condition of the surrounding environment, are the main factors that control how organic matter is transformed. The type and numbers of decomposers in the soil have influence on the decomposition rate and nutrient flows of litter fall. The microbial community, their respiration rate, and specific soil chemical properties indicate the conversion

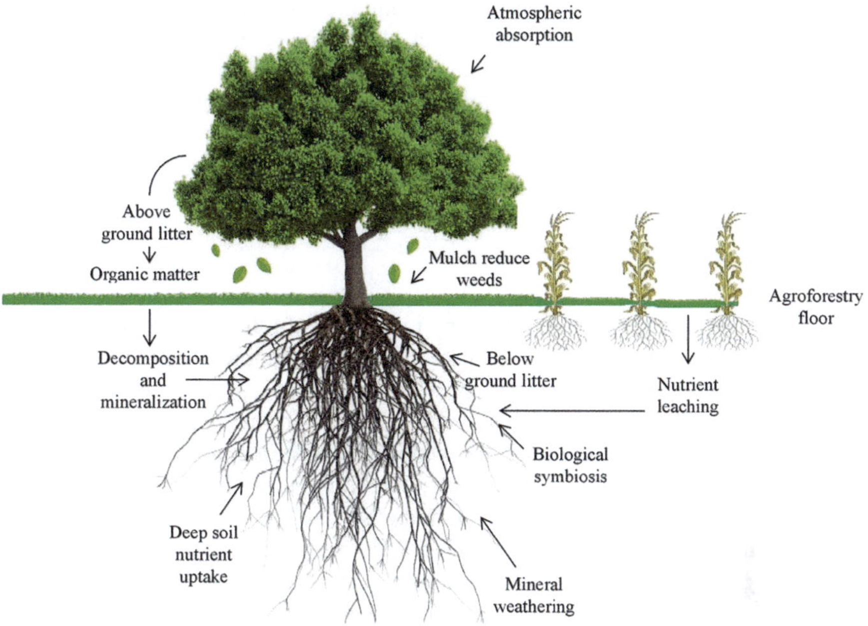

Fig. 22.2 Nutrient cycling in agroforestry system

progress that takes place in agroforestry litters. In addition, abiotic factors and human actions affect the activities of soil microbial enzymes. The nutrients released in the period of litter degradation can be as high as 85% of the total requirement for trees (Zeng et al. 2018). The litter degradation is directly linked with microbial actions that modify the chemical components of litter and control the dynamics of carbon as well as nitrogenin soil. Microorganisms' activities on plant biomass alter the chemical composition of soil organic matter and have an impact on nutrient immobilization. Figure 22.2 illustrates the role of litter degradation plays in recycling of nutrients under agroforestry system.

22.10.1 Carbon Cycle

Microbial biomass consists of less than 3% of the total organic carbon in soils (Zhang et al. 2020). The rate of decomposition of humus in an agroforestry system is low compared to an agricultural land. Carbon compounds can be broke down by the enzymes of microorganisms depending on the composition of litter. The decomposition of leaf litter in agroforestry soils results in large amounts of dissolved organic carbon complexes. Leaching is a major factor in the total carbon losses. Under tree cover less carbon is lost by soil erosion. This suggested that the development of agroforestry systems may slow down decomposition and increase carbon storage. Deforestation can increase the amount of carbon and nitrogen in the soil for a very short period of time, due to the quick absorption of smaller plant parts into the soil facilitating release of the carbon and nutrients in the soil. Table 22.2 presents a list of studies on the impact of tree litter fall on nutrient dynamics in an agroforestry system.

22.10.2 Nitrogen Cycle

The litter sourced nitrogen consumed by microbes is further mobilized. The mineralized nitrogen is back to soil and uptake by plants when microbes die. Only around 1%–2% of the organic nitrogen present in soil is mineralized throughout the process (Salmon et al. 2018). Nitrogen is directly related with the soil organic matter that comprises around 5% of the total nitrogen in soil (Clivot et al. 2017). This organic nitrogen is unable to uptake by plants; therefore, the microbes decompose the organic litter into minor parts through the release of ammonium. The organic nitrogen mineralization in forest covered soils is a regular process and is usually assisted by microbial actions due to the lesser presence of organic nitrogen. Thus, soil nitrogen has limitations to consider as a main nitrogen fraction. Amino acids, polyphenols, and other nitrogenous compounds are broken down into smaller particles with a lesser surface area accessible for enzyme activities, which is one of the main limitations in the decomposition of organic nitrogenous compounds. In addition, the physical absorption of humus by clay particles decreases the active classes of the humus proteins those are inaccessible to microbial proteases. Also, the majority of the soil organic matter located in the soil pores is too small to be obtainable for microorganisms and the arrangement of humic substances are so irregular that there is lesser chance to meet the particular bonds by exact enzymes (Tripathi et al. 2010).

22.10.3 Phosphorous

Phosphorus is second most regulating nutrient after nitrogen. The concentration of phosphorous in litter rises throughout the decomposition process. Though, initially the concentration is reduced because of leaching. Litter fall degradation delivers very low amount of orthophosphate to the plants. Organic acids produced by microbial decomposition of plant litter fall may assemble locally to increase concentrations that can ultimately raise the availability of phosphates to plants. The mineralization of organic phosphorus is an important phenomenon of phosphorous retrieval (Bünemann 2015).

Table 22.2 Effects of tree litter fall on nutrient accumulation under agroforestry systems

Major tree species in agroforestry system	Soil carbon	Nitrogen	Phosphorous	Potassium	Calcium	Magnesium	Sodium	Location	References
Poplar	Positive	Positive	Positive	Positive	Positive	Positive	Positive	Belgium	Pardon et al. (2017)
Rubber	Positive	Positive	Positive	Positive	Positive	Positive		China	Wu et al. (2020)
Coffee–Banana	Positive	Positive	Positive	Positive				Uganda	Zake et al. (2015)
Cacao and Plum		Positive		Positive	Positive	Positive		Cameroon	Munjeb et al. (2018)
Cacao		Positive	Positive	Positive	Positive	Positive		Ghana	Borden et al. (2020)
Bamboo	Positive	Positive	Positive	Positive				India	Nath et al. (2015)
Mahogany		Positive	Positive					Mexico	Falkowski et al. (2016)
Ziziphus	Positive							West Africa	Bado et al. (2021)
Acacia							Negative	India	Behera et al. (2015)
Poplar	Positive	Positive	Positive	Positive				India	Sirohi and Bangarwa (2017)
Multiple	Positive							India	Ramesh et al. (2015)
Multiple		Positive						Nepal	Schwab et al. (2015)
Rubber	Positive		Positive	Positive	Positive	Positive		India	Jessy et al. (2017)
Cacao			Positive	Positive		Positive		Peru	Arévalo-Gardini et al. (2015)
Faidherbia	Soil carbon	Nitrogen	Phosphorous	Potassium				Zambia	Yengwe et al. (2018)

22.10.4 Potassium

The potassium that comes from plant litter generally not accumulates in surface horizons. Though, the quality and quantity of litter decompose may influence its availability to plants over the effect of the left over organic matter on the cation exchange capability of the soil. Potassium and magnesium are vital nutrients for larger vegetation including crops and trees, though, they hardly negatively affect microbial activities and are rapidly released from decompose litter. The existence of tree roots in agroforestry system on the ground reduce leaching losses of plant nutrients through the direct uptake of nutrients. Microorganisms positively affect the nutrient cycling efficiency and other mechanisms in ecosystem. Among the soil living microbes, fungi and bacteria have comparatively higher contribution in the release of nutrients including potassium (Rawat et al. 2016). The microbial population, their respiratory activities, metabolic activities, and soil composites determine the transformation of forest litters under particular soils.

22.11 Carbon/Nitrogen Ratio of the Plant Litter and Its Decomposition

The decomposition of leaf litter can be estimated from the carbon/nitrogen (C/N) ratio. Leaves contain high nutrients are usually decomposed faster than leaves poor in nutrients. The decomposition rate of litter is higher in plants with low C/N ratios and lignin contents. The C/N ratio and nitrogen concentration are very strongly linked with litter decomposition. It is also regarded that the C/P ratio and phosphorous concentration are good forecasters of litter degradation rates (Chen et al. 2016). In addition, the lignin content and the lignin/N ratios in litter are also fair indicators of litter decaying rate and nutrient discharge. Although, all these indicators and their influences on litter decomposition and nutrient release contingent with soil properties and plant species.

22.12 Biological Nitrogen Fixation

Leguminous trees under agroforestry systems improve soil by adding nitrogen through biological fixation. Trees, for example, *Acacia* species have been found to have ability to fix annually 250 kg of nitrogen per hectare. This nitrogen helps symbiotically to the crop growth as well as improves the fertility status of soil. The quantity of nitrogen supplied by the legumes trees uptake up by the first crop is found pretty less and major share is remain in the soil signifying advantage of long-term supply of nitrogen over short-term. Various tree parts such as, leaf, branch, wood, and bark have diverse decay rates which support to allocate the nutrient release over

Table 22.3 Important nitrogen fixing tree species

Common name	Botanical name	Family	Nitrogen fixed (kg N/ha/year)
Whistling pine, Shea	*Casurina equisetifolia*	Casuarinaceae	60–110
Black wattle	*Acacia mearnsii*	Mimosoideae	200
Coral tree, Poro	*Erythrina poeppigiana*	Pipil[onaceae	60
Wild tamarind, Subabul	*Leucaena leucocephala*	Mimosoideae	100–500
Ice cream bean, Inga	*Inga jincicuil*	Mimosoideae	34–50
Maxican lalic	*Gliricidia sepium*	Fabaceae	13
Indian alder	*Alnusnepalensis*	Betulaceae	–

Source: Misra (2011)

time. Some of the major nitrogen fixing tree species are listed in Table 22.3. Biologically, fixation of nitrogen may occur through either symbiotic or nonsymbiotic mechanisms. The symbiotic fixation of nitrogen takes place by the association of nitrogen fixing microbes and tree roots. There are several legume species involved in symbiosis with rhizobium bacteria, though the association of nonleguminous bacteria species like actinomycetes are also important. Nonsymbiotic nitrogen fixation is influenced by free-living soil microbes and plays a considerable aspect in agroforestry ecosystems, which have comparatively lower chance to get nitrogen from outside systems (Soumare et al. 2020). Some of the important nitrogen fixing trees suitable for agroforestry are listed in Table 22.3.

22.13 Summary

The accretion of plant nutrient release through litter fall decomposition is important for the sustainability of ecosystems, as it is one of the main ways to recover nutrients. The nutrient accretion and degradation rate of litter are in equilibrium with the nutrient absorption in an ecosystem, which differ based on the kind of ecosystems. The discharge of plant nutrients via litter fall degradation is a very complex process that includes several biotic and abiotic factors. Though, there is limited information regarding the role of different factors on litter decay and nutrient release in agroforestry systems. Moreover, it is hard to identify the correct rate of litter decomposition, since it is subjective to a number of totally dissimilar factors. There is yet to develop a method to evaluate the rate of litter fall decomposition that includes all the factors. This will also help to define the nutrient release pattern of litter fall in a specific agroforestry system. Besides, it is important to understand litter fall decomposition and nutrient release pattern in the perspective of rising human impacts on biochemical cycles. This chapter discusses on a range of factors that influence the litter decomposition and nutrient release patterns with respect to agroforestry system. It extends our understanding on degradation as well as nutrient release of litter fall under in agroforestry system based on the past studies.

References

Arévalo-Gardini E, Canto M, Alegre J, Loli O, Julca A, Baligar V (2015) Changes in soil physical and chemical properties in long term improved natural and traditional agroforestry management systems of cacao genotypes in Peruvian Amazon. PLoS One 10(7):e0132147

Bado BV, Whitbread A, Manzo MLS (2021) Improving agricultural productivity using agroforestry systems: Performance of millet, cowpea, and ziziphus-based cropping systems in West Africa Sahel. Agric Ecosyst Environ 305:107175

Bargali K, Manral V, Padalia K, Bargali SS, Upadhyay VP (2018) Effect of vegetation type and season on microbial biomass carbon in Central Himalayan forest soils, India. Catena 171:125–135

Behera L, Nayak MR, Patel D, Mehta A, Sinha SK, Gunaga R (2015) Agroforestry practices for physiological amelioration of salt affected soils. J Plant Stress Physiol 1:13–18

Borden KA, Thomas SC, Isaac ME (2020) Variation in fine root traits reveals nutrient-specific acquisition strategies in agroforestry systems. Plant Soil 453(1):139–151

Bünemann EK (2015) Assessment of gross and net mineralization rates of soil organic phosphorus–A review. Soil Biol Biochem 89:82–98

Chandra KK, Bhardwaj AK (2015) Incidence of forest fire in India and its effect on terrestrial ecosystem dynamics, nutrient and microbial status of soil. Int J Agric For 5(2):69–78

Chen Y, Sayer EJ, Li Z, Mo Q, Li Y, Ding Y, Wang J, Lu X, Tang J, Wang F (2016) Nutrient limitation of woody debris decomposition in a tropical forest: contrasting effects of N and P addition. Funct Ecol 30(2):295–304

Chukwuma OB, Rafatullah M, Tajarudin HA, Ismail N (2021) A review on bacterial contribution to lignocellulose breakdown into useful bio-products. Int J Environ Res Public Health 18(11):6001

Clivot H, Mary B, Valé M, Cohan JP, Champolivier L, Piraux F, Laurent F, Justes E (2017) Quantifying in situ and modeling net nitrogen mineralization from soil organic matter in arable cropping systems. Soil Biol Biochem 111:44–59

Cornelissen JH, Grootemaat S, Verheijen LM, Cornwell WK, van Bodegom PM, van der Wal R, Aerts R (2017) Are litter decomposition and fire linked through plant species traits? New Phytol 216(3):653–669

Del Giudice R, Lindo Z (2017) Short-term leaching dynamics of three peatland plant species reveals how shifts in plant communities may affect decomposition processes. Geoderma 285:110–116

Falkowski TB, Diemont SA, Chankin A, Douterlungne D (2016) Lacandon Maya traditional ecological knowledge and rainforest restoration: soil fertility beneath six agroforestry system trees. Ecol Eng 92:210–217

García-Palacios P, Shaw EA, Wall DH, Hättenschwiler S (2016) Temporal dynamics of biotic and abiotic drivers of litter decomposition. Ecol Lett 19(5):554–563

Hsieh IF, Kume T, Lin MY, Cheng CH, Miki T (2016) Characteristics of soil CO_2 efflux under an invasive species, Moso bamboo, in forests of central Taiwan. Trees 30(5):1749–1759

Jessy MD, Joseph P, George S (2017) Possibilities of diverse rubber based agroforestry systems for smallholdings in India. Agrofor Syst 91(3):515–526

Kotowska MM, Leuschner C, Triadiati T, Hertel D (2016) Conversion of tropical lowland forest reduces nutrient return through litterfall, and alters nutrient use efficiency and seasonality of net primary production. Oecologia 180(2):601–618

Kuuskeri J, Häkkinen M, Laine P, Smolander OP, Tamene F, Miettinen S, Nousiainen P, Kemell M, Auvinen P, Lundell T (2016) Time-scale dynamics of proteome and transcriptome of the

white-rot fungus Phlebiaradiata: growth on spruce wood and decay effect on lignocellulose. Biotechnol Biofuels 9(1):1–22
Le Floch A, Jourdes M, Teissedre PL (2015) Polysaccharides and lignin from oak wood used in cooperage: Composition, interest, assays: a review. Carbohydr Res 417:94–102
Liu Y, Wang S, Wang Q, Zhang J (2010) Effects of mixed-species leaf litter on litter decomposition and soil microbial communities in experimental subtropical plantation forest. J Food Agric Environ 8(3&4):1102–1107
Liu D, Ogaya R, Barbeta A, Yang X, Peñuelas J (2015) Contrasting impacts of continuous moderate drought and episodic severe droughts on the aboveground-biomass increment and litterfall of three coexisting Mediterranean woody species. Glob Chang Biol 21(11):4196–4209
Lu J, Breitwieser FP, Thielen P, Salzberg SL (2017) Bracken: estimating species abundance in metagenomics data. Peer J Comput Sci 3:e104
Mazzoli R, Giuffrida MG, Pessione E (2018) Back to the past: "find the guilty bug—microorganisms involved in the biodeterioration of archeological and historical artifacts". Appl Microbiol Biotechnol 102(15):6393–6407
Misra PK (2011) Soil fertility management in agroforestry system. Int J Biotechnol Biochem 7(5): 637–644
Miyauchi S, Hage H, Drula E, Lesage-Meessen L, Berrin JG, Navarro D, Favel A, Chaduli D, Grisel S, Haon M, Piumi F, Levasseur A, Lomascolo A, Ahrendt S, Barry K, LaButti KM, Chevret D, Daum C, Mariette J, Klopp C, Cullen D, de Vries RP, Gathman AC, Hainaut M, Henrissat B, Hildén KS, Kües U, Lilly W, Lipzen A, Mäkelä MR, Martinez AT, Morel-Rouhier M, Morin E, Pangilinan J, Ram AFJ, Wösten HAB, Ruiz-Dueñas FJ, Riley R, Record E, Grigoriev IV, Rosso MN (2020) Conserved white-rot enzymatic mechanism for wood decay in the Basidiomycota genus Pycnoporus. DNA Res. 27(2):dsaa011
Munjeb NL, Kfuban YBP, Louise ATM (2018) Farmers perception of soil and watershed degradation and the assessment of soil nutrients status under agroforestry systems in the Western Highlands of Cameroon: case of Ako sub division. J Soil Sci Environ Manag 9(8):119–126
Nath AJ, Lal R, Das AK (2015) Ethnopedology and soil properties in bamboo (Bambusa sp.) based agroforestry system in North East India. Catena 135:92–99
Nonghuloo IM, Kharbhih S, Suchiang BR, Adhikari D, Upadhaya K, Barik SK (2020) Production, decomposition and nutrient contents of litter in subtropical broad leaved forest surpass those in coniferous forest, Meghalaya. Trop Ecol 61(1):5–12
Pardon P, Reubens B, Reheul D, Mertens J, De Frenne P, Coussement T, Janssens P, Verheyen K (2017) Trees increase soil organic carbon and nutrient availability in temperate agroforestry systems. Agric Ecosyst Environ 247:98–111
Pausch J, Kramer S, Scharroba A, Scheunemann N, Butenschoen O, Kandeler E, Marhan S, Riederer M, Scheu S, Kuzyakov Y, Ruess L (2016) Small but active–Pool size does not matter for carbon incorporation in below-ground food webs. Funct Ecol 30(3):479–489
Ramesh T, Manjaiah KM, Mohopatra KP, Rajasekar K, Ngachan SV (2015) Assessment of soil organic carbon stocks and fractions under different agroforestry systems in subtropical hill agroecosystems of north-east India. Agrofor Syst 89(4):677–690
Rawat J, Sanwal P, Saxena J (2016) Potassium and its role in sustainable agriculture. In: Potassium solubilizing microorganisms for sustainable agriculture. Springer, New Delhi, pp 235–253
Salmon VG, Schädel C, Bracho R, Pegoraro E, Celis G, Mauritz M, Mack MC, Schuur EA (2018) Adding depth to our understanding of nitrogen dynamics in permafrost soils. J Geophys Res Biogeo 123(8):2497–2512
Sarkar S, Das DK (2020) Tree biomass and carbon stock assessment in different agroforestry system trees. Int J Chem Stud 8(4):3858–3861
Schäfer J, Sattler M, Iqbal Y, Lewandowski I, Bunzel M (2019) Characterization of Miscanthus cell wall polymers. GCB Bioenergy 11(1):191–205
Schwab N, Schickhoff U, Fischer E (2015) Transition to agroforestry significantly improves soil quality: a case study in the central mid-hills of Nepal. Agric Ecosyst Environ 205:57–69

Sirohi C, Bangarwa KS (2017) Effect of different spacings of poplar-based agroforestry system on soil chemical properties and nutrient status in Haryana, India. Curr Sci. 113:1403–1407

Soumare A, Diedhiou AG, Thuita M, Hafidi M, Ouhdouch Y, Gopalakrishnan S, Kouisni L (2020) Exploiting biological nitrogen fixation: a route towards a sustainable agriculture. Plan Theory 9(8):1011

Strickland MS, Keiser AD, Bradford MA (2015) Climate history shapes contemporary leaf litter decomposition. Biogeochemistry 122(2):165–174

Tongkaemkaew U, Sukkul J, Sumkhan N, Panklang P, Brauman A, Ismail R (2018) Litterfall, litter decomposition, soil macrofauna, and nutrient contents in rubber monoculture and rubber-based agroforestry plantations. For Soc 2(2):138–149

Tripathi G, Deora R, Singh J (2010) Biological, chemical and biochemical dynamics during litter decomposition at different depths in arable soil. J Ecol Nat Environ 2(3):038–051

Udawatta RP, Anderson SH, Kremer RJ, Garrett HEG (2021) Agroforestry for soil health. North American agroforestry. Springer, Cham, pp 355–386

Ulyshen MD (2016) Wood decomposition as influenced by invertebrates. Biol Rev 91(1):70–85

Van Leeuwen JP, Djukic I, Bloem J, Lehtinen T, Hemerik L, De Ruiter PC, Lair GJ (2017) Effects of land use on soil microbial biomass, activity and community structure at different soil depths in the Danube floodplain. Eur J Soil Biol 79:14–20

Wagg C, Bender SF, Widmer F, Van Der Heijden MG (2014) Soil biodiversity and soil community composition determine ecosystem multifunctionality. Proc Natl Acad Sci 111(14):5266–5270

Werner FA, Homeier J (2015) Is tropical montane forest heterogeneity promoted by a resource-driven feedback cycle? Evidence from nutrient relations, herbivory and litter decomposition along a topographical gradient. Funct Ecol 29(3):430–440

Wu J, Zeng H, Zhao F, Chen C, Liu W, Yang B, Zhang W (2020) Recognizing the role of plant species composition in the modification of soil nutrients and water in rubber agroforestry systems. Sci Total Environ 723:138042

Yan J, Wang L, Hu Y, Tsang YF, Zhang Y, Wu J, Fu X, Sun Y (2018) Plant litter composition selects different soil microbial structures and in turn drives different litter decomposition pattern and soil carbon sequestration capability. Geoderma 319:194–203

Yengwe J, Amalia O, Lungu OI, De Neve S (2018) Quantifying nutrient deposition and yield levels of maize (Zea mays) under Faidherbiaalbida agroforestry system in Zambia. Eur J Agron 99: 148–155

Yue K, García-Palacios P, Parsons SA, Yang W, Peng Y, Tan B, Huang C, Wu F (2018) Assessing the temporal dynamics of aquatic and terrestrial litter decomposition in an alpine forest. Funct Ecol 32(10):2464–2475

Zagyvai-Kiss KA, Kalicz P, Szilágyi J, Gribovszki Z (2019) On the specific water holding capacity of litter for three forest ecosystems in the eastern foothills of the Alps. Agric For Meteorol 278: 107656

Zake J, Pietsch SA, Friedel JK, Zechmeister-Boltenstern S (2015) Can agroforestry improve soil fertility and carbon storage in smallholder banana farming systems? J Plant Nutr Soil Sci 178(2): 237–249

Zeng L, He W, Teng M, Luo X, Yan Z, Huang Z, Zhou Z, Wang P, Xiao W (2018) Effects of mixed leaf litter from predominant afforestation tree species on decomposition rates in the Three Gorges Reservoir, China. Sci Total Environ 639:679–686

Zhang CF, Trofymow JA, Jamieson RC, Meng FR, Gordon R, Bourque CPA (2010) Litter decomposition and nitrogen mineralization from an annual to a monthly model. Ecol Model 221(16):1944–1953

Zhang Q, Jia X, Wei X, Shao M, Li T, Yu Q (2020) Total soil organic carbon increases but becomes more labile after afforestation in China's Loess Plateau. For Ecol Manag 461:117911

Zhou Y, Clark M, Su J, Xiao C (2015) Litter decomposition and soil microbial community composition in three Korean pine (Pinus koraiensis) forests along an altitudinal gradient. Plant Soil 386(1):171–183

Zukswert JM, Prescott CE (2017) Relationships among leaf functional traits, litter traits, and mass loss during early phases of leaf litter decomposition in 12 woody plant species. Oecologia 185(2):305–316

Chapter 23
Agroforestry: A Key Technique for Achieving the Sustainable Development Goals

Sumanta Bhattacharya

Abstract Climate change is the most important issue of the twenty-first century, and it's being discussed in every country. The growth of heat waves, changes in rainfall patterns, the rise of sea levels, desertification, land degradation, drought and floods have all been caused by climate change. Agriculture accounts for 30% of the total greenhouse gas emissions in the world today. As a result of climate change, food shortages and water shortages will become more common. The food problem is a direct result of deforestation and the overuse of natural resources. Agroforestry has emerged as a key component in addressing both the food issue and climate change in the modern world. In the tropical regions of Brazil, Kenya and Indonesia, farmers are paid to grow trees as part of agroforestry programmes. Farming based on agroforestry can provide food security while simultaneously providing a healthy diet. Practicing agroforestry will help recharge groundwater, avoid soil erosion and degradation and lessen the impact of natural disasters. It will reduce poverty, provide food security, produce money and empower tribal communities and rural populations. There is a growing demand for Ayurvedic medicine in India as the country attempts to enhance its forest cover from 23% to 33% through the use of kitchen gardens and agroforestry. This would help the country's commerce industry. Given its location, India's agricultural potential is greatly enhanced by agroforestry. With the majority of its population concentrated in rural areas, India is predominantly an agrarian society, and modern technology plays a crucial role in ensuring food security. In contrast, agroforestry can fill in the gaps where modern technology has not yet reached the agriculture sector. Ecological balance and proper biodiversity can also be achieved through agroforestry. There are around 300 million rural Indians who are completely reliant on forests and other natural resources for their food, clothing and shelter, and agroforestry cultivation in Jharkhand has given tribal communities the power to self-sufficiency. Rural inhabitants in the Indian subcontinent will benefit greatly from community-based farming with agroforestry, which will help reduce rural–urban migration and improve the socioeconomic and food security of the people in rural areas and sustainable use of Land.

S. Bhattacharya (✉)
MAKAUT, Howrah, West Bengal, India

H. S. Jatav et al. (eds.), *Agroforestry to Combat Global Challenges*, Sustainable Development and Biodiversity 36, https://doi.org/10.1007/978-981-99-7282-1_23

Keywords Agroforestry · Climate change · Sustainable use of land · Natural resources · Food security

23.1 Introduction

Agriculture is the earth's most destructive activity and has had the worst effects on the ecosystem. Deforestation for agriculture was the initial cause of environmental degradation, and as population grew and there was a greater need for food, we began to employ various chemicals to hasten the production of food. The green revolution, which concentrated on growing wheat and rice and benefited just a select states, including Punjab, Haryana and Uttar Pradesh because of their superior irrigation systems and land quality, is an example of this following the Independent. India depends on the monsoons for agriculture since it produces crops that need a lot of water. Our lack of high-quality seeds and excessive use of chemical fertilisers render the land unsuited for further development. Seventy per cent of the water is required for irrigation. We barely have any ground water left due to deforestation since trees could naturally recharge ground water, but this is no longer viable as a result of deforestation. The process of climate change and environmental deterioration has caused significant issues for the rural and tribal community. In tribal areas, depending on the location, Jhum cultivation is practised, which involves growing multiple crops simultaneously in the same field. However, climate change has made this impossible to continue with. Prolonged droughts and floods are adding to the problem as changing rainfall patterns and water logging ruin the quality of the land. Increased food security is accompanied with India's biggest water catastrophe. In order to increase food production, the government of India has implemented new laws and programmes such as soil health cards, technologies for reporting temperature and rainfall, the use of drones in the agriculture sector and new machineries and cultivation methods. However, the lack of land reforms since independence and the unequal distribution of resources, which have worsened poverty in the area and a spike in farmer suicide, are the where technology can be used, yet farmers still use traditional farming techniques because of the country's rapid development. There are small- and medium-sized farms; farmers have extremely small holdings. In fact, agriculture is responsible for 30% of the increase in greenhouse gases and is also to blame for increasing river pollution since its waste water, which contains significant amounts of chemical fertilisers, directly enters rivers and affects the marine life there. The globe is returning to the traditional kind of agriculture known as agroforestry, which is environmentally sustainable, may address the dual issues of a food and water shortage and can also increase biodiversity. When agroforestry was used earlier, issues like climate change and environmental damage did not exist. Rising of the food crisis across the world has been presented in Fig. 23.1.

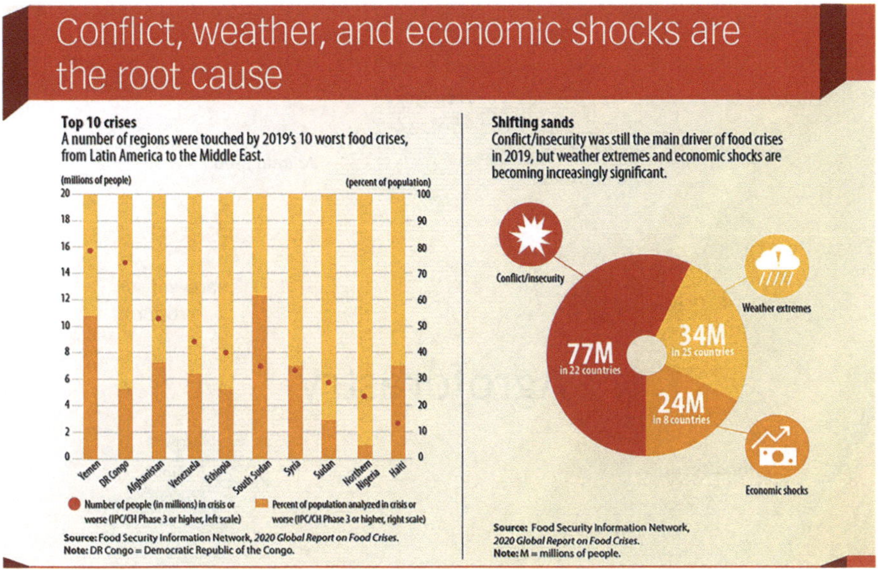

Fig. 23.1 Rising food crisis across the world (Source: Food security Information Network)

23.2 Agroforestry

We need a climate-smart agriculture system that will combat climate change, save the environment, particularly the forest, fisheries and cropland (World Bank 2021). Currently, 75% of the world's population lives in rural areas and depends on agriculture and the forest for their livelihood, which has led to widespread deforestation and resource exploitation. By 2050, our food production needs to expand by 70% in order to keep up with the population growth. In order to implement a smart agriculture system, productivity must be increased, carbon emissions must be reduced and drought, pests and disease vulnerability must be lessened (World Bank 2021). Agroforestry can be defined as growing trees along with the crops which will promote soil health, plant nutrient, crop productivity, tackle climate change, maintain ecological balance and provide animal feed and reduce the use of water for cultivation (Vi Agroforestry n.d.). Agroforestry boosts soil fertility, moisture content and biodiversity, enabling diversified food production and higher yields. With the help of the trees, which also reduce carbon dioxide, agroforestry boosts output while reducing the effects of climate change. The trees give the soil moisture as well as shade, food for the animals, compost and other benefits. A tree used in agroforestry produces more water than it uses. Recreating a natural environment is made easier by agroforestry. Larger harvests, a more favourable environment and enhanced resistance to the effects of climate change are the outcomes. Agroforestry contribution towards sustainability has been presented in Fig. 23.2.

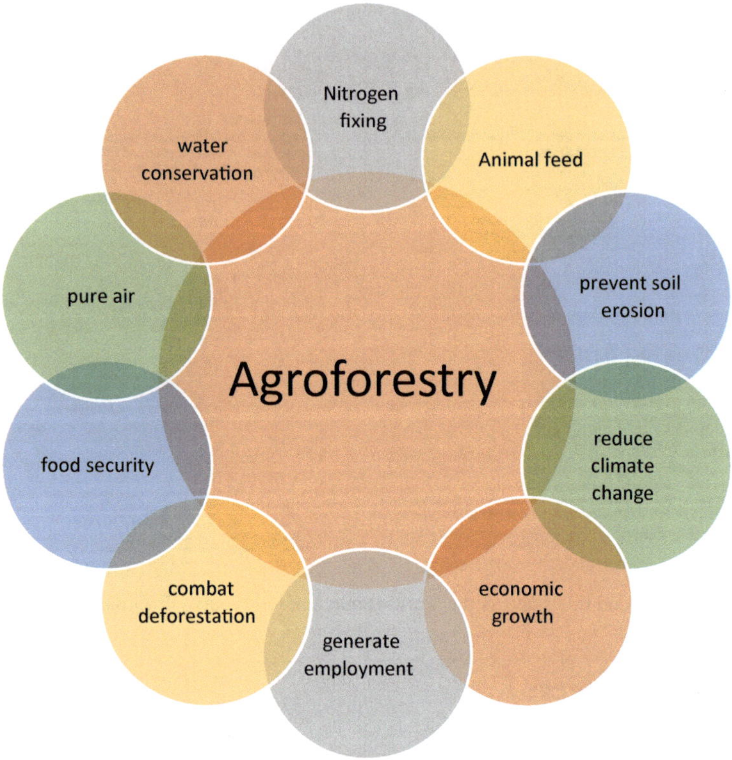

Fig. 23.2 Agroforestry contribution towards sustainability

Especially in the East African region, small farmers are suffering due to climate change, where Vi Agroforestry is providing them with emission profit, contributing to reduce carbon emission. Trees are suffering as firewood and kept throughout the year, different kinds of crops and fruits are grown parallelly and certain trees also have medicinal leaves. Agroforestry can also resolve the problem of water crisis through natural recharge of ground water and prevent soil erosion and natural disasters as trees can control cyclones and floods. Eliminating poverty and increasing the economic activities, trees will provide timber and building materials, reduce pollution from the atmosphere and provide purify air. The whole process is based on Sustainable Land management, with land taken away for development purpose, Agroforestry does the twin work for land management and growing crops, rooftop farming and vertical farming are a product of food insecurity and land management and rise in climate change. South Africa was the first region to start with Agroforestry.

Across the world today, agroforestry has become an important component of cultivating crops, with the change in demand and supply and rapid deforestation, agroforestry is a hope to maintain ecosystem and establish sustainable agriculture practice and sustainable rural development. In India, there is a National Agroforestry

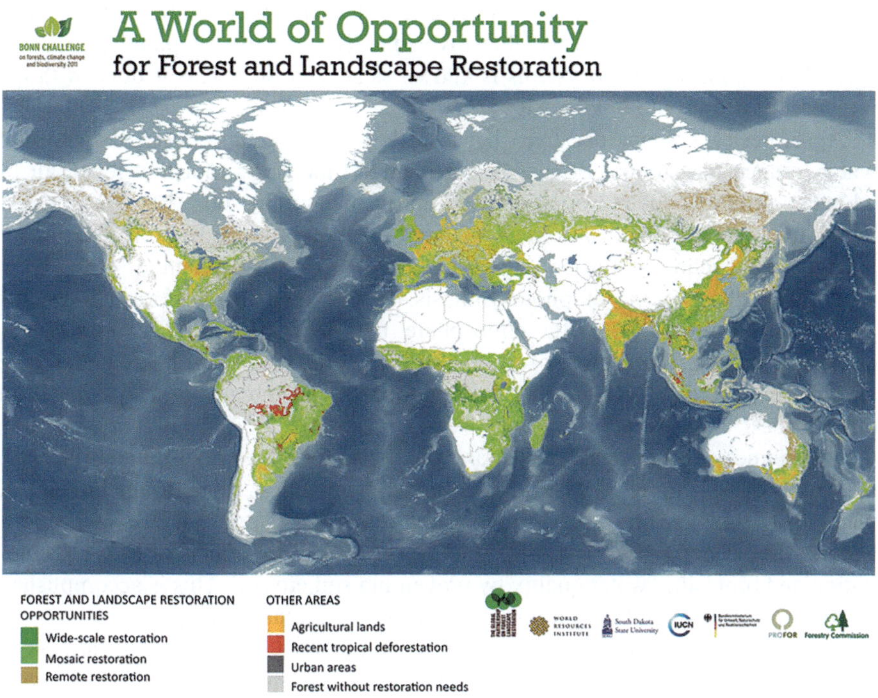

Fig. 23.3 Degraded land across the world (Source: Bonn Challenge)

policy (World Agroforestry 2014). In India, agroforestry is currently implemented on 13.5 million hectares, although there is much more potential. Already, about half of the country's fuel wood and approximately 65% of its timber originate from trees cultivated on farms (Fig. 23.3). It is providing employment to a number of people in the rural region, till now 450 new people have got jobs for the production of Timber. 300 million people depend on forest for their living in rural region.

Agroforestry and sustainable management practices.

Agroforestry can improve the conditions of the tribal community whose land has been taken away for mining and development purpose. Tribal community have been practising agroforestry since ages, there are highly depend on trees for their nutrition and food security where Jhum cultivation is also a kind of agroforestry. Kitchen garden/nutrient gardens mostly practised by rural women are a part of agroforestry. Nutrition literacy is an important part of agroforestry. Agroforestry can lead to nutrition security.

Agroforestry is an important component of sustainable management practises since it involves integrating trees, crops and livestock in a way that is advantageous to all three. This is what makes agroforestry so significant. To build environmentally friendly landscapes, one must draw inspiration from the most successful aspects of agricultural and forestry practises.

A great range of plant and animal life can be found within agroforestry systems, and it is important to preserve this diversity. The practise of planting trees in agriculturally managed ecosystems is known as agroforestry. This is done with the intention of preserving biodiversity.

Agroforestry systems use trees because they mitigate the effects of wind and water, reduce the amount of runoff and maintain the soil's moisture level, all of which contribute to the preservation and improvement of the soil. They contribute to the fertility of the soil in a variety of additional ways, including the fixation of nitrogen, the cycling of nutrients and the deposition of organic matter.

Through the process of photosynthesis, trees grown in agroforestry systems absorb carbon dioxide from the atmosphere, thereby mitigating the consequences of climate change. The capture of carbon through agroforestry has the potential to be an effective strategy for offsetting the emissions of greenhouse gases produced in other places.

Techniques used in agroforestry are beneficial for water management because they reduce the amount of water that is lost to evaporation, increase the amount of water that is replenishing the ground and control the rate at which water runoff occurs. Tree canopies act as natural shields, mitigating the effects of heavy precipitation and protecting water quality by preventing soil erosion. This is accomplished by lowering the amount of sunlight that is able to pass through.

Agroforestry provides economic benefits and resilience by allowing farmers to make money from a larger variety of sources, such as the growing of food crops, lumber, fruits, nuts and medicinal plants. This broadens the farmers' earning potential and increases their financial security. This diversification enhances resistance to the volatility of both the climate and the economy.

Improved food security, nutrition and nutritional diversity are all outcomes of agroforestry systems' commitment to growing a wide range of crops, particularly heirloom varieties, as their primary focus. They provide an alternative to monoculture that is a way of farming that is more robust and sustainable.

A higher standard of living, increased income and reduced costs of living are all possible outcomes for rural communities who use agroforestry practises. The utilisation of trees can enhance the economic stability of farmers because trees give a variety of products that can be marketed and bought and sold.

The establishment of an agroforestry system can provide a number of positive environmental effects, including the production of shade and windbreaks, as well as the regulation of local microclimates. Both crop plants and cattle stand to benefit from these services, which in turn increase the overall productivity and health of the agro-ecosystem.

A sustainable land-use practise known as agroforestry is one that, in addition to boosting agricultural production and improving local ecosystems, also helps to fortify communities. As a result, it is a beneficial instrument for ecologically conscious management as it contributes to striking a balance between agricultural production and environmental protection.

The practise of agroforestry, often known as the 'win-win' strategy, is a type of land management that enables for several uses to be carried out on the same parcel of land. Agriculture has the potential to profit from this land use by generating food and fuel, safeguarding the environment and biodiversity, adapting to or minimising the effects of climate change and conserving the environment. Even though it may appear that implementing this method would call for a significant adjustment to the management philosophy that is now in place, all that is actually required to make it a reality is an increase in the number of trees that are present within the field, either on their own or as part of a certain structural arrangement. The practise of agroforestry, often known as the 'win-win' strategy, is a type of land management that enables for several uses to be carried out on the same parcel of land. Agriculture has the potential to profit from this land use by generating food and fuel, safeguarding the environment and biodiversity, adapting to or minimising the effects of climate change and conserving the environment. It may sound like a system that would require a significant adjustment to the way management is carried out, but all that is required to put it into action is the planting of additional trees in the field. These trees can be planted on their own or as part of a structure such as a shelter belt or buffer strip. One illustration of this would be a shelter belt, often known as a buffer strip. Agroforestry is a form of agriculture that is becoming increasingly common in developing countries, particularly in drier and more arid locations. If the predictions about climate change turn out to be accurate, then British farmers will need to make modifications to their practises in order to keep up with the projected temperature shifts and, at the same time, reduce the environmental damage that their practises may cause. Agroforestry may provide an alternative management strategy for agricultural businesses, which would be beneficial in terms of satisfying their objectives of lowering their impact on the environment and increasing their potential output (Stiles 2017).

23.3 Agroforestry Role in Water Management

By enhancing local water cycles, agroforestry can give farms more control over and access to freshwater. Both the local and global water cycles depend heavily on trees. Through incorporating, it is possible to address water needs and enhance food production systems by planting trees on agricultural land without having a harmful effect on downstream or nearby water users. Climate change and irresponsible land use have consequences.

Agriculture and industry are becoming more of a worldwide problem and a lot of individuals lack access to both food and water insecurity. Around the world, agriculture makes about 70% of freshwater is a factor numerous additional environmental difficulties and consequently, one of the most crucial aspects of modern water and environmental problems (Agroforestry network 2020).

Smallholder farmers are already prevalent throughout much of the world witnessing variations in the patterns of rainfall. This decreased capacity to forecast precipitation and accessibility to both farm productivity and public health are impacted by water.

Climate change requires action to guarantee that everyone has access to sustainable water and sanitation services alteration adaption. Agroforestry provides methods that contribute to greater water security, as well as climate change adaptation and mitigation. Improved fallows, an agroforestry technique, are one instance. Trees have the capacity to increase soil moisture and enrich the soil where they are planted in rotation with cultivated crops while reducing climate change through carbon in the soils sequestration (Agroforestry network 2020).

By increasing infiltration, decreasing runoff and erosion and boosting water availability and quality, agroforestry is an essential component of effective water management.

By decreasing the effects of wind and water, agroforestry systems aid in soil erosion control. Rainfall is absorbed by the tree canopy, slowing the rate at which water runs off the land. Soil erosion is reduced and aggregate development is encouraged because tree roots bind and stabilise the soil. Sedimentation in water bodies can be avoided and aquatic ecosystems can continue to thrive with the help of erosion control measures like these.

Trees in agroforestry systems improve water infiltration into the soil and increase water retention. By breaking up the soil and forming channels, a tree's roots can increase groundwater recharge while decreasing runoff from the surface. Water penetration rates and storage capacity can be greatly improved by incorporating deep-rooted trees into agroforestry systems, such as some agroforestry alley cropping or clavipectoral systems.

Controlling the Flow of Water Agroforestry systems control the flow of water, which is especially useful during times of extreme precipitation. The canopy of trees acts as a sponge, soaking up and dispersing raindrops before they can cause rapid runoff and destructive flash floods. Agroforestry methods reduce soil erosion, shield vital infrastructure and encourage economical water usage because of their ability to control the flow of water.

Trees in agroforestry systems have a significant impact on local climate by moderating wind and water circulation. They block the sun; thus, less water is lost through the soil's surface and evaporation. The tree canopy also mitigates temperature swings, protecting crops from damage caused by high temperatures and decreasing the need for supplemental irrigation. These favourable local weather conditions improve water utilisation efficiency and help conserve water in the larger context.

Reduced runoff from agricultural land into water bodies is one way in which agroforestry systems can improve water quality. Agroforestry systems have tree buffer zones that operate as filters, preventing sediment, fertilisers and agrochemicals from entering groundwater supplies. Trees' root systems help absorb and store nutrients, lowering the likelihood that these substances would seep into nearby water sources and contaminate them.

Agroforestry can be used for river and stream bank restoration, also known as riparian zone restoration. Restoring and stabilising riparian habitats through tree planting helps reduce bank erosion, lowers nutrient runoff and keeps water temperatures down. Improved water quality, protected aquatic habitats and a thriving watershed are all benefits of well-maintained riparian areas.

Agroforestry systems with deep-rooted trees can tap into groundwater and act as a buffer against water scarcity by soaking up extra moisture during dry spells. Certain tree species are more resistant to drought because their thick roots can reach water resources farther underground. Therefore, during times of water scarcity, agroforestry systems help keep water available for crops, livestock and other agricultural activities.

With its combination of forestry and agriculture, agroforestry provides an all-encompassing strategy for water conservation. Agroforestry systems encourage efficient and sustainable use of water resources in agricultural landscapes by decreasing soil erosion, increasing water infiltration and retention, controlling water flow, bettering water quality and adding to water availability.

23.4 Agroforestry and Food Security

Agroforestry system also contributes to increased photosynthetic efficiency of tree species. Enhanced soil fertility and structure have a growing impact on crop productivity. Less soil erosion loss and improved closed-loop nutrient- and organic-cycle management improving the microclimate for the development of agricultural crops. Forest-influenced soils have been shown to produce higher agricultural yields compared to non-forested soils. The Taungya people of Uttar Pradesh's Tarai region are known for their exceptional agricultural yields using just organic methods and no fertiliser. It has been observed that agro forestry in Haryana and western Uttar Pradesh results in grain and wood yields that are almost 20% higher than those obtained from conventional farming methods. The total output of fodder is greater when fodder grasses are cultivated with fodder trees, according to experiments done at IGFRI, Jhansi. Food, animal feed and fuel production all rise when *Leucaena leucocephala* is grown alongside conventional crops and forage grasses in an intercropping arrangement.

Agroforestry-grown nitrogen-fixing trees have a fixing capacity of 50–100 kg N/ha/year. It has been shown that a tree and farm crop production system is more productive in Punjab, Haryana, Uttar Pradesh, Gujarat and some areas of the southern states. Fuel, fodder and small timber are all claimed to be produced and worth far more on degraded fields than the coarse grains typically grown there. One of the most exciting aspects of agroforestry is the role that nitrogen-fixing trees play. As the leaf litter decomposes, humus is formed, nutrients are released and the soil's varied qualities are enhanced; in addition, less fertiliser is required (Prasad n.d.).

When it comes to marginal land, it is most cost-effective to grow trees and fodder crops (including fodder trees). Evidence gathered from Rajasthan's hot, dry and semi-arid regions suggests that the state's marginal lands are not suitable for the production of healthy, abundant crops. Under Haryana, planting trees like Eucalyptus in agroforestry has been proven to be more profitable than pure agriculture. Other types of trees including Prosopis, *Albizia*, *Zizyphus* and Acacia can also be grown in a silvopasture system. In the Tarai region of Uttar Pradesh, *Populus deltoides* doubles farm profits.

Bamboo-based agroforestry can enhance the productivity. It also produces a lot of oxygen and emits more carbon than the tropical and sub-tropical trees (Solomon et al. 2021), making the air clean and balance ecosystem. Practiced in Kerala in India, Nepal and many other developing countries, bamboo-based agriculture can contribute to economic development and employment. In the tropical regions, agroforestry is very common.

In Kenya, only 20% of the land is suitable for growing crops due to soil erosion and lack of rain; due to this, farmers are adopting dryland programme to make their land green and grow crops through agroforestry. Crops like mango, orange, neem and many more are grown in the shades (Nijagi 2021). Many crops especially plantation crops like coffee grow under the shades of trees. Agroforestry provides animal feed which also contributes to the production of better quality milk. Pineapple agroforestry systems (PAFS), which are prevalent in the Indian Eastern Himalayas and other parts of Asia and are typically grown in conjunction with multipurpose trees, can be a sustainable alternative to Jhum cultivation for the North East of India. This practise is traditionally carried out by the ethnic 'Hmar' tribe in southern Assam. In southern Assam, the ethnic 'Hmar' tribe has been growing pineapple for millennia. They currently use the native PAFS for, both for domestic use and to improve economic benefits. They have developed a distinctive agroforestry system by applying indigenous knowledge (PID Delhi 2021). An assortment of commercially significant trees, including *Albizia procera, Parkia timoriana* and *Aquilaria malaccensis*, as well as fruit trees, including mango, papaya, guava, lemon and litchi and with pineapple, caters to both year-round home use and year-round selling. The trees in the higher canopy control light, boost biomass inputs and broaden the range of farms, which improves soil fertility and plant nutrition. The farmers' preferred native fruit trees are preserved thanks to tree-related management techniques. Rubber plants are being introduced by farmers in the older pineapple agroforestry plantations. Today, modern technology and new approaches are used in the agricultural sector to produce crops like through vertical farming, hydroponics, artificial intelligence and machine learning, whereas developing and undeveloped countries are more into agroforestry as digitalisation has not reached it. Agroforestry can protect and preserve natural resources. It is time that the agriculture sector also reduces its carbon emission. There is a long history of agroforestry on the Indian subcontinent. Raising, caring for and loving trees are deeply ingrained in the socioreligious fabric of the people of the subcontinent, trees are heavily integrated into the region's agriculture and livestock production systems (Singh 1987).

The kherji (Prosopis cineraria) and agricultural-crop combination in the hot, arid region meets needs for fodder, small timber and food, while the multi-tier tree-crop combinations in the homegardens of the damp lowlands suit financial and domestic necessities. An good example of a modern but conventional agroforestry system is the combination of *Alnus nepalensis* and *Amomum subulatum* found in the humid sub-temperate regions of Nepal, Bhutan and Sikkim state in India. Other typical instances of widespread agroforestry methods include the purposeful growth of trees on field bunds, their irregular distribution in agricultural fields and the intentional preservation of shade trees in tea and coffee plantations. Similar to this, it is customary to cultivate crops for 2–3 years in newly planted orchards and woods before interplanting shade-tolerant plants like turmeric and ginger.

The greatest number of people can benefit from agroforestry, it is good for the environment, and it does not require the application of modern technology. In order to handle the food crisis, achieve nutrition security and put an end to poverty, hunger and health crises, farmers need to acquire training, and traditional wisdom needs to be put into practise. Several Indian states, including Gujarat, Jharkhand and Maharashtra, have been implementing agroforestry practises as part of their efforts to become more sustainable. Native American groups and tribes have practised this way of life for a very long time.

A large percentage of the goals for sustainable development are significantly reliant on agroforestry. These goals include eliminating hunger and poverty as well as ensuring that everyone has access to clean drinking water and sanitary facilities. The Sustainable Development Goals 2, 3, 6, 13 and 15 all address issues related to climate change and life on land.

23.5 Agroforestry Role in Forest Restoration and Climate Resilience

Agroforestry is the outcome of agricultural practises being integrated with tree planting and management, and it has a substantial impact on the process of forest restoration as well as the climate resilience of the area.

Reforestation is possible through the application of agroforestry practises, which can be applied to reforest damaged or entirely removed areas. A form of farming known as agroforestry involves planting trees in agricultural areas in order to re-establish forest cover, boost biodiversity and help ecosystems recover from damage. The inclusion of tree species that are native to an area is beneficial to agroforestry systems since it aids in the regeneration of the local flora and fauna.

Agroforestry has the potential to be an efficient solution for projects involving reforestation as well as regeneration. Incorporating trees into agricultural holdings allows farmers to make a direct contribution to tree-planting initiatives by increasing the tree density and expanding the amount of land covered in forest. Due to the fact that it enables them to cultivate both trees and food crops at the same time, farmers and landowners can realise financial benefits from practising agroforestry. Because they are helpful in storing carbon, agroforestry systems are an important part of the fight against climate change. Agroforestry methods, in which trees are utilised to combat climate change by absorbing greenhouse gases from the air and storing their own carbon, are gaining in popularity. These methods may be found in more and more agricultural settings. The practise of agroforestry has the potential to be a more

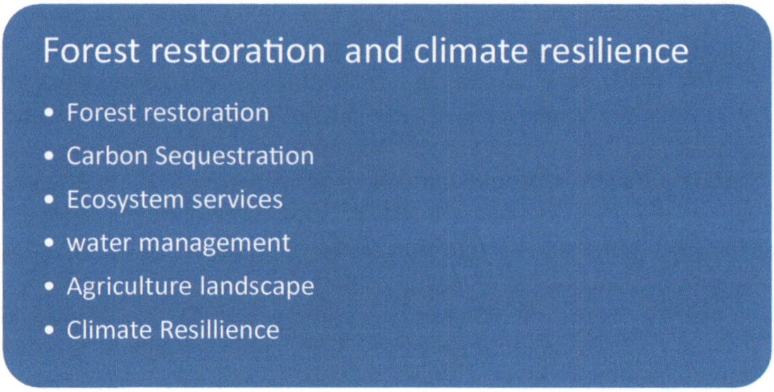

Fig. 23.4 Agroforestry role in forest restoration and climate resilience

successful method of carbon sequestration due to the fact that it combines the benefits that are associated with farming and forestry.

Agroforestry strengthens agricultural systems, making them more resistant to the effects of climate change. The presence of trees in agricultural settings protects crops from the sun and provides shelter from storms, which together make an agroforestry system more resistant to the effects of natural disasters. The tree canopy acts as a windbreak and casts shade on the ground below, both of which contribute to more moderate temperatures and humidity levels. Farmers that practise agroforestry have a more diverse range of income and food sources, which puts them in a better position to withstand the effects of shifting climates.

Because of the method in which they are integrated into landscapes, agroforestry systems can be of assistance in the management of water resources. Having trees in the area can help with water management, as well as the prevention of erosion and the infiltration of water. Their root systems contribute to the reduction of runoff and the improvement of water retention, which is a benefit to the environment. By boosting groundwater levels and retaining surface water during wetter times, agroforestry systems can help lessen the chance of flooding and drought.

Agroforestry systems are able to supply multiple ecological services, which in turn increases the resilience of the landscape (Fig. 23.4). These services include, but are not limited to, the protection of soil, the purification of water, the facilitation of pollination, the improvement of habitat for wildlife and the encouragement of beneficial insects. The capacity of ecosystems to function and recover after disturbances is improved by the practise of agroforestry, which contributes to the fortification of ecosystems. Agroforestry is essential to the process of forest regeneration as well as climate resilience since it combines the benefits of farming with those of trees and the ecosystem services they provide. It presents a holistic perspective of the ecosystem and provides long-term plans for the management of land in the context of climate change.

Agriculture is the main source of revenue and economic growth in rural areas of low-income countries. However, land pressure and climate change are detrimental to agricultural systems in emerging countries, posing a threat to food production. While intensive agricultural methods have been successful in many parts of the world, their promotion has led to a decrease in agricultural output due to degraded soil. Negative feedbacks on climate, food security and on-farm income at local scale result from the numerous environmental implications of agricultural intensification and food production, including negative effects on soil and biodiversity (Mbow et al. 2014).

23.6 Agroforestry and Environment Governance

The two areas of study are intricately intertwined, and law plays an essential part in ensuring the effective use and regulation of agroforestry practises to maximise their potential benefits.

The practises that are known as agroforestry are governed by a set of laws and regulations. In these agreements, the rights and obligations of agroforestry stakeholders including communities, landowners and farmers are laid out in detail. Farmers are one type of agroforestry stakeholder. Legal requirements must be met for 'sustainable land use', 'forest conservation', 'biodiversity protection' and 'tree incorporation into agricultural systems', among other 'green' initiatives.

A stable ownership structure of land is necessary for the widespread implementation of agroforestry practises. If farmers and communities are granted the right to own land and put it to use for agroforestry purposes, then only then can agroforestry be considered a lawful practise. Clear land tenure arrangements, which also encourage sustainable land management, encourage long-term investments in agroforestry by providing stability for land ownership.

It is standard procedure to carry out what is known as an environmental impact assessment (EIA) before to beginning any kind of agroforestry endeavour on a significant scale. EIAs examine the potential environmental, social and financial ramifications of a project in order to guarantee that it will not have a negative impact on the environment, that it will be fair to the community and that it will be profitable.

The management of agroforestry systems, which are crucial for the protection of forests and biodiversity, is significantly impacted by the laws that are in place. They are responsible for the establishment of reserve zones, the establishment of rules for the responsible management of forests and the control of the extraction of timber and other forest items. Legal frameworks provide assistance for agroforestry practises that protect genetic resources and species that are in danger of extinction.

Certification Programmes and Occupational Standards By making the use of sustainable agroforestry practises mandatory, laws and regulations have the potential to increase their use and spread their benefits. By getting the appropriate certifications, such as those for organic farming or sustainable forest management, agroforestry systems can be proven to meet environmental and social requirements through the process of certification verification. Compliance with such criteria may result in gaining access to markets, conducting land management in a more responsible manner and receiving incentives to practise sustainability.

Participation from Stakeholders and Involvement of Stakeholders The participation of stakeholders and other interested parties in the decision-making processes of agroforestry and environmental governance can be enhanced through laws. It is possible for legal frameworks to require the input of the general public, the engagement of local people and the acknowledgement of indigenous and traditional knowledge.

Hearing from a diverse group of people is necessary to ensure that the laws governing agroforestry are effective, fair and in tune with the realities of the local environment.

Because laws provide methods to ensure that norms are obeyed, agroforestry regulations are considered to be enforceable. They spell out the repercussions for disobedience, explain the functions of oversight bodies and offer routes for resolving issues that develop as a result of the practises used in agroforestry. The stringent enforcement of environmental standards for agroforestry practises helps to promote both a clean and safe environment as well as responsible land management.

The legal framework that is provided by legislation in agroforestry and environmental governance is extremely helpful in promoting sustainable practises, ensuring the protection of biodiversity, ensuring the security of land tenure and providing opportunities for stakeholder participation. It is possible to strike a balance between the requirements of agricultural productivity and those of environmental protection through the lawful implementation of agroforestry systems, which requires the establishment of explicit standards and laws.

In order to slow the rate at which biodiversity is being lost and to make sustainable landscape management possible, it is crucial to incorporate protection of natural resources within agricultural practises. Convention for Biological Diversity (CBD) advocates for sustainable agriculture supporting biodiversity and ecosystem functions like connectivity and habitat stability, but governments have instead prioritised expanding the protected area network (Zinngrebe et al. 2020). Changes in environmental governance, which includes all policies and institutions affecting the state of the environment, have profound effects on tree planting and management on farms across the developing world. Multiple facets are undergoing shifts at once. Decentralised multistakeholder committees and local user groups are gradually replacing national forestry agencies as the formal power holders in the field. The use of incentives and market forces to supplement regulatory frameworks is gaining traction in the field of environmental management. Companies are increasingly providing environmental goods and services, such as water, energy and lumber, and protecting biodiversity and watersheds. International agreements and the initiatives of powerful international organisations are increasingly prioritising integrated approaches to ecosystem and landscape management that incorporate local inhabitants as vital partners (Swallow et al. n.d.).

Importers of tropical timber and timber products have been under increasing pressure from the world's largest markets in recent years to provide evidence that their goods come from legal or sustainable sources. There are a number of laws around the world that demand proof of timber's legitimacy, including the Japan Clean Wood Act, the EU Timber Regulation, Australia's Illegal Logging Prohibition

Act and the United States' Lacey Act. In order to better enforce forest laws, tropical timber-producing countries can now access resources made available by a resolution made by the International Tropical Timber Council in November 2001 (ITTO n.d.).

23.7 Role of Agroforestry to Eradicate Poverty and Food Crisis

By fostering long-term and diverse income sources for rural people, agroforestry has the potential to significantly contribute to the fight against poverty.

Farmers have the potential to increase their income through the use of agroforestry techniques. Farmers can increase their revenue diversity by growing and selling lumber, fruits, nuts, medicinal plants and other non-timber forest products by incorporating trees into their agricultural systems. When compared to conventional monoculture crops, agroforestry has the potential to yield higher-value goods, which in turn can lead to higher revenue and more financial security.

Agroforestry can improve rural populations' access to markets and their ability to participate in value chains. Farmers that use tree products in their operations can meet the demands of consumers who want to buy items made in a sustainable manner. Tree-based goods can have their market worth increased through value addition, processing and marketing thanks to agroforestry systems.

Agroforestry helps with food security and better nutrition since it increases the variety of crops grown. A variety of healthy foods can be produced in an agroforestry system because food crops are typically grown alongside trees. Increasing dietary diversity and decreasing reliance on a stable climate can both be achieved through the cultivation of a wide variety of crop and tree species.

Land Management: Agroforestry encourages sustainable land management practises that boost soil fertility, conserve water and strengthen ecosystems. Agroforestry is the practise of incorporating trees into agricultural systems to promote soil health, reduce the risk of soil erosion and increase water penetration. These methods improve farmers' incomes by increasing agricultural output while decreasing the likelihood of crop failures.

When compared to monoculture agriculture, agroforestry systems are more able to withstand the effects of climate change. Agroforestry uses trees to protect crops from the effects of extreme weather by providing shade, windbreaks and microclimate regulation. Farmers' vulnerability to climate-related hazards is mitigated by the diversification of revenue sources afforded by agroforestry, which makes them less reliant on a single crop and more able to adjust to shifting climatic conditions.

Employment and Rural Development: Agroforestry generates jobs in rural regions through on-farm and value-added processes. Communities can benefit from the creation of new jobs that result from the planting, maintenance, harvesting and sale of tree products. Infrastructure upgrades, increased awareness of sustainable land management methods and increased community agency and capability are just few of the ways that agroforestry projects benefit rural advancement.

Agroforestry helps farmers weather market swings and price swings by spreading their revenue among multiple crops instead than relying on a single cash crop. Farmers may expand their market reach and respond to shifting customer preferences by using tree-based products. This resistance to market shocks aids in protecting rural areas from economic downturns and poverty.

In general, agroforestry is a sustainable and diverse method of farming that can boost income, increase food security, facilitate management of natural resources and make farms more resistant to climate change. Agroforestry helps improve the economic and social conditions of rural areas by tackling numerous causes of poverty.

The governments that control and manage over 77% of the world's forests do not respect the rights of indigenous peoples and local populations to the land. The locals who depend on the forests for survival do not get the benefits they should since government goals do not always line up with community needs. The natives in Africa, for instance, do not benefit from the booming forestry and ecotourism sectors. The agricultural practise of agroforestry, in which trees and bushes are grown in and around crop and pastureland, can help solve this issue. In order to avoid the ownership issue and ensure that earnings stay in the community, agroforestry builds on agricultural land currently owned by communities to establish new woods that are not controlled by the government. Although agroforestry systems are on a smaller scale than traditional forests, they still provide many of the same benefits, including increased biodiversity, diversified production and restored soil fertility.

Agroforestry is useful for more than only the environment. Increased food resources and security, enhanced nutrition and higher earnings for farmers are just a few of the ways in which agroforestry can help alleviate worldwide poverty (Quallen 2021).

It is not a coincidence that in some regions of the world natural woods and poverty can be found in close proximity to one another. This is the case for a number of reasons. The natural woods are where humans developed, and even after millennia of settlement, the people who live there have maintained a mainly primitive lifestyle. Many of the people moving from rural areas into wooded areas in quest of extra farmland are economically disadvantaged, as are the areas that they are moving into. For people who are on the socioeconomic periphery of society, for example, work opportunities in forests are often available because of the unequal distribution of land in lowlands. Throughout history, persons on the run from oppression, conflict and war have historically been able to find safety in forests. Agroforestry contribution to reducing poverty has been presented in Fig. 23.5.

There are two ways in which forests can help reduce global poverty. First, they play a crucial role as a safety net, allowing rural residents to either avoid or lessen their exposure to poverty. Second, there is unrealised potential for woods to help some rural residents escape poverty.

Fig. 23.5 Agroforestry contribution to reducing poverty

Many politicians and planners are unaware of these features because the scientific community has not done a good job of explaining the safety net functions of forests.

One explanation for this is that the poorest households' use of woods, whether for sustenance or commerce in local markets, is rarely reflected in national statistics. Some elements of timber resources actually hinder their capacity to aid marginalised people, while the lion's share of timber riches flows to better-off sectors of society. Despite these challenges, if decision-makers realise and act on the promise of forests, they can expand their contribution to poverty alleviation (Sunderlin et al. 2004).

23.8 Agroforestry and the Tribal Community

In addition to improving food security and fostering cultural and ecological resilience, agroforestry can also provide stable income for tribal people by preserving indigenous knowledge and practises.

Livelihood and Income Creation: Agroforestry allows indigenous groups to make a living by growing and selling tree-based goods. Agroforestry systems combine tree planting with crop and livestock husbandry to generate revenue from a wide variety of tree-based and other products. As a result, tribal communities may become more economically secure and less dependent on any one source of revenue.

Protecting Indigenous Knowledge: Agroforestry methods are congruent with indigenous ways of knowing and using land. Tribal communities can keep their indigenous knowledge alive and well by combining traditional methods with contemporary agroforestry practises. This encompasses familiarity with native plant species, agroecological approaches, seed varieties and growing techniques. The restoration of cultural traditions and the sharing of information between generations are two important functions that agroforestry can play in this regard.

Improved nutrition and greater food security are two benefits of agroforestry for indigenous peoples. Agroforestry methods improve food availability and nutritional variety by combining crop cultivation with tree planting. Preserving local food culture and addressing food sovereignty issues are both aided by the incorporation

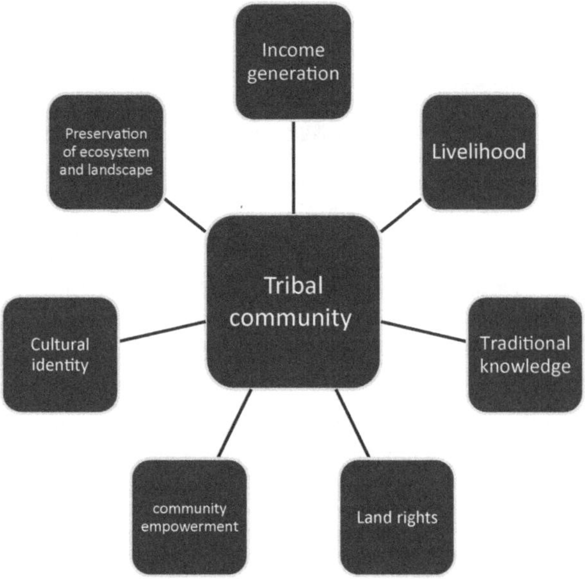

Fig. 23.6 Impact of agroforestry on tribal community

of indigenous food crops and traditional variations. Additionally, a more secure food supply can be maintained through the use of agroforestry systems, which increase resilience to the effects of climate change.

Restoration of degraded land and protection of biodiversity are two ways in which agroforestry increases ecological resilience in a landscape. Agroforestry protects native plant and animal species, especially those of cultural and medical value to indigenous peoples, by combining varied tree species and establishing agroecosystems. Landscape connectivity is improved with the use of agroforestry systems, which help to keep wildlife corridors in good condition.

Tribal communities' cultural identity and resilience can be bolstered by adopting agroforestry practises. The spiritual and cultural relevance of traditional agroforestry systems reflects the intrinsic bond between indigenous peoples and their environments (Fig. 23.6). Agroforestry allows indigenous groups to strengthen their cultural norms, traditional ways of life and relationships to the land. Agroforestry can aid in protecting the rights to and ownership of land for indigenous populations. For agroforestry to be practised in a sustainable manner, it is crucial that indigenous territories and community land rights be recognised and legally protected. Agroforestry can help reduce poverty and promote social justice by giving indigenous tribes more say over their property. Participation, decision-making and community empowerment are all bolstered by agroforestry in indigenous societies. Local leadership may be developed, social cohesiveness can be strengthened, and resources can be reclaimed when tribal communities engage in the planning, implementation and management of agroforestry systems. Through such participation, the tribal people' unique requirements, values and goals can be taken into account while developing agroforestry practices.

When applied with respect for tribal people' cultural and ecological values, agroforestry has the potential to safeguard indigenous knowledge, provide food security, foster social cohesion and fortify ecosystems. It promotes the health and prosperity of indigenous communities by taking an open and collaborative approach that values their unique knowledge, experience and goals.

23.9 Agroforestry and Urban Development

Agroforestry is an approach that should be given more consideration in urban planning because of its potential to assist in the development of more sustainable and resilient urban communities.

It is possible to include agroforestry as a component of green infrastructure in the design of urban landscapes. Planting trees in cities and making use of techniques from the field of agroforestry help to mitigate the heat island effect, purify the air and provide healthier environments in which people may live. Trees that provide shade not only contribute to the aesthetic appeal of a city but also assist reduce the amount of money that is spent on air conditioning. Agroforestry in urban areas can contribute to both food security and self-sufficiency in food production. Community gardens, rooftop gardens and edible landscapes are all examples of urban agroforestry systems that are capable of producing a variety of fruit, vegetables, herbs and other edible plants. This not only lessens the toll that transportation of food takes on the environment, but it also makes it easier for individuals to get their hands on fresh produce grown in their own communities.

The practises of agroforestry can be helpful for managing stormwater in an urban setting. Tree canopies serve as a natural barrier against precipitation, thereby lowering the volume of stormwater flow as well as the rate at which it moves. Tree roots have the ability to remove impurities from the water and improve its quality, in addition to increasing the amount of water that they can absorb. With the assistance of agroforestry systems, capturing and utilising the rainfall that falls in urban areas is made much simpler.

The provision of habitat and food for urban wildlife, such as birds, insects and small animals, is one of the many ways in which urban agroforestry is beneficial. This contributes to the protection of biodiversity. Agroforestry encourages urban biodiversity by re-establishing ecological balance in cities through the use of a range of tree species and the development of urban green spaces. This is accomplished by the use of agroforestry practises such as tree planting and the creation of urban green spaces.

Initiatives to promote urban agroforestry's use have the potential to strengthen community relations and encourage more people to participate in civic life. When local residents, community organisations and schools are involved in agroforestry activities from the very beginning, opportunities for education, the development of skills and the strengthening of community bonds are established. Urban reforestation efforts can instil a sense of pride in one's town as well as a sense of success for having contributed to the betterment of that community.

Both the ability to better regulate the local microclimate and the ability to keep cities cooler in the summer are factors that contribute to the city's ability to withstand the effects of climate change. As natural air conditioners, the trees that are part of agroforestry systems help to reduce the amount of artificial cooling that is required and also contribute to the mitigation of the effects of urban heat islands. Urban agroforestry is an important weapon in the fight against climate change since it may help reduce emissions of greenhouse gases and also contribute to the sequestration of carbon dioxide.

In places that are currently underserved, urban agroforestry has the ability to both boost employment rates and encourage the growth of local businesses. Orchards, nurseries and businesses that add value to tree products can all benefit from this, as can other local businesses. One of the additional benefits of urban agroforestry programmes is the creation of jobs in ancillary businesses such as landscaping, urban forestry and environmental education.

By introducing agroforestry into urban development, improvements can be made to a city's environmental quality, food security, biodiversity and climate resilience, all of which are potential benefits of the practise. Urban agroforestry initiatives not only improve the health of the community, but also the social cohesion and economic prospects of the area. The consequence of these projects is an urban environment that is more habitable and sustainable.

There is growing recognition of the potential contributions of non-traditional forms of GI, such as wastelands and informal green spaces, to CUM and to urban social-ecological systems in general, and this has led to a renewed focus on the role that multifunctional green infrastructure (GI) can play in promoting circular urban metabolism (CUM), reducing the ecological footprint of cities and providing a wide range of services, including biodiversity conservation. Home and communal gardens as well as urban farms are examples of productive urban spaces that might be thought of as an alternative, multipurpose type of GI. These areas can infiltrate stormwater, reduce the effects of urban heat islands, preserve biodiversity, sequester carbon, help form soil and recycle urban wastes (Taylor and Lovell 2021). Role of agroforestry in urban development has been presented Fig. 23.7.

23.10 Agroforestry Impact on Economic Sector

By incorporating trees with crops and cattle, agroforestry systems provide numerous avenues for generating money. Timber, fruits, nuts, medicinal plants, non-timber forest products and animal products are only some of the ways in which farmers might make a living. Farmers' financial security is increased as income risks associated with mono-cropping are mitigated through diversification. Soil fertility, nitrogen cycling and water availability can all be increased through agroforestry, leading to greater agricultural output. By enhancing soil structure, nutrient retention

Fig. 23.7 Role of agroforestry in urban development

and moisture conservation, trees in agroforestry systems boost crop growth and yield. Productivity growth can boost agricultural outputs and, by extension, the economy.

Products from agroforestry can be refined to increase their market value, opening the door for home grown enterprises and regional innovators. Agroforestry goods can benefit from a boost in value and marketability through processing operations like timber milling, fruit processing and herbal medicine manufacture. This helps the economy expand in rural areas by producing jobs and money. Agroforestry systems, from planting and management to processing and selling, all generate employment opportunities along the value chain. Farmers, workers, technicians and processors are all essential members of the agroforestry community. Job creation and reduction in rural unemployment can result from the installation and upkeep of agroforestry systems. By combining tree farming with agricultural practises, agroforestry helps spread awareness of the importance of responsible forest management. With this method, lumber and other forest products will be around for generations to come. By ensuring a steady stream of forest resources and supporting companies that rely on them, sustainable forest management creates economic benefits.

Agroforestry systems can be financially rewarding thanks to carbon credits and payments for ecosystem services. Agroforestry's trees help slow global warming by

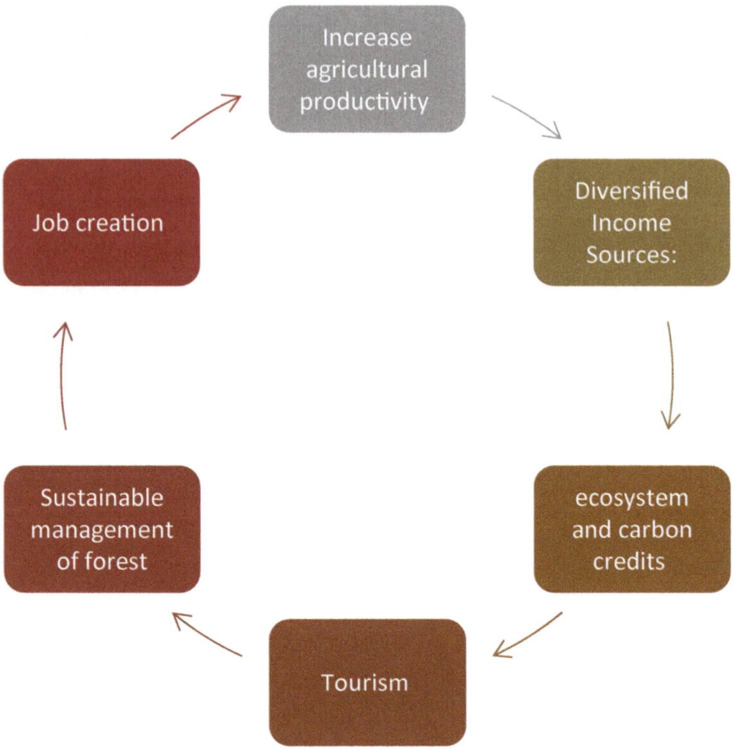

Fig. 23.8 Agroforestry contribution to the economic sector

soaking up carbon dioxide from the air. Carbon offset programmes and compensation for ecosystem services including watershed protection, biodiversity conservation and carbon sequestration are options for farmers and landowners.

The tourism and recreation industries can benefit from agroforestry landscapes due to their potential aesthetic and recreational value. Agroforestry systems, such those that facilitate agroecotourism or farm stays, let sightseers learn about and enjoy the region's rural culture and way of life, as well as its abundant flora and fauna. Having multiple avenues of financial support helps the rural economy thrive.

Farmers benefit from increased resilience to market swings and price instability thanks to agroforestry systems. Farmers can reduce their reliance on any one commodity or market by offering a wider variety of goods and services. Income fluctuations caused by market fluctuations are reduced, and income stability is increased, thanks to this strategy of diversification. Agroforestry has the potential to considerably contribute to the economic sector due to its ability to generate several streams of income, undergo value-added processing, generate new employment opportunities and promote sustainable forest management (Fig. 23.8). In addition to helping farmers make a living, it boosts local economies, encourages entrepreneurship and improves residents' standard of living.

23.11 Conclusion

Ending the food crisis, addressing hunger and achieving a healthy nutrition level can be done through sustainable agriculture and smart agricultural systems. Agroforestry holds great promise for addressing many issues in agriculture, health and climate change. The primary cause of problems in the globe is the exploitation of natural resources. The causes of the global food crisis may range, but environmental deterioration is one of the main ones. In order to stop food waste and produce enough food to meet demand, rural areas must be electrified. Community-based farming and kitchen gardens are assisting in the fight against malnutrition. Around 60 million children worldwide suffer from undernourishment, and 850 million people are experiencing a food crisis. New agricultural techniques for growing crops, such as vertical farming, a type of urban farming, nanotechnology, biotechnology and hydroponics, are now possible thanks to modern technology. However, these techniques are more common in underdeveloped and developing nations, where agroforestry is more prevalent. Agroforestry is a natural strategy that can restore ecological balance, maintain natural resources and increase biodiversity.

References

Agroforestry and water for resilient landscapes – 2020, Agroforestry network

Forest law enforcement, governance and trade – International Tropical Trade Organization (n.d.). https://www.itto.int/sustainable_forest_management/law_enforcement/

Mbow C, Smith P, Skole D, Duguma L, Bustamante M (2014) Achieving mitigation and adaptation to climate change through sustainable agroforestry practices in Africa. Curr Opin Environ Sustain 6:8–14.,ISSN 1877-3435. https://doi.org/10.1016/j.cosust.2013.09.002

Nijagi D (2021, August) Farmers regreen Kenya's drylands with agroforestry and an app, MONGABAY

PID Delhi (2021, August) Traditional Pineapple Agro-forestry Systems can address twin challenges of climate change and biodiversity loss, Ministry of Science & Technology

Prasad K (n.d.) Prospect of agroforestry in India, FAO. https://www.fao.org/3/XII/0931-B5.htm

Quallen B (2021, January) HOW AGROFORESTRY CAN REDUCE GLOBAL POVERTY, The Borgen Project. https://borgenproject.org/agroforestry-can-reduce-global-poverty/

Singh GB (1987) Agroforestry in the Indian subcontinent: past, present and future. World Agroforestry centre Transforming lives and landscapes , International Council for Research in Agroforestry

Solomon T, Moon H, Abebe S, Minale AS (2021) March, Promoting Bamboo -based agroforestry for enhancing ecosystem services from Degraded Lands 16, Agroforestry for Degraded Landscape

Stiles W (2017, September) Agroforestry: An opportunity for sustainability, Sustainable Food Trust. https://sustainablefoodtrust.org/news-views/agroforestry-an-opportunity-for-sustainability/

Sunderlin WD, Angelsen A, Wunder S (2004, April) FORESTS AND POVERTY ALLEVIATION, Center for International Forestry Research

Sustainable Agriculture and Agroforestry, Vi agroforestry (n.d.). https://viagroforestry.org/what-we-do/agroforestry/

Swallow B, Russell D, Fay C (n.d.) Agroforestry and environmental governance, World Agroforestry Centre. Pennsylvania State University

Taylor JR, Lovell ST (2021, August) Designing multifunctional urban agroforestry with people in mind. Urban Agric Reg Food Syst. https://doi.org/10.1002/uar2.20016

World Agroforestry (2014) India leads the way with agroforestry policy. https://www.worldagroforestry.org/news/india-leads-way-agroforestry-policy

World Bank, 2021, April, Climate smart agriculture https://www.worldbank.org/en/topic/climate-smart-agriculture

Zinngrebe Y, Borasino E, Chiputwa B et al (2020) Agroforestry governance for operationalising the landscape approach: connecting conservation and farming actors. Sustain Sci 15:1417–1434. https://doi.org/10.1007/s11625-020-00840-8

Chapter 24
Exploring the Agroforestry Systems for Ecosystem Services: A Synthesis of Current Knowledge and Future Research Directions

Varsha Giri, Tanmaya Kumar Bhoi, Ipsita Samal, J. Komal, and Prasanta Kumar Majhi

Abstract Agroforestry systems (AFS) are land-use systems that integrate trees with crops and/or livestock production. AFS provides multiple ecosystem services (ES) that are crucial for the sustainability of human societies and ecosystems. The promotion of AFS requires a multi-dimensional approach that addresses the social, economic, and environmental dimensions of sustainable development. AFS has been recognized as an important ES provider due to its capacity to enhance biodiversity, soil conservation, carbon sequestration, and water regulation. This book chapter aims to summarize the current state of knowledge on the ES provided by AFS.

Keywords Agroforestry systems · Ecosystem service · Biodiversity · Soil conservation · Carbon sequestration

V. Giri
Silviculture and Forest Management Division, ICFRE-Arid Forestry Research Institute (ICFRE-AFRI), Jodhpur, Rajasthan, India

T. K. Bhoi (✉)
Forest Protection Division, ICFRE-Arid Forestry Research Institute (ICFRE-AFRI), Jodhpur, Rajasthan, India

I. Samal
ICAR-National Research Centre on Litchi, Mushahari, Ramna, Muzaffarpur, Bihar, India

J. Komal
Basic Seed Multiplication and Training Centre, Central Silk Board, Kharsawan, Jharkhand, India

P. K. Majhi
Regional Research and Technology Transfer Station (RRTTS), Odisha University of Agriculture and Technology (OUAT), Keonjhar, Odisha, India

H. S. Jatav et al. (eds.), *Agroforestry to Combat Global Challenges*, Sustainable Development and Biodiversity 36, https://doi.org/10.1007/978-981-99-7282-1_24

24.1 Introduction

Agroforestry systems, which integrate trees with crops and/or livestock on the same land, have been recognized as a sustainable land-use practice that can provide a range of ecosystem services. Ecosystem services are the benefits that humans obtain from nature, such as food, timber, water regulation, climate regulation, and biodiversity conservation. Agroforestry systems have the potential to provide multiple ecosystem services, including improving soil fertility, reducing erosion, and enhancing biodiversity, mitigating climate change, and providing livelihoods for rural communities. Recent studies have highlighted the important role that agroforestry systems can play in providing ecosystem services. For example, a study by Hübner et al. (2021) found that agroforestry systems can significantly increase carbon sequestration compared to conventional agriculture, which can help mitigate climate change. Lelamo (2021) demonstrated that agroforestry systems can improve soil fertility and increase crop productivity, which can enhance food security and rural livelihoods. Muthee et al. (2022) explored the potential of agroforestry systems in providing ecosystem services in sub-Saharan Africa. The study found that agroforestry systems could enhance soil carbon sequestration and nutrient cycling, reduce soil erosion and increase water availability, leading to increased crop yields and improved livelihoods for farmers. Santiago-Freijanes et al. (2021) investigated the role of agroforestry in promoting biodiversity conservation and ecological connectivity in European agricultural landscapes. The study found that agroforestry systems can act as biodiversity hotspots, providing habitat and food for a wide range of species, and enhancing ecological connectivity by facilitating the movement of organisms across fragmented landscapes. Despite these benefits, the potential of agroforestry systems to provide ecosystem services is not yet fully understood, and there is a need for a comprehensive review of the current knowledge in this field. In this review paper, we aim to explore the role of agroforestry systems in providing ecosystem services, focusing on the current state of knowledge and identifying key research gaps. Overall, this review paper will contribute to a better understanding of the potential of agroforestry systems to provide ecosystem services, and will provide insights into how to design and manage these systems to maximize their benefits for both humans and the environment. It is timely and important to revisit the potential of agroforestry systems in providing ecosystem services in light of the growing challenges of climate change and food security (Patel et al. 2020).

24.2 Agroforestry Contribution to Ecosystem Services

Agroforestry has been shown to contribute to a range of ecosystem services, including soil conservation, water quality improvement, carbon sequestration, biodiversity conservation, and climate change mitigation. A global meta-analysis by Donatti et al. (2022) found that agroforestry systems had higher levels of soil organic carbon, soil fertility, and biodiversity than conventional agriculture. The study also

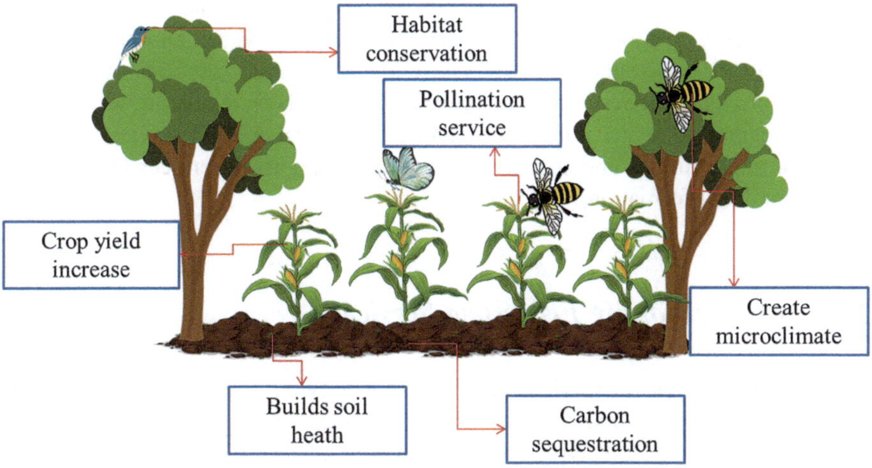

Fig. 24.1 Agroforestry systems promoting diverse tree–crop interactions, nutrient cycling, and biodiversity conservation, resulting in enhanced ecosystem services

found that agroforestry systems could reduce soil erosion and improve water quality. Panwar et al. (2022) examined the potential of agroforestry systems in India to provide multiple ecosystem services. The study found that agroforestry systems could enhance soil fertility, increase biodiversity, and provide carbon sequestration and climate change mitigation benefits. The study also highlighted the potential for agroforestry to improve the livelihoods of small-scale farmers in India. Pradhan et al. (2021) investigated the potential of agroforestry in the Brahmaputra River Basin in India to provide ecosystem services. The study found that agroforestry systems could enhance soil fertility, reduce soil erosion, and provide habitat for biodiversity. The study also highlighted the importance of incorporating local knowledge and practices in the design and implementation of agroforestry systems (Fig. 24.1).

24.3 Classification of Ecosystem Services

Ecosystem services can be broadly classified into four categories: provisioning, regulating, cultural, and supporting services. Obiang Ndong et al. (2020) proposed a new framework for classifying ecosystem services based on their beneficiaries and the underlying ecological processes that support them. The study identified six beneficiary groups: humans, non-human species, ecosystems, the atmosphere, the hydrosphere, and the lithosphere. The study also identified six ecological processes that support ecosystem services: primary productivity, nutrient cycling, water cycling, climate regulation, disturbance regulation, and habitat provision. Pereira et al. (2018) developed a global classification of ecosystem services based on a comprehensive review of existing literature. The study identified 18 categories of ecosystem services, including food provision, water regulation, soil formation, climate regulation, and cultural and spiritual values. Braat and De Groot (2012) proposed a classification of ecosystem services based on the Millennium Ecosystem Assessment (MEA) framework. The study identified four categories of ecosystem services: provisioning services (e.g., food, water, and fiber), regulating services (e.g., climate regulation, water purification, and pollination), cultural services (e.g., recreational opportunities and spiritual values), and supporting services (e.g., nutrient cycling and soil formation).

24.3.1 *Provisioning Ecosystem Services*

Provisioning ecosystem services are those that provide direct benefits to humans, such as food, water, and raw materials. Balvanera et al. (2020) examined the contributions of different types of ecosystems (e.g., forests, grasslands, wetlands) to provisioning ecosystem services globally. The study found that forests and grasslands were the most important ecosystems for food provision, while wetlands were important for water provision. The study also highlighted the need for integrated landscape management approaches to balance trade-offs between different provisioning services. Borelli et al. (2020) assessed the sustainability of provisioning ecosystem services in the context of global food systems. The study found that current food systems were unsustainable due to their negative impacts on ecosystem health and biodiversity. The study proposed a set of interventions to promote more sustainable food systems, including reducing food waste and increasing agroecological practices. Zhang et al. (2021) examined the potential of traditional agroforestry systems in China to provide multiple provisioning ecosystem services. The study found that agroforestry systems had high levels of food and fuel production, as well as soil fertility and biodiversity conservation benefits. The study also highlighted the need for policy support and investment to promote the adoption of agroforestry practices in China (Fig. 24.2).

24.3.1.1 Food and Medicinal Value

Food and medicinal products are important examples of provisioning ecosystem services. A study by Maestre et al. (2020) examined the contributions of different types of ecosystems to food provision in drylands worldwide. The study found that natural ecosystems (e.g., shrublands, grasslands) were more important for food provision than agricultural systems in drylands. The study also highlighted the importance of maintaining the diversity of natural ecosystems for sustainable food production. Langemeyer and Connolly (2020) assessed the contributions of different ecosystem types to medicinal plant provisioning in Europe. The study found that forests were the most important ecosystem type for medicinal plant provisioning, followed by grasslands and wetlands. The study also highlighted the need for integrated conservation and management approaches to ensure the continued provision of medicinal plant resources. Gonçalves et al. (2021) examined the potential of

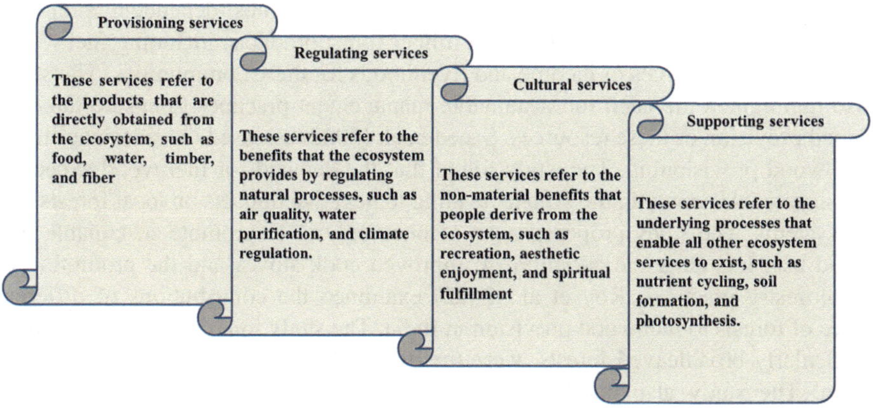

Fig. 24.2 Ecosystem services provided by agroforestry systems. Agroforestry systems can provide multiple ecosystem services, including soil conservation, carbon sequestration, biodiversity conservation, water regulation, and provision of food, fodder, fuel, and other products. These ecosystem services are linked and can have synergistic effects, contributing to the resilience and sustainability of agroecosystems. The type and magnitude of ecosystem services provided by agroforestry systems depend on factors such as the agroforestry design, species selection, management practices, and landscape context

traditional agroforestry systems in Brazil to provide food and medicinal plant resources. The study found that agroforestry systems had high levels of food and medicinal plant production, as well as soil fertility and biodiversity conservation benefits. The study also emphasized the importance of local knowledge and traditional practices in the management of agroforestry systems (Fig. 24.2).

24.3.1.2 Fuel Wood Production

Kim et al. (2021) examined the contributions of different types of ecosystems to fuel wood provisioning in South Korea. The study found that forests were the most important ecosystem type for fuel wood production, followed by agroforestry and grasslands. The study also highlighted the need for sustainable management practices to ensure the continued provision of fuel wood resources. Singh et al. (2021) assessed the potential of traditional agroforestry systems in India to provide fuel wood resources. The study found that agroforestry systems had high levels of fuel wood production, as well as soil fertility and biodiversity conservation benefits. The study also emphasized the importance of local knowledge and traditional practices in the management of agroforestry systems for fuel wood provision. Schaafsma and Bartkowski (2021) examined the impacts of fuel wood extraction on ecosystem services in Tanzania. The study found that fuel wood extraction had negative impacts on forest carbon stocks and biodiversity, but positive impacts on the provision of fuel wood resources. The study highlighted the need for integrated management approaches to balance trade-offs between different ecosystem services. Nandi and Sarkar (2021) examined the potential of non-timber forest products, including fuel wood, to contribute to the livelihoods of forest-dependent communities in India. The study found that non-timber forest products, including fuel wood, were important sources of income and livelihoods for these communities. The study also highlighted the need for sustainable management practices to ensure the continued provision of these resources. Saeed et al. (2022) assessed the sustainability of fuel wood provisioning. The study found that the demand for fuel wood exceeded the sustainable supply in the village, leading to negative impacts on local forests and ecosystems. The study proposed a set of interventions to promote sustainable fuel wood use, including the adoption of improved cook stoves and the promotion of agroforestry practices. Roy et al. (2022) examined the contributions of different types of forests to fuel wood provision in India. The study found that natural forests, particularly broadleaved forests, were the most important sources of fuel wood in India. The study also highlighted the need for sustainable forest management practices to ensure the continued provision of fuel wood resources (Fig. 24.2).

24.3.1.3 Fodder and Feed

Singh et al. (2022) assessed the contributions of different types of ecosystems to fodder production in India. The study found that grasslands and croplands were the most important sources of fodder in India. The study also highlighted the importance of maintaining the diversity of these ecosystems for sustainable fodder production. Tittonell (2021) assessed the global contributions of grasslands to livestock feed production. The study found that grasslands were the most important source of feed for ruminant livestock worldwide. The study also highlighted the need for sustainable management practices to ensure the continued provision of grassland resources for livestock feed production. Jnawali et al. (2021) assessed the contributions of different types of grasslands to livestock feed provision in Nepal. The study found that natural grasslands were the most important sources of livestock feed in the country, providing more than half of the total feed requirements. The study also highlighted the need for sustainability. Raj et al. (2020) examined the role of agroforestry systems in providing feed resources for livestock in India. The study found that agroforestry systems, particularly those with leguminous trees, provided high-quality fodder for livestock and supported the livelihoods of smallholder farmers. The study also highlighted the potential of agroforestry systems to contribute to climate change mitigation and adaptation. Manjunatha et al. (2022) assessed the contribution of natural grasslands to livestock feed provision in the Indian state of Gujarat. The study found that natural grasslands were the most important source of livestock feed in the state, providing more than 80% of the total feed requirements. The study also highlighted the need for sustainable management practices to ensure the continued provision of grassland resources. Fahad et al. (2022) examined the potential of agroforestry systems to provide feed for livestock in India. The study found that agroforestry systems had high levels of feed production, as well as soil fertility and biodiversity conservation benefits. The study also emphasized the importance of promoting agroforestry systems for sustainable livestock production (Fig. 24.2).

24.3.2 Supporting Ecosystem Services

Supporting ecosystem services are those services that are necessary for the functioning of other ecosystem services. Kass (2020) assessed the global distribution of soil biodiversity and its role in supporting ecosystem services. The study found that soil biodiversity is critical for maintaining soil health and productivity, and that its loss could have significant impacts on the provision of ecosystem services. The study highlighted the need for soil conservation and sustainable land management practices (Fig. 24.2). González et al. (2020) assessed the contributions of pollinators to crop production in Mexico. The study found that pollinators play a crucial role in supporting crop production and enhancing crop quality. The study also highlighted the need for pollinator conservation and sustainable land management practices to ensure continued pollination services. Liu et al. (2020) assessed the role of wetlands in supporting water purification services in China. The study found that wetlands are critical for removing pollutants from water and maintaining water quality. The study also highlighted the need for wetland conservation and restoration to ensure the continued provision of water purification services (Fig. 24.2) (Tessema and Nayak 2022).

24.3.2.1 Biodiversity Conservation

Biodiversity conservation is a crucial supporting ecosystem service that plays a vital role in maintaining the productivity and sustainability of ecosystems. In India, biodiversity conservation is crucial for the maintenance of many traditional farming practices, which are dependent on a diversity of plant and animal species (GOI 2021). Studies have shown that the loss of biodiversity can have significant negative impacts on agriculture, human health, and the environment (Majhi et al. 2022; Nayar and Sastry 2020). Globally, biodiversity conservation has been shown to have significant benefits for the environment and human well-being. A study by Kass (2020) found that the loss of biodiversity is a global crisis that threatens human well-being and that biodiversity conservation is essential for achieving sustainable development goals. The Convention on Biological Diversity (CBD), an international treaty signed by over 190 countries, recognizes the critical role of biodiversity conservation in sustainable development. India is a signatory to the CBD and has taken significant steps to conserve its biodiversity. The country has established protected areas, such as national parks and wildlife sanctuaries, and implemented policies and programs to promote sustainable use and conservation of biodiversity (CBD 2021). Jing et al. (2021) assessed the role of biodiversity in supporting ecosystem multi-functionality at a global scale. The study found that biodiversity loss could have significant impacts on the provision of multiple ecosystem services, including nutrient cycling, pollination, and water regulation. The study highlighted the need for conservation efforts to maintain biodiversity and support ecosystem services. Pradhan et al. (2020) assessed the role of plant diversity in supporting ecosystem services in the Indian Himalayas. The study found that plant diversity is critical for supporting multiple ecosystem services, including carbon sequestration, soil health, and water regulation. The study highlighted the need for conservation efforts to maintain plant diversity and support ecosystem services in the region. Wang et al. (2021a, b) assessed the global decline of marine biodiversity and its impacts on supporting ecosystem services. The study found that the loss of marine biodiversity could have significant impacts on the provision of ecosystem services, including fisheries, coastal protection, and carbon sequestration. The study highlighted the need for conservation efforts to maintain marine biodiversity and support ecosystem services. Newbold et al. (2020) assessed the contributions of biodiversity to ecosystem services in terrestrial ecosystems worldwide. The study found that biodiversity plays a key role in supporting many ecosystem services, including carbon storage, soil fertility, and pest regulation. The study also highlighted the need for biodiversity conservation to maintain the provision of these services. Mohan et al. (2021) assessed the contributions of biodiversity to ecosystem services in the Western Ghats region of India. The study found that biodiversity plays a critical role in supporting many ecosystem services, including water regulation, pollination, and soil fertility. The study also highlighted the need for biodiversity conservation to ensure the continued provision of these services. Gustafsson et al. (2020) assessed the economic benefits of biodiversity conservation in European forests. The study found that conserving biodiversity in forest ecosystems can have significant economic benefits, including increased timber production, improved carbon sequestration, and enhanced recreational opportunities. The study highlighted the need for policies that promote biodiversity conservation and sustainable forest management practices (Fig. 24.2).

24.3.2.2 Pollination

Pollination is a critical supporting ecosystem service that is essential for the reproduction of many plant species and the production of many crops. In recent years, there has been growing concern about the decline of pollinators and its potential impacts on ecosystem services and food security. Pollinators play a crucial role in supporting crop production and enhancing crop quality. González et al. (2020) found that pollinators contribute to crop production in Mexico, highlighting the need for pollinator conservation and sustainable land management practices. Pollinators play a critical role in maintaining plant biodiversity. Oliveira et al. (2020) found that pollinators are essential for the reproduction of many plant species and that their loss could have significant impacts on biodiversity. The decline of pollinators could have significant economic impacts on agriculture. Porto et al. (2020) assessed the economic value of pollination services worldwide and found that they contribute significantly to global agricultural production. The conservation of pollinators requires coordinated efforts at the global, national, and local levels. IPBES (2019) assessed the status of pollinators worldwide and highlighted the need for policies and interventions to address the drivers of pollinator decline. Singh et al. (2021) assessed the role of native bees in crop pollination in India. The study found that native bees contribute significantly to crop pollination, particularly for crops such as mango, watermelon, and pumpkin. The study also highlighted the need for conservation of native bee habitats and the promotion of pollinator-friendly agricultural practices to ensure continued pollination services. Khalifa et al. (2021) assessed the contributions of honeybees to crop pollination and honey production in India. The study found that honeybees play a critical role in both crop pollination and honey production, and that their decline could have significant impacts on both agricultural productivity and the livelihoods of beekeepers. The study highlighted the need for pollinator conservation and sustainable beekeeping practices to ensure continued pollination services and honey production. Junqueira et al. (2022) assessed the contributions of wild bees to crop pollination in the United States. The study found that wild bees contribute significantly to crop pollination, particularly for crops such as blueberries, almonds, and cherries. The study also highlighted the need for pollinator-friendly agricultural practices and the conservation of wild bee habitats to ensure continued pollination services. Globally, pollination has been shown to be essential for the production of many food crops, including fruits, vegetables, and nuts. A study by Garibaldi et al. (2022) found that pollinators contribute to the production of 75% of global food crops, including many essential crops such as coffee, cocoa, and almonds. The study also highlighted the significant economic benefits of pollination, estimating that pollination services contribute approximately $235–577 billion annually to the global economy. However, pollinators face numerous threats, including habitat loss, pesticide use, and climate change. In India, studies have shown that the loss of habitat and pesticide use have had significant negative impacts on pollinator populations. A study by Mishra et al. (2019) found that pollinator diversity and abundance have declined in agricultural landscapes in India due to habitat loss and pesticide use (Fig. 24.2).

24.3.2.3 Biomass Production and Soil Fertility Improvement

Biomass production also plays a critical role in soil fertility improvement, with the incorporation of crop residues and other organic materials into the soil contributing to improved soil health and increased crop yields. A study by Gautam et al. (2021) found that the use of biomass-based organic fertilizers significantly increased soil fertility and crop yields in maize and wheat fields in Nepal. In India, studies have shown that agroforestry systems that integrate trees with crops can significantly increase biomass production, improve soil fertility, and provide multiple benefits such as food security, carbon sequestration, and biodiversity conservation. A study by Gupta et al. (2019) found that the integration of trees with crops in agroforestry systems can increase soil organic carbon by up to 60% and improve soil fertility by increasing nutrient availability and water retention. Globally, biomass production and soil fertility improvement have been shown to be essential for sustainable agriculture and food security. A study by Antar et al. (2021) found that improving soil fertility through the use of organic matter can increase crop yields, reduce greenhouse gas emissions, and enhance food security. Biomass production and soil fertility improvement are crucial for sustaining agricultural productivity and reducing greenhouse gas emissions. A study by Lal (2020) estimated that sustainable land management practices, such as conservation agriculture and agroforestry, can sequester up to 3 gigatons of carbon per year and increase soil carbon by up to 1.5 gigatons per year, while also providing multiple benefits such as improved soil fertility, reduced erosion, and increased biodiversity. However, biomass production and soil fertility improvement face numerous challenges, including land degradation, deforestation, and climate change. In India, studies have shown that land degradation and soil erosion are significant challenges in many agricultural landscapes, leading to reduced biomass production and soil fertility. A study by Ghosh et al. (2020) found that soil erosion in the Brahmaputra basin in Northeast India has resulted in significant losses of soil organic carbon and soil nutrients, leading to reduced agricultural productivity. Biomass production is essential for the provision of fuel wood, which is a primary source of energy for many rural households. Studies have shown that sustainable biomass production can help to reduce pressure on forests and improve the livelihoods of rural communities (Sengupta 2022). Soil fertility improvement is also crucial for the production of many food crops and the maintenance of soil health. In India, studies have shown that soil fertility is declining due to factors such as intensive agriculture, soil erosion, and chemical fertilizers. Sustainable land management practices, such as agroforestry and cover cropping, have been shown to improve soil fertility and increase crop yields (Singh et al. 2023a). Sustainable biomass production and soil fertility improvement have been shown to have significant benefits for the environment and human well-being. Schröter et al. (2020) found that sustainable land management practices, including agroforestry and cover cropping, can increase soil fertility and improve the provision of ecosystem services such as food production and carbon sequestration (Fig. 24.2).

24.3.3 Regulating Services

Regulating services are those that help to maintain the balance and functioning of ecosystems, providing benefits such as water purification, climate regulation, and erosion control (Wang et al. 2021a, b). In India, regulating ecosystem services are particularly important due to the country's high population density and dependence on natural resources. For example, the regulation of water flow and quality is essential for agriculture and human consumption, and the regulation of air quality is necessary for human health (Garland et al. 2021) Studies have shown that the loss of regulating ecosystem services can have significant negative impacts on human well-being and the environment. In India, deforestation, land-use changes, and water pollution have contributed to the loss of regulating ecosystem services, leading to negative consequences such as water scarcity, increased air pollution, and soil erosion (Singh and Bhatnagar 2018). Therefore, it is critical to prioritize the conservation and restoration of regulating ecosystem services. Globally, research has highlighted the importance of regulating ecosystem services in mitigating climate change. For example, forests act as carbon sinks, absorbing and storing carbon dioxide from the atmosphere, and wetlands play a vital role in storing carbon and regulating greenhouse gas emissions. However, these ecosystems are under threat from activities such as deforestation, land-use changes, and urbanization, highlighting the need for conservation and restoration efforts (Gomes et al. 2020) (Fig. 24.2).

24.3.3.1 Erosion Control and Soil Conservation

Erosion control and soil conservation are particularly important in India, where soil erosion and land degradation are significant problems due to population growth, unsustainable land-use practices, and climate change (Anantha et al. 2021). Studies have shown that erosion control and soil conservation regulating ecosystem services play a crucial role in maintaining soil fertility and preventing soil erosion, which can lead to nutrient depletion and loss of agricultural productivity. For example, research in India has shown that the use of conservation agriculture techniques such as minimum tillage, mulching, and cover cropping can significantly reduce soil erosion and improve soil fertility (Jat et al. 2021). Globally, erosion control and soil conservation regulating ecosystem services are crucial for maintaining healthy ecosystems and preventing soil degradation, which can have significant negative impacts on food security, water quality, and biodiversity. For example, research has shown that soil erosion can lead to increased sedimentation in rivers and streams, which can negatively impact aquatic ecosystems and reduce water quality (Xiao et al. 2021). Furthermore, erosion control and soil conservation regulating ecosystem services can provide economic benefits, such as increased agricultural productivity and improved water quality. For example, research in India has shown that the use of soil conservation practices can lead to increased crop yields and improved soil quality, leading to increased incomes for farmers (Hossain et al. 2020) (Fig. 24.2).

24.3.3.2 Mitigating Desertification

Desertification is a significant environmental problem worldwide, affecting more than two billion people in over 100 countries. Desertification leads to the loss of soil productivity, reduced biodiversity, and social and economic hardship for affected communities (Sharafatmandrad and KhosraviMashizi 2021). Regulating ecosystem services play important role in mitigating desertification. For example, soil conservation practices such as conservation tillage, terracing, and cover cropping can reduce soil erosion and improve soil health, leading to increased agricultural productivity and reduced vulnerability to desertification (He et al. 2022). Research has also shown that restoring vegetation cover through afforestation and reforestation can help to mitigate desertification. Vegetation cover can help to regulate water cycles, increase soil moisture retention, and reduce soil erosion (de Araujo et al. 2021). Additionally, restored vegetation can provide habitat for biodiversity and ecosystem services such as pollination and pest control. In addition to soil conservation and vegetation restoration, water management is also critical in mitigating desertification. Regulating ecosystem services such as water retention, infiltration, and storage can help to maintain soil moisture and reduce the risk of soil degradation. For example, rainwater harvesting and storage systems can help to retain water for agricultural use during dry periods, reducing the reliance on groundwater resources and reducing the risk of desertification (Yu et al. 2021) (Fig. 24.2).

24.3.3.3 Carbon Sequestration

It is the process by which carbon dioxide (CO_2) is removed from the atmosphere and stored in carbon sinks, such as forests, soil, and oceans. It is a crucial ecosystem service provided by natural systems that helps mitigate the negative impacts of climate change. One recent study explored the potential of natural climate solutions, such as reforestation and forest management, to increase carbon sequestration and mitigate climate change. The study found that natural climate solutions could provide up to one-third of the emissions reductions needed to keep global warming below 2C, while also providing additional benefits, such as biodiversity conservation and sustainable development (Drever et al. 2021). A study was conducted to evaluate the effectiveness of soil carbon sequestration as a climate mitigation strategy. The study found that increasing soil carbon storage by 1 ton per hectare per year could reduce atmospheric CO_2 concentrations by up to 10 parts per million by 2100, which would significantly contribute to achieving the goals of the Paris Agreement (Smith et al. 2020). A study published in the journal *Environmental Research Letters* assessed the carbon sequestration potential of mangrove forests in the Mekong Delta region of Vietnam. The study found that mangrove forests in the region could sequester up to 1.5 million tons of carbon per year, which would provide significant benefits for climate change mitigation and adaptation, as well as local communities that depend on the ecosystem services provided by mangroves (Hauser et al. 2020). Ribeiro et al. (2021) highlight the importance of carbon sequestration as an ecosystem service in the miombo woodlands region of Zimbabwe. The study evaluated the carbon sequestration potential of different land-use systems, including natural forests, degraded forests, grasslands, and croplands. The results showed that natural forests had the highest carbon sequestration potential, followed by degraded forests and grasslands. Croplands, on the other hand, had the lowest carbon sequestration potential. Masson-Delmotte et al. (2022) highlighted the need to increase carbon sequestration in all ecosystems, including forests, grasslands, and wetlands, to limit global warming to 1.5C. The report also emphasized the importance of sustainable land management practices to enhance the capacity of ecosystems to store carbon. Macreadie et al. (2021) assessed the potential of blue carbon ecosystems, such as mangroves, seagrasses, and saltmarshes, to mitigate climate change through carbon sequestration. The review highlighted the significant carbon sequestration potential of blue carbon ecosystems, with estimates suggesting that they could store up to 25 billion tons of carbon dioxide equivalents by 2050 (Fig. 24.2).

24.3.3.4 Control of Weeds, Insect Pest, and Diseases

Weeds are one of the most significant threats to crop production worldwide. Agroforestry systems provide a natural and sustainable approach to weed management. The shade from the trees in agroforestry systems reduces weed growth by reducing the amount of sunlight that reaches the ground, thus reducing weed seed germination and growth (Harms et al. 2020). In addition, tree roots can compete with weeds for nutrients, further reducing weed growth (Singh et al. 2017; Satapathy et al. 2020; Bhoi et al. 2022a, b; Mahanta et al. 2022; Prabhulinga et al. 2022; Samal et al. 2023a, b). Fahad et al. (2022) in Tanzania investigated the effect of agroforestry systems on the control of weeds in maize production. The study found that maize grown in agroforestry systems had lower weed density and biomass compared to maize grown in monoculture systems. This was attributed to the shade provided by the trees, which reduced the amount of light reaching the ground and thus suppressed weed growth. Agroforestry systems can provide disease control by creating a diverse and resilient agroecosystem. A diverse agroecosystem with a mix of crops and trees creates a natural barrier against diseases by reducing the spread of pathogens (LibertAmico et al. 2020). Additionally, some tree species in agroforestry systems have been found to have antifungal and antibacterial properties, further reducing the spread of plant diseases (Mahanta et al. 2023; Majhi et al. 2023; Singh et al. 2023b). Cerda et al. (2020) in Colombia investigated the effect of agroforestry systems on the control of coffee leaf rust disease. The study found that coffee plants grown in agroforestry systems had lower levels of leaf rust disease compared to coffee plants grown in monoculture systems. This was attributed to the presence of shade trees, which created a microclimate that was less favorable for the growth and spread of the leaf rust disease. Agroforestry systems can provide a natural approach to insect pest control by creating habitats for natural enemies of insect pests. Trees in agroforestry systems provide habitat and food sources for natural enemies, such as birds and insects, which prey on insect pests (Kumar et al. 2016). In addition, some tree species in agroforestry systems have been found to have insecticidal properties, further reducing the damage caused by insect pests (Ugwu 2020). Martínez-Sastre et al. (2020) in Spain investigated the effect of agroforestry systems on the control of insect pests in apple orchards. The study found that apple trees grown in agroforestry systems had lower levels of insect pest infestation compared to apple trees grown in monoculture systems. This was attributed to the presence of natural enemies of the insect pests, such as birds and insects, which were attracted by the trees in the agroforestry system and helped to control the pest populations (Fig. 24.2).

24.3.4 Cultural Services

Agroforestry systems provide several ecosystem services, including cultural services. Cultural services are the non-material benefits that people derive from ecosystems, such as aesthetic, spiritual, and recreational values. These services are essential for human well-being and are often overlooked in decision-making processes (Fig. 24.2).

1. **Aesthetic Values:** Agroforestry systems can provide aesthetic values that enhance the visual appeal of the landscape. Trees, crops, and livestock can create a diverse and vibrant landscape that can be attractive to people. Several studies have demonstrated the aesthetic value of agroforestry systems. For example, a study in Uganda showed that farmers perceived their agroforestry systems as more attractive than monoculture systems. Another study in Costa Rica showed that agroforestry systems were perceived as more beautiful than conventional agriculture (Leary et al. 2021) (Fig. 24.2).
2. **Spiritual Values:** Agroforestry systems can also provide spiritual values, such as a connection to nature and cultural heritage. Trees and forests have been associated with spiritual and cultural beliefs in many cultures worldwide. Several studies have documented the spiritual values of agroforestry systems. For example, a study in Ethiopia showed that farmers attributed spiritual values to their agroforestry systems, including the belief that trees were the dwelling place of spirits. Another study in Mexico showed that farmers associated their agroforestry systems with cultural identity and a connection to their ancestors (Sierra-Huelsz et al. 2020) (Fig. 24.2).
3. **Recreational Values:** Agroforestry systems can provide opportunities for recreation activities such as hiking, bird watching, and nature appreciation. The presence of trees and other vegetation in these systems provides habitat for wildlife, making them attractive for recreation activities. Several studies have shown that agroforestry systems can provide recreational opportunities for local communities and tourists (Prihayati and Veriasa 2021) (Fig. 24.2). Various ecosystem services are depicted in Table 24.1.

24.4 Socioeconomic Impact

1. Increased income: Agroforestry systems can supply farmers with a range of products, allowing them to earn more money. Trees, for example, can be harvested for lumber, fruits, or nuts, and livestock can graze on understory vegetation (Mukhlis et al. 2022).
2. Enhanced food security: Agroforestry systems can help to increase food security by providing farmers with a more consistent source of food. During the off-season, trees, for example, can produce fruits and nuts, while cattle can supply milk and meat (Aryal et al. 2023; Kadykalo et al. 2021).
3. Poverty reduction: Agroforestry systems can aid in poverty reduction by providing farmers with a more sustainable way of living. Trees, for example, can provide shade and feed, which can assist to reduce input costs, and agroforestry

Table 24.1 Represent the various ecosystem services provided by agroforestry system

SL. no.	Ecosystem services	Key findings	Reference
Regulating Services			
1	Climate regulation	Agroforestry systems can help regulate local and regional climates by providing shade, reducing wind speed, and increasing humidity.	Fahad et al. (2022)
2	Pest regulation	Agroforestry systems can promote natural pest control by providing habitat for beneficial insects and birds, reducing the need for chemical pesticides.	Monteagudo et al. (2023)
3	Disease regulation	Agroforestry systems can reduce the incidence and severity of plant diseases by promoting healthy soil and diverse crop rotations.	Sollen-Norrlin et al. (2020)
4	Water regulation	Agroforestry systems can improve water management by reducing runoff and improving water infiltration and storage.	Zhu et al. (2020)
5	Soil regulation	Agroforestry systems can promote healthy soil by improving soil structure, increasing soil organic matter, and reducing erosion.	Hartmann et al. (2023)
Cultural services			
6	Aesthetic values	Agroforestry systems can provide aesthetic benefits, such as scenic beauty, wildlife viewing, and recreational opportunities.	Smith et al. (2021)
7	Spiritual and religious values	Agroforestry systems can have cultural and spiritual significance for communities, including providing sacred groves and other cultural practices.	Ormsby and Krishnan (2022)
8	Educational and scientific values	Agroforestry systems can provide opportunities for education and scientific research, such as studying agroforestry practices and their impacts on ecosystems and livelihoods.	Akter et al. (2022)
9	Cultural heritage values	Agroforestry systems can be an important part of cultural heritage, such as traditional agroforestry systems and associated knowledge and practices.	Santoro (2023)
10	Social and community values	Agroforestry systems can provide social and community benefits, such as strengthening local networks and supporting traditional livelihoods.	Meinhold and Darr (2021)
Supporting services			
11	Soil formation	Agroforestry systems can support soil formation and maintenance by promoting nutrient cycling, organic matter accumulation, and soil structure.	Fahad et al. (2022)
12	Nutrient cycling	Agroforestry systems can enhance nutrient cycling by utilizing multiple plant species with complementary nutrient needs and reducing nutrient losses through soil erosion and leaching.	Sileshi et al. (2020)
13	Primary production	Agroforestry systems can support primary production by diversifying crop and tree species, reducing crop failure risks, and increasing crop and tree yields.	Bertsch-Hörmann (2021)

(continued)

Table 24.1 (continued)

SL. no.	Ecosystem services	Key findings	Reference
14	Habitat provision	Agroforestry systems can provide habitat for a range of plant and animal species, supporting biodiversity and ecosystem health.	Egwumah et al. (2022)
15	Genetic resources	Agroforestry systems can contribute to the conservation and management of genetic resources, including crop and tree species with cultural, medicinal, and other values.	Suwardi and Navia (2023)
Provisioning services			
16	Food production	Agroforestry systems can produce a range of food products, including fruits, nuts, vegetables, and livestock, providing diversified and sustainable sources of nutrition.	Damerau et al. (2020)
17	Timber and non-timber forest products	Agroforestry systems can provide timber and non-timber forest products, such as fuel wood, medicinal plants, and ornamental plants, supporting livelihoods and local economies.	Gurung et al. (2021)
18	Water resources	Agroforestry systems can contribute to water resource management by reducing erosion, increasing infiltration and water retention, and improving water quality.	Zhu et al. (2020)
19	Fiber and fodder production	Agroforestry systems can produce fiber and fodder products, such as cotton, bamboo, and forage crops, supporting diverse and sustainable livelihoods.	Santoro et al. (2020)
20	Fuel wood production	Agroforestry systems can provide fuel wood for household and commercial use, reducing pressure on natural forests and supporting sustainable energy sources.	Khadka et al. (2021)

systems can help to save soil and water, lowering the chance of crop failure (Castle et al. 2022).

4. Environmental quality enhancement: Environmental quality can be improved by using agroforestry systems to provide habitat for wildlife, reduce erosion, and conserve water. This can result in a variety of advantages, including enhanced air quality, less floods, and increased resilience to climate change (Akter et al. 2022).
5. Increased social cohesiveness: Agroforestry systems can provide a gathering area for individuals, which can aid in social cohesion. This can result in a variety of advantages, including improved education, lower crime, and a stronger sense of community (Quandt et al. 2023).

24.5 Challenges Associated with Ecosystem Service

1. **Market access:** One of the main challenges associated with ecosystem services in agroforestry systems is market access. Farmers may not have access to markets that are willing to pay for ecosystem services, which can limit their economic benefits from agroforestry systems.
2. **Land tenure:** Another challenge is land tenure, which can limit the adoption of agroforestry systems. Farmers may be reluctant to invest in agroforestry systems if they do not have secure land tenure, as they may not be able to benefit from the long-term ecosystem services provided by these systems (Geressu et al. 2020).
3. **Institutional barriers:** Institutional barriers such as policies, regulations, and governance can also limit the adoption and implementation of agroforestry systems. These barriers can limit the ability of farmers to access financing, technical assistance, and other resources needed to establish and maintain agroforestry systems (Sheppard et al. 2020).

24.6 Opportunities and Ecosystem Service

1. **Payment for ecosystem services:** Payment for ecosystem services (PES) schemes provide opportunities for farmers to earn income from the ecosystem services provided by agroforestry systems. PES schemes can provide financial incentives for farmers to invest in and maintain these systems (Garrett et al. 2021).
2. **Sustainable development goals:** Agroforestry systems can contribute to several sustainable development goals (SDGs), including poverty reduction, food security, climate action, and biodiversity conservation. This provides opportunities for policymakers and stakeholders to support the adoption and scaling up of agroforestry systems (Piemontese et al. 2021).
3. **Climate change mitigation and adaptation:** Agroforestry systems can contribute to climate change mitigation and adaptation by sequestering carbon, reducing greenhouse gas emissions, and increasing resilience to climate change impacts. This provides opportunities for farmers to contribute to global efforts to address climate change (Sharifi 2021).

24.7 Future Prospects

1. **Scaling up:** There is a growing interest in scaling up agroforestry systems to meet the challenges of food security, climate change, and environmental degradation. Scaling up agroforestry systems can provide a range of ecosystem services, including increased carbon sequestration, enhanced biodiversity, and improved soil fertility. This can contribute to the achievement of several sustainable

development goals, including poverty reduction, food security, and climate action (Jurado et al. 2022).
2. **Technological innovations:** Technological innovations such as remote sensing, geographic information systems (GIS), and precision agriculture can enhance the effectiveness and efficiency of agroforestry systems. For example, remote sensing can be used to monitor vegetation cover, while GIS can be used to map ecosystem services at different spatial scales. Precision agriculture can improve crop yields and reduce inputs, leading to more sustainable and profitable agroforestry systems (Plieninger et al. 2020).
3. **Policy support:** Policymakers can play a critical role in supporting the adoption and scaling up of agroforestry systems. Policies that incentivize the adoption of agroforestry systems, such as subsidies or tax breaks, can increase the economic benefits for farmers. Policies that regulate land-use change and deforestation can also support the protection and restoration of forests and other ecosystems (Plieninger et al. 2020).
4. **Partnerships:** Partnerships among governments, civil society, and the private sector can enhance the adoption and scaling up of agroforestry systems. Partnerships can leverage the expertise and resources of different actors to support the development of sustainable and profitable agroforestry systems. For example, partnerships can support the development of value chains that link farmers to markets for agroforestry products and services (Awazi 2022).

24.8 Conclusion

Agroforestry systems offer a wide range of ecosystem services that can benefit both the environment and farmers. These services include carbon sequestration, biodiversity conservation, soil fertility improvement, and more. However, there are challenges and opportunities associated with these services, such as the need for scaling up, technological innovations, policy support, and partnerships. By addressing these challenges and leveraging these opportunities, we can create more sustainable and profitable agroforestry systems that contribute to sustainable development and climate change mitigation and adaptation. The future prospects for ecosystem services in agroforestry systems are promising, and there is growing interest in adopting and scaling up these systems to meet the challenges of food security, climate change, and environmental degradation.

References

Akter R, Hasan MK, Kabir KH, Darr D, Roshni NA (2022) Agroforestry systems and their impact on livelihood improvement of tribal farmers in a tropical moist deciduous forest in Bangladesh. Trees For People 9:100315. https://doi.org/10.1016/j.tfp.2022.100315

Anantha KH, Garg KK, Barron J, Dixit S, Venkataradha A, Singh R, Whitbread AM (2021) Impact of best management practices on sustainable crop production and climate resilience in smallholder farming systems of South Asia. Agric Syst 194:103276. https://doi.org/10.1016/j.agsy.2021.103276

Antar M, Lyu D, Nazari M, Shah A, Zhou X, Smith DL (2021) Biomass for a sustainable bioeconomy: an overview of world biomass production and utilization. Renew Sust Energ Rev 139:110691. https://doi.org/10.1016/j.rser.2020.110691

Aryal K, Maraseni T, Apan A (2023) Transforming agroforestry in contested landscapes: a win-win solution to trade-offs in ecosystem services in Nepal. Sci Total Environ 857:159301

Awazi NP (2022) Agroforestry for climate change adaptation, resilience enhancement and vulnerability attenuation in smallholder farming systems in Cameroon. J Atmos Sci Res. 5(1):1–16. https://doi.org/10.30564/jasr.v5i1.4303

Balvanera P, Siddique I, Dee LE, Paquette A, Isbell F, Gonzalez A et al (2020) Key features and context-dependency of ecosystem service provision across contrasting ecosystems. Environ Res Lett 15(11):114046

Bertsch-Hörmann B (2021) Effects of agroforestry on the carbon dynamics of an agroecological landscape: human appropriation of net primary production in two land use scenarios in the Eisenwurzen region, Austria. Social ecology working paper, 192. Institute of Social Ecology, Vienna

Bhoi TK, Samal I, Majhi PK, Komal J, Mahanta DK, Pradhan AK et al (2022a) Insight into aphid mediated potato virus Y transmission: a molecular to bioinformatics prospective. Front Microbiol 13:1001454. https://doi.org/10.3389/fmicb.2022.1001454

Bhoi TK, Samal I, Mahanta DK, Komal J, Jinger D, Sahoo MR et al (2022b) Understanding how silicon fertilization impacts chemical ecology and multitrophic interactions among plants, insects and beneficial arthropods. SILICON 15:2529–2549. https://doi.org/10.1007/s12633-022-02220-6

Borelli T, Hunter D, Padulosi S, Amaya N, Meldrum G, de Oliveira Beltrame DM (2020) Local solutions for sustainable food systems: the contribution of orphan crops and wild edible species. Agronomy 10(2):231. https://doi.org/10.3390/agronomy10020231

Braat LC, De Groot R (2012) The ecosystem services agenda: bridging the worlds of natural science and economics, conservation and development, and public and private policy. Ecosyst Serv 1:4–15. https://doi.org/10.1016/j.ecoser.2012.07.011

Castle SE, Miller DC, Merten N, Ordonez PJ, Baylis K (2022) Evidence for the impacts of agroforestry on ecosystem services and human well-being in high-income countries: a systematic map. Environ Evid 11(1):1–27

Cerda R, Avelino J, Harvey CA, Gary C, Tixier P, Allinne C (2020) Coffee agroforestry systems capable of reducing disease-induced yield and economic losses while providing multiple ecosystem services. Crop Prot 134:105149. https://doi.org/10.1016/j.cropro.2020.105149

Community-based Distribution (Convention on Biological Diversity) (2021). Retrieved from https://www.cbd.int/convention/. The Convention on Biological Diversity

Damerau K, Davis KF, Godde C, Herrero M, Springmann M, Bhupathiraju SN, Myers SS, Willett W (2020) India has natural resource capacity to achieve nutrition security, reduce health risks and improve environmental sustainability. Nat Food 1(10):631–639. https://doi.org/10.1038/s43016-020-00157-w

de Araujo HFP, Machado CCC, Pareyn FGC, do Nascimento NF, Araújo LDA, de Borges Laís AAP, Santos BA, Beirigo RM, Vasconcellos A, de Dias BO, Alvarado F, da Silva JMC (2021) A sustainable agricultural landscape model for tropical drylands. Land Use Policy 100:104913. https://doi.org/10.1016/j.landusepol.2020.104913

Donatti CI, Andrade A, Cohen-Shacham E, Fedele G, Hou-Jones X, Robyn B (2022) Ensuring that nature-based solutions for climate mitigation address multiple global challenges. One Earth 5(5):493–504. https://doi.org/10.1016/j.oneear.2022.04.010

Drever CR, Cook-Patton SC, Akhter F, Badiou PH, Chmura GL, Davidson SJ (2021) Natural climate solutions for Canada. Sci Adv. 7(23):eabd6034. https://doi.org/10.1126/sciadv.abd6034

Egwumah FA, Ekwugha EU, Ajayi OE (2022) Paramount roles of agroforestry systems as habitat for bird species. Paramount Ecol Resour 9(1):1–7

Fahad S, Chavan SB, Chichaghare AR, Uthappa AR, Kumar M, Kakade V (2022) Agroforestry systems for soil health improvement and maintenance. Sustainability 14(22):14877. https://doi.org/10.3390/su142214877

Garibaldi LA, Gomez Carella DS, Nabaes Jodar DN, Smith MR, Timberlake TP, Myers SS (2022) Exploring connections between pollinator health and human health. Philos Trans R Soc Lond Ser B Biol Sci 377(1853):20210158. https://doi.org/10.1098/rstb.2021.0158

Garland G, Banerjee S, Edlinger A, Miranda Oliveira E, Herzog C, Wittwer R, Philippot L, Maestre FT, van Der Heijden MGA (2021) A closer look at the functions behind ecosystem multifunctionality: a review. J Ecol 109(2):600–613. https://doi.org/10.1111/1365-2745.13511

Garrett LG, Smaill SJ, Addison SL, Clinton PW (2021) Globally relevant lessons from a long-term trial series testing universal hypothesis of the impacts of increasing biomass removal on site productivity and nutrient pools. Forest Ecol Manag 494:119325. https://doi.org/10.1016/j.foreco.2021.119325

Gautam R, Shriwastav CP, Lamichhane S, Baral BR (2021) The residual effect of pre-rice green manuring on a succeeding wheat crop (Triticumaestivum L.) in the rice-wheat cropping system in Banke, Nepal. Int J Agron 2021:1–10. https://doi.org/10.1155/2021/4142743

Geressu R, Siderius C, Harou JJ, Kashaigili J, Pettinotti L, Conway D (2020) Assessing river basin development given water-energy-food-environment interdependencies. Earth Future 8(8): e2019EF001464. https://doi.org/10.1029/2019EF001464

Ghosh A, Das A, Das D, Ray P, Bhattacharyya R, Biswas DR, Biswas SS (2020) Contrasting land use systems and soil organic matter quality and temperature sensitivity in North Eastern India. Soil Tillage Res 199:104573. https://doi.org/10.1016/j.still.2020.104573

Gomes LC, Bianchi FJJA, Cardoso IM, FernandesFilho EI, Schulte RPO (2020) Land use change drives the spatio-temporal variation of ecosystem services and their interactions along an altitudinal gradient in Brazil. Landsc Ecol 35(7):1571–1586. https://doi.org/10.1007/s10980-020-01037-1

Gonçalves CDBQ, Schlindwein MM, Martinelli GDC (2021) Agroforestry systems: A systematic review focusing on traditional indigenous practices, food and nutrition security, economic viability, and the role of women. Sustainability 13(20):11397. https://doi.org/10.3390/su132011397

González AV, Gómez-Silva V, Ramírez MJ, Fontúrbel FE (2020) Meta-analysis of the differential effects of habitat fragmentation and degradation on plant genetic diversity. Conserv Biol 34(3): 711–720. https://doi.org/10.1111/cobi.13422

Government of India (2021) National biodiversity act. Retrieved from https://www.indiacode.nic.in/handle/123456789/2359

Gupta SR, Dagar JC, Jangra R, Gaur A (2019) Tree-based systems for enhancing environmental services of saline environments. Res Dev Saline Agric:461–502

Gurung LJ, Miller KK, Venn S, Bryan BA (2021, November 20) Climate change adaptation for managing non-timber forest products in the Nepalese Himalaya. Sci Total Environ 796:148853. https://doi.org/10.1016/j.scitotenv.2021.148853

Gustafsson L, Bauhus J, Asbeck T, Augustynczik ALD, Basile M, Frey J (2020) Retention as an integrated biodiversity conservation approach for continuous-cover forestry in Europe. Ambio 49(1):85–97. https://doi.org/10.1007/s13280-019-01190-1

Harms NE, Cronin JT, Diaz R, Winston RL (2020) A review of the causes and consequences of geographical variability in weed biological control successes. Biol Control 151:104398. https://doi.org/10.1016/j.biocontrol.2020.104398

Hartmann M, Herzog C, Brunner I, Stierli B, Meyer F, Buchmann N et al (2023) Long-term mitigation of drought changes the functional potential and life-strategies of the forest soil microbiome involved in organic matter decomposition. Front Microbiol 14:1267270. https://doi.org/10.3389/fmicb.2023.1267270

Hauser LT, AnBinh N, Viet Hoa P, Hong Quan N, Timmermans J (2020) Gap-free monitoring of annual mangrove forest dynamics in Ca Mau Province, Vietnamese Mekong delta, using the landsat-7-8 archives and post-classification temporal optimization. Remote Sens 12(22):3729. https://doi.org/10.3390/rs12223729

He C, Niu JR, Xu CT, Han SW, Bai W, Song QL, Dang YP, Zhang HL (2022) Effect of conservation tillage on crop yield and soil organic carbon in Northeast China: A meta-analysis. Soil Use Manag 38(2):1146–1161. https://doi.org/10.1111/sum.12784

Hossain A, Krupnik TJ, Timsina J, Mahboob MG, Chaki AK, Farooq M, Bhatt R, Fahad S, Hasanuzzaman M (2020) Agricultural land degradation: Processes and problems undermining future food security. In: Environment, climate, plant and vegetation growth. Springer, Cham, pp 17–61

Hübner R, Kühnel A, Lu J, Dettmann H, Wang W, Wiesmeier M (2021) Soil carbon sequestration by agroforestry systems in China: A meta-analysis. Agric Ecosyst Environ 315:107437. https://doi.org/10.1016/j.agee.2021.107437

IPBES (2019) Global assessment report on biodiversity and ecosystem services of the Intergovernmental Science-Policy Platform on biodiversity and Ecosystem Services

Jat HS, Datta A, Choudhary M, Sharma PC, Jat ML (2021) Conservation Agriculture: Factors and drivers of adoption and scalable innovative practices in Indo-Gangetic plains of India–a review. Int J Agric Sustain 19(1):40–55. https://doi.org/10.1080/14735903.2020.1817655

Jing X, Prager CM, Borer ET, Gotelli NJ, Gruner DS, He JS et al (2021) Spatial turnover of multiple ecosystem functions is more associated with plant than soil microbial β-diversity. Ecosphere 12 (7):e03644. https://doi.org/10.1002/ecs2.3644

Jnawali SR, Aryal KP, Bhattarai S (2021) Grassland ecosystem services and their contribution to livestock feed provision in Nepal. J Environ Manag 282:111965

Junqueira CN, Pereira RAS, da Silva RC, Alves Cardoso Kobal RO, Araújo TN, Prato A (2022) Do Apis and non-Apis bees provide a similar contribution to crop production with different levels of pollination dependency? A review using meta-analysis. Ecol Entomol 47(1):76–83. https://doi.org/10.1111/een.13092

Jurado JM, López A, Pádua L, Sousa JJ (2022) Remote sensing image fusion on 3D scenarios: a review of applications for agriculture and forestry. Int J Appl Earth Obs Geoinf 112:102856. https://doi.org/10.1016/j.jag.2022.102856

Kadykalo AN, Kelly LA, Berberi A, Reid JL, Findlay CS (2021) Research effort devoted to regulating and supporting ecosystem services by environmental scientists and economists. PLoS One 16(5):e0252463. https://doi.org/10.1371/journal.pone.0252463

Kass MJ (2020) Summary for policymakers of the global assessment report on biodiversity and ecosystem services. Nat Resour Environ 34(3):62–62

Khadka D, Aryal A, Bhatta KP, Dhakal BP, Baral H (2021) Agroforestry systems and their contribution to supplying forest products to communities in the Chure range, Central Nepal. Forests 12(3):358. https://doi.org/10.3390/f12030358

Khalifa SAM, Elshafiey EH, Shetaia AA, El-Wahed AAA, Algethami AF, Musharraf SG (2021) Overview of bee pollination and its economic value for crop production. Insects 12(8):688. https://doi.org/10.3390/insects12080688

Kim JH, Hwang SJ, Kim RH (2021) Ecosystem service assessment of fuelwood provision using a spatially explicit ecological model in South Korea. Forests 12(5):634

Kumar BM, Nair PKR, Kumar MD (2016) Agroforestry as an ecosystem service provider: a framework for sustainable land use in India

Kumar A, Unnithan GJ, Vardhan H, Sharma SK, Gadhwal N, Sinsinwar MS (2022) Bees and pollinators in perspective: a review and a case study. LS Int J Life Sci. 11(1):46–60

Lal R (2020) Soil health and climate change: a review. Soil Sci 185(3–4):65–76

Langemeyer J, Connolly JJT (2020) Weaving notions of justice into urban ecosystem services research and practice. Environ Sci Policy 109:1–14. https://doi.org/10.1016/j.envsci.2020.03.021

Leary J, Grimm K, Aslan C, Mark M, Frey S, Bath-Rosenfeld R (2021) Landowners' socio-cultural valuation of ecosystem services provided by trees in Costa Rican agricultural landscapes. Environ Manag 67(5):974–987. https://doi.org/10.1007/s00267-021-01442-5

Lelamo LL (2021) A review on the indigenous multipurpose agroforestry tree species in Ethiopia: management, their productive and service roles and constraints. Heliyon 7(9):e07874. https://doi.org/10.1016/j.heliyon.2021.e07874

LibertAmico A, Ituarte-Lima C, Elmqvist T (2020) Learning from social–ecological crisis for legal resilience building: multi-scale dynamics in the coffee rust epidemic. Sustain Sci 15(2): 485–501. https://doi.org/10.1007/s11625-019-00703-x

Liu H, Zheng L, Wu J, Liao Y (2020) Past and future ecosystem service trade-offs in Poyang Lake Basin under different land use policy scenarios. Arab J Geosci 13:1–16

Macreadie PI, Costa MDP, Atwood TB, Friess DA, Kelleway JJ, Kennedy H, Lovelock CE, Serrano O, Duarte CM (2021) Blue carbon as a natural climate solution. Nat Rev Earth Environ 2(12):826–839. https://doi.org/10.1038/s43017-021-00224-1

Maestre FT, Eldridge DJ, Soliveres S, Kéfi S, Delgado-Baquerizo M, Bowker MA et al (2020) Structure and functioning of dryland ecosystems in a changing world. Annu Rev Ecol Evol Syst 51:215–237

Mahanta DK, Jangra S, Priti A, Ghosh A, Sharma PK, Iquebal MA et al (2022) Groundnut bud necrosis virus modulates the expression of innate immune, endocytosis, and cuticle development-associated genes to circulate and propagate in its vector, Thrips palmi. Front Microbiol 13:773238. https://doi.org/10.3389/fmicb.2022.773238

Mahanta DK, Samal I, Komal J, Bhoi TK, Majhi PK, Ahmad MA (2023) Understanding anthropogenic climate change, its consequences on insect pests, and techniques in forecasting and monitoring pest dynamics: a contemporary scenario. In: Climate change and insect biodiversity. CRC Press, pp 44–67

Majhi PK, Raza B, Behera PP, Singh SK, Shiv A, Mogali SC et al (2022) Future-proofing plants against climate change: a path to ensure sustainable food systems. In: Biodiversity, functional ecosystems and sustainable food production. Springer International Publishing, Cham, pp 73–116

Majhi PK, Samal I, Bhoi TK, Pattnaik P, Pradhan C, Gupta A, Mahanta DK, Senapati SK (2023) Climate-smart agriculture: an integrated approach to address climate change and food security. In: Climate change and insect biodiversity. CRC Press, pp 221–245

Manjunatha BL, Naorem A, Hajong D, Tewari P (2022) Dynamics of household and regional economy in Banni Grasslands, India: a cross-sectional study. Sustainability 14(18):11236. https://doi.org/10.3390/su141811236

Martínez-Sastre R, García D, Miñarro M, Martín-López B (2020) Farmers' perceptions and knowledge of natural enemies as providers of biological control in cider apple orchards. J Environ Manag 266:110589. https://doi.org/10.1016/j.jenvman.2020.110589

Masson-Delmotte V, Zhai P, Pörtner HO, Roberts D, Skea J, Shukla PR (2022) Sustainable development, and efforts to eradicate poverty. Global Warming of. Cambridge: Cambridge University Press, 1(5)° C: IPCC Special Report on Impacts of Global Warming of 1.5° C above Pre-industrial Levels in Context of Strengthening Response to Climate Change

Meinhold K, Darr D (2021) Using a multi-stakeholder approach to increase value for traditional agroforestry systems: The case of baobab (Adansoniadigitata L.) in Kilifi, Kenya. Agrofor Syst 95(7):1343–1358. https://doi.org/10.1007/s10457-020-00562-x

Mishra SK, Negri MC, Kozak J, Cacho JF, Quinn J, Secchi S et al (2019) Valuation of ecosystem services in alternative bioenergy landscape scenarios. GCB Bioenergy 11(6):748–762. https://doi.org/10.1111/gcbb.12602

Mohan M, Saritha VNK, Rameshan M, Chacko A, Gopikrishna VG (2021) Restoring degraded riparian forest ecosystems of the Western Ghats for ecological sustainability. Restor Ecol 29(4): e13254. https://doi.org/10.1111/rec.13254

Monteagudo N, Benayas JMR, Andivia E, Rebollo S (2023) Avian regulation of crop and forest pests, a meta-analysis. Pest Manag Sci. https://doi.org/10.1002/ps.7421

Mukhlis I, Rizaludin MS, Hidayah I (2022) Understanding socio-economic and environmental impacts of agroforestry on rural communities. Forests 13(4):556

Muthee K, Duguma L, Majale C, Mucheru-Muna M, Wainaina P, Minang P (2022) A quantitative appraisal of selected agroforestry studies in the Sub-Saharan Africa. Heliyon 8(9):e10670. https://doi.org/10.1016/j.heliyon.2022.e10670

Nandi D, Sarkar S (2021) Non-timber forest products based household industries and rural economy—A case study of Jaypur block in Bankura District, pp 505–528. Bengal, India: West. Spatial Modeling in Forest Resources Management: Rural Livelihood and Sustainable Development

Nayar MP, Sastry ARK (2020) India's biodiversity: status, conservation, and threats. Curr Sci 119(4):539–549

Newbold T, Bentley LF, Hill SLL, Edgar MJ, Horton M, Su G, Şekercioğlu ÇH, Collen B, Purvis A (2020) Global effects of land use on biodiversity differ among functional groups. Funct Ecol 34(3):684–693. https://doi.org/10.1111/1365-2435.13500

Obiang Ndong GO, Therond O, Cousin I (2020) Analysis of relationships between ecosystem services: a generic classification and review of the literature. Ecosyst Serv 43:101120. https://doi.org/10.1016/j.ecoser.2020.101120

Oliveira W, Silva JLS, Porto RG, Cruz-Neto O, Tabarelli M, Viana BF, Peres CA, Lopes AV (2020) Plant and pollination blindness: risky business for human food security. Bioscience 70(2):109–110. https://doi.org/10.1093/biosci/biz139

Ormsby AA, Krishnan S (2022) Spiritual values and ecosystem services of sacred Groves in Karnataka, India. In: Religion and nature conservation. Routledge, London, pp 103–114

Panwar P, Mahalingappa DG, Kaushal R, Bhardwaj DR, Chakravarty S, Shukla G (2022) Biomass production and carbon sequestration potential of different agroforestry systems in India: a critical review. Forests 13(8):1274. https://doi.org/10.3390/f13081274

Patel SK, Sharma A, Singh GS (2020) Traditional agricultural practices in India: an approach for environmental sustainability and food security. Energy Ecol Environ 5(4):253–271. https://doi.org/10.1007/s40974-020-00158-2

Pereira HM, Ferrier S, Walters M, Geller GN, Jongman RH, Scholes RJ et al (2018) Essential biodiversity variables. Science 360(6391):1416–1419

Piemontese L, Kamugisha RN, Tukahirwa JMB, Tengberg A, Pedde S, Jaramillo F (2021) Barriers to scaling sustainable land and water management in Uganda: a cross-scale archetype approach. Ecol Soc 26(3). https://doi.org/10.5751/ES-12531-260306

Plieninger T, Muñoz-Rojas J, Buck LE, Scherr SJ (2020) Agroforestry for sustainable landscape management. Sustain Sci 15(5):1255–1266. https://doi.org/10.1007/s11625-020-00836-4

Porto RG, de Almeida RF, Cruz-Neto O, Tabarelli M, Viana BF, Peres CA, Lopes AV (2020) Pollination ecosystem services: a comprehensive review of economic values, research funding and policy actions. Food Secur 12(6):1425–1442. https://doi.org/10.1007/s12571-020-01043-w

Prabhulinga T, Chander S, Arya PS, Bhoi TK, Yele Y (2022) Mechanism and modifications associated with mimicry and camouflage in butterfly. J Entomol Res 46(3):667–672. https://doi.org/10.5958/0974-4576.2022.00115.3

Pradhan R, Manohar KA, Vineeta BC, Sarkar BC, Bhat JA, Shukla G, Chakravarty S (2020) Ecosystem services of urban green sites-A case study from Eastern Himalayan foothills. Trees For People 2:100029. https://doi.org/10.1016/j.tfp.2020.100029

Pradhan NS, Das PJ, Gupta N, Shrestha AB (2021) Sustainable management options for healthy rivers in South Asia: the case of Brahmaputra. Sustainability 13(3):1087. https://doi.org/10.3390/su13031087

Prihayati Y, Veriasa TO (2021) Developing green tourism to create the sustainable landscape: Evidence from Community-based Coffee Tourism (CbCT) in Puncak, Bogor, Indonesia. In: IOPConferenceSeries. IOP Conference Series: Earth and Environmental Science. 879(1): 012027. IOP Publishing. https://doi.org/10.1088/1755-1315/879/1/012027

Quandt A, Neufeldt H, Gorman K (2023) Climate change adaptation through agroforestry: opportunities and gaps. Curr Opin Environ Sustain 60:101244

Raj A, Jhariya MK, Yadav DK, Banerjee A (eds) (2020) Climate change and agroforestry systems: adaptation and mitigation strategies. CRC Press, Boca Raton, FL

Ribeiro NS, Armstrong AH, Fischer R, Kim YS, Shugart HH, Ribeiro-Barros AI et al (2021) Prediction of forest parameters and carbon accounting under different fire regimes in Miombo woodlands, Niassa Special Reserve, Northern Mozambique. Forest Policy Econ 133:102625. https://doi.org/10.1016/j.forpol.2021.102625

Roy M, Sarkar BC, Manohar KA, Shukla G, Vineeta AJ, Nath AJ (2022) Fuelwood species diversity and consumption pattern in the homegardens from foothills of Indian Eastern Himalayas. Agrofor Syst 96(3):453–464. https://doi.org/10.1007/s10457-021-00728-1

Saeed U, Arshad M, Hayat S, Morelli TL, Ali Nawaz MA (2022) Analysis of provisioning ecosystem services and perceptions of climate change for indigenous communities in the Western Himalayan Gurez Valley, Pakistan. Ecosyst Serv 56:101453. https://doi.org/10.1016/j.ecoser.2022.101453

Samal I, Bhoi TK, Majhi PK, Murmu S, Pradhan AK, Kumar D et al (2023a) Combatting insects mediated biotic stress through plant associated endophytic entomopathogenic fungi in horticultural crops Front. Plant Sci 13:1098673. https://doi.org/10.3389/fpls.2022.1098673

Samal I, Bhoi TK, Vyas V, Majhi PK, Mahanta DK, Komal J et al (2023b) Resistance to fungicides in entomopathogenic fungi: underlying mechanisms, consequences, and opportunities for progress. Trop Plant Pathol:1–3. https://doi.org/10.1007/s40858-023-00585-6

Santiago-Freijanes JJ, Mosquera-Losada MR, Rois-Díaz M, Ferreiro-Domínguez N, Pantera A, Aldrey JA, &Rigueiro-Rodríguez, A. (2021) Global and European policies to foster agricultural sustainability: agroforestry. Agrofor Syst 95(5):775–790. https://doi.org/10.1007/s10457-018-0215-9

Santoro A (2023) Traditional oases in northern Africa as multifunctional agroforestry systems: a systematic literature review of the provided Ecosystem Services and of the main vulnerabilities. Agrofor Syst 97(1):81–96. https://doi.org/10.1007/s10457-022-00789-w

Santoro A, Venturi M, Bertani R, Agnoletti M (2020) A review of the role of forests and agroforestry systems in the FAO Globally Important Agricultural Heritage Systems (GIAHS) programme. Forests 11(8):860. https://doi.org/10.3390/f11080860

Satapathy SM, Bhoi TK, Majhi PK, Samal I, Mohapatra S, Dohling PN (2020) Climate smart agriculture (CSA) for sustainable food grain production. Chief Dr. Naresh RK, editor, vol 15, p 1

Schaafsma M, Bartkowski B (2021) Synergies and trade-offs between ecosystem services. In: Life on land. Springer, Cham, pp 1022–1032. https://doi.org/10.1007/978-3-319-95981-8_117

Schröter M, Başak E, Christie M, Church A, Keune H, Osipova E (2020) Indicators for relational values of nature's contributions to good quality of life: the IPBES approach for Europe and Central Asia. Ecosyst People 16(1):50–69. https://doi.org/10.1080/26395916.2019.1703039

SenGupta A (2022) Sustainable development in India with reference to agricultural sector. Available at SSRN 4199047. SSRN Electronic Journal. https://doi.org/10.2139/ssrn.4199047

Sharafatmandrad M, KhosraviMashizi A (2021) Temporal and spatial assessment of supply and demand of the water-yield ecosystem service for water scarcity management in arid to semi-arid ecosystems. Water Resour Manag 35(1):63–82. https://doi.org/10.1007/s11269-020-02706-1

Sharifi A (2021) Co-benefits and synergies between urban climate change mitigation and adaptation measures: a literature review. Sci Total Environ 750:141642. https://doi.org/10.1016/j.scitotenv.2020.141642

Sheppard JP, Bohn Reckziegel R, Borrass L, Chirwa PW, Cuaranhua CJ, Hassler SK (2020) Agroforestry: An appropriate and sustainable response to a changing climate in Southern Africa? Sustainability 12(17):6796. https://doi.org/10.3390/su12176796

Sierra-Huelsz JA, GerezFernández P, LópezBinnqüist C, Guibrunet L, Ellis EA (2020) Traditional ecological knowledge in community forest management: evolution and limitations in Mexican forest law, policy and practice. Forests 11(4):403. https://doi.org/10.3390/f11040403
Sileshi GW, Mafongoya PL, Nath AJ (2020) Agroforestry systems for improving nutrient recycling and soil fertility on degraded lands. Agrofor Degrad Landsc 1:225–253
Singh J, Bhatnagar A (2018) Regulating ecosystem services: a review. J Environ Manag 225:261–275
Singh S, Mishra VK, Bhoi TK (2017) Insect molecular markers and its utility-a review. Int J Agric Environ Biotechnol 10(4):469–479. https://doi.org/10.5958/2230-732X.2017.00058.4
Singh M, Babanna SK, Kumar D, Dwivedi RP, Dev I, Kumar A et al (2021) Valuation of fuelwood from agroforestry systems: a methodological perspective. Agrofor Syst 95(5):977–993. https://doi.org/10.1007/s10457-020-00580-9
Singh DN, Bohra JS, Tyagi V, Singh T, Banjara TR, Gupta G (2022) A review of India's fodder production status and opportunities. Grass Forage Sci 77(1):1–10. https://doi.org/10.1111/gfs.12561
Singh S, Bhoi TK, Khan I, Vyas V, Athulya R, Rathi A et al (2023a) Climate change drivers and soil microbe-plant interactions. In: Climate change and microbiome dynamics: carbon cycle feedbacks. Springer International Publishing, Cham, pp 157–176. https://doi.org/10.1007/978-3-031-21079-2_11
Singh S, Bhoi TK, Vyas V (2023b) Interceding microbial biofertilizers in agroforestry system for enhancing productivity. In: Plant growth promoting microorganisms of arid region. Springer Nature Singapore, Singapore, pp 161–183
Smith P, Soussana JF, Angers D, Schipper L, Chenu C, Rasse DP (2020) How to measure, report and verify soil carbon change to realize the potential of soil carbon sequestration for atmospheric greenhouse gas removal. Glob Chang Biol 26(1):219–241. https://doi.org/10.1111/gcb.14815
Smith MM, Bentrup G, Kellerman T, MacFarland K, Straight R, Ameyaw L (2021) Windbreaks in the United States: a systematic review of producer-reported benefits, challenges, management activities and drivers of adoption. Agric Syst 187:103032. https://doi.org/10.1016/j.agsy.2020.103032
Sollen-Norrlin M, Ghaley BB, Rintoul NLJ (2020) Agroforestry benefits and challenges for adoption in Europe and beyond. Sustainability 12(17):7001. https://doi.org/10.3390/su12177001
Suwardi AB, Navia ZI (2023) Sustainable use and management of wild edible fruit plants: a case study in the Ulu Masen protected forest, West Aceh, Indonesia. J Sustain For 42(8):811–830. https://doi.org/10.1080/10549811.2022.2123355
Tessema SB, Nayak D (2022) Analyzing the perceived prioritized forest ecosystem services under the participatory management: a case of Maksegnit District, Amhara Regional State, Ethiopia. Trees For People 9:100318. https://doi.org/10.1016/j.tfp.2022.100318
Tittonell P (2021) Beyond CO2: Multiple ecosystem services from ecologically intensive grazing landscapes of South America. Front Sustain Food Syst 5:664103. https://doi.org/10.3389/fsufs.2021.664103
Ugwu JA (2020) Insecticidal activity of some botanical extracts against legume flower thrips and legume pod borer on cowpea Vignaunguiculata L. Walp. J Basic Appl Zool 81(1):1–8
Wang B, Zhang Q, Cui F (2021a) Scientific research on ecosystem services and human well-being: a bibliometric analysis. Ecol Indic 125:107449. https://doi.org/10.1016/j.ecolind.2021.107449
Wang J, Soininen J, Heino J (2021b) Ecological indicators for aquatic biodiversity, ecosystem functions, human activities and climate change. Ecol Indic 132:108250. https://doi.org/10.1016/j.ecolind.2021.108250
Xiao L, Li G, Zhao R, Zhang L (2021) Effects of soil conservation measures on wind erosion control in China: a synthesis. Sci Total Environ 778:146308. https://doi.org/10.1016/j.scitotenv.2021.146308

Yu Y, Chen X, Malik I, Wistuba M, Cao Y, Hou D (2021) Spatiotemporal changes in water, land use, and ecosystem services in Central Asia considering climate changes and human activities. J Arid Land 13(9):881–890. https://doi.org/10.1007/s40333-021-0084-3

Zhang Y, Ouyang Z, Tang Z, Wang X, Miao H, Chen J et al (2021) Traditional agroforestry systems contribute to multiple provisioning ecosystem services in China. Ecol Indic 122:107265

Zhu X, Liu W, Chen J, Bruijnzeel LA, Mao Z, Yang X (2020, August) Reductions in water, soil and nutrient losses and pesticide pollution in agroforestry practices: a review of evidence and processes. Plant Soil 453(1–2):45–86. https://doi.org/10.1007/s11104-019-04377-3

Chapter 25
Revitalizing Degraded Soils with Agroforestry Interventions: Opportunities, Challenges, and Future Direction

Vipula Vyas, Tanmaya Kumar Bhoi, Ipsita Samal, Sangeeta Singh, and Deepak Kumar Mahanta

Abstract Soil degradation is a major environmental issue affecting agricultural productivity, food security, and ecosystem services. Agroforestry, a land-use system that integrates trees, crops, and/or livestock in a single management unit, has been recognized as a promising approach for restoring degraded soils. Agroforestry systems provide multiple benefits, including improved soil fertility, increased biodiversity, enhanced ecosystem services, and diversified livelihoods. Agroforestry has been used for restoring degraded mining soils in India, waterlogged soils in Bangladesh, and degraded grasslands in China. Despite the potential of agroforestry intervention for restoring degraded soils, several challenges need to be addressed. These include management complexity, market access, land tenure, and policy issues, as discussed in previous sections. Addressing these challenges will require a concerted effort by stakeholders from different sectors, including farmers, researchers, policymakers, and civil society. The current book chapter provides an overview of the potential of agroforestry intervention for restoring degraded soils and highlights recent research on this topic.

Keywords Soil degradation · Restoration · Agroforestry intervention

V. Vyas · T. K. Bhoi · S. Singh
Forest Protection Division, ICFRE-Arid Forestry Research Institute (ICFRE-AFRI), Jodhpur, Rajasthan, India

I. Samal (✉)
ICAR-National Research Centre on Litchi, Mushahari, Ramna, Muzaffarpur, Bihar, India

D. K. Mahanta
Forest Entomology Discipline, Forest Protection Division, Indian Council of Forestry Research and Education (ICFRE)-Forest Research Institute (ICFRE-FRI), Dehradun, Uttarakhand, India

H. S. Jatav et al. (eds.), *Agroforestry to Combat Global Challenges*, Sustainable Development and Biodiversity 36, https://doi.org/10.1007/978-981-99-7282-1_25

25.1 Introduction

Population growth is a major global concern, as it affects various aspects of human life, including food security, housing, education, and health. The world population has been growing steadily over the past few decades, and this trend is expected to continue in the near future. According to the United Nations (UN 2021), the global population reached 7.9 billion in 2021 and is projected to reach 9.7 billion by 2050. India, which is currently the second-most populous country in the world, is projected to surpass China as the most populous country by 2027. Population growth is a significant driver of soil degradation, as it leads to increased demand for food, fiber, and fuel. This demand often leads to the conversion of natural habitats, deforestation, and intensification of agriculture, which can cause soil degradation. According to a recent report by the Food and Agriculture Organization of the United Nations (FAO), population growth and the resulting increase in demand for food and other resources have led to significant soil degradation worldwide. The report estimates that about 33% of the world's soils are degraded, and this figure is projected to increase to 90% by 2050 if current trends continue (FAO 2020). Degraded soil is soil that has lost its ability to support plant growth or ecosystem functions due to human activities such as intensive agriculture, deforestation, mining, and urbanization. Degraded soil is a major environmental problem that affects food security, biodiversity, and ecosystem services. According to a recent report by the Food and Agriculture Organization (FAO) of the United Nations (FAO 2020), about one-third of the world's soils are degraded, and the problem is getting worse. The report estimates that degraded soil could cost the global economy up to $40 billion annually in lost ecosystem services and crop productivity (Singh et al. 2023a, b, c). Under such circumstances, agroforestry intervention is the only way to sustain the degraded soil in long-term basis. Agroforestry is a sustainable land-use system that combines the benefits of trees and agriculture. It has the potential to restore degraded soils and enhance soil productivity. Agroforestry systems provide multiple benefits, such as improved soil quality, increased biodiversity, and higher crop yields. According to a recent study published in the journal *Land Use Policy*, agroforestry can be an effective way to restore degraded soils. The study examined the impact of agroforestry on soil quality in the Brazilian Cerrado, a region that has been severely degraded by agriculture and grazing. The study found that agroforestry significantly improved soil quality compared to conventional agricultural practices. Agroforestry increased soil organic carbon, nitrogen, and phosphorus content, which improved soil fertility and increased crop productivity. The study also found that agroforestry increased soil water-holding capacity and reduced soil erosion, which helped to conserve soil moisture and reduce soil loss (Menezes et al. 2021). The current book chapter provides an overview of the potential of agroforestry intervention for restoring degraded soils, viz., salt-affected soil, acid soil, waterlogged soil, mine spoils, etc., their mechanism, and their pros and cons are discussed with recent area of research in this topic.

25.2 Agroforestry and their Relation in Degrade Soil Rehabilitation

Agroforestry is a land-use system that combines trees and agriculture in a way that is mutually beneficial. This system can play a crucial role in the rehabilitation of degraded soils by improving soil quality and enhancing ecosystem services. Agroforestry practices can help to prevent soil erosion, increase soil organic matter content, and improve nutrient cycling. According to a recent study, agroforestry can be an effective way to rehabilitate degraded soils. The study evaluated the impact of agroforestry on soil quality in a degraded landscape in northern Ethiopia. The study found that agroforestry increased soil organic matter content, reduced soil erosion, and improved nutrient cycling compared to conventional farming practices (Tadesse et al. 2021). Similarly, another study examined the effect of agroforestry on degraded soil in Indonesia. The study found that agroforestry practices significantly improved soil fertility, reduced soil erosion, and increased crop yields compared to monoculture systems. These studies demonstrate the potential of agroforestry in rehabilitating degraded soils and improving agricultural productivity while enhancing ecosystem services (Nurwijayanto et al. 2020). In Ethiopia, a study found that integrating trees into agricultural landscapes through agroforestry practices improved soil fertility, reduced soil erosion, and increased crop yields. The study reported significant increases in soil organic matter, nitrogen, and phosphorus in agroforestry systems compared to conventional agriculture (Sahle et al. 2022). In the Philippines, a study found that agroforestry interventions improved soil quality, reduced erosion, and increased soil water-holding capacity. The study reported that agroforestry systems had higher levels of soil organic matter, nitrogen, and phosphorus than conventional agriculture, which improved soil fertility and crop productivity (Ota et al. 2020).

25.3 Mechanism of Degrade Soil Restoration Through Agroforestry Intervention

Agroforestry interventions can restore degraded soils through several mechanisms, including:

1. **Improving soil organic matter:** Agroforestry systems can increase soil organic matter through the deposition of leaves, branches, and roots of trees and other vegetation. This improves soil fertility, structure, and water-holding capacity, which can increase crop yields (Nyirenda and Balaka 2021).
2. **Enhancing soil structure:** Agroforestry systems can improve soil structure by reducing soil compaction and increasing soil aggregation, which enhances water infiltration and reduces soil erosion. Trees also help to stabilize soil structure by providing physical support to the soil (Yusnaini et al. 2021).

3. **Reducing soil erosion:** Trees in agroforestry systems can reduce soil erosion by intercepting rainwater, slowing down surface runoff, and reducing the velocity of water flow. Trees also help to anchor the soil and reduce the risk of landslides (Gholamahmadi et al. 2023).
4. **Increasing soil fertility:** Agroforestry systems can increase soil fertility through the cycling of nutrients between trees and crops. Trees can fix atmospheric nitrogen, which can be transferred to the soil through leaf litter and root exudates. Trees also absorb nutrients from deeper soil layers and redistribute them to the upper soil layers, which benefits crops (Sileshi et al. 2020).
5. **Increasing water availability:** Trees in agroforestry systems can increase water availability by reducing evapotranspiration and increasing soil water storage. Trees can also act as windbreaks and reduce water loss due to wind erosion (Gusli et al. 2020) (Fig. 25.1).

A study conducted in Ethiopia found that agroforestry systems increased soil organic matter, improved soil structure and water-holding capacity, and reduced soil erosion. The study also reported that agroforestry systems had higher crop yields than conventional agricultural systems (Fentahun and Gashaw 2014). A study found that agroforestry interventions improved soil structure and reduced soil erosion. The study reported that agroforestry systems had higher levels of soil organic matter and total nitrogen than monoculture agriculture, which improved soil quality and increased crop productivity (Gebrewahid and Meressa 2020). A study conducted in Brazil found that agroforestry systems increased water infiltration and reduced soil compaction. The study also reported that agroforestry systems had higher levels of soil organic matter and nitrogen, which improved soil fertility and increased crop yields (Matos et al. 2022). A study conducted in the state of Gujarat found that agroforestry interventions had a positive impact on soil quality and crop productivity. The study reported that agroforestry systems had higher levels of soil organic carbon, total nitrogen, and available phosphorus than conventional agricultural systems. The study also found that agroforestry systems had higher crop yields and better soil moisture retention (Patel et al. 2020). In the state of Tamil Nadu, a study found that agroforestry interventions improved soil fertility and reduced soil erosion. The study reported that agroforestry systems had higher levels of soil organic carbon, nitrogen, and phosphorus than conventional agricultural systems. The study also found that agroforestry systems had lower soil erosion rates and higher crop yields (Sikka et al. 2014). A study conducted in the state of Jharkhand found that agroforestry interventions increased soil organic matter and improved soil fertility. The study reported that agroforestry systems had higher levels of soil organic carbon, total nitrogen, and available phosphorus than monoculture agriculture. The study also found that agroforestry systems had higher crop yields and better soil moisture retention (Sahoo and Wani 2019) (Fig. 25.2).

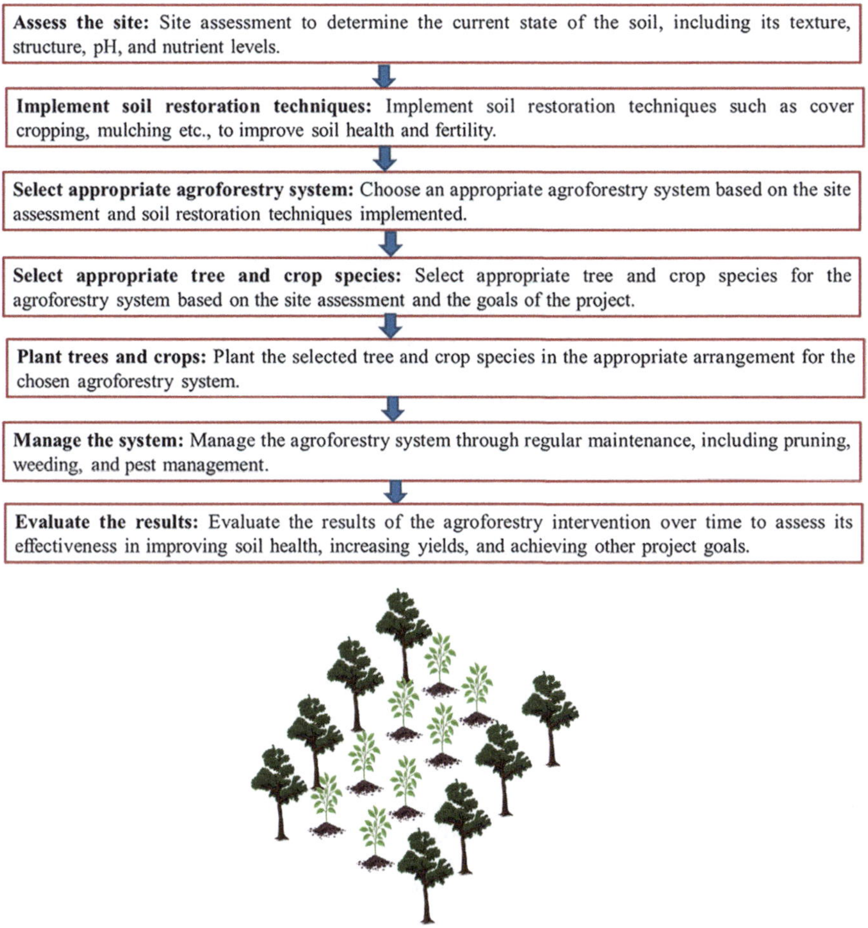

Fig. 25.1 Agroforestry-based climate-resilient soil improvement strategies (ACIS) framework. The ACIS framework is a holistic approach to soil restoration and climate resilience, combining soil restoration techniques with appropriate agroforestry systems, tree and crop species selection, and regular maintenance and monitoring. The framework aims to improve soil health and fertility, increase agricultural productivity, and enhance resilience to climate change. The ACIS framework can be adapted to different agroecological contexts and project goals

25.4 Agroforestry Intervention for Saline Soil Restoration

Saline soils are a major problem in many parts of India, particularly in the arid and semiarid regions. Agroforestry interventions have been used to restore saline soils in some of these regions. A study conducted in the Kachchh district of Gujarat found that agroforestry interventions significantly improved soil quality and reduced soil salinity. The study reported that the agroforestry system had higher levels of soil organic carbon, total nitrogen, available phosphorus, and microbial biomass than the

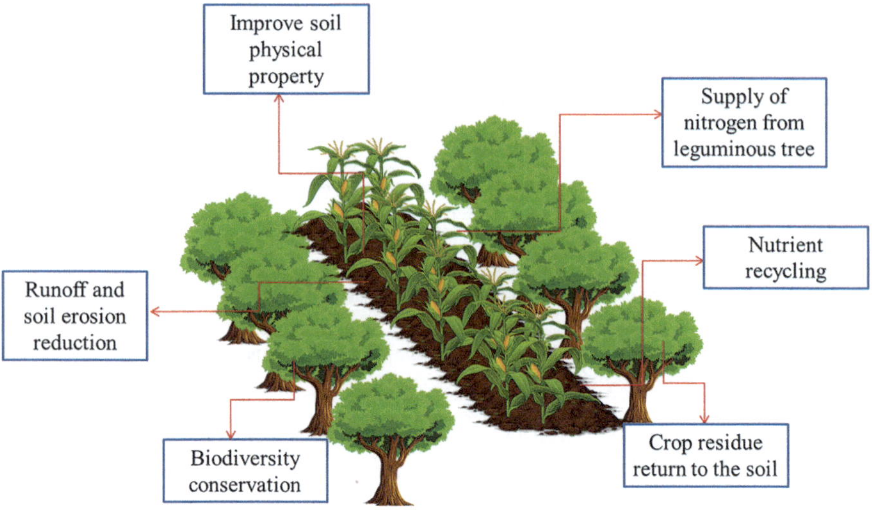

Fig. 25.2 This figure represents the diverse combination of trees, shrubs, and crops are strategically planted to enhance soil fertility and structure. The deep-rooted trees provide stability and prevent erosion, while their fallen leaves and organic matter enrich the soil's nutrient content. The intercropped crops benefit from the shade and protection provided by the tree canopy, reducing water loss and improving overall productivity. This integrated agroforestry approach offers sustainable solutions for rehabilitating degraded soils, promoting biodiversity, and supporting local livelihoods

control (no intervention) and traditional agriculture systems. The study also found that the agroforestry system significantly reduced soil salinity and increased crop yields (Singh 2022). The agroforestry intervention in this study involved planting drought-tolerant tree species (*Prosopis juliflora* and *Acacia tortilis*) in rows with crop cultivation in the interspaces between the rows. This study demonstrates that agroforestry interventions can effectively restore saline soils in India, leading to improved soil quality and increased crop productivity. A study conducted found that agroforestry systems can improve soil quality and crop productivity in saline soils. The study reported that agroforestry systems had higher levels of soil organic carbon, total nitrogen, and available phosphorus than conventional agricultural systems. The study also found that agroforestry systems had higher crop yields and better soil moisture retention in saline soils (Jinger et al. 2023). In the state of Rajasthan, a study found that agroforestry interventions can improve soil quality and reduce salinity levels in saline soils. The study reported that agroforestry systems had higher levels of soil organic matter, total nitrogen, and available phosphorus than monoculture agriculture. The study also found that agroforestry systems had lower salinity levels and higher crop yields (Kumar and Kunhamu 2021). A study conducted in the state of Haryana found that agroforestry interventions can reduce soil salinity and improve crop productivity in saline soils. The study reported that agroforestry systems had higher levels of soil organic carbon, total nitrogen, and available phosphorus than conventional agricultural systems. The study also found that agroforestry systems had higher crop yields and lower soil salinity levels (Kombra et al. 2022). In Iran, a study found that agroforestry interventions can reduce soil salinity and improve crop productivity in saline soils. The study reported that agroforestry systems had higher levels of soil organic matter, total nitrogen, and available phosphorus than monoculture agriculture. The study also found that agroforestry systems had higher crop yields and lower soil salinity levels (Kyrgiakos et al. 2023). A study conducted in the United States found that agroforestry interventions can improve soil quality and reduce salinity levels in saline soils. The study reported that agroforestry systems had higher levels of soil organic carbon, total nitrogen, and available phosphorus than conventional agricultural systems. The study also found that agroforestry systems had lower salinity levels and higher crop yields (Bishaw et al. 2022). The selection of appropriate plant species is critical for the success of agroforestry interventions in saline soil restoration. A study conducted in India evaluated the performance of different agroforestry systems in saline soils and found that the combination of trees (*Casuarina equisetifolia*) and crops (maize and pigeon pea) was the most effective in terms of reducing soil salinity and improving crop productivity. The study also found that the tree species *Eucalyptus tereticornis* was not suitable for saline soil restoration due to its high water consumption (Dev et al. 2020). A study conducted in Pakistan evaluated the performance of different tree species in saline soils and found that the tree species *Prosopis juliflora* and *Acacia nilotica* were the most effective in terms of reducing soil salinity and improving soil quality. The study also found that the tree species *Dalbergia sissoo* and *Acacia modesta* were not suitable for saline soil restoration due to their poor survival rates (Kumar et al. 2022a, b). A study conducted in Iran evaluated the performance of different agroforestry systems in saline soils and found that the combination of trees (*Ailanthus altissima* and *Robinia pseudoacacia*) and crops (barley and clover) was the most effective in terms of reducing soil salinity and improving crop productivity. The study also found that the tree species *Paulownia tomentosa* was not suitable for saline soil restoration due to its high water consumption (Aghajani 2019) (Fig. 25.1).

25.5 Agroforestry Intervention for Acidic Soil Restoration

One recent study that investigated the effectiveness of agroforestry in restoring acid soil was conducted by Härkönen et al. (2023). The study was carried out in Indonesia, where acid soil is a major problem for agricultural production. The researchers evaluated the soil quality and plant growth in a mixed agroforestry system that included teak and rubber trees along with vegetable crops. The results showed that the agroforestry system significantly improved soil quality by increasing soil organic matter, total nitrogen, and available phosphorus. The pH levels also increased, indicating a reduction in soil acidity. In addition, the agroforestry system significantly increased crop yield and plant biomass, suggesting that it can improve the productivity of acid soils. Agroforestry interventions can be effective in restoring acid soils by improving soil quality and increasing soil pH. One study conducted by Sari et al. (2020) in Indonesia found that agroforestry systems incorporating nitrogen-fixing trees, such as *Acacia mangium*, improved soil pH and increased soil organic carbon content. Another study conducted by Wang et al. (2023) in China showed that intercropping with fruit trees, such as apple and pear, improved soil pH and reduced soil acidity compared to conventional monoculture systems. Additionally, a review by Nair et al. (2021a, b) highlighted the potential of agroforestry to restore degraded soils and increase soil fertility, particularly in acid soils. The review identified various agroforestry practices, such as alley cropping, silvopasture, and taungya, that can improve soil quality and increase soil pH in acid soils. Agroforestry interventions can be effective in restoring acid soils in India. One such intervention is the incorporation of leguminous trees and shrubs into agricultural systems. Leguminous plants can fix atmospheric nitrogen and improve soil fertility, leading to increased crop yields and improved soil health. Some examples of leguminous trees and shrubs used in agroforestry systems in India include *Leucaena leucocephala, Acacia auriculiformis*, and *Sesbania sesban.* A recent study by Kanwal and Vishvakarma (2022) examined the effects of agroforestry systems on soil properties and crop yields in acid soils in India. The study found that agroforestry systems led to improvements in soil fertility, with increases in soil organic carbon, total nitrogen, available phosphorus, and exchangeable potassium. The study also found that agroforestry systems led to increased crop yields, with improvements in the yield of crops such as rice, wheat, and maize. Agroforestry systems that incorporate leguminous trees, such as *Acacia auriculiformis, Sesbania* spp., and *Leucaena leucocephala*, have been shown to improve soil fertility and decrease soil acidity (Sileshi et al. 2020). A recent study conducted in India investigated the effect of agroforestry on soil quality in acid soils. The study found that agroforestry systems significantly increased soil organic carbon, total nitrogen, available phosphorus, and exchangeable potassium compared to traditional farming practices (Rathore et al. 2021). Another study conducted in Kerala, India, evaluated the impact of agroforestry on soil pH and nutrient availability. The study found that agroforestry systems that included trees such as *Erythrina variegata, Albizia lebbeck*, and *Acacia auriculiformis* improved soil pH and increased the availability of nutrients such as nitrogen, phosphorus, and potassium (Hasan et al. 2022). Chatterjee et al. (2022) evaluated the impact of agroforestry interventions on soil properties in acid soils of Northeast India. The study found that agroforestry interventions, including the planting of nitrogen-fixing trees and shrubs, significantly improved soil fertility and reduced soil acidity compared to traditional farming practices. Additionally, the study found that agroforestry interventions had positive effects on crop yields, with significant increases in maize, soybean, and rice yields. One study conducted in Brazil investigated the impact of agroforestry on the restoration of acid soils. The researchers found that agroforestry systems, which combined different tree species with crops and/or pasture, significantly increased soil organic matter content, soil pH, and nutrient availability, compared to conventional monoculture systems (Carvalho et al. 2023). Another study conducted in Indonesia showed that agroforestry practices, such as intercropping with leguminous trees and cover crops, improved soil pH and reduced soil acidity in acid soils. The researchers also found that agroforestry increased crop productivity, and enhanced soil microbial activity, which further improved soil health (Alam et al. 2022) (Fig. 25.1).

25.6 Agroforestry Intervention for Waterlogged Soil Restoration

Agroforestry is an effective approach for restoring degraded and waterlogged soil. A recent study conducted in India by Dagar et al. (2022) investigated the effectiveness of agroforestry in restoring waterlogged soils in the Indo-Gangetic Plain. The study found that agroforestry systems that integrated trees with crops and livestock significantly improved soil quality, reduced waterlogging, and enhanced crop yields. The authors concluded that agroforestry can be a viable option for restoring degraded and waterlogged soils in the region. Another recent study by Malobane (2020) in also highlighted the potential of agroforestry for improving waterlogged soils. The study showed that integrating trees into cropland reduced soil compaction and increased soil water-holding capacity, resulting in higher crop yields and reduced nitrate leaching. The authors concluded that agroforestry can be an effective strategy for mitigating the negative impacts of waterlogging on soil health and crop production. Islam et al. (2022) investigated the effects of agroforestry on water-logged soils in the wetlands of the Meghna River Basin in Bangladesh. The study found that agroforestry interventions, such as the planting of tree species like *Acacia auriculiformis* and *Melia azedarach*, significantly improved soil physical properties, including soil moisture, bulk density, and porosity. These improvements led to increased crop yields and reduced soil erosion in the study area. Agroforestry interventions can be effective in restoring waterlogged soil by enhancing drainage, improving soil structure, and increasing organic matter content. One such intervention is the cultivation of vetiver grass (*Chrysopogon zizanioides*), a perennial grass with a deep and extensive root system that can break up compacted soil and enhance water infiltration. A recent study by Malunguja et al. (2022) in India investigated the impact of vetiver grass on soil physicochemical properties and crop productivity in a waterlogged soil. The study found that after 2 years of cultivation, vetiver grass significantly improved soil properties such as bulk density, porosity, and organic matter content. Additionally, the study found that the cultivation of vetiver grass increased crop productivity, especially for rice and wheat crops, by improving soil moisture and nutrient availability. Intervention through planting with suitable tree species that can tolerate waterlogged conditions, such as *Populus deltoides, Eucalyptus tereticornis*, and *Dalbergia sissoo*. These species have been found to improve soil physical properties, increase soil organic carbon content, and enhance soil microbial activity, leading to improved soil fertility and crop productivity. A recent study conducted in India evaluated the impact of agroforestry interventions on waterlogged soil restoration in the Ganga-Yamuna Doab region. The study found that planting *Populus deltoides* in waterlogged soil resulted in a significant improvement in soil physical properties, such as bulk density, porosity, and aggregate stability, as well as an increase in soil organic carbon content and soil microbial biomass. The authors concluded that agroforestry interventions can be an effective and sustainable approach for waterlogged soil restoration in the region (Patel et al. 2020). Gao et al. (2022) study in the Yangtze River Delta region of China, where waterlogging is a common problem in agricultural lands. The researchers compared the effects of three different agroforestry systems (alley cropping, mixed-species planting, and monoculture) with a control treatment (bare fallow) on soil physical properties and biomass production. They found that all three agroforestry systems improved soil physical properties (including soil porosity, bulk density, and water-holding capacity) compared to the control treatment. The mixed-species planting system was the most effective at improving soil physical properties and also resulted in the highest biomass production (Fig. 25.1).

25.7 Agroforestry Intervention for Mining Soil Restoration

Mining activities can have a significant impact on soil health, often leading to degradation and loss of soil productivity. Agroforestry can be an effective intervention for restoring degraded mining soils. A recent study by Grez (2020) investigated the potential of agroforestry for restoring soil quality in mining-affected areas of Mexico. The study found that agroforestry systems with native tree species were effective in improving soil quality parameters such as organic matter, nitrogen, and soil structure. The study also found that agroforestry systems enhanced biodiversity, provided ecosystem services, and increased carbon sequestration in the restored mining soils. The authors concluded that agroforestry can be a sustainable approach for restoring mining-affected soils and improving the livelihoods of local communities. The article by Guan et al. (2021) provides a comprehensive review of the potential of agroforestry for restoring mining soils in China. The authors argue that agroforestry can provide a range of benefits for mine land restoration, including soil improvement, erosion control, biodiversity conservation, and economic benefits for local communities. The article also highlights the challenges of implementing agroforestry in mining landscapes, including land tenure issues, market access, and the need for appropriate policy and institutional frameworks. The authors suggest that successful implementation of agroforestry for mine land restoration requires a multidisciplinary approach involving researchers, policymakers, and local communities. Sengupta (2020) studied the impact of different agroforestry systems on the restoration of mining soil in the coal mining regions of Jharkhand, India. They found that agroforestry systems improved soil quality and plant growth compared to conventional agricultural practices. The study also found that agroforestry systems increased the soil carbon content and improved soil structure, which resulted in better water retention and reduced erosion. The authors concluded that agroforestry can be an effective approach for restoring degraded mining soils in India. The study by Jha et al. (2020) assessed the potential of agroforestry for restoring mined-out areas in India. The authors found that agroforestry can be a cost-effective and sustainable approach for restoring degraded mining lands. They recommend a combination of tree species, crops, and livestock to improve soil fertility, biodiversity, and ecosystem services in the restored areas. The study also highlighted the need for policy support and community participation to promote the adoption of agroforestry for mining soil restoration. Bhardwaj et al. (2023) evaluated the use of agroforestry for restoring mine soil in Singrauli district of Madhya Pradesh, India. The study found that agroforestry significantly improved soil properties such as pH, organic carbon, available nitrogen, phosphorus, and potassium compared to the degraded mine soil. The study also found that agroforestry increased plant biomass and diversity and provided additional benefits such as fuel wood, fodder, and fruits (Fig. 25.1). Reclamation of affected soil through agroforestry intervention has been presented in Table 25.1.

25.8 Advantages of Agroforestry Intervention for Degraded Soil Restoration

1. **Improves soil health:** Agroforestry systems promote the accumulation of organic matter in the soil, which improves soil structure, water retention, and nutrient cycling. This leads to better plant growth and increased soil fertility. A recent study by Berry and Shukla (2021) found that agroforestry systems improved soil quality and increased crop yields compared to conventional agricultural systems.
2. **Provides ecosystem services**: Agroforestry systems provide multiple ecosystem services such as biodiversity conservation, carbon sequestration, and erosion control. Trees in agroforestry systems also provide shade and shelter to crops and livestock, which can improve their health and productivity. A recent study by Hübner et al. (2021) found that agroforestry systems can significantly increase biodiversity and carbon sequestration compared to conventional agricultural systems.
3. **Enhances livelihoods:** Agroforestry systems can provide a range of products, such as timber, fruits, and nontimber forest products, which can generate additional income for farmers. Additionally, agroforestry systems can provide more stable and diversified sources of food and income compared to monoculture systems. A recent study by Awazi (2022) found that agroforestry systems can improve food security and income for smallholder farmers.

Table 25.1 Reclamation of affected soil through agroforestry intervention

SL. no.	Agroforestry system	Tree	Crop plants	Key findings	Reference
Acid soil restoration					
1	Alley Cropping	*Leucaena leucocephala*	Maize, Beans	Increases soil nitrogen levels, improves soil structure, and reduces soil acidity.	Amadu et al. (2021)
2	Taungya Farming	*Gmelina arborea*	Cassava, Yam	Increases soil organic matter and nutrient availability, improves soil water retention, and reduces soil acidity.	Nair et al. (2021a, b)
3	Silvopasture	*Acacia mangium*	Guinea grass	Improves soil physical properties, increases soil organic matter, and reduces soil acidity.	Dibala et al. (2021)
4	Homegardens	*Gliricidia sepium*	Coffee, Banana	Increases soil carbon and nitrogen levels, improves soil fertility and structure, and reduces soil acidity.	Fahad et al. (2022)
5	Agroforestry Parklands	*Faidherbia albida*	Millet, Sorghum	Increases soil nitrogen levels, improves soil water retention, and reduces soil acidity.	Stephen et al. (2020)
Saline soil restoration					
6	Agroforestry in Coastal Saline Soil	*Acacia nilotica*	Rice, Wheat	Improves soil structure, reduces soil salinity and sodicity, and increases soil nutrient availability.	Syed et al. (2021)
7	Salt-Affected Land Agroforestry	*Prosopis juliflora*	Barley, Millet	Reduces soil salinity and improves soil water-holding capacity, nutrient availability, and overall plant growth.	Tomar et al. (2021)
8	Agroforestry for Saline Soil Reclamation	*Casuarina equisetifolia*	Maize, Soybean	Improves soil physical and chemical properties, enhances soil water retention, and increases soil organic matter.	Ondrasek et al. (2022)
9	Alley Cropping in Saline Soils	*Eucalyptus tereticornis*	Sorghum, Sunflower	Improves soil structure and increases soil nutrient availability, leading to increased crop yields and reduced soil salinity.	Singh et al. (2020)
10	Silvopastoral System on Saline Soil	*Populus alba*	Alfalfa	Improves soil quality, increases soil water retention, and enhances plant growth, leading to	Gupta et al. (2020)

(continued)

Table 25.1 (continued)

SL. no.	Agroforestry system	Tree	Crop plants	Key findings	Reference
				improved forage production.	
Waterlogged soil restoration					
11	Agroforestry in Water-logged Soils	*Eucalyptus camaldulensis*	Paddy rice	Improves soil structure, increases soil organic matter, and enhances nutrient cycling, leading to increased crop yields and improved soil drainage.	Singh et al. (2023a, b, c)
12	Agroforestry for Water-logged Soils	*Leucaena leucocephala*	Maize, Mung bean	Improves soil physical and chemical properties, enhances soil water retention and nutrient availability, and increases crop yields.	Kisaka et al. (2023)
13	Agroforestry for Reclamation of Waterlogged Soils	*Melia azedarach*	Wheat, Mustard	Improves soil physical properties, reduces soil salinity, increases soil organic matter, and enhances crop growth and yield.	Kumar et al. (2022a, b)
14	Agroforestry for Water-logged Land Reclamation	*Acacia mangium*	Soybean, Maize	Improves soil drainage and aeration, increases soil organic matter, and enhances crop growth and yield, leading to improved soil fertility.	Das et al. (2020)
15	Agroforestry in Water-logged Soils of Coastal Region	*Casuarina equisetifolia*	Rice, Mung bean	Improves soil physical and chemical properties, enhances nutrient availability, and increases crop yields, leading to improved livelihoods of coastal communities.	Maji et al. (2020)
Mine spoils restoration					
17	Agroforestry for Mine Spoils Reclamation	*Acacia mangium*	Maize, Soybean	Improves soil physical and chemical properties, enhances nutrient availability, and increases crop yields, leading to improved soil fertility and ecosystem services.	Jinger et al. (2023)
18	Agroforestry for Mine Spoils Reclamation	*Eucalyptus camaldulensis*	Paddy rice	Improves soil structure, enhances nutrient cycling, and increases soil organic matter and	da Silva et al. (2022)

(continued)

Table 25.1 (continued)

SL. no.	Agroforestry system	Tree	Crop plants	Key findings	Reference
				microbial activity, leading to improved crop yields and soil health.	
19	Agroforestry for Mine Spoils Reclamation	*Casuarina equisetifolia*	Groundnut, Mustard	Improves soil physical and chemical properties, enhances soil fertility and nutrient availability, and increases crop yields, leading to improved livelihoods of mine-affected communities.	Berry and Shukla (2023)
20	Agroforestry for Mine Spoils Reclamation	*Melia azedarach*	Wheat, Mustard	Improves soil physical and chemical properties, enhances nutrient availability, and increases crop yields, leading to improved soil fertility and ecosystem services.	Samji et al. (2023)
21	Agroforestry for Mine Spoils Reclamation	*Dalbergia sissoo*	Groundnut, Cowpea	Improves soil physical and chemical properties, enhances soil organic matter and nutrient availability, and increases crop yields, leading to improved livelihoods of mine-affected communities.	Jinger et al. (2023)

25.9 Challenges in Adapting Agroforestry Intervention for Degraded Soil Restoration

While agroforestry has many advantages for restoring degraded soils, it also faces several challenges. Some of the challenges are:

1. **Management complexity:** Agroforestry systems are complex and require careful planning, management, and maintenance to ensure their long-term sustainability. Farmers need to have the necessary knowledge and skills to manage the multiple components of agroforestry systems. A recent study by Jahan et al. (2022) found that lack of technical knowledge and training is a major barrier to the adoption of agroforestry systems.
2. **Market access:** Agroforestry systems often produce a mix of products, including trees, crops, and livestock, which can make it difficult for farmers to find markets for their products. This can limit the economic benefits of agroforestry systems and discourage farmers from adopting them. A recent study by Kassa (2021)

found that lack of market access is a major challenge for smallholder farmers practicing agroforestry.

3. **Land tenure and policy issues:** Agroforestry systems require long-term land tenure and supportive policies to ensure their sustainability. In many countries, unclear land tenure systems and policies that favor monoculture agriculture can discourage farmers from adopting agroforestry systems. A recent study by Bettles et al. (2021) found that policy and institutional barriers are significant challenges to the adoption of agroforestry systems.

25.10 Socioeconomic Impact

1. Increased agricultural productivity: Agroforestry procedures such as growing trees and bushes alongside crops, can improve soil fertility, improve nutrient cycling, and boost water retention capacity. This can result in better agricultural productivity, higher crop yields, and improved food security for farmers and local people. Higher yields can also contribute to higher income and better living conditions (Mukhlis et al. 2022).
2. Income diversification: Agroforestry systems frequently incorporate the cultivation of a number of crops and tree species. Farmers can diversify their revenue sources, minimizing their reliance on a particular crop and boosting their resilience to market volatility (Aryal et al. 2023).
3. Increased income: Agroforestry systems can supply farmers with a range of products, allowing them to earn more money. Trees, for example, can be harvested for lumber, fruits, or nuts, and livestock can graze on understory vegetation. A research conducted in Kenya discovered that agroforestry interventions raised household income by 25% on average (Sileshi et al. 2023).
4. Improved food security: Agroforestry systems can help to increase food security by providing farmers with a more consistent source of food. During the off-season, trees, for example, can produce fruits and nuts, while cattle can supply milk and meat. According to a study conducted in India, agroforestry interventions reduced poverty by an average of 15% (Tega and Bojago 2023).
5. Poverty reduction: Agroforestry systems can aid in poverty reduction by providing farmers with a more sustainable way of living. Trees, for example, can provide shade and feed, which can assist to reduce input costs, and agroforestry systems can help to save soil and water, lowering the chance of crop failure. Agroforestry interventions enhanced food security by 10% on average, according to a study conducted in Ethiopia (Belay et al. 2023).
6. Improved environmental quality: Environmental quality can be improved by using agroforestry systems to provide habitat for wildlife, reduce erosion, and conserve water. According to a Chinese study, agroforestry systems can minimize soil erosion by up to 80% (Low et al. 2023).
7. Improved health results: By providing shade, clean air, and a place to gather, agroforestry systems can help to enhance health outcomes. According to a study

conducted in Mexico, agroforestry systems can lower the frequency of respiratory disorders by up to 50% (Lovell et al. 2023).

25.11 Future Direction

Restoration of degraded soil through agroforestry intervention holds great promise for sustainable agriculture and environmental conservation. Agroforestry provides a multifunctional approach to restore soil fertility and biodiversity, improve water retention, and mitigate climate change (Singh et al. 2023a, b, c). Agroforestry systems also contribute to food security, enhance rural livelihoods, and promote sustainable land management practices. The future prospects for agroforestry in restoring degraded soil are promising. The adoption of agroforestry practices is increasing globally, with growing recognition of their potential to restore degraded lands and provide sustainable and resilient agricultural production systems. In addition, agroforestry systems are gaining attention for their potential to contribute to achieving multiple Sustainable Development Goals (SDGs), including zero hunger, climate action, and biodiversity conservation (Majhi et al. 2022). Further research is needed to identify the most effective agroforestry systems for restoring degraded soil in different agroecological zones and socioeconomic contexts. The integration of traditional and indigenous knowledge with scientific research can lead to innovative and context-specific agroforestry systems that restore soil health and support sustainable agriculture. There is also a need for policies and programs that promote the adoption of agroforestry practices, provide technical assistance and financial support, and create enabling environments for scaling up agroforestry interventions.

25.12 Conclusion

In conclusion, agroforestry has emerged as a promising approach for the restoration of degraded soil. Agroforestry systems integrate trees and shrubs with crops and livestock to provide multiple benefits, including soil improvement, erosion control, water retention, biodiversity conservation, and climate change mitigation. Agroforestry systems also contribute to food security, enhance rural livelihoods, and promote sustainable land management practices. Recent studies have shown that agroforestry can be effective in restoring degraded soil, with improvements in soil fertility, erosion control, and crop yields. The future prospects for agroforestry in restoring degraded soil are bright, with growing recognition of the potential of agroforestry to contribute to achieving multiple Sustainable Development Goals, including zero hunger, climate action, and biodiversity conservation. To realize the full potential of agroforestry in restoring degraded soil, there is a need for further research to identify the most effective agroforestry systems for different agroecological zones and socioeconomic contexts. Policies and programs that promote the adoption of agroforestry practices, provide technical assistance and financial support, and create enabling environments for scaling up agroforestry interventions are also essential. Agroforestry has the potential to transform agriculture and contribute to sustainable development by restoring soil health, promoting sustainable land management, and enhancing rural livelihoods. With continued research, investment, and policy support, agroforestry can play a critical role in building a more sustainable and resilient future for our planet and its people.

References

Aghajani H (2019) Soil properties across a chronosequence of Ailanthus altissima in semiarid plantations. Sustain For. 2(1):501

Alam T, Suryanto P, Handayani S, Supriyanta S, Wulandari RA, Anshari A, Aisya AW, Purba AE, Widowati R, Taryono T (2022) Soil quality measurement for sustainable soybean yield agroforestry system under different crop rotation models. Biodivers J Biol Divers. 23(12). https://doi.org/10.13057/biodiv/d231209

Amadu FO, McNamara PE, Davis KE (2021) Soil health and grain yield impacts of climate resilient agriculture projects: evidence from southern Malawi. Agric Syst 193:103230. https://doi.org/10.1016/j.agsy.2021.103230

Aryal K, Maraseni T, Apan A (2023) Transforming agroforestry in contested landscapes: a win-win solution to trade-offs in ecosystem services in Nepal. Sci Total Environ 857(1):159301. https://doi.org/10.1016/j.scitotenv.2022.159301

Awazi NP (2022) Agroforestry for climate change adaptation, resilience enhancement and vulnerability attenuation in smallholder farming systems in Cameroon. J Atmos Sci Res. 5(1). https://doi.org/10.30564/jasr.v5i1.4303

Belay A, Mirzabaev A, Recha JW, Oludhe C, Osano PM, Berhane Z, Olaka LA, Tegegne YT, Demissie T, Mutsami C, Solomon D (2023) Does climate-smart agriculture improve household income and food security? Evidence from Southern Ethiopia. In: Environment, development and sustainability. Springer, Cham, pp 1–2

Berry N, Shukla A (2023) Assessment of growth performance of casuarina equisetifolia clones in tropical region of Jabalpur district of Madhya Pradesh. India. Int J Environ Clim Chang 13 (11):266–271

Bettles J, Battisti DS, Cook-Patton SC, Kroeger T, Spector JT, Wolff NH, Masuda YJ (2021) Agroforestry and non-state actors: a review. Forest Policy Econ 130:102538. https://doi.org/10.1016/j.forpol.2021.102538

Bhardwaj S, Soni R, Gupta SK, Shukla DP (2020) Mercury, arsenic, lead and cadmium in waters of the Singrauli coal mining and power plants industrial zone, Central East India. Environ Monit Assess 192(4):251. https://doi.org/10.1007/s10661-020-8225-2

Bhardwaj DR, Salve A, Kumar J, Kumar A, Sharma P, Kumar D (2023) Biomass production and carbon storage potential of agroforestry land use systems in high hills of north-western Himalaya: an approach towards natural based climatic solution. Biomass Convers Biorefin 28:1–4

Bishaw B, Soolanayakanahally R, Karki U, Hagan E (2022) Agroforestry for sustainable production and resilient landscapes. Agrofor Syst 96(3):447–451. https://doi.org/10.1007/s10457-022-00737-8

Carvalho FEL et al (2023) The interspecific interactions in agroforestry systems enhance leaf water use efficiency and carbon storage in cocoa. Environ Exp Bot 205:105119

Chatterjee D, Kuotsu R, Ray SK, Patra MK, Thirugnanavel A, Kumar R, Borah TR, Chowdhury P, Pongen I, Satapathy BS, Deka BC (2022) Preventing soil degradation in shifting cultivation using integrated farming system models. Arch Agron Soil Sci 68(13):1841–1857

da Silva GOA, Southam G, Gagen EJ (2022) Accelerating soil aggregate formation: A review on microbial processes as the critical step in a post-mining rehabilitation context. Soil Res 61(3): 209–223. https://doi.org/10.1071/SR22092

Dagar S, Gupta (2022) Sustainable land-use systems for rehabilitation of highly sodic lands in the Indo-gangetic plains of North-Western India: Synthesis of long-term field experiments. J Soil Salinity Water Q 14(1):1–14

Das SK, Ghosh GK, Avasthe R (2020) Valorizing biomass to engineered biochar and its impact on soil, plant, water, and microbial dynamics: a review. Biomass Conversion and Biorefinery, pp 1–7

Dev I, Ram A, Ahlawat SP, Palsaniya DR, Singh R, Dhyani SK (2020) Bamboo-based agroforestry system (Dendrocalamusstrictus+ sesame–chickpea) for enhancing productivity in semi-arid tropics of central India. Agrofor Syst 94(5):1725–1739. https://doi.org/10.1007/s10457-020-00492-8

Dibala R, Jose S, Udawatta RP (2021) Silvopasture for food security in a changing climate. Agrofor Ecosyst Serv. 7:173–198

Fahad S, Chavan SB, Chichaghare AR, Uthappa AR, Kumar M, Kakade V, Pradhan A, Jinger D, Rawale G, Yadav DK, Kumar V (2022) Agroforestry systems for soil health improvement and maintenance. Sustainability 14(22):14877

Fentahun T, Gashaw T (2014) Evaluation of land use/land cover changes of Bantneka Watershed, Ethiopia. Civil Environ Res 6(7):90–93

Food and Agriculture Organization of the United Nations (FAO) (2020) Soil pollution: A hidden reality. Retrieved from http://www.fao.org/3/CA0140EN/CA0140EN.pdf. Rome

Gao Y, Wang H, Liu Y, Zhang X, Jing S, Peng Y, Huang D, Li X, Chen S, Lou S, Huang YL, C. (2022) Unexpected high contribution of residential biomass burning to non-methane organic gases (NMOGs) in the Yangtze River delta region of China. J Geophys Res Atmos. 127(4): e2021JD035050. https://doi.org/10.1029/2021JD035050

Gebrewahid Y, Meressa E (2020) Tree species diversity and its relationship with carbon stock in the parkland agroforestry of Northern Ethiopia. Cogent Biol 6(1):1728945

Gholamahmadi B, Jeffery S, Gonzalez-Pelayo O, Prats S, Bastos AC, Keizer JJ, Verheijen FG (2023) Biochar impacts on runoff and soil erosion by water: a systematic global scale meta-analysis. Sci Total Environ 871:161860

Grez (2020) Toward predictable mining rehabilitation strategies: Understanding and overcoming the major belowground barriers for microbial rehabilitation engineers

Guan Y, Zhou W, Bai Z, Cao Y, Wang J (2021) Delimitation of supervision zones based on the soil property characteristics in a reclaimed opencast coal mine dump on the Loess Plateau, China. Sci Total Environ 772:145006. https://doi.org/10.1016/j.scitotenv.2021.145006

Gupta SR, Dagar JC, Teketay D (2020) Agroforestry for rehabilitation of degraded landscapes: achieving livelihood and environmental security. Agrofor Degrad Landsc 1:23–68

Gusli S, Sumeni S, Sabodin R, Muqfi IH, Nur M, Hairiah K, Van UD, Noordwijk M (2020) Soil organic matter, mitigation of and adaptation to climate change in cocoa–based agroforestry systems. Land 9(9):323. https://doi.org/10.3390/land9090323

Härkönen LH, Lepistö A, Sarkkola S, Kortelainen P, Räike A (2023) Reviewing peatland forestry: implications and mitigation measures for freshwater ecosystem browning. For Ecol Manag. 531: 120776. https://doi.org/10.1016/j.foreco.2023.120776

Hasan MK, Islam MT, Akter R, Roshni NA (2022) Effects of reforestation approaches, agroforestry and woodlot, on plant community composition, diversity and soil properties in Madhupur Sal forest, Bangladesh. J Ecol Environ 46:21

Hübner R, Kühnel A, Lu J, Dettmann H, Wang W, Wiesmeier M (2021) Soil carbon sequestration by agroforestry systems in China: a meta-analysis. Agric Ecosyst Environ 315:107437. https://doi.org/10.1016/j.agee.2021.107437

Islam I, Cui S, Hoque MZ, Abdullah HM, Tonny KF, Ahmed M, Ferdush J, Xu L, Ding S (2022) Dynamics of tree outside forest land cover development and ecosystem carbon storage change in eastern coastal Zone, Bangladesh. Land 11(1):76. https://doi.org/10.3390/land11010076

Jahan H, Rahman MW, Islam MS, Rezwan-Al-Ramim A, Tuhin MMUJ, Hossain ME (2022) Adoption of agroforestry practices in Bangladesh as a climate change mitigation option: Investment, drivers, and SWOT analysis perspectives. Environ Chall 7:100509. https://doi.org/10.1016/j.envc.2022.100509

Jha DC, Pal H, Rani M (2020) Significance of grasses in establishment of ecological restoration in mined out degraded land in Jharia coalfield, Dhanbad

Jinger D, Kaushal R, Kumar R, Paramesh V, Verma A, Shukla M, Chavan SB, Kakade V, Dobhal S, Uthappa AR, Roy T (2023) Degraded land rehabilitation through agroforestry in India: achievements, current understanding, and future prospectives. Front Ecol Evol 11:69

Kanwal Y, Vishvakarma SCR (2022) Crop productivity and soil properties under agroforestry system in Kosi watershed of Kumaun Himalaya. Indian J Ecol 49(1):21–30

Kassa (2021) Agroforestry a pathway to climate smart agribusiness: Lessons to smallholder farmers

Kisaka MO, Shisanya C, Cournac L, Manlay JR, Gitari H, Muriuki J (2023) Integrating no-tillage with agroforestry augments soil quality indicators in Kenya's dry-land agroecosystems. Soil Tillage Res 227:105586

Kombra S, Ahlawat KS, Sirohi C, Poonia P, Singh C, Yadav S, Singroha P (2022) Soil moisture status in Eucalypts based agroforestry system in semi-arid region of Haryana. In Biol Forum Int J. 14(1):1526–1529

Kumar BM, Kunhamu TK (2021) Carbon sequestration potential of agroforestry systems in India: a synthesis. In: Agroforestry and ecosystem services. Springer, Cham, pp 389–430

Kumar R, Pande VC, Bhardwaj AK, Dinesh D, Bhatnagar PR, Dobhal S, Sharma S, Verma K (2022a) Long-term impacts of afforestation on biomass production, carbon stock, and climate resilience in a degraded semi-arid ravine ecosystem of India. Ecol Eng 177:106559. https://doi.org/10.1016/j.ecoleng.2022.106559

Kumar R, Singh A, Bhardwaj AK, Kumar A, Yadav RK, Sharma PC (2022b) Reclamation of salt-affected soils in India: progress, emerging challenges, and future strategies. Land Degrad Dev 33(13):2169–2180. https://doi.org/10.1002/ldr.4320

Kyrgiakos LS, Kleftodimos G, Vlontzos G, Pardalos PM (2023) A systematic literature review of data envelopment analysis implementation in agriculture under the prism of sustainability. Oper Res. 23(1):7

Lovell ST, Krishnaswamy K, Lin CH, Meier N, Revord RS, Thomas AL (2023) Nuts and berries from agroforestry systems in temperate regions can form the foundation for a healthier human diet and improved outcomes from diet-related diseases. Agrofor Syst:1–14. https://doi.org/10.1007/s10457-023-00858-8

Low G, Dalhaus T, Meuwissen MPM (2023) Mixed farming and agroforestry systems: a systematic review on value chain implications. Agric Syst 206:103606. https://doi.org/10.1016/j.agsy.2023.103606

Majhi PK, Raza B, Behera PP, Singh SK, Shiv A, Mogali SC, Bhoi TK, Patra B, Behera B (2022) Future-proofing plants against climate change: a path to ensure sustainable food systems. In: Biodiversity, functional ecosystems and sustainable food production. Springer, Cham, pp 73–116

Maji B, Sarangi SK, Mandal UK, Burman D, Balasubramaniam P (2020) Climate change and management options for sustainable soil health and crop production: Eastern coast of India as an example. Adv Agric Res Technol J 4(1):28–41

Malobane ME (2020) Conservation agriculture effects on soil quality and greenhouse gas emission in a sweet sorghum based cropping system in South Africa [Doctoral Dissertation]

Malunguja GK, Thakur B, Devi A (2022) Heavy metal contamination of forest soils by vehicular emissions: ecological risks and effects on tree productivity. Environ Processes 9(1):11. https://doi.org/10.1007/s40710-022-00567-x

Matos PS, Cherubin MR, Damian JM, Rocha FI, Pereira MG, Zonta E (2022) Short-term effects of agroforestry systems on soil health in Southeastern Brazil. Agrofor Syst 96(5–6):897–908. https://doi.org/10.1007/s10457-022-00749-4

Menezes RSC, Sales AT, Primo DC, de Albuquerque ERGM, de Jesus KN, FGC P, da Silva Santana M, dos Santos UJ, Martins JCR, Althoff TD, do Nascimento DM, Gouveia RF, Fernandes MM, Loureiro DC, de Araújo Filho JC, Giongo V, Duda GP, Alves BJR, de Mello Ivo WMP, de Andrade EM, de Sampaio EVSB (2021) Soil and vegetation carbon stocks after land-use changes in a seasonally dry tropical forest. Geoderma 390:114943. https://doi.org/10.1016/j.geoderma.2021.114943

Mukhlis I, Rizaludin MS, Hidayah I (2022) Understanding socio-economic and environmental impacts of agroforestry on rural communities. Forests 13(4):556. https://doi.org/10.3390/f13040556

Nair PR, Kumar BM, Nair VD, Nair PR, Kumar BM, Nair VD (2021a) Shifting cultivation and taungya. In: An introduction to agroforestry, pp 61–86

Nair PR, Kumar BM, Nair VD, Nair PR, Kumar BM, Nair VD (2021b) Global distribution of agroforestry systems. In: An introduction to agroforestry: four decades of scientific developments, pp 45–58

Nurwijayanto A, Náiem M, Syahbudin A, Wahyuono S (2020) Eksplorasiantioksidantumbuhanobat yang berasaldaritamannasionalgunungMerapi Yogyakarta Indonesia. J Tumbuhan Obat Indonesia 13(1):25–31. https://doi.org/10.22435/jtoi.v13i1.1983

Nyirenda H, Balaka V (2021) Conservation agriculture-related practices contribute to maize (Zea mays L.) yield and soil improvement in Central Malawi. Heliyon 7(3):e06636. https://doi.org/10.1016/j.heliyon.2021.e06636

Ondrasek G, Rathod S, Manohara KK, Gireesh C, Anantha MS, Sakhare AS (2022) Salt stress in plants and mitigation approaches. Plants 11(6):717. https://doi.org/10.3390/plants11060717

Ota L, Herbohn J, Gregorio N, Harrison S (2020) Reforestation and smallholder livelihoods in the humid tropics. Land Use Policy 92:104455. https://doi.org/10.1016/j.landusepol.2019.104455

Patel SK, Sharma A, Singh GS (2020) Traditional agricultural practices in India: an approach for environmental sustainability and food security. Energy Ecol Environ 5(4):253–271. https://doi.org/10.1007/s40974-020-00158-2

Rajan A, Boopathy B, Radhakrishnan M, Rao L, Schlüter OK, Tiwari BK (2023) Plasma processing: a sustainable technology in agri-food processing. Sustain Food Technol 1(1): 9–49. https://doi.org/10.1039/D2FB00014H

Rathore AC et al (2021) Biomass, carbon stocks estimation and predictive modeling in mango based land uses on degraded lands in Indian Sub-Himalayas. Agrofor Syst 95:1563–1575

Sahle M, Saito O, Demissew S (2022) Characterization and mapping of enset-based home-garden agroforestry for sustainable landscape management of the Gurагesocioecological landscape in Ethiopia. Environ Sci Pollut Res Int 29(17):24894–24910. https://doi.org/10.1007/s11356-021-17605-0

Sahoo G, Wani AM (2019) Multifunctional agroforestry systems in India for livelihoods. Ann Horticulture 12(2):139–149

Samji A, Eashwarlal K, Shanmugavel S, Kumar S, Warrier RR (2023) Chloroplast genome skimming of a potential agroforestry species Melia dubia. Cav and its comparative phylogenetic analysis with major Meliaceae members. 3 Biotech 13(1):30

Sari RR, Saputra DD, Hairiah K, Rozendaal DMA, Roshetko JM, van Noordwijk M (2020) Gendered species preferences link tree diversity and carbon stocks in cacao agroforest in southeast Sulawesi, Indonesia. Land 9(4):108. https://doi.org/10.3390/land9040108

SenGupta (2020) Study on soil fertility restoration in overburden dumps through biological means in Bastacola opencast mine, Jharia (Jharkhand) coalfields ([Doctoral Dissertation]. Midnapore, West Bengal, India: Vidyasagar University,)

Sikka AK, Madhu M, Chand S, Singh DV, Selvi V, Sundarambal P, Jeevarathanam K, Murgaiah M (2014) Impact analysis of participatory integrated watershed management programme in semi-arid region of Tamil Nadu, India. Indian J Soil Conserv 42(1):98–106

Sileshi M, Nath AJ (2020) Agroforestry systems for improving nutrient recycling and soil fertility on degraded lands. Agrofor Degrad Landsc. 1:225–253

Sileshi GW, Mafongoya PL, Nath AJ (2020) Agroforestry systems for improving nutrient recycling and soil fertility on degraded lands. Agrofor Degrad Landsc 1:225–253

Sileshi GW, Dagar JC, Nath AJ, Kuntashula E (2023) Agroforestry as a climate-smart agriculture: Strategic interventions, current practices and policies. In: Agroforestry for sustainable intensification of agriculture in Asia and Africa. Springer, Singapore, pp 589–640. https://doi.org/10.1007/978-981-19-4602-8_18

Singh G (2022) Role of Prosopis in reclamation of salt-affected soils and soil fertility improvement. In: Prosopis as a heat tolerant nitrogen fixing desert food legume. Academic Press, Cambridge, MA, pp 27–54

Singh RK, Jena K, Chakraborty JP, Sarkar A (2020) Energy and exergy analysis for torrefaction of pigeon pea stalk (cajanus cajan) and eucalyptus (eucalyptus tereticornis). Int J Hydrogen Energy 45(38):18922–18936

Singh A, Singh P, Dhillon GPS, Sharma S, Singh B, Gill RIS (2023a) Differential impacts of soil salinity and water logging on Eucalyptus growth and carbon sequestration under mulched vs. unmulched soils in south-western Punjab, India. Plant Soil 482(1–2):401–425. https://doi.org/10.1007/s11104-022-05700-1

Singh S, Bhoi TK, Vyas V (2023b) Interceding microbial biofertilizers in agroforestry system for enhancing productivity. In: Plant growth promoting microorganisms of arid region. Springer, Singapore, pp 161–183

Singh S, Bhoi TK, Khan I, Vyas V, Athulya R, Rathi A, Samal I (2023c) Climate change drivers and soil microbe-plant interactions. In: Climate change and microbiome dynamics: Carbon cycle feedbacks. Springer, Cham, pp 157–176. https://doi.org/10.1007/978-3-031-21079-2_11

Stephen EA, Evans KD, Akwasi AA (2020) Effects of Faidherbia albida on some important soil fertility indicators on agroforestry parklands in the semi-arid zone of Ghana. Afr J Agric Res 15 (2):256–268

Syed A, Sarwar G, Shah SH, Muhammad S (2021) Soil salinity research in 21st century in Pakistan: Its impact on availability of plant nutrients, growth and yield of crops. Commun Soil Sci Plant Anal 52(3):183–200. https://doi.org/10.1080/00103624.2020.1854294

Tadesse S, Gebretsadik W, Muthuri C, Derero A, Hadgu K, Said H, Dilla A (2021) Crop productivity and tree growth in intercropped agroforestry systems in semi-arid and sub-humid regions of Ethiopia. Agrofor Syst 95(3):487–498. https://doi.org/10.1007/s10457-021-00596-9

Tega M, Bojago E (2023) Farmer's perceptions of agroforestry practices, contributions to rural household farm income, and their determinants in SodoZuria District, Southern Ethiopia. Int J For Res 2023:5439171

Tomar JM, Ahmed A, Bhat JA, Kaushal R, Shukla G, Kumar R (2021, May 26) Potential and opportunities of agroforestry practices in combating land degradation. Agroforestry-small landholder's tool for climate change resiliency and mitigation

United Nations Department of Economic and Social Affairs Population Division (2021) World population prospects 2020: Highlights (ST/ESA/SER.A/423). Retrieved from https://population.un.org/wpp/Publications/Files/WPP2020_Highlights.pdf. New York, United Nations

Wang H, Cao X, Fan W, Deng P, Xu X (2023) Effects of intercropping systems of Phyllostachysedulis and Bletillastriata on soil bacterial community composition and function. Agrofor Syst 97(4):617–630. https://doi.org/10.1007/s10457-023-00814-6

Yusnaini S, Niswati A, Aini SN, Arif MAS, Dewi RP, Rivaie AA (2021) Changes in soil respiration after application of in situ soil amendment and phosphate fertilizer under soybean cultivation at ultisol South Lampung, Indonesia. In IOPConferenceSeries. IOP Conference Series: Earth and Environmental Science. 724(1):012002. IOP Publishing. https://doi.org/10.1088/1755-1315/724/1/012002

Chapter 26
Rice–Fish-Based Agroforestry System: A Climate Smart Way to Reconcile Sustainable Livelihood Options

Suraj Goswami, B. Venkateshwar Reddy, Simran Yadav, Avantika Adhruj, Umakanta Dash, and Arvind Rathore

Abstract Rice–fish-based agroforestry systems are an innovative approach where terrestrial and aquatic ecosystem comined to increasing food production and improving livelihoods while enhancing soil health and reducing negative environmental impacts. In these systems, rice and fish are grown together in the same waterlogged fields, allowing them to share nutrients and optimize land use. This approach can reduce the need for chemical fertilizers and pesticides while decreasing water usage and manual labour. Diversifying crops and incorporating fish, agroecosystems become more resilient to changing environmental conditions such as drought or flooding. Additionally, the approach can provide farmers with a source of high-quality protein and additional income through the sale of surplus fish. This holistic approach can also lead to improved biodiversity and ecosystem services, as well as enhancing the livelihoods of farmers and their communities. Rice-fish-based agroforestry systems are directly linked with SDG 1, 2, 3, 6, 12, 13, and 15, making them a promising practice to achieve the sustainable development goals. By optimizing resource utilization, diversifying crops, and integrating ecological processes, rice–fish agroforestry holds promise for achieving sustainable agricultural intensification while safeguarding natural resources and livelihoods.

S. Goswami (✉) · S. Yadav · A. Adhruj · A. Rathore
Indian Institute of Forest Management, Bhopal, India

B. V. Reddy
Forest college and Research Institute, Hyderabad, India

U. Dash
ARS Scientist (Agroforestry) ICAR-Indian Agricultural Research Institute, Dhemaji, Assam, India

H. S. Jatav et al. (eds.), *Agroforestry to Combat Global Challenges*, Sustainable Development and Biodiversity 36, https://doi.org/10.1007/978-981-99-7282-1_26

26.1 Introduction

Transformation of global food systems is imperative to address environmental degradation and malnutrition. A comprehensive understanding of food systems, encompassing production to consumption and their socioeconomic and environmental outcomes, is crucial. This shift aligns with the Sustainable Development Goals, especially Zero Hunger. Moving beyond productivity-focused approaches, the emphasis should be on sustainability, equity, and resilience to shape future food systems. This paradigm shift is essential to address the complex challenges we face (Freed et al. 2020). Agroecological practices can play a vital role in food system transformation and enhancing resilience to global change. These practices, guided by principles such as leveraging natural processes, local suitability, equity, and systems management, offer diverse solutions for sustainable agriculture (Sinclair et al. 2019; Freed et al. 2020).

Rice is one of the world's most important cereal crops, providing a staple food source for millions of people around the world. With sustainable farming practices and continued innovation in rice cultivation techniques, rice cultivation can be both productive and environmentally responsible. Agroforestry has long been a collective term for land-use systems and practices in which woody perennials are deliberately integrated with crops and/or animals on the same land-management unit, either in a spatial mixture or in temporal sequence (Nair and Garrity 2014).

Rice-based agroforestry is a sustainable and efficient way of using resources and can contribute to food security and rural development. The trees and rice plants complement each other, creating a mutually beneficial relationship. The trees provide shade for the rice, which helps regulate temperature and improve soil moisture, leading to increased yields. The roots of the trees also help to prevent soil erosion and stabilize the soil structure. Additionally, the leaves and branches of the trees provide organic matter, which enriches the soil and provides essential nutrients for the growth of the rice. The trees themselves can also provide a source of income for farmers through the sale of their products, such as fruit, timber, or nontimber forest products. Rice-based agroforestry can also have positive effects on the environment. The trees in the fields can help to reduce greenhouse gas emissions, absorb carbon dioxide from the atmosphere, and provide habitat for wildlife. This system can also help to conserve biodiversity and promote ecosystem health. Furthermore, the integration of trees in rice fields can create a more resilient agricultural system, reducing the impact of extreme weather events, such as floods and droughts (Coche 1967; Wilson and Lovell 2016).

Rice–fish-based agroforestry is a common practice in India, particularly in the eastern and north-eastern regions of the country. These areas are characterized by ample water resources and a long-standing tradition of rice cultivation, making them well-suited for integrating fish farming into agricultural systems. The practice of growing fish in rice fields has ancient origins, and over time, farmers in lowlands have improved rice–fish integration techniques. The aquatic settings available in rice fields are well-utilized for fish farming, generating additional revenue in addition to rice production. Many rice-growing regions worldwide, including China, Bangladesh, Malaysia, Korea, Indonesia, the Philippines, Thailand, and India, use the rice–fish technology. Rice–fish farming is one of the many farming systems suitable for rice ecologies, and it has a particularly high potential in eastern India due to the region's ecology, resource availability, dietary preferences, and socioeconomic and livelihood conditions of small and marginal farmers (Nayak et al. 2020).

According to a study by Nayak et al. (2020), rice–fish-based agroforestry is widely practiced in several regions of India, particularly in the eastern states such as West Bengal, Odisha, and Bihar. The practice has also gained attention in other regions of the country, such as Andhra Pradesh. Rice–fish-based agroforestry is also being implemented in other countries in South Asia, such as Bangladesh and Nepal, and in other parts of Southeast Asia. The potential of this agroforestry system to improve food security and reduce poverty in rural areas has been widely recognized, and it is seen as a means of promoting sustainable agriculture and conserving biodiversity in states such as Assam. The rice–fish farming method is a traditional agricultural practice in the backwaters of Kerala, known locally as "Pokkali" (Arunachalam et al. 2014a, b). The Indian government is also promoting rice–fish-based agroforestry as a means of enhancing food security and promoting sustainable agriculture. The government has implemented various programs and initiatives to encourage the adoption of this type of agroforestry system, including offering training and support to farmers and communities. Overall, the implementation of rice–fish-based agroforestry has the potential to improve food security, promote sustainable agriculture, and increase the income of small and marginal farmers in India.

26.2 History

It is believed that simultaneous rice and fish cultivation dates back more than 2000 years. Chinese archaeologists have discovered ancient clay models of rice fields with miniature objects, such as pieces of fish like the common carp. Tombs from the historic Han era (206 BC–220 AD) have also contained these artifacts (Renkui et al. 1995). Although the original rice–fish systems are thought to have developed in nations like India, Thailand, northern Vietnam, and southern China, the precise location of their origin is uncertain. The most widely accepted opinion holds that the method originated in China due to their advanced aquaculture systems at that time. The concept of rice–fish-based agroforestry can be traced back to ancient times when farmers in Southeast Asia and other regions began to integrate rice cultivation with fish farming. These early systems were based on the idea of using the waste produced by the fish to fertilize the rice fields, and of using the water from the fish ponds to irrigate the rice. Over time, these systems have evolved and become more sophisticated, and today they are widely used as a sustainable and efficient way of producing both rice and fish.

Until the 1980s, the major advantage of rice–fish systems was space optimization and the ability to grow additional animal protein with rice as the main staple food, requiring little upkeep. However, with an increasing population in many nations, the need to maximize available space became more pressing. Starting in the 1980s, the system quickly evolved, adding new species such as the Chinese mitten crab, red swamp crayfish, and softshell turtles. The incorporation of new theories and technology caused the industry to boom. In China, for example, the area used for rice fields increased from 441,027 hectares (1,089,800 acres) to 853,150 ha (2,108,200 acres), and productivity rose sharply (Koohafkan and Altieri 2016).

In recent decades, rice–fish-based agroforestry has gained renewed attention as a way of promoting sustainable agriculture and improving food security. As the global population continues to grow and the demand for food increases, there is a growing recognition of the need for sustainable and efficient agricultural systems that can produce multiple crops and livestock. Rice–fish-based agroforestry is seen as one of the most promising solutions to these challenges, and it is now being promoted and implemented in many countries around the world, including India, China, Indonesia, and the Philippines. In the past, rice field fisheries were the most typical method of producing both rice and fish together (Coche 1967), and they are most prevalent in rainfed and deep-water rice-growing regions (Gregory 1997). For rural communities in low- and middle-income nations across Asia, such as Bangladesh (Dey et al. 2013), Cambodia (Freed et al. 2020), rice field fisheries have been a significant source of food and nutrition security as well as livelihoods. These fisheries get varying degrees of formal recognition and management support, with Cambodia providing the most support. In recent years, rice–fish-based agroforestry has become a popular research topic, with scientists and policymakers around the world exploring the benefits and challenges of these systems and working to develop best practices for their implementation. As a result of this research, we now have a better understanding of the key components of successful rice–fish-based agroforestry systems and the role that these systems can play in promoting sustainable agriculture and reducing poverty and food insecurity.

26.3 Distribution in Global Scenario

The worldwide rice–fish cultivation practices have distinctive agrolandscapes, particularly in tropical and subtropical Asia. China has a long history of rice–fish farming methods, which have also been used in 28 other nations on six different continents, including Africa, Asia, Australia, Europe, North America, and South America (Huat and Tan 1980; Rongquan 1995; Halwart 1998; Suloma and Ogata 2006). Japan, Java, Thailand, India, Vietnam, Bangladesh, Indonesia, Philippines, Malaysia, Nepal, Korea, etc. (Fig. 26.1) are among the nations that use this system. The Philippines, Indonesia, and China all use shallow trenches within their rice fields, whereas India, Thailand, Indonesia, and China use pond refuges next to their rice fields. Deep-water rice fields are another option (Bangladesh). In many parts of

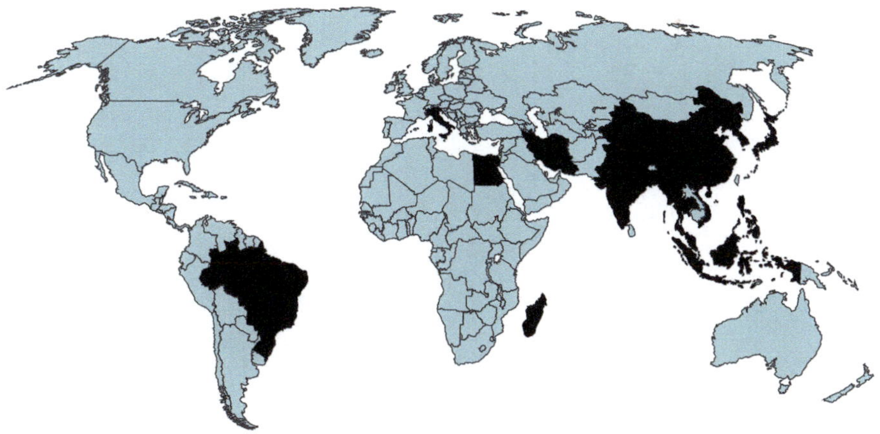

Fig. 26.1 Distribution of rice–fish system across the world

Asia, fish and rice are grown simultaneously, and rice and fish are grown in a rotational pattern. The introduction of an Asian-based Sawah farming system in Africa via an ecotechnology approach has opened a new area for diversification of the rice-based cropping system, with on-farm rice–fish-culture trials recorded (Ofori et al. 2005). It's vital to remember that each nation may see a different level of acceptance and popularity of fish-based agroforestry systems, depending on elements like climate, resource availability, and cultural practices. In addition, continuous research and technology developments continue to support the growth and extension of these systems on a global scale. In general, three types of rice–fish farming field designs predominate: shallow trenches within rice fields (in the Philippines, Indonesia, and China), pond refuges next to rice fields (in Indonesia, India, Thailand, and China), and deep-water rice fields (in Bangladesh) (Mohanty et al. 2008). Carps (found in Indonesia, India, and China), tilapias (found in the Philippines, China, and Thailand), and *Puntius gonionotus* are the most significant fish species found in rice field systems (Thailand). In general, *Cypinus caripo* and *Oreochromis niloticus* are the two species that are most commonly found in rice fields (Mohanty et al. 2008).

26.4 Distribution in India Scenario

In many regions of eastern India, rice–fish integrated farming has a long history. Among them, Zabo farming in Nagaland, Apatani farming in Arunachal Pradesh, Bhasabandha or Bheri system in West Bengal's Sunderbans, and Pokkali system in Kerala are prominent examples. The hill is separated into three sections in this agricultural practice: the upper section is set aside as a forest area, the middle section is used for residential purposes, rainwater collection ponds are built, and the lower section of the hill is used for rice and fish production. Later, the rainwater collected is

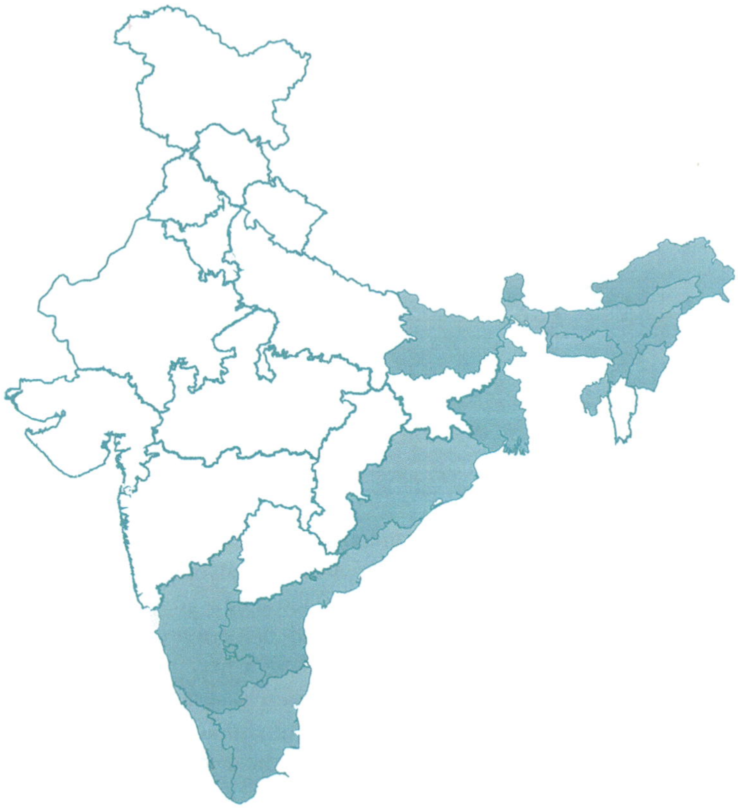

Fig. 26.2 Distribution of rice–fish agroforestry systems across Indian states

effectively used for irrigation and as drinking water for animals (Sathoria and Roy 2022). Assam, Manipur, Tripura, Meghalaya, Sikkim, Karnataka, Goa, Tamil Nadu, Andhra Pradesh, Orissa, and Bihar are among the other Indian states that use this method, (Rani et al. 2019; Mohanty et al. 2008; Sathoria and Roy 2022) also indicated in Fig. 26.2. Although 20 million ha of land in India is ideal for the adoption of integrated rice–fish farming, only 0.23 million ha is currently under rice–fish culture, according to estimates (Mansharamani et al. 2020). The Indian Council of Agriculture Research (ICAR) Cuttack, Odisha, has created models for the Indian scenario in consideration of the expanding population, poverty, and for sustainable environmental conditions in order to promote integrated rice–fish farming (Poonam et al. 2019; Nayak et al. 2020).

Pond refuge and trench systems of rice–fish farming are primarily used in Eastern Indian states, while broad bed-farrows' systems are common in the Andaman Islands (Rani et al. 2019). White fish, such as Danios (*Rasbora*), Barbs (*Puntius*), Snakeskin Gourami (*Trichogaster*), and Half noses, are among the most well-known local fish

Table 26.1 Potential for rice–fish farming in Asia

	Rice field area (million ha)			Rice–fish area (million ha)	
Country	Rainfed	Irrigated	Total	Present	Potential
Bangladesh	9.002	1.227	10.229	n.a.	0.615
China	2.296	30.902	33.198	0.986	5.000
India	26.644	14.349	40.993	n.a.	2.000
Indonesia	3.659	6.230	9.889	0.094	1.570
Korea	0.111	1.118	1.229	<1	0.127
Malaysia	0.220	0.427	0.647	n.a.	0.120
Philippines	1.953	1.473	3.426	1.000	0.181
Thailand	8.065	1.313	9.378	n.a.	0.254
Viet Nam	3.415	2.276	5.691	n.a.	0.326
Total	55.365	59.315	114.680	1.082[a]	10.193
	48%	52%	100%	1%[a]	9%[a]

Source: Lightfoot et al. 1992
[a]Exact figure not available

species (*Xenentodon*). Dark fish such as Sheatfish (*Ompok*), Climbing perch (*Anabas*), Catfish (*Clarias*), Snakehead (*Channa*), and Spiny eels (*Mastacembelus*). It is also possible to harvest other wild oceanic animals including crabs, shrimp, snails, and creepy crawlies (Santhosh 2021).

According to Lightfoot et al. (1992), the highest rice field area in India is 40.993 million hectares, followed by China with 33.198 million hectares, and Bangladesh with 10.229 million hectares. The potential of rice–fish area ranges from 0.12 to 5.0 million hectares in Southeastern Asia. Detailed information about the rice field area and rice–fish area in Southern Asia is presented in Table 26.1.

26.5 Benefits of Rice–Fish-Based Agroforestry System

This type of agroforestry system provides a number of benefits, including:

- Increased food production: By integrating rice cultivation with other agricultural activities, this type of agroforestry system can increase food production and help to improve food security.
- Improved soil health: Rice-based agroforestry systems can help to improve soil health by incorporating the use of organic fertilizers and promoting soil conservation practices.
- Enhanced biodiversity: By integrating a variety of different crops and livestock into the rice fields, this type of agroforestry system can help to promote biodiversity and conserve natural resources.
- Increased income: Rice-based agroforestry systems can provide a source of income for farmers through the sale of a variety of crops and livestock products.

Fig. 26.3 Rice–fish agroforestry system toward sustainable food production

Figure 26.3 illustrates the component of rice, fish, and agroforestry and the benefits of this system.

- Efficient use of resources: Rice–fish agroforestry systems are designed to make efficient use of resources, such as water and nutrients. The waste produced by the fish is used as fertilizer for the rice, and the water from the fish ponds is used to irrigate the rice fields. This helps to conserve water and reduce the need for chemical fertilizers, which can be expensive and can have negative impacts on the environment.
- Climate resilience: Rice-based agroforestry systems can help to increase the resilience of agricultural systems to the impacts of climate change by promoting

Table 26.2 Benefits of rice–fish-based agroforestry system: a triangular overview

Benefits	Social	Economical	Environmental
Food security	Provides a source of protein and income	Increases income from rice and fish sales	Maintains biodiversity
Livelihood	Provides employment opportunities and additional income through surplus produce	Reduces chemical fertilizer use	Improve soil fertility
Good health and well-being	Improved nutrition and well-being	Access to fresh, pesticide free food	Reduced exposure to harmful chemicals and pesticides
Biodiversity	Preservation of traditional knowledge and culture	Increases diversity of crops and fish	Conservation of native and endangered species
Climate resilience	Strengthened community resilience	Buffer against climate-related shocks	Carbon sequestration and mitigation of climate change

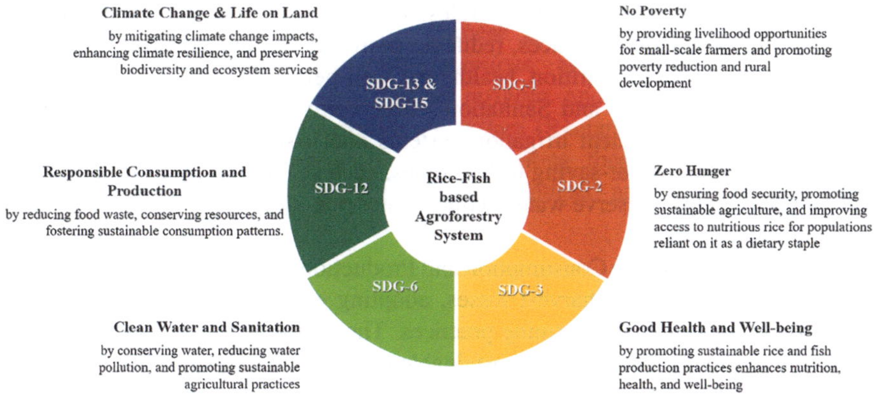

Fig. 26.4 Integration of rice–fish agroforestry system with Sustainable Development Goals

the use of sustainable agricultural practices and incorporating a variety of crops and livestock into the system. Further benefits are mentioned in Table 26.2.

26.6 Rice–Fish-Based Agroforestry Systems and Sustainable Development Goals

Rice, the staple food for more than half of the world's population, plays a vital role in achieving the United Nations Sustainable Development Goals (SDGs) (Fig. 26.4). The SDGs provide a comprehensive framework for addressing global challenges and promoting sustainable development in various sectors, including agriculture, food

26.8 Productivity

Rice–fish-based agroforestry systems have been shown to be highly productive, with potential yields varying depending on location, management practices, and other factors. Several studies have reported that well-managed rice–fish-based agroforestry systems can produce between 4 and 6 tons of rice per hectare, along with several hundred kilograms of fish (Bhattacharyya et al. 2019; Budianta et al. 2018; Chawapun and Sridulyakul 2006). However, it is important to note that productivity can be influenced by a variety of factors, such as soil quality, water availability, and climate conditions.

It is also important to note that the productivity of these systems is not just limited to rice and fish yields. Rice–fish-based agroforestry systems can also contribute to improving soil health, reducing the need for chemical inputs, and promoting biodiversity. For example, in a study conducted in China, the introduction of fish into a rice paddy resulted in a reduction in the need for fertilizer, as the fish helped to break down organic matter and release nutrients into the soil (Chen et al. 2011). Overall, the productivity of a rice–fish-based agroforestry system will depend on a wide range of factors, and it is important to consider these factors when evaluating the potential of these systems in a given location. However, numerous studies have shown that these systems have the potential to be highly productive and sustainable, providing benefits beyond just rice and fish yields.

26.9 Socioeconomic

Rice and fish farming is a multifaceted, inventive method that has many advantages for farmers, the environment, and society. According to studies by Desta et al. (2014) and Saikia and Das (2008), it can increase income and availability of fish for domestic consumption. Additionally, the system promotes biodiversity and reduces the use of fertilizers and pesticides, making it an environmentally friendly and low-cost activity (Ahmed and Garnett 2011; Rothuis et al. 1998). This indigenous farming method also has the potential to significantly improve the livelihood, income, and nutrition of rural people, as noted by Desta et al. (2014) and Noorhosseini-Niyaki and Allahyari (2012). Furthermore, rice and fish are essential foods that improve the health of farmers, with fish being a good source of fatty acids, proteins, vitamins, and minerals (Sathoria and Roy 2022).

In addition to its economic and nutritional benefits, rice and fish farming can also have positive environmental impacts. The system is helpful for restoring soil fertility and preventing soil erosion, as pointed out by Yonghua and Guobin (1998). It also reduces methane emissions by over 30% compared to standard rice farming, which typically contributes between 10% and 20% of the methane that enters the atmosphere (Lu and Li 2006). Furthermore, fish serve as hosts for pests that would otherwise compete with rice for nutrients and eliminate aquatic weeds and algae that spread disease. The water temperature is also kept at a fish-friendly level by shading provided by rice plants in the summer (Kunda et al. 2008). By making the best use of land and resources, integrated rice–fish farming may offer a long-needed solution for sustainable agriculture. The rural youth may find more work opportunities through integrated rice–fish farming, and they may gain from such entrepreneurship activities (Sathoria and Roy 2022). The rice–fish cocultivation also enhances social status, empowers women through work opportunities, and offers prospects for a means of subsistence. Therefore, the benefits of this integrated farming method extend beyond economic and environmental benefits to promote social well-being and empowerment.

26.10 Conclusion

Rice–fish-based agroforestry systems have the potential to address multiple challenges faced by rural communities in developing countries. These systems can increase food production, improve livelihoods, and enhance resource efficiency, among other benefits. Studies have shown that rice–fish-based agroforestry can be an effective tool for poverty reduction and food security, particularly in regions with high rates of malnutrition and poverty. Furthermore, these systems can enhance the resilience of agricultural communities to the impacts of climate change, such as floods and droughts. Overall, scientific evidence supports the potential of rice–fish-based agroforestry systems as a multifaceted approach to address challenges faced by rural communities in developing countries. These systems have shown promising results in increasing food production, improving livelihoods, enhancing resource efficiency, and bolstering the resilience of agricultural communities to climate change impacts. Further research and implementation of these systems can contribute to sustainable development and the achievement of global goals related to poverty reduction, food security, and environmental sustainability. Rice–fish agroforestry systems offer a sustainable and holistic approach to increasing food production, improving livelihoods, and safeguarding the environment. By combining rice and fish cultivation, these systems optimize land use, enhance soil health, and reduce the reliance on chemicals and water. They also enhance resilience to climate change, provide high-quality protein and additional income for farmers, and promote biodiversity and ecosystem services. Overall, rice–fish agroforestry holds promise for achieving sustainable agricultural intensification while benefiting both farmers and the environment.

References

Ahmed N, Garnett ST (2011) Integrated rice-fish farming in Bangladesh: meeting the challenges of food security. Food Secur 3:81–92

Arunachalam A, Subramanian B, Dagar J, Mohan Kumar B (2013) Wetland-based agroforestry. Balancing Between Carbon Sink and Source, Systems. https://doi.org/10.1007/978-81-322-1662-9_10

Arunachalam A, Balasubramanian D, Arunachalam K, Dagar JC, Mohan Kumar B (2014a) Wetland-based agroforestry systems: balancing between carbon sink and source. In: Agroforestry systems in India: livelihood security & ecosystem services. Springer, Cham, pp 333–343

Arunachalam M, Arunachalam K, Arunachalam A, Chinnappa T, Fernandes S (2014b) Backwater rice-fish culture: an eco-friendly and sustainable integrated farming system in Kerala, India. Agric Sci 5(6):562–571

Aryal K, Thapa PS, Lamichhane D (2019) Revisiting agroforestry for building climate resilient communities: a case of package-based integrated agroforestry practices in Nepal. Emerg Sci J 3(5):303–311

Baruah D, Singh ND (2018) Rice-Fish Cultivation of Apatanis: a high altitude farming system in Arunachal Pradesh. J Krishi Vigyan 7(1):187–191

Bhatt BP, Bujarbaruah KM, Patnayak A, Mandal BK, Vinod K, Venkatesh MS, Rajkhowa C, Kumaresan A, Santosh A, Dutta M (2005) Rice based integrated farming system in the North East. In: Agroforestry in North East India: opportunities and challenges. ICAR Research Complex for NEH Region, Umiam, Meghalaya, pp 575–587

Bhattacharyya P, Saha D, Nandy S, Sarker S, Bhattacharyya S (2019) Impact of agroforestry models on soil quality, microbial population, and productivity of rice (Oryza sativa L.)–fish (Catlacatla) integrated system in eastern India. Environ Monitor Assess 191(6):370

Borelli S, Wangpakapattanawong P, Roshetko JM, Widyani N (2017) Agroforestry in rice-production landscapes in South-East Asia: a practical manual. Food and Agriculture Organization of the United Nations and World Agroforestry Centre. https://www.researchgate.net/publication/322339336_Agroforestry_in_rice-production_landscapes_in_South-East_Asia_a_practical_manual

Budianta D, Sukartiko AC, Anggoro S (2018) Evaluation of different tree species in rice–fish farming system for improvement of fish production and soil nutrient availability. Aquacult Rep 11:41–48

CGIAR Report (2017). https://ricecrp.org/wp-content/uploads/2017/03/RICE-and-SDGs.pdf

Chawapun N, Sridulyakul P (2006) Effects of tree legumes on water quality and fish production in rice–fish systems. Aquaculture 253(1–4):394–404

Chen Y, Sha J, Xu P, Wu X (2011) Effects of Chinese mitten crab (Eriocheir sinensis) on ecosystem functioning and water quality in fish–rice paddy fields. Aquaculture 310(3–4):295–301

Coche AG (1967) Fish culture in rice fields a world-wide synthesis. Hydrobiologia 30:1–44

Desta L, Prabha DL, Sreenivasa V, Amede T (2014) Studies on the ecology of the Paddy and Fish co-culture system at DembiGobuMicrowater Shed at Bako, Ethiopia. Int J Fish Aquatic Stud 1: 49–53

Dey MM, Spielman DJ, Haque AM, Rahman MS, Valmonte-Santos R (2013) Change and diversity in smallholder rice–fish systems: recent evidence and policy lessons from Bangladesh. Food Policy 43:108–117

Freed S, Kura Y, Sean V, Mith S, Cohen P, Kim M, Chhy S (2020) Rice field fisheries: wild aquatic species diversity, food provision services and contribution to inland fisheries. Fish Res 229: 105615

Ghosh A (1992) Rice-fish farming development in India: past, present and future. In: Rice-fish research and development in Asia. ICLARM Conf. Proc, pp 27–43

Gregory R (1997) Ricefield fisheries handbook. Cambodia-IRRI-Australia Project. https://www.doc-developpement-durable.org/file/Culture/Culture-plantes-alimentaires/FICHES_PLANTES/riz/Rice%20field_Fisheries_Handbook.pdf

Halwart M (1998) Trends in rice-fish farming. FAN, the FAO Aquaculture Newsletter, No. 18(3) C11

Huat KK, Tan ESP (1980) Review of rice-fish culture in Southeast Asia. In: ICLARM Conference Proceedings (Philippines)

Islam AHMS, Barman BK, Jahan KM (2015) Adoption and impact of integrated rice–fish farming system in Bangladesh. Aquaculture 447:76–85

Joffre O, Sheriff N, Kosal M, Baran E (2012) Integrated rice-fish farming in Cambodia: lessons learned from a participatory action research project. In: Hoanh CT, Szuster BW, Suan-Pheng K, Ismail AM, Noble AD (eds) Tropical deltas and coastal zones: food production, communities and environment at the land-water interface. CABI, Wallingford, pp 433–446

Koohafkan P, Altieri MA (2016) Forgotten agricultural heritage: reconnecting food systems and sustainable development. Earthscan food and agricultural series. Routeledge, London, p 296

Koseki Y (2014) Column: rice-fish culture: the contemporary significance of a traditional practice. Social-Ecological Restoration in Paddy-Dominated Landscapes, pp 165–172

Kunda M, Azim ME, Wahab MA, Dewan S, Roos N, Thilsted SH (2008) Potential of mixed culture of freshwater prawn (Macrobrachiumrosenbergii) and self-recruiting small species mola (Amblypharyngodon mola) in rotational rice–fish/prawn culture systems in Bangladesh. Aquac Res 39(5):506–517

Lightfoot C, Costa-Pierce BA, Bimbao MP, dela Cruz CR (1992) Introduction to rice-fish research and development in Asia. p. 1-10 In CR. dela Cruz C. Lightfoot BA. Costa-Pierce VR. Carangal MP Bimbao, eds. Rice-fish research and development in Asia. Manila, Philippines, International Center for Living Aquatic Resources Management 457 pp.

Lu J, Li X (2006) Review of rice–fish-farming systems in China—one of the globally important ingenious agricultural heritage systems (GIAHS). Aquaculture 260(1–4):106–113

Luo S, Chen X (2018) Rice-fish co-culture system. FAO. http://www.fao.org/3/BU701EN/bu701en.pdf

Mansharamani A, Shrivastava A, Choubey A (2020) Rice-fish farming system in India is in urgent need of conservation and promotion. Available at https://www.downtoearth.org.in/blog/agriculture/rice-fish-farming-system-in-india-is-in-urgent-need-of-conservation-and-promotion-70523

Mariyono J (2023) Sustainable intensification practices of fish-rice co-culture in Java, Indonesia: technical, socio-economic and environmental features. J Agribus Dev Emerg Econ. https://doi.org/10.1108/JADEE-09-2022-0208/full/html

Mohanty RK, Jena SK, Kumar A, Sahoo N, Chowdhury S, R. (2008) Rice-fish culture: an ingenious agricultural heritage system. Res Bull 42:54

Nair, P. K. R., & Garrity, D. (Eds.). (2014). Agroforestry - the future of global land use (1st ed.) Springer Dordrecht. doi: https://doi.org/10.1007/978-94-007-4676-3_13

Nayak PK, Nayak AK, Kumar A, Kumar U, Panda BB, Satapathy BS, Poonam A, Mohapatra SD, Tripathi R, Shahid M, Chatterjee D, Panneerselvam P, Mohanty S, Sunil KD, Pathak H (2020) Rice based Integrated Farming Systems in Eastern India: A Viable Technology for Productivity and Ecological Security

Noorhosseini-Niyaki SA, Allahyari MS (2012) A logistic regression analysis: agro-technical factors impressible from fish farming in rice fields, North of Iran. Int J Agric Manag Dev 2(3):223–227

Ofori J, Abba EK, Otoo E, Wakatsuki T (2005) Rice-fish culture: an option for smallholder sawah rice farmers of the West African lowlands. Ecol Eng 24:235–241

Pathak H, Tripathi R, Jambhulkar NN, Bisen JP, Panda BB (2020) Eco-regional-based rice farming for enhancing productivity, profitability and sustainability

Patra AK (2016) Concept, scope and components of integrated farming system. Root and tuber crops based integrated farming system: a way forward to address climate change and livelihood improvement, p 8

Poonam A, Saha S, Nayak PK, Sinhababu DP, Sahu PK, Satapathy BS, Shahid M, Kumar GAK, Jambhulkar NN, Nedunchezhiyan M, Giri S, Saurabh K, Sangeet K, Nayak AK, Pathak H (2019) Rice-Fish Integrated Farming Systems for Eastern India. NRRI Research Bulletin No. 17, ICAR-National Rice Research Institute, Cuttack, Odisha, p 33+iii

Rani BR, Daniel S, Harshitha M, Rambabu SP (2019) Adoption of integrated rice- fish farming system by farmer of Ramchandrapuram. Int J Curr Microbiol App Sci 8(1):591–597

Rautaray SK, Dash PC, Sinhababu DP (2005) Increasing farm income through rice (Oryza sativa)-fish based integrated farming system in rainfed lowlands of Assam. Indian J Agric Sci 75(2): 79–82

Renkui C, Dashu N, Jianguo W (1995) Rice fish culture in China: the past, present, and future. In: Rice fish culture in China. IDRC, Ottawa, ON

Romanillos R (2010, December) Rice-based agroforestry technology: a strategy in optimizing agricultural productivity and income in marginalized inland valleys in Quezon province, Philippines. In international conference in agroforestry education

Rongquan Z (1995) Scientific and technological development of rice fish culture in China. In: MacKay KT (ed) Rice fish culture in China. Ottawa, ON, IDRC

Rothuis AJ, Nhan DK, Richter CJ, Ollevier F (1998) Rice with fish culture in the semi-deep waters of the Mekong Delta, Vietnam: a socio-economical survey. Aquac Res 29(1):47–57

Saikia SK, Das DN (2008) Rice-fish culture and its potential in rural development: a lesson from Apatani Farmers, Arunachal Pradesh, India. Agric Rural Dev 6:125–131

Santhosh K (2021) Rice field fisheries of Indian states: what types of fishes are found in the rice-fish farming system? J Rice Res 9:265

Sathoria P, Roy B (2022) Sustainable food production through integrated rice-fish farming in India: a brief review. Renew Agric Food Syst:1–9. https://doi.org/10.1017/S1742170522000126

Sinclair F, Wezel A, Mbow C, Chomba S, Robiglio V, Harrison R (2019) The Contribution of Agroecological Approaches to Realizing Climate-Resilient Agriculture. Background paper. Global Commission on Adaptation, Rotterdam

Suloma A, Ogata HY (2006) Future of rice-fish culture, desert aquaculture and feed development in Africa: the case of Egypt as the leading country in Africa. Jpn Agric Res Q 40(4):351–360

Taka T, Tangjang S (2015) Original research article sustainable approach: paddy-cum-fish and bamboo-cum-pine agroforestry in Ziro valley, Arunachal Pradesh, India. J Bioresour 2(1):50–57

Tangjang S, Nair PR (2015) Rice+ fish farming in homesteads: sustainable natural-resource management for subsistence in Arunachal Pradesh, India. J Environ Sci Eng A 4:545–557

Wangpakapattanawong P, Finlayson R, Öborn I, Roshetko JM, Sinclair F, Shono K, Borelli S, Hillbrand A, Conigliaro M (2017) Agroforestry in rice-production landscapes in Southeast Asia: a practical manual. FAO Regional Office for Asia and the Pacific

Wilson MH, Lovell ST (2016) Agroforestry—The next step in sustainable and resilient agriculture. Sustainability 8(6):574

Xu J, World Agroforestry Centre (Kunming) (2011) Participatory agroforestry development in DPR Korea. World Agroforestry Centre

Yadav JSP, Bandyopadhyay AK, Rao KVGK, Sinha TS, Biswai CR (1979) 3 coastal saline soils of India. Bull Cent Soil Salinity Res Inst. (5):1–34. Canning, West Bengal, India

Yassi A, Farid M, Anshori MF, Muchtar H, Syamsuddin R, Adnan A (2023) The integrated Minapadi (Rice-Fish) farming system: compost and local liquid organic fertilizer based on multiple evaluation criteria. Agronomy 13(4):978

Yonghua Z, Guobin D (1998) Benefit analysis and comprehensive evaluation of rice-fish-duck symbiotic model. Chin J Eco-Agric. 6(1):48–51

GPSR Compliance

The European Union's (EU) General Product Safety Regulation (GPSR) is a set of rules that requires consumer products to be safe and our obligations to ensure this.

If you have any concerns about our products, you can contact us on ProductSafety@springernature.com

In case Publisher is established outside the EU, the EU authorized representative is:

Springer Nature Customer Service Center GmbH
Europaplatz 3
69115 Heidelberg, Germany

Batch number: 10725820

Printed by Printforce, the Netherlands